Finite and Locally Finite Groups

NATO ASI Series

Advanced Science Institutes Series

A Series presenting the results of activities sponsored by the NATO Science Committee, which aims at the dissemination of advanced scientific and technological knowledge, with a view to strengthening links between scientific communities.

The Series is published by an international board of publishers in conjunction with the NATO Scientific Affairs Division

A Life Sciences	Plenum Publishing Corporation
B Physics	London and New York
C Mathematical and Physical Sciences	Kluwer Academic Publishers
D Behavioural and Social Sciences	Dordrecht, Boston and London
E Applied Sciences	
F Computer and Systems Sciences	Springer-Verlag
G Ecological Sciences	Berlin, Heidelberg, New York, London,
H Cell Biology	Paris and Tokyo
I Global Environmental Change	

PARTNERSHIP SUB-SERIES

1. Disarmament Technologies	Kluwer Academic Publishers
2. Environment	Springer-Verlag / Kluwer Academic Publishers
3. High Technology	Kluwer Academic Publishers
4. Science and Technology Policy	Kluwer Academic Publishers
5. Computer Networking	Kluwer Academic Publishers

The Partnership Sub-Series incorporates activities undertaken in collaboration with NATO's Cooperation Partners, the countries of the CIS and Central and Eastern Europe, in Priority Areas of concern to those countries.

NATO-PCO-DATA BASE

The electronic index to the NATO ASI Series provides full bibliographical references (with keywords and/or abstracts) to more than 50000 contributions from international scientists published in all sections of the NATO ASI Series.
Access to the NATO-PCO-DATA BASE is possible in two ways:

– via online FILE 128 (NATO-PCO-DATA BASE) hosted by ESRIN,
Via Galileo Galilei, I-00044 Frascati, Italy.

– via CD-ROM "NATO-PCO-DATA BASE" with user-friendly retrieval software in English, French and German (© WTV GmbH and DATAWARE Technologies Inc. 1989).

The CD-ROM can be ordered through any member of the Board of Publishers or through NATO-PCO, Overijse, Belgium.

Series C: Mathematical and Physical Sciences – Vol. 471

Finite and Locally Finite Groups

edited by

B. Hartley †
Department of Mathematics,
University of Manchester,
Manchester, U.K.

G. M. Seitz
Department of Mathematics,
University of Oregon,
Eugene, Oregon, U.S.A.

A. V. Borovik
UMIST,
Manchester, U.K.

and

R. M. Bryant
UMIST,
Manchester, U.K.

Kluwer Academic Publishers

Dordrecht / Boston / London

Published in cooperation with NATO Scientific Affairs Division

Proceedings of the NATO Advanced Study Institute on
Finite and Locally Finite Groups
Istanbul, Turkey
14–27 August, 1994

Library of Congress Cataloging-in-Publication Data

ISBN 0-7923-3669-0

Published by Kluwer Academic Publishers,
P.O. Box 17, 3300 AA Dordrecht, The Netherlands.

Kluwer Academic Publishers incorporates the publishing programmes of
D. Reidel, Martinus Nijhoff, Dr W. Junk and MTP Press.

Sold and distributed in the U.S.A. and Canada
by Kluwer Academic Publishers,
101 Philip Drive, Norwell, MA 02061, U.S.A.

In all other countries, sold and distributed
by Kluwer Academic Publishers Group,
P.O. Box 322, 3300 AH Dordrecht, The Netherlands.

Printed on acid-free paper

TABLE OF CONTENTS

PREFACE

This volume contains the proceedings of the NATO Advanced Study Institute on Finite and Locally Finite Groups held in Istanbul, Turkey, 14–27 August 1994, at which there were about 90 participants from some 16 different countries. The ASI received generous financial support from the Scientific Affairs Division of NATO.

INTRODUCTION

A locally finite group is a group in which every finite set of elements is contained in a finite subgroup. The study of locally finite groups began with Schur's result that a periodic linear group is, in fact, locally finite. The simple locally finite groups are of particular interest. In view of the classification of the finite simple groups and advances in representation theory, it is natural to pursue classification theorems for simple locally finite groups. This was one of the central themes of the Istanbul conference and significant progress is reported herein. The theory of simple locally finite groups intersects many areas of group theory and representation theory, so this served as a focus for several articles in the volume.

Every simple locally finite group has what is known as a *Kegel cover*. This is a collection of pairs $\{(G_i, N_i) \mid i \in I\}$, where I is an index set, each group G_i is finite, $N_i \triangleleft G_i$, and G_i/N_i is simple. Moreover, for each finite subgroup H of G there is an index i such that $H \leqslant G_i$ and $H \cap N_i = 1$. In other words, each finite subgroup of G is contained in one of the simple sections of the cover. If G is countable then the situation is considerably better. Here one can take the index set as the positive integers: $G_1 \leqslant G_2 \leqslant \ldots$, $G = \bigcup G_i$, and $G_i \cap N_{i+1} = 1$ for all i. This is called a *Kegel sequence*.

The existence of Kegel covers clearly indicates how the analysis of simple locally finite groups is closely linked to the study of finite simple groups and their subgroups. The Kegel covers also tie in with representation theory. A result of Mal'cev, based on the theory of ultraproducts, shows that, if each finitely generated subgroup of G has a faithful representation of degree at most n, then the same is true of G. To illustrate the usefulness of this result, assume we are given a Kegel cover such that the simple factors have a faithful representation of bounded degree. Every finite subgroup of G is isomorphic to a subgroup of one of the simple factors, so Mal'cev's result implies G has a nontrivial faithful finite dimensional representation. That is, G is *linear*.

The lead article of Hartley provides a survey of various topics in the theory of simple locally finite groups and serves as a more complete introduction to the rest of the volume. The first section concerns constructions and introduces the linear groups, finitary groups, universal groups of P. Hall, and some recent groups due to Meierfrankenfeld. These topics are each covered individually in subsequent articles. The second section consists of a discussion of Kegel covers, leading to an analysis of how the covers affect the global structure of simple locally finite groups. The final section concerns centralizers, difficulties resulting from the existence of Kegel kernels, and some open problems.

Following the classification of finite simple groups, it was shown in the 1980s that the linear simple locally finite groups are groups of Lie type over locally finite fields. These are the groups that are most closely linked to finite groups. Indeed, such a group is the union of groups of Lie type of a fixed type and the structure and representations of such groups are closely connected with the structure and representation theory of the finite groups of Lie type.

The articles of Liebeck, Saxl, and Seitz are a package aimed at a general understanding of algebraic groups, finite groups of Lie type, subgroups, and representations. The finite groups of Lie type and the simple linear locally finite groups arise naturally from algebraic groups. It has long been recognized that to establish deep results about finite groups of Lie type it is first necessary to understand algebraic groups. Consequently this series of articles begins with Seitz's article on algebraic groups.

The first section of this article proceeds from the definition of algebraic varieties to the classification of simple algebraic groups. This is followed by a discussion of the subgroups of algebraic groups, leading to the determination of the maximal connected subgroups of simple algebraic groups. The final section discusses representation theory, concluding with an application of a new result on abstract homomorphisms to the theory of locally finite groups.

The article of Liebeck presents the general method, based on Lang's theorem, in which one uses information about simple algebraic groups to obtain information about the finite groups of Lie type. Many examples are given. Liebeck also discusses the subgroups of finite simple groups. Included is an outline of a new proof of Aschbacher's reduction theorem on finite classical groups, using descent from algebraic groups. The article concludes with a survey of the current situation on the subgroups of finite exceptional groups, including a reduction theorem analogous to the result for classical groups.

Saxl discusses permutation groups, in particular primitive groups. Included is an outline of the O'Nan-Scott theorem, a reduction theorem for analyzing primitive permutation groups. This result leads one to the study of maximal subgroups of almost simple groups. The article continues with a brief survey of the current state of affairs in the study of maximal subgroups of finite simple groups.

After the linear locally finite groups, the next class of groups to discuss are the *finitary groups*. Let V be a vector space. An element $g \in \mathrm{GL}(V)$ is said to be finitary if $\dim(1 - g)V < \infty$. The collection of all such elements is a group, $\mathrm{FGL}(V)$. A finitary group is a subgroup of this group. Of course, every linear group is finitary. Further examples are fairly easy to construct. Let V have countable basis $v_1, v_2, \ldots$. For each i, let $V_i = \langle v_j \mid 1 \leqslant j \leqslant i \rangle$ and let G_i denote the subgroup of $\mathrm{GL}(V)$ inducing $\mathrm{SL}(V_i)$ on V_i and the identity on $\langle v_{i+1}, v_{i+2}, \ldots \rangle$. Then $G = \bigcup G_i$ is a finitary group, the *stable special linear group on* V. If the base field is locally finite, then G is a simple locally finite group. The special nature of these groups is reflected in the embedding of pairs of terms in the given Kegel sequence—the smaller element of any pair is embedded in the larger in as simple a manner as possible. Another source of examples is provided by alternating groups on an infinite set, using the permutation representation.

The articles of Phillips and Hall discuss these groups in detail. Phillips presents a general survey of finitary groups, not necessarily requiring the group to be locally finite. The article of Hall discusses his recent classification of the finitary simple locally finite groups. Such a group is shown to be similar to those described above: either alternating or a finitary classical group of specific type. The analysis provides a lovely example of how one can work with Kegel covers with the goal of a global classification.

The articles of Meierfrankenfeld and Belyaev each provide further direction for classification results in the theory of locally finite groups. Belyaev discusses the theory of inert subgroups of locally finite groups. These are subgroups for which the intersection of any two conjugates has finite index in each. Nontrivial examples are always present in countable locally finite simple groups which are not linear.

Meierfrankenfeld presents an important new result on the structure of an arbitrary locally finite simple group. Roughly speaking this result shows that a simple locally finite group fits into one of three categories. The first of these consists of the finitary groups. The groups in the next category are those of *alternating type*, which means that there exists a Kegel cover all of whose factors are alternating groups. The third category consists of groups where there is a Kegel cover $\{(G_i, N_i) \mid i \in I\}$ such that, for each i, G_i/N_i is a projective special linear group over a field of characteristic p and $G_i/O_p(G_i)$ is a product of perfect central extensions of classical groups over fields of characteristic p. Groups of the first type have now been classified and Meierfrankenfeld raises the question as to whether groups of the third type are amenable to classification.

The article of Zalesskiĭ concerns the ideal structure of group rings of simple locally finite groups. Connections are explored between various types of groups, their local systems, and the existence of proper ideals of group algebras other than the augmentation ideal. For example, if G is a linear, simple, locally finite group and F a field of characteristic 0, then FG has no proper ideals other than the augmentation ideal.

The study of groups of finite Morley rank lies on the interface between model theory and group theory. Here the outstanding open problem is the Cherlin-Zil'ber conjecture that a simple group of finite Morley rank is an algebraic group over an algebraically closed field. It was shown by S. Thomas that the conjecture holds if the group is also locally finite. The argument makes heavy use of the classification of finite simple groups. In his article, Borovik gives a general discussion of the theory of simple groups of finite Morley rank and indicates how arguments from the classification of finite simple groups, rather than the classification itself, might yield an approach to the conjecture.

Leinen's article gives a lengthy and detailed survey of existentially closed groups in certain classes. Emphasis is placed on classes of nilpotent groups, locally finite groups, and extensions.

The articles of Bryant, Isaacs, and Turull are each concerned with the representation theory of finite groups. Representation theory of finite groups has several connections with the theory of locally finite groups. For example, Mal'cev's theorem ties the representation theory of the finite groups to that of the whole group, and the Kegel kernels often yield representations of the simple factors.

Turull's article discusses length problems in the theory of finite solvable groups. In particular, connections are described between the nilpotence length of a finite solvable group and information about the group of fixed points of a subgroup of the automorphism group. The article includes discussions of fixed-point-free actions and the existence of regular orbits. Issues of this sort have connections with the difficult problem of analyzing centralizers in simple locally finite groups. For example, Hartley's survey article includes an application of Dade's famous result on Carter subgroups of finite solvable groups to the theory of centralizers in locally finite groups.

The article of Bryant concerns the representation theory of $GL_n(K)$, for K an infinite field. There is a natural action on the polynomial algebra $K[x_1, \ldots, x_n]$ in which the homogeneous polynomials of a given degree form an invariant subspace. When K has characteristic 0, these subspaces are each irreducible and afford the appropriate symmetric power of the natural representation. The situation is much more complicated in positive characteristic. Bryant presents an account of results of Doty and Krop which describe the submodule structure. The second part of the article gives a new proof of the result that $K[x_1, \ldots, x_n]$ is asymptotically close to being free when regarded as a module for a finite subgroup of $GL_n(K)$.

Isaacs discusses the modular representation theory of finite solvable groups; in particular, generalizations of the Brauer theory to sets of primes. For a set π of prime numbers, the π-*partial characters* of a finite group G are the restrictions of the ordinary characters of G to the π-elements of G. Isaacs describes an extensive theory of π-partial characters of solvable groups, which can be used to establish results about ordinary characters. Applications are given to the theory of monomial groups and groups of odd order.

The volume concludes with the article of Shalev, which presents a fascinating discussion of recent results in the theory of finite p-groups and their inverse limits, particularly those topics which are closely related to Lie theory. Special topics include a discussion of *powerful p-groups* (p-groups where all p-th powers lie in the derived group) and their connection with the theory of p-adic analytic groups, and also a discussion of Zelmanov's work on the restricted Burnside problem. The article includes a large collection of open problems.

On October 8, 1994, Brian Hartley died suddenly while hiking. The Istanbul conference was from the beginning Brian's idea and he spent an enormous amount of time planning the conference. His vision was a conference where mathematicians from different areas would come together and interact in a nontrivial way. The theory of finite and locally finite groups proved to be a perfect subject for this, and the conference was a huge success. Beyond this, Brian wanted to provide an environment for the education of graduate students—lectures should start at the elementary level and build to recent advances. The format of a NATO Advanced Study Institute was ideal for what he had in mind.

The tone of the conference is clear from the articles in this volume. Brian brought together a collection of participants many of whom rarely see each other. Yet everybody seemed to fit at the conference and attendance at talks rarely dropped off. The special nature of the conference was clear from the beginning and Brian's influence was always present.

This volume is dedicated to the memory of Brian Hartley.

Gary M. Seitz

SIMPLE LOCALLY FINITE GROUPS

B. HARTLEY*
Department of Mathematics
University of Manchester
Manchester M13 9PL
United Kingdom

Abstract. Beginning from basic principles, we outline the current state of affairs in the theory of locally finite simple groups. Particular emphasis is placed on constructions, Kegel sequences, and centralizers.

Key words: simple group, composition series, linear group, finitary group, universal group, Lie type, amalgam, direct limit, embeddings, regular type, diagonal type, absolute simplicity, free extension, Kegel cover, centralizer.

1. Constructions

The classification of finite simple groups (CFSG) was announced in the late 1970s (see for example [12, 13]). To some extent this classification is controversial, in that a proof in the normal sense is not really available. The ingredients of the proof are spread over some thousands of journal pages, and not all parts of the proof are completely without errors. However there seems to be a substantial consensus that the result is correct, and we shall accept it without further question in this article. Our interest is in simple locally finite groups, of which the finite simple groups are of course examples, but we shall mainly be interested in infinite groups of this type. It seems very difficult to make much progress with these without using CFSG, though we could instead assume that all finite non-abelian simple sections of the groups we meet are known. Technically speaking, weaker versions of CFSG also suffice: for example that all finite non-abelian simple groups of large enough order are known, or that there are no more infinite families in a suitable sense. For more on this, see [36].

Having accepted CFSG, it seems at first sight a good idea to try to classify all simple locally finite groups, or perhaps just the countable ones to start with. However this appear to be out of the question, at least in the conventional sense of parametrizing the isomorphism types in some reasonable way. For example, $2^{\aleph_0}$ pairwise non-isomorphic groups can be obtained as unions of towers of finite alternating groups [36]. Instead, it seems that we must look for well behaved families of simple locally finite groups that can be classsified, and try to increase the scope of these as much as possible. At the same time we may ask test questions that will help us to get more insight into what a general simple locally finite group is like. It

---

* At the time of Brian Hartley's tragic death this paper was not yet in its final form. The work on the paper was completed by Richard Phillips.

B. Hartley et al. (eds.), Finite and Locally Finite Groups, 1–44.
© 1995 *Kluwer Academic Publishers. Printed in the Netherlands.*

is desirable to look for new constructions, and test questions may suggest these and lead to new examples.

We will describe some more or less natural examples of simple locally finite groups below, and go on to give some constructions of other examples that may seem less natural but nevertheless exist. In some cases it does not seem easy to understand their properties.

Before doing so, let us say a few words about the rôle of these simple groups in the general theory of locally finite groups. This is not so clear-cut as in the finite case. Every finite group has composition series, usually several of them, but, by the Jordan-Hölder Theorem, they all lead to the same set of composition factors, up to isomorphism. This is the justification for saying that simple groups are the building blocks of finite group theory. Now there is a notion of composition series for a general group, but these series need by no means be of finite length, nor even run in any particular direction. A series of a group G is essentially a totally ordered set $\mathbf{S}$ of subgroups of G with the property that, given any non-identity element x of G, there is a unique largest member V_x of $\mathbf{S}$ among those not containing x, a unique smallest Λ_x among those that do contain x, and $V_x \triangleleft \Lambda_x$; the series is a *composition series* if, for every $1 \neq x$, the factor Λ_x/V_x has only the trivial series consisting of the identity and the group Λ_x/V_x (in other words, the quotient Λ_x/V_x is *absolutely simple* (see Definition 1.28)). For example, we might have

$$\ldots G_{-2} \triangleleft G_{-1} \triangleleft G_0 \triangleleft G_1 \triangleleft \ldots .$$

The terms of this series are indexed by the natural numbers, $\bigcap_{n \leqslant 0} G_n = 1$, and $\bigcup_{n \geqslant 0} G_n = G$. In general, arbitrary ordered sets must be used to index composition series. Since none of the above terms need be normal in G, there is no *a priori* reason why a simple group need not have such a series, and indeed we shall see that it happens. It is the case that every group has a composition series, and indeed that any series can be refined to a composition series, but there is no Jordan-Hölder Theorem in general, and different series can have quite different composition factors. For example we note the following.

Lemma 1.1 *Every locally finite group is a homomorphic image of a group that is both residually and locally finite.*

Proof. Let $\{ G_i \mid i \in I \}$ be the set of finite subgroups of G and define an order $\leqslant$ on I by $i \leqslant j$ if $G_i \leqslant G_j$. The cartesian product $Y = \prod\{ G_i \mid i \in I \}$ is a residually finite group. For each $g \in G$ let $g^* = (g_i)_{i \in I} \in Y$ be given by $g_i = 1$ if $g \notin G_i$ and $g_i = g$ if $g \in G_i$. It is not difficult to show that $B = \langle g^* \mid g \in G \rangle$ is a locally finite subgroup of Y; thus B is both locally finite and residually finite. Now, let E be the set of sequences $(g_i)_{i \in I}$ such that, for some k, $j \geqslant k$ implies that $g_j = 1$. This set E is a normal subgroup of Y and so $D = B \cap E$ is a normal subgroup of B. For the completion of the proof note that the mapping $g \mapsto g^* D$ is an isomorphism from G onto B/D. $\qquad \square$

Now take some countably infinite absolutely simple locally finite group S, for example $\mathrm{PSL}_2(K)$, where K is some infinite algebraic extension of a finite field (the absolute simplicity of $\mathrm{PSL}_2(K)$ follows from Theorem 3 of [2]). Write $S \cong G/H$, where G is locally and residually finite. The proof shows that we can take G to be countable. The series $G > H > 1$ can be refined to a composition series, and in fact the factor G/H cannot be further refined, so S is a composition factor in this composition series. On the other hand, since G is residually finite and countable, it has a series of normal subgroups

$$G = G_0 > G_1 > G_2 > \ldots$$

which have finite index in G and intersect in the identity. The factors of a composition series refining this are clearly all finite. This is quite a striking breakdown of the Jordan-Hölder Theorem. We shall meet this phenomenon again (see Section 1.6). For more about series and composition series of general groups, see [37, 47] though no more than what we have just said is needed for what follows.

Although a locally finite group does not in general have a well defined set of composition factors that can be obtained from a particular series, simple groups often do arise in questions about general locally finite groups. In any case, we will try to show that simple locally finite groups are intriguing objects in their own right.

First, a few more words about the finite simple groups. At the most superficial level, CFSG divides the finite non-abelian simple groups into three kinds:

(1) alternating groups Alt_n ($n \geqslant 5$);

(2) simple groups of Lie type;

(3) 26 sporadic groups.

The sporadic groups seem to have no part to play in the study of infinite simple locally finite groups. The alternating groups are usually among the first groups we meet in a beginning course on group theory, and superficially at least they are familiar.

The finite simple groups of Lie type can be obtained in two ways, from Lie algebras or from algebraic groups. In the first case we begin with a finite dimensional simple Lie algebra L over $\mathbb{C}$ (these are of course classified by Dynkin diagrams) and a finite field k. If the Dynkin diagram of L has a non-trivial symmetry and the field has a suitable automorphism, these may also come into the picture. In L a Chevalley basis is chosen. This is a certain basis with respect to which the multiplication constants are all integers, and so it gives us an integral form $L_{\mathbb{Z}}$ of L. Then $L_k = L_{\mathbb{Z}} \otimes_{\mathbb{Z}} k$ is a finite dimensional Lie algebra over k. The corresponding group is the group of automorphisms of L_k generated by the automorphisms $\exp(\mathrm{ad}(te_r))$, where t ranges over k, and r over the roots of L, and e_r is the element of the Chevalley basis corresponding to r. For a detailed account, see [9]. If the Dynkin diagram has a non-trivial symmetry we may be able to obtain 'twisted versions' which arise as groups of fixed points of automorphisms of the groups just described. In this approach, we are effectively using the adjoint representation of L. If we allow other finite dimensional L-modules and transfer to k via a suitable integral basis, we can obtain quasisimple groups also. For instance we obtain the group $\mathrm{PSL}_{n+1}(k)$ from the adjoint representation of the complex simple Lie algebra of type A_n, which is the Lie algebra of $(n+1) \times (n+1)$ complex matrices of trace zero, and the group $\mathrm{SL}_{n+1}(k)$ from the natural representation of this algebra. Steinberg's notes [49] are a mine of information about this subject.

The second point of view is to begin with a simple algebraic group $\overline{G}$ of adjoint type over the algebraic closure $\overline{\mathbb{F}}_p$ of the field $\mathbb{F}_p$ of p elements, and a Frobenius map σ on $\overline{G}$. A *Frobenius map* on $\overline{G}$ is by definition an algebraic endomorphism of $\overline{G}$ with finite fixed point group. Let $\overline{G}_\sigma$ be this fixed point group. Then $O^{p'}(\overline{G}_\sigma)$, the group generated by the p-elements of $\overline{G}_\sigma$, is a non-abelian simple group in all but a very small number of cases, and these are the simple groups of Lie type. Beginning with groups of non-adjoint type and proceeding similarly produces groups that are in most cases quasisimple. A simple example is to take $\overline{G} = \mathrm{SL}_n(\overline{\mathbb{F}}_p)$ and σ to be the map that raises each matrix entry to the q-th power, where q is a power of p. Then $\overline{G}_\sigma = \mathrm{SL}_n(q)$. The groups $\mathrm{PSL}_{n+1}(\overline{\mathbb{F}}_p)$ are the adjoint groups of type A_n, and give rise to the finite groups $\mathrm{PSL}_{n+1}(q)$, while the groups $\mathrm{SL}_{n+1}(\overline{\mathbb{F}}_p)$ are the universal or simply connected groups of type A_n and give the finite special linear groups as above. See [8] for more detail. This second point of view will be more important to us in this article.

1.1. FINITARY ALTERNATING GROUPS

The finitary alternating groups are probably the most obvious examples of infinite simple locally finite groups, and are natural analogues of the finite alternating groups. Let Ω be any set. The group $\mathrm{Sym}_0(\Omega)$ is the group of all permutations of Ω that move only a finite number of points. If α is such a permutation, then it has a well defined sign, and $\mathrm{Alt}(\Omega)$ is the group of even permutations of Ω. It is trivial to check that $\mathrm{Alt}(\Omega)$ is a locally finite group, which is certainly simple if Ω is infinite. In fact, if Δ is any finite subset of Ω, then $\mathrm{Alt}(\Delta)$ is naturally a subgroup of $\mathrm{Alt}(\Omega)$, and $\mathrm{Alt}(\Omega) = \bigcup_\Delta \mathrm{Alt}(\Delta)$, as Δ ranges over all finite subsets of Ω. Each finite set of elements of $\mathrm{Alt}(\Omega)$ lies in some subgroup $\mathrm{Alt}(\Delta)$. The subgroups $\mathrm{Alt}(\Delta)$ form a *local system of finite subgroups* of $\mathrm{Alt}(\Omega)$ (we give the definition immediately), and here we may restrict to those Δ such that $|\Delta|$ exceeds some given integer.

Definition 1.2   *A local system of subsets of a set S is a family* **F** *of subsets of S such that*

(1) *each element of S belongs to some member of* **F**, *and*

(2) *if $S_1, S_2 \in$* **F**, *then there exists $S_3 \in$* **F** *such that $S_1 \cup S_2 \subseteq S_3$.*

It is trivial to verify the following.

Lemma 1.3   *If a group G has a local system of simple subgroups, then G is simple.*

We shall not say much more about these alternating groups, but refer instead to the articles of Hall [15] and Phillips [44]. We mention only the famous theorem of Wielandt giving a characterization of the infinite alternating groups.

Theorem 1.4 (Wielandt) *Let Ω be an infinite set and let G be an infinite primitive subgroup of $\mathrm{Sym}_0(\Omega)$. Then G is either $\mathrm{Sym}_0(\Omega)$ or $\mathrm{Alt}(\Omega)$.*

For a proof, see [53].

1.2. LINEAR GROUPS

For us, a linear group is simply a subgroup of some $GL(V)$, where V is a finite dimensional vector space over a (commutative) field k. Of course, we shall allow ourselves to think of these as groups of matrices over k. Now the algebraic closure $\overline{\mathbb{F}}_p$ of $\mathbb{F}_p$ is a *locally finite field*, in the sense that every finite set of elements lies in a finite subfield, and of course locally finite fields are precisely the subfields of such algebraic closures. Since a finite number of matrices have only a finite number of entries between them, we see that $GL_n(\overline{\mathbb{F}}_p)$ is a locally finite group and has a local system consisting of groups $GL_n(\mathbb{F}_q)$, where q ranges over powers of p. Thus we see that linear groups over algebraic extensions of finite fields are locally finite groups. Among these are the (linear) algebraic groups over $\overline{\mathbb{F}}_p$, in particular the simple ones, as well as the finite groups of Lie type. (Recall that an algebraic group G is simple as algebraic group if G has no proper non-trivial *connected* normal subgroup. In that case it turns out that G has a finite centre Z and that G/Z is simple as a group. In particular, $SL_n(\overline{\mathbb{F}}_p)$ is a simple algebraic group.) Now we mentioned above that one way to obtain the finite groups of Lie type is from a representation of a simple Lie algebra over the complex field, by choosing a suitable integral basis and transferring the coefficients to a given finite field k. This construction works over any field k (and is explained in this context in [9] and [49]), subject only to the proviso that if we are constructing a twisted form involving a symmetry of the Dynkin diagram then k must have a suitable automorphism with which to combine the symmetry. For example, if we want to construct a Suzuki group $^2B_2$ over k, then k must have an automorphism α such that α^2 is the standard Frobenius map $\lambda \mapsto \lambda^2$ of k. This amounts to saying that k is a union of finite subfields whose orders are odd powers of 2 (we are only talking about locally finite fields here). Thus, if G denotes one of the symbols $A_l, B_l, C_l, D_l, E_6, E_7, E_8, F_4, G_2, {}^2A_l, {}^2D_l, {}^2B_2, {}^3D_4, {}^2E_6, {}^2F_4, {}^2G_2$, and k is any locally finite field, admitting a suitable automorphism if twisting is involved, then we have a group $G(k)$; implicitly we are thinking of starting with the adjoint representation of the Lie algebra. These groups are simple except in a very small number of cases when the rank and the field are very small, and give us a range of examples of infinite simple linear locally finite groups. Each of them is the union of a tower of finite subgroups of the same type over subfields of k. The groups in the infinite families can be identified with classical groups:

$$A_l(k) \cong PSL_{l+1}(k); \ B_l(k) \cong P\Omega_{2l+1}(k); \ C_l(k) \cong PSp_{2l}(k);$$

$$D_l(k) \cong P\Omega_{2l}^+(k); \ {}^2A_l(k) \cong PSU_{l+1}(k); \ {}^2D_l(k) \cong P\Omega_{2l}^-(k).$$

It was shown at more or less the same time by several authors [2, 7, 26, 50] that these are the only infinite simple linear locally finite groups.

Theorem 1.5 *Let G be an infinite simple linear locally finite group. Then G is isomorphic to a group $G(k)$, where G is one of the above symbols and k is an infinite locally finite field.*

For another proof, see [38], and, for a related result, see [19].

The construction of the above groups from algebraic groups is not so convenient as in the finite case.

Lemma 1.6 *Let G be one of the groups $G(k)$ described above, where k is an infinite locally finite field of characteristic p. Then there exists a simple algebraic group $\overline{G}$ of adjoint type over $\overline{\mathbb{F}}_p$, a Frobenius map σ on $\overline{G}$, and a strictly increasing sequence $n(1), n(2), \ldots$ of natural numbers, each dividing the next, such that*

$$G = \bigcup_{i=1}^{\infty} \overline{G}_{\sigma^{n(i)}}. \tag{1}$$

This is not difficult to prove; see [24], [27]. It seems a convenient way of putting these groups into the algebraic group context.

They also have a characterization as the perfect Zariski-dense subgroups of simple algebraic groups of adjoint type over $\overline{\mathbb{F}}_p$ [27].

1.3. FINITARY LINEAR GROUPS

Let V be a vector space of finite or infinite dimension over a field k. An invertible linear map $g : V \rightarrow V$ is called *finitary* if $\dim(g-1)V < \infty$; equivalently, if g acts as the identity on a subspace of V of finite codimension. The set of all these finitary linear maps on V is a subgroup $\mathrm{FGL}(V)$ of $\mathrm{GL}(V)$ called the *finitary general linear group* on V, and a group G is called a *finitary linear group* if G is isomorphic to a subgroup of some $\mathrm{FGL}(V)$. Among these are the groups so far mentioned; linear groups are clearly finitary linear, while, if k is any field, then $\mathrm{Alt}(\Omega)$ acts as a finitary linear group on the permutation module $k\Omega$ having Ω as a basis. As another example, let V be a vector space over k having a countably infinite basis $v_1, v_2, \ldots$. Let $V_n = \sum_{i=1}^{n} kv_i$. Let G_n be the group of all linear transformations that leave V_n invariant, induce an element of $\mathrm{SL}(V_n)$ on it, and fix the basis vectors not in V_n. Let $G = \bigcup_{i=1}^{\infty} G_i$. This is the *stable special linear group* on V (with respect to the given basis). It can also be viewed from a matrix point of view as the direct limit of groups $\mathrm{SL}_n(k)$, where $\mathrm{SL}_n(k)$ is embedded in $\mathrm{SL}_{n+1}(k)$ via

$$g \mapsto \begin{pmatrix} g & 0 \\ 0 & 1 \end{pmatrix}.$$

If k is an infinite locally finite field, then G is an infinite simple locally finite group. If k is finite of order q, and $(n, q-1) = 1$, then $\mathrm{SL}_n(q)$ is simple, and so we see that, like the groups we have considered so far, G above can be written as the union of a tower of finite simple subgroups. However we can carry out the same construction with other classical groups to obtain stable symplectic, unitary and orthogonal groups.

Proposition 1.7 (Serezhkin and Zalesskiĭ [56]) *If k is a finite field of odd order, then the stable symplectic group is an infinite simple locally finite group that cannot be written as the union of a tower of finite simple groups.*

The question of writing simple locally finite groups as unions of towers of finite subgroups will occupy us quite a bit in this article; the above was the first example to show that it is not always possible. It works because if we had

$$\mathrm{Sp}_{2n}(k) \leqslant H \leqslant \mathrm{Sp}_{2m}(k),$$

where H is simple and k is of odd order, then H would clearly be generated by transvections (it is generated by the conjugates of any transvection subgroup from $\mathrm{Sp}_{2n}(k)$). Because of their work on groups generated by transvections, this turns out to be impossible. Bahturin has recently given another rather straightforward example of a simple locally finite group that is not the union of finite simple subgroups.

Finitary groups have come to play an important part in the general theory and are the subject of the articles of Hall and Phillips, and so we say no more about them here.

1.4. UNIVERSAL GROUPS

Hall's universal group was introduced by Philip Hall in 1959 [17]. It is a countably infinite locally finite group U with the following two properties which, as it turns out, characterize it up to isomorphism.

(U1) U contains a copy of every finite group.

(U2) Any two isomorphic finite subgroups of U are conjugate in U.

U can also be characterized by either of the following properties.

EXISTENTIAL CLOSURE. U is countable and existentially closed in the class of locally finite groups.

INJECTIVITY. U is countable locally finite, and if H and K are finite groups with $H \leqslant K$, and $\phi : H \to U$ is an injection, then ϕ can be extended to an injection $K \to U$.

The injectivity property should be viewed as saying that if F is a finite subgroup of U and F can be embedded in another finite group E, paying no attention to the rest of U, then that embedding can be realized inside U. Thus, since every finite group can be embedded in an alternating group, a special linear group, and so on, it follows that every finite subgroup of U is contained in an alternating group, a special linear group, and so on. In particular, U is the union of a tower of finite simple subgroups and is simple.

We shall not enlarge on the subject of existential closure, which is the subject of Leinen's article.

The group U was constructed by Hall as the union of a tower of finite symmetric groups

$$G_1 \leqslant G_2 \leqslant \dots .$$

Given G_n, we take G_{n+1} to be the symmetric group on G_n, and embed G_n in G_{n+1} by the regular representation. The same group results if we use alternating groups instead.

Some other properties of U are as follows.

Proposition 1.8

(1) U *is simple.*

(2) U *contains a copy of every countable locally finite group.*

(3)   (i) *If A is a finite abelian subgroup of U and $C = C_U(A)$, then C/A is an infinite simple group.*

   (ii) *If A is a finite subgroup of U with trivial center, then $C_U(A)$ is isomorphic to U.*

   (iii) *If A is a subgroup of U of prime order and M is a subgroup of U with $C_U(A) \leqslant M < U$ then $M \leqslant N_U(A)$; thus, $N_U(A)$ is a maximal subgroup of U.*

(4) *For every prime p, U contains a copy of every countably infinite locally finite p-group as a maximal p-subgroup.*

From the various parts of the Proposition, parts (1) and (2) were proved by P. Hall in [17]. Part (i) of (3) can be proved using the remarks prior to Theorem 1.12 ($C_U(A)$ is isomorphic to U_A) while (ii) can be proved using the defining conditions (U1) and (U2). Part (iii) has been communicated by U. Meierfrankenfeld and R. E. Phillips. It seems likely that a lot more can be said about maximal subgroups lying above centralizers—see [31] and the subsequent [30] for corresponding results in algebraically closed groups.

Property (4) is a particularly striking illustration of the breakdown of Sylow's Theorem in locally finite groups. It is a special case of a result of Hickin [29, Theorem 4(c)]. We give the proof of this result as it illustrates some useful techniques.

It requires the notion of an *amalgam* of groups: an amalgam $A = (H \cup K)_L$ is the set-theoretic union of two groups H and K that meet in a common subgroup L. It could be defined in a more or less equivalent way as a pair of groups H and K with given monomorphisms of a third group L into H and K. The previous description is more suggestive. An *embedding* of an amalgam A is a group G that contains the groups H and K as subgroups in such a way that they intersect in L precisely. Of course, any amalgam has an embedding, namely in the amalgamated free product $H *_L K$. But it is a well known result of B. H. Neumann, proved by means of permutational products, that every amalgam of finite groups can be embedded in a finite group [41]. The following lemma is a very special case of results in [29].

Lemma 1.9   *Let $A = (H \cup K)_L$ be an amalgam of finite groups and let q be a given prime. Then there exists an embedding of A in a finite group E such that, if $x \in H \smallsetminus L$ and $y \in K \smallsetminus L$, then q divides the order of xy.*

Proof. By the result of B. H. Neumann mentioned, we may embed A in a finite group $G = \langle H, K \rangle$. Write $G = F/N$, where $F = H *_L K$ and N is a normal subgroup of finite index in F. Thus, N is a free group. It is well known that if $x \in H \smallsetminus L$ and $y \in K \smallsetminus L$ then xy has infinite order in $H *_L K$. Thus, for each such pair (x, y), we can choose a non-trivial power $w(x, y)$ in N. Since the set of such pairs is finite and N is residually a finite q-group, we can choose a characteristic subgroup N_1 of finite q-power index, not containing any of the elements $w(x, y)$. The group F/N_1 can then be taken as E. $\square$

Proof of Proposition 1.8(4). Let P be a given countably infinite locally finite p-group and write $P = \bigcup_{i=1}^{\infty} P_i$, where the P_i form a strictly increasing tower of finite subgroups of P. Let $g_1, g_2, \ldots$ be an enumeration of the set of non-trivial p-elements of U, which is countable. The idea is to pick out a subsequence $Q_1, Q_2, \ldots$ of the P_i and construct a sequence $\psi_1, \psi_2, \ldots$ such that ψ_i is an embedding of Q_i in U and the restriction of ψ_{i+1} to Q_i is ψ_i. These will fit together to give an embedding ψ of P in U. The ψ_i will also be required to satisfy the condition that either $g_i \in \psi_i(Q_i)$ or $\langle \psi_i(Q_i), g_i \rangle$ is not a p-group. It is virtually obvious that this implies that $\psi(P)$ is a maximal p-subgroup of U.

To see that the sequence of embeddings $\psi_1, \psi_2, \ldots$ can be constructed, suppose that Q_i and ψ_i have been constructed up to stage n. The method to be given also begins the construction. Choose m such that $|P_m| > |\langle \psi_n(Q_n), g_{n+1} \rangle|$, and put $Q_{n+1} = P_m$. Let θ be some extension of ψ_n to an embedding $Q_{n+1} \to U$. If g_{n+1} belongs to $\theta(Q_{n+1})$, then put $\psi_{n+1} = \theta$. If not, let $H = \theta(Q_{n+1})$, and let $K = \langle \psi_n(Q_n), g_{n+1} \rangle$. We have an amalgam of H and K over their intersection L, which is a proper subgroup of both of them. By Lemma 1.9, we can embed this amalgam in a finite group E in such a way that the group generated by H and K in E is not a p-group. Set-theoretically this may appear a little confusing; H and K are subsets of U but E is not. Nevertheless the identity map on K may be extended to an embedding λ of E in U. Then $\lambda \circ \theta$ can be taken as ψ_{n+1}. □

Universal locally finite groups in general are defined by the properties (U1) and (U2); if $\aleph$ is an uncountable cardinal, then there are $2^{\aleph}$ universal locally finite groups of cardinal $\aleph$ [31, 29]. For more on universal groups, see [36, Chapter 6].

An important generalization of universal groups was given by Hickin [28]; these are universal with respect to the property of having a given periodic abelian group in the centre. Before introducing these let us recall some basic facts about the Schur multiplier of a group. We shall confine ourselves to perfect groups. Thus, let D be any perfect group, finite or infinite. A *stem extension* of D is an extension

$$1 \to Z \to E \to D \to 1 \tag{2}$$

where $Z \leqslant Z(E)$ and E is perfect. Such a stem extension will be called *maximal* if it maps onto any other one by a homomorphism which is unique subject to commuting with the identity map on D. An easy universal argument shows that there can be at most one maximal stem extension of a given perfect group D. A maximal stem extension can be obtained as follows. Let F be a free group for which we have an exact sequence

$$1 \to R \to F \to D \to 1.$$

Then it is not hard to check that

$$1 \to R \cap F'/[F, R] \to F'/[F, R] \to D \to 1 \tag{3}$$

is a maximal stem extension of D. If (2) is a maximal stem extension of D, then E is called the *Schur covering group* of D and Z is called the *Schur multiplier* $M(D)$ of D. We shall only apply these terms to perfect groups in this article. For more details, see [6].

Now the Schur multipliers of all the finite simple groups have been computed and can be found in [14]. By inspection, they all have rank at most 2. On the other hand, Hickin's groups provide examples to show that

Proposition 1.10 *Every countable periodic abelian group A is the Schur multiplier of some countable simple locally finite group.*

This was essentially pointed out by Phillips [42]. Now, since there are $2^{\aleph_0}$ choices for A (for there are $2^{\aleph_0}$ sets of primes, and to each such set S we can associate the direct product of cyclic groups, one for each prime in S), this gives us $2^{\aleph_0}$ non-isomorphic countable simple locally finite groups. Like the stable symplectic group, many of these cannot be written as the union of a tower of finite simple subgroups. More strikingly, we have the following.

Proposition 1.11 *Let A be any countable periodic abelian group of infinite rank (for example, an infinite elementary abelian group), let H be a countable simple locally finite group whose Schur multiplier is isomorphic to A, and let n be a given natural number. Then H cannot be written as the union of a tower of finite groups, each of which is the direct product of at most n simple groups.*

This is because, if such a tower existed, then one of its terms would have to contain the direct product of $2n + 1$ cyclic groups of the same prime order in its Schur multiplier.

Let us now briefly describe Hickin's ideas. We confine ourselves to the countable case for simplicity, though much of what we say is more general. The interested reader is urged to consult Hickin's paper, which contains much more.

Let A be a given countable periodic abelian group. We say that a locally finite group T is a *universal locally finite central extension* of A if

A is contained in the center $Z(T)$ of T, and

A-*injectivity*: whenever $A \leqslant K \leqslant H$ are groups such that $A \leqslant Z(H)$ and $|H : A| < \infty$, and ϕ is a monomorphism of K into T that is the identity on A, then ϕ extends to a monomorphism of H into T.

It follows easily from the injectivity of U that if A is a finite abelian subgroup of U then $C_U(A)$ is a universal locally finite central extension of A. However, Hickin showed that any periodic abelian group A has universal locally finite central extensions, and established a number of remarkable properties of these groups. For simplicity we restrict again to the countable case; if A is a countable periodic abelian group, it turns out that A has a unique (up to isomorphism) countable universal locally finite central extension, which we will denote by U_A. Thus, Hall's group U above is U_1. The group U_A/A will be denoted by H_A.

Theorem 1.12

(1) U_A *is perfect and H_A is simple.*

(2) A *is the Schur multiplier of H_A.*

(3) U_A can be written as the union of a tower of finite subgroups, each of which is the direct product of finite special linear groups (resp. special unitary groups), and H_A can be written as the union of a tower of finite PSLs (resp. PSUs).

(4) If E is any countable locally finite group having A in its centre, then the identity map on A can be extended to an embedding of E in U_A.

Of these, (1) and (4) are due to Hickin and (2) and (3) to Phillips, who also showed that U_A can be constructed as the existentially closed object in a certain inductive class of locally finite groups. If A has infinite rank, then it follows from Proposition 1.11 that, in the towers referred to in (3), the numbers of factors will necessarily be unbounded.

(2) follows in a more or less straightforward way from (3). For U_A is the union of a tower of finite groups each of which has trivial Schur multiplier (the special linear groups can be chosen large enough to avoid any troubles with exceptional multipliers), and so the Schur multiplier of U_A is trivial.

1.5. DIRECT LIMITS OF FINITE SIMPLE GROUPS

Of course, if a group G is the union of a tower of simple subgroups, then G is simple, and, more generally, if G has a local system of simple subgroups, then G is simple. We shall, however, continue to concentrate on countable groups here. Thus, one can construct a countable simple group G simply by taking the union of an arbitrary tower of finite simple groups

$$G_1 \leqslant G_2 \leqslant \ldots, \tag{4}$$

where, having obtained G_n, we choose an embedding of G_n into a finite simple group G_{n+1} and view G_n as a subgroup of G_{n+1} via the chosen embedding. This can be made more formal by using the terminology of direct limits, but it does not really seem necessary in this simple situation. Since we have complete freedom in the choice of embeddings, it is hardly surprising that a very wide variety of groups results from such constructions. Certainly it is easy to obtain $2^{\aleph_0}$ non-isomorphic groups, as we shall soon see. Further there is a serious problem with the isomorphism question; clearly a given group can be obtained in many different ways. We can certainly obtain the same group from any infinite subsequence of the G_i above, and it is easy to see from its injectivity property that Hall's group U can not only be expressed as the union of a tower of alternating groups but also as the union of a tower of subgroups all chosen from any one of the families of simple classical groups, in which we may choose to keep the characteristic fixed, or to vary over different characteristics, as long as we allow the rank parameter to increase without limit. These phenomena make it difficult to imagine how any parametrization of countable simple locally finite groups up to isomorphism could be formulated.

Let us specialize now to the case of unions of finite alternating groups to get some idea of what is going on. Thus, in the tower (4), suppose that $G_i \cong \mathrm{Alt}_{n_i}$. Let $\Omega_i = \{1, 2, \ldots, i\}$. Fixing an isomorphism $G_i \to \mathrm{Alt}_{n_i}$ gives us an action of G_i on Ω_{n_i}, and choosing an embedding of G_i in G_{i+1} amounts to specifying an action of G_i on $\Omega_{n_{i+1}}$. These actions can be specified in any way we like and a simple group will result. Although the situation may seem completely anarchic, we shall see that some order can be perceived. First the following definition seems useful.

Definition 1.13 *Let H be a permutation group on a set Ω, and let k be an integer. Let Δ be the union of k disjoint sets bijective with Ω, and let H act on Δ by acting on each of these subsets as it acts on Ω. This action of H on Δ is uniquely determined up to permutational isomorphism and gives us an embedding of H in $\mathrm{Sym}(\Delta)$ which is unique up to conjugacy in $\mathrm{Sym}(\Delta)$. This embedding will be called* multiplication by k.

Lemma 1.14 *Let p be a prime, let H be a permutation group on a finite set Ω of size n, and multiply H by p as above to obtain an action of H on a set Δ of size np. Then:*

(1) *If x is any element in H, then x has a p-th root in $\mathrm{Sym}(\Delta)$.*

(2) *Let x be an element of H, let r be any natural number, and let $n_\Omega(r)$ and $n_\Delta(r)$ be the numbers of r-cycles of x on Ω and Δ respectively. Then $n_\Delta(r) = pn_\Omega(r)$, and so, if q is a prime different from p, then the powers of q dividing $n_\Omega(r)$ and $n_\Delta(r)$ are the same.*

To prove (1), we note that the number of cycles of any given length of x on Δ is divisible by p. We split Δ into a number of disjoint subsets Λ, on each of which x acts as the product of p cycles of the same length, and note that the p-th power of a transitive cycle on Λ is the product of p disjoint cycles of the same length. (2) is obvious. □

We use this simple idea to construct $2^{\aleph_0}$ non-isomorphic direct limits (4) of alternating groups, as in [36, 6.10]. These will correspond bijectively to non-empty sets of primes. Thus, let π be any non-empty set of primes. Let $p_1, p_2, \ldots$ be a sequence of primes containing each prime in π infinitely often, and no other primes. Let G_1 be some alternating group in its natural representation, and let G_i be embedded in G_{i+1} by multiplication by p_i. Let $G(\pi)$ be the resulting group. The notation is not intended to imply that it depends only on π.

Lemma 1.15

(1) *If $p \in \pi$, then each p-element of $G(\pi)$ is contained in a subgroup of type C_{p^∞} of $G(\pi)$.*

(2) *If $p \notin \pi$, then $G(\pi)$ contains no C_{p^∞}-subgroup.*

Proof. (1) is clear, since whenever $p = p_i$ each p-element of G_i acquires a p-th root in G_{i+1}. On the other hand, suppose that C is a subgroup of type C_{q^∞} of $G(\pi)$, where q is some prime. Let x be an element of order q in C, and suppose that $x \in G_i$. Let n be the number of q-cycles of x on Ω_{n_i}, and q^r the largest power of q that divides n. Let y be an element of C such that $y^{q^{r+1}} = x$, and suppose that $y \in G_m$, where $m \geqslant i$. Then y acts on Ω_{n_m} as the product of a number of disjoint cycles whose lengths are powers of q up to and including q^{r+2}, and thus x acts on Ω_{n_m} as the product of a number k of disjoint q-cycles, where q^{r+1} divides k. It follows from Lemma 1.14 that one at least of $p_i, \ldots, p_{m-1}$ must be q and so $q \in \pi$. □

Corollary 1.16 *There are $2^{\aleph_0}$ pairwise non-isomorphic groups that are unions of towers of finite alternating groups.*

For future reference, we note the following property of $G(\mathbb{P})$, where $\mathbb{P}$ is the set of all primes. I am grateful to A. E. Zalesskiĭ for impelling me towards this result, which is due to E. B. Rabinovich [46].

Proposition 1.17 (E. B. Rabinovich) $G(\mathbb{P})$ *contains Hall's universal group U.*

Proof. We view Hall's group as the union $\bigcup_{i=1}^{\infty} H_i$, where

$$H_1 < H_2 < \dots . \tag{5}$$

Here $H_i \cong \mathrm{Sym}_{m_i}$, which acts on Ω_{m_i}, and H_i acts regularly on $\Omega_{m_{i+1}}$. Thus, $m_{i+1} = m_i!$. Now $G(\mathbb{P})$ is the union of a tower (4). Suppose that, for some i, we have a monomorphism ϕ_i of H_i into some G_j, such that $\phi_i(H_i)$ acts semiregularly on Ω_{n_j}. We will show that ϕ_i can be extended to a monomorphism ϕ_{i+1} of H_{i+1} into some G_k with a similar property. Let $|H_{i+1} : H_i| = r$. Since each prime occurs infinitely often among the primes p_i, we can choose k such that $2r$ divides $p_j p_{j+1} \cdots p_{k-1}$. As a G_j-set, the set Ω_{n_k} is just a union of a number of copies of Ω_{n_j}, say s such copies, where s is divisible by $2r$, and hence, under $\phi_i(H_i)$, Ω_{n_k} splits up into a number of regular orbits, say t, where $2r$ divides t. We group these orbits into sets of size r. Now the regular H_{i+1}-set is also a union of r regular H_i-orbits. Consequently the action of H_i on each union Δ of r regular orbits can be extended to a regular H_{i+1}-action. We allow H_{i+1} to act 'diagonally' on the union of these sets Δ, which is Ω_{n_k}, and so we obtain a monomorphism of H_{i+1} into $\mathrm{Sym}(\Omega_{n_k})$ extending ϕ_i. The image is in $\mathrm{Alt}(\Omega_{n_k})$ since there is an even number of regular H_{i+1}-orbits. □

However, $G(\mathbb{P})$ is not isomorphic to U. For the above argument shows that, for each finite group F, the group $G(\mathbb{P})$ contains a copy of F that acts semiregularly on each sufficiently large Ω_{n_i}; however, each G_i acts naturally on Ω_{n_i} and so each subsequent Ω_{n_j} will break up into a number of natural G_i-orbits. Thus, $G(\mathbb{P})$ contains more than one conjugacy class of subgroups isomorphic to G_i, unlike U.

Embeddings of Diagonal and Regular Type

Definition 1.18 *We refer again to a union (4) of finite alternating groups, and use the notation set up above. The tower (4) is of diagonal type if, for each i, each orbit of G_i on $\Omega_{n_{i+1}}$ is either trivial or natural. (By the natural representation of Alt_n we mean the transitive representation of degree n, which is unique up to isomorphism.) We say that (4) is of regular type if each G_i has at least one regular orbit on $\Omega_{n_{i+1}}$.*

Prototypical situations giving rise to towers of diagonal embeddings are, firstly, the alternating group on a countable set Ω, and, secondly, the groups $G(\pi)$ above, and others like them, obtained by repeated multiplication. Hall's group U gives rise to a tower of regular type. Groups of diagonal type seem to appear first in a paper of Zalesskiĭ [54]. Remarkably, it is essentially true that every tower is of one of these two types. The following is a special case of a substantially more general result proved independently by Praeger and Zalesskiĭ [45, Theorem 1.7].

Theorem 1.19 *Let (4) be a tower of alternating groups. Then either some infinite subtower is of regular type or there exists m such that the tower*

$$G_m \leqslant G_{m+1} \leqslant \cdots$$

is of diagonal type.

Actually the result proved is a little sharper; if the second condition does not hold, then, given r, there exists M such that if $k \geqslant M$ then G_r has a regular orbit on Ω_{n_k}.

Now we see that if G is the union of a tower of finite alternating groups then we may take the tower to be either of regular type or of diagonal type. We will say that G is of regular type or of diagonal type, as the case may be. It is not perhaps immediately clear that a group cannot simultaneously be of both types. But actually this follows from results of Zalesskiĭ on the complex group rings of groups of these types; if G is the union of a tower of regular type then $\mathbb{C}G$ has only three ideals, while if G is the union of a tower of diagonal type then $\mathbb{C}G$ has an infinite number of ideals [54]; see also Zalesskiĭ's article.

Problem 1.20 *Find a group theoretical proof that a group cannot be simultaneously of diagonal and of regular type[1].*

The following was pointed out to me by Ulrich Meierfrankenfeld.

Proposition 1.21 *If G is of regular type, then G involves Hall's group U.*

For a proof of a more general result, see [22]. However, we have seen (Proposition 1.17) that groups of diagonal type can also involve U. In fact most of them do; see Proposition 1.22.

Regarding groups of diagonal type, let t_{i+1} be the number of non-trivial (that is, natural) orbits of G_i on $\Omega_{n_{i+1}}$. Recall that $G_i \cong \mathrm{Alt}_{n_i}$ and Ω_{n_i} is the set on which it naturally acts. There are two possibilities.

Case 1. There exists n such that $t_{i+1} = 1$ for all $i \geqslant n$. In this case we can simply delete the first n terms of the tower and see that $G \cong \mathrm{Alt}(\mathbb{N})$.

Case 2. $t_i \neq 1$ for infinitely many values of i. In this case we may choose a new tower for which $t_i \geqslant 2$ for all i. For each sequence

$$\mathbf{s} = (s_1, s_2, \ldots) \tag{6}$$

---

[1] Editorial Note: such a proof has been given by Meierfrankenfeld.

of integers $s_i \geqslant 2$, let $G(\mathbf{s})$ denote the union of a tower (4), where G_{i+1} is obtained from G_i by multiplying by s_{i+1}, and $G_1 = \text{Alt}_{s_1}$. Another result communicated to me orally by Meierfrankenfeld is the following, which can also be found in more general form in [22].

Proposition 1.22   *Let G be the union of a tower (4) of alternating groups, in which $t_i \geqslant 2$ for infinitely many i. Then G is not isomorphic to $\text{Alt}(\mathbb{N})$, and G involves $G(\mathbf{s})$ for all sequences $\mathbf{s}$ as described above, and hence also involves U.*

Since it is well known that the only infinite simple group involved in $\text{Alt}(\mathbb{N})$ is $\text{Alt}(\mathbb{N})$ itself (an argument can be easily assembled using results discussed in Section 4 of [44]), it follows that $\text{Alt}(\mathbb{N})$ is not isomorphic to any of the groups (4) in which $t_i \geqslant 2$ for infinitely many i. This can be seen in a number of other ways, for example through the complex group rings; it is not clear which is the most direct.

Jilinskiĭ [32, 33] has determined the conditions for two groups $G(\mathbf{s}_1)$ and $G(\mathbf{s}_2)$ to be isomorphic.

A Permutation Representation for a Diagonal Limit of Alternating Groups

Let G be a union of a tower (4) of finite alternating groups of diagonal type, and use the notation set up above. Then we shall see how to form a direct limit of the sets Ω_{n_i} to obtain a set Ω on which G acts. This is just a special case of the general direct limit construction, which we now explain in the appropriate context.

Let Λ be a directed set. This means that Λ is a set with a partial ordering such that if $\lambda, \mu \in \Lambda$ then there exists $\nu \in \Lambda$ such that $\lambda \leqslant \nu$ and $\mu \leqslant \nu$. A simple example is the set of natural numbers, which is the only one that will concern us here. Consider a direct system of permutation groups over Λ. This means that, for each $\lambda \in \Lambda$, we have a permutation group $(G_\lambda, \Delta_\lambda)$, consisting of a group G_λ and a G_λ-set Δ_λ. We also have, for each λ and μ in Λ with $\lambda \leqslant \mu$, a homomorphism of permutation groups $(G_\lambda, \Delta_\lambda) \to (G_\mu, \Delta_\mu)$. This consists of a group homomorphism $\alpha_{\lambda,\mu} : G_\lambda \to G_\mu$ and a map $\beta_{\lambda,\mu} : \Delta_\lambda \to \Delta_\mu$, such that

$$\beta_{\lambda,\mu}(g\delta) = \alpha_{\lambda,\mu}(g)\beta_{\lambda,\mu}(\delta)$$

whenever $\lambda \leqslant \mu$, $g \in G_\lambda$, and $\delta \in \Delta_\lambda$. The maps satisfy the usual consistency conditions but need not be injective here. Then $(G_\lambda, \alpha_{\lambda,\mu})$ is a direct system of groups over Λ and has a direct limit

$$G = \lim_{\longrightarrow} G_\lambda,$$

and $(\Delta_\lambda, \beta_{\lambda,\mu})$ is a direct system of sets over Λ and has a direct limit

$$\Delta = \lim_{\longrightarrow} \Delta_\lambda.$$

The group G acts naturally on the set Δ and we obtain a permutation group

$$(G, \Delta) = \lim_{\longrightarrow}(G_\lambda, \Delta_\lambda),$$

the direct limit of the system of permutation groups. For generalities on direct limits, see [10].

In our situation, suppose we have a tower (4) of finite alternating groups where $G_i \cong \mathrm{Alt}_{n_i}$, and as usual Ω_{n_i} is the natural G_i-set. Then G_i acts on $\Omega_{n_{i+1}}$ via its embedding in G_{i+1}; suppose that G_i has at least one natural orbit on $\Omega_{n_{i+1}}$. Then we have a G_i-map

$$\beta_{i,i+1} : \Omega_{n_i} \to \Omega_{n_{i+1}}.$$

Composing these suitably, we obtain, for each pair of natural numbers i and j with $i \leqslant j$, a G_i-map $\Omega_{n_i} \to \Omega_{n_j}$. Together with the inclusion map $G_i \to G_j$, this gives us a homomorphism of permutation groups $(G_i, \Omega_{n_i}) \to (G_j, \Omega_{n_j})$. Performing the above construction has the effect of reconstituting the group G and providing a set Ω on which G acts; we can regard it as a union of the sets Ω_{n_i}, where

$$\Omega_{n_1} \leqslant \Omega_{n_2} \leqslant \cdots,$$

and G_i acts on Ω_{n_i} as it did initially. Since G_i is transitive on Ω_{n_i}, for it is after all just the natural G_i-set, the group G will be transitive on Ω.

Suppose further that the tower considered is of diagonal type. Then, for each $i \leqslant j$, Ω_{n_j} is a union of a number of natural and trivial G_i-orbits. We obtain the following.

Lemma 1.23 *Suppose the tower* (4) *is of diagonal type. Then G acts transitively on a set Ω with the property that, for each i, Ω is a union of natural and trivial G_i-orbits.*

Corollary 1.24 *No G_i of order greater than 3 has a regular orbit on Ω.*

Permutation representations of this type and the corresponding permutation modules appear to play an important part in the ideal structure of group rings of simple locally finite groups over fields of characteristic zero (see Zalesskiĭ's article [55]).

Definition 1.25 *Let H be a subgroup of a locally finite group G. Then H is said to be* confined *in G if there is a finite subgroup F of G that has no regular orbit on the coset space G/H.*

The statement that F has no regular orbit on G/H is trivially equivalent to the statement that every conjugate of H has non-trivial intersection with F; the conjugates of H are 'confined' near F. The article [55] contains many interesting questions concerning the properties of confined subgroups in locally finite groups.

1.6. MEIERFRANKENFELD'S GROUPS

It was noted above that there is nothing in principle to prevent an infinite simple group from having a non-trivial series. The first examples to exhibit such a phenomenon in practice were constructed by Philip Hall in 1963 [18]. Meierfrankenfeld's groups were constructed in 1991 [39] to show that this phenomenon can also happen in locally finite groups, and they are quite a bit less complicated than Hall's groups. Hall's construction was rather intricate, involving repeated wreath powers. The groups obtained possess non-trivial *ascending* series; an ascending series is a set of subgroups

$$\{ G_\alpha \mid \alpha \leqslant \lambda \}, \tag{7}$$

indexed by ordinals α up to and including some ordinal λ, such that

$$G_0 = 1,$$

$$G_\alpha \lhd G_{\alpha+1} \text{ for } \alpha < \lambda,$$

$$G_\mu = \bigcup_{\alpha<\mu} G_\alpha \text{ for limit ordinals } \mu \leqslant \lambda, \text{ and}$$

$$G_\lambda = G.$$

Each such series has a first and last term and is increasing. The simplest example is of the form

$$1 = G_0 \lhd G_1 \lhd G_2 \lhd \dots$$

indexed by the natural numbers, with $\bigcup G_i = G$. The last term $G_\omega = G$ is often suppressed.

The following provides a little useful background to what follows.

Lemma 1.26 *Let G be an infinite locally finite simple group having an ascending series (7) with $G_1 \neq 1$. Then G_1 is neither locally soluble nor finite.*

Proof. If G_1 is locally soluble and L is the set of finite subgroups of G, then $G_1 \cap F$ is a subnormal soluble subgroup of F for every $F \in L$. From this it follows that $G = \langle (G_1)^G \rangle$ is locally soluble. However, a locally soluble locally finite simple group must be cyclic of prime order ([36, I.B.4] or Lemma 2.4.3 below).

Suppose now that G_1 is finite; replacing G_1 by a minimal subnormal subgroup of itself, we may assume that G_1 is either cyclic of prime order or a non-abelian simple group. In view of the locally soluble case already proved, we may assume that G_1 is non-abelian simple. Note that, if $g \in G$, G_1 and G_1^g are non-abelian simple subnormal subgroups of $\langle G_1, G_1^g \rangle$ and so they commute elementwise unless they are equal. Thus, $G = \langle (G_1)^G \rangle$ is generated by pairwise commuting simple subgroups, and must be the direct product of them. Hence G is finite, contrary to assumption. $\qquad\square$

Question. If G is locally finite, does G_1 necessarily involve an infinite simple group?

As well as ascending series, one can also consider descending series. We will not give the general definition. The simplest example has the form

$$G = G_0 \rhd G_1 \rhd G_2 \rhd \dots,$$

indexed by the natural numbers, with $\bigcap G_i = 1$. If G is simple then it is clear that such a series must contain exactly two terms. Thus, descending series are of no interest to us here. One can also consider series that are neither ascending nor descending, such as

$$\ldots G_{-2} \triangleleft G_{-1} \triangleleft G_0 \triangleleft G_1 \triangleleft G_2 \ldots, \tag{8}$$

indexed by the integers, with $\bigcap G_i = 1$ and $\bigcup G_i = G$. Much more complicated things are possible. The possibility that a simple locally finite group might have a series of some type with finite factors was mentioned by Kargapolov [34]. Meier-frankenfeld's groups turn out to have such a series of the form (8).

For brevity, let us refer to Meierfrankenfeld's groups as Mf-groups. Some of the properties of these strange groups are summarized as follows.

Theorem 1.27   *Let G be any Mf-group. Then:*

(1) *G is a countably infinite simple locally finite group.*

(2) *G has an ascending series (7) with $1 \neq G_1 \neq G$.*

(3) *G has a series (8) with finite factors.*

(4) *Let d be any positive integer. If G is expressed in any way as the union $\bigcup_{i=1}^{\infty} G_i$ of a tower of finite subgroups $G_1 \leqslant G_2 \leqslant \ldots$, then there exists i such that G_i has a perfect subnormal subgroup of defect exactly d.*

(5) *If G has VDA type, then G has a faithful transitive permutation representation in which some finite subgroup has no regular orbit.*

We shall say what it means for G to have VDA type in due course. The phenomenon (5) was met above in the case of diagonal limits of alternating groups. Actually, (2) formally follows from (3), since we can take any of the terms of the series (8) as the first term of a non-trivial ascending series. (4) is relevant to the question of trying to express simple locally finite groups as unions of towers of finite groups that are simple or close to it. Recall that the defect of a subnormal subgroup A of a group B is the smallest number of steps in a chain of subgroups from A to B, with each normal in the next. Thus, (4) says that, in the tower of groups G_i, the subgroups occurring cannot all be simple, or direct products of simple groups, or repeated extensions of such groups and soluble groups a bounded number of times, and so on. Property (4) follows from (2) or (3) because of work of Phillips [43]. Roughly speaking, Phillips' result says that if a group D is the union of a tower of finite subgroups in which the perfect subnormal subgroups have bounded defect then the phenomena (2) and (3) cannot occur in D.

Absolute and Strict Simplicity

Definition 1.28   *A group G is called* strictly simple *if G has no non-trivial ascending series, and* absolutely simple *if G has no non-trivial series at all.*

Thus every absolutely simple group is strictly simple, and every strictly simple group is simple. The Mf-groups are simple but not strictly simple, but it seems that no example is known of a locally finite group that is strictly simple but not absolutely simple. Absolutely simple groups are those groups obtained as composition factors of general groups. It does not seem clear what rôle these concepts play in the context of simple locally finite groups; we refer to Phillips' paper [43] for some discussion.

The Mf-groups are obtained by iterating a construction which we now describe. For more details, see [39] and [22].

Construction

Let H be any transitive permutation group on a finite set X. (This will be the i-th group in our iteration.) Choose finite non-abelian simple groups K and L for which we have a monomorphism $\phi : H \to K$. Let Y and Z be finite sets permuted faithfully and transitively by K and L respectively, and, for notational convenience, suppose that $Z \supseteq \{0, 1\}$. Let W be the wreath product $W = H \wr K \wr L$, which has a natural transitive action on $X \times Y \times Z$. Let $V = H \wr K$. Then $W = V \wr L = \overline{V}L$, where $\overline{V}$, the base group, is a direct product of copies of V indexed by Z. Thus, $\overline{V} = V_0 \times V_1 \times \cdots$. Similarly, $V = \overline{H}K$, where $\overline{H}$ is a direct product of copies of H. We identify H with one of the copies of H in the base group of V_0, and K with the canonical copy of itself in V_1. Then we define $\psi : H \to W$ by $\psi(h) = h\phi(h) \in V_0 \times V_1$ for all $h \in H$.

The following statements are quite easy to verify.

Lemma 1.29   *Let $M = \langle H^W \rangle \prod_{1 \neq z \in Z} V_z$. Then:*

(1) $\overline{V} = MK$, $M \triangleleft \overline{V}$, and $M \cap K = 1$. Thus, $\overline{V}/M$ is simple.

(2) $M \cap \psi(H) = 1$.

(3) *If $1 \neq h \in H$, then the normal closure of $\psi(h)$ in W is $\overline{V}$, which clearly contains $\psi(H)$.*

Since our construction produces another transitive permutation group, it can be repeated. Thus, suppose we begin with any finite non-abelian simple group W_1 acting transitively on a finite set X_1. Having constructed a direct system

$$W_1 \to W_2 \to \ldots \to W_i,$$

in which W_i acts as a transitive permutation group on a set X_i, we construct a finite group W_{i+1} and an embedding $W_i \to W_{i+1}$ by taking W_i to be H in the above construction and putting $W_{i+1} = W$, taking the embedding to be ψ. Let the direct limit of this direct system be G. Identifying the W_i with subgroups of G in the usual way, we have

$$G = \bigcup_{i=1}^{\infty} W_i, \tag{9}$$

$$W_1 \leqslant W_2 \leqslant \ldots . \tag{10}$$

Definition 1.30   *Any group obtained in this way is called an Mf-group.*

Condition (3) of Lemma 1.29 becomes

$$\text{if } 1 \neq w \in W_i, \text{ then } \langle w^{W_{i+1}} \rangle \geqslant W_i. \tag{11}$$

It is very easy to deduce from this that G is simple.

When we carry out our construction, there are many choices for the embedding ϕ of H in the simple group K. One of them is to take K to be the alternating group on H and to take ϕ to be the regular representation. In that case we describe the construction as *regular*. Another is to take $K = \mathrm{Alt}(X)$ and to use the given permutation representation to embed H in K. In that case we call the construction *diagonal*. From now on in these constructions we shall always take the groups L to be alternating groups and we shall begin with an alternating group W_1.

Definition 1.31 *An Mf-group is said to be of visibly regular alternating VRA type if only regular constructions are used in building it up. If only diagonal constructions are used, the group is said to be of visibly diagonal alternating VDA type.*

These are not of course the only possibilities, and the reasons for these names should become clear later.

The following is not too difficult to check.

Lemma 1.32 *Suppose that the construction is diagonal, and allow H to act on $X \times Y \times Z$ via ψ. Then $X \times Y \times Z$ is a union of one point orbits and orbits isomorphic to X.*

This means that, when we use a diagonal construction, the embedding ψ of H in W induces an embedding of permutation groups $(H, X) \rightarrow (W, X \times Y \times Z)$. If G is an Mf-group of VDA type, then G is the union of a tower of subgroups $W_1 \leqslant W_2 \leqslant \cdots$, and there is an associated direct system of permutation groups

$$(W_1, X_1) \rightarrow (W_2, X_2) \rightarrow \cdots \tag{12}$$

such that, if $i \leqslant j$, then the restriction of X_j to W_i is a union of one point orbits and orbits isomorphic to X_i. Notice also that each W_i is a repeated wreath product of alternating groups and X_i is a product of sets acted upon in some way by these alternating groups. Forming the direct limit of the permutation groups (12) as explained in the previous section, we obtain the following.

Proposition 1.33 *Suppose that G is an Mf-group of VDA type. Then G is the union of a tower (10), in which each W_i is a wreath product of finite alternating groups. The group G acts faithfully and transitively on a set X such that, under each W_i, X splits up into a number of trivial orbits and orbits isomorphic to a certain W_i-set X_i. The group W_i acts in product action on X_i, and, in particular, if i is large enough, then W_i has no regular orbit on X.*

This is the last part of Theorem 1.27.

1.7. Free Extensions

It will soon become apparent that in this section there are more questions than answers; in fact we have virtually no answers. We shall give a construction that certainly produces simple locally finite groups, but it seems very difficult to say anything about these groups except that they are simple.

The following result sets the scene.

Lemma 1.34   *Let G be a group. Suppose that G is the union of a tower*

$$G_1 \leqslant G_2 \leqslant \ldots$$

of perfect subgroups G_i. Suppose that, for each i, we have a normal soluble subgroup N_i of G_i, and suppose that $G_i \cap N_{i+1} = 1$ for each i. Suppose further that each G_i/N_i is simple. Then G is simple.

Proof.   Let H be a non-trivial normal subgroup of G. Then there exists i such that $H \cap G_i \neq 1$. Let $j > i$. Then $H \cap G_j \not\leqslant N_j$, and it follows since G_j/N_j is simple that $G_j = (H \cap G_j)N_j$. Hence $G_j/(H \cap G_j)$ is soluble. Since G_j is perfect, we have $G_j \leqslant H$. Since this holds for all $j > i$, we have $H = G$.   □

The following fact will also be useful. We write

$$G = G^{(0)} \geqslant G' = G^{(1)} \geqslant \ldots$$

for the derived series of a group G.

Lemma 1.35   *Let $N \triangleleft G$. Suppose that N is soluble of derived length at most d, and G/N is perfect. Then $G^{(d)}$ is perfect.*

Proof.   Since G/N is perfect, we have $G = G^{(d+1)}N$. Hence $G/G^{(d+1)}$ is soluble of derived length at most d, and it follows that $G^{(d)} \leqslant G^{(d+1)}$.   □

Now begin with an arbitrary perfect finite group G_1. We build up a tower of finite perfect groups satisfying the hypotheses of Lemma 1.34 in a 'free' manner. Suppose we have obtained

$$G_1 \leqslant G_2 \leqslant \ldots \leqslant G_i.$$

Choose an arbitrary finite simple group H containing G_i. Let $H = \langle G_i, X \rangle$ for some finite set X of elements of H. Let E be the free product $G_i * F$, where F is the free group on X. Then we have an obvious exact sequence

$$1 \to R \to E \to H \to 1. \tag{13}$$

Since G_i is mapped injectively, we have $R \cap G_i = 1$. Hence R is free, by the Kurosh Subgroup Theorem. Note that, since it has finite index in E, the group R is of finite rank. Choose a characteristic subgroup M of finite index in R such that R/M is non-trivial and soluble; for example, we might take $M = R^{(d)}R^e$ for some d and e, or $M = \gamma_d(R)R^e$. In particular we might take $M = R'R^e$. Put $H_{i+1} = E/M$. This is a finite group containing (a canonical copy of) G_i, and H_{i+1} has a soluble normal subgroup $L = R/M$, of derived length d say, such that $L \cap G_i = 1$. Put $G_{i+1} = H_{i+1}^{(d)}$ and $N_{i+1} = G_{i+1} \cap L$. Then $G_i \leqslant G_{i+1}$ and $G_i \cap N_{i+1} = 1$. By Lemma 1.35, G_{i+1} is perfect. We can thus construct an infinite tower $G_1 \leqslant G_2 \leqslant \ldots$ satisfying the hypotheses of Lemma 1.34. Hence $G = \bigcup G_i$ is simple, and clearly locally finite.

Let us call the above a *free product construction*, and any resulting group an FP-group. Unfortunately we are not able to prove anything about these groups. Let us ask two contrasting questions.

Problem 1.36

(1) *Is it possible for an FP-group to be the union of a tower of finite simple subgroups?*

(2) *Is it possible to construct an FP-group that has no non-abelian finite simple subgroup?*

Note that in the construction we have required $R/M \neq 1$, so an FP-group cannot actually be constructed as the union of a tower of finite simple subgroups.

A study of these questions seems to require a study of the following problem.

Problem 1.37 *Let $E = H_1 * H_2 * \cdots * H_r * F$, where the H_i are finite groups and F is free of finite rank. Let R be a normal subgroup of E of finite index such that $R \cap H_i = 1$ for $1 \leqslant i \leqslant r$. Let N be a proper characteristic subgroup of R. Let $\overline{E} = E/N$ and $\overline{R} = R/N$. What can be said about the subgroups L of $\overline{E}$ such that $L \cap \overline{R} = 1$?*

It is advantageous to put it in a slightly more general form than appears in the free product construction since, by the Kurosh Subgroup Theorem, every subgroup of finite index in E has the same form as E.

2. Kegel Sequences and Covers

2.1. Kegel Covers and CFSG

We have seen in the previous section that a countable simple locally finite group G need not be the union of a tower of finite simple subgroups. Kegel sequences, or Kegel covers in the uncountable case, provide a substitute for this which is often useful. They were introduced by Otto Kegel [35]; see also [36, 4.5]. The nomenclature is more recent.

Definition 2.1 *Let I be an index set and G a locally finite group. A set*

$$\{ (G_i, N_i) \mid i \in I \} \tag{14}$$

of pairs of subgroups of G is called a sectional cover of G if

$$N_i \triangleleft G_i \quad (i \in I), \tag{15}$$

and

for each finite subgroup F of G there exists
$i \in I$ such that $F \leqslant G_i$ and $F \cap N_i = 1$. $\tag{16}$

(1) alternating groups, labelled Alt;

(2) classical types, labelled

$$A, B, C, D, {}^2A, {}^2D,$$

or alternatively

$$PSL, P\Omega_{odd}, PSp, P\Omega_{even}^+, PSU, P\Omega_{even}^-;$$

(3) exceptional types, labelled E_6, E_7, E_8, F_4, G_2, ${}^2B_2$, ${}^3D_4$, ${}^2E_6$, ${}^2F_4$, ${}^2G_2$;

(4) sporadic groups, of which there are 26.

The groups in the alternating family are labelled by a single numerical parameter. Those in the classical families are labelled by two parameters, a natural number (rank parameter) and a prime power representing the order of a finite field. The groups in each exceptional family are labelled by a field parameter, and each family has a rank label. For consistency, we call n the rank parameter of Alt_n.

Now we can partition our Kegel cover S into a finite number of subsets, one for each of the above families, so that all the factors in the same subset come from the same family. By an obvious extension of Lemma 2.3, at least one of these subsets is a Kegel cover. Since the factors from a Kegel cover have unbounded orders, the sporadic family does not arise from a Kegel cover. Thus we have the following possibilities, according to the type of the Kegel cover or covers obtained. (There is no reason to believe, nor is it true, that only one of these subsets is a Kegel cover.)

(1) G has a Kegel cover of alternating type.

(2) G has a Kegel cover of fixed classical type with unbounded rank parameters.

(3) G has a Kegel cover of fixed classical type with bounded rank parameters.

(4) G has a Kegel cover of fixed exceptional type.

The last two types correspond precisely to the case when G is linear, and then G is known to be of Lie type over a locally finite field.

Theorem 2.6 *Let G be an infinite simple locally finite group. Then the following conditions are equivalent.*

(a) *One of (3) and (4) holds.*

(b) *G is linear.*

(c) *G is of Lie type over a locally finite field.*

(d) *Some finite group is not involved in G.*

We have discussed this already. Note that, in a Kegel cover of type (2), the Weyl groups of the Kegel factors involve alternating groups of unbounded ranks, and hence, in Cases (1) and (2), G involves all finite groups. In types (3) and (4), Mal'cev's Representation Theorem (see below) tells us that G is linear over a field, and then, if q is a prime different from the characteristic of the field, the elementary abelian q-sections of G have bounded rank [36]. Therefore G does not involve most elementary abelian q-groups.

Let us discuss a little further how the proof of this theorem goes. A crucial part is played by Mal'cev's Representation Theorem.

Theorem 2.7 (Mal'cev's Representation Theorem) *Let G be any group and let n be any natural number. Suppose that each finitely generated subgroup of G has a faithful representation of degree at most n over some field (the field in question may depend on the finitely generated subgroup). Then G has a faithful representation of degree at most n over some field.*

The field obtained can be taken to be an ultraproduct of the fields corresponding to the finitely generated subgroups. There are also versions corresponding to projective representations. See [16] for example and also Hall's article. For a proof, see [36, p. 64] and [52, 2.7]. In the situation when G has a Kegel cover of type (3) or (4), since every finite subgroup of G is isomorphic to a subgroup of some Kegel factor in this cover, we find that the hypothesis of Mal'cev's Representation Theorem holds. Hence G is linear over some field. If the field has characteristic zero we find that G is abelian-by-finite, contrary to hypothesis. Hence G is linear in finite characteristic p, say. We find that G is countable [36, 1.L.2], [52, 9.5], so that G has a Kegel sequence, and then that almost all the Kegel kernels are trivial [36, 4.6],[52, 9.32]. Hence (CFSG), G is actually the union of a tower of finite simple subgroups of the same type over various finite fields, and it is then shown that G is itself a group of the same type over a locally finite field.

2.2. KEGEL COVERS FOR NON-LINEAR GROUPS

Let us try further to divide Kegel covers into various kinds according to their factors, and try to say something about which groups have covers of the various kinds. Let G be an infinite simple locally finite group.

We have seen above that G is linear if and only if it has a Kegel cover of type (3) or (4). Presumably the set of factors occurring is more or less unique in this case; this should follow from results on subgroups generated by root subgroups or something like that. It also seems to follow from results in [27]. For if G is a simple locally finite group and is simultaneously a union of finite simple groups of Lie types X_m and Y_n, where X and Y are symbols as above and m and n are rank parameters, then G itself is of type X_m and Y_n, and so the algebraic group of type X_m has a dense subgroup of type Y_n, whence a little more argument seems to give the result on the basis of [27, 2.2]. Thus, we ignore the linear situation from now on.

Next we may consider finitary groups. Presumably here also there must be a more or less unique Kegel cover, though it is not obvious to me how to prove it.

We put this case aside also and,

from now on in this section, G denotes a countable simple locally finite group that is not finitary linear.

Then all Kegel covers for G must have one of types (1) and (2) of the previous subsection. We subdivide them further.

Definition 2.8 *A Kegel sequence of G has type*

(1) RA *if all its factors are alternating and the embeddings are regular,*

(2) DA *if all its factors are alternating and the embeddings are diagonal,*

(3) $\mathrm{CL}(p^\alpha)$ *if all the factors belong to a fixed family of classical groups over a fixed field $\mathbb{F}_{p^\alpha}$,*

(4) $\mathrm{CL}(p^\infty)$ *if all the factors belong to a fixed family of finite classical groups over fields of characteristic p but with unbounded orders,*

(5) $\mathrm{CL}(\infty)$ *if all the factors are classical groups from a fixed family but involving an infinite number of different characteristics.*

Additional commentary on the embeddings implicit in parts (1) and (2) of Definition 2.8 can be found after Proposition 2.9. We follow A. E. Zalesskiĭ in distinguishing the three classical situations. The division of the alternating case into regular and diagonal types follows the lines of the last section but is more technical. Before giving the details, we note that, in general, virtually anything can happen. The following is a direct consequence of the injectivity property of Hall's group U.

Proposition 2.9 *Let $\mathcal{F}$ be any family of finite simple groups with the property that every finite group is contained in some $\mathcal{F}$-group. Then U is the union of a tower of $\mathcal{F}$-groups.*

Now let

$$\{\,(G_i, N_i) \mid i = 1, 2, \ldots\,\} \tag{20}$$

be a Kegel sequence of alternating type. Thus, $G_i/N_i \cong \mathrm{Alt}_{n_i}$, say, which acts on $\Omega_{n_i} = \{\,1, 2, \ldots, n_i\,\}$. Via the isomorphism $G_i \cong G_i N_j/N_j$ $(i < j)$, we have an action of G_i on Ω_{n_j}. We say that (20) is *regular* if, for each $i < j$, G_i has at least one regular orbit on Ω_{n_j}.

The definition of diagonal action is a little more elaborate, and, as given here, is due to Meierfrankenfeld. Let H be a group acting on the right on a set Ω, and let $\mathcal{C}$ be a family of subsets of Ω. Then

$$N_H(\mathcal{C}) = \{\,h \in H \mid Ch \in \mathcal{C} \text{ for all } C \in \mathcal{C}\,\},$$

$$C_H(\mathcal{C}) = \{\,h \in H \mid Ch = C \text{ for all } C \in \mathcal{C}\,\}.$$

When $\mathcal{C}$ is a family of one point sets we obtain the usual setwise and pointwise stabilizers of a subset of Ω. Let Ω and Σ be finite sets and let H be a group acting on Ω. Let $M \lhd H$ with $H/M \cong \mathrm{Alt}(\Sigma)$. We say that H acts *naturally* on Ω with respect to M if H is transitive on Ω and there exists a system of imprimitivity $\mathcal{S}$ for H on Ω such that $C_H(\mathcal{S}) = M$ and $|\mathcal{S}| = |\Sigma|$. In other words, the action of H/M on $\mathcal{S}$ is isomorphic to the natural action of $\mathrm{Alt}(\Sigma)$ on Σ. We say that H acts *essentially* on Ω with respect to M if $C_H(\Omega) \leqslant M$. If it is apparent which normal subgroup we have in mind as M, then we suppress explicit reference to it. It is shown in [39, 1.8] that if H/M is perfect, as it is if $|\Sigma| \geqslant 5$, then, among all subnormal supplements to M in H, there is a unique minimal one, which we denote by R. Clearly $R \lhd H$, and then H acts essentially on Ω if and only if R acts non-trivially.

Finally, we say that the Kegel sequence (20) is of *diagonal type* if, whenever $i < j$, each essential orbit of G_i on Ω_j is natural (with respect to N_i, of course).

Meierfrankenfeld has obtained results which imply the following.

Proposition 2.10 *Let G have a Kegel sequence of alternating type. Then G has a Kegel sequence of either RA or DA type.*

For more details of this result and the next, see [39] and [40]. A similar but slightly different result can be found in [45].

Meierfrankenfeld has given another remarkable dichotomy, based on the following definition.

Definition 2.11   *Let p be a prime. A simple locally finite group is of p-type if G is not finitary and every Kegel cover contains at least one factor which is a classical group in characteristic p.*

Theorem 2.12   (Meierfrankenfeld)   *Let H be a non-finitary simple locally finite group. Then either H is of alternating type (and so, if countable, would be of RA or DA type) or H is of p-type for some prime p. In the latter case, H has a Kegel cover $\{(H_i, M_i) \mid i \in I\}$ such that each factor H_i/M_i is a projective special linear group over a field of characteristic p, and $H_i/O_p(H_i)$ is a central product of perfect central extensions of finite classical groups over fields of characteristic p.*

This remarkable result at first sight reduces us, in studying non-finitary groups, to dealing with Kegel covers whose factors are either alternating groups or projective special linear groups. However this feeling must be tempered by the recollection that, when we pass from a group which may be given to us as the union of a tower of simple groups, say projective symplectic groups for example, and exhibit it as a group of alternating type, the Kegel kernels seem to get out of hand. However it does draw attention to the groups of p-type, which, as Meierfrankenfeld remarks, may be susceptible to being classified. Such a group has the advantage of being covered by finite groups which are more or less known. We will return to this briefly in the next subsection.

The reader may have wondered whether there are in fact any groups of p-type, and, indeed, since the definition involves consideration of all Kegel covers, it is not *a priori* so easy to recognize that a given group is of p-type. The main problem is to know that the group in question is not of alternating type.

Lemma 2.13   (Meierfrankenfeld)   *Let G be a non-finitary simple locally finite group. Suppose that, for some prime p, we have a p-element x in G such that $\langle x^P \rangle$ is soluble for all p-subgroups P of G containing x. Then G is not of alternating type.*

Now we construct an example [16]. It is surely a 'diagonal limit' of projective special linear groups, if this term has any meaning. Let $n_1 \geqslant 2$, let p be an odd prime, and let q be a prime power such that $p \mid (q-1)$. Let $n_i = 2^{i-1}n_1$ for $i \geqslant 2$, and let $H_i = \mathrm{SL}_{n_i}(q)$. We embed H_i in H_{i+1} by the embedding

$$x \mapsto \begin{pmatrix} x & 0 \\ 0 & x \end{pmatrix}.$$

Let $H = \bigcup_{i=1}^{\infty} H_i$, let $Z = Z(H)$, and let $G = H/Z$. Then G is a union of a tower of projective special linear groups over $\mathbb{F}_q$, and so it has a Kegel sequence with these factors. It can also be shown that G is not finitary. If we can show that it is not of alternating type, then, by Lemma 2.13, it must be of q-type. Now let x be any diagonal matrix of order p in H_1 that does not have every p-th root of 1 in $\mathbb{F}_q$ as an eigenvalue. Then the same will be the case when x is considered as an element of any H_i, by the embedding used above. Each Sylow p-subgroup of H_i is a permutational wreath product of the form $C \wr P_i$ where P_i is a Sylow p-subgroup of Sym_{n_i}, and each element of order p outside the base group permutes a set of p linearly independent basis vectors transitively and so has every p-th root of 1 as an eigenvector. It follows that x lies in the base group of each Sylow p-subgroup P of H_i that contains it, and so $\langle x^P \rangle$ is abelian. This is what we need to show that G is of q-type.

Before leaving this section, we mention that A. E. Zalesskiĭ has conjectured unofficially that if G has a Kegel sequence of type $\mathrm{CL}(\infty)$ then it has a great variety of Kegel sequences. As a very special case, he has proved the following (unpublished).

Theorem 2.14 (A. E. Zalesskiĭ) *Let G be a simple locally finite group. Suppose that $G = \bigcup_{i=1}^{\infty} G_i$, where $G_i \cong \mathrm{SL}_{n_i}(q_i)$ and the q_i are powers of distinct primes. Let $\mathcal{F}$ be a set of finite simple groups such that every finite group can be embedded in some $\mathcal{F}$-group. Then G has a Kegel sequence in which all factors come from $\mathcal{F}$.*

The proof depends on results of Gluck bounding the reduced character values of a group of Lie type in terms of the size of the field over which the group is defined.

Zalesskiĭ has suggested that if G has Kegel sequences of two sufficiently different types then it has one of type $\mathrm{CL}(\infty)$. Perhaps we should focus attention by means of the following.

Problem 2.15 *Suppose that G is a simple locally finite group having Kegel sequences of types $\mathrm{CL}(p^{\alpha})$ and $\mathrm{CL}(q^{\beta})$, where p and q are different primes. Does it follow that G has a Kegel sequence of type $\mathrm{CL}(\infty)$?*

2.3. Involvement of Unions of Finite Simple Groups

In view of the fact that unions of towers of finite simple groups are much easier to deal with than groups in which we have to make do with a Kegel sequence or cover, it seems reasonable to try to find groups of the former type that are obtainable somehow from the latter. Recall that a group G is said to *involve* a group H if G contains subgroups $V \lhd U$ such that $U/V \cong H$.

Now an infinite simple locally finite linear group is already a union of a tower of finite simple groups, so certainly involves such a union. We mention the following without much thought.

Problem 2.16 *Does an infinite simple finitary locally finite group always involve an infinite group which is the union of a tower of finite simple groups?*

Regarding the non-finitary groups, the following is true.

Theorem 2.17 *Let G be a non-finitary simple locally finite group. Then G involves Hall's group U.*

We will say a few words about the proof. First, according to Meierfrankenfeld's Theorem 2.12, such a group is either of alternating type or of p-type for some prime p. In the alternating case, the result is essentially due to Meierfrankenfeld (oral communication). He showed that if G is of alternating type then G involves either U or all non-finitary diagonal unions of finite alternating groups; see [22]. As we saw in Proposition 1.22, G involves U in that case as well.

This leaves the case when G is of p-type. To state some results, let us make a provisional definition of what it means for a union of a tower of special linear groups, or quotients thereof, to be diagonal. This is complicated by the fact that these groups may be defined over different fields, though we take the fields to have the same characteristic. Let

$$G = \bigcup_{i=1}^{\infty} G_i, \tag{21}$$

where

$$G_1 \leqslant G_2 \leqslant \ldots . \tag{22}$$

Suppose that G_i is a quotient of $\mathrm{SL}_{n_i}(q_i)$ by a central subgroup, where $q_i = p^{r_i}$. Here p is a fixed prime. Let $S_i = \mathrm{SL}_{n_i}(q_i)$, and fix an epimorphism $S_i \to G_i$. Let V_i be the natural $\mathbb{F}_{q_i} S_i$-module. Let $k = \overline{\mathbb{F}}_p$ and let $\overline{V}_i = V_i \otimes_{\mathbb{F}_{q_i}} k$. The embedding $G_i \to G_{i+1}$ lifts to a unique homomorphism $\sigma_i : S_i \to S_{i+1}$. (We shall assume that no exceptional multipliers are present.) This allows us to view V_{i+1}, and hence $\overline{V}_{i+1}$, as a module over S_i.

Definition 2.18 *Using the notation above, any Frobenius twist of V_i or its contragredient will be called a* fairly natural *module for S_i, and the same for $\overline{V}_i$. The above tower is called* diagonal *if each $\overline{V}_{i+1}$ is a completely reducible kS_i-module and its direct summands are all fairly natural or trivial.*

Theorem 2.19 *Let G be a non-finitary simple locally finite group of p-type. Then G involves a group $H = \bigcup_{i=1}^{\infty} H_i$, where $H_i \cong \mathrm{PSL}_{m_i}(r_i)$, the r_i are powers of p, $m_i \to \infty$, and the tower is diagonal.*

Theorem 2.20 *Let H be a union of diagonal type of groups PSL, over finite fields of the same characteristic, and of unbounded ranks as above. Then H involves U.*

The proof of this is much the same as that of Proposition 1.22. If we have a subgroup L_i of some H_i that acts on $\overline{V}_i$ by permuting the elements of some basis and is isomorphic to some Alt_{n_i}, we can choose a subgroup L_j of some large enough H_j that contains L_i and is isomorphic to some Alt_{n_j}, where the embedding $L_i \to L_j$ is some non-trivial diagonal embedding. In that way we see that H involves some non-finitary diagonal limit of alternating groups and hence, by Proposition 1.22, also involves U.

The proof of Theorem 2.19 involves some careful analysis of the structure of the group, and it seems that much more remains to be said here. We need to set up more notation.

Thus, let G be a countable simple locally finite group of p-type. Then, by Meierfrankenfeld's Theorem 2.12, we can write

$$G = \bigcup_{i=1}^{\infty} G_i, \tag{23}$$

where

$$G_1 \leqslant G_2 \leqslant \ldots \tag{24}$$

is a tower of finite subgroups with the properties given in Theorem 2.12 (with the G_i assuming the rôles of the H_i). Let $P_i = O_p(G_i)$. Then $G_i/P_i = T_i T_i^*$, where T_i is a perfect central extension of $\mathrm{PSL}_{n_i}(q_i)$, the numbers q_i are powers of p, the groups T_i^* are central products of other perfect classical groups over fields of characteristic p, and $[T_i, T_i^*] = 1$. By a theorem of Phillips [43, Theorem 3], G is absolutely simple, and so, by another result of Phillips [42], one can arrange that the inclusion $G_i \to G_{i+1}$ induces an embedding of G_i into T_{i+1} and indeed into each of the simple factors of G_{i+1}. Let $\hat{T}_i = \mathrm{SL}_{n_i}(q_i)$, the covering group of T_i, and in fact let hats denote covering groups generally. Let V_i be the natural $\hat{T}_i$-module as before and let $\overline{V}_i$ be its extension to the algebraic closure. Then we can obtain an action of $\hat{G}_i$ on $\overline{V}_{i+1}$. In this action, $O_p(\hat{G}_i)$ acts trivially on each composition factor and so we obtain an action of $\hat{T}_i$ and $\hat{T}_i^*$ on each such composition factor. By the same token, we can obtain actions of these groups on $\overline{V}_j$, for any $j > i$. Then the claim is the following.

Theorem 2.21 *There is an infinite sequence $i_1 < i_2 < \ldots$ such that, if X is any composition factor of $\hat{G}_{i_r}$ on $\overline{V}_{i_{r+1}}$, then at least one of $\hat{T}_{i_r}$ and $\hat{T}_{i_r}^*$ acts trivially on X, and all non-trivial composition factors of $\hat{T}_{i_r}$ on X are fairly natural.*

3. Centralizers

3.1. GENERALITIES

In studying groups one usually tries to study their subgroup structure, and simple locally finite groups should be no exception to this principle. Important types of subgroups include centralizers of finite subgroups and maximal subgroups, which are of course important objects of study in finite group theory. Rather little is known about maximal subgroups, but see Meierfrankenfeld's article. Other types of subgroups are also coming on the scene, which are not analogous to subgroups studied in finite group theory, but nevertheless seem to have some rôle here. These include *confined subgroups* (see Definition 1.25), which were introduced by Zalesskiĭ in connection with the study of ideals in group rings (see [54, 55]), and *inert subgroups*, which were introduced by Belyaev (see [4]).

Here we shall confine ourselves to centralizers. Quite a number of the results we give have versions which are true for arbitrary locally finite groups, but we shall probably pass over most of them in the interests of brevity.

We may approach the study of centralizers from two directions.

Fix on some particular type of group, and try to make general structural statements about the centralizers of elements, or subgroups, in these groups. For ex-

ample we may take the linear simple locally finite groups, or, more optimistically, the class of all simple locally finite groups.

Study the groups in which some or all centralizers have some particular structure; for example, those in which some centralizer is finite, or some centralizer is linear.

The distinction between these approaches is not completely clear-cut, but it is useful. We will confine ourselves from now on to the case of *countable* groups. Most questions of the type we will consider reduce to that case anyway.

Thus, let G be a countably infinite simple locally finite group, and let

$$\{ (G_i, N_i) \mid i \in I \} \tag{25}$$

be a Kegel sequence of G. Thus, $G = \bigcup_{i=1}^{\infty} G_i$, and the G_i form a tower of finite subgroups of G. Each N_i is a maximal normal subgroup of G_i, and $G_i \cap N_{i+1} = 1$. Suppose we have an element $x \in G$ whose centralizer we wish to study. Clearly we may assume that $x \in G_1$. Now a great deal is known about centralizers in finite simple groups, so, with luck, perhaps we know about the centralizers of x in the Kegel factors, that is, the groups $C_{G_i/N_i}(xN_i)$, which we may denote simply by

$$D_i/N_i = C_{G_i/N_i}(x). \tag{26}$$

However we may want to bear in mind that, for example, the Kegel factors may be linear groups and that x will in general be neither unipotent nor semisimple in these factors. Now trivially we have

$$C_G(x) = \bigcup_{i=1}^{\infty} C_{G_i}(x). \tag{27}$$

Thus, we want to know about the groups $C_{G_i}(x) = C_i$. The kind of things we want to know are probably facts which will survive in some recognizable form through the direct limit (27) as some global statement about $C_G(x)$. We know nothing about the groups N_i, and this is a big problem. Also, although we trivially have

$$C_{G_i}(x)N_i/N_i \leqslant C_{G_i/N_i}(x), \tag{28}$$

in general the connection between these is very tenuous, and we do not get much information about $C_i = C_{G_i}(x)$ from the Kegel factors. This is what I mean by the malevolent influence of the Kegel kernels. Furthermore we now know in general that complicated Kegel kernels cannot be avoided. One way around this is simply to assume that all the N_i are 1. This is what was done in [24], and it seems to me that quite a lot of interesting groups are covered even in this case.

A second possibility is that the Kegel kernels go away of their own accord, after some persuasion. What I mean here is this. Suppose that we are starting with some assumption on $C_G(x)$. We can consider, inside G, the subgroup $N = N_1 N_2 N_3 \cdots$, the product of the Kegel kernels (each of which normalizes the next). This subgroup, which is a repeated semidirect product of finite groups and so is residually finite, is actually an inert subgroup in the sense of Belyaev. Now x acts on N and $C_N(x)$ is a subgroup of $C_G(x)$. Perhaps the structure of $C_G(x)$ forces N to have some specific structure which more or less gets rid of it.

Probably this is enough of these generalities, so let us illustrate with a few cases. We will not try to give the most general results here.

Theorem 3.1 [24] *Let G be a simple locally finite group containing an element x with finite centralizer. Then G is finite.*

See [20] for a more general result.

Sketch of proof. We use the notation above. Let $|C_G(x)| = n$ and let x have order r. Then $|C_{G_i}(x)| \leqslant n$, and so $|C_{G_i/N_i}(x)| \leqslant n$ (for any group Y and $x \in Y$, let $\mathrm{cl}_Y(x)$ denote the set of conjugates of x in Y; then, for a normal subgroup N of Y, $|\mathrm{cl}_Y(x)| \leqslant |N||\mathrm{cl}_{Y/N}(xN)|$ and it follows that, for finite Y, $|C_{Y/N}(xN)| \leqslant |C_Y(x)|$). This is one situation when the Kegel kernels do no harm. Now there are only a finite number of finite simple groups containing an element whose order divides a given integer r and the order of whose centralizer is bounded by a given integer n. If $r = 2$ this is the Brauer-Fowler Theorem; in general it can be deduced from CFSG. Hence we find that the groups G_i/N_i have bounded order, and this implies that G is finite. $\square$

Theorem 3.2 *Let G contain an element x whose centralizer is a Černikov group. Then G is finite.*

Recall that a group X is a Černikov group if X contains a normal subgroup X^0 of finite index that is a direct product of a finite number of groups of type C_{p^∞}, for possibly several primes p. For us the point is that a Černikov group satisfies the descending chain condition on subgroups.

Sketch of proof. Let $C = C_G(x)$, and use the above notation again. We consider $C_N(x)$. Put $N_r^* = \langle N_i \mid r \leqslant i \rangle$. Then $\bigcap_{r=1}^\infty N_r^* = 1$, and so there exists r such that $C_N(x) \cap N_r^* = 1$. We may discard $G_1, \ldots, G_{r-1}$ and assume that $C_N(x) = 1$. Now a finite group admitting a fixed point free automorphism is soluble, and, applying this to the finite subgroups of N, we find that N is locally soluble. Also every finite subgroup of N lies in an x-invariant finite subgroup of N. Let F be such a finite x-invariant subgroup. Then $C_F(x) = 1$, and so, if $L = F\langle x \rangle$, then $\langle x \rangle$ is a Carter subgroup of L. Thus, by a well known result of Dade [11], or the earlier work of Shamash and Shult [48], it follows that the Fitting height of F is bounded in terms of the order of x. This tells us that N has a finite series with locally nilpotent factors, and, in particular, if $N \neq 1$, that $P = O_p(N) \neq 1$ for some prime p. Let i be such that $P_i = P \cap G_i \neq 1$. Then if $j > i$ we have $P_j = P \cap G_j \not\leqslant N_j$. Now $[P_j, N_j] \leqslant P \cap N_j \leqslant O_p(N_j) = Q_j$, say. Let $C_j = C_{G_j}(N_j/Q_j)$; then $C_j \lhd G_j$, and we have just seen that $C_j \not\leqslant N_j$. Since G_j/N_j is simple, we find that $G_j = C_j N_j$. But N_j is soluble, and, assuming that $G_j = G_j'$, which we can by Corollary 2.5, we find that $G_j = C_j$, that is, G_j/Q_j is quasisimple.

The Kegel kernels have not exactly gone away but they are in disarray. It is possible to continue along these lines and obtain the result, but better methods leading to better results are possible [5], and the above is just an illustration. □

Next suppose that G contains an element of prime order p whose centralizer is elementary abelian. In this case we should expect at least that G is linear. If the G_i are simple then it is a matter of showing that they cannot be either all alternating or all from a family of classical groups of unbounded rank, and that is easy to do on the basis of results in the literature. But suppose for example that the N_i are all non-trivial p-groups. Then it is not clear that we can deduce anything about the groups $C_{G_i/N_i}(x)$. Here the Kegel kernels are a very serious obstacle. Further we have that x acts on the locally finite p-group N with elementary abelian centralizer, and it is not clear what that tells us either. This case seems rather resistant.

But these arguments focus attention on the following problem.

Problem 3.3 *Study the action of finite groups on residually finite groups. Suppose that a finite group F acts on a residually finite group R. What can be said about the structure of R in terms of the structure of $C_R(F)$? What about the case when $C_R(F) = 1$? In the above context it would be reasonable to assume that R is a repeated semidirect product of F-invariant finite groups.*

Some examples for the case $C_R(F) = 1$ can be found in [1]; it is not to be expected that R has bounded Fitting height.

3.2. UNIONS OF TOWERS OF SIMPLE GROUPS

This case was studied at some length in [24] and the results have been extended in [25], by taking account of results of Wall [51]. The ideas can be seen by considering the case when G is a union of a tower

$$G_1 \leqslant G_2 \leqslant \ldots \tag{29}$$

of finite alternating groups. Let x be an element of order r, and take $x \in G_1$. Now if x is an element of order r in a finite symmetric group $S = \mathrm{Sym}(\Omega)$ then we can partition Ω as $\Omega = \bigcup_{d|r} \Omega_d$, where x acts on Ω_d as a product of e_d cycles of length d. Some of these sets Ω_d may in general be empty. Now $C_S(x)$ leaves each Ω_d invariant and permutes the d-cycles on Ω_d. They can be permuted in any way and it is not difficult to show, as is well known, that $C_S(x) = \prod_{d|r} Z_d \wr \mathrm{Sym}_{e_d}$, where Z_d denotes a cyclic group of order d. The convention is that the corresponding factor is omitted if some Ω_d is empty. The point is that the number of non-abelian composition factors of $C_{\mathrm{Sym}(\Omega)}(x)$ is at most the number of divisors of r, a number that depends only on the order r of x. The same applies to the centralizer of x in $\mathrm{Alt}(\Omega)$.

Returning to the tower (29), we see that the number of non-abelian composition factors of $C_{G_i}(x)$ is bounded independently of x. A slightly stronger version of this can be given which turns out to imply that $C_G(x)$ has a finite series in which some factors are simple and the others soluble.

On the other hand the non-abelian composition factors in the groups $C_i = C_{G_i}(x)$ all occur in a single layer as a direct product. That feature need not be preserved in the group G, where centralizers can involve non-split extensions of infinite simple groups [24].

By more elaborate versions of the above argument, taking into account the various possibilities for simple groups in a tower (29), the following can be obtained.

Theorem 3.4 [25] *Let G be an infinite simple locally finite group and $x \in G$. Suppose that G has a local system of finite simple subgroups. Then $C_G(x)$ has a finite series in which each factor is either simple or soluble. If G is not linear, then some simple factor of $C_G(x)$ is not linear.*

We ask the following.

Problem 3.5 *Suppose that G is simple locally finite and is the union of a tower of finite groups, each of which is a direct product of simple groups. What can be said about the structure of the centralizers of the elements of G?*

If the number of direct factors is bounded, this is not much different from the above case. But Hickin's groups tell us that the number need not be bounded.

3.3. OCCURRENCE OF STRANGE CENTRALIZERS

We mentioned the possibility of looking for general statements about centralizers of elements in arbitrary simple locally finite groups. In linear groups the situation seems quite well under control and not much different from that for algebraic groups. The situation for semisimple elements is reasonably well documented in [24] and more general elements will probably be dealt with to some extent in [25]. Let us leave aside the finitary case. For non-finitary groups one might feel that the groups are very complex and this complexity should be reflected in the centralizers. However there are many ways in which this might be made precise and it is not clear that any of them is actually true. Roughly speaking, Theorem 3.4 tells us that if G has a local system of finite simple subgroups then the centralizer of every element has only a finite number of non-abelian composition factors. Actually the way in which the theorem is phrased makes sure that the centralizer of each element has a well defined set of non-abelian composition factors. It is natural to wonder whether centralizers in all simple locally finite groups have this structure, but Meierfrankenfeld's groups furnish counterexamples (see below). A next shot might be to suspect that the centralizer of every element involves an infinite simple group, which is non-linear if G is. In Meierfrankenfeld's groups the centralizer of every element, and so of every finite subgroup, is residually finite, and so has a composition series with finite factors. In some sense they are moral counterexamples. But we remarked that a residually and locally finite group can involve an infinite simple group, and this phenomenon occurs in Mf-groups of VDA type. Thus they are not actual counterexamples. In the other Mf-groups we (or at least I) do not know whether centralizers involve infinite simple groups. The situation is not at all clear. Here is the theorem.

Theorem 3.6 *Let G be an Mf-group and F a non-trivial finite subgroup of G. Then:*

(1) $C_G(F)$ *is residually finite.*

(2) *If G is of* VDA *type, then $C_G(F)$ involves every group $G(\mathbf{s})$ (cf. Proposition 1.22) and hence involves Hall's group U.*

It could still presumably be the case that every composition factor of $C_G(F)$ is finite, since the infinite simple groups obtained do not seem to be part of a composition series of $C_G(F)$.

We have noted that we cannot exclude the possibility that there exists a non-linear simple locally finite group with an elementary abelian centralizer. At least the following gives some limit on how far this kind of thing can go. It shows, for example, that there will always be plenty of centralizers that are not locally soluble.

Lemma 3.7   *Let G be a non-linear simple locally finite group and S a finite simple group. Then there is a non-trivial element $x \in G$ such that $C_G(x)$ involves S.*

Proof.   Choose a prime p that does not divide $|S|$, and let C be a cyclic group of order p. Then $S \times C$ is involved in G (Theorem 2.6). Therefore there exist finite subgroups $K \triangleleft H \leqslant G$ such that $H/K = S_0/K \times C_0/K$, where $S_0/K \cong S$ and $C_0/K \cong C$. We can take K to be nilpotent. It is not too hard to prove then by induction that there exist subgroups S_1 and C_1 of H such that $[S_1, C_1] = 1$, $S_1 K = S_0$ and $C_1 K = C_0$. Thus $S_1 \leqslant C_G(C_1)$ and S_1 has S as a homomorphic image.   □

The following, however, seems to be open. The first part of it seems to have been raised explicitly first in a seminar at Middle East Technical University.

Problem 3.8   *Does there exist a non-linear infinite simple locally finite group in which the centralizer of every non-trivial element is almost soluble, that is, has a soluble subgroup of finite index? More explicitly, do some or all of the* FP *(free product) groups mentioned in Subsection 1.7 have this property?*

Studying this for FP-groups seems inevitably to lead to the following.

Problem 3.9   *Let $E = H_1 * H_2 * \cdots * H_r * F$, where the H_i are finite groups and F is free. Let R be a normal subgroup of finite index in E such that $R \cap H_i = 1$ for $1 \leqslant i \leqslant r$, and let N be a proper characteristic subgroup of R. Let $\overline{E} = E/N$ and $\overline{R} = R/N$. Let x be an element of $\overline{E}$ outside $\overline{R}$. Let $C = C_{\overline{E}}(x)$. What can be said about $C\overline{R}/\overline{R}$? What about the case $N = R'R^n$, where $n = |E/R|$?*

What we have in mind is that this image of the centralizer should be small unless x is somehow closely related to one of the H_i, and then it should somehow be controlled in terms of the relevant H_i. As a very special case we mention the following.

Lemma 3.10   *Let F be a free non-abelian group, let R be a normal subgroup of finite index in F, and let p be a prime. Let $\overline{F} = F/R'R^p$, and let x be a p-element of $\overline{F}$ outside $\overline{R} = R/R'R^p$. Let $C = C_{\overline{F}}(x)$. Then $C\overline{R}/\overline{R}$ has a normal p-complement with cyclic quotient.*

For example, if $p = 2$, we see that the centralizer of every 2-element of $\overline{F}$ outside $\overline{R}$ is soluble (using the Feit-Thompson Theorem).

This seems to be something of a jungle.

3.4. Finite Centralizers

It seems to be an open question whether a non-linear simple locally finite group can contain a finite subgroup with trivial centralizer. We will discuss what seems to be known about this. It follows from the classification of such groups that a simple linear locally finite group does contain such a subgroup, and indeed it follows without the classification, because centralizers in a linear group are Zariski closed and so the minimal condition on centralizers holds.

Note the following.

Lemma 3.11 *Let G be a locally finite group with $Z(G) = 1$. Then the following two statements are equivalent.*

(1) *G contains a finite subgroup with trivial centralizer.*

(2) *G contains a finite subgroup with finite centralizer.*

For suppose that F is a finite subgroup of G such that $C_G(F) = C$, a finite group. Since $Z(G) = 1$, we can choose, for each $1 \neq c \in C$, an element g_c such that $[c, g_c] \neq 1$. Let $E = \langle F, g_c \mid 1 \neq c \in C \rangle$. Then E is a finite subgroup of G with trivial centralizer. □

A problem which is related to the question of finite subgroups with trivial centralizer is the following.

Problem 3.12 *Let G be a countably infinite locally finite group. Is it true that $|\operatorname{Aut} G| = 2^{\aleph_0}$?*

This seems rather unlikely but it does not seem easy to find a counterexample. Where can we look for automorphisms? Inner automorphisms will not give us enough. We might next try locally inner automorphisms. An automorphism ϕ of a group H is *locally inner* if, given any finite subset X of G, there exists $g \in G$ such that $\phi(x) = x^g$ for all $x \in X$. The set of all locally inner automorphisms of G is a subgroup $\operatorname{Linn} G$ of $\operatorname{Aut} G$. The following is well known.

Lemma 3.13 *Let G be a countable locally finite group.*

(1) *Suppose that $C_G(F) \nleq Z(G)$ is satisfied for all finite subgroups F of G. Then $|\operatorname{Linn} G| = 2^{\aleph_0}$.*

(2) *Suppose that there is a finite subgroup F of G such that $C_G(F) = Z(G)$. Then $\operatorname{Linn} G = \operatorname{Inn} G$.*

Proof. (1) Write $G = \bigcup_{i=1}^{\infty} G_i$, where the G_i form a strictly increasing tower of finite subgroups of G. We build up a tower of finite subgroups H_i of G such that $H_i \geqslant G_i$ and each inner automorphism of H_i can be extended to an inner automorphism of H_{i+1} in two different ways. Suppose that H_i has been obtained. Choose an element $x \in C_G(H_i) \smallsetminus Z(G)$, let y be an element of G such that $[x, y] \neq 1$, and let $H_{i+1} = \langle H_i, G_{i+1}, x, y \rangle$. Let $t \in H_i$. Then t and xt induce the same inner automorphism of H_i, but $y^t \neq y^{xt}$. Now it is easy to build up $2^{\aleph_0}$ locally inner automorphisms of G by successively extending from each term of the tower (H_i) to the next.

(2) Let ϕ be a locally inner automorphism of G, let F be the given finite subgroup, and let x be an element of G such that $\phi(f) = f^x$ for all $f \in F$. Let $g \in G$. Then there exists an element y of G such that $\phi(f) = f^y$ for all $f \in \langle F, g \rangle$. It follows that $yx^{-1} \in C_G(F) = Z(G)$, and so $\phi(g) = g^y = g^x$. In other words, ϕ is the inner automorphism of G induced by x.   □

Thus, if we have a simple locally finite group G containing a finite subgroup with trivial centralizer, then the group of locally inner automorphisms is countable, and, unless the group is linear or otherwise well under control, it is likely to be difficult to know where to look for other automorphisms.

For the rest of this subsection, suppose that G is a countably infinite simple locally finite group containing a finite subgroup F such that $C_G(F) = 1$. What can we do? It is easy to see that F cannot be nilpotent, but here we wish to avoid any structural assumption on F. Let $\{ (G_i, N_i) \mid i \in I \}$ be a Kegel sequence. We can take $F \leqslant G_1$. Then F acts on the residually finite group $N = N_1 N_2 \cdots$ and $C_N(F) = 1$. We are back to actions of finite groups on residually finite ones and we cannot do anything about this at present. So let's assume that $N_i = 1$ for all i. Thus

$$G_1 \leqslant G_2 \leqslant \ldots \qquad (30)$$

and the G_i are finite simple. We can assume that the tower is of a type corresponding to Alt, $\mathrm{CL}(p^\alpha)$, $\mathrm{CL}(p^\infty)$, or $\mathrm{CL}(\infty)$.

Lemma 3.14 *Let F be a subgroup of $A = \mathrm{Alt}_n$ such that $|C_A(F)| \leqslant m$. Then n is bounded in terms of m and $|F|$.*

This is an easy exercise. It just depends on the fact that if n is very large then F will have many isomorphic orbits on the natural A-set. Hence G is not of type Alt.

Now, if G is of type $\mathrm{CL}(\infty)$, then we may delete the terms of (30) that are classical groups over fields of characteristic dividing $|F|$, and suppose that none are such. Then a roughly similar argument to the above also leads to a contradiction. It involves proving a lemma similar to Lemma 3.14, where A is replaced by a classical group over a field of characteristic prime to $|F|$. See [21].

Suppose then that G is of type $\mathrm{CL}(q)$, where q is a power of a prime p. For definiteness take $G_i \cong \mathrm{PSL}_{n_i}(q)$. Thus we have a fixed group F that can be embedded in infinitely many groups $\mathrm{PSL}_n(q)$, where q is fixed, as a group with trivial centralizer. Unfortunately this can happen. It is shown in [21, Proposition C2] that if p is any prime, k is any field of characteristic p, and L is any finite group such that $O_p(L) = 1$, then there exists a finite group H of the form $H = AL$, where $A \lhd H$ and $A \cap L = 1$, that can be embedded in infinitely many groups $\mathrm{PSL}_n(k)$ as a subgroup with trivial centralizer.

This construction has some connections with modular representation theory. It turns out that H is also isomorphic to a subgroup of $\mathrm{SL}_n(k)$, and that the centralizer of this subgroup consists of the scalar matrices. Let V_n be the natural $\mathrm{SL}_n(k)$-module. Then V_n is a kH-module, and $\mathrm{End}_{kH} V_n = k$. In particular V_n is an indecomposable kH-module; we have indecomposable kH-modules of unbounded finite dimensions. This phenomenon was of course known, but in our case we have the stronger condition that the endomorphism rings of these modules consist only of scalars.

This situation is intrinsically bound up with our counterexample. But in that situation we could replace F by the simple group G_1. Maybe it is reasonable to ask

Problem 3.15   *Let F be a finite simple group, or perhaps a finite group such that $O_p(F) = 1$. Suppose that $F \leqslant \mathrm{PSL}_n(q) = M$, where q is a power of p and $C_M(F) = 1$. Does it follow that n is bounded in terms of $|F|$? We are particularly interested in the case when F is also a projective special linear group over $\mathbb{F}_q$. What about the algebraic group situation, that is, say, $G_i \cong \mathrm{PSL}_{n_i}(\overline{\mathbb{F}}_p)$, and G_1 is a closed subgroup of G_2 with $C_{G_2}(G_1) = 1$? Does it follow that n_2 is bounded in terms of n_1?*

It may also be noted that, in the situation of the counterexample, one can find an infinite semidirect product $N = N_1 N_2 \cdots$ normalized by F; this brings us back yet again to actions of finite groups on residually finite ones.

3.5. LINEAR CENTRALIZERS

We have previously noted that we have not been able to say anything about the structure of a simple locally finite group containing an element x of prime order p whose centralizer is elementary abelian. Now every elementary abelian group is linear (being isomorphic to the additive group of some field), and so this is in one sense a very basic case of the following problem.

Problem 3.16   *Let G be a simple locally finite group containing an element with linear centralizer. Does it follow that G is linear?*

However the case just mentioned is that of a unipotent centralizer, and if we assume instead that the unipotent radical of the centralizer is finite we can answer the above affirmatively. More generally we can obtain some rather strong results for locally finite groups containing elements with such centralizers, without assuming simplicity. The results in this section are due to V. V. Belyaev and the author, [5] and [23].

Let L be any locally finite linear group over a field of characteristic $p \geqslant 0$, and suppose that $O_p(L)$ is finite. This condition is taken to be automatically satisfied if $p = 0$. Then it is not difficult to show, using the classification of simple locally finite linear groups and other things, that L has a normal subgroup J of finite index such that

$$J = TS_1 \cdots S_r, \tag{31}$$

where T is an abelian group of finite rank (that is, a direct product of a finite number of locally cyclic groups, a sort of 'torus', in general involving infinitely many primes), each S_i is an infinite quasisimple linear group, and so of Lie type over an algebraic extension of a finite field, and the factors commute elementwise. When $p = 0$, we take r to be zero. Thus, we shall consider locally finite groups containing an element or finite subgroup whose centralizer has a subgroup of finite index of the form (31). We shall not assume that all the S_i are defined in the same characteristic.

Definition 3.17 *We say that a locally finite group G is* of linear type *if G contains a normal subgroup N such that N is a direct product of a finite number of simple linear groups (some of which may be finite) and $C_G(N) = 1$.*

Clearly, a simple group of linear type is linear. We write $S(G)$ for the locally soluble radical of a locally finite group G, and $F(G)$ for the Hirsch-Plotkin radical of G (the unique maximal normal locally nilpotent subgroup).

Theorem 3.18 *Let G be a locally finite group containing a finite nilpotent subgroup F. Suppose that $C_G(F)$ contains a subgroup of finite index of the form (31). Then:*

(1) *If F is cyclic and $F(G) = 1$, then G is of linear type.*

(2) *If T is Černikov and $F(G) = 1$, then G is of linear type.*

(3) *If T is finite and F is cyclic, then $G/S(G)$ is of linear type, and $S(G)$ has a finite series with locally nilpotent factors.*

Corollary 3.19 *With the same hypotheses, if G is simple and either F is cyclic or T is Černikov, then G is linear.*

Presumably the result should remain true assuming only that G is simple, without the subsidiary assumptions on F and T. Indeed it should be possible to combine the three parts of Theorem 3.18 into one, but we seem to be a long way from that at the present.

We shall say a few words about (3) and then concentrate on (1) and (2). (3) is basically a result in the theory of automorphisms of finite soluble groups.

Proposition 3.20 *Let A be a finite cyclic group acting on a finite soluble group S, and let $C = C_S(A)$. Then the Fitting height of S is bounded in terms of $|A|$ and $|C|$.*

The point here is that it is not assumed that the orders of A and S are coprime. The result can be viewed as a generalization of the work of Shamash and Shult on finite soluble groups with cyclic Carter subgroups, which is the case $C = 1$. However the bound obtained on the Fitting height is astronomical. Applying this to the action of F on the finite F-invariant subgroups of $S(G)$ is what tells us that $S(G)$ has finite 'Fitting height'. One then has to do some work to make sure that the hypotheses are inherited by $G/S(G)$ (and the structure just obtained for $S(G)$ is used here), and then either of the first two parts can be applied.

Passing to (1) and (2), let $C = C_G(F)$. The proofs revolve around a certain type of finite 'inseparable' subgroup of C analogous to those considered by Belyaev [3].

Definition 3.21 *Let H be a locally finite group, X a finite subgroup of H, and π a set of primes. Then X is called π-inseparable if, whenever Y is a subgroup of H normalized by X and such that $X \cap Y = 1$, we have that Y is an abelian π'-group of finite rank. If the hypothesis on Y implies that $Y = 1$, we say that X is inseparable.*

These subgroups Y are the X-signalizers.

Lemma 3.22 *Suppose that L is a group having a subgroup J of finite index with the structure (31), and let π be any finite set of primes. Then L has a finite π-inseparable subgroup. Further, if T is Černikov, then L has a finite inseparable subgroup.*

We shall not discuss the proof of this. It uses standard methods. The main point is that, if L is a group of Lie type over an infinite locally finite field, and we take a subgroup L_0 of L of the same type over a suitable finite subfield, then L_0 is contained in every non-trivial subgroup of L that it normalizes, so L_0 is certainly inseparable. Now consider what this says about our problem. We concentrate on Case (2) of Theorem 3.18 for definiteness. We have a locally finite group containing a finite nilpotent subgroup F such that $C_G(F) = C$ satisfies the hypotheses of Lemma 3.22. Then C contains a finite inseparable subgroup R. Let N be any subgroup of G normalized by F and R and such that $N \cap R = 1$. Then F acts on N and $C_N(F) = C \cap N = 1$, since $(C \cap N) \cap R = 1$. Next we prove

Lemma 3.23 *Let H be a finite group admitting a finite nilpotent group Q of automorphisms such that $C_H(Q) = 1$. Then H is a soluble group and the Fitting height of H is bounded by a function of the composition length of Q.*

The statement that H is soluble is new, as far as we know. It uses CFSG and is a generalization of the fact that a finite group admitting a fixed point free automorphism is soluble. The bound on the Fitting height then follows from Dade's work on Carter subgroups and Fitting heights [11]. Thus, in particular, if $N \triangleleft G$ and $N \cap R = 1$, we find that N is locally soluble and of finite 'Fitting height', and, since $F(G) = 1$ by assumption, we find that $N = 1$. Therefore every non-trivial normal subgroup of G has non-trivial intersection with R. It follows from this that G has a minimal normal subgroup. To finite group theorists this may come as no surprise, but in infinite group theory this is real progress. The reason why G has a minimal normal subgroup is as follows. Choose a non-trivial normal subgroup N of G such that $N \cap R$ is as small as possible, recalling that R is finite. Let $N \cap R = N_0$. Let

$$M = \bigcap \{\, K \triangleleft G \mid K \cap R = N_0 \,\}.$$

It is an easy exercise to prove that M is a minimal normal subgroup of G. With a bit more work along the same lines we can prove that a minimal normal subgroup of G is a direct product of simple non-abelian groups. These are permuted by F, and we reduce quickly to the case when

$$G = SF, \tag{32}$$

where S is a simple non-abelian normal subgroup. Our task then is to prove that S is linear. This follows from the following lemma.

Lemma 3.24 *Let S be any non-linear simple locally finite group and let E be a finite subgroup of Aut S. (We identify S with Inn S.) Then there exists a subgroup N of S such that*

(1) *N is a repeated semidirect product of finite groups*

$$N = N_1 N_2 \cdots,$$

where N_i normalizes N_j if $i \leqslant j$, each group N_i is E-invariant, and

$$N_i \cap N_{i+1} N_{i+2} \cdots = 1,$$

and

(2) *if n_i is the product of the orders of the non-abelian composition factors of N_i then $n_i \to \infty$ as $i \to \infty$.*

In the situation (32), take $E = \langle F, R \rangle$, and suppose that S is non-linear. We obtain the group N as in Lemma 3.24. There exists i such that $R \cap N_i N_{i+1} \cdots = 1$. Then F acts fixed point freely on $N_i N_{i+1} \cdots$, by the inseparable nature of R. Hence, by Lemma 3.23, N_j is soluble if $j \geqslant i$. But this contradicts Lemma 3.24(2) and concludes the proof.

The proof of Theorem 3.18(1) is broadly similar. The extra work comes in Lemma 3.23. There, although we know that Q is cyclic, we only have that $C_H(Q)$ is an abelian π'-group of bounded rank. We can take π to be the set of prime divisors of $|Q|$. The conclusion is that $S(H)$ has bounded Fitting height and $F^*(H/S(H))$ is a direct product of a bounded number of simple groups which are either sporadic, alternating of bounded degree, or of Lie type and bounded rank. To combine Cases (1) and (2) seems to need a version of this for an arbitrary nilpotent group Q; this would be an extension of Dade's work on Carter subgroups and Fitting heights.

References

1. S. D. Bell and B. Hartley, A note on fixed-point-free actions of finite groups, *Quart. J. Math. Oxford* (2) 41 (1990), 127–130.
2. V. V. Belyaev, Locally finite Chevalley groups, in *Investigations in Group Theory*, Urals Scientific Centre, Sverdlovsk, 1984, pp. 39–50 (in Russian).
3. V. V. Belyaev, Locally finite groups containing a finite inseparable subgroup, *Sibirsk Mat. Zh.* 34 (1993), 23–41 (in Russian).
4. V. V. Belyaev, Inert subgroups in simple locally finite groups, *these Proceedings*.
5. V. V. Belyaev and B. Hartley, to appear.
6. F. Beyl and J. Tappe, *Group Extensions, Representations and the Schur Multiplier*, Lecture Notes in Math. 958, Springer-Verlag, Berlin, 1982.
7. A. V. Borovik, Periodic linear groups of odd characteristic, *Soviet Math. Dokl.* 26 (1982), 484–486.
8. R. Carter, *Simple Groups of Lie Type*, Wiley-Interscience, 1972 .
9. R. W. Carter, *Finite Groups of Lie Type: Conjugacy Classes and Complex Characters*, Wiley, London, 1985.
10. P. M. Cohn, *Universal Algebra*, Harper and Row, New York, 1965.
11. E. C. Dade, Carter subgroups and Fitting heights of finite solvable groups, *Illinois J. Math.* 13 (1969), 449–514.
12. D. Gorenstein, *The Classification of Finite Simple Groups*, Plenum Press, New York, 1983.
13. D. Gorenstein, *Finite Simple Groups; an Introduction to their Classification*, Plenum Press, New York, 1982.
14. R. Griess, Schur multipliers of the known finite simple groups II, *Proc. Symp. Pure Math.* 37 (1980), 279–282.
15. J. I. Hall, Locally finite simple groups of finitary transformations, *these Proceedings*.

16. J. I. Hall and B. Hartley, A group theoretical characterization of simple, locally finite, finitary linear groups, *Arch. Math.* 60 (1993), 108–114.
17. P. Hall, Some constructions for locally finite groups, *J. London Math. Soc.* 34 (1959), 305–319.
18. P. Hall, On non-strictly simple groups, *Proc. Cambridge Philos. Soc.* 59 (1963), 531–553.
19. B. Hartley, Fixed points of automorphisms of certain locally finite groups and Chevalley groups, *J. London Math. Soc.* (2) 37 (1988), 421–436.
20. B. Hartley, A theorem of Brauer-Fowler type and centralizers in locally finite groups, *Pacific J. Math.* 152 (1992), 101–117.
21. B. Hartley, Centralizing properties in simple locally finite groups and large finite classical groups, *J. Austral. Math. Soc.* (Ser. A) 49 (1990), 502–513.
22. B. Hartley, On some simple locally finite groups constructed by Meierfrankenfeld, Preprint No. 1994/01, Manchester Centre for Pure Mathematics, 1994 (obtainable from Department of Mathematics, University of Manchester).
23. B. Hartley, to appear.
24. B. Hartley and M. Kuzucuoğlu, Centralizers of elements in locally finite simple groups, *Proc. London Math. Soc.* (3) 62 (1991), 301–324.
25. B. Hartley and M. Kuzucuoğlu, to appear.
26. B. Hartley and G. Shute, Monomorphisms and direct limits of finite groups of Lie type, *Quart. J. Math. Oxford* (2) 35 (1984), 49–71.
27. B. Hartley and A. E. Zalesskiĭ, On simple periodic linear groups—dense subgroups, permutation representations and induced modules, *Israel J. Math.* 82 (1993), 299–327.
28. K. Hickin, Universal locally finite central extensions of groups, *Proc. London Math. Soc.* (3) 52 (1986), 53–72.
29. K. Hickin, Complete universal locally finite groups, *Trans. Amer. Math. Soc.* 239 (1978), 213–227.
30. K. Hickin, A. c. groups, extensions, maximal subgroups, and automorphisms, *Trans. Amer. Math. Soc.* 290 (1985), 457–481.
31. K. K. Hickin and A. Macintyre, Algebraically closed groups: embeddings and centralizers, in *Word Problems II*, North-Holland, New York, 1980, 141–155.
32. A. G. Jilinskiĭ, Coherent systems of finite type of inductive systems of non-diagonal embeddings of simple Lie algebras, *Doklady AN Belorus. SSR* 36 (1992), 9–13 (in Russian).
33. A. G. Jilinskiĭ, Classification of diagonal algebras, unpublished.
34. M. I. Kargapolov, Locally finite groups having a normal system with finite factors, *Sibirsk. Mat. Zh.* 11 (1961), 853–873 (in Russian).
35. O. H. Kegel, Über einfache, lokal endliche Gruppen, *Math. Z.* 95 (1967), 169–195.
36. O. H. Kegel and B. A. F. Wehrfritz, *Locally Finite Groups*, North-Holland, Amsterdam, 1973.
37. A. G. Kurosh, *Theory of Groups*, Vol II, trans. K. A. Hirsch, Chelsea, New York, 1970.
38. M. W. Liebeck and G. M. Seitz, Subgroups generated by root elements in groups of Lie type, *Ann. Math.* 139 (1994), 293–361.
39. U. Meierfrankenfeld, Non-finitary, locally finite, simple groups, preprint, Michigan State University, 1991.
40. U. Meierfrankenfeld, Non-finitary locally finite simple groups, *these Proceedings*.
41. B. H. Neumann, Permutational products of groups, *J. Austral. Math. Soc.* 1 (1959/60), 299–310.
42. R. E. Phillips, Existentially closed locally finite central extensions, multipliers and local systems, *Math. Z.* 187 (1984), 383–392.
43. R. E. Phillips, On absolutely simple locally finite groups, *Rend. Sem. Mat. Padova* 79 (1988), 213–220.
44. R. E. Phillips, Finitary linear groups: a survey, *these Proceedings*.
45. C. Praeger and A. E. Zalesskiĭ, Orbit lengths of permutation groups, and group rings of locally finite simple groups of alternating type, *Proc London Math. Soc.* (3) 70 (1995), 313–335.
46. E. B. Rabinovich, Inductive limits of symmetric groups and universal groups, *Visti Acad. Navuk Belorussian SSR* (ser. fiz.-mat. navuk) no. 5 (1981), 39–42 (in Russian).
47. D. J. S. Robinson, *Finiteness Conditions and Generalized Soluble Groups*, Part I, Springer-Verlag, Berlin, 1972.

48. J. Shamash and E. Shult, On groups with cyclic Carter subgroups, *J. Algebra* 11 (1969), 564–597.

49. R. Steinberg, *Lectures on Chevalley Groups,* Mimeographed notes, Yale University, 1967.

50. S. Thomas, The classification of the simple periodic linear groups, *Arch. Math.* 41 (1983), 103–116.

51. G. E. Wall, On the conjugacy classes in the unitary, symplectic and orthogonal groups, *J. Austral. Math. Soc.* 3 (1963), 1–63.

52. B. A. F. Wehrfritz, *Infinite Linear Groups,* Springer-Verlag, Berlin, 1973.

53. H. Wielandt, *Unendliche Permutationsgruppen,* Lecture notes prepared by Adolf Mader, Math. Inst. der Univ. Tübingen, 1960.

54. A. E. Zalesskiĭ, Group rings of inductive limits of alternating groups, *Algebra and Analysis* 2 (1990), 132–149 (in Russian). English translation: *Leningrad Math. J.* 2 (1991), 1287–1303.

55. A. E. Zalesskiĭ, Group rings of simple locally finite groups, *these Proceedings.*

56. A. E. Zalesskiĭ and V. N. Serezhkin, Finite linear groups generated by reflections, *Math. USSR Izv.* 17 (1981), 477–503.

ALGEBRAIC GROUPS

G. M. SEITZ
Department of Mathematics
University of Oregon
Eugene, Oregon 97403
USA
E-mail: seitz@math.uoregon.edu

Abstract. The notes discuss material on the theory of algebraic groups which is essential for a detailed study of the subgroup structure of algebraic groups, finite groups of Lie type, and certain locally finite groups.

The first section covers the general theory of algebraic groups, starting from the definition of algebraic variety. The interaction of various layers of structures (the coordinate ring, associated Lie algebra, Zariski topology) is discussed. Key points of the basic theory are mentioned (Jordan decomposition, reductive and semisimple groups, big cell, commutator relations), progressing towards the classification of simple algebraic groups.

The second section concerns the subgroup theory of simple algebraic groups. This begins with a study of subsystem groups and parabolic subgroups. Included is a brief discussion of modules occurring within parabolic subgroups and connections with unipotent classes. The section concludes with a discussion of the determination of the maximal closed connected subgroups of simple algebraic groups.

The final section is an overview of the representation theory of simple algebraic groups. This begins with a discussion of high weight modules and proceeds to the parametrization of irreducible modules by high weights. Included is a discussion of Weyl modules and consequences for extension theory. The section on tensor product theorems includes results of Steinberg, Borel-Tits, and a recent result of the author on abstract homomorphisms. The section concludes with applications to the theory of locally finite groups.

Key words: algebraic varieties, algebraic groups, reductive groups, subgroups, representations.

1. Introduction

Our goal here is to provide a brief introduction to the structure and representations of simple algebraic groups. This is a very large subject, so we limit ourselves to a few topics that are particularly significant for the analysis of the subgroup structure of algebraic groups and associated finite groups of Lie type. We indicate some of the key results and illustrate the ideas with examples, but omit essentially all proofs. The interested reader should consult some of the references for more complete accounts of the subject.

The material is divided into three main sections. The first covers fundamentals of algebraic varieties, algebraic groups, and associated Lie algebras, up to the classification of simple algebraic groups. The second section contains a number of results on the subgroup structure of simple algebraic groups. This includes material on subsystem subgroups, internal modules within parabolic subgroups, semisimple and unipotent classes, groups containing long root elements, and a brief discussion of maximal subgroups. The final section concerns representation theory. We give the

B. Hartley et al. (eds.), Finite and Locally Finite Groups, 45–70.
© 1995 *Kluwer Academic Publishers. Printed in the Netherlands.*

basic theory of high weight modules and describe the parametrization of irreducible modules and their connection with Weyl modules. A fundamental theorem of Steinberg on tensor products is stated followed by results of Borel-Tits and the author on abstract representations. We close with an application to the theory of locally finite groups.

2. Preliminary Results

In this first section we describe basic material from the study of algebraic varieties and algebraic groups. The theory of algebraic groups is particularly rich because of the interplay between several types of structures. We will try to illustrate connections between the various structures. In particular, we introduce the Lie algebra of an algebraic group and describe results leading to the classification of the simple algebraic groups. The material in this section is standard and we use [4, 23] as general references.

2.1. ALGEBRAIC VARIETIES

Let K be an algebraically closed field and consider $K[x_1, \ldots, x_n]$ as a ring of polynomial functions on K^n. An *affine algebraic variety* is the zero set of a set of polynomials. The zero set of a set of polynomials is the same as the zero set of the ideal generated by the polynomials, so we work entirely with ideals.

There is a map $I \mapsto V(I)$ sending an ideal to the subset of K^n annihilated by it. Conversely, there is a map $X \mapsto I(X)$ sending a subset of K^n to the ideal annihilating it. For an ideal I, we have $V(I) = V(\sqrt{I})$ and the Hilbert Nullstellensatz implies that $I(V(I)) = \sqrt{I}$. Consequently, we focus on zero sets of radical ideals. The correspondence between radical ideals and the zero sets of these ideals is bijective.

Given a radical ideal I, the quotient ring $K[x_1, \ldots, x_n]/I$ can be viewed as a ring of functions on $V(I)$. Consequently, we now regard an affine algebraic variety as a pair $(V, K[V])$, consisting of a subset V of K^n together with a ring of functions on V, without nilpotent elements. The ring $K[V]$ is called the *coordinate ring* of V. We often call the variety V, suppressing the coordinate ring, but this ring is essential to the study as it connects the geometrical and algebraic points of view.

This can all be done in a coordinate free way. An affine algebraic variety is defined to be a set with a finitely generated algebra of functions satisfying certain conditions. However, for our purposes we can work with subsets of K^n. Given affine algebraic varieties $V \subseteq K^m$ and $W \subseteq K^n$, then $V \times W \subseteq K^{m+n}$ and it is an algebraic variety if we set $K[V \times W] = K[V] \otimes K[W]$. There are a few things to check here and some identifications have been made.

The *Zariski topology* on K^n is defined in terms of the above concepts. The closed subsets are just the zero sets of ideals of polynomials. Each affine algebraic variety inherits the induced topology. If $(V, K[V])$ is an affine algebraic variety, then $K[V]$ is Noetherian, so V satisfies the descending chain condition on closed sets and hence the ascending chain condition for open sets. We say that V is a *Noetherian topological space*. As a topological space, V is *quasicompact*, where this is just compactness in the ordinary sense, but without the Hausdorff axiom.

An affine algebraic variety V is *irreducible* if it cannot be expressed as the union of two proper closed subsets. For example, K^n is irreducible. Indeed, suppose $K^n = C_1 \cup C_2$, with each C_i a proper closed subset of K^n. Then, $I(C_1) \cap I(C_2)$ annihilates all of K^n and hence is 0. This intersection contains the product of the ideals, forcing one of the ideals to be 0. This is a contradiction. More generally, the same argument shows

Proposition 2.1 [23, 1.4] *V is irreducible if and only if $K[V]$ is an integral domain.*

If V is an irreducible variety, then we define $\dim(V)$ to be the transcendence degree of the quotient field of $K[V]$. So, for example, $\dim(K^n) = n$. This notion of dimension behaves well with respect to subvarieties.

Proposition 2.2 [23, 3.2] *Let I_1 and I_2 be irreducible varieties such that $I_1 \subseteq I_2$. Then $\dim(I_1) \leqslant \dim(I_2)$. Moreover, the dimensions coincide if and only if the varieties do.*

An arbitrary affine variety is the union of its finitely many maximal irreducible subsets, called *irreducible components*. Dimension of such a variety is defined to be the maximum of the dimensions of its irreducible components.

Let $(V, K[V])$ and $(W, K[W])$ be affine algebraic varieties. Let $\varphi : V \to W$ be a function. For $f \in K[W]$, $\varphi^*(f) = f \circ \varphi$ is a K-valued function on V. We say φ is a *morphism* provided $\varphi^*(f) \in K[V]$ for all $f \in K[W]$. So a morphism φ gives rise to an algebra homomorphism $\varphi^* : K[W] \to K[V]$, called the *comorphism* of φ.

An *isomorphism* between algebraic varieties is a morphism which is bijective as a function and for which the comorphism is an isomorphism. This is a strong condition and the following simple example is instructive. Let $V = K^1 = K$, where K is a field of positive characteristic p. Consider the *Frobenius morphism*, $F : c \mapsto c^p$. Then F is an automorphism of the field K, but not an automorphism of K viewed as algebraic variety. This is because the comorphism is the map $f(x) \mapsto f(x^p)$ on the coordinate ring $K[x]$, and hence is not surjective.

So far we have ignored an important part of the theory. For many purposes the notion of affine algebraic variety is inadequate and it is necessary to introduce a more general notion of algebraic variety. For example, it is often necessary to do this in considering group actions. Roughly speaking one considers Noetherian topological spaces having a covering by open sets, each of which is an affine algebraic variety. Consistency is obtained by further imposing a sheaf of functions which restricts to the coordinate ring for each of the affine open sets.

An important example is afforded by *projective space*, P^n, the set of all elements in $K^{n+1} \setminus \{0\}$, with elements identified if they are scalar multiples of each other. One can also think of P^n as the set of 1-spaces of K^{n+1}. The required topology on P^n is obtained by defining a closed set to be the zero set of a collection of homogeneous polynomials.

For each $1 \leqslant i \leqslant n+1$, let A_i denote the subset of P^n consisting of all elements of P^n having nonzero entry in the i-th coordinate. An element of A_i can be identified with an element of K^n by first normalizing to get a 1 in the i-th coordinate and then deleting the i-th coordinate. In this way we see that P^n can be covered by the A_i, so has a covering by open sets, each a copy of K^n.

2.2. ALGEBRAIC GROUPS

We can now define the notion of an *affine algebraic group*. This is a group G which also has the structure of an affine algebraic variety in such a way that the multiplication map $m : G \times G \to G$ and the inverse map $i : G \to G$ are morphisms of algebraic varieties.

Proposition 2.3 [23, 8.6] *An affine algebraic group is isomorphic to a closed subgroup of* $\mathrm{GL}_n(K)$ *for some integer* n.

An example is in order. Let $G = \mathrm{SL}_n(K)$. View this as a subvariety of $\mathrm{M}_n(K)$ (which can be regarded as K^{n^2}), so that the coordinate ring of G is $K[x_{ij}]/I$ where I is the ideal generated by the polynomial $\det - 1$. So what is the comorphism of m? To ease notation, we identify x_{ij} with its restriction to G. Then $m^*(x_{ij}) = x_{ij} \circ m$. Applying this to a pair of elements of G we find that

$$m^*(x_{ij}) = \sum_k x_{ik} \otimes x_{kj},$$

so that m^* does indeed map

$$K[G] \to K[G \times G] = K[G] \otimes K[G].$$

Similarly for the inverse map.

It is clear that $\mathrm{SL}_n(K)$ is a closed subvariety of $\mathrm{M}_n(K)$. However, $\mathrm{GL}_n(K)$ is another matter. Indeed, it is an open set—the complement of $V(\det)$. However, it still has the the structure of an algebraic variety. Indeed, view $\mathrm{GL}_n(K)$ as the subgroup of $\mathrm{M}_{n+1}(K)$, consisting of determinant 1 matrices which are block diagonal with blocks of size $n \times n$ and 1×1. This can be generalized to show that, if V is an algebraic variety and if $f \in K[V]$, then the set of nonzeros of f has the structure of an affine algebraic variety.

If G is an algebraic group, then the irreducible component containing the identity is a closed normal subgroup. We denote this group by G°. The other irreducible components are just the (finitely many) cosets of G°. An algebraic group is irreducible if and only if it is connected in the Zariski topology. So when working with algebraic groups we use the words *irreducible* and *connected* interchangeably. Thus G is connected (i.e. irreducible) if and only if it contains no proper closed subgroup of finite index.

A *homomorphism* from one algebraic group to another is a morphism of algebraic varieties which is also a homomorphism of groups. The image of an algebraic group under a homomorphism is closed in the target group. Moreover, the image of a connected group is connected.

The usual group theoretic constructions produce closed subgroups. For example, the centralizer of any subset is closed, the normalizer of a closed subgroup is again closed, and the commutator of closed groups is closed. Moreover, the following result can be used to construct connected algebraic groups.

Theorem 2.1 [23, 7.5] *Suppose G is an algebraic group and $\{X_i \mid i \in I\}$ is a family of closed connected subgroups of G. Then $\langle X_i \mid i \in I \rangle$ is a closed connected subgroup of G.*

For example, using the known fact that $\mathrm{SL}_n(K)$ is generated by its root groups we see that $\mathrm{SL}_n(K)$ is connected. Indeed, each root subgroup is isomorphic to the additive group of K. A similar argument shows that any Chevalley group constructed from a representation of a complex simple Lie algebra and then realized over K has the structure of a connected algebraic group.

Given an algebraic group G there is an action of G on its coordinate ring as follows. For $g \in G$ and $f \in K[G]$, define a function λ_g with domain $K[G]$ so that, for $f \in K[G]$ and $x \in G$, $\lambda_g(f)(x) = f(g^{-1}x)$. It turns out that $\lambda_g(f) \in K[G]$, so this gives a group action of G on $K[G]$. Similarly, there is a right action.

The action of G on $K[G]$ is *locally finite* in the sense that any finite subset of functions is contained in a finite dimensional G-invariant subspace of $K[G]$. Hence, $K[G]$ gives rise to a large number of finite dimensional representations of G. These are *rational representations* in the sense that they are morphisms of G to $\mathrm{GL}(V)$ for suitable finite dimensional vector spaces V.

Let G be an algebraic group and H a closed subgroup. Then there is a quotient structure on the coset space G/H, giving the space the structure of an algebraic variety on which G acts morphically. In the special case where $H \trianglelefteq G$, then the quotient group has the structure of an affine algebraic group. Indeed, the coordinate ring consists of those elements of $K[G]$ which are constant on cosets of H. However, for subgroups that are not normal, the quotient may fail to be affine.

2.3. LIE ALGEBRAS

We add still another layer of structure by assigning a Lie algebra to each affine algebraic group. One way to do this is to set $L(G) = \mathrm{Der}_K(K[G])^G$, the Lie algebra of G-invariant derivations of $K[G]$.

Consider the simplest case, where $G = K$, the additive group of the field. Here $K[G] = K[x]$, the polynomial ring in one indeterminate. Let $0 \neq \delta \in \mathrm{Der}_K(K[G])$. Then δ is G-invariant if and only if, for all $c \in K$, $\lambda_c \circ \delta = \delta \circ \lambda_c$. One checks that this forces $\delta(x)$ to be a constant polynomial. Normalizing, we may take $\delta(x) = 1$. Hence, $L(K) = \langle d/dx \rangle$.

The following is a general result on dimensions.

Proposition 2.4 [4, I.3.6] *If G is an algebraic group, then*

$$\dim(G) = \dim(G^\circ) = \dim(L(G)).$$

For computational purposes it is often easier to work with another formulation of the Lie algebra of an algebraic group. For G an algebraic group, let $T(G)_1$ denote the *tangent space at the identity*. An element γ of $T(G)_1$ is a function from $K[G]$ to K such that, $\gamma(fg) = \gamma(f) \cdot g(1) + f(1) \cdot \gamma(g)$, for all $f, g \in K[G]$.

There is a K-linear map $L(G) \to T(G)_1$ obtained by evaluation at 1. That is, for $\delta \in L(G)$ we obtain an element $\gamma \in T(G)_1$ such that, for all $f \in K[G], \gamma(f) = \delta(f)(1)$.

It turns out that this map is a vector space isomorphism. We can then transport the Lie algebra structure of $L(G)$ to $T(G)_1$. We often identify the Lie algebra and the tangent space. For example, we find that the Lie algebra of $\mathrm{GL}_n(K)$ is $\mathfrak{gl}_n(K)$, where a matrix is applied to a coordinate function x_{ij} by taking the (i, j)-entry of the matrix and is applied to arbitrary functions by using the derivation property and evaluation at the identity.

Viewing the Lie algebra of an algebraic group as the tangent space to the identity also permits the following definition. Suppose $\varphi : G \to G'$ is a morphism of algebraic groups. There is an associated Lie algebra homomorphism $\partial\varphi : L(G) \to L(G')$, given by $\partial\varphi(\gamma) = \gamma \circ \varphi^*$. This homomorphism is called the *differential of φ*. In particular, taking $G' = \mathrm{GL}(V)$ we see that a rational representation of an algebraic group gives rise to a representation of Lie algebras.

A great deal of information about the structure and embeddings of algebraic groups is reflected in the Lie algebra structure. But there are subtle points. Indeed, consider the Frobenius morphism F of K given by $c \mapsto c^p$. The comorphism F^* is a homomorphism of $K[x]$ and we have already mentioned that $F^*(f(x)) = f(x^p)$. As above, $L(K)$ is generated by $\gamma = d/dx$. But

$$\partial F(\gamma)(f(x)) = \gamma(F^*(f(x))) = \gamma(f(x^p)) = 0,$$

so $\partial F = 0$.

The above example of the Frobenius morphism of the field suggests that the rank of the differential of a morphism may be relevant to whether or not the morphism is actually an isomorphism. The following is a positive result in this direction.

Theorem 2.2 [48, 4.3.4(ii)] *Let $\varphi : G \to G'$ be a homomorphism of algebraic groups. Then φ is an isomorphism if and only if it is an isomorphism of abstract groups and $\partial\varphi : L(G) \to L(G')$ is an isomorphism.*

The Lie algebra of an affine algebraic group affords a representation of the group as follows. For each element g of G, let $\varphi_g : G \to G$ be conjugation by g. Then $\partial\varphi_g : L(G) \to L(G)$ is a linear transformation. The map $G \to \mathrm{GL}(L(G))$, sending g to $\partial\varphi_g$, is a representation, called Ad. The differential of Ad is a homomorphism $L(G) \to \mathfrak{gl}(L(G))$. It turns out that this is just the adjoint representation of $L(G)$, arising from the Lie algebra bracket.

Proposition 2.5 [23, 10.4] *If G is an algebraic group, then $\partial\mathrm{Ad} = \mathrm{ad}$.*

2.4. JORDAN DECOMPOSITION

At this point we have set up the main ingredients of the basic theory and we are in a position to discuss some group theoretical results. The first topic is the *Jordan decomposition*. Given an algebraic group G, an element $g \in G$ is said to be *semisimple* (resp. *unipotent*) if the element λ_g induces a semisimple (resp. unipotent) element on each finite dimensional G-invariant subspace of $K[G]$.

Proposition 2.6 [23, 15.3] *If $g \in G$, then there are unique elements g_{s} and g_{u} in G such that g_{s} is semisimple, g_{u} is unipotent, and $g = g_{\mathrm{s}} \cdot g_{\mathrm{u}} = g_{\mathrm{u}} \cdot g_{\mathrm{s}}$. Moreover, this decomposition is invariant under all morphisms of algebraic groups.*

In the special case where $K = \overline{\mathbb{F}}_p$, for p a prime, G is a locally finite group and the Jordan decomposition is just the usual way of associating to an element of a finite group its p-part and p'-part.

We can define analogs of p'-groups and p-groups as follows. A *torus* is a connected abelian group consisting entirely of semisimple elements. Tori can be diagonalized in any rational representation. As an algebraic group, a torus is isomorphic to a direct product of copies of the multiplicative group of the base field. At the other extreme, an algebraic group is said to be *unipotent* if it consists entirely of unipotent elements. The following is a group theoretic version of Engel's theorem for Lie algebras.

Proposition 2.7 [23, 17.5] *Let G be a unipotent algebraic group. Then G is nilpotent. Moreover, if $\varphi : G \to \mathrm{GL}_n(K)$ is a morphism of algebraic groups, then $\varphi(G)$ is conjugate to a subgroup of the group of lower unitriangular matrices.*

There is a similar result for solvable groups, although it is easy to see by examples that here we must assume that the solvable group is connected. The result is the Lie-Kolchin theorem.

Theorem 2.3 [23, 17.6] *Assume G is a connected solvable algebraic group. If $\varphi : G \to \mathrm{GL}_n(K)$ is a morphism of algebraic groups, then G fixes a flag of the underlying vector space. Hence, $\varphi(G)$ is conjugate to a subgroup of the group of lower triangular matrices.*

Let G be a connected algebraic group. The product of two closed connected unipotent normal subgroups of G is another such subgroup. Consequently, G contains a unique largest connected unipotent normal subgroup, $R_u(G)$, called the *unipotent radical* of G. We say G is *reductive* if $R_u(G) = 1$. Similarly, there is a unique largest closed connected solvable normal subgroup of G, $R(G)$, called the *radical* of G. A *semisimple group* is one for which the radical is trivial.

2.5. CLASSIFICATION

The notion of semisimplicity is analogous to the corresponding notion for complex Lie algebras. Eventually, it is shown that a semisimple group is the commuting product of closed simple groups. However, the terminology may be a little confusing. A connected group is said to be *simple* if it has no proper normal subgroup which is closed and connected. It turns out that such a group has finite center, with quotient group simple as an abstract group. A reductive group is the commuting product of a semisimple group with a torus.

Using the classification of semisimple Lie algebras as a model, one might expect that the maximal closed, connected, solvable subgroups play a prominent role in the classification of semisimple algebraic groups. One of the key points in the classification of semisimple complex Lie algebras is establishing the conjugacy of the maximal solvable subalgebras and the conjugacy of the Cartan subalgebras. The situation for algebraic groups is entirely analogous, but much more complicated. The following result is a corollary of a fixed point theorem of Borel.

Theorem 2.4 [23, 21.3] *Any two maximal closed, connected, solvable subgroups of an algebraic group are conjugate.*

This theorem is proved by showing that, if B is a maximal, closed, connected solvable subgroup, then the quotient variety G/B is *complete*, which means that it satisfies certain properties similar to those of projective varieties. In particular, this variety is not affine except in the trivial case where $G = B$. The result is established by showing that a connected solvable group acting on a complete variety has a fixed point.

The above result is fundamental to the classification of simple algebraic groups. The maximal closed connected solvable subgroups are called *Borel subgroups*. If B is a Borel subgroup of the connected group G, then $B = U \cdot T$, where $U = R_u(B)$ and T is a maximal torus of G. Using Theorem 2.4 and other arguments it is shown that all maximal tori are conjugates of T.

The factorization of B is not surprising. Indeed, if we embed G in a group $\mathrm{GL}_n(K)$ and apply Proposition 1.9, then we conclude that B is isomorphic to a subgroup of the lower triangular group. Clearly the latter has such a factorization.

In the following we give an outline of ideas involved in the classification of semisimple algebraic groups. Let G be a semisimple algebraic group with Borel subgroup B and maximal torus T. Define $\mathrm{rank}(G) = \dim(T)$. It is shown that any torus has reductive centralizer. If one considers a torus of codimension 1 in T, then either this torus has centralizer equal to T or it is the commuting product of the torus and a simple group of rank 1. Moreover, it is shown, and this is a key result, that simple rank 1 groups are quotients of $\mathrm{SL}_2(K)$ by a finite kernel.

The key idea here is this. If G is a simple group of rank 1 then the variety G/B is shown to be isomorphic to P^1. This gives an action of G on P^1, and the result follows from the fact that the automorphism group of P^1 is $\mathrm{PGL}_2(K)$.

Once the rank 1 groups have been determined one proceeds to construct the root system associated with the maximal torus. The nontrivial T-invariant unipotent subgroups of the rank 1 groups arising within centralizers of tori of codimension 1 in T are called the *root subgroups* of G. Their Lie algebras are T-invariant 1-spaces of $L(G)$ which play the role of the usual root spaces of semisimple Lie algebras over the complex numbers. So the roots arise as characters of T. By passing back and forth between the Lie algebra and the group it is eventually shown that G has a (B,N)-*pair*, with B a Borel subgroup in the usual sense.

It is shown that $N_G(T)/T = W$, the *Weyl group*, is a finite group generated by reflections associated to a root system Σ. We may take it that U is the product of root subgroups U_α corresponding to positive roots α. Also, $G = \bigcup BwB$, the union over elements of W (well-defined as $T \leqslant B$). The Weyl group has a unique element w_0 interchanging the sets of positive and negative roots. The corresponding double coset Bw_0B has special properties. Its translate $w_0^{-1}Bw_0B$ is called the *big cell*. The big cell can be expressed U^-TU, where U^- is the product of all root subgroups corresponding to negative roots.

Proposition 2.8 [4, 14.4] *Let G be a semisimple algebraic group. The big cell is an open dense subset of G. Moreover, the product map is an isomorphism of varieties*

$$\prod_{\alpha>0} U_{-\alpha} \times T \times \prod_{\alpha>0} U_\alpha \longrightarrow U^-TU.$$

In particular, $\dim(G) = 2 \cdot |\Sigma| + \dim(T)$.

The subgroups of G containing B are the *parabolic subgroups*. There is one for each subset of a base, Π, of Σ. Further results regarding the structure of these groups will be presented in the next section.

The root groups are 1-dimensional unipotent groups and hence isomorphic to the additive group of K. For each root α, fix an isomorphism $c \mapsto U_\alpha(c)$. One can then establish the Chevalley commutator relations. Note the polynomial nature of the expressions, which, of course, should be expected in the theory of algebraic groups.

Proposition 2.9 [23, 32.5] *Let α and β be independent roots. There are elements $c_{i,j} \in K$ such that, for all $c, d \in K$,*

$$[U_\alpha(c), U_\beta(d)] = \prod_{i,j>0} U_{i\alpha+j\beta}(c_{i,j}c^i d^j),$$

where the product is taken over all roots of the form $i\alpha + j\beta$ (in some fixed order).

Once the root system has been constructed and the commutator relations established we are in a position to proceed with the classification of semisimple algebraic groups. For this we require more than just the root system. Fix a maximal torus T of G and set $X(T) = \mathrm{Hom}(T, K^\#)$, the character group of T. It turns out that $X(T) \cong \mathbb{Z}^n$, where $n = \dim(T)$. Then $X(T)$ is a lattice in $X(T)_\mathbb{Q} = X(T) \otimes_\mathbb{Z} \mathbb{Q}$, containing the root lattice, $\mathbb{Z}\Sigma$. One can choose a positive definite inner product, $(\ ,\ )$, on $X(T)_\mathbb{Q}$ which is invariant under the Weyl group.

A *weight* is defined to be an element λ of $X(T)_\mathbb{Q}$ such that

$$\langle \lambda, \alpha \rangle = 2(\lambda, \alpha)(\alpha, \alpha)^{-1} \in \mathbb{Z},$$

for all roots α. Let Λ be the set of weights, another lattice in $X(T)_\mathbb{Q}$. We have $\mathbb{Z}\Sigma \leqslant X(T) \leqslant \Lambda$. The lattices $\mathbb{Z}\Sigma$ and Λ depend only on the root system, and the quotient $\Lambda/\mathbb{Z}\Sigma$ is a finite group, called the *fundamental group*. Following [4], we call the pair consisting of the root system and the lattice $X(T)$ the *diagram of G*. There is a notion of isomorphism of diagrams: an isomorphism of root systems that extends to the weight lattice and sends one intermediate lattice to the other.

Theorem 2.5 [4, 24.1] *Two semisimple algebraic groups are isomorphic if and only if their diagrams are isomorphic. Moreover, each diagram has an associated semisimple algebraic group.*

The existence part of the classification asserts that in fact there is a semisimple algebraic group corresponding to each root system and subgroup of the fundamental group. This is independent of the characteristic.

Construction can be done as in Steinberg [51]. Special terminology exists for the extreme cases. If $X(T) = \Lambda$, then G is said to be *simply connected*, while, if $X(T) = \mathbb{Z}\Sigma$, then G is said to be *adjoint*. It is shown in [51] that one Chevalley group maps to another (of the same type) if its corresponding subgroup of the fundamental group contains the other. Consequently, the simply connected group maps onto all others.

There are subtle points about the classification theorem which can be illustrated by an example. Consider the groups $SL_2(K)$ and $PSL_2(K) = SL_2(K)Z/Z$, where Z is the group consisting of the scalars in $GL_2(K)$. There is a natural surjection $\pi : SL_2(K) \to PSL_2(K)$. When $char(K) = 2$ this is an isomorphism of abstract groups. However, it is not an isomorphism of algebraic groups. Indeed, the comorphism is not surjective. If T is a maximal torus of $SL_2(K)$, then $X(T) = \Lambda$, while its image has character group just the root lattice.

The distinction can also be detected at the level of Lie algebras. Of course $L(SL_2(K)) = \mathfrak{sl}_2$. If $char(K) = 2$ this Lie algebra has a 1-dimensional center with quotient a Frobenius twist of the usual module for $SL_2(K)$. On the other hand, $L(PSL_2(K))$ has a submodule isomorphic to the twist of the usual module, with quotient the trivial module. One checks that $\partial\pi$ has kernel equal to $Z(\mathfrak{sl}_2)$.

3. Subgroup Structure of Algebraic Groups

In this section we describe some general results on the subgroup structure of algebraic groups. Let G be a simple algebraic group over the algebraically closed field K of characteristic p. Let T be a maximal torus of G and Σ the corresponding root system. So G has a system of T-root subgroups, one root group for each root in Σ. The easiest subgroups to describe are those which are closely tied to the root system, so we begin with these.

3.1. Subsystem Subgroups

In working with simple algebraic groups one often encounters subgroups containing (or normalized by) a maximal torus. For example, each semisimple element of G is conjugate to an element of T, so centralizers of semisimple elements contain maximal tori. It turns out that the groups normalized by a maximal torus are completely determined by the root system. Semisimple subgroups normalized by a maximal torus are called *subsystem subgroups*.

A subset Δ of the root system Σ of G is said to be *closed* if it satisfies the following two conditions: (1) $\alpha \in \Delta$ if and only if $-\alpha \in \Delta$; (2) if $\alpha, \beta \in \Delta$ and $\alpha + \beta \in \Sigma$, then $\alpha + \beta \in \Delta$. Let Δ be closed. Then Δ is itself a root system and using the Bruhat decomposition we find that $G(\Delta) = \langle U_\alpha \mid \alpha \in \Delta \rangle$ is a T-invariant semisimple subgroup having maximal torus $T \cap G(\Delta)$ and corresponding root system Δ.

The group $G(\Delta)$ is a subsystem subgroup of G and, except for some special cases which occur only in characteristic 2 and 3, these are the only subsystem subgroups. The subsystem subgroups are well understood. We will see later that these subgroups are important in understanding other subgroups as well.

A lovely algorithm due to Borel and de Siebenthal [5] determines all subsystems of Σ. The procedure is as follows. Start from the Dynkin diagram and form the extended Dynkin diagram by adjoining the negative of the highest root. Remove any collection of nodes and repeat the process with each of the components of the resulting graph. All diagrams obtained in this way are Dynkin diagrams of subsystems. There is one conjugacy class of subsystem groups for each such subsystem. In [9] one can find tables of subsystems for each of the exceptional groups.

For example, we indicate below the Dynkin diagram for E_8 and next to it the extended Dynkin diagram. An extra node has been added on the right.

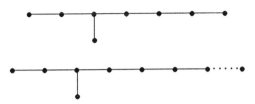

Removing appropriate nodes it is clear that we can construct subsystems of type D_8, A_8, $A_4 \times A_4$, etc. Consequently, there are subsystem subgroups of G corresponding to each of these types.

When the root system is not simply laced and the characteristic is 2 or 3 (the latter only for G_2), there are certain degeneracies in the Chevalley commutator relations which give rise to subsystem subgroups not of the form $G(\Delta)$ for Δ a closed subsystem. We illustrate this by example. If $\Sigma = C_2$, then the collection of all short roots forms a collection closed under negatives, but not closed in the usual sense. There is a corresponding subgroup $A_1 \times A_1 < \mathrm{Sp}_4$, which contains a maximal torus and does not correspond to a closed subsystem. For these groups the set Δ is closed under negatives but not under sums if the roots involved correspond to a degeneracy in the commutator relations. It is not difficult to list all the possibilities for Δ.

The next result describes all closed connected subgroups of G which contain T.

Proposition 3.1 [41, 2.5] *Let $T \leqslant X$ be a closed connected subgroup of G. Then $X = R_{\mathrm{u}}(X)DZ$, where D is a semisimple subsystem subgroup, Z is a torus commuting with D and contained in T, and $R_{\mathrm{u}}(X)$ is a product of T-root subgroups.*

Each semisimple element is conjugate in G to an element of T. So we can apply the above to study centralizers of semisimple elements. Let $t \in T$. Then $C_G(t)^\circ$ is a connected group. From the action of T on root groups it is clear that the set of root groups centralized by t is closed under negatives. It follows that $C_G(t)^\circ$ is reductive, hence a product of a (possibly trivial) subsystem group and a central torus. Further, $C_G(t)$ is known to be connected if G is simply connected. In general, the component group, $C_G(t)/C_G(t)^\circ$, is isomorphic to a section of the fundamental group.

One further comment is in order in connection with semisimple elements. As indicated above, all semisimple classes are represented in T. Moreover, G-conjugacy among elements of T is determined by the Weyl group, $N_G(T)/T$.

3.2. PARABOLIC SUBGROUPS

Let D be a closed subgroup of G which is not reductive. These subgroups have a normal unipotent subgroup. Moreover, a theorem of Borel and Tits [6] shows that any such group is contained in a canonical parabolic subgroup. So it is natural to study the parabolic subgroups in some detail. It turns out that there are some beautiful features of their internal structure. To make the statements uniform we rule out the cases where (Σ, p) is $(B_n, 2)$, $(C_n, 2)$, $(F_4, 2)$, $(G_2, 2)$ or $(G_2, 3)$.

Fix a parabolic subgroup $P = QL$, where $Q = R_u(P)$ and L is the Levi factor containing T. The structure of L is not a problem. It is the product of a subsystem group corresponding to removing some nodes from the Dynkin diagram (the extended diagram is not needed here) and a central torus. We wish to discuss the action of L on Q. It turns out that Q has a filtration by particularly nice modules for L. To discuss these we need a little terminology taken from [1].

Let Π be a base of Σ and for convenience take P to be the opposite of the standard parabolic subgroup. That is, there is a subset J of Π such that L corresponds to the subsystem Σ_J with base J, while Q is the product of all root subgroups $U_{-\alpha}$ for roots $\alpha \in \Sigma^+ \setminus \Sigma_J^+$.

Let $\beta \in \Sigma^+ \setminus \Sigma_J^+$. Then we can write $\beta = \beta_J + \beta_{J'}$, where

$$\beta_J = \sum_{\alpha_j \in J} c_j \alpha_j, \quad \beta_{J'} = \sum_{\alpha_j \notin J} d_j \alpha_j.$$

Set $\text{level}(\beta) = \sum d_j$ and $\text{shape}(\beta) = \beta_{J'}$. Let

$$Q(i) = \prod_{\text{level}(\beta) \geqslant i} U_{-\beta}.$$

We will not need this, but in fact $Q(i)$ is the i-th term of the descending central series of Q. In any case, we have a descending series $Q = Q(1) \geqslant Q(2) \geqslant \cdots$ of normal subgroups of P.

Fix i and consider

$$Q(i)/Q(i+1) \cong \prod_{\text{level}(\beta)=i} U_{-\beta}.$$

Then

$$Q(i)/Q(i+1) \cong \prod_S V_S,$$

where S ranges over the set of shapes of roots having level i and V_S corresponds to the direct product of all U_β for β a positive root of shape S.

The theorem to follow combines results in [1] and [38].

Theorem 3.1 *Let $P = QL$ be a parabolic subgroup of G.*

(1) *For each i, $Q(i)/Q(i+1)$ has the structure of a completely reducible rational L-module and each V_S is an irreducible submodule.*

(2) *For each shape S, L has finitely many orbits on V_S. In particular, L has an open dense orbit on V_S.*

(3) *For $S \neq S'$, $V_S \not\cong V_{S'}$.*

We remark that it is shown in [1] that similar results hold for finite groups of Lie type, including twisted groups. Also, another result of Richardson (see Theorem 5.2.1 of [10]) shows that there always exists an open dense orbit of P on Q.

It may be instructive to give an example of how this all plays out. Consider the case where $G = E_8$ and P is the parabolic subgroup with Levi factor of type

A_7. Let α be the fundamental root which is not contained in the root system for L. Then $Q(4) = 1$ and each of the quotients $Q(i)/Q(i+1)$ is irreducible. Specifically, $Q(1)/Q(2) = \wedge^3 W$, $Q(2)/Q(3) = \wedge^2 W^*$ ($\cong \wedge^6 W$), and $Q(3) = W$, where W is the usual module for $\mathrm{SL}_8(K)$.

A number of other interesting modules arise within parabolic subgroups. In particular, the usual 27-dimensional modules for E_6 occur, as does the 56-dimensional module for E_7. To get these, take $G = E_7$ or $G = E_8$, respectively, with P a maximal parabolic subgroup with Levi factor of type E_6 or E_7, respectively. In the first case, the unipotent radical, Q, is abelian and affords an irreducible 27-dimensional module for E_6. In the second case, $Q(1)/Q(2)$ is irreducible and affords a 56-dimensional module for E_7.

In the next section we will discuss the representation theory of simple algebraic groups. We will see that irreducible modules are parametrized by what are known as *high weights*. These are certain elements of the character group, $X(T)$. It is quite easy to specify the high weights for the modules V_S in the theorem (see the comments following Theorem 4.2).

Modules for simple algebraic groups having finitely many orbits are very rare. One naturally wonders about the number of orbits on these modules. In one special case there is precise information. This is the case where Q is abelian. Here P is a maximal parabolic subgroup. There is just one irreducible module in the filtration, so that Q itself is an irreducible module for L.

Theorem 3.2 [40, 2.10] *Let P be a parabolic subgroup such that $Q = R_u(P)$ is abelian. Then:*

(1) *The number of orbits of L on Q equals $|P \backslash G / P|$.*

(2) *If Q^- denotes the unipotent radical of the opposite parabolic subgroup, then $PwP \cap Q^-$ is an L-orbit, for each element w of the Weyl group.*

A nice description for the orbits was given in [39]. Let α be the simple root not in the root system of L. Consider the maximum number of orthogonal roots having coefficient of α equal to 1. This is the number of nonidentity L-orbits on Q and representatives are given in a natural way. Namely, if a maximal set is $\{\beta_1, \ldots, \beta_s\}$, then orbit representatives are given by $U_{-\beta_1}(1)$, $U_{-\beta_1}(1) \cdot U_{-\beta_2}(1)$, $\ldots$, $U_{-\beta_1}(1) \cdots U_{-\beta_s}(1)$.

It is still unclear how many orbits there are in the general case, but there is a conjecture that should be mentioned. If V_S is one of the internal modules, then one can construct a certain subsystem group D such that there is a maximal parabolic subgroup $P_D = Q_D L_D$ of D such that $L_D = L$ and $Q_D/Q'_D \cong V_S$ as L-modules. Consequently, for purposes of determining the number of L-orbits on V_S we may assume that $V_S = Q/Q'$.

Conjecture. *If P is a maximal parabolic subgroup of G, then the number of L-orbits of Q/Q' is bounded above by $|P \backslash G / P|$.*

For classical groups this has been verified by Röhrle (see (2.15) of [40]).

3.3. Unipotent Elements

The Jordan decomposition shows that every element of a simple algebraic group is the product of commuting semisimple and unipotent elements. We have commented on the classes and centralizers of semisimple elements. In this section we discuss a few results concerning unipotent elements. We begin with the following result of Lusztig.

Theorem 3.3 [32] *The number of unipotent conjugacy classes in a simple algebraic group is finite.*

This should not come as a surprise. For consider the conjugacy classes of unipotent elements in the group $GL_n(K)$. Using the Jordan form for matrices we see that unipotent classes are in bijective correspondence with partitions of n.

With the Jordan form as guide, we indicate two extreme cases. We say that a nonidentity unipotent element is a *long root element* if it is conjugate to an element of a root subgroup of G, corresponding to a long root of Σ. In the case of $GL_n(K)$ these correspond to transvections—elements with a single nontrivial Jordan block of size 2. At the other extreme are the *regular* unipotent elements. These correspond to unipotent matrices in $GL_n(K)$ which are a single Jordan block. To get such an element take a nontrivial element from each of the root groups for a set of fundamental roots and multiply them.

There are several ways to describe conjugacy classes of unipotent elements. For groups over algebraically closed fields of characteristic 0, there is a method going back to Dynkin [17]. Each unipotent class corresponds to a certain labelled Dynkin diagram, with the labels being 0, 1, or 2. For classical groups there is a version of the Jordan form which takes into account the underlying form. See [49] for details. A general approach for large primes was given by Bala-Carter ([2], [3]) and it was shown by Pommerening ([35], [36]) that the same classification holds for all good primes.

For exceptional groups there are results listing all conjugacy classes of unipotent elements. For small primes there are occasionally extra conjugacy classes. See Mizuno ([33], [34]) for groups of type E_6, E_7, and E_8; Shoji and Shinoda ([45], [46]) for groups of type F_4, and Enomoto [18] for groups of type G_2.

Analysis of the centralizers is sometimes complicated, mainly because the centralizers are not reductive. The following results describe the centralizers in the two extreme cases mentioned above and give some additional properties of these elements. The first result can be found in (1.1) and (1.2) of [29].

Proposition 3.2 *Let $u \in G$ be a long root element contained in the long root subgroup U. Then $P = N_G(U)$ is a parabolic subgroup of G and $C_G(u) = C_G(U) = P'$. Moreover, for $g \in G$, $\langle U, U^g \rangle$ is either a unipotent group or a subsystem group of type A_1.*

The situation for regular unipotent elements is probably less well-known, even in the case where $G = GL_n(K)$. See [49], III, §1.

Proposition 3.3 *Let $u \in G$ be a regular unipotent element. Then $C_G(u) = A \times Z(G)$, where A is an abelian unipotent group with $\dim(A) = \text{rank}(G)$. Moreover, u is contained in a unique Borel subgroup of G.*

Over the years there have been a number of papers concerned with subgroups of a particular group which contain some special type of element. The case where the element is a long root element is perhaps the case most studied. Kantor [26] studied subgroups of finite classical groups generated by long root subgroups and Cooperstein ([11]–[15]) carried out a similar analysis for finite exceptional groups. In both cases the results were in the form of lists. In an attempt to get more conceptual results, Liebeck and the author [29] studied subgroups of simple algebraic groups containing long root elements. The main result is the following.

Theorem 3.4 [29] *Let X be a simple, closed connected subgroup of G such that $|U \cap X| > 2$, where U is a long root subgroup of G. There is a subsystem subgroup Y of G such that $X \leqslant Y$ and one of the following holds:*

(1) $X = Y$;

(2) $X = Y_\tau$, *where τ is a graph automorphism of Y;*

(3) $X = G_2$ *and* $Y = B_3$;

(4) $X = C_4$, $Y = E_6$, *and* $p = 2$.

Note that, when $\operatorname{char}(K) \neq 2$, the hypothesis just says that X contains a long root element of G.

The above theorem provides a good example of how the subsystem subgroups help in analyzing the subgroup structure of G. The embedding of X in G is determined in two steps. We first find an appropriate subsystem subgroup. These groups we know quite well. Then we identify X within the subsystem subgroup.

The connection with subsystem groups can be useful in several respects. For example, suppose one wishes to find $C_G(X)$, with X as in Theorem 3.4. Of course, this centralizer contains $C_G(Y)$. The latter centralizer is easy to find. It is essentially another subsystem group, corresponding to the subsystem of all roots orthogonal to the system determining Y. Moreover, it is shown in [29] that usually $C_G(X)$ is reductive of rank equal to the rank of $C_G(Y)$. This makes the process of finding $C_G(X)$ fairly easy.

Theorem 3.4 can also be applied to study the analogous problem for finite groups of Lie type. Roughly speaking, the finite groups of Lie type occur as fixed points of certain morphisms of the simple algebraic groups over fields of positive characteristic. Let $G(q)$ be such a group, arising from the algebraic group G. If $X \leqslant G(q)$ is a quasisimple group containing at least two nonidentity elements of a long root group U of G, then it is shown that $\langle X, U \rangle$ is a closed connected simple subgroup of G and that X arises from this group by taking fixed points under a suitable morphism.

One can also study subgroups containing other types of unipotent elements. If K is algebraically closed of characteristic 0, then it can be shown that every unipotent element is contained in a closed subgroup of type A_1. When the unipotent element is regular, then this is often the only proper reductive subgroup containing the unipotent element.

How would we construct such a subgroup? Consider the case of $\operatorname{GL}_n(K)$. Consider the representation of $\operatorname{SL}_2(K)$ on homogeneous polynomials of degree $n - 1$. This is an irreducible representation, with image $\operatorname{SL}_2(K)$ if n is even and $\operatorname{PSL}_2(K)$ if n is odd. It is easily checked that a nonidentity unipotent element of the image is a regular unipotent element of $GL_n(K)$. By taking direct sums of such representations one can show that any unipotent element of $GL_n(K)$ is contained in a subgroup of type A_1. The same can be shown for other types of groups, as well.

However, when the characteristic is $p > 0$, then even these subgroups may not be available. Indeed, if we take a regular unipotent element in $GL_n(K)$, where $n > p$, then the unipotent element has order greater than p and hence cannot lie in any subgroup of type A_1.

The following result of Testerman shows that elements of order p usually are in groups of type A_1.

Theorem 3.5 [53] *Let G be a simple algebraic group over a closed field of characteristic $p > 0$ and assume p is a good prime. Each element of order p is contained in a closed connected subgroup of type A_1.*

The above theorem does not determine all groups of type A_1 containing a fixed unipotent element. For certain special cases, for example when the unipotent element is a root element or a regular element, there exist results describing all overgroups of type A_1. But no completely general result exists to date. However, a recent result of Lawther and Testerman [28] obtains strong information on exceptional groups, given mild restrictions on the characteristic.

3.4. MAXIMAL SUBGROUPS

In this section we will briefly describe some results on maximal subgroups of simple algebraic groups. Further information will be given in the lectures of Liebeck. Let G be a simple algebraic group over an algebraically closed field K. Then G has an associated root system and we distinguish cases according to whether this system has classical or exceptional type.

First consider the case where G is a classical group. Say $G = I(V)$ is one of the groups $SL(V)$, $SO(V)$, or $Sp(V)$, where V is a finite dimensional vector space over K. Equip V with the trivial form in the first case and a nondegenerate bilinear form in the latter two cases.

The following results appear in [43]. The first result is an easy reduction theorem.

Proposition 3.4 *Let G be $SL(V)$, $SO(V)$, or $Sp(V)$, and let M be maximal among closed connected subgroups of G. Then one of the following holds:*

(1) *M stabilizes a proper subspace of V;*

(2) *$V = V_1 \otimes V_2$ and $M = I(V_1) \cdot I(V_2)$;*

(3) *M is simple and $V|_M$ is tensor indecomposable.*

It is an easy matter to determine when groups as in parts (1) and (2) of the above result are in fact maximal. A much deeper problem is to determine when groups of type (3) are maximal. The problem can be stated as follows. Let X be a simple closed subgroup of $GL(V)$. Assume that $V|_X$ is irreducible and tensor indecomposable. When is X maximal among closed connected subgroups of one of the classical groups $SL(V)$, $SO(V)$, or $Sp(V)$? Well, if not, then there exists a closed connected subgroup Y of $GL(V)$ such that $X < Y$ and Y is not one of the groups $SL(V), SO(V)$, or $Sp(V)$.

A remarkable thing occurs. It turns out that this is a very rare event.

Theorem 3.6 *All triples* (X, Y, V) *are known explicitly.*

Let K have characteristic p. Dynkin [16] and [17] settled the case $p = 0$. For $p > 0$, this is the main content of [43] or [52], according to whether Y is of classical or exceptional type.

These are all extremely long, complicated papers, but in the end we have an explicit list of possible configurations, giving the precise embedding of X in Y together with the precise representations of both groups on V. We illustrate with one example. Consider the group $X = E_6$. It is well known that X has representations of degree 27. Pick one of these which is restricted. This notion will be defined in the next section, but here it just amounts to the assertion that the differential is nontrivial. Then we have X embedded in Y, where $Y = SL_{27}(K)$. Now take $V = \wedge^4(V)$. For $p \neq 2, 3$ this gives one of the configurations of the theorem.

Theorem (3.6) is as complete an answer to the analysis of the closed, connected, maximal subgroups of classical algebraic groups as one can expect.

Recent results of Ford [19] provide information if one drops the connectivity assumption on M.

Now assume G is of exceptional type. Here the rank of G is bounded so it makes sense to ask for a complete list of the maximal subgroups. To avoid certain technicalities involving the characteristic of K, we state a weakened form of the main result.

Theorem 3.7 ([17], [44]) *Let G be a simple algebraic group of exceptional type. Assume that either $p = 0$ or $p > 7$. Let M be a maximal closed connected subgroup of G. Then M is a parabolic subgroup or a subsystem subgroup, or one of the following holds for G and M:*

(1) $G = G_2$ *and* $M = A_1$;

(2) $G = F_4$ *and* $M = A_1$ *or* $A_1 \cdot G_2$;

(3) $G = E_6$ *and* $M = A_2, G_2, F_4, C_4,$ *or* $A_2 \cdot G_2$;

(4) $G = E_7$ *and* $M = A_1$ *(two classes)*, $A_2, A_1 \cdot A_1, A_1 \cdot G_2, A_1 \cdot F_4,$ *or* $G_2 \cdot C_3$;

(5) $G = E_8$ *and* $M = A_1$ *(three classes)*, $B_2, A_1 \cdot A_2,$ *or* $G_2 \cdot F_4$.

The maximal subgroups of type A_1 do not exist unless the characteristic of K is 0 or suitably large. They correspond to certain conjugacy classes of unipotent elements.

There is a result quite similar to the above which drops the connectivity assumption. This result determines all maximal closed subgroups of exceptional groups having positive dimension. It is described in the article by Liebeck. See [30] for a precise statement. In recent work [31] Liebeck and Seitz have determined all closed, simple subgroups of exceptional groups.

4. Representations of Algebraic Groups

In this section we discuss some standard results on representations of algebraic groups, leading up to some new results on abstract homomorphisms with applications to the theory of locally finite groups.

Throughout this section we take G to be a semisimple algebraic group over an algebraically closed field K of characteristic p. Take G to be simply connected. Fix a Borel subgroup B of G and a maximal torus T of B. Then $B = UT$, where $U = R_u(B)$. Let $\Pi = \{\alpha_1, \ldots, \alpha_n\}$ be a base of the root system Σ of G.

4.1. High Weight Modules

Let $G \to \mathrm{GL}(V)$ be a rational representation. The action of T on V can be diagonalized, so this yields a decomposition

$$V = \bigoplus_\mu V_\mu,$$

where V_μ denotes the weight space of V corresponding to the weight $\mu \in X(T)$. The action of the Weyl group on T induces an action on $X(T)$. Consequently, the weights μ for which $V_\mu \neq 0$ fall into orbits under the action of the Weyl group.

The Lie-Kolchin theorem (Theorem (2.3)) shows that a Borel subgroup B of G stabilizes a 1-space of V. The action of B on this 1-space affords a character, say λ. As $U = B'$, λ can be identified with a weight of T.

Let $\langle v_\lambda \rangle$ be a B-invariant 1-space with weight λ and consider the module $M = \langle Gv_\lambda \rangle$. The big cell $U^-TU = U^-B$ is dense in G, so that $M = \langle U^-Bv_\lambda \rangle = \langle U^-v_\lambda \rangle$. Now

$$U^- = \prod_{\alpha > 0} U_{-\alpha}.$$

Moreover, for each positive root α and each weight μ, one has (see 27.2 of [23])

$$U_{-\alpha} V_\mu \subseteq \bigoplus_{c \geqslant 0} V_{\mu - c\alpha}.$$

It follows from the preceding remarks that all weights μ of M are less than λ in the sense that they are obtained from λ by subtracting positive roots. We write $\mu \ll \lambda$. For this reason M is called a *high weight module* with associated high weight λ. By definition this means that M is a cyclic module generated by a 1-space fixed by B and affording the weight λ. Clearly, λ is uniquely determined among weights of M. A nonzero vector in V_λ is called a *maximal vector*.

When V is irreducible for G, then $M = V$ and we have the following theorem.

Theorem 4.1 [23, 31.3] *Let G be a semisimple algebraic group and V an irreducible rational representation of G.*

(1) *V contains a maximal vector, v^+, of weight λ which spans V_λ.*

(2) *If μ is any weight of V, then $\mu \ll \lambda$.*

(3) *V is uniquely determined, up to isomorphism, by λ.*

With V as above we write $V = V_G(\lambda)$. We will have completely parametrized irreducible representations of semisimple algebraic groups once we determine precisely which weights λ occur as high weights of irreducible rational representations. A weight λ is said to be *dominant* if $\langle \lambda, \alpha \rangle \geqslant 0$ for all positive roots $\alpha \in \Sigma$. It turns out that λ is the high weight of an irreducible rational representation of G if and only if λ is dominant.

If λ is a dominant weight, then so is $-w_0\lambda$ and we have the following result relating the high weights of an irreducible module and its dual.

Proposition 4.1 [23, 31.6] $V_G(-w_0\lambda) \cong V_G(\lambda)^*$.

We have seen in Section 1 that the representation of G yields a corresponding representation of $L(G)$ on V. If $p = 0$, it turns out that this is an irreducible representation. So here one can exploit the well-known Weyl degree formula (see, for example, 24.3 of [24]).

Theorem 4.2 *Let G be a semisimple algebraic group over an algebraically closed field of characteristic 0. If λ is a dominant weight, then*

$$\dim(V_G(\lambda)) = \frac{\prod_{\alpha > 0}\langle \lambda + \rho, \alpha \rangle}{\prod_{\alpha > 0}\langle \rho, \alpha \rangle}.$$

In the theorem ρ denotes the half sum of all positive roots. When $p > 0$ things are very much more complicated. Indeed, a dimension formula for irreducible representations remains an open problem.

For $1 \leqslant i \leqslant n$, define a dominant weight, ω_i, by the equation $\langle \omega_i, \alpha_j \rangle = \delta_{ij}$, for all j. The weights $\{\omega_1, \ldots, \omega_n\}$ are called the *fundamental dominant weights*. The fundamental dominant weights form a $\mathbb{Z}$-basis of the additive group of all weights. If λ is a dominant weight then $\lambda = \sum_i c_i\omega_i$, with $c_i \geqslant 0$ for each i. Therefore, the collection of irreducible rational representations is in bijective correspondence with n-tuples $(c_1, \ldots, c_n)$ of nonnegative integers.

Some examples are in order. If $G = \mathrm{SL}_n(K)$, then $V = V_G(\omega_1)$ is the usual n-dimensional module for G (or its dual) and $V_G(\omega_i) = \wedge^i V$. On the other hand, $V_G(d\omega_1) = S^d V$, provided either $p = 0$ or $p > 0$ with $d < p$.

What are the high weights of the modules which occur within parabolic subgroups? Fix a parabolic subgroup $P = QL$ and a shape S. Let V_S be the corresponding module. So we can regard $V_S = \prod_\beta U_{-\beta}$, where β ranges over all positive roots of shape S. Each of the root groups is a 1-dimensional T-invariant subspace. Also, $T \cap L$ is a maximal torus of L. So the weights within this module have the form $-\beta|_{T\cap L}$, where β ranges over all positive roots of shape S. It is shown in [1] that all roots of a given length of shape S are conjugate under the Weyl group of L. Consequently, this Weyl group has at most two orbits on weight spaces and each weight space is 1-dimensional.

To describe the high weight of V_S we need one more notion. If β has shape S, then write β as an integral combination of fundamental roots. The height of β is defined to be the sum of the coefficients.

Proposition 4.2 [1, Theorem 2] *Fix a shape S. There is a unique root β of minimal height and shape S. Also, $-\beta|_{T\cap L}$ affords the high weight of V_S.*

We illustrate with the example considered earlier. Let P be the parabolic of E_8 with Levi factor of type A_7. Let $\Pi = \{\alpha_1, \ldots, \alpha_8\}$ and let $\alpha = \alpha_2$ be the simple root not in the root system of L. Then the possible shapes are α, 2α, and 3α. To find the corresponding high weights we must find the positive roots of minimal height having the corresponding shape. These are easily seen to be $\alpha = \alpha_2$, $\alpha_1 + 2\alpha_2 + 2\alpha_3 + 3\alpha_4 + 2\alpha_5 + \alpha_6$, and $\alpha_1 + 3\alpha_2 + 3\alpha_3 + 5\alpha_4 + 4\alpha_5 + 3\alpha_6 + 2\alpha_7 + \alpha_8$. Upon restriction to $T \cap L$ we find that these weights restrict to the fundamental dominant weights λ_3, λ_6, λ_1 of A_7, in the usual ordering of simple roots. Hence, we have composition factors as previously indicated.

4.2. IRREDUCIBLE MODULES AND WEYL MODULES

We have already mentioned that when $p > 0$ there does not presently exist a formula for $\dim(V_G(\lambda))$ like the formula given in 4.2. However, there is a module closely related to the irreducible module having precisely the dimension given by 4.2 and which plays an important role in representation theory.

For the rest of the section assume $p > 0$.

Fix a dominant weight λ. Start with a semisimple complex Lie algebra and an irreducible representation, V, of high weight λ. Using the Kostant $\mathbb{Z}$-form of the enveloping algebra of the Lie algebra and a suitable lattice we construct a Chevalley group $G(K)$ over K and a module V_K for $G(K)$ such that $\dim(V_K) = \dim(V)$ (see [51]). Moreover, Theorem 1.1 shows that $G(K)$ is a connected algebraic group, a subgroup of $\mathrm{GL}(V_K)$. Also, V_K is a module of high weight λ.

We can now map G morphically to $G(K)$ and obtain a rational representation of G on $V(K)$. The structure of the module depends on the lattice chosen. If the lattice is generated under the action of the $\mathbb{Z}$-form of the enveloping algebra by a maximal vector, then the resulting module, $W_G(\lambda)$, is called a *Weyl module*.

There is also an intrinsic construction of Weyl modules. Let λ be a dominant weight and $\langle v \rangle$ a 1-dimensional module for B affording λ upon restriction to T. Now $G \times B$ acts on $K[G]$, with G acting on the left and B on the right. This yields an action of $G \times B$ on $K[G] \otimes_K \langle v \rangle$. Set $H(\lambda) = (K[G] \otimes_K \langle v \rangle)^B$, where B denotes the direct factor of $G \times B$. One checks that $H(\lambda) = D(\lambda) \otimes_K \langle v \rangle$, where $D(\lambda) = \{f \in K[G] \mid f(x) = \lambda(b)f(xb),\ x \in G,\ b \in B\}$.

It turns out that $H(\lambda)$ has a simple socle isomorphic to $V_G(\lambda)$. The Weyl module is related to the dual of this module (see Proposition 4.1 above). Indeed, $H(-\omega_0\lambda)^*$ is isomorphic to $W_G(\lambda)$ (see Part II, 2.12 of [25]).

The following result indicates some fundamental properties of Weyl modules. Proofs can be found in sections 2.13, 2.14, 5.11 of [25].

Theorem 4.3 *Let λ be a dominant weight.*

(1) $W_G(\lambda)$ *has a unique maximal submodule M such that $W_G(\lambda)/M \cong V_G(\lambda)$.*

(2) $\dim(W_G(\lambda))$ *is given by the Weyl dimension formula (see Theorem 4.2).*

(3) $W_G(\lambda) = \langle Gv_\lambda \rangle$ *and is universal among high weight modules of weight λ.*

The assertion in part (3) of the above theorem means that there is a surjective homomorphism from $W_G(\lambda)$ to any cyclic G-module generated by a maximal vector of weight λ. Using this we obtain the following result showing that extensions of simple modules can be analyzed within Weyl modules.

Proposition 4.3 *Let G be a semisimple algebraic group and let V be an indecomposable extension of $V_G(\mu)$ by $V_G(\lambda)$. Either $\mu \ll \lambda$ or $\lambda \ll \mu$. In the first case V is an image of the Weyl module $W_G(\lambda)$ and in the second case V^* is an image of $W_G(-w_0\mu)$.*

The mystery about Weyl modules concerns the nature of the maximal submodule. The structure and size of this module can vary wildly and hence our information about the irreducibles is far from complete, in spite of the nice parametrization by dominant weights.

Given a weight λ, the dimension of $V_G(\lambda)$ can be much smaller than that of the corresponding Weyl module. For example, suppose $p > 0$, $G = \mathrm{SL}_n(K)$, and $\lambda = p\lambda_1$. Then $\dim(V_G(p\lambda_1)) = n$, whereas $\dim(W_G(p\lambda_1)) = (n+p)!/(n!)(p!)$.

In the remainder of this subsection we indicate a few significant results on irreducible modules. The first is a result of Smith that is useful for inductive arguments.

Proposition 4.4 [47] *Let λ be a dominant weight and $P = QL$ a standard parabolic subgroup of the semisimple group G. Then the fixed point space $V_G(\lambda)^Q$ is irreducible upon restriction to L, with high weight λ.*

A dominant weight $\lambda = \sum_i c_i \lambda_i$ is said to be *restricted* if each $c_i < p$. We will see in the next subsection that the dimensions of all irreducible modules will be known once we know the dimensions of all restricted irreducibles. It is the restricted modules for which the differential gives an irreducible representation of $L(G)$.

In passing from the Weyl module to its irreducible quotient the dimension of weight spaces may decrease. However, the following result of Premet shows that weight spaces rarely disappear entirely.

Theorem 4.4 [37] *Let G be a semisimple algebraic group. If the Dynkin diagram for G has a double bond, assume $p > 2$, and, if G has a normal subgroup of type G_2, assume $p > 3$. Then, for all restricted dominant weights λ, the sets of weights in $V_G(\lambda)$ and $W_G(\lambda)$ agree.*

We close this subsection with a recent result of Kleshchev concerning the restriction of irreducible representations to particular subgroups. The result was used to resolve a conjecture of Jantzen-Seitz (see also [20]) on the modular representations of symmetric groups. It seems likely that this result will have other applications.

Theorem 4.5 [27] *Consider one of the following configurations for a simple group G and subgroup X: $(\mathrm{SL}_n(K), \mathrm{GL}_{n-1}(K))$, (B_n, D_n), (D_{n+1}, B_n). Let V be an irreducible rational restricted module for G. Then the socle and cosocle of $V|_X$ are multiplicity-free with all simple constituents being restricted.*

4.3. TENSOR PRODUCT THEOREMS

In this section we describe three results in increasing generality, starting with Steinberg's famous tensor product theorem. This result shows that irreducible rational representations can be decomposed into the tensor product of twists of restricted representations. To ease exposition we state some of the results a little less generally than possible.

We need a little terminology. Suppose K is a field and σ is a field endomorphism of K. Let $G(K)$ be a universal Chevalley group defined over K. So, if K is algebraically closed, then $G(K)$ is a simply connected algebraic group. Fixing a system of root subgroups we can define an endomorphism of $G(K)$, which we also call σ. This endomorphism induces the map $U_\alpha(c) \rightarrow U_\alpha(c^\sigma)$ on each root subgroup.

Given a representation $G(K) \rightarrow \mathrm{GL}(V)$ for V a finite dimensional space over K, we obtain a new representation $G(K) \rightarrow G(K) \rightarrow \mathrm{GL}(V)$, where the first homomorphism is the field twist determined by σ. We use the notation V^σ for the new representation. In particular, if K has characteristic $p > 0$, let F denote the Frobenius morphism $c \mapsto c^p$ of K and also the corresponding endomorphism of $G(K)$.

Let G be a semisimple algebraic group over a field K of characteristic $p > 0$. If λ is a dominant weight, then it has a p-adic expansion $\lambda = \lambda_0 + p\lambda_1 + \cdots + p^r \lambda_r$, where each of $\lambda_0, \ldots, \lambda_r$ is a restricted dominant weight. We can now state the Steinberg tensor product theorem.

Theorem 4.6 [50] *Let G be a simply connected semisimple algebraic group and let λ be a dominant weight with p-adic expansion $\lambda = \lambda_0 + p\lambda_1 + \cdots + p^r \lambda_r$, where each of $\lambda_0, \ldots, \lambda_r$ is restricted. Then*

$$V_G(\lambda) \cong V_G(\lambda_0) \otimes V_G(\lambda_1)^F \otimes \cdots \otimes V_G(\lambda_r)^{F^r}.$$

As mentioned earlier, this theorem reduces many questions concerning irreducible representations to the study of the restricted ones. It is also shown in [50] that each irreducible representation of a finite group of Lie type over an algebraically closed field in the natural characteristic lifts to an irreducible representation of a corresponding algebraic group. Consequently, the above theorem yields a similar theorem for irreducible representations of finite groups of Lie type in the defining characteristic.

We next describe a similar result for *abstract* irreducible representations. Let $G = G(K)$ be a simple algebraic group over K and consider a group homomorphism $G \rightarrow \mathrm{GL}(V)$, where V is a finite dimensional vector space over the algebraically closed field K and where we impose no assumption of rationality. So this is a representation, but not necessarily a homomorphism of algebraic groups.

We can even be a bit more general by taking V to be a vector space over some other algebraically closed field L of the same characteristic. In this case let $G(L)$ denote a simple, simply connected algebraic group over L which has the same type as $G = G(K)$. As above, for each field morphism $\sigma : K \rightarrow L$, there is a corresponding group homomorphism $\sigma : G(K) \rightarrow G(L)$.

If V is a rational representation of $G(L)$ and σ a field morphism from K to L, we write V^σ for the representation $G(K) \rightarrow G(L) \rightarrow \mathrm{GL}(V)$, where the first map is the field morphism corresponding to σ and the second is the given representation.

The following is a fundamental result of Borel-Tits.

Theorem 4.7 [7] *Let G be a simply connected, simple algebraic group over K and let $G \rightarrow \mathrm{GL}(V)$ be an abstract irreducible representation, where V is finite* *dimensional over an algebraically closed field L of the same characteristic. Then*

$$V \cong V_1^{\sigma_1} \otimes \cdots \otimes V_r^{\sigma_r},$$

where each V_i is an irreducible rational representation of $G(L)$ and $\sigma_1, \ldots, \sigma_r$ are field morphisms from K to L.

One can view the above theorem in the following way. An irreducible abstract representation $G \to \mathrm{GL}(V)$ can be factored $G \to G(L) \times \cdots \times G(L) \to \mathrm{GL}(V)$, where the first map is the product of field morphisms and the second is an irreducible rational representation. The Steinberg tensor product theorem can also be formulated in this way.

In recent work ([42]), a similar result has been established for arbitrary abstract homomorphisms of algebraic groups.

Theorem 4.8 [42, Theorem 1] *Let $\varphi : G(K) \to Y(L)$ be an abstract homomorphism, where $G(K)$ is a simply connected simple algebraic group over K and $Y(L)$ is an algebraic group over the algebraically closed field L. If $\mathrm{char}(K) = 0$, assume that a maximal torus of $G(K)$ is mapped into a torus of $Y(L)$. Then the following hold:*

(1) *φ can be factored*

$$G(K) \to G(L) \times \cdots \times G(L) \to Y(L),$$

where the first map is a product of field morphisms and the second is a rational homomorphism.

(2) *The Zariski closure of $\varphi(G(K))$ is a commuting product of images of $G(L)$.*

Both Theorem 4.7 and the theorem to follow hold, more generally, for universal Chevalley groups over perfect fields. For related results see the paper of Tits [54]. The above theorem has various applications. Let $G = G(K)$ be a simple algebraic group with K an algebraically closed field of characteristic $p > 0$. One consequence shows that extensions of abstract irreducible G-modules can be reduced to extensions of irreducible rational modules. Another application shows that, with essentially one exception, an abstract representation of G is rational if and only if each composition factor is rational (the exception is where G has type C_n with $p = 2$). See [42] for details. In the next section we give an application of a variation of the above theorem to the theory of locally finite groups.

4.4. LOCALLY FINITE GROUPS

In this last section we describe some recent connections between the theory of algebraic groups and the theory of locally finite groups. Let p be a prime and $K = \overline{\mathbb{F}}_p$. Then K is a locally finite field and simple algebraic groups over K are quasisimple, locally finite groups.

The following result is due to Hartley and Zalesskiĭ. To state it we need some notation. If k is an infinite subfield of K, then we write $G(k)$ to indicate a Chevalley group or twisted version (including Suzuki and Ree groups) over k of the same type as $G = G(K)$.

Theorem 4.9 [22] *Let $G = G(K)$ be a simple algebraic group over K.*

(1) *If M is a dense subgroup of G, then there is an infinite subfield k of K such that $G(k) \leqslant M \leqslant N_G(G(k))$.*

(2) *All maximal subgroups of G are closed.*

This result shows that abstractly maximal subgroups are necessarily closed. Hence, information is obtained by applying results from the theory of algebraic groups.

The proof of the above result proceeds by first reducing to the case where M is almost simple. That is, M is contained in the automorphism group of a simple group, say X. Then X is a simple infinite locally finite group having a faithful finite dimensional representation (e.g. X acts on $L(G)$). Such groups have been classified, using the classification of finite simple groups (see [21]). The result is that X is of Lie type over an infinite subfield k of K. Further arguments with representation theory show that X necessarily has the same type as G.

The above result and its proof suggest the following problem. Given an infinite quasisimple subgroup, say X, of G, what can be said about its closure? A priori, it is not clear that any useful information can be obtained, since the closure might have a large unipotent subgroup. On the other hand, Theorem 4.9 implies that rarely is the closure the whole group. As above, we have $X = X(k)$ for some infinite subfield k of K.

The following result is a special case of a result in [42]. It reduces the analysis of arbitrary infinite quasisimple subgroups of G to the study of closed, semisimple subgroups.

Theorem 4.10 [42, Theorem 7] *Let k be an infinite subfield of $K = \overline{\mathbb{F}}_p$ and assume $X = X(k)$ is a simple subgroup of $G(K)$, a semisimple algebraic group over K. Then the Zariski closure of X is a commuting product of simple algebraic groups of type $X(K)$.*

The point of view here is that to study the embedding of X in G we first study the embedding of X in its Zariski closure and then the embedding of the closure in G. The former is easy, as the proof shows that each projection map is a natural embedding composed with a field twist. Information about the latter embedding belongs to the realm of the subgroup structure of simple algebraic groups. At this point the reader should consult the article of Liebeck.

References

1. H. Azad, M. Barry and G. Seitz, On the structure of parabolic subgroups, *Comm. Alg.* **18** (1990), 551–562.

2. P. Bala and R. Carter, Classes of unipotent elements in simple algebraic groups, I, *Proc. Cambr. Phil. Soc.* **79** (1976), 401–425.

3. P. Bala and R. Carter, Classes of unipotent elements in simple algebraic groups, II, *Proc. Cambr. Phil. Soc.* **80** (1976), 1–17.

4. A. Borel, *Linear Algebraic Groups*, Springer-Verlag, New York, 1991.

5. A. Borel and J. de Siebenthal, Les sous-groupes fermés de rang maximum des groupes de Lie clos, *Comment. Math. Helv.* 23 (1949), 200–221.

6. A. Borel and J. Tits, Éléments unipotents es sousgroupes paraboliques de groupes réductifs, I, *Invent. Math.* 12 (1971), 95–104.

7. A. Borel and J. Tits, Homomorphisms 'abstract' de groupes algébriques simples, *Ann. Math.* 98 (1973), 499–571.

8. A. Borel et al., *Seminar on Algebraic Groups and Related Finite Groups*, Lec. Notes in Math. 131, Springer-Verlag, Berlin-New York, 1970.

9. R. Carter, Conjugacy classes in the Weyl group, *Comp. Math.* 25 (1972), 1–59.

10. R. Carter, *Finite Groups of Lie Type: Conjugacy Classes and Complex Characters*, Wiley, 1985.

11. B. Cooperstein, Subgroups of the group $E_6(q)$ which are generated by root subgroups, *J. Algebra* 46 (1977), 355–388.

12. B. Cooperstein, The geometry of root subgroups in exceptional groups, *Geom. Dedicata* 8 (1979), 317–381.

13. B. Cooperstein, Subgroups of exceptional groups of Lie type generated by long root elements, I. Odd characteristic, *J. Algebra* 70 (1981), 270–282.

14. B. Cooperstein, Subgroups of exceptional groups of Lie type generated by long root elements, II. Even characteristic, *J. Algebra* 70 (1981), 283–298.

15. B. Cooperstein, The geometry of root subgroups in exceptional groups II, *Geom. Dedicata* 15 (1983), 1–45.

16. E. Dynkin, Maximal subgroups of the classical groups, *Amer. Math. Soc. Translations* 6 (1957), 245–378.

17. E. Dynkin, Semisimple subalgebras of semisimple Lie algebras, *Amer. Math. Soc. Translations* 6 (1957), 111–244.

18. H. Enomoto, The conjugacy classes of Chevalley groups of type (G_2), *J. Fac. Sci. Univ. Tokyo* 16 (1969), 497–512.

19. B. Ford, *Overgroups of Irreducible Linear Groups*, Ph.D thesis, University of Oregon, 1993.

20. B. Ford, Irreducible restrictions of representations of the symmetric groups, to appear.

21. B. Hartley and G. Shute, Monomorphisms and direct limits of finite groups of Lie type, *Quart. J. Math.* 35 (1984), 49–71.

22. B. Hartley and A. Zalesskiĭ, On simple periodic linear groups—dense subgroups, permutation representations, and induced modules, *Israel J. Math.* 82 (1993), 299–327.

23. J. Humphreys, *Linear Algebraic Groups*, Springer-Verlag, New York, 1972.

24. J. Humphreys, *Introduction to Lie Algebras and Representation Theory*, Springer-Verlag, New York, 1972.

25. J. Jantzen, *Representations of Algebraic Groups*, Academic Press, 1987.

26. W. Kantor, Subgroups of classical groups generated by long root elements, *Trans. Amer. Math. Soc.* 248 (1979), 347–379.

27. A. Kleshchev, On restrictions of irreducible modular representations of semisimple algebraic groups and symmetric groups to some natural subgroups, to appear.

28. R. Lawther and D. Testerman, to appear.

29. M. Liebeck and G. Seitz, Subgroups generated by root elements in groups of Lie type, *Ann. Math.*, to appear.

30. M. Liebeck and G. Seitz, Maximal subgroups of exceptional groups of Lie type, finite and algebraic, *Geom. Dedicata* 35 (1990), 353–387.

31. M. Liebeck and G. Seitz, Reductive subgroups of exceptional algebraic groups, to appear.

32. G. Lusztig, On the finiteness of the number of unipotent classes, *Invent. Math.* 34 (1976), 201–213.

33. K. Mizuno, The conjugate classes of unipotent elements of the Chevalley groups of type E_6, *J. Fac. Sci. Univ. Tokyo* 24 (1977), 525–563.

34. K. Mizuno, The conjugate classes of unipotent elements of the Chevalley groups of type E_7 and E_8, *Tokyo J. Math.* 3 (1980), 391–461.

35. K. Pommerening, Über die unipotenten Klassen reduktiver Gruppen, *J. Algebra* 49 (1977), 525–536.

36. K. Pommerening, Über die unipotenten Klassen reduktiver Gruppen, II, *J. Algebra* 65 (1980), 373–398.

37. A. Premet, Weights of infinitesimally irreducible representations of Chevalley groups over a

field of prime characteristic, *Math. USSR Sbornik* 61 (1988), 167–183.

38. R. Richardson, Conjugacy classes in Lie algebras and algebraic groups, *Indag. Math.* 88 (1985), 337–344.
39. R. Richardson, G. Röhrle and R. Steinberg, Parabolic subgroups with abelian unipotent radical, *Invent. Math.* 110 (1992), 649–671.
40. G. Röhrle, On the structure of parabolic subgroups in algebraic groups, *J. Algebra* 157 (1993), 80–115.
41. G. Seitz, The root subgroups for maximal tori in finite groups of Lie type, *Pacific J. Math.* 106 (1983), 153–244.
42. G. Seitz, Abstract homomorphisms of algebraic groups, to appear in *Proc. London Math. Soc.*
43. G. Seitz, The maximal subgroups of classical algebraic groups, *Mem. Amer. Math. Soc.* 365 (1987), 1–286.
44. G. Seitz, Maximal subgroups of exceptional algebraic groups, *Mem. Amer. Math. Soc.* 441 (1991), 1–197.
45. T. Shoji, The conjugacy classes of Chevalley groups of type (F_4) over finite fields of characteristic $p \neq 2$, *J. Fac. Sci. Univ. Tokyo* 21 (1975), 1–17.
46. K. Shinoda, The conjugacy classes of Chevalley groups of type (F_4) over finite fields of characteristic 2, *J. Fac. Sci. Univ. Tokyo* 21 (1974), 133–159.
47. S. Smith, Irreducible modules and parabolic subgroups, *J. Algebra* 75 (1982), 286–289.
48. T. A. Springer, *Linear Algebraic Groups*, Birkhauser, Boston, 1981.
49. T. A. Springer and R. Steinberg, Conjugacy classes, in *Seminar on Algebraic Groups and Related Finite Groups*, Lec. Note Math. 131, Springer-Verlag, Berlin-New York, 1970.
50. R. Steinberg, Representations of algebraic groups, *Nagoya Math. J.* 22 (1963), 33–56.
51. R. Steinberg, *Lectures on Chevalley Groups*, lecture notes, Yale University.
52. D. Testerman, Irreducible subgroups of exceptional algebraic groups, *Mem. Amer. Math. Soc.* 390 (1988), 1–190.
53. D. Testerman, Overgroups of unipotent elements in simple groups of Lie type, finite and algebraic, to appear.
54. J. Tits, Homomorphismes 'abstract' de groupes de Lie, *Symposia Math.* XIII, Istituto Nazionale di Alta Math., Bologna, 1974, pp. 479–499.

SUBGROUPS OF SIMPLE ALGEBRAIC GROUPS AND OF RELATED FINITE AND LOCALLY FINITE GROUPS OF LIE TYPE

M. W. LIEBECK
Imperial College
London SW7 2BZ
England
E-mail: m.liebeck@ic.ac.uk

Abstract. This is a survey of some recent results on the subgroup structure of simple algebraic groups and of related finite and locally finite groups of Lie type. The first section contains basic material on simple algebraic groups, their automorphisms and Frobenius morphisms, and shows how every finite or locally finite group of Lie type can be realised as the fixed point group G_σ or G_ϕ of a suitable automorphism σ or ϕ of a simple algebraic group G. The other two sections deal with classical groups and exceptional groups. Both sections start with results on the closed subgroups of G, and then show how these results can be used to deduce corresponding results about the subgroups of G_σ and G_ϕ.

Key words: algebraic groups, groups of Lie type, maximal subgroups, closed subgroups.

Introduction

The purpose of this article is to discuss some recent results concerning the subgroup structure of simple algebraic groups and of related finite and locally finite groups of Lie type. For simple algebraic groups G, the results are of the following form. When G is a classical group with associated vector space V, a certain collection $\mathcal{C}$ of natural 'geometric' subgroups of G is defined; the main result on subgroups of G (Theorem 2.2) states that any closed subgroup either lies in a member of $\mathcal{C}$, or is essentially a simple group acting irreducibly on V. For G of exceptional type, the main results (Theorems 3.1, 3.2 and 3.8) describe all maximal closed subgroups and all semisimple closed subgroups of G, subject to some mild characteristic restrictions.

The finite groups of Lie type arise as fixed point groups G_σ of so-called Frobenius morphisms σ of G (see Section 1 for discussion of these). We shall discuss how the results mentioned above on subgroups of G can be used to deduce analogous results on subgroups of the finite groups G_σ. For G classical, this process yields a new proof of a well known theorem of Aschbacher [1] on subgroups of finite classical groups. And for G exceptional, we shall see that our knowledge of various important classes of subgroups of G_σ, such as maximal subgroups and semisimple subgroups, is nearing completion except in some small characteristics.

The groups of Lie type over infinite locally finite fields form an important class of locally finite simple groups (see the articles of Hartley and others in these Proceedings). Each of these groups can be realised as the fixed point group G_ϕ of a suitable automorphism ϕ of a simple algebraic group G, and we shall see that our

B. Hartley et al. (eds.), Finite and Locally Finite Groups, 71–96.

results on subgroups of G can also be used to obtain analogous results on subgroups of the locally finite groups G_ϕ.

The article has three sections. In the first, we present some basic material on simple algebraic groups, their automorphisms and Frobenius morphisms. We also discuss the fundamental Lang-Steinberg theorem concerning Frobenius morphisms, which is the basic tool for relating the subgroups of G with those of the finite groups G_σ. The second and third sections contain the results discussed above on subgroups of classical and exceptional groups, respectively.

1. Frobenius Morphisms and the Lang-Steinberg Theorem

We begin this section with a brief dicussion of the types of simple algebraic groups and their automorphisms. We then define Frobenius morphisms and finite groups of Lie type, and state the Lang-Steinberg theorem. Some applications of this result are given, showing how to relate subgroups of simple algebraic groups with those of the corresponding finite groups of Lie type. The section concludes with a discussion of representations of finite groups of Lie type, and a few words about locally finite groups.

Many of the results presented in this section are due to Steinberg [29, 30] and to Springer and Steinberg [28]. We shall follow the approaches of these papers, and also refer freely to the article of Seitz in these Proceedings.

1.1. SIMPLE ALGEBRAIC GROUPS

We summarise some material which is discussed in more detail in Seitz's article. Let K be an algebraically closed field of characteristic p. The simple algebraic groups over K were classified by Chevalley [8], and fall into the following families:

$$\text{classical types}\ :\ A_n(K), \quad B_n(K), \quad C_n(K), \quad D_n(K)$$
$$(\text{examples}\ :\ \mathrm{SL}_{n+1}(K),\ \mathrm{SO}_{2n+1}(K),\ \mathrm{Sp}_{2n}(K),\ \mathrm{SO}_{2n}(K)),$$
$$\text{exceptional types}\ :\ G_2(K), \quad F_4(K), \quad E_6(K), \quad E_7(K), \quad E_8(K).$$

For each fixed type (e.g. $A_n(K)$ with n fixed), there may be several pairwise non-isomorphic simple algebraic groups, which can usually (but not always) be distinguished by their centres. For example, when $p = 0$ or p does not divide $n + 1$, the groups of type $A_n(K)$ are $\mathrm{SL}_{n+1}(K)$ and its quotients by subgroups of the group of scalar matrices; the group with the largest centre, $\mathrm{SL}_{n+1}(K)$, is the *simply connected* group of type $A_n(K)$, while the group with trivial centre, $\mathrm{PSL}_{n+1}(K)$, is the *adjoint* group of this type. (When p divides $n + 1$, more subtle considerations are required to distinguish the different groups of type $A_n(K)$, as discussed in Seitz's article, Theorem 2.5.)

For the rest of this section, let G be a simple algebraic group over K. Fix a maximal torus T of G. A root subgroup of G is a 1-dimensional T-invariant unipotent subgroup $U_\alpha = \{U_\alpha(c) \mid c \in K\}$, where $\alpha \in X(T) = \mathrm{Hom}(T, K^*)$ and, for $t \in T, c \in K$,

$$U_\alpha(c)^t = U_\alpha(\alpha(t)c).$$

(For example, if $G = \mathrm{SL}_{n+1}(K)$ we can take T to be the subgroup of all diagonal matrices in G; the root subgroups are then of the form $\{I + cE_{ij} \mid c \in K\}$ for $i \neq j$, where E_{ij} is the matrix with 1 in the ij-entry and zeros elsewhere.) The root subgroups U_α generate G, and the collection of roots α forms the root system $\Sigma(G)$ of G. The Weyl group $W(G) = N_G(T)/T$ is a finite group generated by reflections corresponding to roots in $\Sigma(G)$. Associated with the root system $\Sigma(G)$ is the Dynkin diagram of G, which has nodes labelled by a set of fundamental roots $\alpha_1, \ldots, \alpha_n$, and edges determined by the inner products among these roots. If α_0 is the highest root in $\Sigma(G)$—that is, $\alpha_0 = \sum c_i \alpha_i$ with $\sum c_i$ maximal—then the adjoining of $-\alpha_0$ to the Dynkin diagram (with appropriate edges) gives the extended Dynkin diagram of G. Full descriptions of root systems and diagrams (and many other related things) can be found in [4, p. 250]. For example, here are the Dynkin diagram and extended Dynkin diagram of type E_8.

1.2. Automorphisms

Continue to assume that G is a simple algebraic group over K. We now describe various automorphisms of G as an abstract group. To distinguish those automorphisms which respect the structure of G as an algebraic group, we make the following definition.

Definition An automorphism τ of G is called an *algebraic automorphism* if τ and τ^{-1} are both morphisms $G \to G$. Write $\mathrm{Aut_{alg}}(G)$ for the group of all algebraic automorphisms of G, and $\mathrm{Aut_{abs}}(G)$ for the group of all automorphisms of G as abstract group.

A well known result of Steinberg [29, Theorem 30] describes these automorphism groups. In order to state the result, we list some examples of automorphisms of G.

(1) *Inner automorphisms.* These are clearly algebraic automorphisms.

(2) *Graph automorphisms* of types A_n, D_n, D_4, E_6. For these types, the Dynkin diagram possesses a symmetry of order 2,2,3, or 2 respectively, which extends linearly to a permutation ρ of the root system $\Sigma(G)$. The graph automorphism corresponding to the symmetry ρ is given by

$$U_\alpha(c) \mapsto U_{\rho(\alpha)}(\epsilon_\alpha c) \quad (\alpha \in \Sigma(G), c \in K),$$

where each ϵ_α is ± 1 (see [29, p. 156]). These graph automorphisms are algebraic.

(3) *Field automorphisms.* For each automorphism ϕ of the field K, there is a corresponding field automorphism of G given by

$$U_\alpha(c) \mapsto U_\alpha(c^\phi) \quad (\alpha \in \Sigma(G), c \in K).$$

Nontrivial field automorphisms are not algebraic, as their inverses are not morphisms. Note that the field automorphisms which are morphisms are those of the form $U_\alpha(c) \mapsto U_\alpha(c^q)$, where q is a power of p. We shall write σ_q for this particular field automorphism.

(4) *More graph automorphisms* of types $B_2\,(p = 2)$, $F_4\,(p = 2)$, $G_2\,(p = 3)$. For these types the Dynkin diagram has a symmetry of order 2, extending to a permutation ρ of $\Sigma(G)$ which interchanges long and short roots. This time the corresponding graph automorphism of G is given by

$$U_\alpha(c) \mapsto U_{\rho(\alpha)}(\epsilon_\alpha c^{p(\alpha)}),$$

where $p(\alpha) = 1$ if α is a long root, $p(\alpha) = p$ if α is a short root, and $\epsilon_\alpha = \pm 1$ (see [29, p. 156]).

These graph automorphisms are not algebraic, as their inverses are not morphisms.

Theorem 1.1 [29, Theorem 30]

(i) $\mathrm{Aut}_{\mathrm{alg}}(G)$ *is generated by inner automorphisms and graph automorphisms of type* (2) *above.*

(ii) $\mathrm{Aut}_{\mathrm{abs}}(G)$ *is generated by inner, graph and field automorphisms as in* (1)−(4) *above.*

1.3. FROBENIUS MORPHISMS

For an automorphism σ of the simple algebraic group G, denote by G_σ the fixed point subgroup $\{g \in G \mid g^\sigma = g\}$. When σ is a morphism, there is a remarkable dichotomy concerning the fixed point groups G_σ:

Theorem 1.2 [30, 10.13] *Suppose* $\sigma \in \mathrm{Aut}_{\mathrm{abs}}(G)$ *and* σ *is a morphism* $G \to G$. *Then exactly one of the following statements holds:*

(i) σ *is an algebraic automorphism;*

(ii) G_σ *is finite.*

This implies, for example, that, for any $x \in G$, the centralizer $C_G(x)$ is infinite.

Definition If $\sigma \in \mathrm{Aut}_{\mathrm{abs}}(G)$ is a morphism such that G_σ is finite, then σ is called a *Frobenius morphism* of G.

Examples 1.3 Let $G = \mathrm{SL}_n(K)$ with K of characteristic $p > 0$, and let q be a power of p.

(a) If $\sigma = \sigma_q$, the field automorphism sending (a_{ij}) to (a_{ij}^q) for all matrices (a_{ij}) in G, then G_σ is the finite group $\mathrm{SL}_n(q)$.

(b) Let $\bar{G}$ be the adjoint group $GZ/Z = \mathrm{PSL}_n(K)$ (where Z is the group of nonzero scalar matrices), and again let $\sigma = \sigma_q$, the field automorphism sending $(a_{ij})Z$ to $(a_{ij}^q)Z$. As K is algebraically closed, for every $x \in \mathrm{GL}_n(K)$ there exist $g \in G$, $z \in Z$ such that $x = gz$. It follows quickly that $\bar{G}_\sigma = \mathrm{PGL}_n(q)$. Notice that the corresponding finite simple group $\mathrm{PSL}_n(q)$ is not in general equal to $\bar{G}_\sigma$, but is equal to $O^{p'}(\bar{G}_\sigma)$.

(c) Now let σ be the automorphism sending (a_{ij}) to $(a_{ij}^q)^{-T}$ (where $-T$ is the inverse-transpose map). Then G_σ is the finite unitary group $\mathrm{SU}_n(q)$.

The precise form of all Frobenius morphisms is given by the next result, which is again due to Steinberg. For notational convenience, we denote by σ_1 the identity automorphism of G.

Theorem 1.4 [30, §11] *Every Frobenius morphism of G is G-conjugate to either σ_q or $\tau\sigma_q$, where τ is a graph automorphism as in (2) or (4) above, and σ_q is a field automorphism as in (3) with q a positive power of p (except for $\tau\sigma_q$ with τ as in (4), when $q = 1$ is also allowed).*

Corollary 1.5 *The finite groups G_σ, where σ is a Frobenius morphism of G, are as follows:*

(i) *$\sigma = \sigma_q$, G_σ a Chevalley group $\mathrm{A}_n(q)$, $\mathrm{B}_n(q)$, $\mathrm{C}_n(q)$, $\mathrm{D}_n(q)$, $\mathrm{G}_2(q)$, $\mathrm{F}_4(q)$, $\mathrm{E}_6(q)$, $\mathrm{E}_7(q)$, $\mathrm{E}_8(q)$;*

(ii) *$\sigma = \tau\sigma_q$, G_σ a twisted group $^2\mathrm{A}_n(q)$, $^2\mathrm{D}_n(q)$, $^3\mathrm{D}_4(q)$, $^2\mathrm{E}_6(q)$, $^2\mathrm{B}_2(pq^2)$, $^2\mathrm{F}_4(pq^2)$, $^2\mathrm{G}_2(pq^2)$.*

When G is of adjoint type, $O^{p'}(G_\sigma)$ is usually simple (see Example 1.3(b)); these finite simple groups are the *simple groups of Lie type*.

Theorem 1.6 [29, Theorem 30] *Let $L = O^{p'}(G_\sigma)$ be a finite simple group of Lie type. Then every automorphism of L lifts to a morphism of G commuting with σ. Thus $\mathrm{Aut}(L)$ is generated by G_σ (acting by conjugation), together with the restrictions to L of field and graph automorphisms of G leaving L invariant.*

1.4. THE LANG-STEINBERG THEOREM

This is a basic result which underlies much of the interplay between the finite and algebraic groups of Lie type.

Theorem 1.7 [30, 10.1] *Let H be a connected linear algebraic group, and suppose that $\sigma : H \to H$ is a surjective homomorphism of algebraic groups, such that H_σ is finite. Then the map*

$$h \mapsto h^{-1}h^\sigma$$

from H to H is surjective.

For example, the result applies when $H = G$, a simple algebraic group, and σ is a Frobenius morphism of G. Theorem 1.7 was proved for field automorphisms $\sigma = \sigma_q$ by Lang [14], and in full generality by Steinberg.

Whenever a group H as in Theorem 1.7 acts on a set S on which σ also acts, the Lang-Steinberg theorem applies to determine the H_σ-orbits on the fixed point set S_σ (Proposition 1.8 below gives the precise result). We shall give applications of this when S is a conjugacy class of elements or of subgroups of a simple algebraic group G.

To state the proposition, we need a definition.

Definition Let X be a group, and $\sigma : X \to X$ a homomorphism. Denote by $H^1(\sigma, X)$ the set of equivalence classes of X under the equivalence relation

$$x \sim y \quad \Leftrightarrow \quad y = z^{-1}xz^\sigma \text{ for some } z \in X.$$

For example, if σ is the identity on X then $H^1(\sigma, X)$ is the set of conjugacy classes of X.

Proposition 1.8 [28, I, 2.7] *Let H, σ be as in Theorem 1.7. Suppose that H acts transitively on a set S, and that σ also acts on S in such a way that $(sh)^\sigma = s^\sigma h^\sigma$ for all $s \in S$, $h \in H$. Then the following hold.*

(i) *S contains an element fixed by σ.*

(ii) *Fix an element $s_0 \in S_\sigma$, and assume that $X = H_{s_0}$ is a closed subgroup of H. Then there is a bijective correspondence between the set of H_σ-orbits on S_σ and the set $H^1(\sigma, X/X^\circ)$.*

Note that in conclusion (ii) the group X/X° is of course finite. The proof of Proposition 1.8 is a fairly straightforward application of Theorem 1.7. Here is a sketch.

For part (i), fix $s \in S$ and observe that, by the transitivity of H, there exists $h \in H$ with $s^\sigma = sh$. By Theorem 1.7, we have $h = k^{-1}k^\sigma$ for some $k \in H$; then sk^{-1} is fixed by σ.

Now consider (ii). We show first that the natural map $H^1(\sigma, X) \to H^1(\sigma, X/X^\circ)$ is bijective. It is clearly surjective. If $x, y \in X$ represent the same element of $H^1(\sigma, X/X^\circ)$, then, replacing y by an element equivalent to it in G, we have $x \equiv y \bmod X^\circ$. By Theorem 1.7, $y = g^{-1}g^\sigma$ for some $g \in G$. Then σ fixes the connected group $gX^\circ g^{-1}$; applying Theorem 1.7 to this group, we see that

$$gxy^{-1}g^{-1} = (gzg^{-1})^{-1}(gzg^{-1})^\sigma$$

for some $z \in X^\circ$. This gives $x = z^{-1}yz^\sigma$, so $x \sim y$. This proves the injectivity of the map.

Now define a map from the set of H_σ-orbits on S_σ to $H^1(\sigma, X)$ as follows. Take an orbit representative $s_0 h \in S_\sigma$ (where $h \in H$); then $s_0 h^\sigma = s_0 h$, hence $h^\sigma h^{-1}$ belongs to $H_{s_0} = X$, and we map the orbit of $s_0 h$ to the equivalence class of $h^\sigma h^{-1}$. It is not hard to check that this map is a well-defined bijection, completing the proof of Proposition 1.8.

We now give a number of applications of Proposition 1.8 in the case where $H = G$, a simple algebraic group, and σ is a Frobenius morphism of G.

Example 1.9 Let $x \in G$ and suppose that the conjugacy class x^G of x in G is σ-stable (i.e. x^σ is conjugate to x). Taking $S = x^G$ in Proposition 1.8, we deduce:

(i) $x^G \cap G_\sigma \neq \emptyset$, and

(ii) the G_σ-classes in $x^G \cap G_\sigma$ are in bijective correspondence with the set

$$H^1(\sigma, C_G(x)/C_G(x)^\circ).$$

As discussed in Seitz's article, a great deal is known about conjugacy classes and centralizers in G. For example, when x is a semisimple element, $C_G(x)$ is a reductive group containing a maximal torus of G; moreover $C_G(x)/C_G(x)^\circ$ is isomorphic to a subgroup of $\ker \pi$, where π is the natural map $\hat{G} \to G$, $\hat{G}$ being the simply connected group of the same type as G (see [28, II, 4.1, 4.4]). Notice also that if $(C_G(x)^\circ)'$ is nontrivial then it is a semisimple subsystem subgroup of G; discussion of these and their fixed point groups can be found in Example 1.12 below.

Example 1.10 Let D be a closed σ-stable subgroup of G, and define the set

$$S = \{D^g \mid g \in G\}.$$

By Proposition 1.8, there is a bijective correspondence between the G_σ-classes of σ-stable subgroups in S and the set $H^1(\sigma, N_G(D)/N_G(D)^\circ)$.

In the next five examples we apply 1.10 to various families of subgroups of G.

Example 1.11 *Parabolic subgroups.* The parabolic subgroups of the finite group G_σ are defined to be the subgroups P_σ, where P is a σ-stable parabolic subgroup of G (see Subsection 3.2 of Seitz's article for discussion of these). It is a basic fact that $N_G(P) = P = P^\circ$ for a parabolic subgroup P. Therefore, by 1.10, each σ-stable G-class of parabolic subgroups of G gives just one G_σ-class of parabolic subgroups of G_σ.

Example 1.12 *Subsystem subgroups.* Let Δ be a closed subsystem of the root system $\Sigma(G)$, and define $D = \langle U_\alpha \mid \alpha \in \Delta \rangle$, a *subsystem subgroup* of G (see Subsection 3.1 of Seitz's article). Assume that D is σ-stable (this is automatic if $\sigma = \sigma_q$ (notation as in 1.4), and amounts to Δ being stable under the permutation ρ of $\Sigma(G)$ corresponding to τ when $\sigma = \tau\sigma_q$). To apply 1.10, we need to know the group $N_G(D)/N_G(D)^\circ$. In fact this is determined by the action of the Weyl group $W = W(G)$ on the root system: we have

$$N_G(D)/N_G(D)^\circ \cong N_W(\Delta)/W(\Delta)W(\Delta^\perp),$$

where $W(\Delta)$ is the group generated by the reflections corresponding to the roots in Δ, and $\Delta^\perp$ is the set of roots orthogonal to all those in Δ (see [5, §2]). Essentially, this means that $N_G(D)/N_G(D)^\circ$ is the group of symmetries induced by W on the Dynkin diagram of D.

If D^g is a σ-stable G-conjugate of D, we call $(D^g)_\sigma$ a subsystem subgroup of G_σ. Using the information given above, it is possible to work out all classes of subsystem subgroups of G_σ. There are many such calculations in the literature. For example, as remarked in Example 1.9, centralizers of semisimple elements give rise to subsystem subgroups, and the corresponding centralizers in G_σ are found in [5, 6, 10, 11], among other papers. Also, in [22], those subsystem subgroups whose normalizers are maximal in G_σ are determined.

The method of Example 1.12 plays an important role in some of our results in Sections 2 and 3, and we illustrate it in the next three examples.

Remark As noted in the discussion of subsystem subgroups in Seitz's article (Subsection 3.1), when $\Sigma(G)$ has roots of different lengths and p is 2 or 3 (the latter only for $G = G_2$), one has to extend the definition of a subsystem subgroup slightly, to include a few subgroups corresponding to non-closed subsystems.

Example 1.13 Let $G = \mathrm{SL}_6$ (type A_5), and let $D = (\mathrm{SL}_2)^3$, a subsystem subgroup of type $(A_1)^3$. Here $N_G(D)/N_G(D)^\circ \cong \mathrm{Sym}_3$, permuting the factors SL_2. Take $\sigma = \sigma_q$, so $G_\sigma = \mathrm{SL}_6(q)$. As σ acts trivially on W, $H^1(\sigma, N_G(D)/N_G(D)^\circ)$ is just the set of conjugacy classes in Sym_3, with representatives 1, (12) and (123). If D^g is a σ-stable conjugate of D, then $g^\sigma g^{-1} \in N_G(D)$, hence maps to a conjugate w of one of the above representatives. The action of σ on D^g is like that of $w\sigma$ on D; thus the three G_σ-classes of subsystem subgroups of G_σ arising from conjugates of D are as follows:

(1) $w = 1$: $D_\sigma = \mathrm{SL}_2(q)^3$;

(2) $w = (12)$: σ-action on D^g is $(g_1, g_2, g_3) \mapsto (g_2^{(q)}, g_1^{(q)}, g_3^{(q)})$ for $g_i \in \mathrm{SL}_2$, so

$$(D^g)_\sigma = \mathrm{SL}_2(q^2) \times \mathrm{SL}_2(q);$$

(3) $w = (123)$: σ-action is $(g_1, g_2, g_3) \mapsto (g_3^{(q)}, g_1^{(q)}, g_2^{(q)})$, and

$$(D^g)_\sigma = \mathrm{SL}_2(q^3).$$

Example 1.14 Take $G = \mathrm{Sp}_{2n}$ (type C_n) and $D = \mathrm{SL}_n$, a subsystem subgroup of type A_{n-1} (the subsystem being that spanned by the short fundamental roots). Then $N_G(D)/N_G(D)^\circ = \langle w \rangle \cong C_2$, where w induces a graph automorphism on D. Thus we obtain two classes of subsystem subgroups $\mathrm{SL}_n(q)$ and $\mathrm{SU}_n(q)$ of $G_\sigma = \mathrm{Sp}_{2n}(q)$.

Example 1.15 Here is a slightly more complicated example, taken from [22] (see also [12, p. 35, nos. 36, 37]). Let $G = \mathrm{E}_8$, and let D be a subsystem subgroup of type $A_4 A_4$ (this subsystem is obtained by deleting the node α_5 from the extended Dynkin diagram of type E_8 given in Subsection 1.1). It turns out that

$$N_G(D)/N_G(D)^\circ = \langle w \rangle \cong C_4,$$

where w interchanges the two factors A_4 and w^2 induces a graph automorphism on each factor. Thus, in $G_\sigma = \mathrm{E}_8(q)$, there are four classes of subsystem subgroups $(D^g)_\sigma$. The structures of these subgroups are as follows:

$$\text{corresponding to } 1 \ : \quad D_\sigma \cong (\mathrm{SL}_5(q) \circ \mathrm{SL}_5(q)).(5, q-1),$$
$$\text{corresponding to } w^2 \ : \quad (D^g)_\sigma \cong (\mathrm{SU}_5(q) \circ \mathrm{SU}_5(q)).(5, q+1),$$
$$\text{corresponding to } w \ : \quad (D^g)_\sigma \cong \mathrm{SU}_5(q^2),$$
$$\text{corresponding to } w^3 \ : \quad (D^g)_\sigma \cong \mathrm{PGU}_5(q^2).$$

Example 1.16 *Maximal tori.* By 1.10, the G_σ-classes of σ-stable maximal tori of G are in bijective correspondence with $H^1(\sigma, N_G(T)/T) = H^1(\sigma, W)$, where W is the Weyl group of G. For example, if $G = \mathrm{SL}_n$ and $\sigma = \sigma_q$ then the classes of maximal tori $(T^g)_\sigma$ in $G_\sigma = \mathrm{SL}_n(q)$ correspond with conjugacy classes in $W = \mathrm{Sym}_n$. In similar fashion to Example 1.13, we see that a maximal torus in the G_σ-class corresponding to the W-class containing permutations of cycle-type $(a, b, \ldots, c)$ has order $(q^a - 1)(q^b - 1) \cdots (q^c - 1)/(q - 1)$.

The classes and orders of maximal tori in all the finite groups G_σ are known, although there remain some questions about their structure; see [7, §10] for the cases where G_σ is of untwisted type, and [11] for a summary of the twisted cases.

1.5. Elements of Prime Order

We now discuss briefly elements of prime order in the simple algebraic group G and the centralizers of these elements. The discussion will be used in some of the proof sketches given in Sections 2 and 3.

Let $x \in G$ have prime order r, and assume that $r \neq p$, so that x is a semisimple element. As remarked in Example 1.9, if $C = C_G(x)^\circ$ then C is a reductive group and the group C', if nontrivial, is a subsystem subgroup. The next result shows that the possible subsystems arising are very restricted. Recall from Subsection 1.1 the definition of the highest root $\alpha_0 = \sum c_i \alpha_i$ in the root system $\Sigma(G)$.

Theorem 1.17 [12, 14.1] *Let $x \in G$ have prime order $r \neq p$ and let C' be the subsystem subgroup $(C_G(x)^\circ)'$. If Δ is the Dynkin diagram of the root system of C', then one of the following holds:*

(i) *Δ is obtained by deleting nodes from the Dynkin diagram of G;*

(ii) *Δ is obtained from the extended Dynkin diagram of G by deleting one node α_i, where $r = c_i$, the coefficient of α_i in the highest root α_0.*

For example, when $G = \mathrm{E}_8$ we have

$$\alpha_0 = \begin{array}{ccccccc} 2 & 4 & 6 & 5 & 4 & 3 & 2 \\ & & & 3 & & & \end{array}$$

where the node α_i of the E_8 Dynkin diagram is labelled by its coefficient c_i. Referring to the extended Dynkin diagram shown in Subsection 1.1, we see that in E_8 we have the following classes of elements as in Theorem 1.17(ii):

for $p \neq 2$, elements of order 2 with centralizers of type D_8 and $\mathrm{A}_1\mathrm{E}_7$;
for $p \neq 3$, elements of order 3 with centralizers of type A_8 and $\mathrm{A}_2\mathrm{E}_6$;
for $p \neq 5$, elements of order 5 with centralizer of type $\mathrm{A}_4\mathrm{A}_4$.

Definition The primes dividing the coefficients c_i of the highest root α_0 are called *bad* primes for G.

The bad primes are as follows (see [28, I, §4]):

$$G = A_n : \text{none},$$
$$G = B_n, C_n, D_n : \text{only } 2,$$
$$G = G_2, F_4, E_6, E_7 : 2 \text{ and } 3,$$
$$G = E_8 : 2, 3 \text{ and } 5.$$

Corollary 1.18 *Suppose $x \in G$ has prime order r, and r is not a bad prime for G. Then $Z(C_G(x))$ is infinite, except possibly when G is of type A_n; in this case, $Z(C_G(x)^\circ)$ is infinite.*

Here is a sketch of the proof of the corollary. Assume first that $r \neq p$. When G is not of type A_n, $C_G(x)$ is necessarily connected, by [28, II, 4.6]; and in all cases, by Theorem 1.17, $C_G(x)^\circ$ is a Levi subgroup, so $Z(C_G(x)^\circ)$ is infinite. Now assume that $r = p$, and let $\hat{G}$ be the simply connected cover of G. By [28, III, 3.12] there is a $\hat{G}$-equivariant bijection $e \mapsto v(e)$ from the nilpotent elements of the Lie algebra $L(\hat{G})$ to the unipotent elements of $\hat{G}$. If $0 \neq e \in L(\hat{G})$ is nilpotent, then $C_{\hat{G}}(e) = C_{\hat{G}}(\lambda e)$ for any $\lambda \in K^*$, whence $C_{\hat{G}}(v(e)) = C_{\hat{G}}(v(\lambda e))$. It follows that $Z(C_{\hat{G}}(v(e)))$ is infinite, and hence $Z(C_G(x))$ is infinite.

1.6. NON-REDUCTIVE SUBGROUPS

Suppose that M is a closed connected subgroup of the simple algebraic group G and M is not reductive; in other words, $R_u(M) \neq 1$, where $R_u(M)$ is the unipotent radical of M (that is, the largest connected unipotent normal subgroup of M). Write $Q_1 = R_u(M)$, and define a chain of connected unipotent subgroups

$$Q_1 \leqslant Q_2 \leqslant \cdots \leqslant Q_i \leqslant \cdots$$

as follows:

$$Q_2 = R_u(N_G(Q_1)^\circ), Q_3 = R_u(N_G(Q_2)^\circ), \ldots, Q_i = R_u(N_G(Q_{i-1})^\circ).$$

Evidently this chain becomes stationary, at a subgroup Q_r with $Q_r = R_u(N_G(Q_r)^\circ)$. A result of Borel and Tits [2, 2.3] says that, for such a subgroup Q_r, the normalizer $N_G(Q_r)$ must be a parabolic subgroup of G. Since

$$M \leqslant N_G(Q_1) \leqslant N_G(Q_2) \leqslant \cdots \leqslant N_G(Q_r),$$

this means that M lies in a parabolic subgroup. Also [2, 3.12] gives a similar conclusion for subgroups of the finite groups G_σ; in this case, for a subgroup M of G_σ, instead of $R_u(M)$ we consider $O_p(M)$, where, as always, p is the underlying characteristic. To summarise:

Theorem 1.19 [2]

(i) *If M is a closed connected subgroup of G, then either M is reductive or M lies in a parabolic subgroup of G.*

(ii) *Let σ be a Frobenius morphism of G, and let U be a nontrivial p-subgroup of G_σ. Then $N_{G_\sigma}(U)$ lies in a parabolic subgroup of G_σ.*

1.7. Representations

In this section assume that the simple algebraic group G over K is simply connected. Section 4 of Seitz's article gives basic material on representations of G, and shows in particular how the finite-dimensional rational KG-modules are parametrized by dominant weights, the KG-module corresponding to the weight λ being denoted by $V_G(\lambda)$. Each dominant weight λ is of the form $\sum c_i \omega_i$, where the ω_i are the fundamental dominant weights and the c_i are non-negative integers.

Now let σ be a Frobenius morphism of G. By 1.4, we may assume that σ is equal to σ_q or $\tau \sigma_q$ for some power q of p. A remarkable result of Steinberg states that every irreducible KG_σ-module is the restriction of some irreducible KG-module $V_G(\lambda)$, and gives a parametrization of these modules:

Theorem 1.20 [30, 13.1]

(i) *If $\sigma = \sigma_q$, define*
$$\Lambda = \{\sum c_i \omega_i \mid 0 \leqslant c_i \leqslant q - 1\}.$$

Then the restricted modules $V_G(\lambda) \downarrow G_\sigma$ ($\lambda \in \Lambda$) form a complete set of non-isomorphic irreducible KG_σ-modules.

(ii) *If $\sigma = \tau \sigma_q$, sending $U_\alpha(c)$ to $U_{\rho(\alpha)}(\epsilon_\alpha c^{q(\alpha)})$ for $\alpha \in \Sigma(G)$, $c \in K$, define*
$$\Lambda = \{\sum c_i \omega_i \mid 0 \leqslant c_i \leqslant q(\alpha_i) - 1\}$$

(where the α_i form a basis of fundamental roots dual to the ω_i). Then the conclusion of (i) holds for this set Λ.

1.8. Locally Finite Groups

Assume now that $p > 0$ and $K = \overline{\mathbb{F}}_p$, the algebraic closure of $\mathbb{F}_p$. Then K is a locally finite field, so the simple algebraic group G over K is a locally finite group.

Let k be an infinite subfield of K. Then there is an automorphism ψ of K with fixed field k, and ψ gives a field automorphism $U_\alpha(c) \mapsto U_\alpha(c^\psi)$ of G, which we also denote by ψ. The fixed point group G_ψ is $G(k)$, an untwisted group of the same type as G over k.

Now suppose that G has a graph automorphism τ, as in (2) or (4) in Subsection 1.2. Then provided k satisfies suitable conditions, the fixed point group $G_{\tau\psi}$ is a twisted group of the same type as G over k. For example, when τ is as in (2) of Subsection 1.2, the condition is that k possess an extension of degree $|\tau|$, namely $K_{\psi|\tau|}$. (For a discussion of all this, see [29, §11].) We summarise in the next result, taken from [29, Lemma 64 and Theorem 34].

Theorem 1.21 *Every infinite locally finite simple group of Lie type is of the form $G_\phi/Z(G)_\phi$, for some simply connected simple algebraic group G, and some automorphism $\phi = \psi$ or $\tau\psi$ of G as described above.*

In order to perform calculations like those in Examples 1.10–1.16 for subgroups of G_ϕ, it would be desirable to have a version of the Lang-Steinberg theorem for the automorphism ϕ. Unfortunately, no such result is true in general. However, it is possible to salvage something useful; the next result is a weak analogue of 1.10.

Let $\phi = \psi$ or $\tau\psi$ as above, and exclude the cases where G_ϕ is a twisted Suzuki or Ree group (of type $^2B_2$, $^2F_4$ or $^2G_2$). Define an automorphism ϕ_p of G as follows: $\phi_p = \sigma_p$ if $\phi = \psi$, and $\phi_p = \tau\sigma_p$ if $\phi = \tau\psi$.

Proposition 1.22 [21] *Suppose that D is a closed subgroup of G which is both ϕ-stable and ϕ_p-stable. Let $S = \{D^g \mid g \in G\}$. Then there is an injection from the set of G_ϕ-classes of ϕ-stable subgroups in S to the set $H^1(\phi, N_G(D)/N_G(D)^\circ)$.*

For instance, consider $G = \mathrm{SL}_2(K)$, and let D be a maximal torus of G. Take $\phi = \psi$, $\phi_p = \sigma_p$. Then $N_G(D)/N_G(D)^\circ = \langle w \rangle \cong C_2$, and

$$H^1(\phi, N_G(D)/N_G(D)^\circ) = \{1, w\}.$$

The element w is in the image of the injection in Proposition 1.22 (hence corresponds to a ϕ-stable conjugate of D) if and only if K_{ψ^2} has degree 2 over $k = K_\psi$.

2. Subgroups of Classical Groups

Throughout this section, let G be a simple algebraic group of classical type. We shall describe results on the subgroup structure of G and of the related finite and locally finite classical groups G_σ and G_ϕ (where σ is a Frobenius morphism and ϕ is an automorphism as in Subsection 1.8). Sketches of some of the proofs will be given; details can be found in [20, 21].

We take G to be one of the groups $\mathrm{SL}(V)$, $\mathrm{Sp}(V)$ and $\mathrm{SO}(V)$, where V is a finite-dimensional vector space over the algebraically closed field K of characteristic p. The main results take the following form. We define a collection $\mathcal{C}$ of 'natural' subgroups of G (or G_σ or G_ϕ). Then any closed subgroup of G (or of G_σ or G_ϕ) either lies in a member of $\mathcal{C}$ or is essentially a simple group acting irreducibly on V. We call such a result a 'reduction theorem', since for many purposes it reduces the study of subgroups to the study of irreducible representations of simple groups. The reduction theorems for the finite groups G_σ and the locally finite groups G_ϕ are deduced from the Reduction Theorem for G. For G_σ, this gives a new proof of a well known result of Aschbacher [1] on subgroups of finite classical groups.

2.1. ALGEBRAIC GROUPS

To ease notation, write

$$G = \mathrm{Cl}(V)$$

to indicate that G is one of the classical groups $\mathrm{SL}(V)$, $\mathrm{Sp}(V)$, or $\mathrm{SO}(V)$. Let $n = \dim V$.

We begin with an elementary reduction theorem for connected subgroups of G.

Proposition 2.1 *Let H be a closed connected subgroup of G. Then one of the following holds:*

(i) $H \leqslant G_U$, *the stabilizer in G of a proper nonzero subspace U of V;*

(ii) $V = V_1 \otimes V_2$ and H lies in a subgroup of G of the form $\mathrm{Cl}(V_1) \circ \mathrm{Cl}(V_2)$ acting naturally on the tensor product;

(iii) H is simple, and acts irreducibly and tensor-indecomposably on V.

Remarks 1. When $G = \mathrm{Sp}(V)$ or $\mathrm{SO}(V)$, the subspace U in (i) can be taken to be totally singular or non-degenerate, or, in the case where $(G, p) = (\mathrm{SO}(V), 2)$, nonsingular of dimension 1. Also, the precise tensor product subgroups occurring in (ii) are of the following types:

$$\mathrm{SL} \otimes \mathrm{SL} < \mathrm{SL}, \quad \mathrm{Sp} \otimes \mathrm{SO} < \mathrm{Sp}, \quad \mathrm{Sp} \otimes \mathrm{Sp} < \mathrm{SO}, \quad \mathrm{SO} \otimes \mathrm{SO} < \mathrm{SO}.$$

2. The subgroups H in (iii) are usually maximal in one of the groups $\mathrm{Cl}(V)$, with an explicit list of exceptions (see Theorem 3.6 of Seitz's article).

Here is a sketch proof of Proposition 2.1. If H is non-reductive, then the Borel-Tits Theorem 1.19 implies that $H \leqslant P$ for some parabolic subgroup P; and $P \leqslant G_U$ for some (totally singular) subspace U, giving conclusion (i). Thus we may assume that H is reductive; we may also take it that H is irreducible on V (otherwise (i) holds again). Then $Z(H)$ acts as scalars on V, hence is finite. It follows that H is semisimple, say $H = H_1 \cdots H_t$, a commuting product of simple groups H_i. If $t > 1$ then H preserves a tensor decomposition $V = V_1 \otimes \cdots \otimes V_t$, and it is not hard to see that (ii) holds. And if $t = 1$ then H is simple and we obtain (iii).

Proposition 2.1 only covers connected subgroups, and hence is not applicable in some important situations—in particular it will not apply to the study of finite subgroups. What is required is the generalization given in Theorem 2.2 below, a reduction theorem for arbitrary closed subgroups of G.

In order to state the result, we need to define a number of classes of subgroups M of the classical group $G = \mathrm{Cl}(V)$.

CLASS $\mathcal{C}_1$. *Subspace stabilizers*: $M = G_U$, where U is a proper nonzero subspace of V; moreover, U is totally singular or non-degenerate, or, in the case where $(G, p) = (\mathrm{SO}(V), 2)$, nonsingular of dimension 1.

CLASS $\mathcal{C}_2$. *Stabilizers of orthogonal decompositions*: here $V = \bigoplus_1^t V_i$, where $t > 1$, the subspaces V_i are mutually orthogonal and isometric, and $M = G_{\{V_1, \ldots, V_t\}}$; in general, $M \cong \mathrm{Cl}(V_1) \wr \mathrm{Sym}_t$ (where Sym_t denotes the symmetric group of degree t).

CLASS $\mathcal{C}_3$. *Stabilizers of totally singular decompositions*: here $G = \mathrm{Sp}(V)$ or $\mathrm{SO}(V)$, $V = W \oplus W'$ where W and W' are maximal totally singular subspaces, and $M = G_{\{W, W'\}}$; if $\dim V = 2m$, then $M \cong \mathrm{GL}_m.2$ or GL_m (the latter only when $G = \mathrm{SO}(V)$ and m is odd).

CLASS $\mathcal{C}_4$. *Tensor product subgroups*: either

(i) $V = V_1 \otimes V_2$ and $M = \mathrm{Cl}(V_1) \circ \mathrm{Cl}(V_2)$ as in Proposition 2.1(ii), or

(ii) $V = \bigotimes_1^t V_i$ with $t > 1$, the V_i mutually isometric, and $M = N_G(\prod \mathrm{Cl}(V_i))$, where $\prod \mathrm{Cl}(V_i)$ acts naturally on the tensor product; in general, we have

$$M = (\prod \mathrm{Cl}(V_i)).\mathrm{Sym}_t.$$

CLASS $\mathcal{C}_5$. *Finite local subgroups*: let r be a prime different from p, and let R be an extraspecial r-group of order r^{1+2m}, or, when $r = 2$, a central product of such a group with a cyclic group of order 4. Every faithful irreducible representation of R has degree r^m. Take $n = \dim V$ to be r^m, and embed R as a subgroup of $G = \mathrm{Cl}(V)$ via one of these representations, where $\mathrm{Cl}(V)$ is $\mathrm{SL}(V)$ if r is odd or $R = C_4 \circ 2^{1+2m}$, and $\mathrm{Cl}(V)$ is $\mathrm{Sp}(V)$ or $\mathrm{SO}(V)$ otherwise. Then the members of the class $\mathcal{C}_5$ are the subgroups $M = N_G(R)$ with R as above. If $Z = Z(G)$, then $\bar{M} = M/Z \cong C_{\mathrm{Aut}(R)}(Z(R))$. Thus the subgroups in the class are as follows:

$$G = \mathrm{SL}_{r^m}, \quad \bar{M} = r^{2m}.\mathrm{Sp}_{2m}(r),$$
$$G = \mathrm{Sp}_{2m}, \quad \bar{M} = 2^{2m}.\mathrm{O}_{2m}^-(2),$$
$$G = \mathrm{SO}_{2m}, \quad \bar{M} = 2^{2m}.\mathrm{O}_{2m}^+(2).$$

Definition We write $\mathcal{C}(G)$ for the collection $\mathcal{C}_1 \cup \cdots \cup \mathcal{C}_5$ of subgroups of G defined above.

We can now state the Reduction Theorem for G. If $p > 0$, let σ be a Frobenius morphism of G such that G_σ is a finite classical group; for simplicity in the statement, exclude the case where G_σ is a unitary group.

Theorem 2.2 [20] *Let $G = \mathrm{Cl}(V)$ be a classical algebraic group, and let H be a closed subgroup of G. Then one of the following holds:*

(i) *H is contained in a member of $\mathcal{C}(G)$, which is σ-stable if H is σ-stable;*

(ii) *modulo scalars, H is almost simple, and $E(H)$ acts irreducibly on V.*

Remarks 1. In (ii), $E(H)$ is the unique quasisimple normal subgroup of H. Of course H may be finite or infinite; if it is infinite, the conclusion of (ii) can be strengthened to say that $E(H)$ also acts tensor-indecomposably on V.

2. If $G = \mathrm{SL}(V)$ and G_σ is a finite unitary group, the statement of Theorem 2.2 remains true, with one proviso: to make the σ-stability conclusion in (i) valid, the collection $\mathcal{C}_1$ of subgroups must be extended to include stabilizers of pairs of subspaces U, W of V such that $\dim U + \dim W = \dim V$ and either $U \subseteq W$ or $U \cap W = 0$.

3. If $K = \bar{\mathbb{F}}_p$, the algebraic closure of $\mathbb{F}_p$, and ϕ is an automorphism of G such that G_ϕ is a locally finite classical group (as discussed in Subsection 1.8), the statement of Theorem 2.2 remains true with ϕ replacing σ in conclusion (i), subject to the proviso of the previous remark when G_ϕ is unitary.

We briefly sketch some of the ideas involved in the proof of Theorem 2.2. As we are assuming that G_σ is not a unitary group, we may take σ to act on the space V

in a way which is compatible with its action on G. Suppose first that H is infinite. As in the proof of Proposition 2.1, we may assume that H° is reductive. Also, H (and σ) fixes the socle of the restriction $V{\downarrow}H^\circ$, so we may take it that this socle is equal to V. If $V{\downarrow}H^\circ$ is not homogeneous, then H (and σ) permute the homogeneous components, and one sees that H lies in a (σ-stable) subgroup in $\mathcal{C}_1, \mathcal{C}_2$ or $\mathcal{C}_3$. And if $V{\downarrow}H^\circ$ is homogeneous but reducible, an argument shows that H lies in a (σ-stable) tensor product subgroup as in $\mathcal{C}_4(\mathrm{i})$. Finally, if H° is irreducible on V and is a product of more than one simple group, then H lies in a (σ-stable) tensor product subgroup in $\mathcal{C}_4$. This gives Theorem 2.2 when H is infinite.

Now assume H is finite. If H normalizes a (σ-stable) closed connected subgroup D of G such that $1 < D < G$ and D is not simple, then the previous paragraph applies to give the result. Thus we suppose that H normalizes no such subgroup D.

We now work in $\bar{G} = G/Z$, where Z is the group of scalars in G. Denote $\bar{H} = HZ/Z$. Assume that $\bar{H}$ is an r-local subgroup for some prime r. Then $r \neq p$ by Theorem 1.19. Now the supposition at the end of the previous paragraph quickly forces H to lie in one of the finite local subgroups in $\mathcal{C}_5$ (or in $\mathcal{C}_2$ in the case where each V_i is an orthogonal space of dimension 1).

We are left with the case where $\bar{H}$ is non-local, so that

$$F^*(\bar{H}) = F_1 \times \cdots \times F_t,$$

a direct product of non-abelian simple groups F_i. Suppose that $t > 1$. At this point we apply an argument due to Borovik [3]. For each i, let $CC(F_i)$ be the double centralizer $C_{\bar{G}}(C_{\bar{G}}(F_i))$ of F_i in $\bar{G}$. Then $\prod CC(F_i)$ is a commuting product of subgroups of $\bar{G}$ which is σ-stable, proper and nontrivial. Therefore, by assumption, $CC(F_i)$ is finite for all i. But, if we choose $x \in F_i$ of prime order at least 5, then $CC(F_i)$ contains $Z(C_{\bar{G}}(x))$, which is infinite by 1.18 if $G \neq \mathrm{SL}(V)$; when $G = \mathrm{SL}(V)$, we apply the same argument with $\prod CC^\circ(F_i)$, where

$$CC^\circ(F_i) = C_{\bar{G}}(C_{\bar{G}}(F_i)^\circ)^\circ \geqslant Z(C_G(x)^\circ)^\circ,$$

which is infinite. This contradiction forces $t = 1$; in other words, $\bar{H}$ is almost simple. An easy argument now shows that $E(H)$ is irreducible on V, giving (ii) of Theorem 2.2.

2.2. Finite Groups

Continue to assume that $G = \mathrm{Cl}(V) = \mathrm{SL}(V), \mathrm{Sp}(V)$ or $\mathrm{SO}(V)$, and let σ be a Frobenius morphism of G such that $G_\sigma = \mathrm{SL}_n(q), \mathrm{Sp}_n(q)$ or $\mathrm{SO}_n^\pm(q)$; write $G_\sigma = \mathrm{Cl}_n(q)$. (As above, we have omitted the finite unitary groups to avoid a few minor complications in statements.) As remarked at the beginning of the proof of Theorem 2.2, we can take σ to act on V, and the natural module for G_σ is then $V_\sigma = V_n(q)$. We now outline how to use the Reduction Theorem 2.2 for G to obtain a reduction theorem for the finite groups G_σ; details can be found in [20].

Let H be a subgroup of G_σ. By Theorem 2.2, either

(i) H lies in a σ-stable subgroup $M \in \mathcal{C}(G)$, or

(ii) modulo scalars, H is almost simple, and $E(H)$ is absolutely irreducible on V_σ.

If (i) holds then $H \leqslant M_\sigma$. Thus, to obtain a reduction theorem for G_σ, we need to consider the G_σ-classes of subgroups M_σ; for this we use the Lang-Steinberg theorem, as illustrated in Examples 1.10–1.16.

If M lies in $\mathcal{C}_1$, then $M = G_U$ for some subspace U of V and we may take it that U is σ-stable. Therefore M_σ stabilizes the subspace U_σ of V_σ, and H lies in a member of the following class of subgroups of G_σ.

CLASS $\mathcal{C}_1^\sigma$. Stabilizers in G_σ of subspaces of V_σ.

Now let $M \in \mathcal{C}_2$, so $M = G_{\{V_1,\ldots,V_t\}}$ where $V = \bigoplus V_i$. Assume that $\mathrm{Cl}(V_i)$ is insoluble (i.e. not of type GL_1, O_1 or O_2). Then $M^\circ = \prod \mathrm{Cl}(V_i)$ and σ permutes the factors $\mathrm{Cl}(V_i)$. Moreover, the group of permutations induced by M_σ on the factors is a subgroup of $C_{\mathrm{Sym}_t}(\sigma)$. If $\langle\sigma\rangle$ induces a nontrivial imprimitive subgroup of Sym_t, then M_σ permutes the sums of subspaces corresponding to each block of imprimitivity. Hence, replacing M by another member of $\mathcal{C}_2$ if necessary, we may assume that either $\langle\sigma\rangle$ induces a trivial group or t is prime and $\langle\sigma\rangle$ induces a transitive group. In the first case, in general $M_\sigma \cong \mathrm{Cl}_m(q)\wr\mathrm{Sym}_t$ (where $n = mt$); in the second, roughly speaking, $M_\sigma \cong \mathrm{Cl}_m(q^t).t$ (compare Example 1.13). A similar argument gives the same conclusion when $\mathrm{Cl}(V_i)$ is soluble. Thus we obtain the following classes of subgroups of G_σ.

CLASS $\mathcal{C}_2^\sigma$. Stabilizers in G_σ of orthogonal decompositions of V_σ (type $\mathrm{Cl}_m(q)\wr\mathrm{Sym}_t$, where $n = mt$).

CLASS $\mathcal{C}_2^{\prime\sigma}$. Extension field subgroups (type $\mathrm{Cl}_m(q^t).t$ with $n = mt$ and t prime).

Next, suppose that $M \in \mathcal{C}_3$, so M is the stabilizer of a totally singular decomposition $V = W \oplus W'$. Here $M = \mathrm{GL}_{\frac{n}{2}}.2$ (or $\mathrm{GL}_{\frac{n}{2}}$), and we see as in Example 1.14 that M_σ is $\mathrm{GL}_{\frac{n}{2}}(q).2$ or $\mathrm{GU}_{\frac{n}{2}}(q).2$ (or possibly a subgroup of small index in one of these); thus H lies in a member of the following class.

CLASS $\mathcal{C}_3^\sigma$. Subgroups of type $\mathrm{GL}_{\frac{n}{2}}(q).2$ and $\mathrm{GU}_{\frac{n}{2}}(q).2$, where G_σ is symplectic or orthogonal.

Now consider $M \in \mathcal{C}_4$, a tensor product subgroup. If $M = N_G(\prod \mathrm{Cl}(V_i))$, where $V = \bigotimes V_i$ and the V_i are isometric, we argue as in the case where $M \in \mathcal{C}_2$ that either σ acts trivially on the factors or t is prime and $\langle\sigma\rangle$ acts transitively. Thus H lies in one of the following subgroups.

CLASS $\mathcal{C}_4^\sigma$. Tensor product subgroups of type $\mathrm{Cl}_{n_1}(q) \circ \mathrm{Cl}_{n_2}(q)$ $(n = n_1 n_2)$, or of type $N_{G_\sigma}(\prod_1^t \mathrm{Cl}_m(q))$ $(n = m^t)$.

CLASS $\mathcal{C}_4^{\prime\sigma}$. Subgroups of type $\mathrm{Cl}_m(q^t).t$ with $n = m^t$.

Finally, suppose that $M \in \mathcal{C}_5$, so M is a finite local subgroup $N_G(R)$. In this case we argue that if $M_\sigma < M$ then M lies in a member of $\mathcal{C}_1 \cup \cdots \cup \mathcal{C}_4$. Hence we may take it that $M = M_\sigma$, and define the class

CLASS $C_5'^\sigma$. Local subgroups $N_{G_\sigma}(R) = N_G(R)$ as in C_5.

We have now outlined the proof of the Reduction Theorem for finite classical groups:

Theorem 2.3 [1] *If G_σ is a finite classical group as above, and H is a subgroup of G_σ, then either*

(i) *H lies in a member of $C_1^\sigma \cup \cdots \cup C_5^\sigma$, or*

(ii) *modulo scalars, H is almost simple, and $E(H)$ is absolutely irreducible on V_σ.*

Remarks 1. A suitably adapted proof gives a similar result for the finite unitary groups (which were excluded from the above discussion).

2. In fact, Aschbacher's result [1] is more detailed than that stated in Theorem 2.3. His result involves two further classes of subgroups: C_6^σ, a class of subgroups whose representations on V_σ are realised over a proper subfield of $\mathbb{F}_q$; and C_7^σ, the class of classical groups on V_σ which lie in G_σ. His conclusion in (ii) is correspondingly more refined—for example, the representation of $E(H)$ can be taken not to be realised over a proper subfield. In [20] we use methods similar to those outlined above to obtain the more refined result.

Also, Aschbacher deals not only with the groups G_σ, but with almost all groups whose socle is the associated classical finite simple group. Our methods also handle these groups.

3. The subgroups in $C_4'^\sigma$ do not occur in [1], and indeed satisfy both conclusions (i) and (ii) of 2.3. These subgroups play a major role in the extension of Aschbacher's theorem proved in [24, Corollary 6]. They act on V_σ as follows. Let $W = V_m(q^t)$ be the natural module for $C = \mathrm{Cl}_m(q^t)$, and let $\bar{V}$ be the $\mathbb{F}_{q^t}C$-module

$$W \otimes W^{(q)} \otimes \cdots \otimes W^{(q^{t-1})}$$

(where the superscript (q) denotes a twist of the module W by the Frobenius q-power map). Then $\bar{V} \cong \bar{V}^{(q)}$, so $\bar{V}$ is realised over $\mathbb{F}_q$, and we take V_σ to be the corresponding $\mathbb{F}_q C$-module. In particular, C is not absolutely tensor indecomposable on V_σ. It should be possible to refine the conclusion (ii) in Theorem 2.3 to say that, at least if $E(H)$ is of Lie type in characteristic p, then $E(H)$ is absolutely tensor indecomposable on V_σ.

4. There is a considerable literature concerning the question of the maximality of subgroups H as in (ii) of Theorem 2.3; the article of Saxl in these Proceedings (see Section 3) contains a discussion of this question.

2.3. LOCALLY FINITE GROUPS

Assume now that $G = \mathrm{Cl}(V)$, where the natural module V is a finite-dimensional vector space over $K = \bar{\mathbb{F}}_p$, the algebraic closure of $\mathbb{F}_p$. Let k be an infinite subfield of K. As discussed in Subsection 1.8, the locally finite classical groups over k arise as fixed point groups G_ϕ of suitable automorphisms ϕ of G.

Let H be a subgroup of G_ϕ. If H is dense in G, then a result of Hartley and Zalesskii [13] implies that H is a group of the same type as G (possibly twisted) over a subfield of k (see 4.9 of Seitz's article). So assume now that H is not dense. Then $\bar{H}$, the closure of H, is a proper ϕ-stable subgroup of G. We now apply Theorem 2.2: either $\bar{H}$ lies in a ϕ-stable member of $\mathcal{C}(G)$ or $\bar{H}$ is almost simple and irreducible on V (see Remark 3 after the statement of Theorem 2.2). In the latter case, another application of [13] shows that $E(H)$ is a quasisimple group of Lie type over a subfield of k.

Thus, to obtain a reduction theorem for subgroups of G_ϕ, it remains to work out the G_ϕ-classes of fixed point groups M_ϕ for ϕ-stable subgroups M in $\mathcal{C}(G)$. This is done using Proposition 1.22 in the same way as in Subsection 2.2. The upshot is a reduction theorem (Theorem 2.4 below) with the following very similar list of subgroups.

CLASS $\mathcal{C}_1^\phi$. Stabilizers of subspaces.

CLASS $\mathcal{C}_2^\phi$. Stabilizers of orthogonal decompositions, type $\mathrm{Cl}_m(k) \wr \mathrm{Sym}_t$ (where $n = mt$).

CLASS $\mathcal{C}_2'^\phi$. Extension field subgroups, type $\mathrm{Cl}_m(\tilde{k}).t$ (where $n = mt$ and $\tilde{k}$ is an extension of k of degree t, if this exists).

CLASS $\mathcal{C}_3^\phi$. Subgroups $\mathrm{GL}_{\frac{n}{2}}(k).2$, $\mathrm{GU}_{\frac{n}{2}}(k).2$ for $G = \mathrm{Sp}, \mathrm{SO}$ (the unitary subgroup only exists if k has a degree 2 extension).

CLASS $\mathcal{C}_4^\phi$. Tensor product subgroups of type $\mathrm{Cl}_{n_1}(k) \circ \mathrm{Cl}_{n_2}(k)$ (where $n = n_1 n_2$), or $N_{G_\phi}(\prod_1^t \mathrm{Cl}_m(k))$ (where $n = m^t$).

CLASS $\mathcal{C}_4'^\phi$. Subgroups of type $\mathrm{Cl}_m(\tilde{k}).t$ (where $n = m^t$ and $\tilde{k}$ is an extension of k of degree t, if this exists).

Theorem 2.4 [21] *Let $G_\phi = G(k)$ be a locally finite non-unitary classical group over k. If H is a subgroup of G_ϕ, then one of the following holds:*

(i) *$H \leqslant N(G(l))$, where $G(l)$ is a subgroup of the same type as G over a proper subfield l of k;*

(ii) *H lies in a member of $\mathcal{C}_1^\phi \cup \cdots \cup \mathcal{C}_4'^\phi$;*

(iii) *modulo scalars, $H = H(k)$ is an almost simple group of Lie type over k, acting irreducibly on V.*

As before, there is a similar result for the locally finite unitary groups. For details see [21].

3. Subgroups of Exceptional Groups

Throughout this section, let G be a simple algebraic group of exceptional type G_2, F_4, E_6, E_7 or E_8 over the algebraically closed field K of characteristic p. As usual, denote by σ a Frobenius morphism of G. Under some mild assumptions on p, there is a complete description of the maximal subgroups and of the closed connected semisimple subgroups of G (see Theorems 3.1, 3.2 and 3.8 below). For the finite groups G_σ, one expects that, in general, maximal or semisimple subgroups are just fixed point groups of such (σ-stable) subgroups of G. This is now known to be true in almost all cases (see Corollary 3.7 and Theorem 3.12). In the last section we describe analogous results for the locally finite exceptional groups G_ϕ.

3.1. MAXIMAL SUBGROUPS OF G

As for the classical groups in Subsection 2.1, we begin with a result which determines the maximal closed connected subgroups of G under some assumptions on the characteristic. This result of Seitz [25] underlies the entire body of work presented in this section.

Theorem 3.1 [25, Theorem 1] *Let M be a maximal closed connected subgroup of G. If M is simple, assume that either $p = 0$ or $p > 7$. Then either*

(i) *M is a parabolic subgroup or a subsystem subgroup, or*

(ii) *G and M are as follows (one $\mathrm{Aut}_{\mathrm{alg}}(G)$-class for each M listed, except when otherwise stated):*

$$G = G_2 : M = A_1,$$
$$G = F_4 : M = A_1, G_2 \text{ or } A_1 G_2,$$
$$G = E_6 : M = A_2, G_2, F_4, C_4 \text{ or } A_2 G_2,$$
$$G = E_7 : M = A_1 \text{ (two classes)}, A_2, A_1 A_1, A_1 G_2, A_1 F_4 \text{ or } G_2 C_3,$$
$$G = E_8 : M = A_1 \text{ (three classes)}, B_2, A_1 A_2 \text{ or } G_2 F_4.$$

Remark The assumptions on p in [25, Theorem 1] are in fact much weaker than the assumption $p = 0$ or $p > 7$ in Theorem 3.1. Roughly speaking they are as follows:

if $M = A_1$, assume $p = 0$ or $p > 7$;
if M has rank 2, assume $p = 0$ or $p > 5$;
if M has rank 3 or $(G, M) = (E_8, B_4)$, assume $p \neq 2$.

In all other cases, no assumption on p is made. This remark applies to all the other results in this section in which characteristic assumptions are made.

As in Subsection 2.1, for applications one needs to strengthen Theorem 3.1 in two ways: firstly, by dropping the assumption of connectedness on M; and secondly, by keeping track of σ-stability, where σ is a Frobenius morphism of G. This is achieved in the next result, which determines the maximal closed subgroups M of G of positive dimension. Surprisingly, it turns out that, when M° does not contain a maximal torus of G, there are just three classes for which M is maximal closed but M° is not maximal connected:

Theorem 3.2 [16, Theorem 1] *Suppose that M is a maximal closed (σ-stable) subgroup of G with $M^\circ \neq 1$. If M° is simple, assume that $p = 0$ or $p > 7$. Then one of the following holds:*

(i) *M° contains a maximal torus of G;*

(ii) M° is as in (ii) of Theorem 3.1;

(iii) $G = \mathrm{E}_7$ and $M = (2^2 \times \mathrm{D}_4).\mathrm{Sym}_3$ (where $M^\circ = \mathrm{D}_4 < \mathrm{A}_7$ subsystem);

(iv) $G = \mathrm{E}_8$ and $M = \mathrm{A}_1 \times \mathrm{Sym}_5$ (where $M^\circ = \mathrm{A}_1 < \mathrm{A}_4 \mathrm{A}_4$ subsystem);

(v) $G = \mathrm{E}_8$ and $M = (\mathrm{A}_1 \mathrm{G}_2 \mathrm{G}_2).2$ (where $M^\circ = \mathrm{A}_1 \mathrm{G}_2 \mathrm{G}_2 < \mathrm{F}_4 \mathrm{G}_2 < G$).

Note that the subgroups M° in (i) are easily determined: either M° is parabolic or it is reductive of the form $M^\circ = DT$, where $D = (M^\circ)'$ is a subsystem subgroup (possibly trivial) and $T = Z(M^\circ)^\circ$ is a torus (possibly trivial).

3.2. MAXIMAL SUBGROUPS OF G_σ: REDUCTION THEOREM

Now assume that $p > 0$ and σ is a Frobenius morphism of the exceptional algebraic group G, and take G to be of adjoint type. We outline first the proof of a reduction theorem for maximal subgroups of G_σ (see Theorem 3.4 below). The flow of the proof is quite similar to that of the reduction theorem 2.2 for finite classical groups.

Let H be a maximal subgroup of G_σ. If H normalizes a σ-stable closed connected subgroup D with $1 < D < G$, then HD lies in a maximal closed σ-stable subgroup M of G with $M^\circ \neq 1$. The possibilities for M are given by Theorem 3.2, and we conclude from the maximality of H that $H = M_\sigma$.

Now assume that H normalizes no such subgroup D. Suppose first that H is local, so that $H = N_{G_\sigma}(E)$ for some elementary abelian r-subgroup E of G_σ (where r is prime). We may take it that $r \neq p$, by Theorem 1.19. Some rather interesting local subgroups emerge in this situation:

Theorem 3.3 [9, 3] *Under the above hypotheses, one of the following holds (one G_σ-class of subgroups in each case):*

$$G = \mathrm{G}_2 : E = 2^3, \ H = N_G(E) = 2^3.\mathrm{SL}_3(2),$$
$$G = \mathrm{F}_4 \text{ or } \mathrm{E}_6 : E = 3^3, \ H = N_G(E) = 3^3.\mathrm{SL}_3(3) \text{ or } 3^{3+3}.\mathrm{SL}_3(3),$$
$$G = \mathrm{E}_8 : E = 2^5 \text{ or } 5^3, \ H = N_G(E) = 2^{5+10}.\mathrm{SL}_5(2) \text{ or } 5^3.\mathrm{SL}_3(5).$$

Definition We call the subgroups $N_G(E)$ in the conclusion of Theorem 3.3 *exotic local* subgroups.

Returning to the argument, we now assume that our maximal subgroup H of G_σ is non-local. Thus $F^*(H) = F_1 \times \cdots \times F_t$, where the F_i are non-abelian simple groups. When $t = 1$, H is almost simple, which is our desired conclusion. So assume that $t > 1$.

At this point we apply Borovik's double centralizer argument, as in the proof of the Reduction Theorem 2.2 for classical groups. The subgroup $\prod CC(F_i)$ is σ-stable and is normalized by H; hence it must be finite. Consequently $Z(C_G(x))$ must be finite for all $x \in F_i$. By 1.18, this implies that $|F_i|$ can only be divisible by primes which are bad for G. Hence $G = \mathrm{E}_8$ and each F_i is a simple $\{2,3,5\}$-group. The only such simple groups are Alt_5, Alt_6 and $\mathrm{U}_4(2)$ (where Alt_k denotes the alternating group of degree k). Further argument now forces

$$G = \mathrm{E}_8, \ F^*(H) = \mathrm{Alt}_5 \times \mathrm{Alt}_6.$$

Somewhat amazingly, there *is* such a subgroup H in $G_\sigma = E_8(q)$ for suitable q, as was first shown by Borovik [3]. This subgroup is closely tied up with the subgroup $A_1 \times \mathrm{Sym}_5$ of Theorem 3.2(iv).

We have now sketched the proof of our Reduction Theorem:

Theorem 3.4 [16, Theorem 2] *Let H be a maximal subgroup of the finite exceptional group G_σ. Then one of the following holds:*

(i) H *is almost simple;*

(ii) $H = M_\sigma$, *where M is a maximal σ-stable closed subgroup with $M^\circ \ne 1$, given by Theorem 3.2;*

(iii) H *is an exotic local subgroup;*

(iv) $G = E_8$, $p > 5$ *and* $H = (\mathrm{Alt}_5 \times \mathrm{Alt}_6).2^2$.

Remarks 1. No assumptions on the characteristic p are required for this result; this is essentially because Theorem 3.2 only needs to be used for non-simple subgroups. A less explicit version of Theorem 3.4 was given by Borovik [3].

2. Theorem 2 of [16] also handles maximal subgroups of all groups whose socle is the simple group $O^{p'}(G_\sigma)$.

3. The subgroups M_σ in (ii) can be worked out using the Lang-Steinberg theorem, as in Examples 1.10–1.16. When M° is a subsystem subgroup or a maximal torus, the maximal subgroups M_σ are given explicitly in [22]; the other cases are listed in [16].

3.3. ALMOST SIMPLE MAXIMAL SUBGROUPS

In view of the Reduction Theorem 3.4, it remains to study the maximal subgroups of G_σ which are almost simple.

Let H be an almost simple subgroup of G_σ (not assumed maximal, for the time being), and let $X = F^*(H)$, a finite simple group. When X is alternating, sporadic, or of Lie type in p'-characteristic, there is a bounded number of possible isomorphism types for X, since G has a representation of degree at most 248; see [22, §4A] for further discussion of this matter.

We now focus attention on the 'generic' case—that in which X is of Lie type in characteristic p; say $X = X(q)$, a group of Lie type over $\mathbb{F}_q$, where $q = p^a$. In order to apply the results in Subsection 3.1 on subgroups of G, one would like to show that, in general, X lies in a proper closed connected subgroup of G. For large characteristics, this was proved by Seitz and Testerman [27]; indeed, they showed that either X is contained in a parabolic subgroup or it lies in a connected simple subgroup of the same type as X. Recently, in [19], we have obtained a similar result which is free of characteristic assumptions; however the result does require an assumption on q, where $X = X(q)$ as above. In order to specify this assumption, we need a definition.

Definition Let T be a maximal torus of G, and $X(T)$ the character group of T. For a sublattice L of the lattice $X(T)$, let $t(L)$ be the exponent of the torsion subgroup of $X(T)/L$. For $\alpha, \beta \in \Sigma(G)$, call the element $\alpha - \beta$ of $X(T)$ a *root difference*. Define

$$t(G) = \max\{t(L) \mid L \text{ a sublattice of } X(T) \text{ generated by root differences}\}.$$

In view of the next result, it would be of interest to know the values of $t(G)$ for the exceptional groups G.

Theorem 3.5 [19] *Let $X = X(q)$ be a simple group of Lie type in characteristic p, and suppose that $X < G$. Assume that*

$$q > 16 \;\; if \;\; X \neq L_2(q), \, {}^2B_2(q), \, {}^2G_2(q),$$

and

$$q > t(G) \;\; if \;\; X = L_2(q), \, {}^2B_2(q) \;\; or \;\; {}^2G_2(q).$$

Then there is a closed connected subgroup $\bar{X}$ of G containing X such that every X-invariant subspace of the Lie algebra $L(G)$ is also $\bar{X}$-invariant. Further, if $X \leqslant G_\sigma$, then $\bar{X}$ is σ-stable.

The main point of this is that usually X will act reducibly on $L(G)$, so $\bar{X}$ will be a proper subgroup of G, at which point the results in Subsection 3.1 apply. The upshot for maximal subgroups is the following result.

Corollary 3.6 [19] *Suppose that H is a maximal subgroup of G_σ, and suppose $F^*(H) = X(q)$ with q as in Theorem 3.5. Then one of the following holds:*

(i) *$X(q)$ has the same type as G (possibly twisted);*
(ii) *$X(q) = O^{p'}(\bar{X}_\sigma)$ for some maximal closed σ-stable subgroup $\bar{X}$ of G with $\bar{X}° \neq 1$ (given by Theorem 3.2 if $\bar{X}$ is non-simple or if $p > 7$).*

Remarks 1. There is no assumption on p. The result holds, more generally, for maximal subgroups of all groups with socle $O^{p'}(G_\sigma)$.

2. The subgroups satisfying (i) are just subgroups of the same type as G, possibly twisted, over subfields of $\mathbb{F}_{q_1}$, where $G_\sigma = G(q_1)$; they are unique up to G_σ-conjugacy, by [17, 5.1]. We call these *subfield subgroups* and *twisted subgroups*.

3. The conclusion of (ii) implies that q must be a power of q_1 (where $G_\sigma = G(q_1)$). In fact, the only unknown subgroups $X(q)$ occurring in (ii) must have $q = q_1$ and $p \leqslant 7$ (since $\bar{X}$ must be simple in an unknown case).

As a consequence of the discussion in this section and the previous one, we see that, in general, the maximal subgroups of the finite exceptional groups are just fixed point groups of maximal closed subgroups of the corresponding algebraic groups. To be specific:

Corollary 3.7 [19] *There is a constant c such that, if H is a maximal subgroup of G_σ with $|H| > c$, then either*

(i) *H is the normalizer of a subfield or twisted subgroup (see Remark 2 after Corollary 3.6), or*
(ii) *$H = \bar{X}_\sigma$ for some maximal closed σ-stable subgroup $\bar{X}$ of G of positive dimension.*

3.4. THE ULTIMATE

One of the ultimate aims in our study of subgroups is to determine all closed semisimple subgroups of G (and of G_σ), where by a semisimple group we mean a commuting product of simple groups. This aim has now largely been achieved, under the usual assumptions on the characteristic. The result for G (Theorem 3.8 below) says that every closed connected semisimple subgroup of G is embedded in an explicitly described way in a subsystem subgroup; and the result for G_σ (Theorem 3.12) provides, for each 'generic' simple subgroup of G_σ, a connected σ-stable semisimple subgroup of G containing it.

In order to state the results in detail, we need a definition.

Definition Let $Y = Y_1 \cdots Y_t$ be a semisimple connected algebraic group, with each Y_i simple. We say that a closed connected semisimple subgroup X of Y is *essentially embedded* in Y if the following hold for all i:

(i) if Y_i is a classical group with natural module V_i, then either the projection of X in Y_i is irreducible on V_i or $Y_i = D_n$ and the projection lies in a natural subgroup $B_r B_{n-r-1}$ of Y_i, irreducible in each factor with inequivalent representations;

(ii) if Y_i is of exceptional type, then the projection of X in Y_i is either Y_i or a maximal connected subgroup of Y_i not containing a maximal torus (hence given in Theorem 3.1(ii) for $p = 0$ or $p > 7$).

Theorem 3.8 [18, Theorem 5] *Let X be a closed connected semisimple subgroup of the exceptional group G, and suppose that $p = 0$ or $p > 7$. Assume that X has no factor of type A_1. Then there is a subsystem subgroup Y of G such that X is essentially embedded in Y.*

When X has a factor A_1, there is a similar result which is somewhat more complicated to state; we refer the reader to [18, Theorem 7]. As a consequence, when $p = 0$ there are only finitely many conjugacy classes of closed connected semisimple subgroups in G, whereas there are infinitely many when $p > 0$.

The classical factors of subsystem groups are of small rank (at most 8), so their essentially embedded closed connected semisimple subgroups can be determined, using some representation theory. Using Theorem 3.8, the conjugacy classes and centralizers in G of all simple connected subgroups of rank at least 2 are explicitly listed in [18, §8]; the same is done for subgroups of type A_1 in [15]. We illustrate the process with an easy example.

Example 3.9 Suppose we wish to find all conjugacy classes of subgroups of type A_3 in $G = E_8$ (assuming $p = 0$ or $p > 7$—actually $p \neq 2$ is all that is assumed for this case in [18, Theorem 5]). Let $A_3 \cong X < E_8$. By Theorem 3.8, X is essentially embedded in a subsystem subgroup Y of G. Since A_3 does not occur in the conclusion of Theorem 3.1(ii), Y has no factors of exceptional type; so Y is a product of classical groups. Now the nontrivial irreducible representations of A_3 have dimensions 4, 6, 15 and higher (embedding A_3 in A_3, D_3, B_7 respectively). Consideration of the subsystems of E_8 shows that Y must be a subsystem group of type A_3, $A_3 A_3$, A_5 or D_8; there is just one G-class of subsystems of each of the types A_3, A_5, D_8, and two classes of type A_3A_3. The embedding of X in Y is uniquely determined up to conjugacy for $Y = A_3$, A_5, D_8; and for $Y = A_3A_3$, it is determined up to a twist by a field automorphism in one of the projections. This covers all possibilities for X.

Using Theorem 3.8, one can work out centralizers of reductive subgroups of G, and restrictions of the Lie algebra $L(G)$ to reductive subgroups. Several interesting consequences emerge, such as the following two results.

Corollary 3.10 [18, Theorem 2] *Let X be a closed connected reductive subgroup of G, and assume that $p = 0$ or $p > 7$. Then $C_G(X)^\circ$ is reductive.*

Corollary 3.11 [18, Theorem 3], [15] *Suppose that X is a closed connected simple subgroup of G, and $p = 0$ or $p > 7$. Then $C_{L(G)}(X) = L(C_G(X))$.*

We now move on to subgroups of the finite groups G_σ. Here the key is Theorem 3.5, which leads to the following result.

Theorem 3.12 [19] *Let $X = X(q) < G$, with q as in Theorem 3.5, and assume that $p > 7$. Then*

   (i) *X lies in a closed connected simple subgroup $\tilde{X}$ of G of the same type as X, and*

   (ii) *if also $X < G_\sigma$ then X lies in a closed connected semisimple σ-stable subgroup $\bar{X} = X_1 \cdots X_t$ of G, where each factor X_i is simple of the same type as X, and σ acts transitively on the X_i; moreover, $C_G(X)^\circ = C_G(\bar{X})^\circ$.*

Of course, the embeddings of the subgroups $\tilde{X}$ and $\bar{X}$ in the conclusion are given by Theorem 3.8, so Theorem 3.12(ii) determines the embeddings of $X(q)$ in G_σ. For $X = X(q)$ of Lie rank more than $\frac{1}{2}\text{rank}(G)$, the conclusion of Theorem 3.12 is obtained in [23, Theorem 2] with no assumption on p, and assuming only that $q > 2$.

3.5. LOCALLY FINITE GROUPS

We conclude by describing some analogues of the results already discussed for subgroups of the infinite locally finite groups of exceptional Lie type. Recall from Subsection 1.8 that these are fixed point groups of the form G_ϕ for suitable automorphisms ϕ of the exceptional simple algebraic group G. In the discussion to follow, we exclude the case where G_ϕ is a Ree group of type $^2G_2$ or $^2F_4$, and assume that $p > 7$.

First we consider maximal subgroups. Let H be a maximal subgroup of G_ϕ. Arguing as in the second paragraph of Subsection 2.3, we may assume that the closure $\bar{H}$ of H in G is a proper ϕ-stable subgroup of G. In [21] it is shown that the conclusion of Theorem 3.2 holds for maximal closed ϕ-stable subgroups of G. Hence we see that $H = M_\phi$ for some such subgroup M of G, given by Theorem 3.2(i)–(v). Further argument shows that each such M has a G-conjugate which is both ϕ-stable and ϕ_p-stable (where ϕ_p is as in Proposition 1.22). Hence the G_ϕ-classes of maximal subgroups can be worked out using Proposition 1.22.

Now we consider arbitrary infinite simple subgroups X of G_ϕ. The classification of locally finite simple linear groups (see Hartley's article) implies that $X = X(l)$, a simple group of Lie type over a locally finite field l of characteristic p. At this point, a result of Seitz [26] (see 4.10 of Seitz's article) implies that the closure $\bar{X}$ of X in G is a commuting product of simple groups of the same type as X. The embedding of $\bar{X}$ in G is given by Theorem 3.8. Moreover, using Theorem 3.8 we show in [21] that $\bar{X}$ has a conjugate which is ϕ-stable and ϕ_p-stable, so Proposition 1.22 again applies to determine the G_ϕ-classes of simple subgroups X.

References

1. M. Aschbacher, On the maximal subgroups of the finite classical groups, *Invent. Math.* 76 (1984), 469–514.

2. A. Borel and J. Tits, Éléments unipotents et sousgroupes paraboliques de groupes réductifs, *Invent. Math.* 12 (1971), 95–104.

3. A. Borovik, The structure of finite subgroups of simple algebraic groups, *Algebra and Logic* 28 (1989), 249–279 (in Russian).

4. N. Bourbaki, *Groupes et Algèbres de Lie* (Chapters 4, 5 and 6), Hermann, Paris, 1968.

5. R. W. Carter, Centralizers of semisimple elements in finite groups of Lie type, *Proc. London Math. Soc.* 37 (1978), 491–507.

6. R. W. Carter, Centralizers of semisimple elements in the finite classical groups, *Proc. London Math. Soc.* 42 (1981), 1–41.

7. R. W. Carter, Conjugacy classes in the Weyl group, *Compositio Math.* 25 (1972), 1–59.

8. C. Chevalley, *Seminaire Chevalley, Vols. I,II: Classifications des Groupes de Lie Algébriques*, Paris, 1956–8.

9. A. M. Cohen, M. W. Liebeck, J. Saxl and G. M. Seitz, The local maximal subgroups of exceptional groups of Lie type, finite and algebraic, *Proc. London Math. Soc.* 64 (1992), 21–48.

10. D. I. Deriziotis, The centralizers of semisimple elements of the Chevalley groups E_7 and E_8, *Tokyo J. Math.* 6 (1983), 191–216.

11. D. I. Deriziotis and M. W. Liebeck, Centralizers of semisimple elements in finite twisted groups of Lie type, *J. London Math. Soc.* 31 (1985), 48–54.

12. D. Gorenstein and R. Lyons, The local structure of finite groups of characteristic 2 type, *Memoirs Amer. Math. Soc.* 276 (1983).

13. B. Hartley and A. E. Zalesskii, On simple periodic linear groups—dense subgroups, permutation representations and induced modules, *Israel J. Math.* 82 (1993), 299–327.

14. S. Lang, Algebraic groups over finite fields, *Amer. J. Math.* 78 (1956), 555–563.

15. R. Lawther and D. M. Testerman, A_1 subgroups of exceptional algebraic groups, to appear.

16. M. W. Liebeck and G. M. Seitz, Maximal subgroups of exceptional groups of Lie type, finite and algebraic, *Geom. Dedicata* 36 (1990), 353–387.

17. M. W. Liebeck and G. M. Seitz, Subgroups generated by root elements in groups of Lie type, *Ann. Math.* 139 (1994), 293–361.

18. M. W. Liebeck and G. M. Seitz, Reductive subgroups of exceptional algebraic groups, to appear.

19. M. W. Liebeck and G. M. Seitz, Finite subgroups of exceptional algebraic groups, to appear.

20. M. W. Liebeck and G. M. Seitz, On subgroups of algebraic and finite classical groups, to appear.

21. M. W. Liebeck and G. M. Seitz, Subgroups of locally finite groups of Lie type, to appear.

22. M. W. Liebeck, J. Saxl and G. M. Seitz, Subgroups of maximal rank in finite exceptional groups of Lie type, *Proc. London Math. Soc.* 65 (1992), 297–325.

23. M. W. Liebeck, J. Saxl and D. M. Testerman, Subgroups of large rank in groups of Lie type, *Proc. London Math. Soc.*, to appear.

24. G. M. Seitz, Representations and maximal subgroups of finite groups of Lie type, *Geom. Dedicata* 25 (1988), 391–406.

25. G. M. Seitz, Maximal subgroups of exceptional algebraic groups, *Memoirs Amer. Math. Soc.*

441 (1991).

26. G. M. Seitz, Abstract homomorphisms of algebraic groups, to appear in *J. London Math. Soc.*

27. G. M. Seitz and D. M. Testerman, Extending morphisms from finite to algebraic groups, *J. Algebra* 131 (1990), 559–574.

28. T. A. Springer and R. Steinberg, Conjugacy classes, in *Seminar on Algebraic Groups and Related Topics* (ed. A. Borel et al.), Lect. Notes Math. 131, Springer, Berlin, 1970, pp. 168–266.

29. R. Steinberg, *Lectures on Chevalley Groups*, Yale University Lecture Notes, 1968.

30. R. Steinberg, Endomorphisms of linear algebraic groups, *Memoirs Amer. Math. Soc.* 80 (1968).

FINITE SIMPLE GROUPS AND PERMUTATION GROUPS

J. SAXL
DPMMS
Cambridge University
16 Mill Lane
Cambridge CB2 1SB
England
E-mail: saxl@pmms.cam.ac.uk

Abstract. This paper surveys some results in the area of maximal subgroups of the finite simple groups and their automorphism groups. The first two sections are concerned with maximal subgroups of the alternating and symmetric groups. We outline the Reduction Theorem and discuss the maximality of primitive groups in the corresponding symmetric groups. In the third section we consider the corresponding questions for the classical groups. The next section contains a brief survey of knowledge of the maximal subgroups of the other almost simple groups. Finally, the last section outlines some recent work on primitive permutation groups of special degrees.

Key words: finite simple groups, maximal subgroups, permutation groups, groups of Lie type.

Introduction

This article is concerned with maximal subgroups of the finite simple groups and their automorphism groups. At the outset we concentrate on subgroups of the symmetric groups. In the first section, we outline a version of the Reduction Theorem for subgroups of symmetric groups. This was first stated by O'Nan and Scott [51]; since then such results have been obtained for other families of almost simple groups ([2, 38]). The idea is as follows: given a group X in a family of almost simple groups (e.g. the symmetric groups in Section 1), construct a collection $\mathcal{C} = \mathcal{C}(X)$ of 'natural' subgroups of X such that, for any subgroup G of X (not containing the socle $F^*(X)$), either G is contained in a member of $\mathcal{C}$, or G is almost simple and 'nice'.

The definitions of 'nice' and 'natural' depend on the family of almost simple groups which we are investigating. For example, in dealing with the symmetric groups, nice means primitive as a permutation group, whereas in classical groups the definition involves absolute irreducibility (and more) on the underlying space. The groups in $\mathcal{C}$ should be 'natural' enough to be well understood. The Reduction Theorem then allows us to concentrate on those subgroups which are almost simple (and nice); we are in a position to invoke the classification of finite simple groups to investigate these. In Section 2 we concentrate on the primitive permutation actions of almost simple groups and ask the question of when such an action leads to a maximal subgroup of the corresponding symmetric group. This was answered in [31]: the subgroup obtained is usually maximal, with an explicit list of exceptions. We are then led to study maximal subgroups of the other almost simple groups. In Section 3

B. Hartley et al. (eds.), Finite and Locally Finite Groups, 97–110.
© *1995 Kluwer Academic Publishers. Printed in the Netherlands.*

we discuss the subgroup structure of classical groups, using an analogous approach. The next section contains a brief survey of knowledge of maximal subgroups in the other almost simple groups. Finally, in Section 5 we are concerned with primitive permutation groups of special degrees n (such as n odd, or prime to 3, or congruent to 2 mod 4).

1. Reduction Theorem for Symmetric Groups

Throughout this section, we consider $X = \mathrm{Sym}_n$, acting naturally on the set Ω of n points. (Similar statements can be made for the alternating groups Alt_n.)

We define a family $\mathcal{C}$ of 'natural' subgroups (in general maximal) in X such that, whenever G is a subgroup of X (not containing the socle $F^*(X)$), either G is contained in a member of $\mathcal{C}$, or G is almost simple and primitive on Ω.

The collection $\mathcal{C}$ consists of five classes. The first two take care of groups G which are not primitive on Ω.

(1) The intransitive subgroups $\mathrm{Sym}_k \times \mathrm{Sym}_{n-k}$, for $1 \leqslant k \leqslant n/2$; if G is a subgroup intransitive on Ω then it has an orbit of size $k \leqslant n/2$, and so G is contained in the full stabilizer $\mathrm{Sym}_k \times \mathrm{Sym}_{n-k}$ of this orbit in Sym_n.

(2) The imprimitive subgroups $\mathrm{Sym}_k \wr \mathrm{Sym}_{n/k}$, for k any proper divisor of n; if G is transitive but imprimitive on Ω then it preserves a nontrivial partition of Ω into l blocks of size k (with $n = kl$); then G is contained in the full subgroup preserving this partition, that is, $G \leqslant \mathrm{Sym}_k \wr \mathrm{Sym}_l$.

The next three classes arise from examples of groups which are primitive on Ω but are not almost simple.

(3) The full affine group $\mathrm{AGL}_d(p)$ of degree $n = p^d$: here Ω has the additional structure of a vector space $V = V_d(p)$ of dimension d over the field of p elements, and $\mathrm{AGL}_d(p)$ is the semidirect product $\mathrm{GL}_d(p).V$, acting on V by

$$(A, v) : x \mapsto xA + v.$$

It has an elementary abelian normal subgroup consisting of translations, and the stabilizer of the point 0 is the linear group $\mathrm{GL}_d(p)$.

(4) The full primitive wreath product $\mathrm{W}(m, k)$ of degree $n = m^k$ (with $m, k > 1$): as an abstract group, this is again the wreath product $\mathrm{Sym}_m \wr \mathrm{Sym}_k$, but the action is on the set Ω of k-tuples with entries in a given m-set—so the stabilizer of a point is $\mathrm{Sym}_{m-1} \wr \mathrm{Sym}_k$.

(5) The full diagonal group $\mathrm{D}(T, k)$ of degree $n = |T|^k$, where $k > 1$ and T is a non-abelian simple group: here T^k acts on the set Ω of right cosets of a diagonal subgroup $D = \{(t, \ldots, t) \mid t \in T\}$ in T^k; this action is not primitive for $k > 2$, but the group $\mathrm{D}(T, k) = N_{\mathrm{Sym}(\Omega)}(T^k)$ is primitive. Moreover, $F^*(\mathrm{D}(T, k)) = T^k$

and $D(T, k)/T^k \cong \mathrm{Sym}_k \times \mathrm{Out}T$. The action of $D(T, k)$ can be described as:

$$D(x_1, \ldots, x_k)(g_1, \ldots, g_k) = D(x_1 g_1, \ldots, x_k g_k) \quad \text{for } g_i \in T,$$
$$D(x_1, \ldots, x_k)\alpha = D(x_1\alpha, \ldots, x_k\alpha) \quad \text{for } \alpha \in \mathrm{Aut}T,$$
$$D(x_1, \ldots, x_k)\pi = D(x_{1\pi^{-1}}, \ldots, x_{k\pi^{-1}}) \quad \text{for } \pi \in \mathrm{Sym}_k.$$

In the special case $k = 2$, identifying Ω with T, we obtain as a subgroup the well-known primitive action of $T \times T$ on T by $(t_1, t_2) : x \mapsto t_1^{-1} x t_2$ —so the first (respectively second) coordinate acts by left (respectively right) regular action on Ω.

These five classes (closed under conjugation in X) taken together give us a class $\mathcal{C}$ of subgroups suitable for our purpose. To summarize, we put the groups in $\mathcal{C}$ in a table (some of the conditions for inclusion are tightened slightly for reasons of maximality).

Groups in $\mathcal{C}(\mathrm{Sym}_n)$	conditions	degree n
$\mathrm{Sym}_k \times \mathrm{Sym}_l$	$1 \leqslant k < n/2$	$n = k + l$
$\mathrm{Sym}_k \wr \mathrm{Sym}_l$	$k, l > 1$	$n = kl$
$\mathrm{AGL}_d(p)$	p prime	$n = p^d$
$W(m, k)$	$k > 1$, $m \geqslant 5$	$n = m^k$
$D(T, k)$	$k > 1$, T nonabelian simple	$n = |T|^{k-1}$

We are now ready for the Reduction Theorem; this was first stated by O'Nan and Scott—see [51, Appendix].

Theorem 1.1 *If $G < \mathrm{Sym}_n$ then either G is contained in a member of $\mathcal{C}$ or G is an almost simple primitive group.*

Corollary 1.2 *If G is a maximal subgroup of Sym_n then either $G \in \mathcal{C}$ or G is an almost simple primitive group.*

We proceed to sketch a proof of the Reduction Theorem. The proof we outline uses some ideas of Buekenhout, Cameron, Scott and others.

First we state a well-known easy observation about groups with a regular normal subgroup.

Lemma 1.3 *If G on Ω has a regular normal subgroup N, then we can identify N and Ω so that the actions of the stabilizer G_α on N by conjugation and on Ω are isomorphic. The centralizer of N in $\mathrm{Sym}(\Omega)$ is isomorphic to N (in fact it equals N if N is abelian, and it is N in its left regular action in general).*

Here is a proof. For a fixed $\alpha \in \Omega$, the mapping $\phi : n \mapsto \alpha n$ is a bijection from N to Ω, and $(n^g)\phi = (n\phi)g$ for $g \in G_\alpha$, so the actions of G_α on N by conjugation and on Ω are isomorphic. Using this identification, the action of N becomes the right regular action of N. It is easy to check that the centralizer of any transitive subgroup is semiregular; in particular, this is true of the centralizer C of N. On the other hand, the left and right regular actions on N commute, so C contains all the permutations of N of the form $\lambda_c : n \mapsto c^{-1}n$ for $c \in N$. Thus C is regular, and in fact equals N in its left regular action.

Now we go back to the proof of Theorem 1.1. We may assume that G is primitive on Ω. Let N be a minimal normal subgroup of G (note that this is not necessarily unique—however it follows from the argument below that there are at most two such, isomorphic to each other). Then $N = T_1 \times \cdots \times T_r$ with each T_i simple, isomorphic to some T. Note that N is transitive on Ω since it is normal in G, so $G = G_\alpha N$. If N is abelian, then N is a (multiplicative) vector space of dimension r over F_p, the field of p elements; identifying N and Ω as in Lemma 1.3, N acts as a group of translations and G_α as a group of non-singular linear transformations, so $G \leqslant \mathrm{AGL}_r(p)$. So assume that T is a non-abelian simple group. Suppose that $r = 1$. If there is no other minimal normal subgroup of G then G is almost simple. If, on the other hand, N_2 is another minimal normal subgroup then N, N_2 centralize each other, so both are regular and N, N_2 act by right, left regular action of T—so $G < \mathrm{D}(T, 2)$. So assume that $r > 1$. The next lemma helps with deciding whether $G \leqslant \mathrm{W}(m, k)$ for some m and k.

Lemma 1.4   *If $N = U_1 \times \cdots \times U_k$ with $k > 1$ and G acts transitively on the set $\{U_1, \ldots, U_k\}$, and if $Y_1 = U_2 \times \cdots \times U_k$ is intransitive on Ω with m orbits, then $G \leqslant \mathrm{W}(m, k)$.*

For, write $Y_i = \prod_{j \neq i} U_j$; then G acts on the set $\{Y_1, \ldots, Y_k\}$, and is transitive there. Let $\mathcal{H}_i$ be the set of Y_i-orbits on Ω. Then each $\mathcal{H}_i$ has the same size m. Write $\mathcal{H}_i = \{H_{i1}, \ldots, H_{im}\}$ and let Γ be the set of subsets of Ω given by the intersections $H_{1i_1} \cap H_{2i_2} \cap \cdots \cap H_{ki_k}$ as $i_1, \ldots, i_k$ vary. Note that the elements of Γ are certainly disjoint and partition Ω, and that Γ is closed under the action of G. As each U_i is transitive on $\mathcal{H}_i$ while fixing each element of $\mathcal{H}_j$ for $j \neq i$, we deduce that each element of Γ is non-empty, whence Γ is a system of imprimitivity for G on Ω. The primitivity of G on Ω implies that each element of Γ is a singleton. Hence the map $\omega \mapsto (i_1, \ldots, i_k)$ where $\{\omega\} = H_{1i_1} \cap H_{2i_2} \cap \cdots \cap H_{ki_k}$ gives a bijection between Ω and the set of k-tuples from $\{1, \ldots, m\}$. This imposes a Cartesian structure on Ω, which is certainly G-invariant, and so $G \leqslant \mathrm{W}(m, k)$.

To continue with the proof of Theorem 1.1, we shall assume that the situation in Lemma 1.4 does not arise. In particular we see then that $X_i = \prod_{j \neq i} T_j$ is transitive on Ω for each i. It follows that $N = N_\alpha X_i$ and hence each natural projection $p_i : N \to T_i$ remains surjective when restricted to N_α; in other words, N_α is a subdirect product. It is a general fact (easy to prove inductively [51, Lemma p. 328]) about subdirect products of simple groups that there is a partition $\{I_1, \ldots, I_k\}$ of the set of indices such that if $U_i = \prod_{j \in I_i} T_j$ then the intersection $D_i = N_\alpha \cap U_i$ is a diagonal subgroup of U_i and $N_\alpha = D_1 \times \cdots \times D_k$. Now G_α acts on the set $\{D_1, \ldots, D_k\}$, since $N_\alpha \lhd G_\alpha$ and the D_i are the simple normal subgroups of N_α. Moreover, this action of G_α is transitive, as can be seen from the maximality of G_α in G. Since $G = G_\alpha N$, we deduce that G acts on the set $\{U_1, \ldots, U_k\}$ and is transitive there. Also, if $k > 1$ then $U_2 \times \cdots \times U_k$ is intransitive on Ω; this is however contrary to our assumption that the situation of Lemma 1.4 does not arise. Thus $k = 1$, so N_α is a diagonal subgroup of N and we deduce that $G \leqslant D(T, r)$.

Remarks 1.5 (A) The parameters of each of the groups determine the group up to conjugacy in Sym_n. For example, the class of $D(T, k)$ is determined by the parameters (T, k). The groups in $\mathcal{C}$ are maximal in Sym_n (though there are precisely five cases, all with $n \leqslant 23$, where the intersection with Alt_n is no longer maximal in Alt_n); this is shown in [31], using the methods discussed in Section 2 below.

(B) The argument of the first part of the proof of Theorem 1.1 can be modified to show the well-known important fact that the socle of a finite primitive permutation group is characteristically simple; in fact, the socle is either minimal normal or is the product of two minimal normal subgroups, isomorphic to each other.

(C) There is another important stronger version of the Reduction Theorem, not much harder to prove (see e.g. [30]). This gives a fairly explicit description of all the primitive permutation groups G which are not almost simple. For example, an affine permutation group is any permutation group with an elementary abelian regular normal subgroup N; the primitivity of G is equivalent to the linear action of G_α on N being irreducible. A permutation group in diagonal action is a primitive subgroup of $D(T, k)$ containing T^k. The main change is needed in class $\mathcal{C}_4$ which has to be replaced by two classes:

(4a) groups in wreath product action: the degree is $n = m^k$ and there is a primitive permutation group G_0 of degree m, almost simple or of diagonal type, such that G is contained in $G_0 \wr \mathrm{Sym}_k$ in product action of degree m^k and such that the socle of G is $F^*(G_0)^k$;

(4b) groups in twisted wreath action: this is characterized by the existence of a unique regular normal subgroup which is nonabelian.

The groups in class (4b) are rather rare; the smallest example has degree 60^6. (In fact they were omitted by mistake in the early papers [51, 10].) They are investigated in much detail in the recent paper of Baddeley [7].

Using this version of the Theorem, many specific problems about finite permutation groups can be reduced to questions about primitive almost simple groups (and hence to questions about maximal subgroups of almost simple groups).

(D) There are versions of the Theorem for infinite permutation groups. One is in the recent paper [44] of Macpherson and Praeger. There is also a version for groups of finite Morley rank by Macpherson (announced in his lecture in Istanbul).

2. Maximality in Symmetric Groups

We start with an easy example. Consider the action of degree $n = 10$ of $H = \mathrm{Sym}_5$ on the set of unordered pairs in $\{1, 2, 3, 4, 5\}$. This leads to an embedding of Sym_5 in Sym_{10}, and there is a unique conjugacy class of such subgroups. These subgroups are however not maximal in Sym_{10}: there is an action of $G = \mathrm{Sym}_6$ on the set of partitions of the set $\{1, 2, 3, 4, 5, 6\}$ into two equal parts, leading to an embedding $\mathrm{Sym}_6 \leqslant \mathrm{Sym}_{10}$, and any primitive subgroup Sym_5 of Sym_{10} is contained in such a Sym_6.

In general, let H be an almost simple primitive group. An important natural question is whether H is maximal in Sym_n (or more precisely in $\mathrm{Alt}_n H$). This was answered fully in [31]. The key idea is quite simple. Assume that $H < G < \mathrm{Alt}_n H$ with $F^*(G) \not\leqslant H$; since we wish to find all such embeddings, we may assume that H is maximal in G. Since G is primitive, the stabilizer G_α of any point is maximal in G. Since H is transitive, $G = G_\alpha H$; thus we obtain a 'maximal' factorization of G. Moreover, $H_\alpha = H \cap G_\alpha$ is maximal in H. Using the Reduction Theorem of Section 1, we are led to study maximal factorizations of almost simple groups.

Theorem 2.1 [15, 32] *Let G be an almost simple group, and let A, B be subgroups of G not containing $F^*(G)$, both maximal in G. If $G = AB$ then the triple (G, A, B) is explicitly known.*

The list of such factorizations with G classical is rather long, though in most cases there is a geometric setting for the examples: if V is the underlying vector space, there is a subspace W of V and an automorphism α of $F^*(G)$ such that either A^α or B^α is the full stabilizer in G of W. If G is an exceptional group of Lie type, examples are rare and all factorizations (not just the maximal ones) are determined in [15]—there are only three families here, two with $F^*(G) = G_2(q)$ in characteristic 3 (with $\mathrm{SL}_3(q) \triangleleft A$, and B normalizing either $\mathrm{SU}_3(q)$ or $^2G_2(q)$) and one with $F^*(G) = F_4(q)$ in characteristic 2 (with $\mathrm{Sp}_8(q) \triangleleft A$ and $^3D_4(q) \triangleleft B$).

We remark that a corresponding result for algebraic groups appears in [36], where we determine all factorizations of simple algebraic groups over an algebraically closed field as a product of two closed subgroups under the assumption that if one of the factors is non-reductive then it is a full parabolic subgroup.

Theorem 2.1 is then used to resolve our original question of maximality:

Theorem 2.2 [31] *If H is an almost simple primitive permutation group of degree n, and if G is a group with $H < G < \mathrm{Alt}_n H$ and $F^*(G) \not\leqslant H$, then (H, G, n) is explicitly known.*

The list of examples, while not short, is quite manageable; in fact, there are 21 infinite families of examples and 58 exceptional triples. Some of the examples are well-known: for example, we have the well-known embeddings

$$G_2(q) < \mathrm{Sp}_6(q) < \mathrm{L}_6(q) < \mathrm{Alt}_n$$

with $n = (q^6 - 1)/(q - 1)$ for every q a power of 2. Secondly, the factorization $\mathrm{U}_6(2) = \mathrm{U}_5(2)\mathrm{M}_{22}$ with the intersection $\mathrm{U}_5(2) \cap \mathrm{M}_{22} = \mathrm{L}_2(11)$ maximal in both factors (of index 672 and 20736, respectively) leads to the two examples

$$\mathrm{M}_{22} < \mathrm{U}_6(2) < \mathrm{Alt}_{672}$$

and

$$\mathrm{U}_5(2) < \mathrm{U}_6(2) < \mathrm{Alt}_{20736}.$$

As a final example, given any r, there exists a chain $G_1 < \ldots < G_{r+1} < \mathrm{Alt}_n$ with $G_i = \mathrm{Sp}_{2^i}(2^{2^{r-i+1}})$ of degree $n = 2^{2^r-1}(2^{2^r} \pm 1)$.

We pause to summarize how far we have got at this stage in our quest for all maximal subgroups of the symmetric (and alternating) groups.

The Reduction Theorem leads us to consider the almost simple primitive groups. Theorem 2.2 assures us that these give usually maximal subgroups of the relevant symmetric group. Thus, usually, if G is an almost simple group and G_1 is a maximal subgroup of G (not containing $F^*(G)$) of index n, then G (or more precisely the normalizer in Alt_n or Sym_n of the socle $F^*(G)$) is a maximal subgroup of Sym_n or Alt_n. We need therefore consider the maximal subgroups of all the almost simple groups. Note that if G is alternating or symmetric of degree m then m is (much) smaller than n; proceeding inductively, it will suffice to deal with all the other types of almost simple groups. We proceed to outline our current state of knowledge in the next two sections.

3. Reduction Theorem and Maximality in Classical Groups

In this section we consider the maximal subgroups of the finite classical groups. We follow the approach outlined above for symmetric groups.

The Reduction Theorem for the finite classical groups was proved by Aschbacher in [2]. It is discussed in some detail in Section 2 of Liebeck's article (together with a new proof), so we shall be rather brief.

Let L be a finite simple classical group, and let G be a group wich satisfies $L \triangleleft G \leqslant \mathrm{Aut}(L)$. Let $V = V_d(q)$ be the underlying vector space associated naturally with L. Aschbacher's collection $\mathcal{C} = \mathcal{C}(G)$ consists of various 'geometric' subgroups of G. Roughly speaking, they are:

(a) stabilizers in G of subspaces of V, totally isotropic or non-degenerate;

(b) stabilizers of certain direct sum decompositions of V into subspaces of equal dimensions;

(c) stabilizers of certain tensor product decompositions of V;

(d) subgroups (not containing $F^*(G)$) which are classical groups on V over subfields or extension fields; and

(e) normalizers of r-subgroups of symplectic type in an irreducible representation on V.

We remark that the definition of $\mathcal{C}(G)$ has to be adjusted if L is one of $\mathrm{L}_d(q)$, $\mathrm{Sp}_4(q)$ with q even or $\mathrm{P}\Omega_8^+(q)$, if G contains elements in $\mathrm{Aut}(L) \setminus \mathrm{P\Gamma L}_d(q)$; in the first two of these cases, this is done in [2]; in the third case all maximal subgroups are known from the work of Kleidman [19].

Theorem 3.1 [2] *Let $L \triangleleft G \leqslant \mathrm{Aut}(L)$, with L a simple classical group. If H is any subgroup of G (not containing L), then either*

H is contained in a member of $C(G)$

or

H is almost simple, and the (projective) representation of $F^*(H)$ on V is absolutely irreducible and cannot be realized over a proper subfield of F_q.

The groups in C with their structure and conjugacy classes are described in detail in the book [22] of Kleidman and Liebeck, where the question of maximality of these groups is also solved completely.

The next step in the investigation is to decide on maximality of the almost simple subgroups. This is very much harder than the corresponding problem for the symmetric group in Section 2. It will be convenient to write $\mathrm{Cl}(V)$ for the classical subgroup of $\mathrm{SL}_d(q)$ corresponding to G. The task is to determine all triples $H < K < \mathrm{Cl}(V)$ with both H, K almost simple modulo scalars, with $E(H)$, $E(K)$ absolutely irreducible on V, and with $E(K) \not\leq H$.

The problem splits naturally into cases according to the *characteristic triples* (r, s, p), where r is the defining characteristic of H (defined to be 1, respectively -1, if H' is alternating, respectively sporadic, modulo scalars), s is the defining characteristic of K, and p is the characteristic of V.

Most of the effort has been concentrated on the 'generic' case, where $r = p$, that is the defining characteristics of H and of G are equal. The next remarkable theorem is the result of the work of a number of authors over a number of years.

Theorem 3.2   *The triples $H < K < \mathrm{Cl}(V)$, with $E(H)$ quasisimple of Lie type in characteristic p absolutely irreducible on V, with $E(K) \not\leq H$ and with K not a classical group on V, are known, with only finitely many unresolved questions remaining.*

If K is not also of Lie type of characteristic p, the examples are listed in [34]; there are finitely many undecided questions. This brings us to the heart of the problem, where K is also of Lie type in characteristic p. The aim is to lift to a correponding situation in algebraic groups over the algebraic closure of the prime field. This is possible by [55], unless K is an exceptional group and p is fairly small. The recent results in [40] (see section 3.3 of Liebeck's article) enable us to lift also in this remaining case, provided that H is maximal in K and the defining field of H is large enough. At that stage the pivotal results of Seitz [52] and Testerman [57] (discussed in Section 3 of Seitz's article) give us the result.

Another case which has received much attention over the last few years is the one where the characteristic triple is $(1, 1, p)$. In particular, the difficult special case with $H = \mathrm{Sym}_{n-1}$ embedded naturally in $K = \mathrm{Sym}_n$ has been resolved recently by Kleshchev [27] and Ford, completing the work of Jantzen and Seitz [17]:

Theorem 3.3   *Let V be the irreducible module V^λ of Sym_n corresponding to the p-regular partition $\lambda = (\lambda_1^{a_1}, \ldots, \lambda_r^{a_r})$ of n with $\lambda_1 > \ldots > \lambda_r > 0$. Then the restriction of V to Sym_{n-1} is irreducible if and only if $\lambda_i - \lambda_{i-1} + a_i + a_{i-1}$ is 0 modulo p for each $i \geqslant 2$.*

The special case $p = 2$ was first conjectured by Benson [8]. The general conjecture was stated in [17]; the 'if' part was proved there, and so was a special case of the converse. Kleshchev's proof uses representation theory of algebraic groups. In particular he proves a remarkable branching rule for linear groups (this is mentioned in Section 4 of Seitz's article). For new improved proofs see [13] and [28].

If $H = \mathrm{Sym}_m$ with other values of m, one can reduce to two possible families of candidates: n is either the number of k-subsets (with $k < m/2$) or k-partitions (into k equal parts) of $\{1, \ldots, m\}$. As far as I know, little is known at present if K is Alt_n or its covering group. (The corresponding problems in the case where V is a complex space are resolved completely in [49] and [25].)

For the other cases, information is less complete. Seitz [54] has a result in the case (r, r, p), where the middle group is classical.

Theorem 3.4 [54] *Let $H < K < \mathrm{Cl}(V)$, with H and K quasisimple of Lie type of equal characteristic r, $r \neq p$. Assume that H is not contained in a geometric subgroup of $\mathrm{Cl}(V)$, and that the defining field of H has at least four elements. Assume also that K is a classical group, and K is minimal among quasisimple overgroups of H. Then $(H/Z(H), K/Z(K))$ is one of $(\mathrm{PSp}_{2n}(s), \mathrm{L}_{2n}(s))$, $(\mathrm{PSp}_{2n}(s), \mathrm{U}_{2n}(s))$, $(\mathrm{P}\Omega_{n-1}(s), \mathrm{P}\Omega_n(s))$, $(\mathrm{PSp}_{2n}(s^i), \mathrm{PSp}_{2ni}(s))$ and $(\mathrm{G}_2(s), \mathrm{P}\Omega_7(s))$.*

The precise nature of V for the examples above remains undecided.

Finally, the announcements [45] and [16] contain preliminary progress reports in some subcases of the cases (r, p, p) and $(1, r, p)$, with $1 < r \neq p$.

To summarize, much has been done and much remains open. It is possible that eventually we shall know all the triples $H < K < \mathrm{Cl}(V)$ with H almost simple modulo scalars and $E(K) \not\leq H$. At that stage, we shall know that any other (projective) modular irreducible representation of an almost simple group leads to a maximal subgroup of the corresponding classical group. Thus we need to understand the modular representation theory of all the finite simple groups. This area is wide open at present.

4. Maximal Subgroups of other Almost Simple Groups

To start with, if $F^*(G)$ is one of the 26 sporadic simple groups, the situation is pretty satisfactory. The classification of the maximal subgroups is complete in 24 of the 26 cases. Much information is in the Atlas [6]. A useful list of references in 21 of the cases is in [23, p. 383]. The next three (Fi_{23}, Fi'_{24} and Th) have been dealt with since in [24, 43, 42]. The two remaining groups are the Monster and the Baby Monster. For these two groups, much information, including all the maximal odd locals, is in Wilson's work [58, 59, 60]. Meierfrankenfeld [47] has announced the classification of maximal 2-locals in the Monster, and there is hope that his techniques may deal also with the corresponding problem for the Baby Monster.

Subgroups of exceptional groups of Lie type are discussed extensively in Section 3 of Liebeck's article. Recent progress here is most impressive ([38, 40, 41]). The results in [38] (together with [11] and [35]) enable us to concentrate on maximal subgroups which are almost simple. (This is in fact another occurrence of the Reduction Theorem, this time for the family of exceptional groups; admittedly, the corresponding family $\mathcal{C}$ consists of groups rather less obviously natural in this case.) If $F^*(H)$ is not of Lie type in the same characteristic, the modular representation theory limits the number of possibilities to a finite list. If on the other hand H is of Lie type in the same defining characteristic then, under a very mild assumption on the size of the field over which H is defined, [40] allows us to translate the question into one about maximal connected closed subgroups of exceptional algebraic groups over an algebraically closed field. That was solved in the fundamental paper of Seitz [53], under the assumption of the defining characteristic being large enough.

Finally we remark that the maximal subgroups have been classified by entirely different methods for all the families of almost simple exceptional groups of Lie type of rank at most two ([56, 12, 4, 20, 21, 46]), as well as for $\mathrm{F}_4(2)$ and $\mathrm{E}_6(2)$ ([48, 26]).

5. Special Degrees

In this last section we discuss some fairly complete results on primitive permutation groups one can obtain by putting some mild restrictions on the degrees. The first result in this direction, the odd degree theorem, is now more than ten years old.

Theorem 5.1 [18, 33] *The primitive permutation groups of odd degree are classified.*

As an immediate corollary, we have:

Theorem 5.2 *The maximal subgroups of all the alternating and symmetric groups of odd degrees are known.*

We now summarize the list in Theorem 5.1.

Firstly, the stronger Reduction Theorem (cf. Remark 1.5C) leads to two classes: affine primitive groups, and wreath products of almost simple examples (that is, those in which there is an almost simple primitive group G_0 of odd degree m such that $G \leqslant G_0 \wr \mathrm{Sym}_k$ in product action of degree $n = m^k$ and $F^*(G) = F^*(G_0)^k$—cf. Remark 1.5B). These yield the maximal subgroups $\mathrm{AGL}_d(p)$ for $n = p^d$ and $\mathrm{W}(m,k)$ for $n = m^k$ in Theorem 5.2.

Secondly, the almost simple groups G are listed according to the possibilities for their socle $L = F^*(G)$:

(a) if $L = \mathrm{Alt}_c$, the action of L is on a set of subsets or partitions of $\{1, \ldots, c\}$, except for $n = 15$ and $c \in \{7, 8\}$;

(b) if L is sporadic, the actions are known from [3];

(c) if L is of Lie type of even characteristic, then the action is parabolic [9];

(d) if L is of Lie type of odd characteristic, then one of three cases holds:

   (i) G_1 is a subfield subgroup of G (that is, $L = L(q)$ and $L_1 = L(q_0)$, where $q = q_0^c$ for some odd c);

(ii) G is classical and either G_1 is a certain subgroup in $\mathcal{C}$ or the dimension is at most 8 and the action is as given in a short explicit list;

(iii) G is exceptional, and a short list follows for each family; for example, for $L = E_8(q)$, G_1 is the normalizer in G of one of the subsystem subgroups $2^4 L_2(q)^8$, $2P\Omega_{16}^+(q)$, $2^2 P\Omega_8^+(q)^2$, or of a torus T^ϵ of order $(q - \epsilon)^8$, where $\epsilon = \pm 1$ and $q \equiv \epsilon \bmod 4$.

The proof of Theorem 5.1 relies on Aschbacher's work [1]. In particular, the stabilizer of a point contains a Sylow 2-subgroup of G. If G is a group of Lie type in odd characteristic then G_1 contains classical involutions (cf. [1]), and we can use [1] to identify it.

Some variations on this theme are currently under investigation in [50]. In particular, we have a theorem for degree prime to 3:

Theorem 5.3 [50] *The primitive permutation groups of degree prime to 3 are classified.*

This has the immediate corollary:

Theorem 5.4 *All the maximal subgroups of* Alt_n *and* Sym_n *with* $3 \nmid n$ *are known.*

The list in Theorem 5.3 is rather similar to that in Theorem 5.1.

Firstly, there are three families of groups not almost simple: affine groups of characteristic prime to 3; diagonal or twisted wreath groups involving the Suzuki groups $^2B_2$; and wreaths of smaller examples, either almost simple or diagonal. Secondly, there is a list of almost simple group actions: this now includes all the primitive actions of the Suzuki groups, but apart from that it is not very different from the list above; for example, if $L = E_8(q)$ with $3 \nmid q$ then G_1 is the normalizer in G of $\mathrm{SL}_3^\epsilon(q) \circ E_6^\epsilon(q)$ or T^ϵ, with $q \equiv \epsilon \bmod 3$, or a subfield subgroup $E_8(q_0)$ with $q = q_0^\epsilon$. Here SL_3^ϵ stands for SL_3, respectively SU_3, if ϵ is $+1$, respectively -1, and similarly E_6^ϵ stands for E_6, respectively $^2E_6$.

A few words about the proof. The stabilizer G_1 contains a Sylow 3-subgroup of G. We reduce quickly to the case where G is an almost simple group of Lie type of characteristic not 3 (other than $^2B_2$). Now G (and hence G_1) contains an elementary abelian 3-subgroup of large rank. A further reduction leads to G_1 being also almost simple of Lie type of the same characteristic. The Lie rank of G_1 has to be large, and we are in position to apply the results of [37] to identify it.

Another class currently under investigation in [50] is that of degrees $n \equiv 2 \bmod 4$. Here the list will be much shorter. Firstly, all the primitive groups which arise are almost simple. Secondly, here is the list for exceptional groups of Lie type in odd characteristic:

$L = E_7$, $G_1 = P_1$, $q \equiv 1(4)$;
L is G_2 or $^3D_4$, G_1 is parabolic or $N(\mathrm{SL}_3^\epsilon(q))$.

We conclude by giving an 'asymptotic' result, which explains partly the shape of the lists above.

Lemma 5.5 *Let p^m be a prime power greater than 1. There exist constants c_0, d_0 such that, if $p^m \nmid n$ and n is the degree of a faithful primitive permutation action of G,*

(i) *if G is Alt_c or Sym_c with $c \geqslant c_0$ then the action is on a set of subsets or partitions of $\{1, \ldots, c\}$;*

(ii) *if G is a classical group of dimension $d \geqslant d_0$ then the action is a geometric action (i.e. the stabilizer lies in the Aschbacher family C).*

Proof.   Let H be the stabilizer of a point in the action of G. Let P, Q be Sylow p-subgroups of G, H, respectively, with $Q \leqslant P$. Since $p^m \nmid n$, we have $|P : Q| \leqslant p^{m-1}$.

(i) We claim that $c_0 = p^m + p$ (or $c_0 = 9$ for $p = 2$ and $m \leqslant 2$) will do. To simplify, we shall assume that p is odd. Consider a set of $p^{m-1} + 1$ disjoint p-cycles in P; at least two of these lie in the same coset of Q in P. Hence Q contains an element x which is a p-cycle or a double p-cycle. If H is a primitive subgroup of Sym_c, we get a contradiction from a classical result of Jordan. So H is intransitive or imprimitive on $\{1, \ldots, c\}$, and, since it is maximal in G, the result follows.

(ii) Suppose that G is classical of dimension d over the field F_r. If $p \mid r$, we can use Liebeck's bounds [29] (in fact much more can be proved), so assume that r is prime to p. Then $p \mid r^k - 1$ for some $k \leqslant p - 1$; take such k minimal. Assume that $d \geqslant (p^{m-1} + 1)(2k + 1)$. Then, by the argument of (i), H contains an element of order p with at most two irreducible blocks, each of size k, and fixing a large subspace. Taking d_0 large enough, the result follows from [14, Theorem 2].

References

1. M. Aschbacher, A characterization of Chevalley groups over fields of odd order, *Ann. of Math.* 106 (1977), 353–468.
2. M. Aschbacher, On the maximal subgroups of the finite classical groups, *Invent. Math.* 76 (1984), 469–514.
3. M. Aschbacher, Overgroups of Sylow subgroups in sporadic groups, *Memoirs Amer. Math. Soc.* 343 (1986).
4. M. Aschbacher, Chevalley groups of type G_2 as the group of a trilinear form, *J. Algebra* 109 (1987), 193–259.
5. M. Aschbacher and L. L. Scott, Maximal subgroups of finite groups, *J. Algebra* 92 (1985), 44–80.
6. J. H. Conway, R. T. Curtis, S. P. Norton, R. A. Parker and R. A. Wilson, *An Atlas of Finite Groups*, Oxford University Press, 1985.
7. R. W. Baddeley, Primitive permutation groups with a regular non-abelian normal subgroup, Proc. London Math. Soc. 67 (1993), 547–595.
8. D. Benson, Some remarks on the decomposition numbers of the symmetric groups, in *The Santa Cruz Conference on Finite Groups* (eds. B. Cooperstein and G. Mason), *Proc. Symp. Pure Math.* 37 (1980), 381–394.
9. A. Borel and J. Tits, Elements unipotents et sousgroupes paraboliques de groupes reductifs, *Invent. Math.* 12 (1971), 95–104.
10. P. J. Cameron, Finite permutation groups and finite simple groups, *Bull. London Math. Soc.* 13 (1981), 1–22.
11. A. M. Cohen, M. W. Liebeck, J. Saxl and G. M. Seitz, The local maximal subgroups of exceptional groups of Lie type, finite and algebraic, *Proc. London Math. Soc.* 64 (1992), 21–48.
12. B. N. Cooperstein, Maximal subgroups of $G_2(2^n)$, *J. Algebra* 70 (1981), 23–36.
13. B. Ford, Irreducible restrictions of representations of the symmetric groups, *J. London Math. Soc.*, to appear.

14. J. I. Hall, M. W. Liebeck and G. M. Seitz, Generators for finite simple groups, with applications to linear groups, *Quart. J. Math.* 43 (1992), 441–458.
15. C. Hering, M. W. Liebeck and J. Saxl, The factorizations of the finite exceptional groups of Lie type, *J. Algebra* 106 (1987), 517–527.
16. W. J. Hussen, On the maximality of symmetric and alternating groups in the classical groups, *Abstracts of Papers Presented to the Amer. Math. Soc.* (1994), 891–20–53.
17. J. C. Jantzen and G. M. Seitz, On the representation theory of the symmetric groups, *Proc. London Math. Soc.* 65 (1992), 475–504.
18. W. M. Kantor, Primitive permutation groups of odd degree, and an application to finite projective planes, *J. Algebra* 106 (1987), 15–45.
19. P. B. Kleidman, The maximal subgroups of the finite 8-dimensional orthogonal groups $P\Omega_8^+(q)$ and of their automorphism groups, *J. Algebra* 110 (1987), 173–242.
20. P. B. Kleidman, The maximal subgroups of the Chevalley groups $G_2(q)$ with q odd, of the Ree groups $^2G_2(q)$, and of their automorphism groups, *J. Algebra* 117 (1988), 30–71.
21. P. B. Kleidman, The maximal subgroups of the Steinberg triality groups $^3D_4(q)$ and of their automorphism groups, *J. Algebra* 115 (1988), 182–199.
22. P. B. Kleidman and M. W. Liebeck, *The Subgroup Structure of the Finite Classical Groups*, Cambridge University Press, 1990.
23. P. B. Kleidman and M. W. Liebeck, A survey of the maximal subgroups of the finite simple groups, *Geom. Ded.* 25 (1988), 375–389.
24. P. B. Kleidman, R. A. Parker and R. A. Wilson, The maximal subgroups of the Fischer group Fi_{23}, *J. London Math. Soc.* 39 (1989), 89–101.
25. P. B. Kleidman and D. B. Wales, The projective characters of the symmetric groups that remain irreducible on subgroups, *J. Algebra* 138 (1991), 440–478.
26. P. B. Kleidman and R. A. Wilson, The maximal subgroups of $E_6(2)$ and $\mathrm{Aut}(E_6(2))$, *Proc. London Math. Soc.* 60 (1990), 266–294.
27. A. S. Kleshchev, On restrictions of irreducible modular representations of semisimple algebraic groups and symmetric groups to some natural subgroups, *Proc. London Math. Soc.* 69 (1994), 515–540.
28. A. S. Kleshchev, Branching rules for modular representations of symmetric groups, I, *J. Algebra*, to appear; II, *J. reine angew. Math.* 459 (1995), 163–212.
29. M. W. Liebeck, On the orders of maximal subgroups of the finite classical groups, *Proc. London Math. Soc.* 50 (1985), 426–446.
30. M. W. Liebeck, C. E. Praeger and J. Saxl, On the O'Nan-Scott reduction theorem for finite primitive permutation groups, *J. Australian Math. Soc.* 44 (1988), 389–396.
31. M. W. Liebeck, C. E. Praeger and J. Saxl, A classification of the maximal subgroups of the alternating and symmetric groups, *J. Algebra* 111 (1987), 365–383.
32. M. W. Liebeck, C. E. Praeger and J. Saxl, The maximal factorizations of the finite simple groups and their automorphism groups, *Memoirs Amer. Math. Soc.* 432 (1990).
33. M. W. Liebeck and J. Saxl, The primitive permutation groups of odd degree, *J. London Math. Soc.* 31 (1985), 250–264.
34. M. W. Liebeck, J. Saxl and G. M. Seitz, On the overgroups of irreducible subgroups of the finite classical groups, *Proc. London Math. Soc.* 55 (1987), 507–537.
35. M. W. Liebeck, J. Saxl and G. M. Seitz, Subgroups of maximal rank in finite exceptional groups of Lie type, *Proc. London Math. Soc.* 65 (1992), 297–325.
36. M. W. Liebeck, J. Saxl and G. M. Seitz, Factorizations of simple algebraic groups, *Trans. Amer. Math. Soc.*, to appear.
37. M. W. Liebeck, J. Saxl and D. M. Testerman, Simple subgroups of large rank in groups of Lie type, *Proc. London Math. Soc.*, to appear.
38. M. W. Liebeck and G. M. Seitz, Maximal subgroups of exceptional groups of Lie type, finite and algebraic, *Geom. Dedicata* 36 (1990), 353–387.
39. M. W. Liebeck and G. M. Seitz, On subgroups of algebraic and finite classical groups, to appear.
40. M. W. Liebeck and G. M. Seitz, Finite subgroups of exceptional algebraic groups, to appear.
41. M. W. Liebeck and G. M. Seitz, Reductive subgroups of exceptional algebraic groups, to appear.
42. S. A. Linton, The maximal subgroups of the Thompson group, *J. London Math. Soc.* 39 (1989), 79–88.

43. S. A. Linton and R. A. Wilson, The maximal subgroups of the Fischer groups Fi_{24} and Fi'_{24}, *Proc. London Math. Soc.* 63 (1991), 113–164.

44. H. D. Macpherson and C. E. Praeger, Infinitary versions of the O'Nan-Scott Theorem, *Proc. London Math. Soc.* 68 (1994), 518–540.

45. K. Magaard, On the maximality of irreducible cross characteristically embedded classical groups, *Abstracts of Papers Presented to the Amer. Math. Soc.* (1994), 891-20-88.

46. G. Malle, The maximal subgroups of $^2F_4(q^2)$, *J. Algebra* 139 (1991), 52–69.

47. U. Meierfrankenfeld, Maximal 2-locals of the monster, *Abstracts of Papers Presented to the Amer. Math. Soc.* (1994), 891-20-87.

48. S. P. Norton and R. A. Wilson, The maximal subgroups of $F_4(2)$ and its automorphism group, *Comm. in Algebra* 17 (1989), 2809–2824.

49. J. Saxl, The complex characters of the symmetric groups that remain irreducible in subgroups, *J. Algebra* 111 (1987), 210–219.

50. J. Saxl, Primitive permutation groups of special degrees, in preparation.

51. L. L. Scott, Representations in characteristic p, in *The Santa Cruz Conference on Finite Groups* (eds. B. Cooperstein and G. Mason), *Proc. Symp. Pure Math.* 37 (1980), 319–331.

52. G. M. Seitz, The maximal subgroups of classical algebraic groups, *Memoirs Amer. Math. Soc.* 365 (1987).

53. G. M. Seitz, Maximal subgroups of exceptional algebraic groups, *Memoirs Amer. Math. Soc.* 441 (1991).

54. G. M. Seitz, Cross-characteristic embeddings of finite groups of Lie type, *Proc. London Math. Soc.* 60 (1990), 166–200.

55. G. M. Seitz and D. M. Testerman, Extending morphisms from finite to algebraic groups, *J. Algebra* 131 (1990), 559–574.

56. M. Suzuki, On a class of doubly transitive groups, *Annals of Math.* 75 (1962), 105–145.

57. D. M. Testerman, Irreducible subgroups of exceptional algebraic groups, *Memoirs Amer. Math. Soc.* 390 (1988).

58. R. A. Wilson, Some subgroups of the Baby Monster, *Invent. Math.* 89 (1987), 197–218.

59. R. A. Wilson, The odd local subgroups of the Monster, *J. Australian Math. Soc.* 44 (1988), 1–16.

60. R. A. Wilson, Is J_1 a subgroup of the Monster?, *Bull. London Math. Soc.* 18 (1986), 349–350.

FINITARY LINEAR GROUPS: A SURVEY

R. E. PHILLIPS
Department of Mathematics
Michigan State University
East Lansing, MI 48824
U.S.A.
E-mail: rphillips@math.msu.edu

Abstract. This is a survey article outlining the current state of research in the area of finitary linear groups. While many aspects of the subject are covered, special emphasis is placed on foundations, locally solvable groups, and locally finite groups. Generally, proofs are not included.

Key words: finitary linear group, finite dimensional group, finitary permutation, primitive and imprimitive groups, Tits alternative, unipotent, locally solvable, locally finite, local systems and countable local systems.

1. Introduction

Let V be a vector space over a field $\mathcal{K}$; an element $g \in \mathrm{GL}_{\mathcal{K}}(V)$ is said to be *finitary* if $[V, g] = \{v(g - 1) \mid v \in V\}$ is finite dimensional. Note that g is finitary if and only if

$$(1.1) \qquad C_V(g) = \{v \in V \mid vg = v\} \text{ has finite codimension in } V.$$

The set $\mathrm{FGL}_{\mathcal{K}}(V)$ of all finitary transformations is a normal subgroup of $\mathrm{GL}_{\mathcal{K}}(V)$ and a group G is *finitary linear* if G is isomorphic to a subgroup of $\mathrm{FGL}_{\mathcal{K}}(V)$. More generally, a *finitary representation* (or $\mathcal{K}$-finitary representation) of a group G is a homomorphism $\alpha : G \to \mathrm{FGL}_{\mathcal{K}}(V)$ where V is a vector space over $\mathcal{K}$. The group G will be called *finite dimensional* if G has a faithful representation in $\mathrm{FGL}_{\mathcal{K}}(V)$ for some finite dimensional V. Much of the current work on finitary groups is devoted to determining

(1.2) which groups G (in some specified class $\mathcal{W}$) have faithful finitary linear representations.

This line of research is an extension of the program of determining which groups G are finite dimensional (see, for example, [69, Chapter 2]). In the sequel, we attempt to give an exposition of the results and ongoing research in the area of finitary linear groups. Generally, proofs are not given, and, when there are proofs, frequently they amount to little more than a sketch. Rough guidelines for the inclusion of arguments are (i) the material is not easily available in published form, and (ii) the import of certain ideas dictates further commentary. Many of the critical entries listed in the References are not yet in print (several not even submitted) which illustrates the

B. Hartley et al. (eds.), Finite and Locally Finite Groups, 111–146.

degree of current activity in the subject. An attempt has been made to discuss all aspects of this topic; omissions are likely due to ignorance rather than choice.

2. Examples

(i) *Subgroups of* $GL(n, \mathcal{K})$ which will henceforth be called *finite dimensional groups*.

(ii) *Groups of Finitary Permutations*. Considerable detail on these groups will be given in the next section. For the discussion here, suppose that g is a permutation on a set Ω; then $supp(g)$, the *support* of g, is the set of all elements $\omega \in \Omega$ such that $\omega g \neq \omega$. The permutation g is called *finitary* if $supp(g)$ is finite. The set of all finitary permutations on Ω is a normal subgroup of $Sym(\Omega)$ and is denoted $FSym(\Omega)$. Primary references for finitary permutation groups are [63, Chapter 11] and [79]. A permutation group G acting on Ω is called finitary if $G \subseteq FSym(\Omega)$.

If $G \subseteq FSym(\Omega)$, and $\mathcal{K}$ is a field, the permutation module V for G over $\mathcal{K}$ is the $\mathcal{K}$-space V with basis $\{v_\omega \mid \omega \in \Omega\}$ with G-action defined by $g : v_\omega \mapsto v_{\omega g}$. Evidently, if $G \subseteq FSym(\Omega)$ and V is the permutation module of G over $\mathcal{K}$, then $G \subseteq FGL_\mathcal{K}(V)$—thus, G is a finitary linear group.

(iii) *Wreath Products*. Let $S = H \ Wr_\Omega \ (G, \Omega)$ where (G, Ω) is a group of finitary permutations. Recall that the base group B of S is a direct product of the form $B = Dr\{H(\omega) \mid \omega \in \Omega\}$ where $H(\omega)$ is a copy of H (under the isomorphism $h \mapsto h(\omega)$). The equation $h(\omega)^g = h(\omega g)$ $(h \in H, g \in G)$ and its coordinatewise extension defines an action of G on B and the split extension BG is the restricted permutational wreath product of H by G. Let A be an n-dimensional $\mathcal{K}$-space on which H acts and for $\omega \in \Omega$ let $A(\omega)$ be a copy of A (under the isomorphism $a \mapsto a(\omega)$): also put $V = \bigoplus\{A(\omega) \mid \omega \in \Omega\}$. Then B acts coordinatewise on V and G permutes the direct factors of V via $A(\omega)g = A(\omega g)$. It is easy to check that the action just described defines a faithful finitary representation of S on V. Thus, S is a finitary linear group.

We note that if $1 \neq H$ is an irreducible subgroup of $GL(n, \mathcal{K})$ and G acts transitively on Ω then S acts irreducibly on V. Finally, the group S is a prototype of an imprimitive irreducible group (see §7).

(iv) *Stable Groups*. Let V be a $\mathcal{K}$-space with basis $\mathcal{B}$ and put

$$stab(V, \mathcal{B}) = \{g \in GL_\mathcal{K}(V) \mid vg = v \text{ for all but a finite number of } v \in V\}.$$

Then $G = stab(V, \mathcal{B})$ is a finitary linear group. Note that, if $\mathcal{C}$ is another basis of V, elements of G need not fix all but a finite number of vectors in $\mathcal{C}$—the point is that this is a highly basis dependent concept.

We shall see later (in (5.7)) that for every countable subgroup G of $FGL_\mathcal{K}(V)$ there is a countable independent subset $\mathcal{B}$ of V such that $G \subseteq stab(W, \mathcal{B})$, where W is the span of $\mathcal{B}$. To see that this fact can not hold in general, note that, if V has countably infinite dimension and $\mathcal{K}$ is a countable field, then, for any basis $\mathcal{B}$ of V, $stab(V, \mathcal{B})$ is countable while $FGL_\mathcal{K}(V)$ is uncountable. Thus, there is no basis $\mathcal{B}$ of V for which $FGL_\mathcal{K}(V) = stab(V, \mathcal{B})$.

(v) *Free Groups.* The following observation is due to U. Meierfrankenfeld [41, Example 8.1]. Let I be a set with $|I| > 2$ and F a free group with generators $\{f_i \mid i \in I\}$ and M a free Abelian group with free generators $\{v_i \mid i \in I\}$. Further, let $\{z_i \mid i \in I\}$ be a set of integers with $|z_i| \geqslant 6$ for all $i \in I$. Then

$$v_i f_i = v_i, \quad \text{and, if } j \neq i, \ v_j f_i = v_j + z_i v_i,$$

defines a faithful representation of F in $\mathrm{Aut}(M)$, and, for any $f \in F$, $[M, f]$ lies in a finitely generated subgroup of M; thus M is a finitary $\mathbb{Z}F$-module. If $\mathcal{K}$ is a field with $z_i \neq 0$ for all $i \in I$, then $V = \mathcal{K} \otimes \mathbb{Z}M$ is a finitary and irreducible $\mathcal{K}F$-module; further, V is a faithful $\mathcal{K}F$-module if $\mathrm{char}(\mathcal{K}) = 0$.

Thus, for fields $\mathcal{K}$ of characteristic 0, free groups F of rank at least 2 have faithful, irreducible, finitary representations of degree $|I|$. Similar constructions can be carried out in characteristic $p > 0$ provided $\mathcal{K}$ has a sufficient number of independent transcendental elements. Finally, note that any irreducible representation of F must be primitive (primitive groups are discussed in §8).

3. History

The concept of a finitary linear transformation seems to have been introduced by Dieudonné [15] in an attempt to use the determinant function for linear transformations on infinite dimensional spaces. The idea reappears in Rosenberg [62] as an essential feature in the classification of normal subgroups of general linear groups. Structural properties of finitary linear groups are undertaken by Zalesskiĭ [80], [81]. The theory of finitary permutation groups seems to have been developed independently of the linear groups; key contributions in the theory of finitary permutation groups can be found in [27], [28], [47], [48], [49], [64], [65], [63, Chapter 11], [67], [68], [78], [79], [4], [5], and [7].

There are many occurrences of examples in the infinite groups literature which are finitary linear groups—notably in the work of McLain [37], [38], [39], [40], [61, pp. 14-18]. These examples were generally designed to respond to questions raised from a group theoretic (or ring theoretic) viewpoint, and they were not studied from the perspective of linear groups. Applications of the concept of 'finitary linear' can be found in [51, Lemmas 2.9 and 2.10], [50], [52], [6] and [13]. Initial use of the term 'finitary linear' is due to J. I. Hall [17], and this paper initiated the current interest in finitary linear groups. Intrinsic properties of finitary linear groups are developed in [11], [53], [54], [18], [41], [43], [46], [31], [1], [2], [3], [4] ,[5], [7]. Within the context of locally finite groups (especially locally finite simple groups), the study of these groups seems to have been especially worthwhile [17], [19], [22], [42], [53], [54], [56], [34], [5], [6]. One aspect of the study of finitary groups is the extension of the highly developed theory of periodic finite dimensional groups (see [79, Chapter 9]). Other thrusts include work on unipotent groups [35], [36], [76] and generalized solvable and nilpotent groups [46], [70]. Finitary groups over division rings are taken up in the series of papers [71], [72], [73], [74], [76], [77].

4. Groups of Finitary Permutations

In addition to the examples provided in the previous section, groups of finitary permutations will play an extremely important role in the general structure of finitary linear groups, and for this reason we discuss here some of the primary features of these groups.

(4.1) *Alternating Groups.* If g is a finitary permutation on a set Ω, g is assigned a parity according to the parity of g restricted to its support. Thus, every finitary permutation on Ω is even or odd and the alternating group $\text{Alt}(\Omega)$ is the group of even permutations on Ω. $\text{Alt}(\Omega)$ is a normal subgroup of index 2 in $\text{FSym}(\Omega)$, and it is routine to show that $\text{Alt}(\Omega)$ is a simple locally finite group. As in the finite case, $\text{Alt}(\Omega)$ is generated by 3-cycles. Also, if the set Ω is infinite, $\text{Alt}(\Omega)$ is n-transitive for every positive integer n (that is, highly transitive). Further, $\text{Alt}(\Omega)$ acts primitively on Ω, and a remarkable theorem of Wielandt asserts that every primitive group of finitary permutations on the infinite set Ω must contain $\text{Alt}(\Omega)$. Specifically,

(4.1.1) **Theorem** [79, Satz 9.2] *If Ω is an infinite set and G is a primitive subgroup of* $\text{FSym}(\Omega)$ *then either* $G = \text{FSym}(\Omega)$ *or* $G = \text{Alt}(\Omega)$.

The proof of (4.1.1) makes essential use of a classical theorem of Jordan on the structure of finite primitive groups containing elements of small degree. We note that, for an infinite set Ω, $\text{Alt}(\Omega)$ contains a copy of every finite group. However, $\text{Alt}(\Omega)$ does not contain a subgroup of type p^∞ for any prime p [32, p. 187] which perhaps is the first indication of the restrictive nature of this group. Indeed, the work of [17], [22], [42] make a strong case that $\text{Alt}(\Omega)$, along with the other finitary linear, locally finite, simple groups, are extremely restricted in nature.

Since Wielandt's theorem plays a central role in what follows, we indicate some of the ideas behind the proof; several of these concepts are of independent interest.

(4.1.2) Let $H \triangleleft G$; G has the *conjugacy centralizer property* (ccp) with respect to H if for every finitely generated subgroup F of G there is an $h \in H$ such that $[F, F^h] = 1$. It is not difficult to show (as is done in [49, p. 10] and later in [53, p. 419]) that, if G has the ccp relative to $H \triangleleft G$, then

(i) $G' \subseteq H$, and

(ii) if $T \triangleleft H$ then $T \triangleleft G$.

Now every infinite transitive group G of finitary permutations has the ccp with respect to every transitive normal subgroup [49, p. 10] and this fact, combined with earlier comments, gives considerable information on the positioning of transitive normal subgroups in an infinite transitive group. Before collecting all of this information we note that the derived group G' of every infinite transitive group of finitary permutations is also transitive; thus item (ii) above applies to G'. In summary:

(iii) If $G \subseteq \text{FSym}(\Omega)$ where Ω is infinite and G is transitive, then G' is the minimal subnormal transitive subgroup of G. Further, every normal subgroup of G' is normal in G.

(4.1.3) If Ω is infinite and $G \subseteq \mathrm{FSym}(\Omega)$ is primitive, then every normal subgroup of G must be transitive. It follows from (4.1.2) that G' is simple.

The facts presented in (4.1.2) and (4.1.3) illustrate why Wielandt's theorem should be anticipated. The remainder of the proof of (4.1.1) involves identifying the simple group G' as an alternating group. An outline of the proof of (4.1.1) is given in [12, pp. 38–39].

(4.2) *The Dichotomy.* Thanks to the work of P. M. Neumann [48] a surprising amount of information is available regarding the structure of infinite transitive groups of finitary permutations.

Let Ω be an infinite set and G a transitive subgroup of $\mathrm{FSym}(\Omega)$. Since the group G is finitary, G- blocks must be finite sets. Further, if $\Delta_1 \subseteq \Delta_2 \subseteq \cdots \subseteq \Delta_n \subseteq \cdots$ is an ascending union of G-blocks, $\Sigma = \bigcup\{\Delta_i \mid i \geqslant 1\}$, and $g \in G$, then either $\Sigma g = \Sigma$ or $\Sigma g \cap \Sigma = \emptyset$. Thus, we must have $\Sigma = \Omega$. It follows that

(4.2.1) either

(i) Ω contains a maximal G-block, or

(ii) Ω is the ascending union of a countable number of finite G-blocks.

The groups of type (i) are called *almost primitive* (since G acts primitively on the block system defined by a maximal block), and those of type (ii) are called *totally imprimitive.*

We list several of the more important properties of totally imprimitive groups in

(4.3) Theorem *If G is a totally imprimitive group of finitary permutations on the infinite set Ω, both of the following properties obtain.*

(i) *Ω and G are both countable.*

(ii) *There is an ascending chain $N_1 \subseteq N_2 \subseteq \cdots \subseteq N_k \subseteq \cdots$ of normal subgroups of G such that*

(a) *$\bigcup\{N_k \mid k \geqslant 1\} = G$;*

(b) *each N_k is a subdirect power of a finite group (and so is an FC-group).*

Corresponding properties for the 'almost primitive' groups are given in

(4.4) Theorem *If G is almost primitive on an infinite set Ω, then G has a normal subgroup N satisfying:*

(a) *N is a subdirect power of a finite group, and*

(b) *there is an infinite set Δ such that G/N is isomorphic to either $\mathrm{Alt}(\Delta)$ or $\mathrm{FSym}(\Delta)$.*

Note the rather amazing fact:

(4.5) if G is an uncountable transitive group of finitary permutations, then G is 'almost primitive'. Thus, G has a chief factor isomorphic to an infinite alternating group.

(4.6) *Generalized Wreath Products.* It is helpful to keep in mind the following construction of totally imprimitive finitary permutation groups. In its full generality, the construction is due to P. Hall [23] (see also [61, pp. 18–22]); we require a highly specialized case.

Let $\mathbb{N}$ be the set of positive integers and (G_n, Ω_n), $n \in \mathbb{N}$, be a set of finite transitive permutation groups. For each n, fix an element 1_n in Ω_n, and let Ω be the set of all sequences $\{\omega_n \mid n \geqslant 1\} = \{\omega_n\}$ where $\omega_n \in \Omega_n$ and $\omega_n = 1_n$ for almost all n. The set Ω is called the *direct product* of the sets Ω_n. We will identify Ω_n with the n-th coordinate set of Ω.

If $g \in G_k$, g acts on Ω by

$$\{\omega_n\}g = \begin{cases} \{\omega_n\} & \text{if } \omega_t \neq 1_t \text{ for some } t > k \\ \{\omega_1, \ldots, \omega_{k-1}, \omega_k g, 1_{k+1}, \ldots\} & \text{if } \omega_t = 1_t \text{ for all } t > k. \end{cases}$$

Since the sets Ω_k are finite, $G_k \subseteq \mathrm{FSym}(\Omega)$; thus, $G = \langle G_k \mid k \geqslant 1 \rangle \subseteq \mathrm{FSym}(\Omega)$. The group G is called the *wreath product* of the permutation groups (G_n, Ω_n) and is denoted

(4.6.1) $G = \mathrm{Wr}\{(G_n, \Omega_n) \mid n \geqslant 1\}.$

As is shown in [61, p. 19], the group G is transitive on Ω, and for any finite subset $\{n_1, \ldots, n_s\}$ of $\mathbb{N}$, $\langle G_{n_1}, \ldots, G_{n_s} \rangle$ is isomorphic to the iterated wreath product

$$(\cdots(((G_{n_1}, \Omega_{n_1}) \wr (G_{n_2}, \Omega_{n_2})) \wr (G_{n_3}, \Omega_{n_3})) \cdots) \wr (G_{n_s}, \Omega_{n_s}).$$

Further, for any integer k, $G \cong \langle G_1, \ldots, G_k \rangle \mathrm{Wr}_\Lambda \langle G_k, G_{k+1}, \ldots \rangle$, where Λ is the direct product of the Ω_i, $i \geqslant k$.

It is easy to see that each of the sets $\Omega_1 \times \Omega_2 \times \cdots \times \Omega_k$ is a G-block and it follows from this that G is totally imprimitive. It has been shown by P. M. Neumann in [49, pp. 14–17] that every infinite totally imprimitive group can be embedded as a transitive subgroup of a group of the form (4.6.1). Thus, this wreath product construction provides a type of universal totally imprimitive group.

There are rather large classes of groups Λ with the property that

(4.7) every finitary, transitive permutation representation of a Λ-group has finite degree.

Examples of such classes Λ are the classes *solvable*, FC, *hypercentral*, *poly -* FC *or Abelian'*, and *any variety of groups* [48], [78]. To illustrate the type of arguments used, we here give a proof of the relatively simple solvable case.

(4.8) **Theorem** *Let G be a solvable, transitive group of finitary permutations on a set Ω. Then $|\Omega|$ is finite.*

Proof. We induct on the derived length of G. If G is Abelian, then G is also regular and in this case we have $|\Omega| = |\mathrm{supp}(g)|$ for every $g \in G$; thus Ω is a finite set. In the general case, induction implies that the derived group G' must have finite orbits. Since G' is a normal subgroup of G the set Σ of these orbits forms a system of imprimitivity for G. Further, if π is the representation of G on Σ, then $G' \subseteq \ker(\pi)$. Thus G/G' acts transitively and finitarily on Σ and the $n = 1$ case shows that Σ is finite. Thus Ω is a finite union of finite sets and so is finite.

(4.9) *Residual Properties.* In [49] it is shown that if $G \subseteq \mathrm{FSym}(\Omega)$ is residually finite (solvable) then all G-orbits are finite. Thus, G is a subdirect product of finite groups. The work of Belyaev [4], [5], [7] also deals with residual properties of finitary permutation groups. In particular, it is shown in [4, Theorem 6.1] that any residually finite image of $G \subseteq \mathrm{FSym}(\Omega)$ is locally normal (locally normal means that x^G is finite for all $x \in G$).

5. Basic Ideas and Concepts

(5.1) *The Degree Function.* If $G \subseteq \mathrm{FGL}_K(V)$ and $g \in G$, the *degree* of g, $\deg(g)$, is the dimension of $[V, g]$. Likewise, for a subgroup H of G, $\deg(H) = \dim[V, H]$. This concept is fundamental in much of what follows. Elementary results regarding degrees may be found in [53, p. 412]; we list several here in

(5.1.1) **Lemma** *If $G \subseteq \mathrm{FGL}_K(V)$ and $g, h \in G$, then*

   (i) $\deg(g) = \deg(g^h)$;

   (ii) $\deg(g^{-1}) = \deg(g)$;

   (iii) *for all integers n,* $\deg(g^n) \leqslant \deg(g)$;

   (iv) $\deg(gh) \leqslant \deg(g) + \deg(h)$;

   (v) $\deg[g, h] \leqslant 2\min\{\deg(g), \deg(h)\}$.

The finitely generated subgroups of a finitary linear group G play a critical role in the study of G; before noting the relevant properties, we recall some definitions in

(5.2) If H is a non-empty subset of $\mathrm{FGL}_K(V)$ and W is a non-empty subset of V, then $\langle WH \rangle$ is the K-span of all vectors of the form wh, $w \in W$, $h \in \langle H \rangle$, while $[W, H]$ is the K-span of all vectors of the form $w(h - 1)$, $w \in W$, $h \in \langle H \rangle$. Both $\langle WH \rangle$ and $[W, H]$ are $\langle H \rangle$-subspaces of V.

Returning to the finitely generated groups, we have

(5.3) **Lemma** [46] *Suppose that $G \subseteq \mathrm{FGL}_K(V)$ is generated by the finite set T. Then*

   (i) $\dim(V/C_V(G))$ *is finite;*

   (ii) $\dim[V, G] \leqslant \sum\{\dim[V, t] \mid t \in T\}$ *and so $[V, G]$ has finite dimension;*

   (iii) *if W is a finite dimensional subspace of V, $\langle WG \rangle$ is finite dimensional;*

(iv) *there is a finite dimensional G-subspace* $X = X_G$ *of* V *satisfying*

   (a) $[V, G] \subseteq X$, *and*
   (b) $V = X + C_V(X)$.

Note that G acts faithfully on any X satisfying (a) and (b).

For any finitely generated $G \subseteq \mathrm{FGL}_\mathcal{K}(V)$ the *degree* of G, $\deg(G)$, is the dimension of $[V, G]$. Since G does not always act faithfully on $[V, G]$, a different degree function (associated with a faithful representation) is occasionally useful.

(5.3.1) Lemma [31] *Let G be a finitely generated subgroup of $\mathrm{FGL}_\mathcal{K}(V)$ and define $\delta(G) = \dim(X)$ where X is a minimal subspace satisfying the properties (a) and (b) of (5.3)(iv) (such an X is not unique, but its dimension is unique). The integer $\delta(G)$ will be called the* faithful degree *of G. We note the easily verified properties*

   (i) $\deg(G) \leqslant \delta(G)$, *and*
   (ii) *if $G \subseteq H$ where H is a finitely generated subgroup of $\mathrm{FGL}_\mathcal{K}(V)$, then we have $\deg(G) \leqslant \deg(H)$ and $\delta(G) \leqslant \delta(H)$.*

The import of part (iv) of (5.3) is that the finitely generated subgroups of $\mathrm{FGL}_\mathcal{K}(V)$ are finite dimensional groups; i.e., finitary linear groups are locally finite dimensional (in a very strong sense). Applying a well known theorem of Schur [14, p. 252] we immediately arrive at the fact that *periodic finitary linear groups are locally finite.* We are also in a position to define the determinant of a $g \in \mathrm{FGL}_\mathcal{K}(V)$. For such a g, let

$$\det(g) = \det(g|_{[V,g]}),$$

and note that for any $X_{(g)}$ satisfying (iv) of (5.3) we have $\det(g) = \det(g|_{X_{(g)}})$. Of course, the usual rules of determinants carry over from the finite dimensional case. In view of this development, $\mathrm{SL}_\mathcal{K}(V) = \{g \in \mathrm{FGL}_\mathcal{K}(V) \mid \det(g) = 1\}$ is a normal subgroup of $\mathrm{FGL}_\mathcal{K}(V)$ and $\mathrm{FGL}_\mathcal{K}(V)/\mathrm{SL}_\mathcal{K}(V)$ is isomorphic with $\mathcal{K}^*$, the multiplicative group of $\mathcal{K}$. If $\dim(V)$ is infinite, $\mathrm{SL}_\mathcal{K}(V)$ is a simple group; if $\mathcal{K}$ is a locally finite field (that is, a subfield of the algebraic closure of the Galois field $\mathbb{F}_p$, p a prime), $\mathrm{SL}_\mathcal{K}(V)$ is a (uncountable) locally finite simple group.

(5.4) An additional consequence of (5.3) and (5.3.1) is a characterization of finitary linear groups in terms of abstract degree functions. In particular, if $G \subseteq \mathrm{FGL}_\mathcal{K}(V)$ and $\mathcal{S}$ is the set of finitely generated subgroups of G, we have the function $\delta : \mathcal{S} \to \mathbb{N}$ satisfying the following two conditions.

   (a) For each group $D \in \mathcal{S}$, there is a field $\mathcal{K}_D$ and a faithful $\mathcal{K}_D$-representation $\psi_D : D \to \mathrm{GL}_{\mathcal{K}_D}(X_D)$ where $\dim(X_D) = \delta(D)$.
   (b) If $D, E \in \mathcal{S}$ with $D \subseteq E$, then $\dim(D\psi_E) = \delta(D)$.

(As presented the field $\mathcal{K}_D$ is, of course, $\mathcal{K}$; the reason for the more general setup will become apparent in the next paragraph.)

It is shown by Kegel and Schmidt in [31] that any group G with a function $\delta : \mathcal{S} \to \mathbb{N}$ satisfying (a) and (b) is, in fact, a finitary linear group over some field $\mathcal{F}$.

The faithful finitary representation of such a G is constructed using ultraproducts (see §13 and the Appendix of [19]).

(5.5) The strong local linearity of finitary linear groups also allows the use of *Jordan decompositions*. Recall that, if $\mathcal{K}$ is algebraically closed, every $g \in \mathrm{GL}(n, \mathcal{K})$ has a unique factorization:

(5.5.1) $g = g_d g_n$ where g_d is diagonalizable, g_n is nilpotent and $g_d g_n = g_n g_d$ (multiplicative Jordan decomposition—see [69, Chapter 7]).

If $g \in \mathrm{FGL}(V, \mathcal{K})$, there are finitary linear transformations g_n and g_d satisfying (5.5.1) [70]; such factorizations are of great utility in studying nilpotence criteria in finitary groups.

(5.6) *Tits Alternative for Finitary Linear Groups.* The Tits alternative for subgroups of $\mathrm{GL}(n, \mathcal{K})$ asserts that, if $H \subseteq \mathrm{GL}(n, \mathcal{K})$ is finitely generated, then either H is 'solvable-by-finite' or H contains non-cyclic free subgroups [69, pp. 145]. To develop a corresponding alternative for finitary linear groups, let

$\mathcal{WF}$ be the class of finitary linear groups G that do not contain non-cyclic free subgroups ($\mathcal{WF}$ indicates 'without free').

Suppose that $G \in \mathcal{WF}$ and let $\mathcal{S}$ be the set of finitely generated subgroups of G. If $H \in \mathcal{S}$, let $S = S_H$ be the maximum solvable normal subgroup of H—this exists since H is a finite dimensional group. Since $G \in \mathcal{WF}$, S has finite index in G, and the fact that the Zariski closure of a solvable group is solvable implies that S is closed. It follows from this that H°, the connected component of 1 in H, is contained in S. Thus, H° is solvable, has finite index in H, and is triangularizable (since it is connected).

Consider $D, E \in \mathcal{S}$ with $D \subseteq E$; since $D \cap E^\circ$ is a closed subgroup of finite index in D, we have $D^\circ \subseteq E^\circ$. Thus, if $Y = \bigcup\{D^\circ \mid D \in \mathcal{S}\}$, Y is a normal, locally solvable subgroup of G with G/Y locally finite. Further, it is easy to see that Y' is locally nilpotent. We conclude that the following is true.

(5.6.1) **Theorem** *If $G \in \mathcal{WF}$, then there is a normal subgroup Y of G such that Y is locally solvable, G/Y is locally finite, and Y' is locally nilpotent (in fact, Y' is unipotent—see §6.3).*

Note the similarity between (5.6.1) and the classical results of Mal'cev-Zassenhaus [69, Chapter 3]. Arguments of the type preliminary to (5.6.1) can be found in [70] and [57]. Much of what follows will be done in the context of $\mathcal{WF}$-groups with the primary focus on the locally solvable and locally finite sections of such a group. We also note

(5.6.2) the normal subgroup Y of (5.6.1) can be chosen in such a way that G/Y is a finitary linear group.

The proof of (5.6.2) involves ideas not yet presented; more will be said on this point in §8.

(5.7) As an additional consequence of (5.3)(iv) we note that a simple basis extension argument applied on the finite dimensional level shows that for a countable $G \subseteq \mathrm{FGL}_\mathcal{K}(V)$ there is an independent set $\mathcal{B}$ of V such that $G \subseteq \mathrm{stab}(W, \mathcal{B})$, where W is the span of $\mathcal{B}$—cf. §2, Example 4.

6. General Structure; Irreducibility and Unipotence

For $G \subseteq \mathrm{FGL}_\mathcal{K}(V)$, a G-*system* of subspaces of V is a complete, linearly ordered set $\mathcal{L}$ of G-subspaces of V with $0, V \in \mathcal{L}$; a *jump* in $\mathcal{L}$ is a pair (B, T) of elements of $\mathcal{L}$ such that B is properly contained in T and no element of $\mathcal{L}$ lies strictly between B and T. The system $\mathcal{L}$ is a G-*composition system* if, for every jump (B, T) in $\mathcal{L}$, T/B is an irreducible G-module; in this case, T/B is called a G-*composition factor* of V. Standard Zorn's lemma type arguments show that G-composition systems always exist; more generally, if $\mathcal{H}$ is any chain of G-submodules of V, there is always a G-composition system $\mathcal{L}$ with $\mathcal{H} \subseteq \mathcal{L}$. A G-composition factor T/B is trivial if $[T, G] \subseteq B$. Incidentally, up to this point, these concepts all make sense for any group (or ring) of transformations on V, but we will continue to assume our group is finitary linear.

(6.1) *Jordan-Hölder theory.* A surprising Jordan-Hölder theory has been developed for finitary linear groups by Meierfrankenfeld [41] and Wehrfritz [75]. Let G be the trivial group and $\mathcal{K}$ a field. The $\mathcal{K}G$-space $V = \prod\{\mathcal{K}v_i \mid 1 \leqslant i \leqslant \infty\}$ ($\prod$ denotes the cartesian product) has dimension $2^{\aleph_0}$ and has G-composition systems $\mathcal{L}$ and $\mathcal{H}$ such that the set of jumps in $\mathcal{L}$ has cardinal $\aleph_0$ and the set of jumps in $\mathcal{H}$ has cardinal $2^{\aleph_0}$. Evidently all the composition factors are trivial G-modules. The interesting fact is that, if one ignores the trivial composition factors, the Jordan-Hölder theory carries over exactly as in the classical case. In particular,

(6.1.1) **Theorem** *Let $G \subseteq \mathrm{FGL}_\mathcal{K}(V)$ and $\mathcal{L}$ and $\mathcal{H}$ be two G-composition systems of V. Then there is a 1-1 correspondence between the non-trivial composition factors of $\mathcal{L}$ and $\mathcal{H}$ such that the corresponding factors are isomorphic as $\mathcal{K}G$-modules.*

An interesting consequence of (6.1.1) is that, if $\mathcal{L}$ is a G-composition series of V with all factors finite dimensional, then all G-composition factors of V are finite dimensional.

(6.2) *The Completely Reducible Representation* [53, pp. 421–422]. Let $\mathcal{L}$ be a G-composition system for $G \subseteq \mathrm{FGL}_\mathcal{K}(V)$ and let $\{V_i \mid i \in I\}$ be the set of G-composition factors of $\mathcal{L}$. For each $i \in I$, let $\psi_i : G \to \mathrm{GL}_\mathcal{K}(V_i)$ be the representation of G on the composition factor V_i. Then ψ_i is an irreducible and finitary representation of G; i.e., $G\psi_i \subseteq \mathrm{FGL}_\mathcal{K}(V_i)$. Further, for $g \in G$, $g\psi_i = 1$ for all but a finite number of i.

Put $G_i = G\psi_i$; then the map $\psi : G \to \mathrm{Dr}\{G_i \mid i \in I\}$ defined by $(g\psi)_i = g\psi_i$ is a finitary representation of G on the space $V = \bigoplus\{V_i \mid i \in I\}$ (with coordinatewise action). Evidently, this representation is completely reducible. Thus, modulo $\ker(\psi)$, considerable information about the structure of G is determined by the irreducible images G_i of G.

The group $U = \ker(\psi)$ acts trivially on all the composition factors V_i, a feature we can take as a defining condition of a *unipotent* group. Since unipotent groups are of considerable importance in the theory of finitary linear groups, we discuss this topic here in considerable detail.

(6.3) *Unipotent Groups.* The primary reference here is [46]; many of the same ideas appear in [4, Section 4]. An element $x \in \mathrm{FGL}_{\mathcal{K}}(V)$ is called *unipotent* if there is a non-negative integer $n = n(x)$ such that $(x - 1)^n = 0$. A subgroup S of $\mathrm{FGL}_{\mathcal{K}}(V)$ is *unipotent* if each of its elements is unipotent. The fundamental properties of unipotent groups follow in

(6.3.1) Theorem *Let $G \subseteq \mathrm{FGL}_{\mathcal{K}}(V)$.*

(i) *If G is unipotent, then G is locally nilpotent; if $\mathrm{char}(\mathcal{K}) = p > 0$, G is a locally finite p-group, and, if $\mathrm{char}(\mathcal{K}) = 0$, G is torsion free.*

(ii) *If p is a prime, $\mathrm{char}(\mathcal{K}) = p$, and G is a p-group, then G is unipotent.*

(iii) *G is unipotent if and only if V has a G-composition system with all composition factors trivial G-modules.*

(iv) *If G is irreducible and H is a normal unipotent subgroup of G, then $H = 1$.*

(v) *$\mathrm{unip}(G) = \langle H \text{ asc } G \mid H \text{ is unipotent}\rangle$ is a normal unipotent subgroup of G; thus, an irreducible subgroup of $\mathrm{FGL}_{\mathcal{K}}(V)$ has no non-trivial ascendant unipotent subgroups. Note that H asc G means that there is a well-ordered ascending series of the shape $H = H_0 \triangleleft H_1 \triangleleft \cdots \triangleleft H_\alpha \triangleleft \cdots$ where the union of all of the H_α is G.*

(vi) *If $x \in G$ and $U = \mathrm{unip}(x^G)$, then U is nilpotent of class not exceeding $2(\deg(x))$; if $U = x^G$, then U is nilpotent of class not exceeding $\deg(x)$; thus, in a unipotent group, the normal closure of every element is nilpotent.*

Note that it follows from part (iii) that the group $\ker(\psi)$ of (6.2) is a unipotent group; using (iv) of (6.3.1), it is not difficult to show that

$$\text{if } \psi \text{ is as in (6.2), then } \ker(\psi) = \mathrm{unip}(G).$$

Thus every finitary linear group is 'unipotent-by-a subdirect product of irreducible finitary groups'. Immediate refinements are possible if conditions are imposed on the field or the group. For example, if $G \subseteq \mathrm{FGL}_{\mathcal{K}}(V)$ where $\mathrm{char}(\mathcal{K}) = 0$ and G is periodic, then G has no non-trivial unipotent subgroups, and so G is a subdirect product of irreducible groups. Similar comments obtain if $\mathrm{char}(\mathcal{K}) = p > 0$ and G is a periodic p'-group.

The question of which groups have faithful, finitary unipotent representations is of interest. Condition (vi) of (6.3.1) places obvious restraints on unipotent groups, as is illustrated in

(6.3.2) Every unipotent finitary linear group is generated by its normal nilpotent subgroups; i.e., unipotent groups are Fitting groups.

The information embodied in (6.3.2) is by no means the entire story. For example, *if G is a p^∞-group, G has no faithful unipotent representation in any characteristic.* To see this, note first that, if G is a unipotent subgroup of $\mathrm{FGL}_\mathcal{K}(V)$, then $\mathrm{char}(\mathcal{K}) = p$. Further, if $[V, G, G] = 0$, an easy argument shows that G has exponent not exceeding p. Thus, there is an $x \in G$ such that $[V, x]$ is a non-trivial G-module. Since $[V, x]$ is a finite dimensional G-subspace of V, some non-trivial image of G acts faithfully and unipotently on $[V, x]$. Since all non-trivial images of G are isomorphic to G, we may assume that G acts faithfully on $[V, x]$; i.e., G is a unipotent subgroup of $\mathrm{GL}(n, \mathcal{K})$ where $n = \dim[V, x]$. However, unipotent groups in finite dimension and positive characteristic have finite exponent and this contradiction completes the argument. This observation is due to O. Puglisi. In [35, Lemma 2], the following much stronger fact is proved.

(6.3.3) Theorem *If* $\mathrm{char}(\mathcal{K}) = p > 0$, *G is a unipotent subgroup of* $\mathrm{FGL}_\mathcal{K}(V)$, *and* $x \in G$ *with* $|x| \geqslant p^2$, *then* x *has finite height in* G; *it is possible for elements of order* p *to have infinite height.*

In the finite dimensional case, unipotent groups are unitriangularizable—note that the concept of triangular requires an ordering of a basis of the vector space. Generalizations of triangular representations are discussed in the following.

(6.3.4) *Unipotent Groups as Unitriangular Groups.* In [36] it is shown that every countable unipotent group has a stable and unitriangular representation. In particular, if $G \subseteq \mathrm{FGL}_\mathcal{K}(V)$ is countable and unipotent then there is a $\mathcal{K}$-space W and an ordered basis Λ of W such that G is simultaneously stable and unitriangular with respect to Λ. It follows that G is embeddable in a McLain group (indexed by the ordered set of rational numbers).

In [35], Leinen has also taken up the study of nilpotent (and hypercentral) unipotent groups (over skew fields), while the paper [36] investigates existentially closed unipotent groups and universality properties of unipotent groups; considerable information on unipotent groups is developed in [76, Section 2] (again over skew fields).

The group $\mathrm{unip}(G)$ of part (v) of (6.3.1) will be called the *unipotent radical* of the finitary linear group G. Now that the concept of unipotence has been developed, we observe that the group Y' of (5.6.1) is locally an upper triangular group, and so is unipotent. Thus, (5.6.1) can be slightly rephrased as '*every* $\mathcal{WF}$-*group is unipotent-by-Abelian-by-locally finite*'. Versions of this type of result have been obtained by Zalesskiĭ in [82], [83] for locally solvable subgroups of the multiplicative group of locally finite algebras.

(6.4) *Extension of the Base Field.* If the group G is an irreducible subgroup of $\mathrm{FGL}_\mathcal{K}(V)$ then $\mathcal{D} = C_{\mathrm{End}_\mathcal{K}(V)}(G)$ is a division ring with $\mathcal{K}$ contained in the center of $\mathcal{D}$. Note that, for any $g \in G$, $\mathcal{D}$ acts faithfully on the finite dimensional space $[V, g]$. It follows that $\mathcal{D}$ is finite dimensional over $\mathcal{K}$. If the field $\mathcal{K}$ is algebraically closed, we obtain the same result as in the finite dimensional case.

(6.4.1) Theorem *If G is an irreducible subgroup of $\mathrm{FGL}_{\mathcal{K}}(V)$ and $\mathcal{K}$ is algebraically closed, then $\mathcal{D} = \mathcal{K}$. Here, $\mathcal{D}$ is as above.*

If $\mathcal{K} \subseteq \mathcal{F}$ are fields and $G \subseteq \mathrm{FGL}_{\mathcal{K}}(V)$, then G acts on the $\mathcal{F}$-space $V^{\mathcal{F}} = \mathcal{F} \otimes_{\mathcal{K}} V$ via linear extension of $(1 \otimes v)g = 1 \otimes vg$, $g \in G$, and with this action $G \subseteq \mathrm{FGL}_{\mathcal{F}}(V^{\mathcal{F}})$. Regarding absolute irreducibility, the criteria present in the finite dimensional case persist for finitary groups.

(6.4.2) Theorem [33] *If $G \subseteq \mathrm{FGL}_{\mathcal{K}}(V)$, the following three conditions are equivalent.*

(a) *G is absolutely irreducible.*

(b) *If $\mathcal{F}$ is the algebraic closure of $\mathcal{K}$, $V^{\mathcal{F}}$ is irreducible as an $\mathcal{F}G$-module.*

(c) *V is an irreducible $\mathcal{K}G$-module and $C_{\mathrm{End}_{\mathcal{K}}(V)}(G) = \mathcal{K}$.*

Using the results in (6.4.2), irreducible subgroups can be shifted to absolutely irreducible subgroups provided we accept a finite field extension.

(6.4.3) Theorem [33] *Suppose that G is an irreducible subgroup of $\mathrm{FGL}_{\mathcal{K}}(V)$. Then there is a finite field extension $\mathcal{F}$ of $\mathcal{K}$ such that V is absolutely irreducible as an $\mathcal{F}G$-module. The finite index $[\mathcal{F} : \mathcal{K}]$ divides the $\mathcal{K}$-dimension of every finite dimensional $\mathcal{F}$-subspace of V. In particular, $[\mathcal{F} : \mathcal{K}]$ divides $\dim_{\mathcal{K}}[V, g]$ for all $g \in G$.*

A byproduct of either (6.4.2) or (6.4.3) is that an irreducible finitary linear group over an algebraically closed field is absolutely irreducible.

We also note a very useful consequence of (6.4.2)(c).

(6.4.4) Theorem *If $G \subseteq \mathrm{FGL}_{\mathcal{K}}(V)$ is absolutely irreducible, then $\mathcal{K}G$ (that is, the $\mathcal{K}$-span of G in $\mathrm{End}_{\mathcal{K}}(V)$) is a dense ring of linear transformations. Thus, for any $n \geq 1$, if $\{v_1, \ldots, v_n\}$ is an independent subset of V and $u_1, \ldots, u_n$ are in V, then there is an $h \in \mathcal{K}G$ such that $v_i h = u_i$ for $1 \leq i \leq n$.*

7. Imprimitive Groups

For $G \subseteq \mathrm{FGL}_{\mathcal{K}}(V)$, a *$G$-system of imprimitivity* of V is a set of subspaces V_i of V such that $V = \bigoplus \{V_i \mid i \in I\}$ and, for all $g \in G$, $V_i g = V_j$ for some $j \in I$. If $\mathcal{P} = \{V_i \mid i \in I\}$ is a G-system of imprimitivity then $\pi : G \to \mathrm{Sym}(\mathcal{P})$ defined by $V_i(g\pi) = V_i g$ is a permutation representation of G. In fact,

(7.1) π maps G into the finitary permutations $F\mathrm{Sym}(\mathcal{P})$, and, if the space V_i is moved by any element of G, $\dim(V_i)$ must be finite.

To obtain the proof of (7.1), as well as other results, note that, if $V_{i_0} \in \mathrm{supp}(g\pi)$, then $\dim(V_{i_0}(g - 1)) = \dim(V_{i_0})$. Thus,

(7.2) $\dim(V_{i_0}) \leqslant \deg(g)$; in particular, $\dim(V_{i_0})$ is finite.

(7.3) For any g-orbit Ω of $\mathcal{P}$, let n_Ω be the common dimension of the spaces $V_i \in \Omega$. It is not difficult to show that

$$\sum \{n_\Omega(|\Omega| - 1) \mid \Omega \text{ is a } g\text{-orbit}\} \leqslant \deg(g).$$

The proof of (7.1) follows easily from these observations. In the special case where G is irreducible, $G\pi$ must be transitive; we collect the above results specialized to the irreducible case in

(7.4) **Theorem** *Suppose $G \subseteq \mathrm{FGL}_K(V)$ is irreducible, $\mathcal{P} = \{V_i \mid i \in I\}$ is a G-system of imprimitivity, and π is the permutation representation of G into $\mathrm{Sym}(\mathcal{P})$. Then*

(i) *$G\pi$ is a transitive subgroup of $\mathrm{FSym}(\mathcal{P})$;*
(ii) *the V_i are all finite dimensional (of the same dimension);*
(iii) *$\ker(\pi)$ is a subdirect product of finite dimensional groups and $G/\ker(\pi)$ is a transitive group of finitary permutations.*

As with finite dimensional groups, systems of imprimitivity arise when studying reducible normal subgroups of irreducible groups. The salient features of this Clifford theory are noted in

(7.5) **Theorem** [53, pp. 423–425] *If G is an irreducible subgroup of $\mathrm{FGL}_K(V)$ where $\dim(V)$ is infinite, $1 \neq H \lhd G$, and $D \neq V$ is an irreducible H-submodule of V, then the set $\mathcal{D} = \{D_i \mid i \in I\}$ of H-homogeneous components determined by D is a G-system of imprimitivity of V. Further, each D_i is finite dimensional (and, in particular, D is finite dimensional) and, if π denotes the representation of G on $\mathcal{D}$, then $HC_G(H) \subseteq \ker(\pi)$.*

Note that the irreducibility of G implies that D is a non-trivial H-module. Also each D_i is a direct sum of isomorphic (non-trivial) H-modules of the form Dg, $g \in G$. If $D_i = \bigoplus \{D_{i,j} \mid j \in J\}$ where the $D_{i,j}$ are H-isomorphic with D, and $[D_i, h] \neq 1$ for some $h \in H$, then $1 \neq \bigoplus \{[D_{i,j}, h] \mid j \in J\}$ and, since the $D_{i,j}$ are all H-isomorphic, we must have $[D_{i,j}, h] \neq 1$ for all $j \in J$. It follows that the set J is finite. To see that D must be finite dimensional note that, if $g \in G$ is such that $Dg \neq D$, then $Dg + D = Dg \oplus D$ (since both are irreducible H-modules) and it now follows that $\dim(D(g - 1)) = \dim(D)$; thus, $\dim(D) \leqslant \deg(g)$ which implies that $\dim(D)$ is finite. The remaining parts of (7.5) follow as in the finite dimensional case (with liberal use of Zorn's lemma).

A remarkable strengthening of (7.5) is due to U. Meierfrankenfeld [41] (for a special case see [53, p. 427]).

(7.6) **Theorem** *If G is an irreducible subgroup of $\mathrm{FGL}_K(V)$ and H is an ascendant subgroup of G, then V contains an irreducible H-submodule D. Moreover, if W is an H-submodule of V, D can be chosen so that $D \subseteq W$. It follows from (7.5) that, if H is reducible, then V contains a finite dimensional irreducible H-submodule.*

This result is by no means obvious, and it illustrates some of the difficulties encountered in infinite dimensional representation theory that do not arise in the finite dimensional case. The content of (7.6) gives a partial answer to the more general question:

(7.7) If $G \subseteq \mathrm{FGL}_{\mathcal{K}}(V)$, does V contain an irreducible G-submodule?

The answer to (7.7) is, in general, negative; an easy example is the following McLain type group.

(7.7.1) Let $\mathcal{K}$ be the field of p elements (p a prime) and V be the $\mathcal{K}$ space with basis $\{v_n \mid n \in \mathbb{Z}\}$. Let G be the subgroup of $\mathrm{FGL}_{\mathcal{K}}(V)$ generated by the transvections $\{\tau_{m,n} \mid m < n\}$ where

$$\tau_{m,n} : v_m \mapsto v_n + v_m \text{ and } \tau_{m,n} : v_t \mapsto v_t \text{ if } t \neq m.$$

Suppose that V contains an irreducible G-submodule W. Since the group G is unipotent, G acts trivially on W. Thus, for every $v \in W$ and $g \in G$, $vg = v$. Write $v = a_1 v_{n_1} + a_2 v_{n_2} + \cdots + a_t v_{n_t}$ where $n_1 < n_2 < \cdots < n_t$ and the a_i are non-zero elements of the field $\mathcal{K}$. Choose $s \in \mathbb{Z}$ with $s > n_t$; an easy computation shows that $v\tau_{n_1,s} \neq v$. From this contradiction, conclude that V contains no irreducible submodule W.

A more elaborate discussion of the question (7.7) will be presented in §12. Note that Meierfrankenfeld's theorem gives an affirmative answer to (7.7) for those groups which are ascendant subgroups of irreducible groups.

There are variants of Clifford's theorem which do not require that G be irreducible and this approach can occasionally be used to answer questions of the type (7.7).

(7.8) **Theorem** [53, pp. 423–425] *Suppose that $G \subseteq \mathrm{FGL}_{\mathcal{K}}(V)$, $H \lhd G$, and D is a non-trivial irreducible H-submodule of V. Let $L = \langle DG \rangle$ be the G-submodule generated by D. Then L decomposes as in (7.5) into a system of G-imprimitivity. Further, L has both the acc and dcc on G-submodules. Thus, V contains an irreducible G-submodule.*

8. Applications of §7 to Some Classes of Groups—Locally Solvable Groups

The techniques of the previous section can be used to obtain precise information about the finitary representations of certain classes of groups. One striking feature that will emerge is the paucity of primitive irreducible representations of $\mathcal{WF}$-groups in infinite dimensions; as usual, a *primitive* representation is an irreducible representation that is not imprimitive. As an illustration of this phenomenon, we note (and discuss in detail later) that, if $G \subseteq \mathrm{FGL}_{\mathcal{K}}(V)$ is a primitive $\mathcal{WF}$-group, $\dim(V)$ is infinite, and $\mathrm{char}(K) = 0$, then G is isomorphic to either $\mathrm{FSym}(\Omega)$ or $\mathrm{Alt}(\Omega)$ for some infinite set Ω.

A starting point for this development is a determination of some groups, all of whose irreducible finitary representations are finite dimensional. Rather large classes of such groups are discussed in [11] (at one point in [11], critical use is made of [21]).

(8.1) Theorem *Suppose that G has a subnormal series $1 = H_0 \lhd H_1 \lhd \cdots \lhd H_n = G$ where, for $1 \leqslant i \leqslant n$, H_i/H_{i-1} is either abelian, finite, or a finite dimensional locally finite simple group of Lie type. Then*

(i) *every transitive finitary permutation representation of G has finite degree;*

(ii) *every irreducible finitary linear representation is finite dimensional;*

(iii) *if $G \subseteq \mathrm{FGL}_\kappa(V)$ then V contains an irreducible G-submodule.*

Further, every subdirect product of groups satisfying the defining conditions above also satisfies the conclusions (i), (ii) *and* (iii).

It may be instructive to present the proof of a relatively easy case of (8.1), and we do this in

(8.1.1) Theorem *If G is a solvable group, then every irreducible finitary linear representation of G has finite dimension* (*cf.* (4.7)).

Proof. Since every image of G is solvable, we may assume that G is an irreducible subgroup of $\mathrm{FGL}_\kappa(V)$. We use induction on the derived length of G. If G is Abelian and $x \in G$, then $[V, x]$ is a finite dimensional G-submodule of V and this forces $\dim(V) < \infty$. For the general case, suppose that $\dim(V)$ is infinite; by induction, G' is a reducible normal subgroup of G, and (7.6) implies that V contains an irreducible G'-submodule D. It follows from (7.4) and (7.5) that G/G' has a transitive, finitary, permutation representation on the system of imprimitivity $\mathcal{D} = \{D_i \mid i \in I\}$ determined by D and G'. Since the finitary transitive representations of G/G' have finite degree (see (4.7)) the set $\mathcal{D}$ must be finite. Since the D_i are finite dimensional, $\dim(V)$ is finite, and this completes the proof.

We note that the existence of the irreducible G'-submodule D can be proved by induction, thus avoiding the appeal to (7.6), and this approach may be used in all cases of (8.1). We also point out that the class of groups defined in (8.1) includes the class of *periodic finite dimensional groups* [11, p. 384].

(8.2) *Locally Solvable Groups.* The next layer of results pertain to the representation theory of locally solvable groups. One of the basic theorems in this theory is

(8.2.1) Theorem [46] *If $G \subseteq \mathrm{FGL}_\kappa(V)$ and $x \in G$ is such that x^G is locally solvable, then x^G is solvable and there is a function $\mu : \mathbb{N} \to \mathbb{N}$ such that the derived length of x^G is bounded by $\mu(\deg(x))$.*

The following (8.2.2), (8.2.3) and (8.2.5) give additional information about the locally solvable subgroups of a finitary linear group; the first requires only local linearity, while the other two depend on (8.2.1).

(8.2.2) Theorem *Let $G \subseteq \mathrm{FGL}_\kappa(V)$. Then*

(a) $\mathrm{ls}(G) = \langle H \lhd G \mid H$ *is locally solvable*$\rangle$ *is a locally solvable normal subgroup of G; here $\mathrm{ls}(G)$ is called the* locally solvable radical *of G.*

(b) *If $H \lhd G$ and both H and G/H are locally solvable, then G is locally solvable.*

(c) $G/\mathrm{ls}(G)$ *has only the trivial locally solvable normal subgroup.*

In general, abstract groups need not have a unique maximal locally solvable normal subgroup ([23] or [61, p. 91]), nor is the class of locally solvable groups closed under extensions. For the proof of part (a) of (8.2.2) it suffices to show that if H and K are two normal locally solvable subgroups of G then HK is also locally solvable. To this end, if $T = \langle h_1 k_1, \ldots, h_n k_n \rangle$ is a finitely generated subgroup of HK then

$$T \subseteq \langle h_1, \ldots, h_n, k_1, \ldots, k_n \rangle = \langle h_1, \ldots, h_n \rangle^{\langle k_1, \ldots, k_n \rangle} \langle k_1, \ldots, k_n \rangle = S.$$

The group S is finitely generated and so is finite dimensional (by (5.3)). Since $Y = \langle h_1, \ldots, h_n \rangle^{\langle k_1, \ldots, k_n \rangle}$ is locally solvable, Y must be solvable and it follows that S is an extension of a solvable group by a solvable group and so is solvable. For part (b), suppose that both H and G/H are locally solvable and let X be a finitely generated subgroup of G. Then X is a finite dimensional group and so the locally solvable $H \cap X$ must be solvable. Since $X/H \cap X$ is solvable, X must be solvable. Part (c) is an easy consequence of the other two parts.

An additional application of (8.2.1) considerably reduces the scope of primitive groups.

(8.2.3) Theorem *Let* $G \subseteq \mathrm{FGL}_{\mathcal{K}}(V)$.

(a) *If* $\mathrm{ls}(G) \neq 1$, G *is irreducible, and* $\dim(V)$ *is infinite, then* G *is imprimitive.*

(b) *Every infinite dimensional, primitive* $\mathcal{WF}$-*group* G *is locally finite and has no non-trivial locally solvable normal subgroups.*

Proof. Let G be an irreducible subgroup of $\mathrm{FGL}_{\mathcal{K}}(V)$ where $\dim(V)$ is infinite. Suppose that $Y = \mathrm{ls}(G) \neq 1$ and let $1 \neq y \in Y$. Then (8.2.1) implies that $A = y^G$ is solvable and (8.1) shows that A must be reducible. The imprimitivity of G now follows from (7.5) and (7.6), and part (a) follows. Part (b) is easily obtained from part (a) and (5.6.1).

The introductory remarks of this section regarding the primitive groups in characteristic 0 can now be put in a more precise context.

(8.2.4) Theorem *If* G *is a primitive* $\mathcal{WF}$-*group in* $\mathrm{FGL}_{\mathcal{K}}(V)$ *where* $\dim(V)$ *is infinite and* $\mathrm{char}(\mathcal{K}) = 0$, *then* G *is either* $\mathrm{FSym}(\Omega)$ *or* $\mathrm{Alt}(\Omega)$ *for some infinite set* Ω.

Proof. From (8.2.3) we see that G is locally finite. Furthermore, it has been shown in [53, Proposition 3] and [17] that an infinite dimensional primitive group in characteristic 0 is either $\mathrm{FSym}(\Omega)$ or $\mathrm{Alt}(\Omega)$. More precisely, J. Hall proved in [17] that, if $\mathrm{char}(\mathcal{K}) = 0$, $\dim(V)$ is infinite, and G is a periodic simple subgroup of $\mathrm{FGL}_{\mathcal{K}}(V)$, then there is an infinite set Ω such that G is isomorphic to $\mathrm{Alt}(\Omega)$; further, V is the natural module $[\mathcal{K}\Omega, G]$. In [53] it is shown that, if $\mathrm{char}(\mathcal{K}) = 0$, $\dim(V)$ is infinite, and G is a periodic irreducible primitive subgroup of $\mathrm{FGL}_{\mathcal{K}}(V)$, then the derived group G' is infinite and simple. Thus, by Hall's result, $G' \cong \mathrm{Alt}(\Omega)$; it is also proved in [53] that if $G \neq G'$ we must have $G \cong \mathrm{FSym}(\Omega)$ and that, again, the module V must be the natural module.

In the proof of (8.2.4), we note that both Hall and Phillips use the classification of finite simple groups to prove the relevant results. Thus, considerably more machinery is used than in the proof of the permutation group analogue (4.1.1). We remark also that, in the paper [2], Belyaev has arrived at the same result with different methods.

There are several additional properties of finitary locally solvable groups which can be easily deduced from (8.2.1).

(8.2.5) Theorem *Let G be a locally solvable and irreducible subgroup of $\mathrm{FGL}_{\mathcal{K}}(V)$ and suppose that $\dim(V)$ is infinite. Then G is imprimitive, $\dim(V) = \aleph_0$, and $|G| \leqslant \max\{|\mathcal{K}|, \aleph_0\}$. Thus, in uncountable dimensions, locally solvable groups must be reducible.*

Proof. The imprimitivity of G follows from the lines of argument given in (8.2.3). If $\mathcal{D}$ is a G-system of imprimitivity of V, and π is the permutation representation of G on $\mathcal{D}$, then $R = G/\ker(\pi)$ is a transitive group of finitary permutations (by (7.4) and (7.5)). Since R contains no infinite alternating sections, R must be totally imprimitive (see (4.2)). Thus, R is countable and $\mathcal{D}$ is countable. Since $\mathcal{D}$ consists of finite dimensional subspaces whose sum is V, V has countably infinite dimension. Now $\ker(\pi)$ is a subdirect product of finite dimensional solvable groups. If the field $\mathcal{K}$ is infinite, the order of the countable subdirect product can not exceed $|\mathcal{K}|$. If $\mathcal{K}$ is a finite field, the order of $\ker(\pi)$ is bounded by $\aleph_0$. In all cases, $|G| \leqslant \max\{|\mathcal{K}|, \aleph_0\}$.

There are other large classes of groups G all of whose finitary representations are imprimitive, notably periodic p'-groups over fields of characteristic $p > 0$ [53, Theorem A]. Further discussion of these groups will be taken up in §10.

(8.3) *Irreducible Normal Subgroups in Imprimitive Groups.* Let G be an imprimitive subgroup of $\mathrm{FGL}_{\mathcal{K}}(V)$ where $\dim(V)$ is infinite, and suppose that H is an irreducible normal subgroup of G. Using ideas borrowed from the permutation group situation, it is not difficult to show that G has the conjugacy centralizer property with respect to H. We now use (4.2.2) to conclude that $G' \subseteq H$. A little additional argument establishes

(8.3.1) Theorem *If $G \subseteq \mathrm{FGL}_{\mathcal{K}}(V)$ is imprimitive and $\dim(V)$ is infinite, then G' is the minimal subnormal irreducible subgroup of G. Further, if $T \lhd G'$, then $T \lhd G$.*

An interesting consequence of the results of this section is

(8.4) Corollary *If $G \subseteq \mathrm{FGL}_{\mathcal{K}}(V)$ is irreducible, then either G is locally solvable or $\mathrm{ls}(G)$ is solvable.*

Proof. If $\dim(V)$ is finite, $\mathrm{ls}(G)$ is solvable, so from this point on we assume that $\dim(V)$ is infinite. Let $Y = \mathrm{ls}(G)$; if $Y = 1$, there is nothing to prove. If $Y \neq 1$,

(8.2.5) implies that G is imprimitive. If Y is irreducible, use (8.3.1) and conclude that $G' \subseteq Y$. We now apply (8.2.2)(b) and deduce that G is locally solvable. If Y is reducible, then $Y \subseteq \ker(\pi)$ where π is the permutation representation of G on the system of imprimitivity forced by Y. Since $\ker(\pi)$ is a subdirect product of linear groups of fixed degree, Y must be solvable.

We now have sufficient machinery to sketch a proof of (5.6.2). The argument is due to U. Meierfrankenfeld.

(8.5) Theorem   *Let $G \subseteq \mathrm{FGL}_{\mathcal{K}}(V)$ and $Y = \mathrm{ls}(G)$; then G/Y has a faithful $\mathcal{K}$-finitary representation.*

Proof.   Suppose first that G is irreducible. If $\dim(V)$ is finite then Y is a closed subgroup of G and so G/Y has a faithful finite dimensional $\mathcal{K}$-representation. We assume now that $\dim(V)$ is infinite. Now apply (8.4) and deduce that Y is solvable. Since Y is reducible, (7.5) and (7.6) imply that V has a system of imprimitivity $\mathcal{D} = \{D_i \mid i \in I\}$ of Y-homogeneous components determined by an irreducible Y-module D. Recall that the subspaces D_i are finite dimensional and that, if π is the permutation representation of G on $\mathcal{D}$, $\ker(\pi)$ fixes each of the subspaces D_i.

Fix $i \in I$, put $W = D_i$, and define $N = N_G(W)$ and $C = C_G(W)$; then N/C is isomorphic to a subgroup of $\mathrm{GL}_{\mathcal{K}}(W)$; i.e., there is a finite dimensional representation $\mu : N \to \mathrm{GL}_{\mathcal{K}}(W)$ with $\ker(\mu) = C$. Note that the induced representation μ^G is equivalent to the representation of G on V and so μ^G is a finitary representation. Let S/C be the maximum solvable normal subgroup of the finite dimensional group N/C. Since S/C is closed in N/C, $(N/C)/(S/C)$ has a faithful finite dimensional $\mathcal{K}$-representation. It follows that there is a finite dimensional $\mathcal{K}$-space M and a representation $\psi : N \to \mathrm{GL}_{\mathcal{K}}(M)$ with $\ker(\psi) = S$. The representation of interest is the induced representation $\alpha = \psi^G : G \to \mathrm{GL}_{\mathcal{K}}(A)$, where A is the induced module $A = M^G$ (see [14, pp. 73–76] for relevant concepts). The representation α is actually a finitary representation. We will not prove this here, but we note that this is forced by the fact that μ^G is a finitary representation.

The remaining part of the proof is to show that $\ker(\alpha) = Y$. We know that $\ker(\alpha) = \bigcap\{S^x \mid x \in G\}$ and the group $E = \bigcap\{YS^x \mid x \in G\}$ has solvable action on every D_i, $i \in I$. It follows that E is a subdirect product of solvable groups whose derived lengths are bounded by an integer d. Thus, E is solvable and the maximality of Y implies that $E = Y$. Since $Y(\bigcap\{S^x \mid x \in G\}) \subseteq E$ we have $Y = \bigcap\{S^x \mid x \in G\} = \ker(\alpha)$, as desired.

Passage from the irreducible case to the general case is not difficult, and we do not present this part of the proof here.

In [70], Wehrfritz takes up the study of generalized nilpotent subgroups of $\mathrm{FGL}_{\mathcal{K}}(V)$ with particular emphasis on the relationships between such subgroups as the Fitting subgroup, the Hirsch-Plotkin radical, the set of right Engel elements, the set of left Engel elements, and the hypercenter. These results extend the work in Chapter 8 of [69]. For a sampling of the results of [70] we note

(8.6)   If $G \subseteq \mathrm{FGL}_{\mathcal{K}}(V)$ and $U = \mathrm{unip}(G)$, then, modulo U, G has upper central height at most $\omega 2$.

As a final remark in this section, we mention a nice result of Belyaev on locally finite p-groups.

(8.7) Theorem [4, p. 277] *Let p be a prime and G be an irreducible p-subgroup of $\mathrm{FGL}_{\mathcal{K}}(V)$, where $p \neq \mathrm{char}(\mathcal{K})$. Then there is a normal abelian subgroup A of G such that G/A is isomorphic to a group of finitary permutations.*

The approach to proving (8.7) is to let A be a maximal Abelian normal subgroup of G (which is non-trivial by (8.2.1)) and then show that $A = \ker(\pi)$ where π is the permutation representation on the relevant system of imprimitivity.

Locally nilpotent groups of finitary permutations are discussed in [27], [28], and [67].

9. Groups Generated by Elements of Small Degree

Both of the results (6.3.1)(vi) and (8.2.1) suggest that subgroups of the form x^G, $x \in G \subseteq \mathrm{FGL}_{\mathcal{K}}(V)$, have much in common with finite dimensional groups. We pursue this idea by first considering finite dimensional groups generated by elements of bounded degrees.

(9.1) Theorem [54] *Let t be a positive integer and G an irreducible $\mathcal{WF}$-subgroup of $\mathrm{GL}(n, \mathcal{K})$ generated by a set X of elements of degree t or less.*

  (i) *If G is primitive and $n > 4t^2$, then G has a unique component H, H is normal in G, V is an irreducible H-module, and $C_G(H) \subseteq \zeta(G)$.*

  (ii) *If G is imprimitive and $n > 6t^2$, then there is an integer $e > 6t$ and a normal subgroup D of G such that $\mathrm{Alt}_e \subseteq G/D \subseteq \mathrm{Sym}_e$ and D is a subdirect product of subgroups of $\mathrm{GL}(t, \mathcal{K})$.*

Observe that a consequence of (9.1) is that there is a bound on the linear degrees of irreducible solvable groups generated by elements of small degree. Also, the hypothesis that G is a $\mathcal{WF}$-group can not be dropped since non-cyclic irreducible free subgroups of $\mathrm{GL}(n, \mathcal{K})$ can be generated by elements of degree 1 (see §2).

An obvious example of a subgroup generated by elements of fixed degree is a group of the form x^G. Using (9.1) together with (6.3.1)(vi), the CFSG (that is, the classification of finite simple groups), and the fact that solvable images of periodic linear groups have derived lengths bounded by the linear degree of the group, we are able to establish

(9.1.1) Theorem *If G is a $\mathcal{WF}$-group contained in $\mathrm{GL}(n, \mathcal{K})$ and $x \in G$, there is a function $\tau : \mathbb{N} \to \mathbb{N}$ such that every locally solvable subnormal section of x^G is solvable of derived length bounded by $\tau(\deg(x))$—here a subnormal section is a quotient group H/K where H is a subnormal subgroup of G.*

It follows that derived lengths of images of x^G and the derived lengths of solvable normal subgroups of x^G are bounded by a function of the degree of x (and independent of n). In particular, if x^G is solvable, its derived length is bounded by a function of the degree of x. This fact enables us to push up this result to finitary linear groups; we illustrate this in the following theorem.

(9.2) Theorem *If G is a $\mathcal{WF}$-group in $\mathrm{FGL}_{\mathcal{K}}(V)$, and $x \in G$, then there is a function $\mu : \mathbb{N} \to \mathbb{N}$ such that every locally solvable section of x^G is solvable of derived length bounded by $\mu(\deg(x))$. In particular, $\mathrm{ls}(x^G)$ is a solvable group and every locally solvable image of x^G is solvable.*

Note that (9.2) provides a generalization of (8.2.1). To relate the importance of the types of bounds appearing in (9.1.1) we present the proof of (9.2).

Proof of (9.2). Let G be as in (9.2), $x \in G$, $Y = x^G$ and let $\mathcal{S}$ be the set of finitely generated subgroups H of G with $x \in H$. Suppose that T is a subnormal subgroup of Y, and that $S \triangleleft T$ with T/S locally solvable. As noted in (5.3), $H \in \mathcal{S}$ implies that H is a finite dimensional group; thus, if $R = x^H$, $(T \cap R)/(S \cap R)$ is a locally solvable subnormal section of the finite dimensional group R. From (9.1.1) we deduce that $(T \cap R)^{\tau(\deg(x))} \subseteq S \cap R$. Here, the exponent denotes the appropriate term of the derived series of $T \cap R$. Since the groups $\{T \cap R = T \cap x^H \mid H \in \mathcal{S}\}$ form a local system of T, we see that $T^{\tau(\deg(x))} \subseteq S$. Thus, T/S is solvable of derived length bounded by $\tau(\deg(x))$.

If $G \subseteq \mathrm{FGL}_{\mathcal{K}}(V)$ and $x \in G$, put $H = x^G$ and let $\mathcal{L}$ be a G-composition series for V. Since $\deg(x)$ is finite, there are only a finite number of G-composition factors V_i (at most $\deg(x)$ of them) of $\mathcal{L}$ on which H acts non-trivially. Moreover,

(9.3) [46, p. 36] there is a finite G-series $0 = V_0 \subseteq V_1 \subseteq V_2 \subseteq \cdots \subseteq V_n = V$ such that the factors are either irreducible G-modules or trivial H-modules. These two types of factors are interlaced in the series. It follows that $H/\mathrm{unip}(H)$ acts on a direct sum of a finite number of irreducible G-spaces. If $H = G$ (in particular, if G is simple), then there are only a finite number of isomorphism types of $\mathcal{K}G$-modules that occur as sections of V.

The Jordan-Hölder theorem (6.1) is being used in the last assertion of (9.3). The result (9.1) is similar in spirit to many of the theorems in [21]. In the same paper, the authors also identify the modules for some of the relevant groups.

10. Locally Finite Groups

Perhaps the most decisive results to date in the theory of finitary linear groups are in the area of locally finite groups—recall that periodic and locally finite mean the same thing in the context of finitary linear groups. Especially noteworthy is the manner in which the development of locally finite, finitary linear simple groups has affected the general theory of locally finite simple groups (see [24], [42], [19] in these Proceedings and [3]). Nearly all of the results in this area use CFSG. The initial breakthrough was proved by J. I. Hall in [17]; this result has been mentioned earlier (cf. the introductory remarks in §8), but is important enough to merit further commentary.

(10.1) Theorem *Suppose G is a locally finite simple subgroup of $\mathrm{FGL}_{\mathcal{K}}(V)$ where* $\mathrm{char}(\mathcal{K}) = 0$ *and* $\dim(V)$ *is infinite. Then G is an infinite alternating group.*

In the same paper, the possible finitary representations of an infinite alternating group are determined (see also [2]). Substantial information from the representation theory of finite groups of Lie type is used in the various proofs found in [17]. Using many of Hall's ideas, the author established in [53] the following result.

(10.2) Theorem *If G is an irreducible, primitive subgroup of $\mathrm{FGL}_{\mathcal{K}}(V)$ where* $\mathrm{char}(\mathcal{K}) = 0$ *and* $\dim(V)$ *is infinite, then G is either an infinite alternating group or an infinite finitary symmetric group. Further, the irreducible module V must be the natural permutation module.*

As noted earlier, these results provide a complete list of all irreducible, primitive $\mathcal{WF}$-groups in characteristic 0. In [53], a study of locally finite p'-groups in characteristic $p > 0$ is also undertaken. The primary result here is

(10.3) Theorem *If G is a locally finite irreducible p'-subgroup of $\mathrm{FGL}_{\mathcal{K}}(V)$ where* $\mathrm{char}(\mathcal{K}) = p > 0$, *then V is an imprimitive G-module.*

The simple, locally finite, finitary groups in prime characteristic have been determined by J. I. Hall [19], and will be discussed in detail in another entry in these Proceedings. Prior to discussing this theorem, we note some of the locally finite simple subgroups of $\mathrm{FGL}_{\mathcal{K}}(V)$. For the rest of this paragraph, $\mathcal{K}$ will be a locally finite field of characteristic $p > 0$ and the dimension of V will be infinite. The simple locally finite groups in the finite dimensional case have been determined for some time, and we will not discuss them here (see [19] and [24]). The alternating group has been discussed earlier; it appears as a locally finite simple subgroup of $\mathrm{FGL}_{\mathcal{K}}(V)$ in any characteristic. We have also observed earlier (see §5) that $\mathrm{SL}_{\mathcal{K}}(V)$ is a locally finite simple group. If a, h, q are respectively non-degenerate alternating, Hermitian, and quadratic forms on V, the corresponding finitary symmetry groups are denoted $\mathrm{FSp}_{\mathcal{K}}(V, a)$, $\mathrm{FU}_{\mathcal{K}}(V, h)$, and $\mathrm{FO}_{\mathcal{K}}(V, q)$ (the existence of h requires a quadratic extension of $\mathcal{K}$). In all three cases, the derived groups are simple. Of course, the isomorphism types of these groups will generally depend on the similarity class of the form. We note that, in uncountable dimensions, there are a multitude of inequivalent alternating forms [20]. The finitary symplectic groups have been studied in [66]; several of the types of finitary classical groups are discussed in [13]. An additional class of simple groups is obtained from special transvection groups. If $\alpha \in V^*$ (the dual space of V) and $x \in \ker(\alpha)$, the map $t(\alpha, x) \in \mathrm{GL}_{\mathcal{K}}(V)$ defined by $t(\alpha, x) : v \mapsto v + ((v)\alpha)x$ is a finitary transformation on V, and, if $W \subseteq V^*$, $T(W, V) = \langle t(\alpha, x) \mid \alpha \in W, x \in \ker(\alpha) \rangle$ is a locally finite subgroup of $\mathrm{FGL}_{\mathcal{K}}(V)$. In general, $T(W, V)$ is not a simple group (unipotent groups are frequently of this shape), but, if W is a sufficiently large subspace of V^*, the group will be simple. A sufficient condition for simplicity of $T(W, V)$ is $\mathrm{ann}_V(W) = \{v \in V \mid (v)\alpha = 0 \text{ for all } \alpha \in W\} = 0$. An informative discussion of these transvection groups can be found in [13]. The following theorem shows that the groups discussed above form a complete list of locally finite, simple subgroups of $\mathrm{FGL}_\mathcal{K}(V)$.

(10.4) Theorem [19] *Let G be a locally finite, infinite, non-'finite dimensional', simple subgroup of $\mathrm{FGL}_\mathcal{K}(V)$, where $\dim(V)$ is infinite. Then G is either alternating, the derived subgroup of $\mathrm{FSp}_\mathcal{K}(V, a)$, $\mathrm{FU}_\mathcal{K}(V, h)$, or $\mathrm{FO}_\mathcal{K}(V, q)$, or a group of type $T(W, V)$, where $W \subseteq V^*$ and $\mathrm{ann}_V(W) = 0$.*

The list of locally finite, infinite, finitary linear, simple groups is very concise and manageable. The possible finitary modules for the classical groups in (10.4) have been determined by Meierfrankenfeld in [43]. This classification of the simple groups (and their finitary modules) together with general structure theorems for locally finite, finitary groups (later in this section) provide the framework for a theory structured very much like locally finite, finite dimensional groups.

Many of the results of this and prior sections can be put in a more restrictive setting with the aid of a theorem of Leinen [34].

(10.5) Theorem *Let $\mathcal{K}$ be an algebraically closed field and G a locally finite, irreducible subgroup of $\mathrm{FGL}_\mathcal{K}(V)$. Let $\mathcal{P}$ be the subfield of $\mathcal{K}$ generated by the m-th roots of 1 for all positive integers m that occur as orders of elements of G. Then G is equivalent to a finitary representation over the field $\mathcal{P}$; i.e., there is an $x \in \mathrm{GL}_\mathcal{K}(V)$ such that $G^x \subseteq \mathrm{FGL}_\mathcal{P}(V)$ (which means that there is a basis Λ of V such that with respect to the basis Λ the matrix entries of the elements of G lie in the subfield $\mathcal{P}$). If $\mathcal{K}$ has positive characteristic, the field $\mathcal{P}$ is locally finite; thus, in this case, an irreducible, locally finite, finitary linear group can be represented over a locally finite field.*

The result (10.5) generalizes to locally finite, finitary linear groups a well known theorem of Winter for finite dimensional groups and Brauer for finite groups (see [69, p. 113]). It provides the additional lever (for absolutely irreducible locally finite groups) of working over a countable field. Immediate improvement of some of our prior results is possible.

(10.6) Theorem [34, Theorem D] *If $G \subseteq \mathrm{FGL}_\mathcal{K}(V)$ is an irreducible, locally finite, locally solvable group, then G is countable (compare with (8.2.5)).*

Without giving all the details, the method of proof is as follows. First, we may assume the field $\mathcal{K}$ is algebraically closed; (6.4.2) and (6.4.3) assert that this is possible. The finite dimensional version can be easily handled by Winter's theorem cited above. In the general case, we may assume that V is an imprimitive G-module of countably infinite dimension. Let $\{D(\omega) \mid \omega \in \Omega\}$ be the system of imprimitivity. If N is the stabilizer of the system of imprimitivity, $G/N = Y$ is a countably infinite group of finitary permutations on Ω. The group G is embedded in a wreath product $S = H \, \mathrm{Wr}_\Omega(H, \Omega) \subseteq \mathrm{FGL}_\mathcal{K}(V)$ of the type discussed in (iii) of §2. The base group $B = \mathrm{Dr}\{H(\omega) \mid \omega \in \Omega\}$ acts on $V = \bigoplus\{D(\omega) \mid \omega \in \Omega\}$ coordinatewise and Y permutes the direct factors according to the action of Y on Ω. The finite dimensional groups $H(\omega)$ act irreducibly on $D(\omega)$ and so there is a basis $\mathcal{B}(\omega)$ of $D(\omega)$ such that the respective matrix entries of $H(\omega)$ are in the field $\mathcal{P}$ of (10.5). The union of the bases $\mathcal{B}(\omega)$, $\omega \in \Omega$, provides a basis $\mathcal{B}$ of V which is stable for S. Relative to $\mathcal{B}$, the matrix entries are in the countable field $\mathcal{P}$. Since Ω is a countable set, S is countable and it follows that G is countable.

Note that we have used only the finite dimensional version of (10.5). The foregoing type of argument can be applied to any imprimitive group to obtain results of the following type.

(10.7) Theorem [34]

   (a) *If $G \subseteq \mathrm{FGL}_{\mathcal{K}}(V)$ is locally finite, irreducible, and imprimitive, then G has a stable representation in $\mathrm{FGL}_{\mathcal{P}}(V)$ where $\mathcal{P}$ is as in (10.5). Thus*

$$|G| \leqslant \max\{\aleph_0, \dim(V)\}.$$

   (b) *If $G \subseteq \mathrm{FGL}_{\mathcal{K}}(V)$ is a locally finite p'-group and $\mathrm{char}(\mathcal{K}) = p$, then*

$$|G| \leqslant \max\{\aleph_0, \dim(V)\}.$$

If G is irreducible, G is countable.

In contrast to (10.7), primitive locally finite groups in countable dimensions need not be countable; for example, if $\mathcal{K}$ is an infinite locally finite field, and V is a $\mathcal{K}$-space of dimension $\aleph_0$, $\mathrm{FGL}_{\mathcal{K}}(V)$ is an uncountable, primitive, locally finite group. However, Leinen's work does provide bounds (in terms of $\dim(V)$) on the orders of all finitary linear, locally finite groups.

(10.8) Theorem [34] *If $G \subseteq \mathrm{FGL}_{\mathcal{K}}(V)$ is locally finite, then*

$$|G/\mathrm{unip}(G)| \leqslant \max\{\aleph_0, 2^{\dim(V)}\}.$$

(10.9) *Kegel Covers in Finitary Groups.* Relevant information about a locally finite simple group is frequently obtained by locating a Kegel cover that is amenable to study (see [24] and [42] for a discussion of Kegel covers and sequences). This approach is implicit in the proof of (10.4) [19]. Independently of Hall's work, both Meierfrankenfeld [44] and Belyaev [1, Theorem B] have proved the following result.

(10.9.1) Theorem *Let G be a locally finite simple subgroup of $\mathrm{FGL}_{\mathcal{K}}(V)$ where $\mathrm{char}(\mathcal{K}) = p \geqslant 0$. Then G has a Kegel cover $\mathcal{L} = \{(G_i, M_i) \mid i \in I\}$ where the G_i are perfect and, for every i, $G_i/O_p(G_i)$ is a quasi-simple group. Here, $O_p(G_i)$ is the largest normal p-subgroup of G_i if $p > 0$, and $O_p(G_i) = \{1\}$ if $p = 0$.*

It follows from (10.9.1) together with Theorem 3 of [55] that every locally finite, simple, finitary linear group is *absolutely simple*. The existence of a Kegel cover (or sequence) in a perfect group does not imply that the group is simple [32, p. 116]. Whether or not this is the case in a finitary group is an open question.

(10.10) Let G be a perfect, locally finite, finitary linear group with a Kegel cover. Does it follow that G is simple?

With respect to (10.10), the author has proved the following related result in [56].

(10.10.1) Suppose $x \in G \subseteq \mathrm{FGL}_K(V)$, $\dim(V)$ is infinite, and $H = x^G$ is irreducible and has a Kegel cover. Then H has a simple normal subgroup M with H/M solvable. Thus, if H is perfect, H is simple.

A far-reaching structure theorem for locally finite, finitary groups has been developed by both Meierfrankenfeld [44] and Belyaev [1]. First, we look at the version in [1].

(10.11) **Theorem** *Let $1 \neq G \subseteq \mathrm{FGL}_K(V)$ be a locally finite group with $\mathrm{ls}(G) = 1$. Then G has a minimal normal subgroup $M \neq 1$; further, if M is a minimal normal subgroup of G then either*

(i) *M is an infinite simple group which is not finite dimensional, or*

(ii) *M is a direct product of isomorphic finite dimensional groups.*

The hard work in the proof of (10.11) is establishing that G has a non-trivial simple subnormal subgroup R. Once this is established, the group R^G generated by the conjugates of R in G yields a minimal normal subgroup. Even at this point there are some surprises. In particular, it follows from (10.11) that any non-'finite dimensional' subnormal simple subgroup must be normal. To see this, let R be a non-Abelian simple subnormal subgroup of G which is not finite dimensional and suppose there is a $g \in G$ such that $R^g \neq R$. Then $\langle R, R^g \rangle$ is a direct product $R \times R^g$ and so R and R^g centralize each other. If $x \in R^g$, $[V, x]$ is a finite dimensional R-space and, since R is not a finite dimensional group and R is simple, $[V, x, R] = 0$. Thus, $[V, R^g] = [V, R]g \subseteq C_V(R)$. Since $[V, R] \cap [V, R]g$ has finite co-dimension in $[V, R]$, R must have a faithful finite dimensional representation on a quotient of $[V, R]$; this contradiction shows that $R \triangleleft G$.

A more elaborate version of (10.11) has also been developed in [44]. Before we state this result recall (see (8.5)) that in a finitary linear group G, $G/\mathrm{ls}(G)$ is again a finitary linear group.

(10.11.1) **Theorem** *Let $G \subseteq \mathrm{FGL}_K(V)$ be locally finite and put $G^0 = G/\mathrm{ls}(G)$.*

(a) *G^0 has the properties of (10.11). Moreover, if R is a simple subnormal subgroup of G^0 and $x \in G^0$, then x normalizes all but a finite number of the conjugates R^g in G^0; in fact the number of conjugates moved by x can not exceed $\deg_G(x)$. Put somewhat differently, G^0 acts finitarily on the conjugates of R.*

(b) *Let $\ln(G)$ be the Hirsch-Plotkin radical of G, that is,*

$$\ln(G) = \langle H \triangleleft G \mid H \text{ is locally nilpotent} \rangle,$$

let $E(G) = \langle S \lhd \lhd G \mid S$ is quasi-simple$\rangle$ *and put* $F^*(G) = \mathrm{ln}(G)E(G)$; *here* $F^*(G)$ *is the analogue of the* generalized Fitting subgroup *in a finite group.*

Then $C_G(F^*(G)) \subseteq F^*(G)$, *and so* $G/F^*(G)$ *acts faithfully on* $F^*(G)$.

(c) $C_{G^0}(E(G^0)) \subseteq E(G^0)$.

We note that the component theory of (10.11.1) applies to locally finite subgroups of $\mathrm{GL}(n, \mathcal{K})$, a fact that does not seem to be recorded in the literature. The content of (10.11) and (10.11.1) can be used to obtain some information about primitive, locally finite, finitary groups.

(10.12) If $G \subseteq \mathrm{FGL}_{\mathcal{K}}(V)$ is locally finite and primitive and $\dim(V)$ is infinite, then G has a unique simple component M and $C_G(M) = 1$. Thus $G/M \subseteq \mathrm{Out}(M)$.

Proof. Since G is primitive, $\mathrm{ls}(G) = 1$. Thus, G has a minimal normal subgroup $M \neq 1$ as in (10.11); note that M is irreducible. If $g \in C_G(M)$, $[V, g]$ is a finite dimensional M-module. Thus, $C_G(M) = 1$ and it follows that M is the unique minimal normal subgroup of G. If M is not simple, M contains a simple normal subgroup $A \neq 1$ which is finite dimensional. Since A is reducible (see (8.1)), V has a system of imprimitivity $\mathcal{D}$ normalized by $AC_M(A) = M$. Thus, M is reducible and from this contradiction we deduce that M is simple. The remaining assertions follow from (10.11.1).

The result (10.12) together with (8.2.3) suggests the following conjecture.

(10.12.1) Conjecture *If* $G \subseteq \mathrm{FGL}_{\mathcal{K}}(V)$, G *is locally finite and primitive, and* $\dim(V)$ *is infinite, then* G' *is an infinite simple group.*

A weak form of (10.12.1) has been proved by the author in [56].

(10.12.2) Theorem *Suppose that* $G \subseteq \mathrm{FGL}_{\mathcal{K}}(V)$ *is locally finite, primitive, and that* $\dim(V)$ *is infinite. Then* G *has a normal simple subgroup* M *such that* G/M *is solvable of derived length less than* 6. *As in* (10.12) *the group* G/M *is a group of outer automorphisms of* M.

(10.13) *More on Simple Groups.* Criteria have been developed by Hall-Hartley [22] and Belyaev [3] which can be used to identify finitary groups within the class of locally finite simple groups. We begin with a discussion of Belyaev's criteria. Many of the ideas in Belyaev [3] rely on a rather detailed knowledge of residually finite finitary linear groups, a topic we discuss later in §11. In [3] the notion of an *inert* subgroup is introduced: a subgroup H of a group G is inert if, for every $g \in G$, the intersection $H \cap H^g$ has finite index in H. An inert subgroup H is *proper* if H is neither G nor a finite subgroup of G. Inert subgroups arise naturally in a countable locally finite simple group in the following manner. Let $\{(G_i, M_i) \mid i \geqslant 1\}$ be a Kegel sequence of the countable locally finite simple G and $H = \langle M_i \mid i \geqslant 1 \rangle$; the subgroups $H_n = \langle M_i \mid i > n \rangle$ form a descending chain of normal subgroups of finite index in H. Further, for $n > 1$, H is a split extension $H = H_n \langle M_i \mid 1 \leqslant i \leqslant n \rangle$. In particular, H is residually finite. The subgroup H is also inert since, if $g \in G_n$, H_n

is contained in the intersection $H \cap H^g$.

General properties of inert subgroups in locally finite simple groups are developed in [9]. Here we confine our comments to those aspects of inert subgroups which relate to finitary linear groups. The first result connecting the concept of inert with simplicity appears in [8].

(10.13.1) Theorem *A locally finite simple group is finite dimensional if and only if it has no proper inert subgroups.*

The criteria for finitary linear groups are

(10.13.2) Theorem [3, Theorem 1] *If G is a countable locally finite simple group, the following three statements are equivalent.*

(i) *G is finitary linear.*

(ii) *There exists a prime p such that, for every proper inert subgroup H of G, $H/O_p(H)$ is locally normal (locally normal means that the normal closure of every element is finite).*

(iii) *If H is a proper inert subgroup with $\mathrm{ls}(H) = 1$, then H is locally normal.*

Similar criteria are given in the same paper which rely only on the residually finite subgroups of G.

(10.13.3) Theorem [3, Theorem 6] *If G is a locally finite simple group, the following three statements are equivalent.*

(i) *G is finitary linear.*

(ii) *If H is a countable residually finite subgroup of G with $\mathrm{ls}(H) = 1$, then H contains a non-trivial finite normal subgroup.*

(iii) *If H is a countable residually finite subgroup of G with $\mathrm{ls}(H) = 1$, then H is locally normal.*

A more detailed discussion on the role of inert subgroups in locally finite simple groups is given in [10]. A very different condition has been given by J. I. Hall and B. Hartley in [22]. This condition exploits the information provided in (8.2.1).

(10.13.4) Theorem *The locally finite simple group G is finitary if and only if there is an involution $x \in G$ and a positive integer t such that, for every 2-subgroup P of G with $x \in P$, x^P is solvable of derived length not exceeding t.*

Note that the necessity of the condition in (10.13.4) is implied by (8.2.1).

(10.14) *Sylow Theory.* While finitary linear p-groups (p a prime) have a reasonably well understood structure (see (8.6) and (8.7)), little has been done with the analysis of a general Sylow theory. The difficulties implicit in such a theory can be noted in the infinite alternating group $G = \mathrm{Alt}(\Omega)$ on a countably infinite set Ω, where, for any prime p, there are $2^{\aleph_0}$ isomorphism types of Sylow p-subgroups [27]—incidentally, there is only one isomorphism type of *transitive* Sylow p-subgroup in this group. In the papers [29], [30], E. G. Kosman has studied special types of Sylow p-subgroups of countable stable linear groups over a finite field. It would be interesting to know the answer to the following question.

(10.14.1) Let G be a locally finite, irreducible subgroup of $\mathrm{FGL}_\mathcal{K}(V)$ and p a prime. Is there a unique conjugacy class of irreducible Sylow p-subgroups of G?

11. Residual Properties

We begin the discussion with a review of the situation in finitary permutation groups (see (4.9)). In [49, Theorem 2] it is shown that a group $G \subseteq \mathrm{FSym}(\Omega)$ which is residually finite (residually solvable) has all orbits finite. Thus, such a group is a subdirect product of finite (finite solvable) groups. Similar theorems for finitary linear groups have been proved in [11], [4], [57], and [7, Section 5]. We begin with a more general form of Theorem A of [11].

(11.1) **Theorem** *Suppose that $G \subseteq \mathrm{FGL}_\mathcal{K}(V)$ satisfies either of the following two conditions:*

(i) *G is residually solvable and locally solvable;*

(ii) *G is residually finite and locally finite, such that either $\mathrm{char}(\mathcal{K}) = 0$ or $\mathrm{char}(\mathcal{K}) = p > 0$ and G is a p'-group.*

Then

(a) *every G-composition factor of V is finite dimensional;*

(b) *$G/\mathrm{unip}(G)$ is a subdirect product of finite dimensional groups;*

(c) *if the hypothesis (i) holds, then $G/\mathrm{unip}(G)$ is a subdirect product of solvable groups.*

The substantive part of (11.1) is the conclusion (a); parts (b) and (c) follow easily from (6.2). If the hypothesis (ii) holds, $\mathrm{unip}(G) = 1$ and so, in this case, G is a subdirect product of finite dimensional groups. Using the classification (10.4) it has been shown in [59] that the conclusions of (11.1) remain valid if the hypothesis (ii) is replaced by (ii)' G is residually finite and locally finite. A consequence of (11.1) is

(11.2) **Theorem** *If G satisfies the hypothesis (i) of (11.1), then $G/\mathrm{unip}(G)$ is also a residually solvable group.*

In [57] a study is undertaken of when homomorphic images of finitary linear groups with certain residual properties have the same residual property. A sample of this type of result is given in

(11.3) **Theorem** *Suppose that $G \subseteq \mathrm{FGL}_\mathcal{K}(V)$ is a $\mathcal{WF}$-group and is residually $\mathcal{X}$, where $\mathcal{X}$ is any class of groups. Then, if H is either the locally solvable radical of G or the Hirsch-Plotkin radical of G, G/H is also residually $\mathcal{X}$. If, in addition, the group G is locally finite, then $G/O_q(G)$ is residually $\mathcal{X}$ for every prime q.*

Suppose now that G is a p-subgroup (p a prime) of $\mathrm{FGL}_{\mathcal{K}}(V)$ which is residually finite, and that $\mathrm{char}(\mathcal{K}) \neq p$. Then (11.1) implies that G is a subdirect product of finite dimensional p-groups. Thus, if $x \in G$, x^G is a finite dimensional, residually finite p-group. Since x^G is a Cernikov group, x^G must be finite. Thus, G *is an FC-group* (and so, locally normal) (see [4, Theorem 8.1]). Similar remarks obtain if the locally finite group G involves only finitely many primes Π and the characteristic of $\mathcal{K}$ is co-prime to the primes in Π [7, Section 5].

We close this section by noting that there appears to be nothing very special about residually finite unipotent groups (see [11, p. 391]).

12. Finitary Modules

As we have seen in the foregoing sections, considerable information is available about the structure of finitary linear groups (especially $\mathcal{WF}$-groups) and the irreducible modules for these groups. This does not seem to be the case for general finitary modules. Repeating the question raised in (7.7), we do not have criteria on subgroups G of $\mathrm{FGL}_{\mathcal{K}}(V)$ that shed light on whether or not V contains an irreducible G-submodule. The following recent result of Meierfrankenfeld [45] seems to be the most incisive theorem available.

(12.1) Theorem *Suppose that $G \subseteq \mathrm{FGL}_{\mathcal{K}}(V)$ and that G has a local system $\mathcal{L}$ of subgroups such that $H \in \mathcal{L}$ implies that V is a completely reducible $\mathcal{K}H$-module. Then $[V, G]$ is a completely reducible $\mathcal{K}G$-module.*

Obvious consequences of (12.1) are the cases where G is locally finite and either $\mathrm{char}(\mathcal{K}) = 0$ or $\mathrm{char}(\mathcal{K}) = p > 0$ and G is a p'-group. Observe that, even in the cases noted, V need not be a completely reducible $\mathcal{K}G$-module (in characteristic 0, and with $\dim(V)$ infinite, the permutation module for $\mathrm{Alt}(\Omega)$ does not split over the natural module). Another relevant result in [45] is

(12.2) Theorem *Suppose that G is a locally finite subgroup of $\mathrm{FGL}_{\mathcal{K}}(V)$ and the group H is an extension of V by G. If there is a local system $\mathcal{L}$ of finite subgroups of G such that, for all $L \in \mathcal{L}$, the first cohomology group $H^1(L, V)$ is finite dimensional, then H splits over V.*

There are other types of groups for which something can be said about the module structure—notably the groups $H = x^G$ of §9. An immediate consequence of the module structure outlined in (9.3) is a positive answer to (7.7) for groups of this shape.

(12.3) Theorem *If $H = x^G$ and $G \subseteq \mathrm{FGL}_{\mathcal{K}}(V)$, then V contains an irreducible H-module.*

For the proof of (12.3), use the module structure indicated in (9.3); if V_1/V_0 is a trivial H-module, V_1 contains a one dimensional H-submodule, while, if V_1 is an irreducible G-module, the assertion follows from (7.6). The difficulties encountered in producing an irreducible G-submodule from the given information on H arise in the case where the only H-submodules of V are trivial. In this case, the approach indicated in (7.8) breaks down.

13. Ultraproducts and Countable Local Theorems

Generalizations of Mal'cev's representation theorem [32, p. 64] have proved very useful in the study of finitary linear groups; the reader is referred to the appendix of [19] for the basic algebraic constructions using ultraproducts. Our comments will be simplified by first establishing some notation. If $\mathcal{F} = \{G_i \mid i \in I\}$ is a local system of subgroups of a group G, we introduce an order $\leqslant$ on the set I by defining $i \leqslant j$ if and only if $G_i \subseteq G_j$; $\mathrm{uf}(\mathcal{F})$ will denote an ultrafilter on I containing all subsets of the form $S_i = \{j \in I \mid i \leqslant j\}$ where $i \in I$. The ultrafilter $\mathrm{uf}(\mathcal{F})$ is the type of ultrafilter used in the Mal'cev representation theorem.

One illustration of the power of ultraproduct techniques is the following observation in [18].

(13.1) Let G be a group, $x \in G$, and $\mathcal{F} = \{G_i \mid i \in I\}$ a local system of subgroups of G with $x \in G_i$ for all i. Suppose in addition that there is a family of representations $\psi_i : G_i \to \mathrm{GL}_{\mathcal{K}_i}(V_i)$ and an integer n such that, for all i, $\dim[V_i, x\psi_i] \leqslant n$. Let $\mathcal{E} = \mathrm{uf}(\mathcal{F})$, $\mathcal{K} = \prod(\mathcal{K}_i)_{\mathcal{E}}$, $V = \prod(V_i)_{\mathcal{E}}$ and $\psi : G \to \mathrm{FGL}_{\mathcal{K}}(V)$ the Mal'cev representation of G. Then $\dim[V, x\psi] \leqslant n$.

The following local theorem is a very nice consequence of (13.1).

(13.2) [18] Suppose that the group G has a countably directed cover of simple, finitary linear groups. Then G is simple and finitary linear.

A *countably directed cover* of a group G is a set $\mathcal{L}$ of subgroups of G satisfying (i) for every countable subgroup H of G there is a $Y \in \mathcal{L}$ such that $H \subseteq Y$, and (ii) if $\{H_i \mid i \in I\}$ is a countable subset of $\mathcal{L}$, there is an $L \in \mathcal{L}$ such that $\langle H_i \mid i \in I \rangle \subseteq L$. Thus, a countably directed cover of G is a special type of local system of G.

The conclusion that G in (13.2) is simple is routine; the more interesting aspect is that G is finitary. In the proof, (13.1) is used in the following way; let $\mathcal{C}$ be a countably directed cover of G consisting of simple finitary linear groups and fix $1 \neq x \in G$. Then there is a positive integer n and a countably directed cover $\mathcal{D} \subseteq \mathcal{C}$ such that, for all $H \in \mathcal{D}$, $\deg_H(x) \leqslant n$ (we are not claiming that the existence of $\mathcal{D}$ is obvious—additional argument is required). It follows from (13.1) that x has finite degree in the representation ψ of (13.1). Thus, $x \in \mathrm{FGL}_{\mathcal{K}}(V)$, and it follows that $G = x^G \subseteq \mathrm{FGL}_{\mathcal{K}}(V)$, as desired.

There are no known analogues of (13.2) for general finitary linear groups—indeed it is not true that groups with countably directed covers of finitary linear groups are themselves finitary linear. An example is provided by the group $G = \mathrm{tor}(\prod\{C_{p^n} \mid n \geqslant 1\})$. G has the property that every countable subgroup H of G is a direct product of cyclic groups, but G is not a direct product of cyclic groups. It is a fact that the countable subgroups H are finitary linear in any characteristic— the characteristic p case requires [35, Theorem B]. It also follows from [35] (and indeed is noted in the same paper) that G is not finitary linear in characteristic p. If $q \neq p$ (q a prime or $q = 0$) suppose that G is finitary linear in characteristic q. Since G is residually finite, G is a subdirect product of finite groups (see §11, especially (11.1)) and it follows that G is a direct product of cyclic groups. From this contradiction, we conclude that G *is not a finitary linear group*.

The countable subgroups in this example are (highly) reducible and this type of local theorem is still open for irreducible groups; a specialized version of this question, as yet unresolved, is

(13.3) Let $\mathcal{K}$ be a field and suppose that G has a countably directed cover of irreducible $\mathcal{K}$-finitary linear groups. Is G a $\mathcal{K}$-finitary linear group?

Arguments similar to that given for (13.2) have also been used by Belyaev in [4, Section 5] to prove countable type local theorems for groups of finitary permutations. Both J. I. Hall (with attribution to K. Hickin) and Simon Thomas (unpublished) have proved that *a group G which has a countably directed cover of alternating groups is an alternating group*. Hall derives this fact from a key part of the proof of (10.1); this result can also be proved using a permutation group analogue of (13.1).

The frequency with which the representation theorem (13.1) is used in proofs of results in this survey is somewhat surprising; for example, (13.1) or a variant is used in the proofs of (5.4), (10.4), and (10.5). Frequently, (13.1) can be applied to obtain finitary representations that are in some sense an improvement over a given finitary representation. Another result where ultraproduct techniques give interesting information is

(13.4) **Theorem** *Let G be a countable irreducible subgroup of $\mathrm{FGL}_{\mathcal{K}}(V)$ and suppose that $G_1 \subseteq G_2 \subseteq \cdots \subseteq G_n \cdots$ are finitely generated subgroups of G such that $\bigcup G_n = G$. Fix a non-zero vector $v \in V$ and, for each n, put $L_n = \langle vG_n \rangle$ and let M_n be a maximal G_n-submodule of L_n. If there is an integer d such that $\dim(L_n/M_n) \leqslant d$ for all n, then there is a field $\mathcal{K}^*$ (an ultraproduct of copies of $\mathcal{K}$) and a faithful representation $\psi : G \to \mathrm{GL}(d, \mathcal{K}^*)$. Thus, if G is not a finite dimensional group, the dimensions $\dim(L_n/M_n)$ are unbounded.*

We remark that (13.4) is used in the proofs of (10.10.1) and (10.12.2).

In [58], Puglisi has recently used ultraproduct techniques to study finitary representations of free products.

(13.5) **Theorem** *Suppose that, for $i \in I$, $G_i \subseteq \mathrm{FGL}_{\mathcal{K}_i}(V_i)$ and the $\mathcal{K}_i$ all have the same characteristic $p \geqslant 0$. Then there is a field $\mathcal{K}$ of characteristic p and a $\mathcal{K}$-space V such that the free product $G = * \{G_i \mid i \in I\} \subseteq \mathrm{FGL}_{\mathcal{K}}(V)$. Further, V has only a finite number of non-trivial G-composition factors, and the action of G on these factors is primitive.*

As a final result in this section we have a somewhat different type of countable local theorem (ultraproducts are not used in the arguments).

(13.6) **Theorem** [26] *Let $\{(G_i, \Omega_i) \mid i \in \mathbb{N}\}$ be a set of finite transitive permu-*

tation groups, and let $G = \mathrm{Wr}\{(G_n, \Omega_n) \mid n \geqslant 1\}$ be the wreath product discussed in (4.8.1). Now let Σ be the class of all groups of this shape. If the group G has a countably directed cover of Σ-groups, then the group G is countable (and so is a Σ-group). Thus, there are no uncountable groups with countably directed covers of Σ-groups.

The result (13.6) is of particular interest in the special case where the groups in the countable directed cover $\mathcal{C}$ of Σ-groups are all isomorphic to a fixed group Y. It implies that there are no uncountable locally finite models in the $\infty - \omega$ theory of Y. It is interesting to contrast this result with groups T isomorphic to wreath products of finite groups over the negative integers, where there are large numbers of isomorphism types of groups G with countable directed covers of groups isomorphic with T [25, p. 799]. Aside from the groups in (13.6), the author is unaware of other locally finite groups X for which there are no uncountable groups with countably directed covers of groups isomorphic to X. In particular, the local theorem (13.6) is unresolved for arbitrary totally imprimitive groups of finitary permutations.

14. General Problems and New Areas

This final section will include a smattering of comments and suggestions for further work in the subject. If problems have been explicitly stated earlier, they will not be mentioned again here.

(14.1) *Homomorphic Images of Finitary Linear Groups.* Since free groups of countably infinite rank are finitary linear groups (even finite dimensional groups) every countable group is a homomorphic image of a finitary linear group. Thus, if anything is to be said about images, the class of groups under discussion will have to be trimmed down considerably. A candidate for such a class is the class of locally finite, finitary linear groups (or the $\mathcal{WF}$-groups). A striking feature of finite dimensional, locally finite groups is that many structural properties of such groups carry over to homomorphic images (such as the conjugacy of Sylow subgroups—see [69, Chapter 9]). It would be interesting to carry this philosophy over to the locally finite, finitary linear groups. Some results of this type appear in this survey, notably in (9.2). Also, *a simple image of a locally finite finitary group is necessarily finitary linear* (see (10.13.4)). A typical type of question that one may ask in this context is whether a residually solvable image of a locally finite, finitary linear G is necessarily a subdirect product of solvable groups (cf. (11.1))?

A second type of problem in the 'image' theory is a determination of which homomorphic images of $\mathcal{WF}$-groups are finitary linear. There appears to be no satisfactory analogue of the Zariski topology for infinite dimensional V; nonetheless, there are results (such as (6.3.2) and (8.5)) that suggest a relationship between the Zariski theory in the finite dimensional groups and images of infinite dimensional groups ($\mathrm{unip}(G)$ and $\mathrm{ls}(G)$ are both closed in the finite dimensional case, and, in the general case, the images modulo these subgroups are finitary linear). An investigation of the infinite dimensional analogues of other canonical closed subgroups in the finite dimensional case would seem to be in order. In particular, if H is normal in a finitary linear group G, and $H = C_G(X)$ for some subset X of G, is G/H finitary linear?

(14.2) *Cofinite Algebras.* Let V be a vector space over $\mathcal{K}$ and

$$\mathrm{CF}_{\mathcal{K}}(V) = \{g \in \mathrm{End}_{\mathcal{K}}(V) \mid V/\ker(g) \text{ has finite dimension}\}.$$

The algebra $\mathrm{CF}_{\mathcal{K}}(V)$ is a $\mathcal{K}$-subalgebra of $\mathrm{End}_{\mathcal{K}}(V)$ and is also a 2-sided ideal in $\mathrm{End}_{\mathcal{K}}(V)$. This algebra will be called the *cofinite algebra* of V over $\mathcal{K}$. Note that, if $g, h \in \mathrm{CF}_{\mathcal{K}}(V)$, then $[g, h] = gh - hg \in \mathrm{CF}_{\mathcal{K}}(V)$. Thus, $\mathrm{CF}_{\mathcal{K}}(V)$ also carries a Lie algebra structure; we denote the Lie algebra by $\mathfrak{cf}_{\mathcal{K}}(V)$. One interesting feature of these algebras comes from the following observation: if $\mathbb{C}$ is the field of complex numbers and $g \in \mathrm{CF}_{\mathbb{C}}(V)$, then $\exp(g) = \sum_0^{\infty} \frac{g^n}{n!}$ is defined and is an element of $\mathrm{FGL}_{\mathbb{C}}(V)$; i.e., the exponential maps the cofinite algebra into the finitary linear group. Also, for any field $\mathcal{K}$, any invertible element of $1 + \mathrm{CF}_{\mathcal{K}}(V)$ is an element of $\mathrm{FGL}_{\mathcal{K}}(V)$.

Current information about these algebras is rudimentary, although it is relatively easy to show that they are locally finite dimensional in the sense of (5.3). It seems likely that the rich theory of finite dimensional algebras can be used to obtain significant information about these (possibly) infinite dimensional algebras. One expects techniques to be similar to those used in the study of finitary linear groups. It is not out of the question that the simple Lie algebras contained in $\mathfrak{cf}_{\mathcal{K}}(V)$ can be classified in the same spirit as the classification of the finite dimensional complex simple Lie algebras.

Other subalgebras that arise in the theory of finitary linear groups are the subalgebras $\mathcal{K}G$ of $\mathrm{End}_{\mathcal{K}}(V)$ spanned by the group elements $g \in G \subseteq \mathrm{FGL}_{\mathcal{K}}(V)$ (see (6.4.2) and (6.4.4)). Algebras of this type seem not to have been studied in any systematic way. Properties such as density, irreducibility, etc., are closely related to the corresponding group theoretic concepts, and they merit further investigation in this context.

(14.3) *Changing Field Characteristics.* We say that a locally finite, finitary linear group $G \subseteq \mathrm{FGL}_{\mathcal{K}}(V)$ is *non-modular* if either $\mathrm{char}(\mathcal{K}) = 0$ or G is a p'-group and $\mathrm{char}(\mathcal{K}) = p$. It is a fact that, if the G above is finite (irreducible) and $\dim(V)$ is finite, then G is isomorphic to a subgroup (irreducible subgroup) of $\mathrm{GL}(n, \mathbb{C})$ where $\mathbb{C}$ is the field of complex numbers and $n = \dim(V)$ [16, p. 62]. If G is not finite and $\dim(V) = n$ is finite, then $\mathcal{K}G = \mathcal{K}H$ for some finite subgroup H of G, and so, again, $G \subseteq \mathrm{GL}(n, \mathbb{C})$. The question here is whether this same phenomenon is true in the infinite dimensional situation for finitary linear groups. In particular, we have the following question.

(14.3.1) **Question** *If $G \subseteq \mathrm{FGL}_{\mathcal{K}}(V)$ is non-modular, is there a $\mathbb{C}$-space W such that $G \subseteq \mathrm{FGL}_{\mathbb{C}}(W)$? Assuming this is true, does G irreducible in $\mathrm{FGL}_{\mathcal{K}}(V)$ imply that G is irreducible in $\mathrm{FGL}_{\mathbb{C}}(W)$?*

(14.4) *Maximal Subgroups.* Very little is known about the structure of maximal subgroups of locally finite groups. For example, it has only recently been established that countable locally finite simple groups have maximal subgroups [42, Theorem B]. How complicated these maximal subgroups are is open to speculation. It seems reasonable to conjecture that *any maximal subgroup of a non-'finite dimensional' locally finite simple group contains an infinite simple section.* The stature of this conjecture has not been examined for finitary linear groups.

(14.4.1) Question *If G is a locally finite, finitary linear simple group which is not finite dimensional and M is a maximal subgroup of G, does M contain infinite simple sections?*

The classification (10.4) will likely play a role in the study of this conjecture. Aside from simple groups, there are other interesting questions on maximal subgroups. For example, there is

(14.4.2) Question *If G is a locally finite, finitary linear group which is not finite dimensional and G has a maximal subgroup M which is locally nilpotent, is the group G locally solvable?*

References

1. V. V. Belyaev, Semisimple periodic groups of finitary transformations, *Algebra and Logic* **32** (1) (1993), 17–33.
2. V. V. Belyaev, Finitary linear representations of infinite symmetric and alternating groups, *Algebra and Logic* **32** (6) (1993), 319–327.
3. V. V. Belyaev, Local characterizations of periodic simple groups of finitary transformations, *Algebra and Logic* **32** (3) (1993), 231–250.
4. V. V. Belyaev, The structure of p-groups of finitary transformations, *Algebra and Logic* **31** (5) (1992), 271–284.
5. V. V. Belyaev, Local characterizations of infinite alternating and Lie type groups, *Algebra and Logic* **31** (4) (1992), 221–234.
6. V. V. Belyaev, Locally finite simple groups as a product of two inert subgroups, *Algebra and Logic* **31** (4) (1992), 216–221.
7. V. V. Belyaev, Locally solvable groups of finitary transformations, *Algebra and Logic* **31** (6) (1992), 331–342.
8. V. V. Belyaev, Locally finite subgroups with a finite inseparable subgroup, *Siberian Math. J.* **34** (1993), 218–232.
9. V. V. Belyaev, Inert subgroups in infinite simple groups, *Siberian Math. J.* **34** (1993), 606–611.
10. V. V. Belyaev, Inert subgroups in infinite locally finite groups, *these Proceedings*.
11. B. Bruno and R. E. Phillips, Residual properties of finitary linear groups, *J. Algebra* **166** (1994), 379–392.
12. P. Cameron, *Oligomorphic Permutation Groups*, London Math. Soc. Lecture Note Series, 152, Cambridge University Press, Cambridge, New York, 1990.
13. P. Cameron and J. I. Hall, Some groups generated by transvection subgroups, *J. Algebra* **140** (1991), 184–209.
14. C. Curtis and I. Reiner, *Representation Theory of Finite Groups and Associative Algebras*, Interscience, New York, 1962.
15. J. Dieudonné, Les determinants sur un corps non commutatif, *Bull. Math. Soc. France* **71** (1943), 27–45.
16. J. D. Dixon, *The Structure of Linear Groups*, Van Nostrand Reinhold Co., New York, 1971.
17. J. I. Hall, Infinite alternating groups as finitary linear transformations groups, *J. Algebra* **119** (1988), 337–359.
18. J. I. Hall, Finitary linear groups and elements of finite degree, *Arch. Math.* **50** (1988), 315–318.
19. J. I. Hall, Locally finite simple groups of finitary transformations, *these Proceedings*.
20. J. I. Hall, The number of trace valued forms and extraspecial groups, *J. London Math. Soc.* (2) **37** (1988), 1–13.
21. J. I. Hall, M. Liebeck and G. Seitz, Generators for finite simple groups, with applications to

linear groups, *Quart. J. Math.* 43 (2) (1992), 441–458.

22. J. I. Hall and B. Hartley, A group theoretical characterization of simple, locally finite, finitary linear groups, *Arch. Math.* 60 (1993), 108–114.

23. P. Hall, Wreath powers and characteristically simple groups, *Proc. Cambridge Philos. Soc.* 58 (1962), 49–71.

24. B. Hartley, Simple locally finite groups, *these Proceedings.*

25. K. Hickin, Some applications of tree limits to groups, Part 1, *Trans. Amer. Math. Soc.* 305 (2) (1988), 797–839.

26. K. Hickin and R. E. Phillips, Some countably stunted finitary permutation groups, to appear.

27. I. D. Ivanjuta, Sylow *p*-subgroups of the countable symmetric group, *Ukrain. Mat. Zh.* 15 (1963), 240–249 (in Russian).

28. I. D. Ivanjuta, On complete sets of Sylow subgroups of a countable symmetric group, *Ukrain. Mat. Zh.* 18 (1966), 112–115 (in Russian).

29. E. G. Kosman, Geometrical characterizations of Sylow *p*-subgroups of a bounded linear group, *Ukranian Math. J.* 40 (1988), 337–343.

30. E. G. Kosman, Constructions of Sylow subgroups of a bounded linear group, *Ukranian Math. J.* 39 (1987), 144–149.

31. O. H. Kegel and D. Schmidt, Existentially closed finitary linear groups, in *Groups St Andrews 1989*, eds. C. M. Campbell, E. F. Robertson, Cambridge University Press, LMS Lecture Note Series 159, 1991, pp. 355–362.

32. O. H. Kegel and B. A. F. Wehrfritz, *Locally Finite Groups*, North-Holland, Amsterdam, London, 1973.

33. F. Leinen, Absolute irreducibility for finitary linear groups, to appear in *Rend. Sem. Mat. Univ. Padova.*

34. F. Leinen, Irreducible representations of periodic finitary linear groups, to appear in *J. Algebra.*

35. F. Leinen, Hypercentral unipotent finitary skew linear groups, *Comm. Algebra* 22 (3) (1994), 939–949.

36. F. Leinen and O. Puglisi, Unipotent finitary linear groups, *J. London Math. Soc.* 48 (2) (1993), 59-76.

37. D. H. McLain. A characteristically simple group, *Proc. Cambridge Phil. Soc.* 50 (1954), 641–642.

38. D. H. McLain, On locally nilpotent groups, *Proc. Cambridge Phil. Soc.* 52 (1956), 5–11.

39. D. H. McLain, *A Class of Locally Nilpotent Groups*, Ph.D. dissertation, Cambridge University, 1956.

40. D. H. McLain, Finiteness conditions in locally soluble groups, *J. London Math. Soc.* 34 (1959), 101–107.

41. U. Meierfrankenfeld, Ascending subgroups of finitary linear groups, to appear in *J. London Math. Soc.*

42. U. Meierfrankenfeld, Non finitary locally finite simple groups, *these Proceedings.*

43. U. Meierfrankenfeld, A characterization of the natural module for some classical groups, to appear.

44. U. Meierfrankenfeld, A simple subnormal subgroup for locally finite, finitary linear groups, to appear.

45. U. Meierfrankenfeld, A note on the cohomology of finitary modules, to appear.

46. U. Meierfrankenfeld, R. E. Phillips, and O. Puglisi, Locally solvable finitary linear groups, *J. London Math. Soc.* 47 (1993), 31–40.

47. A. S. Mikles and R. I. Tyskevic, The transitive subgroups of a bounded symmetric group, *Veski Akad. Nauk BSSR, Ser. Fiz-Mat. Nauk* 6 (1971), 39–45 (in Russian).

48. P. M. Neumann, The lawlessness of groups of finitary permutations, *Arch. Math.* 26 (1975), 561–566.

49. P. M. Neumann, The structure of finitary permutations groups, *Arch. Math.* 27 (1976), 3–17.

50. D. S. Passman, Semiprimitivity of group algebras of locally finite groups, II, to appear.

51. D. S. Passman, The Jacobson radical of a group ring, *Proc. London Math. Soc.* 38 (3) (1979), 169–192.

52. D. S. Passman and A. E. Zalesskiĭ, Semiprimitivity of group algebras of locally finite simple groups, *Proc. London Math. Soc.* 67 (3) (1993), 243–276.

53. R. E. Phillips, The structure of groups of finitary transformations, *J. Algebra* 119 (1988),

400–448.

54. R. E. Phillips, Finitary linear groups generated by elements of small degree, to appear.

55. R. E. Phillips, On absolutely simple locally finite groups, *Rend. Sem. Mat. Padova* 79 (1988), 213–220.

56. R. E. Phillips, Primitive, locally finite, finitary linear groups, to appear.

57. O. Puglisi, Homomorphic images of finitary linear groups, *Arch. Math.* 60 (1993), 497–504.

58. O. Puglisi, Free products of finitary linear groups, to appear.

59. A. Radford, *Residually Finite, Locally Finite, Finitary Groups*, Ph.D. thesis, Michigan State University, 1995.

60. D. J. S. Robinson, *Finiteness Conditions and Generalized Solvable Groups, Part I*, Springer-Verlag, Berlin, Heidelberg, New York, 1972.

61. D. J. S. Robinson, *Finiteness Conditions and Generalized Solvable Groups, Part II*, Springer-Verlag, Berlin, Heidelberg, New York, 1972.

62. A. Rosenberg, The structure of the infinite general linear group, *Ann. Math.* 68 (1958), 278–294.

63. W. R. Scott, *Group Theory*, Prentice Hall, Englewood Cliffs, 1964.

64. D. Segal, Normal subgroups of finitary permutation groups, *Math. Z.* 140 (1974), 81–85.

65. D. Segal, A note on finitary permutation groups, *Arch. Math.* 25 (1974), 470–471.

66. V. N. Serezhkin and A. E. Zalesskiĭ, Finite linear groups generated by reflections, *Math. USSR-Izvestia* 17 (1981), 477–503.

67. D. A. Suprenenko, On locally nilpotent subgroups of infinite symmetric groups, *Dokl. Akad. Nauk SSSR* 167 (1966), 302–304 (in Russian); *Soviet Math. Doklady* 7 (1966), 392–394.

68. D. A. Suprenenko, Indecomposibility of transitive subgroups of the group SF(X), *Dokl. Akad. Nauk SSSR* 186 (1969), 779-780 (in Russian); *Soviet Math. Doklady* 10 (1969), 674–676.

69. B. A. F. Wehrfritz, *Infinite Linear Groups*, Springer-Verlag, Berlin, Heidelberg, New York, 1973.

70. B. A. F. Wehrfritz, Nilpotence in finitary linear groups, *Michigan Math. J.* 40 (1993), 419–432.

71. B. A. F. Wehrfritz, Locally solvable finitary skew linear groups, *J. Algebra* 160 (1993), 226–241.

72. B. A. F. Wehrfritz, Nilpotence in finitary skew linear groups, *J. Pure Appl. Algebra* 83 (1992), 27–41.

73. B. A. F. Wehrfritz, Algebras generated by locally nilpotent finitary skew linear groups, *J. Pure Appl. Algebra* 88 (1993), 305–316.

74. B. A. F. Wehrfritz, Irreducible locally nilpotent finitary skew linear groups, to appear.

75. B. A. F. Wehrfritz, A Jordan Hölder theorem for finitary linear groups, to appear.

76. B. A. F. Wehrfritz, Locally soluble primitive finitary skew linear groups, to appear.

77. B. A. F. Wehrfritz, Generalized soluble primitive skew linear groups, to appear.

78. J. Wiegold, Groups of finitary permutations, *Arch. Math.* 25 (1974), 466–469.

79. H. Wielandt, *Unendliche Permutationsgruppen*, Lecture Notes: Mathematisches Institut der Universität, Tübingen, 1960.

80. A. E. Zalesskiĭ, Groups of bounded transformations of a vector space, *Dokl. Akad. Nauk BSSR* 13 (1969), 485–488 (in Russian).

81. A. E. Zalesskiĭ, Groups of bounded automorphisms, *Dokl. Akad. Nauk BSSR* 19 (1975), 681–684 (in Russian).

82. A. E. Zalesskiĭ, The structure of certain classes of matrix groups over skewfields, *Sibirsk. Mat. Zh.* 8 (1967), 978–988 (in Russian).

83. A. E. Zalesskiĭ, Solvable subgroups of the multiplicative group of a locally finite algebra, *Mat. Sb.* 61 (103) (1963), 408–417 (in Russian).

LOCALLY FINITE SIMPLE GROUPS
OF FINITARY LINEAR TRANSFORMATIONS

J. I. HALL*
Department of Mathematics
Michigan State University
East Lansing
Michigan 48824
USA
E-mail: jhall@math.msu.edu

Abstract. The classification of the groups of the title is a natural successor to the classifications of finite simple groups and of locally finite linear simple groups. The statement of the classification is given along with an explanation of the examples. The proof is then discussed.

Key words: simple group, locally finite group, finitary linear group.

1. Introduction

Let $_K V$ be a left vector space over the field K. The element $g \in \mathrm{GL}_K(V)$ is *finitary* if $V(g-1) = [V, g]$ has finite K-dimension. This dimension is the *degree* of g on V, $\deg_V g = \dim_K[V, g]$. The invertible finitary linear transformations of V form a normal subgroup $\mathrm{FGL}_K(V)$ of $\mathrm{GL}_K(V)$, and any subgroup of $\mathrm{FGL}_K(V)$ is called a *finitary linear group*, obvious examples being the linear groups, subgroups of $\mathrm{GL}_K(V)$ for finite dimensional V. We stretch this terminology further by referring to any group which is isomorphic to a finitary linear group as a *finitary group*. Similarly, a *linear group* is any group which is isomorphic to a subgroup of $\mathrm{GL}_K(V)$, for some finite dimensional V.

The classification of locally finite simple groups of finitary linear transformations to be discussed here is a natural successor to CFSG, the classification of finite simple groups [8], and the work BBHST of Belyaev [2], Borovik [6], Hartley and Shute [14], and Thomas [29], which classified the locally finite simple groups that are linear.

Theorem 1.1 (CFSG: Classification of finite simple groups) *Each finite simple group is isomorphic to one of the following:*

(1) *an alternating group* Alt_n;

(2) *a classical linear group* $\mathrm{PSp}_n(q)$, $\mathrm{PSU}_n(q)$, $\mathrm{P\Omega}_n^\epsilon(q)$, *or* $\mathrm{PSL}_n(q)$;

(3) *an exceptional group of Lie type* $\mathrm{E}_6(q)$, $\mathrm{E}_7(q)$, $\mathrm{E}_8(q)$, $\mathrm{F}_4(q)$, $\mathrm{G}_2(q)$, $^2\mathrm{B}_2(q)$, $^3\mathrm{D}_4(q)$, $^2\mathrm{E}_6(q)$, $^2\mathrm{F}_4(q)$, *or* $^2\mathrm{G}_2(q)$;

(4) *one of 26 sporadic groups;*

(5) *a cyclic group of prime order.*

---

* Partial support provided by the NSA.

B. Hartley et al. (eds.), Finite and Locally Finite Groups, 147–188.
© 1995 *Kluwer Academic Publishers. Printed in the Netherlands.*

Theorem 1.2 (BBHST: Belyaev, Borovik, Hartley, Shute, Thomas) *Each locally finite simple group which is not finite but has a faithful representation as a linear group in finite dimension over a field is isomorphic to a group $\Phi(K)$ of Lie type, where K is an infinite, locally finite field, that is, an infinite subfield of $\overline{\mathbb{F}}_p$, for some prime p.*

Theorem 1.3 *Each locally finite simple group which is not linear in finite dimension but has a faithful representation as a finitary linear group over a field is isomorphic to one of the following:*

(1) *an alternating group* Alt_Ω *with Ω infinite;*

(2) *a finitary symplectic group* $\mathrm{FSp}_K(V, s)$;

(3) *a finitary special unitary group* $\mathrm{FSU}_K(V, u)$;

(4) *a finitary orthogonal group* $\mathrm{F}\Omega_K(V, q)$;

(5) *a special transvection group* $\mathrm{T}_K(W, V)$.

Here K is a (possibly finite) subfield of $\overline{\mathbb{F}}_p$, for some prime p; the forms s, u, and q are nondegenerate on the infinite dimensional K-space V; and W is a subspace of the dual $V^\star$ whose annihilator in V is trivial: $0 = \{v \in V \mid vW = 0\}$.

This paper is devoted to a discussion of Theorem 1.3 and its proof [12]. The characteristic 0 case of the theorem is to be found in [10], the only examples being the alternating groups Alt_Ω. See Subsection 4.3 below and its Theorems 4.6 and 4.7.

The second section contains a detailed discussion of the examples in the conclusion to the theorem since such intimate knowledge is needed for their reconstruction in the proof. Particular attention is paid to the root elements of each group, these being special elements of small degree from which the underlying geometry can be recovered. Typical examples of root elements are the 3-cycles of an alternating group and the transvections of a special linear group. The third section contains in turn three subsections and introduces the general tools which are most important in the proof. Its first subsection deals with Kegel covers and their properties, the most crucial being the fact that every locally finite simple group can be glued together out of finite simple groups in an appropriate manner. This observation is due to Otto Kegel. Together with the classification of finite simple groups, it accounts for the recent activity and progress in the classification theory of locally finite simple groups.

The second part of Section 3 presents a theorem which is a linear version of an old result of Jordan, who proved that a primitive permutation group of finite degree which contains an element of small support must be alternating or symmetric. An irreducible finitary group in infinite dimension can be thought of as a group generated by elements of small degree. The linear version of Jordan's theorem says that a primitive linear group in finite dimension which is generated by elements of small degree comes from a short list of groups each of which is highly geometric. The third part of Section 3 deals with ultraproducts and specifically with a theorem, proved using ultraproducts, which allows us to sew together local geometries, obtained from the subgroups of a Kegel cover, into a global geometry which will ultimately emerge as the classical geometry of the group at hand. This theorem is motivated by Mal'cev's Representation Theorem. (Since ultraproducts and the representation theorem may not be familiar to many people, we have provided an introductory appendix.)

The fourth section then contains the actual discussion of the proof of Theorem 1.3. The basic outline is very simple. The general theory of Kegel covers tells us where to look for the internal subgroup geometry of the group. Our theorem of Jordan type then implies that this internal geometry is as expected. Using ultraproducts, we can build from the internal geometry an external geometry with the desired local properties. We then identify this external geometry as the expected classical geometry.

It now appears that, as hoped in [10, 11], the classification of locally finite simple groups which are finitary linear has a natural place in the general theory of locally finite simple groups. The nonfinitary groups exhibit certain 'universal' behavior not seen in the finitary groups. See the present articles by Belyaev [5] and Meierfrankenfeld [21] for precise statements and discussion along these lines. The classification has also been used by Passman in his study of the semiprimitivity problem for group algebras of locally finite groups [24, 25].

Our general references for group theory and the geometry of the classical groups are the books of Aschbacher and Taylor [1, 28]. For an introduction to finitary linear groups, there is no better reference than the present article by Phillips [27].

2. Examples

We discuss in some detail the conclusions to Theorem 1.3, that is, the infinite dimensional finitary classical groups including the infinite alternating groups. We pay special attention to their generation by special kinds of elements of small degree— transvections, Siegel elements, and 3-cycles—which encode group theoretically much of the group's natural geometry.

2.1. Alternating Groups

The group Sym_Ω is the group of all permutations of the point set Ω. We usually write Sym_n for Sym_Ω when $\Omega = \{1, 2, \ldots, n\}$. The *support* of the permutation $g \in \mathrm{Sym}_\Omega$ is the subset of those points in Ω which are moved by g rather than fixed. The *finitary symmetric group*, FSym_Ω, is the normal subgroup of Sym_Ω containing all permutations with finite support. (Other notation exists; this is the group $\mathrm{Sym}_0(\Omega)$ of [15].) Clearly if Ω is finite then $\mathrm{FSym}_\Omega = \mathrm{Sym}_\Omega$, but for infinite Ω this is false. It is the case of infinite Ω which interests us here.

The finitary symmetric group FSym_Ω is that subgroup of Sym_Ω which is generated by the class of 2-cycles and can be thought of as the union (or direct limit) of its finite subgroups Sym_Δ, as Δ ranges over the finite subsets of Ω. In particular, it is locally finite. If it were possible to find an odd number of 2-cycles whose product was the identity, this would be achieved within some subgroup Sym_Δ with Δ finite. Since this can not be the case, the members of FSym_Ω can be divided into odd and even elements, as in the finite case. The normal subgroup of Sym_Ω consisting of all even finitary permutations is the *alternating group*, Alt_Ω. It is the union of its finite alternating subgroups Alt_Δ. In particular, it is locally finite and simple.

For any field K, consider the permutation module $K\Omega$ for Alt_Ω (and FSym_Ω). If the support of an element g contains t points of Ω in s distinct orbits, then the degree of g on $K\Omega$ is $t-s$. Thus the module $K\Omega$ gives rise to a finitary linear representation of Alt_Ω. (In particular, Alt_Ω is finitary in all characteristics, including 0.) On the other hand, when Ω is infinite Alt_Ω contains p-groups of unbounded class for all primes p, and so can not be linear in any characteristic. We thus have our first example of a locally finite, finitary linear simple group which is not linear in any finite dimension.

For alternating groups, the *root elements* will be the 3-cycles. These special elements of order 3 have minimal support (size 3) and minimal degree (equal to 2) in Alt_Ω. The following proposition should be thought of as saying that the geometry of the underlying set Ω is embedded group theoretically in Alt_Ω via the class of 3-cycles.

Proposition 2.1 *A subgroup of Alt_Ω which is generated by 3-cycles is a direct product of natural, alternating subgroups Alt_Σ, for $\Sigma \subseteq \Omega$.*

Proof. Let $G \leqslant \mathrm{Alt}_\Omega$ be generated by 3-cycles, and let Σ be an orbit of G on Ω. We need to show that G contains Alt_Σ.

For $\alpha, \omega \in \Sigma$, there is a set of 3-cycles $t_1, t_2, \ldots, t_n$ in G (for some n) with α in the support of t_1, ω in the support of t_n, and each pair t_i, t_{i+1} having supports with nontrivial intersection. Within $\langle t_{n-1}, t_n \rangle \leqslant \mathrm{Alt}_5$, we can find a 3-cycle t_{n-1}^* whose support contains ω and meets the support of t_{n-2} nontrivially. This shortens the path from α to ω; and, proceeding in this manner, we find a 3-cycle t of G which has both α and ω in its support. If β is a third member of the orbit Σ, then similarly we find a 3-cycle s of G whose support contains β and ω. Finally, in $\langle t, s \rangle \leqslant \mathrm{Alt}_5$, there is a 3-cycle of G whose support is α, β, ω. Thus all 3-cycles with support from Σ belong to G, as desired. $\square$

Corollary 2.2 (See [10, (4.7)] and [13, Theorem 3]) *Let $g \in \mathrm{Alt}_n$ be an element of odd prime order p which is composed of s disjoint p-cycles, where $2sp - 1 \leqslant n$. Then Alt_n is generated by $2 + \lceil (n-2)/s(p-1) \rceil$ conjugates of g. (Here $\lceil x \rceil$ denotes the smallest integer which is greater than or equal to the real number x.)*

Proof. An easy induction shows that, for each $t \geqslant 2$, there are t conjugates of g in Alt_n which generate a subgroup transitive on anything up to $ts(p-1) + 1 \leqslant n$ points. Therefore a subgroup of Alt_n with orbits of length $n - 1$ and 1 can be generated by $\lceil (n-2)/s(p-1) \rceil$ conjugates of g, one of which can be g itself. One more generates with these a doubly transitive hence primitive subgroup of Alt_n. Now choose a conjugate h of g whose support intersects that of g in a set of size 1. (This is possible as $n \geqslant 2sp - 1$.) Then $[g, h]$ is a 3-cycle. By the proposition, the $2 + \lceil (n-2)/s(p-1) \rceil$ conjugates of g now selected must generate all of Alt_n. $\square$

2.2. SPECIAL TRANSVECTION GROUPS

The alternating and finitary symmetric groups mentioned above are finitary linear over any field and are locally finite, but in general a finitary group need not be locally finite. Indeed a slight modification to an old result of Schur says that a finitary linear group is locally finite if and only if it is periodic. (See [10, 26, 27].)

Any subgroup of $\mathrm{FGL}_K(V)$ will be locally finite if the field K is locally finite, that is, a subfield of the algebraic closure of some field of prime order, $K \leqslant \overline{\mathbb{F}}_p$. To see this, first note that any finite set Σ of such transformations generates a group which acts faithfully on a finite dimensional subspace of V, and the matrices for Σ in this action have only finitely many entries. These entries therefore generate a finite subfield F of locally finite K. That is, $\langle \Sigma \rangle$ is a subgroup of $\mathrm{GL}_n(F)$, for some finite $F \leqslant K$, and so is finite.

A transvection of $\mathrm{GL}_K(V)$ is a nonidentity element which is as close as possible to being the identity. More precisely, a *transvection* t has $(t-1)^2 = 0$ with the range $V(t-1)$ of dimension 1. Choose a representative x of this range, $V(t-1) = \langle x \rangle$. Then $v \mapsto v(t-1) = \alpha x$ gives a linear functional $\varphi : v \mapsto \alpha$ such that $x\varphi = 0$ (since $(t-1)^2 = 0$).

The pair x, φ completely determines t; and, conversely, for any pair $x \in V$ and $\varphi \in V^\star$ (the dual) with $x\varphi = 0$, we have a transvection $t = t_{\varphi,x}$ given by

$$v.t_{\varphi,x} = v + (v\varphi)x \,.$$

The *K-transvection subgroup* $T(\langle \varphi \rangle, \langle x \rangle)$ is then the subgroup composed of the identity plus all transvections $t_{\varphi,\alpha x} = t_{\varphi\alpha,x}$, as α runs through the nonzero elements of the field K, and is isomorphic to the additive group of K. For $\mathrm{SL}_K(V)$, the transvections are the *root elements* and the K-transvection subgroups are the corresponding *root subgroups*.

The transvection t is finitary of degree 1 (by definition) and is unipotent since $(t-1)^2 = 0$. In particular, if V has finite dimension then t has determinant 1. As it has degree 1, the element t can act nontrivially in at most one H-composition factor in V whenever $t \in H \leqslant \mathrm{GL}_K(V)$. In particular, if $H = \langle t^H \rangle$, then H has at most one nontrivial composition factor in V and so is unipotent-by-irreducible. The next lemma contains a related and important geometric property of transvections.

Lemma 2.3   *The transvection $t_{\varphi,x}$ leaves invariant the subspace $W \leqslant V$ if and only if either $x \in W$ or $W \leqslant \ker \varphi$.*

Proof.   If W is not in $\ker \varphi$, there is a $w \in W$ with $w.t_{\varphi,x} \neq w$, whence W contains

$$w.t_{\varphi,x} - w = w + w\varphi.x - w = \alpha x \neq 0 \,. \quad \square$$

If $V = K^n$ is spanned by the canonical basis $e_1, \ldots, e_n$ with dual basis $e_1^\star, \ldots, e_n^\star$, then the transvection $t_{e_j^\star, \alpha e_j}$ is just the usual elementary matrix $I + \alpha e_{i,j}$, where $e_{i,j}$ is a matrix unit. Gaussian elimination proves:

Theorem 2.4   *If $\dim_K V$ is finite, then $\mathrm{SL}_K(V)$ is generated by its transvections.*

For general V we define the *finitary special linear group* $\mathrm{FSL}_K(V)$ to be that subgroup of $\mathrm{GL}_K(V)$ which is generated by all the transvections $t_{\varphi,x}$, with $\varphi \in V^\star$, $x \in V$, and $x\varphi = 0$. Because it is true in finite dimensions, we always have $\mathrm{FSL}_K(V) = \mathrm{FGL}_K(V)'$, the derived group, excepting the usual small cases. (It is possible to define a determinant function on $\mathrm{FGL}_K(V)$ because finitary transformations have only finitely many eigenvalues not 1. The theorem then implies that the kernel of the determinant homomorphism is $\mathrm{FSL}_K(V)$.[1]

If V has finite dimension, then $V \cong V^\star$ and $\mathrm{FSL}_K(V) = \mathrm{SL}_K(V)$. When V has infinite dimension then $V^\star$ has uncountably infinite dimension; and, in particular, $\mathrm{FSL}_K(V)$ is uncountable. There is another finitary counterpart to the special linear group which remains countable for countable V, the stable special linear group $\mathrm{SL}^0_\infty(K)$. This is best introduced in terms of matrices. Every $k \times k$ matrix A_k can be extended to a $(k+1) \times (k+1)$ matrix A_{k+1} by placing A_k in the upper lefthand corner of A_{k+1} and then bordering A_k with 0's in A_{k+1} except for a new diagonal 1:

$$A_k \longrightarrow A_{k+1} = \left(\begin{array}{c|c} A_k & 0 \\ \hline 0 & 1 \end{array} \right) .$$

This gives us natural embeddings

$$\mathrm{GL}_1(K) \to \mathrm{GL}_2(K) \to \mathrm{GL}_3(K) \to \mathrm{GL}_4(K) \to \cdots \to \mathrm{GL}_k(K) \to \cdots .$$

The union of these groups is then the *stable linear group* $\mathrm{GL}^0_\infty(K)$ and is countable when K is, since it is the ascending union of countable groups.

The stable linear group has a natural finitary action on the K-space V spanned by $\mathcal{B} = \{e_1, e_2, \ldots, e_k, \ldots\}$, where $\mathcal{B}_k = \{e_1, e_2, \ldots, e_k\}$ is the standard basis for the natural module of $\mathrm{GL}_k(K)$, for each k. Its derived subgroup (and determinant 1 subgroup) is the *stable special linear group* $\mathrm{SL}^0_\infty(K) = \mathrm{GL}^0_\infty(K)'$ and is the corresponding ascending union of the subgroups $\mathrm{SL}_k(K)$.

If we think of the elements of $\mathrm{GL}_K(V)$ as infinite matrices with respect to the basis $\mathcal{B}$, then $\mathrm{GL}^0_\infty(K)$ is that finitary subgroup of matrices which differ from the identity only within a finite number of rows and columns. In contrast, if A is an arbitrary matrix of the finitary linear group $\mathrm{FGL}_K(V)$, then $A - I$ will have only a finite number of nonzero columns, but there may be infinitely many rows in which A differs from the identity.

We can unify and generalize our two infinite dimensional versions of the special linear group, $\mathrm{FSL}_K(V)$ and $\mathrm{SL}^0_\infty(K)$, by first realizing that both are generated by K-transvection subgroups. By definition, $\mathrm{FSL}_K(V)$ is generated by all transvections, while the theorem and our construction show that the stable group is generated by the various elementary matrix transvections $I + \alpha e_{i,j}$ (Gaussian elimination again).

Let U be a K-subspace of V and W a K-subspace of the dual $V^\star$. Then the *special K-transvection group* $\mathrm{T}_K(W, U)$ is defined as

$$\mathrm{T}_K(W, U) = \langle t_{\varphi,x} \mid \varphi \in W, x \in U, x\varphi = 0 \rangle ,$$

the subgroup of $\mathrm{GL}_K(V)$ generated by all the transvections $t_{\varphi,x}$ where the eligible pairs φ, x are restricted to W and U. Clearly such a group is finitary. In fact it is a subgroup of $\mathrm{FSL}_K(V) = \mathrm{T}(V^\star, V)$. On the other hand $\mathrm{SL}^0_\infty(K) = \mathrm{T}(W, V)$, where

---

[1] Although any unipotent element could lay claim to having determinant 1, the determinant function really only makes sense in the finitary context; so our notation is somewhat redundant. We might better use $\mathrm{SL}_K(V)$ in place of $\mathrm{FSL}_K(V)$, just as we use Alt_Ω over the redundant FAlt_Ω. We nevertheless prefer the notation $\mathrm{FSL}_K(V)$.

W is the subspace of $V^\star$ spanned by $\mathcal{B}^\star = \{e_1^\star, e_2^\star, \ldots, e_k^\star, \ldots\}$, the dual of the basis $\mathcal{B} = \{e_1, e_2, \ldots, e_k, \ldots\}$.

There is a certain amount of symmetry here between W and U. A transvection on V is also a transvection in its natural action on $V^\star$, so in some respects it may be better to think of the transvection group G as acting on the product $_KU \times W_K$ respecting the natural pairing $p\colon U \times W \to K$ given by $p(u, w) = u.w$. The members of $\mathrm{T}(W, U)$ act as isometries of $U \times W$ equipped with this pairing in the sense that, for all $u \in U$, $w \in W$, and $g \in \mathrm{T}(W, U)$,

$$p(u, w) = u.w = ug.g^{-1}w = ug.wg = p(ug, wg),$$

using the natural right action of $\mathrm{GL}_K(V)$ on $V^\star$. The group $\mathrm{T}(W, U)$ is faithful on both U and W if and only if the pairing p is nondegenerate on both sides, that is, if $\mathrm{ann}_U W = 0$ and $\mathrm{ann}_W U = 0$. Here by definition

$$\mathrm{ann}_U W = \{\, u \in U \mid u.w = 0, \text{ for all } w \in W \,\},$$

and similarly for $\mathrm{ann}_W U$.

For $G = \mathrm{T}(W, V)$ we are guaranteed $\mathrm{ann}_W V = 0$ since $W \leqslant V^\star$. Consider the case where $\mathrm{ann}_V W$ is also 0, so that G is irreducible in its action on V by Lemma 2.3. If $\dim_K V = n$ is finite, then the only possible such choice for W is the complete dual $V^\star$; and $\mathrm{T}_K(W, V) = \mathrm{SL}_K(V) \cong \mathrm{SL}_n(K)$. Assume now that $\dim_K V$ is infinite, where (as we have seen) there are various distinct choices for W with $\mathrm{ann}_V W = 0$.

Let Σ be a finite subset of G, so that $[W, \Sigma] = W_0$ and $[V, \Sigma] = V_0$ both have finite dimension. Since the pairing p is nondegenerate, there are finite dimensional W_1 and V_1, with $W_0 \leqslant W_1 \leqslant W$ and $V_0 \leqslant V_1 \leqslant V$, for which the restriction of the pairing p is nondegenerate. Thus $\langle \Sigma \rangle \leqslant \mathrm{T}(W_1, V_1) \leqslant G$. Nondegeneracy then guarantees that W_1 and V_1 have the same finite dimension, k say, and that $\mathrm{T}(W_1, V_1)$ is isomorphic to $\mathrm{SL}_k(K)$.

Therefore $G = \mathrm{T}(W, V)$ with $\mathrm{ann}_V W = 0$ has every finitely generated subgroup contained in a quasisimple subgroup. This forces G itself to be simple or possibly quasisimple. As G is irreducible on V, any central element is multiplication by a member of the division ring $\mathrm{Hom}_K(V, V)$. When V has infinite dimension, such a nontrivial multiplication is not finitary whereas G is. We conclude that in this case G is simple. Indeed, if K is locally finite, then G is locally finite, simple, and finitary. (It can not be linear, because it has alternating sections of unbounded degree.)

In Proposition 2.1 we saw that the geometry of the alternating group can be reclaimed from its class of 3-cycles, its root elements. A similar statement is true here for groups generated by K-transvection subgroups, that is, root subgroups.

Theorem 2.5 *Let $G \leqslant \mathrm{GL}_K(V)$ be an irreducible group generated by the conjugacy class T^G of K-transvection subgroups with $|T| = |K| > 2$. Then G is either*

(1) $\mathrm{FSp}_K(V, s)$, *for s a nondegenerate symplectic form, or*
(2) $\mathrm{T}_K(W, V)$, *for some $W \leqslant V^\star$ with $\mathrm{ann}_V W = 0$.*

Here $\mathrm{FSp}_K(V, s)$ is a finitary symplectic group, as discussed in the next subsection.

This theorem is from [7]. (There the reducible case and the case $|T| = |K| = 2$ are also handled, but the results are more complicated. For instance, the 2-cycles of the finitary symmetric group are transvections on a natural module in characteristic 2.) For finite dimensional V, the theorem is due to McLaughlin [20]. Related results appear in Zalesskiĭ and Serezhkin [32]. Kantor [16] considers the more general situation of indecomposable subgroups of orthogonal groups which are generated by Siegel elements; see the subsection on orthogonal groups below and also [19].

We have already observed that, for $_K V$ of countable dimension, the stable group $\mathrm{SL}_\infty^0(K)$ can be realized as a group $T(W, V)$ with W of countable dimension and $\mathrm{ann}_V W = 0$. The converse holds. If W is a subspace of V^* with $\mathrm{ann}_V W = 0$ then the dimension of W must be infinite, and such a W of countable dimension has a basis which is dual to some basis of V. (See [7] for this and other remarks on the groups $T(W, V)$.) For W of countable dimension with $\mathrm{ann}_V W = 0$, the group $T(W, V)$ is thus isomorphic to $\mathrm{SL}_\infty^0(K)$. Indeed, any two such subgroups are conjugate in $\mathrm{GL}_K(V)$. Therefore, although the initial construction of the stable group appears to be very basis dependent, it actually has a very natural and basis-free definition as a minimal irreducible group among the $T(W, V)$. In particular, any group $T(W, V)$ which is countable and irreducible on V must be a stable linear group $\mathrm{SL}_\infty^0(K)$ for a suitable choice of basis $\mathcal{B}$.

2.3. FINITARY SYMPLECTIC GROUPS

We now consider the K-space V endowed with the *symplectic* (or alternating) form $s: V \times V \to K$. That is, s satisfies the following, for all $\alpha, \beta \in K$ and $x, y, z \in V$:

1. $s(\alpha x + \beta y, z) = \alpha s(x, z) + \beta s(y, z)$;
2. $s(x, y) = -s(y, x)$;
3. $s(x, x) = 0$.

In this case, the pair $(_K V, s)$ is called a *symplectic space* (although we sometimes abuse the terminology by referring to V itself as the symplectic space). The *radical* of the symplectic space is its subspace

$$\mathrm{rad}\,(_K V, s) = \{v \in V \mid s(v, w) = 0, \text{ for all } w \in V\}.$$

The symplectic space is *nondegenerate* if its radical is 0, and otherwise it is *degenerate*.

The linear transformation $g \in \mathrm{GL}_K(V)$ is an *isometry* of the symplectic space $(_K V, s)$ provided that, for all $x, y \in V$,

$$s(x, y) = s(xg, yg).$$

The subgroup of $\mathrm{GL}_K(V)$ consisting of all isometries is the *symplectic group*, denoted by $\mathrm{Sp}_K(V, s)$. The *finitary symplectic group* is then the group of isometries which are finitary:

$$\mathrm{FSp}_K(V, s) = \mathrm{Sp}_K(V, s) \cap \mathrm{FGL}_K(V).$$

For a given V of uncountable dimension, there are many fundamentally different nondegenerate symplectic forms and so different groups. The countable dimension case mimics that of finite dimension, in that there is a unique nondegenerate form up to similarity, and so a unique nondegenerate symplectic group up to isomorphism. (See [7].)

The null axiom $s(x, x) = 0$ for symplectic spaces is an immediate consequence of the preceding alternating axiom $s(x, y) = -s(y, x)$ when K has characteristic other than 2. Symplectic spaces in characteristic 2 have other exotic properties. In particular, the symplectic groups in characteristic 2 also arise as orthogonal groups (see below); and it will on occasion be convenient to restrict consideration of the symplectic case to characteristics other than 2, leaving the treatment of characteristic 2 to the orthogonal case.

Let $(_K V, s)$ be a nondegenerate symplectic space. For each vector $x \in V$, the map $\delta_x \colon V \to K$ given by mapping v to $s(v, x)$ is a K-linear functional, an element of $V^\star$. The map $\delta \colon x \mapsto \delta_x$ is then a canonical K-isomorphism of V into $V^\star$. If the transvection $t_{\varphi, x}$ is an isometry of (V, s), then it is an easy exercise to see that

$$x^\perp = \{v \in V \mid s(v, x) = 0\},$$

the kernel of the functional δ_x, must also be the kernel of the functional φ. That is, $t_{\varphi, x}$ is a symplectic isometry if and only if $\varphi = \delta_x \alpha$, for some $\alpha \in K$. The symplectic *root elements* are then the *symplectic transvections* $t_{\delta_x \alpha, x} = t_{\delta_x, \alpha x}$ which we write as

$$t_{x, \alpha x} \colon v \to v + s(v, x)\alpha x.$$

The corresponding *root subgroup* T_x is the K-transvection subgroup consisting of the identity and the elements $t_{x, \alpha x}$, as α runs through the nonzero elements of K. Again it is isomorphic to the additive group of K.

Theorem 2.6 (See [28, 8.5, 8.8]) *For a finite dimensional and nondegenerate symplectic space $(_K V, s)$, the symplectic group $\mathrm{Sp}_K(V, s)$ is generated by its root elements, the symplectic transvections. The symplectic group is quasisimple, except for certain small, finite exceptions, and has center $\{\pm 1\}$.*

From this we easily conclude:

Corollary 2.7 *For a nondegenerate and infinite dimensional symplectic space $(_K V, s)$, the finitary symplectic group $\mathrm{FSp}_K(V, s)$ is simple and generated by its root elements, the symplectic transvections.*

We have already anticipated this corollary in Theorem 2.5 where finitary symplectic groups and the groups $\mathrm{T}(W, V)$ were characterized as those irreducible groups generated by K-transvection subgroups. Unlike the special linear case, for a given nondegenerate symplectic space $(_K V, s)$ we do not have several different finitary analogues, but only the one. This is because the 'duality' δ identifies a canonical subspace $W = \delta(V)$ of the dual $V^\star$ for which $\mathrm{FSp}(V, s) \leqslant \mathrm{T}(W, V)$. In particular, for V of countable dimension the finitary symplectic group is isomorphic to the stable symplectic group, which can be constructed by embedding $\mathrm{Sp}_{2k}(K)$ in the upper lefthand corner of $\mathrm{Sp}_{2k+2}(K)$ as before. (Extend the nondegenerate symplectic space $V_{2k} = K^{2k}$ to $V_{2k+2} = K^{2k+2}$ by adding on a perpendicular direct summand which is nondegenerate of dimension 2.)

Our next result is a symplectic relative of Corollary 2.2. It states that, in a strong sense, the minimum possible number of root element (transvection) generators is in general enough.

Theorem 2.8   *Let $G = \mathrm{Sp}(V, s) \cong \mathrm{Sp}_{2n}(q)$, where q is odd but $(n, q) \neq (1, 9)$. For t a transvection, consider a subgroup $H = \langle t^H \rangle$ with $\dim[V, H] = k < 2n$. Then there is a set $\Sigma = \{t_1, \ldots, t_{2n-k}\}$ of $2n - k$ distinct G-conjugates of t such that $G = \langle H, \Sigma \rangle$.*

In the exceptional cases where q is even or $(n, q) = (1, 9)$, there is a suitable Σ of size $2n - k + 1$.

Proof. The case $n = 1$ only asks how many transvections are required to generate $\mathrm{SL}_2(q)$. The answer [16, Theorem 4.9] is two, if q is odd but not 9, and three otherwise.

In the nonexceptional general case (q odd but not 9), we observe that from the case $n = 1$ there is a t_1 with $H_1 = \langle H, t_1 \rangle$ satisfying $\dim[V, H_1] = k + 1$ and such that H_1 contains a subgroup $\mathrm{Sp}_2(q)$ and hence a full K-transvection subgroup. Now the result follows easily from Theorem 2.5 and indeed from McLaughlin's original result [20].

For $\mathrm{Sp}_4(9)$, the result should be checked by hand, and then an argument as in the preceding paragraph takes over for $(n, q) = (\geqslant 3, 9)$. Similarly, for q even, the previous argument works as well, provided we are willing to accept one additional generator.   □

For instance, taking $H = \langle t \rangle$, we learn that G can always be generated by $2n$ distinct transvections except for $(n, q) = (1, 9)$ and q even where $2n + 1$ suffice. Any number of transvections smaller than $2n$ would generate a subgroup whose commutator was proper in V and so could not be all of G. Although we have not proven it, in characteristic 2 the additional transvection generator is always needed. This is associated with the fact that symplectic groups in characteristic 2 are also orthogonal groups, as will be discussed further below. In $\mathrm{Sp}_2(9) \cong \mathrm{Alt}_6$ transvections are 3-cycles, and three are required for generation.

For the finite classical groups of other types, similar theorems hold. Theorem 2.5 and Proposition 2.11 include special cases.

2.4. FINITARY UNITARY GROUPS

Let σ be an automorphism of order 2 of the field K. We provide the K-space V with a *unitary* (or hermitian) form $u \colon V \times V \to K$. That is, u satisfies the following, for all $\alpha, \beta \in K$ and $x, y, z \in V$:

1. $u(\alpha x + \beta y, z) = \alpha u(x, z) + \beta u(y, z)$;
2. $u(x, y) = u(y, x)^\sigma$.

In this case, the pair $(_K V, u)$ is called a *unitary space* (briefly, V is a unitary space). The *radical* of the unitary space is, as before,

$$\mathrm{rad}\,(_K V, u) = \{v \in V \mid u(v, w) = 0, \text{ for all } w \in V\}.$$

The unitary space is *nondegenerate* if its radical is 0, and otherwise it is *degenerate*.

The linear transformation $g \in \mathrm{GL}_K(V)$ is an *isometry* of the unitary space $(_K V, u)$ provided that, for all $x, y \in V$,

$$u(x, y) = u(xg, yg).$$

The subgroup of $\mathrm{GL}_K(V)$ consisting of all isometries is the *general unitary group* $\mathrm{GU}_K(V, u)$. The *finitary unitary group* is then the group of isometries which are finitary:

$$\mathrm{FGU}_K(V, u) = \mathrm{GU}_K(V, u) \cap \mathrm{FGL}_K(V).$$

The notion of unitary space which we study is not the most general. It is not necessary to restrict attention to fields rather than division rings. More seriously, we shall only consider those unitary spaces which contain isotropic vectors. A nonzero vector x is called *isotropic* if $u(x, x) = 0$. A complex space endowed with its usual inner product provides an example of a unitary space which has no isotropic vectors. Such a space is called *anisotropic*. The isometry groups of anisotropic spaces behave in general differently from those of our unitary spaces. We are interested primarily in groups and spaces over finite or locally finite fields, and over such a field any anisotropic space must have dimension 1.

Let $({}_K V, u)$ be a nondegenerate unitary space. For $\alpha \in K$ and $x \in V$, the transvection

$$t_{x,\alpha x} : v \to v + u(v, x)\alpha x$$

is an isometry if and only if x is isotropic and $\alpha^\sigma = -\alpha$ (α is called *skew*). These *unitary transvections* are our *root elements* in this case. For a fixed isotropic 1-space $\langle x \rangle$, the corresponding *root subgroup* is the subgroup consisting of the identity and the elements $t_{x,\alpha x}$, as α runs through the nonzero skew elements of K. The root subgroup is isomorphic to the additive group of $K_0 = \{\beta \,|\, \beta = \beta^\sigma\}$, the degree 2 subfield of K composed of elements fixed by the automorphism σ.

In finite dimensions, a unitary isometry need not have determinant 1, so we are also interested in the *special unitary group*

$$\mathrm{SU}_K(V, u) = \mathrm{GU}_K(V, u) \cap \mathrm{SL}_K(V)$$

or, more generally, in the *finitary special unitary group*

$$\mathrm{FSU}_K(V, u) = \mathrm{GU}_K(V, u) \cap \mathrm{FSL}_K(V).$$

Theorem 2.9 (See [28, 10.23]) *Let $({}_K V, u)$ be a finite dimensional and nondegenerate unitary space which contains an isotropic vector. Then, except for certain small, finite exceptions, the special unitary group $\mathrm{SU}_K(V, u)$ is quasisimple and is generated by its root elements, the unitary transvections.*

Corollary 2.10 *If $({}_K V, u)$ is a nondegenerate and infinite dimensional unitary space which contains an isotropic vector, then the finitary special unitary group $\mathrm{FSU}_K(V, u)$ is simple and is generated by its root elements, the unitary transvections.*

As before, subgroups of unitary groups which are generated by transvections must have a very restricted structure. For instance, the minimum conceivable number of transvection generators for the full special unitary group can be achieved in most cases.

Proposition 2.11 (See [16, Theorem 4.9]) *If finite $\dim_K V \geqslant 5$, K is finite, and u is nondegenerate, then $\mathrm{SU}_K(V, u)$ is generated by $\dim_K V$ unitary transvections.*

A more general result like Theorem 2.8 is true here as well.

For the finite group $\mathrm{SU}_K(U, u)$ of the proposition, the field K must be $\mathbb{F}_{r^2}$, for some prime power r; and the form u is uniquely determined up to isometry. The group is thus unique up to isomorphism and is often denoted by $\mathrm{SU}_n(r^2)$ or by $\mathrm{SU}_n(r)$. For the purposes of this paper we prefer and use exclusively the first of these two differing pieces of notation. In particular the natural module for $\mathrm{SU}_n(q)$ is defined over $\mathbb{F}_q$.

2.5. FINITARY ORTHOGONAL GROUPS

On the K-space V the *orthogonal space* (or, sometimes, quadratic space) $(_KV, q)$ with *quadratic* form $q\colon V \to K$ and associated *orthogonal* (or symmetric bilinear) form $b\colon V \times V \to K$ satisfies the following, for all $\alpha, \beta \in K$ and $x, y, z \in V$:

1. $b(\alpha x + \beta y, z) = \alpha b(x, z) + \beta b(y, z)$;
2. $b(x, y) = b(y, x)$;
3. $q(\alpha x) = \alpha^2 q(x)$;
4. $q(x + y) - q(x) - q(y) = b(x, y)$.

Clearly q uniquely determines b, the *polar form* of q. Conversely we have

$$2q(x) = q(2x) - q(x) - q(x) = b(x, x).$$

Therefore, if the characteristic of K is not 2, the orthogonal form b uniquely determines q. In characteristic 2 the polar form of q is revealed as symplectic since $2q(x) = 0$. Conversely, starting with any nonzero symplectic b and a map q defined on a basis, we can extend q to a quadratic form with b as it polar form by using (3) and (4) above. In particular, in characteristic 2 any symplectic b will be the polar form for more than one quadratic form q.

The *radical* of the orthogonal space $(_KV, q)$ with polar form b is

$$\mathrm{rad}\,(_KV, q) = \{v \in V \mid q(v) = 0,\ b(v, w) = 0,\ \text{for all } w \in V\}.$$

The orthogonal space is *nondegenerate* if its radical is 0, and otherwise it is *degenerate*. In characteristic not 2, the radical of the quadratic form q equals the radical of its polar form b (with the obvious definition). In characteristic 2, the form b is symplectic on V and $\mathrm{rad}\,(_KV, q)$ could be strictly smaller than the symplectic radical $\mathrm{rad}\,(_KV, b)$. If the symplectic radical is 0, then the space $(_KV, q)$ is *nondefective*; otherwise it is *defective*.

The axioms imply that the restriction of a nondegenerate but defective quadratic form q to the symplectic radical is a σ-semilinear map whose image is a K^2-subspace of K, where K^2 is the subfield of squares in K. In the cases of interest to us, K^2 will always be equal to K, and the symplectic radical will have K-dimension 1 (or 0).

The linear transformation $g \in \mathrm{GL}_K(V)$ is an *isometry* of the orthogonal space $(_KV, q)$ provided that, for all $x \in V$,

$$q(x) = q(xg).$$

The subgroup of $\mathrm{GL}_K(V)$ consisting of all isometries is the *general orthogonal group* $\mathrm{GO}_K(V, q)$. The *finitary general orthogonal group* is then the group of isometries which are finitary:

$$\mathrm{FGO}_K(V, q) = \mathrm{GO}_K(V, q) \cap \mathrm{FGL}_K(V).$$

We may again pass to the special orthogonal group of determinant 1 isometries, but here this subgroup need not be perfect. We are more interested in the derived group

$$\mathrm{F}\Omega_K(V, q) = \mathrm{FGO}_K(V, q)'$$

which we shall call the *finitary orthogonal group*. (For V of finite dimension we write $\Omega_K(V, q) = \mathrm{GO}_K(V, q)'$. In certain small, finite cases, this notation may go against convention.)

In characteristic other than 2, a nondegenerate orthogonal space has no isometries which are transvections; in characteristic 2, a K-transvection subgroup meets the general orthogonal group in a subgroup of order at most 2. Let $({}_K V, q)$ be a nondegenerate orthogonal space. The subspace U of V is *totally singular* if the restriction of q to U is identically 0, so that U is its own radical. In an orthogonal group, a *root element* s is a unipotent element with $(s - 1)^2 = 0$ for which the range $V(s - 1)$ is totally singular of dimension 2. We shall refer to these orthogonal root elements as *Siegel elements*. (In [28, Chap. 11] these are the Siegel transformations of type II.) The *root subgroups* correspond to the various totally singular 2-spaces and are isomorphic to the additive group of K. It is possible to give a precise formula for the action of a Siegel element just as we have done earlier for transvections, but we shall not require the exact description.

Let $({}_K V, q)$ be a nondefective orthogonal space in characteristic 2 with polar symplectic form b. Choose $x \in V$ with $q(x) \neq 0$. Furthermore, let W be the subspace $x^\perp = \{v \in V \mid b(v, x) = 0\}$ and p the restriction of q to W. The orthogonal space $({}_K W, p)$ is then nondegenerate but defective with symplectic radical $\langle x \rangle$. The space $\bar{W} = W / \langle x \rangle$ is a nondegenerate symplectic space with form $\bar{b}$ induced by the restriction of b to W. Any symplectic isometry of the space $\bar{W}$ then lifts uniquely to an orthogonal isometry of W. Therefore the nondegenerate symplectic space $({}_K \bar{W}, \bar{b})$ and its group can be realized as a nondegenerate orthogonal space $({}_K W, p)$ and its group. In fact any symplectic group in characteristic 2 can be 'made orthogonal' by this process. As mentioned above, we often prefer to think of symplectic groups in characteristic 2 as orthogonal groups. There is the possibility of confusion here because there are two different types of root elements involved—the symplectic transvections and the orthogonal Siegel elements. Each transvection on symplectic $({}_K \bar{W}, \bar{b})$ lifts to a transvection on orthogonal $({}_K W, p)$, but two different transvections of the symplectic root subgroup $T_{\bar{z}}$ necessarily lift to t_{φ_1, z_1} and t_{φ_2, z_2} with $\langle z_1 \rangle \neq \langle z_2 \rangle$.

Theorem 2.12 (See [28, 8.8, 11.9, 11.48]) *Let $({}_K V, q)$ be a finite dimensional and nondegenerate orthogonal space which has dimension at least 5 and contains totally singular 2-spaces. In characteristic 2, assume further that the dimension is at least 6 and that $K = K^2$. The orthogonal group $\Omega_K(V, q)$ is quasisimple and, in particular, is generated by its root elements, the Siegel elements.*

Corollary 2.13 *If $({}_K V, q)$ is a nondegenerate and infinite dimensional orthogonal space which contains a totally singular 2-space and also, in the characteristic 2 case,* has $K = K^2$, then the finitary orthogonal group $F\Omega_K(V, q)$ is simple and generated by its root elements, the Siegel elements.

For the finite and nondegenerate orthogonal groups, there are results about subgroups generated by root elements similar to Theorems 2.5 and 2.8 and Proposition 2.11 above. In particular, the minimum conceivable number of root element generators (namely, one-half the dimension) is always nearly enough. See also [16, 19].

3. Tools

In this section we discuss the main tools of our proof—Kegel covers, a linear theorem of Jordan type, and representations constructed via ultraproducts.

3.1. KEGEL COVERS

Let G be a simple group and let X be a countable (or finite) subset of G. It will be no disadvantage to assume that X is a subgroup, since every countable subset of a group generates a countable subgroup. The simplicity of G is in fact a local property, in the sense that it can be checked within the finitely generated subgroups of G:

The group G is simple if and only if, for each ordered pair $x, y \in G$, there is a finite set $g_1, \ldots, g_k$ (for some k which depends upon x and y) with $x = \prod_{i=1}^{k} y^{\pm g_i}$.

With this in mind, we try to build $X = X_0$ into a simple group. For each pair x, y from X_0, find $g_1, \ldots, g_k$ as described, and let X_1 be the subgroup generated by X_0 together with all the g_i required for the various pairs x, y. As X_0 is countable, the number of pairs is countable, as is the total number of the various g_i; and so, ultimately, the subgroup X_1 is itself countable. We have designed X_1 to verify simplicity relative to all pairs x, y from its subgroup X_0, but in the process many more elements have presumably been introduced. We continue in the same manner, but now starting from the subgroup X_1. Choose in turn each of the countable number of pairs $x, y \in X_1$; find suitable g_i for each pair; and define the new subgroup X_2 to be generated by X_1 together with all the new g_i. Now we have a countable X_2 which verifies simplicity relative to each pair x, y from its subgroup X_1. This is a machine whose crank we can keep on turning, at each new stage i producing a countable subgroup X_i which verifies simplicity relative to all pairs from its subgroup X_{i-1}. If we now set $H = \bigcup_i X_i$, then the subgroup H of G verifies simplicity relative to each pair of its elements; and so H itself is simple. It contains the original subgroup X and, being the ascending union of countable subgroups, is countable itself. We have proven:

Theorem 3.1 (P. Hall) *In a simple group, every countable subset is contained in a countable simple subgroup.*

This argument, which is due to Phillip Hall, is in fact a special case of the downward Löwenheim-Skolem argument of model theory. There are several places in the theory of locally finite simple groups where ideas and methods from model theory appear to great advantage. Another important instance is found in the third subsection (and the appendix).

Hall's result is for arbitrary simple groups, whereas we are mainly concerned with locally finite groups which are simple. We thus ask whether something stronger is true in this smaller class. We can hope that, in a locally finite simple group, every finite subgroup is contained in a finite simple subgroup. This is false; see Corollary 3.8 for the counterexample discovered by Zalesskiĭ and Serezhkin. A slightly weaker statement is true, as was first proven by Kegel [18, 4.3]. The subgroup X is said to be *covered* by the section H/M if X is a subgroup of H which meets the normal subgroup M trivially. Such a covering leads to an isomorphic embedding of X in the section H/M.

Proposition 3.2 *In a locally finite simple group, every finite subgroup is covered by a simple section of a finite subgroup.*

Proof. The proof of Hall's result can be attempted. At each stage the subgroup X_i is finitely generated and so finite, but the proof falls down at the last moment. The ascending union of finite subgroups still will likely be countably infinite rather than finite.

The answer is to stop along the way. First, consider the embedding of X in X_1. Since every nontrivial element of X is in the X_1-normal closure of each other element of X, there is a X_1-chief factor P/Q which covers X. It is tempting to choose $H = P$ and M a maximal normal subgroup of P containing Q, but this may not be good enough. The section P/Q is a direct product of isomorphic finite simple groups, but certain elements of X might project trivially onto the direct factor S selected by our choice of M. That is, if the element x of X projects nontrivially onto S, it is still possible for $y \in X$ to project trivially onto S and have $x \in \langle y^{g_1}, \ldots, y^{g_k} \rangle$ since the g_i, while all belonging to X_1, are not necessarily in its subgroup P.

To remedy the problem, we go one step further, considering the embedding of X_1 (and its subgroup X) in X_2. Again there is a chief factor $\tilde{P}/\tilde{Q}$ of X_2 which covers X_1 and is a direct product of isomorphic simple groups. Consider again our pair x, y from the original X, and let $\tilde{S}$ now be a simple direct factor of $\tilde{P}/\tilde{Q}$ onto which x projects nontrivially. Since all the g_i from the previous paragraph are in X_1, they are in $\tilde{P}$ and are covered by the section. If y projected trivially onto $\tilde{S}$, then x could not be in its $\tilde{P}$-normal closure, which is the case since $X_1 \leqslant \tilde{P}$. We conclude that y must also project nontrivially onto $\tilde{S}$. That is, once one element of X projects nontrivially onto $\tilde{S}$, they all must. If we set $H = \tilde{P}$ and take M to be that maximal normal subgroup of H containing $\tilde{Q}$ which picks up all the factors of $\tilde{P}/\tilde{Q}$ except $\tilde{S}$, then X is covered by the simple section $H/M \cong \tilde{S}$ of finite $\tilde{P}$, as required. $\square$

A *sectional cover* of a group G is a set $\mathcal{C} = \{(G_i, N_i) \mid i \in I\}$ of pairs of subgroups such that, for each $i \in I$, N_i is normal in G_i, and, for every finitely generated subgroup X of G, there is an i with X covered by the section G_i/N_i. (If each N_i equals 1, then we speak of a *subgroup cover*.) The previous proposition then says that the locally finite simple group G has a sectional cover by simple sections G_i/N_i of finite subgroups G_i. Such a sectional cover is called a *Kegel cover* after Otto Kegel, who first proved their existence. (See [18, 4.3] and also [15, 21].) The subgroups N_i are the *Kegel kernels*, and the simple factors G_i/N_i are the *Kegel quotients*. If G is finite, then any Kegel cover contains the trivial cover $\{(G, 1)\}$.

The Kegel cover $\mathcal{C}$ of G is a particular type of *local system* for G, which is to say, G is the union of its subgroups G_i, and for any two G_i and G_j there is a third G_k with $\langle G_i, G_j \rangle \leqslant G_k$. A Kegel cover has a stronger property: we can choose k so that G_k not only contains the subgroup $\langle G_i, G_j \rangle$ but the simple section G_k/N_k covers $\langle G_i, G_j \rangle$. Once such containment and covering statements are true about pairs of subgroups G_i, G_j, they actually hold for all finite collections of subgroups G_i; there is a k such that the section G_k/N_k covers the subgroup $\langle G_1, G_2, \ldots, G_n \rangle$.

The group G is the direct limit of the members of any local system. Indeed the study of covers and local systems really stems from a desire to refine the trivial observation that any group is the direct limit of its set of finitely generated subgroups. The next result is actually a combinatorial result about partial orderings in which finite sets have upper bounds, but it is used frequently and crucially to replace a large Kegel cover by a smaller and more manageable one.

Lemma 3.3 (Coloring Argument) *Let G be a locally finite simple group, and suppose that the pairs of the Kegel cover $\mathcal{K} = \{(G_i, N_i) \mid i \in I\}$ are colored with a finite set $1, \ldots, n$ of colors. Then $\mathcal{K}$ contains a monochromatic subcover. That is, if $\mathcal{K}_j$ is the set of pairs from $\mathcal{K}$ with color j, for $1 \leqslant j \leqslant n$, then there is a color j for which $\mathcal{K}_j$ is itself a Kegel cover of G.*

Proof. Otherwise, for each j, there is a finite subgroup X_j of G which is not covered by any section which is colored by j. The subgroup $X = \langle X_1, \ldots, X_j, \ldots, X_n \rangle$ is therefore not covered by a section with any of the colors $1, 2, \ldots, n$. As X is generated by a finite number of finite groups, it is finite itself. Therefore some section of the Kegel cover $\mathcal{K}$ covers X, a contradiction which proves the lemma. □

A trivial example of the coloring argument is nevertheless instructive. Choose a nonidentity element g in G and color the sections (that is, pairs) of the Kegel cover $\{(G_i, N_i) \mid i \in I\}$ with two colors, one indicating that $g \notin G_i \setminus N_i$ and the second color indicating $g \in G_i \setminus N_i$. One of the color classes must be a subcover, but by definition the first can not, since it nowhere covers $\langle g \rangle$. We have proven that the sections which cover a specific element are themselves a Kegel cover. (This would have worked with any finite X in place of g. Less evident applications of the argument appear later.)

Proposition 3.4 *Let G be a locally finite simple group with Kegel cover*

$$\mathcal{K} = \{(G_i, N_i) \mid i \in I\},$$

and choose a nonidentity element g in G. For each $j \in J = \{j \in I \mid g \in G_j \setminus N_j\}$, set $H_j = \langle g^{G_j} \rangle$ and $M_j = H_j \cap N_j$. Then $\mathcal{K}_g = \{(H_j, M_j) \mid j \in J\}$ is a Kegel cover whose collection of Kegel quotients is contained in that of the original cover.

Proof. As G is the direct limit of the G_j, the normal closures H_j are also directed. If H is the subgroup of G which is the direct limit of the H_j, then H must be normal in G and nontrivial since $g \in H$. Therefore $H = G$. Certainly

$$H_j/M_j = H_j/H_j \cap N_j \cong H_j N_j/N_j = G_j/N_j.$$

It remains to observe that if G_i/N_i covers G_j, then H_i/M_i covers H_j. □

Proposition 3.2 tells us that Kegel covers exist. Proposition 3.4 and the coloring argument of Lemma 3.3 then allow us to begin searching for covers which are in some sense nice. The next observation [21, 2.8] will give us further opportunity to specialize our covers.

Lemma 3.5 *Let N be normal in the finite group G with G/N perfect. Then G has a unique subgroup H which is minimal subject to being normal in G and a supplement to N.*

Proof. Let $\bar{G} = G/N$. If H_1 and H_2 are normal supplements to N, then

$$\overline{H_1 \cap H_2} \geqslant \overline{[H_1, H_2]} = [\bar{H_1}, \bar{H_2}] = [\bar{G}, \bar{G}] = \bar{G}.$$ □

In this case, we call $H = H(G, N)$ the *heart* of the pair (G, N). If

$$\mathcal{S} = \{(G_i, N_i) \mid i \in I\}$$

is a Kegel cover of the group G, then the *heart* of $\mathcal{S}$ is

$$\mathcal{H} = \{(H(G_i, N_i), N_i \cap H(G_i, N_i)) \mid i \in I\}.$$

The quotients of $\mathcal{H}$ are identical to those of $\mathcal{S}$.

For $G = \text{Alt}_\Omega$, let the *canonical cover* of G be

$$\mathcal{CC}(\text{Alt}_\Omega) = \{(\text{Alt}_\Delta, 1) \mid \Delta \subseteq \Omega, |\Delta| < \infty\}.$$

Notice that this is actually a cover by simple subgroups, not just simple sections.

Proposition 3.6 *Let $\mathcal{S}$ be a Kegel cover of $G = \text{Alt}_\Omega$, for some infinite Ω, and let $\mathcal{H}$ be the heart of $\mathcal{S}$. Then $\mathcal{H} \cap \mathcal{CC}(G)$ is a Kegel cover of G, and $\mathcal{H} \smallsetminus \mathcal{CC}(G)$ is not a cover.*

Proof. Let t be a 3-cycle of G, and let $\mathcal{S}_t = \{(H_i, M_i) \mid i \in I\}$ be the Kegel cover produced as in Proposition 3.4 by taking the normal closure of t for each section G_i/N_i of $\mathcal{S}$ which covers t. Each $H_i = \langle t^{G_i} \rangle$ is a normal supplement to N_i in G_i, so H_i contains the heart of (G_i, N_i).

By Proposition 2.1 the group H_i is a direct product of natural finite alternating subgroups of G. As M_i is normal in H_i with H_i/M_i simple, M_i can only be a direct product of all but one of the direct factors. Let Alt_{Δ_i} be the missing factor. By construction, H_i is normal in G_i, and so $M_i = H_i \cap N_i$ is also. This forces the lone missing factor Alt_{Δ_i} to be normal itself. The 3-cycle t is in H_i but not in N_i, so Alt_{Δ_i} must be the normal closure of t in G_i and is clearly a simple complement to N_i in G_i. That is, $H_i = \text{Alt}_{\Delta_i}$ is the heart of (G_i, N_i); $M_i = 1$; and the cover $\mathcal{S}_t$ is contained in $\mathcal{H} \cap \mathcal{CC}(G)$. On the other hand, $\mathcal{H} \smallsetminus \mathcal{CC}(G)$ does not cover t and so can not be a cover. □

The proposition is a recast version of [10, (8.1)] with the elementary proof promised there. Among other things, it provides an elementary proof of Hartley's observation [15] that the infinite alternating groups can not be written as a nonnatural limit of diagonally embedded finite alternating groups.

Each of the simple groups G discussed in Section 2 has a Kegel cover, $\mathcal{CC}(G)$, which is a canonical cover in that every cover can be viewed as a modified version of this particular one. We present only those for the symplectic groups.

Let s be a nondegenerate symplectic form on the vector space V, infinite dimensional over the locally finite field K; and let $G = \mathrm{FSp}_K(V, s)$ be the associated finitary symplectic group. If $\{(G_i, N_i) \mid i \in I\}$ is a Kegel cover of G, then G is the direct limit of its finite subgroups G_i, and V is the direct limit of its finite dimensional subspaces $[V, G_i]$. As s is nondegenerate, every finite dimensional subspace is contained in one for which the restriction of s is nondegenerate. This implies that G has a cover by finite dimensional quasisimple symplectic subgroups. For the alternating groups, the canonical cover was composed of simple groups; but here nothing like that is true. Although, as just seen, quasisimple covers exist, not every cover reduces to a quasisimple one. This is essentially because the space V, although nondegenerate, can nevertheless be written as a direct limit of finite dimensional subspaces, each of which is degenerate. The associated subgroup cover is then not quasisimple but, instead, looks to be composed of arbitrary finite dimensional 'parabolic subgroups', groups which are unipotent-by-quasisimple. Such subgroups do provide a canonical cover.

Choose a finite subset $X = \{u_1, \ldots, u_n\}$ of V, and set

$$K_X = \langle s(u_i, u_j) \mid 1 \leqslant i, j \leqslant n \rangle,$$

that subfield of K generated by the entries in the Gram matrix of the set X. The subfield K_X is finite, since X is finite and K is locally finite. Next choose a finite subfield F of K which contains K_X. Set $FX = \sum_{i=1}^{n} Fu_i$, an F-subspace of V. We then let $G_{X,F} = \langle t_{u,\alpha u} \mid u \in FX - \mathrm{rad}\, FX, \alpha \in F - 0 \rangle$.

The group $G_{X,F}$ can be viewed as being generated by full transvection subgroups over F, and so is a unipotent normal subgroup P extended by a symplectic group over F (see Lemma 2.3 and Theorem 2.5). In particular, $G_{X,F}$ is unipotent-by-quasisimple. Its solvable radical $N_{X,F}$ is the unipotent radical P extended by the subgroup of symplectic scalars $\{\pm 1\}$ (which is 1 in characteristic 2). We then have the canonical cover

$$\mathcal{CC}(G) = \{(G_{X,F}, N_{X,F}) \mid X \subset V, |X| < \infty, K_X \leqslant F \leqslant K, |F| < \infty\},$$

and the counterpart of the previous result is valid.

Proposition 3.7   *Let $\mathcal{S}$ be a Kegel cover of $G = \mathrm{FSp}_K(V, s)$, and let $\mathcal{H}$ be the heart of $\mathcal{S}$. Then $\mathcal{H} \cap \mathcal{CC}(G)$ is a Kegel cover of G, and $\mathcal{H} \smallsetminus \mathcal{CC}(G)$ is not a cover.*

As a corollary we have the result of Zalesskiĭ and Serezhkin [32] mentioned earlier.

Corollary 3.8   *A stable symplectic group over a finite field of odd order does not have a cover by finite simple subgroups.*

More generally, a finitary symplectic group in infinite dimension over a locally finite field of odd characteristic does not have a cover by finite simple subgroups. Indeed in odd characteristic the group $G_{X,F}$ is never simple; even when its unipotent radical is trivial, it has a central subgroup of order 2.

Belyaev [3] and Meierfrankenfeld [23] have proven by elementary methods that every locally finite simple group which is finitary has a Kegel cover whose members are unipotent-by-quasisimple, as is the case with $CC(G)$. It is easy to see that the subset $CC_{qs}(G)$ composed of those $G_{X,F}$ which are quasisimple (corresponding to FX which are nondegenerate for the restriction of s) is still a cover. Indeed from Theorem 1.3 we may conclude that every locally finite, finitary simple group has a quasisimple cover; but we do not have an elementary proof of this.

As mentioned before, for each classical finitary group there is a similarly defined canonical cover and a corresponding proposition stating that every Kegel cover is at heart canonical. This responds in a precise manner to Hartley's remark [15] that, for finitary locally finite simple groups, Kegel covers should be essentially unique. All the proofs are basically the same. Once a member of the cover has root elements outside its kernel, its heart must be of a very restricted form.

3.2. A Theorem of Jordan Type

A group which has a faithful representation in finite dimension will in general have many fundamentally different ones coming from, for instance, tensor, symmetric, and exterior powers. This is not the case for finitary groups which are not finite dimensional. Each simple group in the conclusion of Theorem 1.3 has only one basic finitary representation. All others come from this one via direct sums, 'duality', or playing with the field [4, 10, 22].

We can think of an infinite dimensional, finitary group as being generated by 'elements of small degree'. Jordan [9, Théorème II] proved that a finite, primitive subgroup of Sym_n which contains an element of small support is either Alt_n or Sym_n. This idea was extended by Wielandt [30] who proved that a primitive subgroup of Sym_Ω, for infinite Ω, which contains a finitary permutation must in fact contain all of Alt_Ω. In our classification we need a theorem from [13] providing a result similar to Jordan's for finite, primitive linear groups. This theorem is really all of the classification of finite simple groups, CFSG, that is required for the proof of Theorem 1.3. (More is of course needed for Theorem 1.2, BBHST; and most conceivable applications of Theorem 1.3 would also require BBHST.)

Theorem 3.9 *Let H be a perfect finite subgroup of $\text{GL}_K(V)$, for K algebraically closed, with H primitive on V. Assume that H is generated by elements whose degree on V is less than $\sqrt{n}/12$, where $n = \dim_K V$. Then either*

(1) *H is Alt_n and V is a natural module, or*

(2) *H is a quasisimple classical group in the same characteristic as K, and V is a nearly natural module.*

If the finite group G is a quasisimple classical group $\text{Cl}_n(q)$ or simple $\text{PCl}_n(q)$, then a *natural* module for G is the module $\mathbb{F}_q^n$ for its defining projective representation (or any twist of this module via automorphisms). A *nearly natural* module is a natural module tensored up to a (possibly) larger field. If G is an alternating or symmetric group on Ω, then a *natural* module for G is the nontrivial irreducible factor in the permutation module $K\Omega$, for any field K.

The irreducible, imprimitive case of the theorem can be handled as well.

Proposition 3.10   *Let H, V, and K be as in Theorem 3.9, except that the representation of H on V is assumed to be irreducible but not primitive. Let $\Omega = \{V_i \mid i \in I\}$ be a maximal block system. Then in its action on Ω the group H induces Alt_Ω, and so H is the extension of a subdirect product of $|I|$ copies of $L \leqslant \mathrm{GL}_K(V_i)$ by Alt_Ω.*

Proof. The permutation action of H on Ω must involve a characteristic 0 solution to the primitive case. Therefore, by Theorem 3.9 (or Jordan's theorem), H induces Alt_Ω on Ω.   □

A representation and module as in the proposition will be called *generalized monomial*. The representation and module are *monomial* if all the V_i have dimension 1, so that L is cyclic. Any permutation module $K\Omega$ is monomial for Alt_Ω, and on occasion we shall abuse notation mildly by including the natural module (and any other nontrivial section of the permutation module) as a monomial module.

It has been observed by Phillips [27, (9.1)] that a stronger version of the proposition (with essentially the same proof) is valid in the broader context of finitary groups which are irreducible but imprimitive and are generated by elements of small degree. There finiteness plays no role and local finiteness is replaced by the weaker assumption that there are no noncyclic free subgroups. For a careful discussion of imprimitive modules, see Phillips' article. They arise naturally in the study of general finitary groups; for instance, an irreducible, locally solvable finitary group is either finite dimensional or imprimitive in countable dimension [27, (8.2.5)].

3.3. ULTRAPRODUCTS

In the previous two subsections we have seen that each group G from Theorem 1.3 is highly geometric. The linear algebra and geometry of G is essentially unique, as seen in the previous subsection, and precisely dictates the internal subgroup structure of G, as seen in the first subsection. In classifying these groups we seek to reverse the process by rebuilding the geometry out of the subgroup structure. We do this using the ultraproduct construction for groups and representations. From a suitably chosen Kegel cover we are able to fabricate the natural module and its geometry. There are few references concerning ultraproducts which are elementary and readily available (but see [18, pp. 64–67]). Therefore in an appendix we provide a primer on their use in the context of interest to us, the representation theory of groups.

Let $\{(G_i, N_i) \mid i \in I\}$ be a Kegel cover of the infinite locally finite simple group G. Order the index set I by declaring $i < j$ if and only if $G_i < G_j$ with $G_i \cap N_j = 1$. This ordering has the property that $i \leqslant k \geqslant j$ implies $\langle G_i, G_j \rangle \leqslant G_k$. Next let $\mathcal{F}$ be an ultrafilter generated by the directed set $(I, \leqslant)$, as described in the appendix. For each $i \in I$, let $(\varphi_i, c_i) \colon G_i \to \mathrm{GL}_{F_i}(W_i)$ be a projective representation of G_i whose kernel is N_i. Let $W = \prod_{\mathcal{F}} W_i$ be the ultraproduct vector space over the ultraproduct field $F = \prod_{\mathcal{F}} F_i$.

Theorem 3.11   *The ultraproduct provides a faithful projective representation*

$$(\varphi, c): G \to \mathrm{GL}_F(W).$$

Furthermore:

(1) *If, for each $i \in I$, the cocycle c_i is trivial, then c is trivial; that is, if each φ_i is a representation, then φ is a representation.*

(2) *If, for each $i \in I$, G_i in its action on W_i leaves invariant a nondegenerate form of type Cl, then G in its action on W leaves invariant a nondegenerate form of type Cl.*

(3) *If, for each $i \in I$, the space W_i has F_i-dimension less than k, then W has F-dimension less than k and G is a finite dimensional linear group.*

(4) *Let $1 \neq g \in G$. If, for each $i \in I$, the commutator $[W_i, \varphi_i(\langle g \rangle \cap G_i)]$ has F_i-dimension less than k, then $[W, \varphi(g)]$ has dimension less than k. In particular, G is a finitary linear group on W which leaves invariant a nondegenerate form of type Cl if each $\varphi_i(G_i)$ does.*

Proof.   Most of this is immediate from the remarks and results presented in the appendix, particularly Theorem D.3.

For the final remark of parts (3) and (4), we must show that the existence of a faithful projective representation $(\varphi, c): G \to \mathrm{FGL}_F(W)$ implies the existence of a corresponding genuine representation $\tilde{\varphi}$. For (3) this is immediate; there is a faithful and finite dimensional representation $\tilde{\varphi}: G \to \mathrm{GL}(W^* \otimes W)$. In the situation of (4) this particular representation is not finitary when W has infinite dimension, and something else must be done.

Consider now (4), and assume that W has infinite dimension. Let Z be the group of scalars in $\mathrm{GL}_F(W)$, so that Z is isomorphic to the multiplicative group of F. As φ is a projective representation, $\varphi(G).Z$ is a subgroup of $\mathrm{GL}_F(W)$. Here $\varphi(G).Z/Z \cong G$ is simple. As $\varphi(G).Z \cap \mathrm{FGL}_F(W).Z \geqslant \langle g, Z \rangle$, we must have $\varphi(G).Z \leqslant \mathrm{FGL}_F(W).Z$. As W has infinite dimension, $Z \cap \mathrm{FGL}_F(W) = 1$; so $\mathrm{FGL}_F(W).Z = \mathrm{FGL}_F(W) \times Z$. Therefore we can construct the desired representation $\tilde{\varphi}$ as the map φ followed by projection onto $\mathrm{FGL}_F(W)$, so that the image $\tilde{\varphi}(G)$ equals $\varphi(G).Z \cap \mathrm{FGL}_F(W)$.

If $\varphi(G)$ leaves invariant a nondegenerate form of type Cl, then $\varphi(G).Z$ acts on a 1-space of such forms. Its perfect subgroup $\tilde{\varphi}(G)$ therefore leaves each of these forms invariant.   □

Theorem 3.11 provides us with two valuable corollaries.

Corollary 3.12   *A locally finite simple group G which has a sectional cover composed of sporadic, cyclic, exceptional Lie type groups, or classical groups of bounded dimension, has a faithful representation as a linear group in finite dimension.*

Proof.   For the groups belonging to these classes, there is a k such that each group has a faithful representation on a vector space of dimension at most k. Therefore G has a faithful representation in dimension at most k by part (3) of the theorem.   □

In particular BBHST (Theorem 1.2) applies to say that G is either finite or of Lie type over a locally finite field.

Now choose $1 \neq g \in G$. If $g \in G_i$ and G_i/N_i is an alternating or classical group, let (φ_i, c_i) be a natural representation chosen to minimize $\dim_{F_i} \varphi_i(g)$, the *natural degree* of g in G_i/N_i. (If $g \notin G_i$, take the natural degree of g in G_i/N_i to be 0.) An immediate consequence of part (4) of the theorem is then:

Corollary 3.13 *A locally finite simple group G which has a sectional cover composed of alternating groups or classical groups of unbounded dimension in which the natural degrees of the element $g \neq 1$ are bounded has a faithful representation as a finitary linear group.*

4. Around a Proof

We discuss a proof of Theorem 1.3. We first show that a group as hypothesized resembles the conclusions internally, then we reconstruct its associated geometry and identify it externally.

4.1. THE ATTACK

Let G be a locally finite simple group which is a subgroup of $\mathrm{FGL}_K(V)$ but has no faithful representation as a linear group in finite dimension. Assume that G has the Kegel cover $\mathcal{K} = \{(G_\lambda, N_\lambda) \mid \lambda \in \Lambda\}$.

Our approach has five basic steps:

Step A.   Reconstruct the unipotent-by-quasisimple cover '$\mathcal{H} \cap \mathcal{CC}(G)$'.

Step B.   Find root elements.

Step C.   Find a quasisimple cover '$\mathcal{CC}_{\mathrm{qs}}(G)$'.

Step D.   Rebuild the natural module and its geometry.

Step E.   Identify G as the isometry group generated by all eligible root elements.

We begin by restricting the cover $\mathcal{K}$ under consideration.

Choose an arbitrary but fixed nonidentity element g of G, and set $d = \deg_V g = \dim_K[V, g]$. As mentioned after Lemma 3.3, the members of $\mathcal{K}$ with g in $G_\lambda \setminus N_\lambda$ form a subcover. Discarding the unnecessary members of $\mathcal{K}$, we may assume that $g \in G_\lambda \setminus N_\lambda$, for all λ.

We next further prune $\mathcal{K}$ in a typical application of the coloring argument, Lemma 3.3. Choose a constant κ. The exact value of κ is not crucial, but we want it to be large compared to d, that is, $\kappa \gg d$. In particular, to use Theorem 3.9 we need $\kappa > 144d^2$. Color the members (G_λ, N_λ) of the cover $\mathcal{K}$ with six colors, according to the isomorphism type of the simple quotient G_λ/N_λ:

1.   alternating Alt_{n_λ} with $\kappa < n_\lambda$;

2.   symplectic PSp_{n_λ} in odd characteristic with $\kappa < n_\lambda$;

3.   unitary PSU_{n_λ} with $\kappa < n_\lambda$;

4.   orthogonal $\mathrm{P}\Omega_{n_\lambda}$ with $\kappa < n_\lambda$;

5.   linear PSL_{n_λ} with $\kappa < n_\lambda$;

6.   cyclic, sporadic, exceptional Lie type, classical of degree n_λ with $\kappa \geqslant n_\lambda$, or alternating of degree n_λ with $\kappa \geqslant n_\lambda$.

Take note that, as discussed earlier in Subsections 2.3 and 2.5, we are including symplectic groups in even characteristic under the heading of orthogonal groups.

By the classification of finite simple groups, Theorem 1.1 (CFSG), we have assigned each member of the cover a color. Therefore by Lemma 3.3 there is a monochromatic subcover, one color class which is itself still a Kegel cover $\mathcal{S}$. As G is finitary but not linear, Corollary 3.12 says that the last color, class **6**, does not give a cover.

In summary, we have a locally finite simple group $G \leqslant \mathrm{FGL}_K(V)$ which is not linear, and an element $g \neq 1$ of G with $\deg_V g = \dim_K[V, g] = d$. The group G has a Kegel cover $\mathcal{S} = \{(G_i, N_i) \mid i \in I\}$ with $g \in G_i \smallsetminus N_i$, for all i; and the simple quotient G_i/N_i is of a fixed type, alternating or classical, one of **1** through **5**. The degree of G_i/N_i is n_i with $n_i > \kappa$, for a fixed constant κ $(\gg d)$. Indeed the set $\{n_i \mid i \in I\}$ is unbounded (again by Corollary 3.12).

In making these coloring arguments, we appear to have used the full strength of the classification of finite simple groups, CFSG. In fact it is possible to use Theorem 3.9 to show directly that nonlinear G must be covered by alternating or classical groups. Therefore, as mentioned before, Theorem 3.9 contains as much of the classification as required for the proof of Theorem 1.3. On the other hand, most envisioned applications would involve at least further appeal to Theorem 1.2 whose proof requires more serious use of the classification.

4.2. A Unipotent-by-Quasisimple Cover (Step A)

We first aim to prove that, starting from any Kegel cover, we can find one which looks like one of those for the groups of Theorem 1.3. These covers are parabolic for classical groups and natural for alternating groups. The nearly natural modules for classical groups and generalized monomial modules for alternating groups were defined and discussed in Subsection 3.2.

Proposition 4.1   *For each i, there is a G_i-composition factor V_i in V such that $\ker_{G_i} V_i \leqslant N_i$ and V_i is a nearly natural module for classical G_i/N_i or generalized monomial for alternating G_i/N_i.*

In particular, for a classical cover $\mathcal{S}$, the defining characteristic of the group G_i/N_i is the same as that of V. If V has characteristic 0, then $\mathcal{S}$ is an alternating cover.

Proof.   The heart of (G_i, N_i) can not be in the kernel of every G_i-composition factor since the heart is perfect. Thus there is some G_i-composition factor V_i whose kernel is in N_i. If the action on V_i is primitive, then the proposition comes from Theorem 3.9. In the imprimitive case we find that V_i is a generalized monomial module for G_i, as in Proposition 3.10. $\square$

For each i, now choose a genuinely natural module W_i over the field F_i.

G_i/N_i	W_i	F_i
Alt_{n_i}	$\mathbb{Q}^{n_i-1}$	$\mathbb{Q}$
$\mathrm{PCl}_{n_i}(q_i)$	$\mathbb{F}_{q_i}^{n_i}$	$\mathbb{F}_{q_i}$

Here PCl indicates one of the projective classical groups. For each i, next choose a (projective) natural representation

$$\varphi_i \colon G_i \to \mathrm{GL}_{F_i}(W_i),$$

taking care to minimize $d_i = \deg_{W_i} \varphi_i(g)$. (For alternating covers, we have a genuine representation; the associated cocycle can be taken to be trivial.)

Corollary 4.2   *For the projective representation $\varphi_i \colon G_i \to \mathrm{GL}_{F_i}(W_i)$ we have*

$$d_i = \deg_{W_i} \varphi_i(g) = \dim_{F_i}[W_i, \varphi_i(g)] \leqslant \dim[V_i, g] \leqslant d.$$

Proof.   By Proposition 4.1 the commutator $[W_i, \varphi_i(g)]$ has dimension at most that of $[V_i, g]$ which is a section of $[V, g]$ of dimension d.   □

As in Theorem 3.11 and Corollary 3.13, the simple group G acts on $\prod_{\mathcal{F}} W_i$, the ultraproduct of the spaces W_i and a vector space over the field $\prod_{\mathcal{F}} F_i$. It is convenient to extend this field ultraproduct to its algebraic closure F and the vector space ultraproduct to W, its tensor product with the field F. By the results mentioned and Corollary 4.2, $\dim_F[W, g] = d_0 \leqslant d$. Thus G acts as a subgroup of $\mathrm{FGL}_F(W)$.

At this point it may be worth standing back and considering what we have accomplished; starting with a finitary group, we have proved that it is indeed a finitary group. This does not sound like much, but in fact we have made a great deal of progress. While the original representation had no additional structure, the new representation has been constructed according to a precise recipe and so has many built-in properties. For instance, again by Theorem 3.11, if S has classical type PCl, then G acts on W leaving invariant a form of the same type as the members of S. Our original space might have been highly irreducible, say, a direct sum of finitely many copies of some faithful finitary module. As is the case with the natural modules for the groups we seek, our new space must be essentially irreducible since each G_i has a unique nontrivial composition factor within it (see Theorem 4.3 below).

Consider now the implications of Proposition 4.1 and its corollary for the finitary action of G on W. We learn that each G_i has in W a composition factor whose kernel is contained in N_i and which is nearly natural for G_i/N_i. In this factor the element g has degree at most d_0. But from our choice of the φ_i, in any nearly natural representation, g has degree at least d_0. Therefore the degree is exactly d_0; there is a unique G_i-composition factor in W on which g acts nontrivially; and this factor is nearly natural for G_i/N_i. If G_i/N_i is alternating then this factor is monomial, admitting some quotient $Z_m^{n_i-1}.\mathrm{Alt}_{n_i}$ of G_i, since the restriction on the degree of g forces any associated block of imprimitivity to have dimension 1.

We pull together some of the properties of our new finitary representation for G. As G is faithful on W, we may identify G with its image in $\mathrm{FGL}_F(W)$.

Theorem 4.3 *The locally finite simple group G is a subgroup of $\mathrm{FGL}_F(W)$ with F algebraically closed. For a fixed $1 \neq g \in G$, we have $d_0 = \dim_F[W, g] \ll \kappa$. Let $S = \{(G_i, N_i) \,|\, i \in I\}$ be a Kegel cover for G having one of the types **1** through **5**, and such that $g \in G_i \setminus N_i$, for all $i \in I$.*

(1) *In its action on W, each G_i has a unique nontrivial composition factor.*

(2) *If the cover S has alternating type as in **1**, then F has characteristic 0 and the composition factor of (1) is monomial.*

(3) *If the cover S has classical type as in one of **2** through **5**, then F has the same characteristic p as each classical group G_i/N_i, the composition factor of (1) is nearly natural of degree n_i, and the set $\{n_i \,|\, i \in I\}$ is unbounded. In cases **2**, **3**, and **4**, G leaves invariant a nondegenerate form of the same type as that for the classical groups G_i/N_i.*

For each $i \in I$, let $H_i = \langle g^{G_i} \rangle$ and $M_i = H_i \cap N_i$.

Proposition 4.4 *$\mathcal{H} = \{(H_i, M_i) \,|\, i \in I\}$ is a Kegel cover of G whose factors are the same as those of S. If the cover S is of classical type, then the subgroups H_i are unipotent-by-quasisimple in their action on W. If S is alternating, then the H_i are abelian-by-simple.*

Proof. The first sentence is immediate from Proposition 3.4.

As g acts nontrivially on only one G_i-composition factor in W, the same is true of H_i. Therefore the kernel of the action on this factor is a unipotent normal subgroup, and modulo this normal subgroup we get the action described above. In the case of an alternating cover, our space W has characteristic 0; so a unipotent group is torsion free. In a locally finite group such a group can only be trivial, so the monomial quotient group must be all of H_i. $\square$

It is not hard to go one step further. If we select those members of $\mathcal{H}$ with H_i perfect, then we actually have a Kegel cover of G which is contained in the heart of S.

Corollary 4.2 provides us with a converse to Theorem 3.11(4). In that theorem we saw that having an element of bounded representation degree in Kegel quotients forces a locally finite simple group to be finitary. Now Corollary 4.2 says that, in a finitary, locally finite simple group, elements must have bounded representation degree in their Kegel quotients. A little more can be squeezed out of these observations.

Theorem 4.5 *A simple section of a finitary locally finite group is itself finitary.*

To prove this, first find a Kegel cover for the simple section S. Pull the members of this cover back to finite subgroups of the parent locally finite group. The arguments of Proposition 4.1 and its corollary go through to prove that elements of the simple section S have bounded representation degree in the Kegel quotients of the section. Therefore by Theorem 3.11(4) the simple group S is finitary.

4.3. Alternating Groups

In this section we complete a proof of the following two theorems.

Theorem 4.6   *A locally finite simple group which is infinite and finitary and has a Kegel cover all of whose quotients are alternating groups is isomorphic to an alternating group* Alt_Ω, *for some infinite set* Ω.

Theorem 4.7   *A locally finite simple group which is infinite and finitary in characteristic* 0 *is isomorphic to an alternating group* Alt_Ω, *for some infinite set* Ω.

These theorems can be found in [10]. (See also [4].) The two results are seen to be equivalent, using what we have proven earlier. Indeed, a Kegel cover of a characteristic 0 group must have alternating type by Proposition 4.1. Conversely, in the previous section we constructed a faithful finitary representation in characteristic 0 for any locally finite simple and finitary group that possesses a Kegel cover of alternating type.

We isolate the alternating case from the classical because we are able to give a nearly complete proof. Although the alternating case is a little different and a little easier than the classical case, the proof given here is a good introduction to the more difficult arguments. Indeed some of the results in this section have slightly easier proofs if we use more permutation group theory, but we stay with proofs similar in spirit to those of the general case. These are, in any event, not overly deep or complex.

Assume then that we have an infinite, locally finite simple G which is finitary and has a Kegel cover whose quotients are alternating groups as in Theorem 4.3(2). Thus $G \leqslant \mathrm{FGL}_F(W)$ where the algebraically closed field F has characteristic 0. As in Proposition 4.4, G has the Kegel cover $\mathcal{H} = \{(H_i, M_i) \mid i \in I\}$, where $H_i/M_i \cong \mathrm{Alt}_{n_i}$. By Maschke's theorem, $W = [W, H_i] \oplus C_W(H_i)$. The module $[W, H_i]$ is monomial for H_i. Its dimension is $n_i - \epsilon$, and the base group $M_i \cong Z_m^{n_i-1}$ acts diagonally. (Here ϵ is 1 or 0 as $m = 1$ or $m \neq 1$.)

The element g was originally chosen as an arbitrary nonidentity element. Let us now assume that it was chosen to have odd prime order (as certainly was possible). In each representation

$$H_i \cong M_i.\mathrm{Alt}_{n_i} \xrightarrow{\varphi_i} \mathrm{GL}_{F_i}(W_i) \cong \mathrm{GL}(\mathbb{Q}^{n_i-1}),$$

we have $\ker \varphi_i = M_i$ and $d_0 = \deg g = \dim[W_i, g] = s(p-1) \ll n_i$, where the element $\varphi_i(g)$ is represented by s distinct p-cycles in the natural permutation representation of $\varphi_i(H_i) \cong \mathrm{Alt}_{n_i}$.

We also have the injections

$$\eta_{i,j} : H_i \to H_j/M_j \cong \mathrm{Alt}_{n_j}$$

available to us, furnished by the Kegel cover for every $j > i$. Without further information, this faithful permutation representation of H_i could have many different types; but we also know that the element g is always represented by s distinct p-cycles and no more. (We have $d_0 = s(p-1)$; consider φ_j and the representation of H_j on W.) This additional knowledge is enough for us to kill off the Kegel kernels M_i, and to identify the injections $\eta_{i,j}$ explicitly.

Proposition 4.8 *Let the monomial group $A \cong Z_m^{n-1}.\mathrm{Alt}_n$ be a subgroup of Alt_Δ, and let the base group of A be $B \cong Z_m^{n-1}$. Further assume that the element g of $A \smallsetminus B$ has odd prime order and is represented by s distinct p-cycles both in Alt_Δ and in the natural representation of $A/B \cong \mathrm{Alt}_n$. If $n \gg sp$, then there is a subgroup $C \cong \mathrm{Alt}_n$ which contains g and complements B in A.*

Proof. As a consequence of Corollary 2.2, $\bar{A} = A/B \cong \mathrm{Alt}_n$ is generated by at most $e = 3 + (n-2)/s(p-1)$ conjugates of $\bar{g}$, a p-element composed of s distinct p-cycles. (One of these can be taken to be $\bar{g}$ itself.) For each such conjugate $\bar{h}_j$ of $\bar{g}$, choose a preimage h_j in A which is a conjugate of g. (Take care to lift $\bar{g}$ to g.) Let C be the subgroup of A generated by the various h_j. As $\bar{C} = \bar{A}$, we have $A = BC$; so C is isomorphic to $Z_k^{n-1}.\mathrm{Alt}_n$, for some divisor k of m. We claim that $k = 1$. Indeed C moves at most $esp = (3 + (n-2)/s(p-1))sp$ points of Δ, a number only slightly larger than n and a good deal less than $2n$ since p is odd and $n \gg sp$. On the other hand, it is an easy exercise to prove that any nontrivial elementary abelian q-group Z_q^{n-1} has no faithful permutation representation of degree less than $(n-1)q$, which is at least of the order of $2n$. Thus no prime divisor q of k exists; and $k = 1$, as claimed. □

Lemma 4.9 *Root elements, that is, 3-cycles, exist in G. More precisely, there is an element t of order 3 and a subset J of I, such that $\mathcal{H}_J = \{(H_j, M_j) \mid i \in J\} \subseteq \mathcal{H}$ is a subcover and t acts as a 3-cycle in $H_j/M_j \cong \mathrm{Alt}_{\Omega_j}$, for each $j \in J$.*

Proof. Choose $i \in I$ with $n_i \gg sp$, as in Proposition 4.8, for $H_i/M_i \cong \mathrm{Alt}_{n_i}$. Restricting the monomial module $[W, H_i]$ to the subgroup $C_i \cong \mathrm{Alt}_{\Omega_i}$ guaranteed by the proposition, we have a natural module. In particular, a 3-cycle t of C_i has degree 2 on $[W, H_i]$ and so degree 2 on W. But then it also has degree 2 on $[W, H_j]$, a monomial module for H_j, for every $j \geq i$. The image of t in $\mathrm{Alt}_{\Omega_j} \cong H_j/M_j$ is thus also a 3-cycle, for all $j \in J = \{j \mid j \geq i\}$. □

Lemma 4.10 *For all $j \in J$, we have $M_j = 1$ and $H_j \cong \mathrm{Alt}_{\Omega_j}$.*

Proof. For each $j \in J$ and any $j < k \in J$, the monomial group H_j is embedded isomorphically in $H_k/M_k \cong \mathrm{Alt}_{\Omega_k}$. Its image contains 3-cycles of H_k/M_k by Lemma 4.9. Therefore, by Proposition 2.1, we must have $M_j = 1$ and $H_j \cong \mathrm{Alt}_{\Omega_j}$. □

Lemma 4.11 *For all $i, j \in J$, the injection $\eta_{i,j}$ of $H_i \cong \mathrm{Alt}_{\Omega_i}$ with $|\Omega_i| = n_i$ into $H_j \cong \mathrm{Alt}_{\Omega_j}$ with $|\Omega_j| = n_j$ is induced by a unique injection of sets $\eta_{i,j}^*: \Omega_i \to \Omega_j$.*

Proof. By Lemma 4.9, in this embedding 3-cycles go to 3-cycles; so this is an immediate consequence of Proposition 2.1. □

Now let $\Omega = \underrightarrow{\lim}\, \Omega_j$ be the direct limit of the finite sets Ω_j with respect to the injections $\eta_{i,j}^*$.

Proposition 4.12 *The group G is isomorphic to Alt_Ω.*

Proof. The element t of Lemma 4.9 is a 3-cycle in each $H_j \cong \mathrm{Alt}_{\Omega_j}$, for $j \in J$; so it acts on Ω as a 3-cycle. For any triple α, β, γ from Ω, there is an i with $\alpha, \beta, \gamma \in \Omega_i$. Therefore the subgroup $H_i \cong \mathrm{Alt}_{\Omega_i}$ of G contains the 3-cycle (α, β, γ) of Alt_Ω. All possible 3-cycles are in G, so the simple group G is all of Alt_Ω.   □

4.4. ROOT ELEMENTS (STEP B)

The previous subsection handles the alternating case (2) of Theorem 4.3, so from now on we may assume we are in the classical case (3), with a Kegel cover having one of the types **2** through **5**.

In the alternating case we found root elements by first splitting each Kegel quotient off its kernel and then using the structure of the natural module. The same is done in the classical case, starting with the following proposition which was pointed out by Ulrich Meierfrankenfeld. (The statement $E = O^p(E)$ says that E has no nontrivial homomorphic image which is a p-group.)

Proposition 4.13   *Suppose that the finite group $E = O^p(E)$ acts on the finite dimensional K-vector space U in characteristic p with a unique nontrivial composition factor. Assume also that $E = E_0 O_p(E)$ for $E_0 \leqslant E$ implies $E = E_0$. Then $[E, O_p(E)] \leqslant C_E(U)$.*

Proof. We proceed by induction on $\dim_K U$. Set $Q = [E, O_p(E)]$.

First assume that $[U, Q]$ is not trivial as a KE-module. Then, since $E = O^p(E)$, the unique nontrivial composition factor is in $[U, Q] = [U, E]$. As Q itself is unipotent, we also have $[U, Q] < U$.

Let Y be an E-invariant hyperplane of U which contains $[U, Q] = [U, E]$. By induction we have $Y \leqslant C_U(Q)$. In particular, the action of Q on U is quadratic: $[U, Q, Q] = 0$. Choose $x \in U \smallsetminus Y$, so that $U = Kx \oplus Y$ and $[U, Q] = [x, Q]$ as K-space. By quadratic action the set $W = \{[x, q] \mid q \in Q\}$ is an $\mathbb{F}_p Q$-submodule of U. Indeed it is an $\mathbb{F}_p E$-submodule since $[U, E] \leqslant C_U(Q)$, so that, for $q \in Q$ and $e \in E$,

$$[x, q]^e = [x^e, q^e] = [x + [x, e], q^e] = [x, q^e] + [[x, e], q^e] = [x, q^e] \in W.$$

Consider now the KE-module $\bar{U} = U/C_U(E)$. The image of $[U, E]$ is an irreducible KE-submodule $\bar{T}$ with $\bar{T} = K\bar{W}$, because $[U, E] = [x, Q]$ as a K-space. As an $\mathbb{F}_p E$-module (of possibly infinite dimension), $\bar{T}$ has a nonzero irreducible submodule $\bar{W}_0$ within finite $\bar{W}$. Thus $\bar{T} = K\bar{W} = K\bar{W}_0$ is a sum of $\mathbb{F}_p E$-irreducibles and so is completely reducible. Therefore $\bar{W}$ is complemented in $\bar{T}$; there is a $\mathbb{F}_p E$-submodule $\bar{Z}$ of $\bar{T}$ with $\bar{T} = \bar{W} \oplus \bar{Z}$.

For an arbitrary $e \in E$, we have

$$\bar{x}^e = \bar{x} + [\bar{x}, e] = \bar{x} + (\bar{w} + \bar{z}) = \bar{x} + [\bar{x}, q] + \bar{z} = \bar{x}^q + \bar{z},$$

where $\bar{z}$ is in $\bar{Z}$ and $\bar{w}$ is in $\bar{W}$, so that $\bar{w} = [\bar{x}, q]$, for some $q \in Q$. Therefore $(\bar{x} + \bar{Z})^e = (\bar{x} + \bar{Z})^q$, and generally

$$(\bar{x} + \bar{Z})^E = (\bar{x} + \bar{Z})^Q.$$

By a Frattini argument, $E = QN_E(\bar{x} + \bar{Z})$; so by assumption $E = N_E(\bar{x} + \bar{Z})$. That is, for each $e \in E$, we have $[\bar{x}, e] \in \bar{Z}$. In particular, for each $q \in Q$, this gives $[\bar{x}, q] \in \bar{Z}$; but already $[\bar{x}, q] \in \bar{W}$. Therefore $[\bar{x}, q] \in \bar{W} \cap \bar{Z} = \bar{0}$. We conclude that $[\bar{x}, Q] = \bar{0}$, which is not true since $\bar{W}$ is nonzero. The contradiction shows that this case can not occur, and therefore $[U, Q]$ must be trivial as a KE-module.

Dually, $U/C_U(Q)$ is a trivial E-module. Therefore

$$[U, Q, E] = [E, U, Q] = 0\,,$$

whence $[Q, E, U] = 0$ by the Three Subgroups Lemma $[1, (8.7)]$. As $E = O^p(E)$, we have $[Q, E] = [O_p(E), E, E] = [O_p(E), E] = Q$; that is, Q is trivial on U. $\square$

Corollary 4.14   *For all i, H_i splits over $O_p(H_i)$.*

Proof.   Let E be a minimal supplement to $O_p(H_i)$ in H_i. In particular E is perfect. As E is finite, there is in W an E-invariant finite dimensional subspace U with $W = U \oplus C$, for some $C \leqslant C_W(E)$. Now E is faithful in its action on U and satisfies all the hypotheses of the proposition. Therefore perfect E intersects $O_p(H_i)$ only in a central p-subgroup. The Schur multiplier of the classical group H_i/M_i in characteristic p has trivial p-part $[8, \text{p. } 302]$, so this intersection is trivial. $\square$

Proposition 4.15   *Let H be a classical group $\mathrm{Cl}_n(q)$, for $n \gg 0$, and let U be an extension of a trivial KH-module Z by a nearly natural KH-module. Then either*

(i) $U = Z \oplus [U, H]$, or

(ii) $H \cong \mathrm{Sp}_{2m}(q)$ with q even, and U is a nondegenerate but defective nearly natural module for $H \cong \Omega_{2m+1}(q)$ with symplectic radical of dimension 1.

Proof.   Since the dual of a nearly natural module is also nearly natural, the proposition is equivalent $[1]$ to the cohomological statement that $H^1(H, \mathbb{F}_q^n)$ is 0 but for the exceptional case (ii) where it has dimension 1. As such, the result is reasonably well-known; see, for instance, $[17, \text{Theorem } 2.14]$ or $[19, \S1]$.

In fact, this result can be proven in an elementary fashion using generation results like Theorem 2.8 and Proposition 2.11. Consider the case in which H is a unitary group with $n \geqslant 5$. We may assume that Z has dimension 1.

As $\bar{Y} = Y/Z$ is nearly natural, a transvection t of H has commutator of dimension 1 or 2 on Y, hence centralizer $C_Y(t)$ of codimension 1 or 2. Clearly $C_Y(t)$ contains Z, so $\bar{W} = \overline{C_Y(t)}$ has the same codimension 1 or 2 in $\bar{Y}$. The subspace $\bar{W}$ is invariant under $C_H(t)$. As $n \geqslant 5$, it can only be the hyperplane $C_{\bar{Y}}(t)$. Therefore $C_Y(t)$ has codimension 1, and t is a transvection on Y.

By Proposition 2.11, the group H is generated by n of its transvections. Therefore $[U, H]$ has dimension at most n. But this commutator must cover the nearly natural quotient U/Z of dimension n. We conclude that $[U, H]$ has dimension exactly n, and $U = Z \oplus [U, H]$, as claimed.

This handles the unitary case, and the argument for the special linear case is essentially identical. (In large enough dimension n, these groups are generated by n transvections $[16, \text{Theorem } 4.9]$.) For H symplectic in odd characteristic, the result is trivial since $Z(H) = \{\pm 1\}$ and $[U, H] = [U, Z(H)]$ intersects Z trivially. For H symplectic in characteristic 2 and of dimension $2n > 2$, the argument given above for the unitary groups can be adapted. Transvections again act as transvections, but now (see Theorem 2.8) we need $2n + 1$ transvections to generate H. The argument of the previous paragraph then suggests the (real) possibility of a nonsplit extension with a trivial submodule of dimension 1, but larger trivial submodules can be ruled out. Similar arguments for the orthogonal groups require the use of Siegel elements and so are messier.   □

Theorem 4.16   *G contains root elements in its action on W.*

Proof.   Choose an i, and let L be a complement to $O_p(H_i)$ in H_i. (If L is symplectic in characteristic 2 so that $L \cong \mathrm{Sp}_{2n}(q) \cong \Omega_{2n+1}(q)$, then replace L by an orthogonal subgroup $L_0 \cong \Omega_{2n}^\epsilon(q)$.) In W there is a unique L-composition factor and that is nearly natural. Therefore $[W, L]$ is an extension of a trivial KL-module Z by this nearly natural composition factor. By the previous proposition, we have $[W, L] = Z \oplus [W, L, L]$; but L is perfect, so $[W, L] = [W, L, L]$ is nearly natural for L and $Z = 0$. Thus $W = [W, L] \oplus C_W(L)$. In the nonorthogonal cases, a transvection of L on its natural module is also a transvection on W. In the orthogonal case, a Siegel element s of L will also have commutator dimension 2 on W. As L is irreducible on nearly natural $[W, L]$, the L-invariant forms on this module are unique up to scalar multiples. Therefore this dimension 2 commutator which is singular for L is singular in the orthogonal geometry on W, and s is a Siegel element on W. Thus in all cases, the root elements of L become root elements of G on W.   □

4.5. The Rest

Now that we have root elements, the end of the proof is in sight, although some of the arguments are still rather delicate in nature. We present this part in less detail.

4.5.1. Quasisimple Complements (Step C)

Unlike the alternating case, for the classical groups there are Kegel covers which are not just modified quasisimple covers. To pave the way for the geometric reconstruction to come, we replace our given Kegel cover by a related quasisimple cover. If we think of our cover as having parabolic type, what we now wish to do is restrict our attention to Levi subgroups. From the geometric viewpoint, we want only to consider subspaces with trivial radical.

Originally the element g was chosen as an arbitrary nonidentity element. It is now convenient to go back and rechoose so that it is a root element of G. This is possible by Theorem 4.16, and allows us to assume that each H is generated by root elements.

For each $i \in I$, let

$$\mathcal{L}_i = \{L \leqslant H_i \,|\, L \cap O_p(H_i) = 1,\ L \text{ quasisimple of type Cl}\}.$$

Then set $\mathcal{L} = \bigcup_i \mathcal{L}_i$.

Proposition 4.17   *$\mathcal{L}$ is a quasisimple cover of G.*

Consider the case in which $\mathrm{Cl} = \mathrm{Sp}$ and F has odd characteristic. For each $i \in I$, we need to find a $j \in I$ and an $L \in \mathcal{L}_j$ with $H_i \leqslant L$. If H_i itself is quasisimple, there is nothing to prove; so we may assume that this is not the case.

Choose a j with $H_i \leqslant H_j$ but $H_i \cap M_j = 1$. The symplectic space W is nondegenerate, and in particular $C_W(G) = 0$. Thus we may choose our j so that additionally $[W, H_i] \cap C_{[W,H_j]}(H_j) = 0$. (Some argument is needed here. This is actually a 'Kegel cover' property for the irreducible G-module W.) Therefore $[W, H_i]$ is embedded isometrically in the nearly natural module $\bar{X} = [W, H_j]/C_{[W,H_j]}(H_j)$ for the symplectic group $\bar{H} = H_j/O_p(H_j) \cong \mathrm{Sp}_{2n}(q)$. Set $k = \dim[W, H_i] = \dim[\bar{X}, \bar{H}_i]$. As H_i is not quasisimple, we have $k < 2n$.

By Theorem 2.8, there is a set $\bar{\Sigma}$ of $2n - k$ transvections $\bar{t}$ of $\bar{H}$ such that $\bar{H} = \langle \bar{H}_i, \bar{\Sigma} \rangle$. We lift each individual transvection $\bar{t}$ of $\bar{\Sigma}$ to a transvection t of H_j and call the corresponding set of $2n - k$ transvections Σ. Consider $L = \langle H_i, \Sigma \rangle$. By construction $\bar{L} = \bar{H}$ with $\bar{X} = [\bar{X}, \bar{L}]$ a nearly natural module of dimension $2n$. However

$$\dim[W, L] \leqslant \dim[W, H_i] + \dim[W, \Sigma] \leqslant k + (2n - k) = 2n.$$

We conclude that $[W, L]$ of dimension $2n$ is a nearly natural module for the quasisimple group $L \cong \bar{H} \cong \mathrm{Sp}_{2n}(q)$. In particular $H_i \leqslant L \in \mathcal{L}_j$, as required.

4.5.2. *Reconstructing the Space (Step D)*

In Theorem 4.3 the space $_K V$ was replaced by the new, essentially irreducible G-module $_F W$; but this new space is still not ideal. The field F may not be the natural one for the group G. Indeed since F is the algebraic closure of an ultraproduct of finite fields it typically has large transcendence degree. Now we construct another new space on which G acts, this one defined over a direct limit of finite fields, which is thus locally finite.

We begin with a result of Hartley and Shute:

Theorem 4.18 [14, Theorem C] *Let $\mathbb{F}_{q_1}$ and $\mathbb{F}_{q_2}$ be two finite fields, and let $G_1 \cong \Phi(\mathbb{F}_{q_1})$ and $G_2 \cong \Phi(\mathbb{F}_{q_2})$ be two quasisimple Lie type groups of the same type Φ. Then G_1 is isomorphic to a subgroup of G_2 if and only if $\mathbb{F}_{q_1}$ is isomorphic to a subfield of $\mathbb{F}_{q_2}$. In this case, any two subgroups of G_2 isomorphic to G_1 are conjugate via an automorphism of G_2.*

We are particularly interested in the case where each G_i is a classical group, $\mathrm{Cl}_n(\mathbb{F}_{q_i})$. There is a canonical subgroup of the matrix group G_2 isomorphic to G_1, namely the corresponding matrix subgroup over the subfield $\mathbb{F}_{q_1}$ of $\mathbb{F}_{q_2}$. The embedding of G_1 as this subgroup is induced by the canonical $\mathbb{F}_{q_1}$-linear map of $X_1 \cong \mathbb{F}_{q_1}^n$ into $X_2 \cong \mathbb{F}_{q_2}^n$. The space X_2, as $\mathbb{F}_{q_1} G_1$-module, is a direct sum of $|\mathbb{F}_{q_1} : \mathbb{F}_{q_2}|$ isomorphic copies of X_1, permuted among themselves by $\mathbb{F}_{q_2}$ scalar multiplication.

In the cases $\mathrm{Cl} \in \{\mathrm{Sp}, \mathrm{SU}, \Omega\}$, each automorphism of G_2 comes from conjugation by a matrix of $\mathrm{GL}_{\mathbb{F}_{q_2}}(X_2)$ which normalizes G_2 followed by an automorphism of the field $\mathbb{F}_{q_2}$ (if we assume that $n > 8$ when $\mathrm{Cl} = \Omega$). Thus, in these cases, Theorem 4.18 can in part be rephrased in the geometric form:

Theorem 4.19   *Suppose that η is an isomorphism of the quasisimple classical group $G_1 \cong \text{Cl}_n(\mathbb{F}_{q_1})$ into $G_2 \cong \text{Cl}_n(\mathbb{F}_{q_2})$, where $\text{Cl} \in \{\text{Sp}, \text{SU}, \Omega\}$ and $n > 8$. Then there is an isomorphism σ of $\mathbb{F}_{q_1}$ into $\mathbb{F}_{q_2}$, and a σ-semilinear injective isometry η^* of $X_1 \cong \mathbb{F}_{q_1}^n$ into $X_2 \cong \mathbb{F}_{q_2}^n$ which induces η. The map η^* is uniquely determined up to scalar multiplication by a member of $\mathbb{F}_{q_2}$.*

A new proof (and extension) of Hartley and Shute's Theorem 4.18 has been given by Liebeck and Seitz [19]. The basic observation is that root elements of G_1 must be mapped to root elements of G_2. To convince yourself of this, think of how the group of lower unitriangular matrices over the subfield $\mathbb{F}_{q_1}$ fits into that over the field $\mathbb{F}_{q_2}$. The center of the small subgroup (Sylow in $\text{SL}_n(\mathbb{F}_{q_1})$) falls into the center of the large subgroup (Sylow in $\text{SL}_n(\mathbb{F}_{q_2})$). But these centers are in both cases composed of transvections, root elements in the case $\text{Cl} = \text{SL}$. For any abstract embedding η of $\text{SL}_n(\mathbb{F}_{q_1})$ into $\text{SL}_n(\mathbb{F}_{q_2})$ consideration of nilpotency class shows that the center of the (small) Sylow group of the image of η must lie in the center of the large Sylow subgroup. Therefore root elements are taken by η to root elements.

The apparently small change of allowing G_1 and G_2 to have different dimensions produces drastic results. Concerning an arbitrary injection η of $G_1 \cong \text{Cl}_{n_1}(\mathbb{F}_{q_1})$ into $G_2 \cong \text{Cl}_{n_2}(\mathbb{F}_{q_2})$, we can say almost nothing. Indeed, from any pair of fields $\mathbb{F}_{q_1}$ and $\mathbb{F}_{q_2}$ and any degree k permutation representation of G_1, we can construct an embedding of G_1 in $\text{Cl}_k(\mathbb{F}_{q_2})$. If we now require that η respect root elements (rather than proving it along the way), then order is restored:

Theorem 4.20   *Suppose that η is an isomorphism of the quasisimple classical group $G_1 \cong \text{Cl}_{n_1}(\mathbb{F}_{q_1})$ into $G_2 \cong \text{Cl}_{n_2}(\mathbb{F}_{q_2})$, where $\text{Cl} \in \{\text{Sp}, \text{SU}, \Omega\}$ and $n_1 > 8$. Assume additionally that, for the root element g of G_1, the element $\eta(g)$ is a root element of G_2. Then there is an isomorphism σ of $\mathbb{F}_{q_1}$ into $\mathbb{F}_{q_2}$, and a σ-semilinear injective isometry η^* of $X_1 \cong \mathbb{F}_{q_1}^{n_1}$ into $X_2 \cong \mathbb{F}_{q_2}^{n_2}$ which induces η. The map η^* is uniquely determined up to scalar multiplication by a member of $\mathbb{F}_{q_2}$.*

The result can be proven using the earlier result, or it can be given a direct proof from first principles since root elements carry so much information about the related geometry. As mentioned before, a classical group is generated by what is essentially the smallest possible number of root elements (see Theorem 2.8 and Proposition 2.11). Since η respects root elements, we find that $Y = [X_2, \eta(G_1)]$ has dimension essentially n_1 and so must look something like the natural module X_1 tensored up to $\mathbb{F}_{q_2}$. This is the beginning of the construction of the semilinear map. Theorem 4.19 gives the action of $\eta(G_1)$ on the nondegenerate and nearly natural module Y, and the only way of extending this action to all of X_2 is by making $\eta(G_1)$ trivial on $Y^\perp$.

Some care must be taken when the n_i are both odd, the q_i are even, and $\text{Cl} = \Omega$. In this case neither X_1 nor X_2 is irreducible, and there are two fundamentally different embeddings, depending upon whether or not the symplectic radical of X_1 is mapped into that of X_2 or not.

We are now in a position to reconstruct the natural space on which our group G acts in the cases $\text{Cl} \in \{\text{Sp}, \text{SU}, \Omega\}$. For each $L \in \mathcal{L}$, we have $L \cong \text{Cl}_{n_L}(\mathbb{F}_{q_L})$. Let $X_L \cong \mathbb{F}_{q_L}^{n_L}$ be a natural $\mathbb{F}_{q_L}$ L-module. To create our G-space X, it will be helpful to fix some member $L \in \mathcal{L}$ and prune $\mathcal{L}$ to the quasisimple cover $\mathcal{Q} = \{Q \in \mathcal{L} \mid L \leqslant Q\}$. By the previous theorem, for each $Q \in \mathcal{Q}$, there is a field automorphism $\sigma_{L,Q}$ and a $\sigma_{L,Q}$-semilinear map $\eta^* \colon X_L \to X_Q$ which induces the inclusion $L \leqslant Q$. This map η^* is unique up to scalar multiplication by some member of $\mathbb{F}_{q_Q}$. For each $Q \in \mathcal{Q}$, choose and fix one of these multiples $\eta^*_{L,Q}$. Next, for each pair $P, Q \in \mathcal{Q}$ with $P \leqslant Q$, there is a collection of maps ρ which are $\sigma_{P,Q}$-semilinear maps from X_P to X_Q, inducing the inclusion $P \leqslant Q$. Pairwise these maps ρ differ only by a scalar multiple from $\mathbb{F}_{q_Q}$. In particular, their images are $\mathbb{F}_{q_P}$-subspaces of X_Q with trivial pairwise intersections. On the other hand, the restriction of the inclusion $P \leqslant Q$ from P to its subgroup L is the inclusion $L \leqslant Q$. Therefore exactly one of the maps ρ will satisfy

$$\eta^*_{L,P} \rho = \eta^*_{L,Q}.$$

In this case, we set $\eta^*_{P,Q} = \rho$.

The injective maps $\eta^*_{P,Q}$ and $\sigma_{P,Q}$, for all $P, Q \in \mathcal{Q}$ with $P \leqslant Q$, allow us to construct the vector space $X = \varinjlim X_Q$ over the field $E = \varinjlim \mathbb{F}_{q_Q}$. We then see that X is a finitary EG-module. We can further construct an invariant nondegenerate form of type Cl on X as the direct limit of invariant forms on the X_Q, having used a L-invariant nondegenerate form to normalize as before. (Indeed a small amount of additional argument would show that E is a subfield of F and that X is isometric to an E-subspace of W.) In these cases, this completes the reconstruction of Step D.

There are two related reasons why the case Cl $=$ SL must be treated differently from that of the other classical groups. First, the result corresponding to Theorem 4.19 must allow for the transpose-inverse automorphism; geometrically, we must allow for dualities as well as semilinear maps. The second difficulty arises because, in the proof of the result corresponding to Theorem 4.20, there is no canonical complement $Z^{\perp}$ to $Z = [X_2, \eta(G_1)]$ available for unique extension of the action on Z. The answer to both problems is to consider G_i as acting on the direct product of the spaces X_i and $Y_i = X_i^*$ preserving the natural nondegenerate pairing $p \colon X_i \times Y_i \to \mathbb{F}_{q_i}$ given by $p(x, y) = xy$.

Theorem 4.21 *Suppose that η is an isomorphism of the quasisimple classical group $G_1 \cong \mathrm{SL}_{n_1}(\mathbb{F}_{q_1})$ into $G_2 \cong \mathrm{SL}_{n_2}(\mathbb{F}_{q_2})$ with $n_1 > 2$. Assume additionally that, for the root element (transvection) g of G_1, the element $\eta(g)$ is a root element of G_2. Set $X_i = \mathbb{F}_{q_i}^{n_i}$ and $Y_i = X_i^*$.*

Then there is an isomorphism σ of $\mathbb{F}_{q_1}$ into $\mathbb{F}_{q_2}$, and a σ-semilinear injective isometry $\eta^* \colon X_1 \times Y_1 \to X_2 \times Y_2$ which induces η and such that either

(a) *$\eta^*(X_1) \leqslant X_2$ and $\eta^*(Y_1) \leqslant Y_2$, or*

(b) *$\eta^*(X_1) \leqslant Y_2$ and $\eta^*(Y_1) \leqslant X_2$.*

In any event the map η^* is uniquely determined up to scalar multiplication by a member of $\mathbb{F}_{q_2}$.

Part (b) describes dualities as opposed to semilinear maps. The proof of uniqueness involves the observation that the set of transvections fixing a 1-space of X_1 must either fix a 1-space of X_2 (case (a)) or a hyperplane of X_2 (case (b)), but does not fix both since $n_1 > 2$. A canonical complement to $[X_2, \eta(G_1)]$ in X_2 is now provided by $[Y_2, \eta(G_1)]^{\perp}$ (and similarly with the roles of X_2 and Y_2 reversed).

To find paired spaces on which G acts, we once again prune $\mathcal{L}$ down to its subcover $\mathcal{Q}$ of all members containing some fixed L. This and the theorem allow us to construct as before field maps $\sigma_{L,Q}$ and semilinear isometries $\eta_{L,Q}^*$. Here we must choose, for all Q including L, paired spaces X_Q and Y_Q admitting Q. We make our choices so that always $\eta_{L,Q}^*$ takes X_L into X_Q and Y_L into Y_Q. Continuing as before, for each pair $P, Q \in \mathcal{Q}$ with $P \leqslant Q$, we find a scalar collection of $\sigma_{P,Q}$-semilinear maps from $X_P \times Y_P$ to $X_Q \times Y_Q$, each of which induces the inclusion $P \leqslant Q$. Exactly one of these possible choices for $\eta_{P,Q}^*$ will make true the equation

$$\eta_{L,P}^* \eta_{P,Q}^* = \eta_{L,Q}^*,$$

and this is the choice we make.

Our original choices for L guarantee that always $\eta_{P,Q}^*$ takes X_P into X_Q and Y_P into Y_Q. Thus we can define the pair of spaces $X = \varinjlim X_Q$ and $Y = \varinjlim Y_Q$ over the field $E = \varinjlim \mathbb{F}_{q_Q}$. On $X \times Y$ there is a G-invariant nondegenerate pairing which is the direct limit of those on the various $X_Q \times Y_Q$. Our reconstruction is complete for the groups of type SL as well.

4.5.3. *Identifying the Group (Step E)*

At the close of the previous subsection we had identified our group G as a subgroup of one of $\mathrm{FSp}_E(X, s)$, $\mathrm{FSU}_E(X, u)$, $\mathrm{F\Omega}_E(X, q)$, or $\mathrm{T}_E(Y, X)$. It remains to prove that we have equality. This is now largely a matter of book-keeping. Using the construction of X as a direct limit (and Y in the final case), we can reveal the large finitary group as a direct limit itself, and that can then be seen to be isomorphic to G, the direct limit of the Q in $\mathcal{Q}$. Alternatively we can prove that, in its action on X (and Y), the group G contains every root element of the larger group. At that point results like Theorem 2.5 prove the equality. The proof of Theorem 1.3 is then complete.

References

1. M. G. Aschbacher, *Finite Group Theory*, Cambridge University Press, Cambridge, 1986.
2. V. V. Belyaev, Locally finite Chevalley groups, in *Studies in Group Theory*, Urals Scientific Centre of the Academy of Sciences of USSR, Sverdlovsk, 1984, pp. 39–50 (in Russian).
3. V. V. Belyaev, Semisimple periodic groups of finitary transformations, *Algebra i Logika* **32** (1993), 17–33 (English translation 8–16).
4. V. V. Belyaev, Finitary linear representations of infinite symmetric and alternating groups, *Algebra i Logika* **32** (1993), 591–606 (English translation 319–327).
5. V. V. Belyaev, Inert subgroups in simple locally finite groups, *these Proceedings*.
6. A. V. Borovik, Periodic linear groups of odd characteristic, *Dokl. Akad. Nauk. SSSR* **266** (1982), 1289–1291.
7. P. J. Cameron and J. I. Hall, Some groups generated by transvection subgroups, *J. Algebra* **140** (1991), 184–209.
8. D. Gorenstein, *Finite Simple Groups: an Introduction to their Classification*, Plenum, New York, 1982.
9. C. Jordan, Théorèmes sur les groupes primitifs, *J. Math. Pure Appl.* **16** (1871), 383–408.
10. J. I. Hall, Infinite alternating groups as finitary linear transformation groups, *J. Algebra* **119** (1988), 337–359.

11. J. I. Hall, Finitary linear transformation groups and elements of finite local degree, *Arch. Math.* 50 (1988), 315–318.

12. J. I. Hall, Periodic finite simple groups of finitary linear transformations, in preparation.

13. J. I. Hall, M. W. Liebeck, G. M. Seitz, Generators for finite simple groups, with applications to linear groups, *Quart. J. Math.* (2) 43 (1992), 441–458.

14. B. Hartley and G. Shute, Monomorphisms and direct limits of finite groups of Lie type, *Quart. J. Math.* (2) 35 (1984), 49–71.

15. B. Hartley, Simple locally finite groups, *these Proceedings.*

16. W. M. Kantor, Subgroups of classical groups generated by long root elements, *Trans. Amer. Math. Soc.* 248 (1979), 347–379.

17. W. M. Kantor and R. A. Liebler, The rank 3 permutation representations of the finite classical groups, *Trans. Amer. Math. Soc.* 271 (1982), 1–69.

18. O. H. Kegel and B. A. F. Wehrfritz, *Locally Finite Groups*, North-Holland, Amsterdam, 1973.

19. M. W. Liebeck, G. M. Seitz, Groups generated by long root elements, preprint, 1993.

20. J. McLaughlin, Some groups generated by transvections, *Arch. Math. Basel* 18 (1967), 364–368.

21. U. Meierfrankenfeld, Non finitary simple locally finite groups, *these Proceedings.*

22. U. Meierfrankenfeld, A characterization of the natural module for some classical groups, *Geom. Ded.*, to appear.

23. U. Meierfrankenfeld, A simple subnormal subgroup for locally finite, finitary linear groups, preprint, 1992.

24. D. S. Passman, Semiprimitivity of group algebras of locally finite groups II, preprint, April 1994.

25. D. S. Passman, The semiprimitivity problem for group algebras of locally finite groups, preprint, October 1994.

26. R. E. Phillips, The structure of groups of finitary transformations, *J. Algebra* 119 (1988), 337–359.

27. R. E. Phillips, Finitary linear groups: a survey, *these Proceedings.*

28. D. E. Taylor, *The Geometry of the Classical Groups*, Heldermann Verlag, Berlin, 1992.

29. S. Thomas, The classification of the simple periodic linear groups, *Arch. Math.* 41 (1983), 103–116.

30. H. Wielandt, *Unendliche Permutationsgruppen*, Lecture Notes: Mathematisches Institut der Universität, Tübingen, 1960.

31. A. E. Zalesskiĭ, Group rings of locally finite groups, *these Proceedings.*

32. A. E. Zalesskiĭ and V. N. Serezhkin, Finite linear groups generated by reflections, *Math. USSR-Izv.* 17 (1981), 477–503.

Appendix

A Primer on Ultraproducts of Groups

A. Introduction

Let G be a group and $\{N_i \mid i \in I\}$ be a set of normal subgroups. When $\bigcap_i N_i = 1$, there is a natural embedding of G in the Cartesian product $\prod_{i \in I} \bar{G}_i = \prod_i \bar{G}_i$ of the groups $G/N_i = \bar{G}_i$. This allows us to transfer many properties of the groups $\bar{G}_i$ to the group G. For instance, if each $\bar{G}_i$ has exponent bounded by e, then so does G. Of particular interest is the ability to combine modules for the individual $\bar{G}_i$ together into one large module for G.

Of course many groups G do not have this rich normal structure. Indeed in the main body of the paper we are primarily interested in simple groups G. What every group does have is subgroups, so it would be highly desirable to write G as some sort of sub-Cartesian product of certain of its subgroups G_i. This we do using the ultraproduct of groups. Ultraproducts allow us to patch together global properties of G out of local properties of the G_i.

The reader may notice that the proofs presented here have a repetitive nature. This is because the general ultraproduct is a model theoretic construct which transfers first order properties from the coordinate objects to a global object that is a quotient of the Cartesian product. As such, ultraproducts have many interesting applications outside of the group theoretic realm, but we shall not discuss them here.

We would prefer not to require information on all subgroups of G, but if not all then how large a set of subgroups $\{G_i \,|\, i \in I\}$ (where I is some indexing set) is needed? Before the condition $\bigcap_i N_i = 1$ was enough, but a normal subgroup N_i arrives with more luggage than an arbitrary subgroup G_i. Certainly we must have $\bigcup_i G_i = G$, but now this is not enough. Equations such as $g \cdot h = k$ must be verifiable entirely within some coordinate, so there must be an i for which all three of g, h, k belong to G_i. (For instance, if g and h are in no common G_i, then their natural images in $\prod_i G_i$ commute.) In fact we want still more. The motivating observation is that every set is the direct limit of its finite subsets, whence every group is the direct limit of its finitely generated subgroups. A suitable set $\{G_i \,|\, i \in I\}$ of subgroups will be one which has G as its direct limit. As defined in the main body, a set of subgroups $\{G_i \,|\, i \in I\}$ of G is a *local system* for G if $\bigcup_i G_i = G$ and, for each pair i, j from I, there is a k with $\langle G_i, G_j \rangle \leqslant G_k$. As desired, the group G is the direct limit of the members of any local system $\mathcal{C}$; so all the information describing G is to be found in the members of $\mathcal{C}$ together with their containment relations.

Starting with a local system $\mathcal{C} = \{G_i \,|\, i \in I\}$, we attempt to reconstruct G within $\prod_i G_i$. There is a natural 'diagonal embedding' which injects G as a set into $\prod_i G_i$ via $g \mapsto g^{\mathcal{C}}$:

$$g_i^{\mathcal{C}} = g \text{ if } g \in G_i,$$
$$g_i^{\mathcal{C}} = 1 \text{ if } g \notin G_i.$$

Unfortunately we will not, in general, have

$$g^{\mathcal{C}} h^{\mathcal{C}} = (gh)^{\mathcal{C}},$$

since the two sides differ at any coordinate i for which G_i contains exactly one of g and h. Still the equality is true 'almost everywhere', and this is the essence of the ultraproduct construction.

B. Filters, Ultrafilters, and Ultraproducts

Let I be any nonempty set. A *filter* $\mathcal{F}$ on I is a set of subsets of I which satisfies two axioms:

(1) if $A, B \in \mathcal{F}$, then $A \cap B \in \mathcal{F}$;
(2) if $A \in \mathcal{F}$ and $A \subseteq B$, then $B \in \mathcal{F}$.

If the empty set is in $\mathcal{F}$, then by (2) the filter $\mathcal{F}$ must in fact be the complete power set 2^I. To avoid 2^I, the *trivial filter*, the axiom $\emptyset \notin \mathcal{F}$ is sometimes included; but we do not make this assumption.

An example of a nontrivial filter on I is the *principal filter* $\mathcal{F}_a$ composed of all subsets of I which contain the element $a \in I$. For infinite I, the *cofinite filter* composed of all cofinite subsets of I is also nontrivial. (A subset is cofinite if its complement is finite.)

We say that the set I is *directed* by the partial order $\leqslant$ if, for every pair i, j of elements of I, there is a $k \in I$ with $i \leqslant k \geqslant j$. In this case define

$$\mathcal{F}(i) = \{a \in I \mid i \leqslant a\}.$$

The nontrivial filter generated by the directed set $(I, \leqslant)$ is then

$$\mathcal{F}_{(I,\leqslant)} = \{A \mid A \supseteq \mathcal{F}(i), \text{ for some } i \in I\}.$$

If the filter $\mathcal{F}$ on I contains A and B with $A \cap B = \varnothing$, then $\mathcal{F}$ is the trivial filter 2^I. A filter which instead satisfies

(**3**) for all $A \subseteq I$, $A \in \mathcal{F}$ if and only if $I \smallsetminus A \notin \mathcal{F}$

is called an *ultrafilter* and is a maximal nontrivial filter. Principal filters are ultrafilters, but the cofinite filter is not. In general the filter generated by a directed set is not an ultrafilter. (*Exercise*: all ultrafilters on finite I are principal.)

The union of an ascending chain of nontrivial filters on I is itself a nontrivial filter, so by Zorn's lemma every nontrivial filter is contained in an ultrafilter. Thus completing the cofinite filter gives us a nonprincipal ultrafilter on each infinite set I. (*Exercise*: a nonprincipal ultrafilter contains the cofinite filter.) We mildly abuse terminology by saying that any ultrafilter containing $\mathcal{F}_{(I,\leqslant)}$ is an ultrafilter generated by the directed set $(I, \leqslant)$.

The following property of ultrafilters is used often.

Lemma B.1 *If $\mathcal{F}$ is an ultrafilter on I, then for any finite coloring of I there is exactly one color class which belongs to $\mathcal{F}$.*

The case of a 2-coloring is just the axiom (**3**), and the lemma follows by induction.

Now suppose that $\{G_i \mid i \in I\}$ is a set of sets (typically the underlying sets of a collection of groups, rings, fields, etc.) and that $\mathcal{F}$ is an ultrafilter on I. On the Cartesian product $\prod_i G_i$ define an equivalence relation $\sim_{\mathcal{F}}$ by

$$(x_i)_{i \in I} \sim_{\mathcal{F}} (y_i)_{i \in I} \text{ iff } \{i \in I \mid x_i = y_i\} \in \mathcal{F}.$$

The *ultraproduct* $\prod_{\mathcal{F}} G_i$ is then the quotient of the set $\prod_i G_i$ by the equivalence relation $\sim_{\mathcal{F}}$. As there may be many different ultrafilters, there are also different ultraproducts of the same set of sets. (*Exercise*: what happens if $\mathcal{F}$ is principal?)

As an instance of Lemma B.1 we have

Lemma B.2 *If each set G_i has finite cardinality at most q, then $\prod_{\mathcal{F}} G_i$ has cardinality at most q.*

Proof. For each i, color the individual members of G_i with the colors $1, 2, \ldots, q$, at most one element of G_i receiving any given color. The coordinate positions of each element $g = (g_i)_{i \in I}$ in the Cartesian product are q-colored. By Lemma B.1 the element g is equivalent with respect to $\mathcal{F}$ to exactly one of the monochromatic elements, of which there are at most q. $\square$

A more difficult exercise is to prove that an ultraproduct of arbitrary finite sets is either finite or uncountably infinite.

Extra structure on the G_i can be transferred to the ultraproduct.

Proposition B.3   *An ultraproduct of groups is a group.*

Proof. Since the Cartesian product $\Gamma = \prod_i G_i$ is a group, it is enough to prove that the multiplication of Γ induces a well-defined multiplication on the ultraproduct $\Gamma_{\mathcal{F}} = \prod_{\mathcal{F}} G_i$. The associativity, identity, and inverses of $\Gamma_{\mathcal{F}}$ will then be naturally induced by those of Γ.

Let $g_1 = (g_{i1})_{i \in I}$ and $g_2 = (g_{i2})_{i \in I}$ be a pair of elements from Γ which are equivalent in $\Gamma_{\mathcal{F}}$, and let h_1 and h_2 be a second such pair. Equivalence implies that the subsets $J_g = \{j \in I \,|\, g_{j1} = g_{j2}\}$ and $J_h = \{j \in I \,|\, h_{j1} = h_{j2}\}$ both belong to $\mathcal{F}$, so by **(1)** their intersection $J_g \cap J_h$ does as well. This intersection is certainly contained in $J = \{j \in I \,|\, g_{j1}h_{j1} = g_{j2}h_{j2}\}$, so $J \in \mathcal{F}$ by **(2)**. That is, g_1h_1 and g_2h_2 are equivalent with respect to $\mathcal{F}$; and multiplication in $\Gamma_{\mathcal{F}}$ is indeed well-defined. $\square$

A similar result is

Proposition B.4   *An ultraproduct of fields is a field.*

Proof. This is a little more subtle. A Cartesian product of groups is a group, whereas a Cartesian product of fields is only a commutative ring with identity. Everything proceeds as with groups, except we must additionally check that in the ultraproduct we can invert nonzero elements. Indeed an element g is nonzero if and only if (it is represented by elements for which) the set of coordinate positions i with $g_i = 0_i$ is not in $\mathcal{F}$. But then the set of positions with $g_i \neq 0_i$ is in $\mathcal{F}$ by **(3)**. The element with g_i^{-1} in those positions and 0_i elsewhere is (a representative of) an inverse for g in the ultraproduct. $\square$

We shall henceforth blur the distinction between an element of the ultraproduct and the elements of the Cartesian product which represent it. For the calculation of Proposition B.3 it was enough that $\mathcal{F}$ be a filter, whereas Proposition B.4 uses the full strength of the ultrafilter definition.

A typical consequence of Lemma B.2 is that an ultraproduct of bounded finite fields is finite. An arbitrary ultraproduct of finite fields is either finite or has uncountable transcendence degree, by the remark which follows Lemma B.2. (*Exercise:* (i) prove that if, for some prime p, we have $\{i \in I \,|\, \mathrm{char}\, F_i = p\} \in \mathcal{F}$, then the ultraproduct of fields $\prod_{\mathcal{F}} F_i$ has characteristic p; (ii) prove that if there is no p for which (i) holds, then $\prod_{\mathcal{F}} F_i$ has characteristic 0.)

The forming of ultraproducts commutes with the taking of products. We leave as an *Exercise* the proof of

Lemma B.5 *There is a natural isomorphism between the products $\prod_{\mathcal{F}} A_i \times \prod_{\mathcal{F}} B_i$ and $\prod_{\mathcal{F}}(A_i \times B_i)$.*

To prove that group multiplication in the Cartesian product $\prod_i G_i$ induces a well-defined multiplication for the ultraproduct $\prod_{\mathcal{F}} G_i$, we needed to show that the individual coordinate multiplication functions $G_i \times G_i \to G_i$ induce a well-defined global function. Suppose more generally that, for each $i \in I$, we have a set map $\varphi_i \colon A_i \to Z_i$, where A_i and Z_i are arbitrary sets. The φ_i become the coordinate functions of a map $\varphi \colon \prod_i A_i \to \prod_i Z_i$. Assume $x = (x_i)_{i \in I}$ and $y = (y_i)_{i \in I}$ are equivalent in $\prod_i A_i$ with respect to the ultrafilter $\mathcal{F}$, and set $J = \{i \in I \mid x_i = y_i\}$, $J \in \mathcal{F}$. Then $K = \{i \in I \mid \varphi_i(x_i) = \varphi_i(y_i)\}$ contains J and so also belongs to $\mathcal{F}$. That is, in $\prod_i Z_i$ the two elements $\varphi(x)$ and $\varphi(y)$ are equivalent with respect to $\mathcal{F}$. Therefore the map $\varphi \colon \prod_i A_i \to \prod_i Z_i$ induces a well-defined map from $\prod_{\mathcal{F}} A_i$ to $\prod_{\mathcal{F}} Z_i$. We denote this map by $\varphi_{\mathcal{F}}$.

As with group multiplication, many properties of algebraic objects are described in terms of the behavior of certain maps. For instance, the fact that the abelian group M_i is a module for the ring R_i is equivalent to certain properties of the function $\varphi_i \colon M_i \times R_i \to M_i$ given by $\varphi_i(m, r) = mr$. The coordinate maps φ_i can now be sewn together to produce an ultraproduct map $\varphi_{\mathcal{F}}$, and properties which can be verified in individual coordinates are lifted to the full ultraproduct. We discover that the abelian group $M = \prod_{\mathcal{F}} M_i$ is a module for the ring $R = \prod_{\mathcal{F}} R_i$.

We can iterate this observation. If each V_i is a vector space over the field F_i, then $V = \prod_{\mathcal{F}} V_i$ is a vector space over $F = \prod_{\mathcal{F}} F_i$. If indeed V_i is an $F_i G_i$-module, for some group G_i, then V is an FH-module where $H = \prod_{\mathcal{F}} G_i$. We restate this in the language of representation theory.

Theorem B.6 *Let I be an index set and let $\mathcal{F}$ be an ultrafilter on I. For each $i \in I$, let $\varphi_i \colon G_i \to \mathrm{GL}_{F_i}(V_i)$ be a representation. Then $\varphi_{\mathcal{F}} \colon \prod_{\mathcal{F}} G_i \to \mathrm{GL}_F(V)$ is a representation, where $F = \prod_{\mathcal{F}} F_i$ and $V = \prod_{\mathcal{F}} V_i$.* $\qquad\qquad\Box$

C. Mal'cev's Representation Theorem

We are now ready to combine remarks from the previous two sections. Let G be a group and $\mathcal{C} = \{G_i \mid i \in I\}$ a local system for G. Let the index set I be given a direct ordering so that $i \leqslant k \geqslant j$ implies $\langle G_i, G_j \rangle \leqslant G_k$. We say that such a directed set is *compatible* with the local system $\mathcal{C}$. There may be many compatible ways of directing I. The most obvious is '$i \leqslant k$ if and only if $G_i \leqslant G_k$', but a slightly different order is used for the proof of Theorem 3.11 in the main body of the paper. Let $\mathcal{F}$ be an ultrafilter generated by the directed set $(I, \leqslant)$.

Theorem C.1 *The injection $g \mapsto g^{\mathcal{C}}$ provides an isomorphism of G into $\prod_{\mathcal{F}} G_i$.*

Proof. The ultraproduct formalizes our earlier statement that $g^{\mathcal{C}} h^{\mathcal{C}}$ and $(gh)^{\mathcal{C}}$ agree 'almost everywhere'. They agree on a member J of $\mathcal{F}$ and so are equal in the ultraproduct. Indeed, if $g \in G_a$ and $h \in G_b$, then $(g^{\mathcal{C}})_i = g$ for $i \in \mathcal{F}(a) \in \mathcal{F}$ and $(h^{\mathcal{C}})_i = h$ for $i \in \mathcal{F}(b) \in \mathcal{F}$. Therefore $(g^{\mathcal{C}})_i(h^{\mathcal{C}})_i = gh$ for $i \in \mathcal{F}(a) \cap \mathcal{F}(b) \in \mathcal{F}$. On the other hand, for $i \in \mathcal{F}(a) \cap \mathcal{F}(b)$ we have $g \in G_i \geqslant G_a$ and $h \in G_i \geqslant G_b$. Therefore $gh \in G_i$, and we have $(gh)_i^{\mathcal{C}} = gh$ for every $i \in \mathcal{F}(a) \cap \mathcal{F}(b) \in \mathcal{F}$. We may take $J = \mathcal{F}(a) \cap \mathcal{F}(b)$. $\qquad\qquad\qquad\qquad\qquad\qquad\qquad\qquad\qquad\qquad\qquad\qquad$ □

Combining this theorem with Theorem B.6, we have

Theorem C.2 *For each $i \in I$, let $\varphi_i \colon G_i \to \mathrm{GL}_{F_i}(V_i)$ be a representation. Then $\Phi_{\mathcal{F}} \colon G \to \mathrm{GL}_F(V)$ is a representation, where $F = \prod_{\mathcal{F}} F_i$, $V = \prod_{\mathcal{F}} V_i$, and $\Phi_{\mathcal{F}}$ is the restriction of $\varphi_{\mathcal{F}}$ to G, $\Phi_{\mathcal{F}} = \varphi_{\mathcal{F}}|_G$.* $\qquad\qquad\qquad\qquad\qquad\qquad$ □

Exercise: the element $g \in G$ belongs to $\ker(\Phi_{\mathcal{F}})$, the kernel of $\Phi_{\mathcal{F}}$, precisely when $\{i \in I \mid g \in \ker(\varphi_i)\} \in \mathcal{F}$.

We have already seen that the basic defining properties of the coordinate algebraic objects are transferred to the ultraproduct. The rules which check the existence of an inverse in a field or define the action of a module are stated in terms of finite subsets of the object under concern and their relationships. Their validity in individual coordinates breeds their validity in the ultraproduct. A more specialized case of this is:

Theorem C.3 *Let $\mathcal{F}$ be an ultrafilter on the index set I, and let k be an integer. If, for each $i \in I$, the dimension of the vector space V_i over the field F_i is k, then the dimension of $V = \prod_{\mathcal{F}} V_i$ over $F = \prod_{\mathcal{F}} F_i$ is k.*

Proof. Subsets of the Cartesian product $\prod_i V_i$ of size n larger than k are linearly dependent over the ring $\prod_i F_i$. Consider $v_j = (v_{ij})_{i \in I} \in \prod_i V_i$, for $1 \leqslant j \leqslant n$. Choose $\alpha_j = (\alpha_{ij})_{i \in I}$ so that, for each i, $\sum_j \alpha_{ij} v_{ij} = 0$ is a nontrivial linear dependence. We claim that not every α_j is equivalent to 0 with respect to $\mathcal{F}$. Let $I_j = \{i \in I \mid \alpha_{ij} = 0\}$. If each I_j is in $\mathcal{F}$, then $\bigcap_j I_j$ is nonempty by (1) and, for i in this intersection, $\sum_j \alpha_{ij} v_{ij} = 0$ is a trivial linear dependence, against construction. This proves the claim. Therefore any n elements v_j of V are linearly dependent over F, and $\dim_F V$ is at most k.

Now let $v_j = (v_{ij})_{i \in I}$, for $1 \leqslant j \leqslant k$, be elements of $\prod_i V_i$ such that the set $\{v_{ij} \mid 1 \leqslant j \leqslant k\}$ is linearly independent, for each i. We claim that the v_j represent F-linearly independent elements of the ultraproduct V.

Suppose that

$$\sum_{j=1}^{k} \alpha_j v_j = 0_V \,,$$

for elements $\alpha_j = (\alpha_{ij})_{i \in I} \in F$. That is, $\sum_{j=1}^{k} \alpha_j v_j \sim_{\mathcal{F}} 0 \in \prod_i V_i$, for elements $\alpha_j = (\alpha_{ij})_{i \in I} \in \prod_{i \in I} F_i$. Thus, for some $K \in \mathcal{F}$ and all $i \in K$, $\sum_{j=1}^{k} \alpha_{ij} v_{ij} = 0 \in V_i$. By linear independence, $\alpha_{ij} = 0 \in F_i$, for all $i \in K$ and all $1 \leqslant j \leqslant k$. That is, α_j and 0 as elements of $\prod_i V_i$ agree in the coordinate positions of $K \in \mathcal{F}$. They are therefore equal in the ultraproduct; and $\alpha_j = 0_F$, for $1 \leqslant j \leqslant k$, as required. $\qquad$ □

Generally in the ultraproduct construction, any first order sentence will hold globally provided it holds often enough locally, that is, on some member of $\mathcal{F}$. See, for instance, the exercises which follow Proposition B.4 and Theorem C.2. (*Exercise*: prove the previous theorem under the weaker hypothesis $\{i \in I \mid \dim_{F_i} V_i = k\} \in \mathcal{F}$.)

Clearly a group which is linear in dimension k has local systems of subgroups with each subgroup linear in dimension k. Take, for instance, the system of all finitely generated subgroups. As a consequence of Theorems C.2 and C.3, we now have Mal'cev's surprising converse.

Theorem C.4 (Mal'cev's Representation Theorem) *Let G be a group with a local system of subgroups each of which has a faithful representations in dimension k. Then G has a faithful representation in dimension k.* □

D. Refinements

While Mal'cev's Representation Theorem C.4 is striking and beautiful, much of its power stems from our ability to extend and refine it. Here we offer certain modifications and additions of specific use to us. The proofs are left as *Exercises*.

Suppose that G is a group with local system $\{G_i \mid i \in I\}$ and, for each $i \in I$, let $\varphi_i \colon G_i \to \mathrm{GL}_{F_i}(W_i)$ be a map. In general, for a map $\rho \colon A \to \mathrm{GL}(X)$, the commutator $[X, A]$ is that subspace of X which is spanned by the images of all maps $\rho(a) - 1$, as a runs through A.

Give I a direct ordering which is compatible with the local system, and let $\mathcal{F}$ be an ultrafilter generated by the directed set $(I, \leqslant)$. Define $W = \prod_{\mathcal{F}} W_i$, a vector space over the field $F = \prod_{\mathcal{F}} F_i$. We identify G with its image in $\prod_{\mathcal{F}} G_i$, as described in Theorem C.1. Thus, as seen above, the map $\Phi_{\mathcal{F}} = (\prod_{\mathcal{F}} \varphi_i)|_G$ takes G into $\mathrm{GL}_F(W)$.

D.1. Controlling Subgroups

Let $H \leqslant G$ be a subgroup of G. If, for each $i \in I$, we set $H_i = H \cap G_i$, then $\{H_i \mid i \in I\}$ is a local system for H.

Theorem D.1 *Suppose that, for each $i \in I$, there is a subspace U_i with the properties $[W_i, H_i] \leqslant U_i \leqslant W_i$ and $\dim_{F_i} U_i = k$. Then $[W, H] \leqslant U = \prod_{\mathcal{F}} U_i$, an F-subspace of W having dimension k.*

The case $G = H$ and $W_i = U_i$ gives Mal'cev's Theorem again. If instead $H = \langle g \rangle$, then we learn that any g of bounded local degree acts as a finitary linear transformation from $\mathrm{FGL}_F(W)$, even when W itself has infinite dimension.

D.2. Invariant Forms

Many of the classical groups are isometry groups of certain forms. The ultraproduct of isometry groups becomes an isometry group in its action on the ultraproduct.

Proposition D.2 *If, on each W_i, there is a quadratic or sesquilinear form of some specific type which is left invariant by $\varphi_i(G_i)$, then on W there is a $\varphi(G)$-invariant form of the same type.*

So, for instance, if each φ_i has its image in $\mathrm{Sp}_{F_i}(W_i, f_i)$, then the form $f_{\mathcal{F}} = \prod_{\mathcal{F}} f_i$ is a symplectic form from $W \times W$ to F and is $\varphi(G)$-invariant. Furthermore, f is nondegenerate if the individual f_i are.

D.3. Projective Representations

In our applications we need a representation theorem which starts with projective representations φ_i, that is, homomorphisms into projective groups $\mathrm{PGL}_{F_i}(V_i)$, since the natural representations of the classical simple groups are projective representations. We define projective representation in a different but equivalent form. The map $\varphi: G \to \mathrm{GL}_F(V)$ with associated cocycle $c: G \times G \to F$ is a *projective representation* provided, for all $g, h \in G$,

$$\varphi(g)\varphi(h) = c(g, h)\varphi(gh).$$

Thus a projective representation whose cocycle is identically 1 is a representation in the usual sense, a 'genuine' representation. As a consequence of this definition, the cocycle c is characterized by the property

$$c(g, h)c(gh, k) = c(g, hk)c(h, k), \text{ for all } g, h, k \in G.$$

(*Exercise:* verify this.) The *kernel* of the projective representation φ is given by

$$\ker(\varphi) = \{g \in G \mid \varphi(g) \text{ is scalar on } V\}.$$

The more general theory of projective representations can be developed in terms of modules with action 'twisted' by the cocycle c. (Beware: this is not the study of what are usually called projective modules.) If we had done that earlier, then we would have found in place of Theorem B.6 the result that an ultraproduct of projective representations is a projective representation whose associated cocycle is the ultraproduct of the coordinate cocycles. We then find a projective version of Theorem C.2 which can be further refined by the results of the previous two subsections.

Theorem D.3   *For each $i \in I$, let $(\varphi_i, c_i): G_i \to \mathrm{GL}_{F_i}(V_i)$ be a projective representation. Then $(\Phi_{\mathcal{F}}, c_{\mathcal{F}}): G \to \mathrm{GL}_F(V)$ is a projective representation, where $c_{\mathcal{F}} = \prod_{\mathcal{F}} c_i$, $F = \prod_{\mathcal{F}} F_i$, $V = \prod_{\mathcal{F}} V_i$, and $\Phi_{\mathcal{F}} = (\prod_{\mathcal{F}} \varphi_i)|_G$. The element $g \in G$ is in $\ker(\Phi_{\mathcal{F}})$ if and only if $\{i \in I \mid g \in \ker(\varphi_i)\} \in \mathcal{F}$.*

   (1) *If, for each $i \in I$, the dimension $\dim_{F_i} W_i$ is at most k, then $\dim_F W$ is at most k.*

   (2) *If, for some $g \in G$ and each $i \in I$, the dimension $\dim_{F_i}[W_i, \varphi_i(\langle g \rangle \cap G_i)]$ is at most k, then $\dim_F[W, \Phi_{\mathcal{F}}(g)]$ is at most k.*

   (3) *If each W_i has a $\varphi_i(G_i)$-invariant nondegenerate form of type Cl, then on W there is a $\Phi_{\mathcal{F}}(G)$-invariant nondegenerate form of type Cl.*

Since the ultrafilter $\mathcal{F}$ is generated by the compatible directed set $(I, \leqslant)$, the statement of (2) can be simplified to: if, for some $g \in G$ and each i with $g \in G_i$, the dimension $\dim_{F_i}[W_i, \varphi_i(g)]$ is at most k, then $\dim_F[W, \Phi_{\mathcal{F}}(g)]$ is at most k.

NON-FINITARY LOCALLY FINITE SIMPLE GROUPS

U. MEIERFRANKENFELD
Department of Mathematics
Michigan State University
East Lansing, MI 48824
USA
E-mail: meier@math.msu.edu

Abstract. In this paper we prove and discuss consequences of the following theorem.
  Let G be a locally finite simple group. Then one of the following holds:
  (a) G is finitary.
  (b) G is of alternating type.
  (c) There exists a prime p and a Kegel cover $\{(H_i, M_i) \mid i \in I\}$ such that G is of p-type and, for all i in I, H_i/M_i is a projective special linear group.

Key words: locally finite group, Kegel cover, finitary group, group of alternating type, group of p-type.

1. Introduction

Define a LFS-group to be an infinite, locally finite, simple group. This paper is a contribution to the general theory of LFS-groups. Recall that a group G is finitary if there exist a field K and a faithful KG-module V such that $[V, g]$ is finite dimensional for all g in G. The infinite alternating groups and all finitary classical groups defined over locally finite fields provide examples of finitary LFS-groups and it has been conjectured and almost proved by J. I. Hall that every finitary non-linear LFS-group is of that kind. In contrast, although many examples exist, not much is known about general non-finitary LFS-groups. One purpose of this paper is to demonstrate that the division of LFS-groups into finitary and non-finitary groups is natural and allows us to obtain a considerable amount of information about the non-finitary LFS-groups.

Fundamental to the study of locally finite groups is the concept of Kegel covers. Let G be a locally finite group. A set of pairs $\{(H_i, M_i) \mid i \in I\}$ is called a *Kegel cover* for G if, for all i in I, H_i is a finite subgroup of G and M_i is a maximal normal subgroup of H_i, such that for each finite subgroup H of G there exists $i \in I$ with $H \leqslant H_i$ and $H \cap M_i = 1$. The groups H_i/M_i, $i \in I$, are called the factors of the Kegel cover.

It has been proven in [5, 4.3, p. 113] that every LFS-group has a Kegel cover. Using Kegel covers many questions about LFS-groups can be transferred to questions about finite simple groups, which in turn may be answered using the classification of finite simple groups. Define a LFS-group to be of alternating type if it possesses a Kegel cover all of whose factors are alternating groups. If p is a prime, define a LFS-group G to be of p-type if G is non-finitary and every Kegel cover for G has a

B. Hartley et al. (eds.), *Finite and Locally Finite Groups*, 189–212.

factor which is isomorphic to a classical group defined over a field in characteristic p. In Theorem 3.3 we prove

Theorem A   *Let G be a LFS-group. Then one of the following holds:*

(a) *G is finitary.*

(b) *G is of alternating type.*

(c) *There exists a prime p and a Kegel cover $\{(H_i, M_i) \mid i \in I\}$ for G such that G is of p-type and, for all $i \in I$, $H_i/O_p(H_i)$ is the central product of perfect central extensions of classical groups defined over a field in characteristic p and H_i/M_i is a projective special linear group.*

Using Theorem A many questions about non-finitary LFS-groups can now be transferred to questions about alternating and projective special linear groups. As an example we prove in Section 4:

Theorem B   *Let G be a countable, non-finitary LFS-group and H a finite subgroup of G. Then H is contained in a maximal subgroup of G. In particular, G has maximal subgroups.*

In Theorem 5.5 we give an affirmative answer to Questions 1 and 2 raised in J. Hall's and B. Hartley's paper [1] on a characterization of finitary LFS-groups. In Section 6 we provide examples of countable LFS-groups which are not absolutely simple. We remark that, by [7], every Kegel cover for a countable LFS-group which is not absolutely simple has a fairly complicated structure. In particular, no such group has a Kegel cover where each of the H_i is a central product of quasisimple groups and, using Theorem A, any such group is of alternating type.

We hope that Theorem A will draw attention to LFS-groups of p-type. In [1, Proposition 1] non-finitary LFS-groups G have been constructed which have an element x of order q, q an odd prime, such that $\langle x^S \rangle$ is abelian for every q-subgroup S of G containing x. It follows from Theorem 5.5 that no such group is of alternating type and thus we obtain examples of LFS-groups of p-type. It seems plausible that the restricted structure of the Kegel covers for LFS-groups of p-type provided by Theorem A might lead to a classification of LFS-groups of p-type. It also seems worthwhile to examine the $\mathbb{F}_p$-module arising from the above Kegel cover via the ultrafilter construction in [5, 1L Appendix]. It might be possible to characterize LFS-groups of p-type in terms of that module.

We remark that Theorem 3.4(b) has been proven independently and with different methods by C. Praeger and A. Zalesskiĭ in [8, Theorem 1.7].

We finish the introduction with a list of some of the notations used throughout. Let K be a field, V a vector space over K and q a symplectic, orthogonal or unitary form on V. Then $O(V, q)$ is the largest subgroup of $\mathrm{GL}_K(V)$ preserving q, whereas $\Omega(V, q) = O(V, q)'$ and $\mathrm{P}\Omega(V, q) = \Omega(V, q)/Z(\Omega(V, q))$. A singular subspace of V is a K-subspace of V on which q vanishes. $\mathcal{F}$ is the class of finite simple groups isomorphic to $\mathrm{P}\Omega(V, q)$ for some finite field K, some finite dimensional vector space V over K and some non-degenerate symplectic, orthogonal or unitary form q on V. Also, $\mathcal{L}$ is the class of finite simple groups isomorphic to $\mathrm{PSL}_K(V)$ for some finite field K and some finite dimensional vector space V over K. Furthermore, $\mathcal{C} = \mathcal{L} \cup \mathcal{F}$ and $\mathcal{A}$ is the class of finite simple alternating groups. If p is a prime, then $\mathcal{C}(p)$,

$\mathcal{F}(p)$ and $\mathcal{L}(p)$ are defined similarly except that the defining fields are assumed to be in characteristic p. If x is acting on V, then

$$\deg_V(x) = \dim_K[V, x], \ \mathrm{pdeg}_V(x) = \min\{\deg_V(kx) \mid 0 \neq k \in K\}$$

and

$$\mathrm{pdeg}_V(Z(\mathrm{GL}_K(V))x) = \mathrm{pdeg}_V(x).$$

If Ω is a set and x acts on Ω, $\mathrm{pdeg}_\Omega(x)$ and $\deg_\Omega(x)$ are both equal to the number of elements in Ω not fixed by x. Also, if $G = \mathrm{PSL}_K(V)$, $\mathrm{P\Omega}(V, q)$ or $\mathrm{Alt}(\Omega)$, and x induces an inner automorphism y on G, then $\mathrm{pdeg}_G(x) = \mathrm{pdeg}_V(y)$ or $\mathrm{pdeg}_G(x) = \mathrm{pdeg}_\Omega(y)$, respectively.

Let G be a locally finite group. A set of pairs $\{(H_i, M_i) \mid i \in I\}$ is called a sectional cover for G if, for all i in I, H_i is a finite subgroup of G and M_i is a normal subgroup of H_i, and if, for each finite subgroup H of G, there exists i in I with $H \leqslant H_i$ and $H \cap M_i = 1$. The groups H_i/M_i, $i \in I$, are called the factors of the sectional cover. A set of subgroups $\{H_i \mid i \in I\}$ of G is called a sectional cover for G if each finite subgroup of G is contained in one of the H_i, i.e. if $\{(H_i, 1) \mid i \in I\}$ is a sectional cover for G. A Kegel sequence for G is a Kegel cover $\{(H_i, M_i) \mid i \geqslant 1\}$ for G such that, for all i, $H_i \leqslant H_{i+1}$ and $H_i \cap M_{i+1} = 1$.

For a finite group H let $w(H)$ be the number of isomorphism types of transitive permutation representations for H or, equivalently, the number of conjugacy classes of subgroups of H.

Let G be a group acting on a set Ω. Let I be a subset of Ω. Then

$$N_G(I) = \{g \in G \mid i^g \in I \text{ for all } i \in I\}$$

and

$$C_G(I) = \{g \in G \mid i^g = i \text{ for all } i \in I\}.$$

Let Δ be a set of subsets of Ω. Then

$$N_G(\Delta) = \{g \in G \mid D^g \in \Delta \text{ for all } D \in \Delta\}$$

and

$$C_G(\Delta) = \{g \in G \mid D^g = D \text{ for all } D \in \Delta\}.$$

Suppose that $G/M \cong \mathrm{Alt}(\Sigma)$ for some set Σ and some normal subgroup M of G. Let t be a positive integer with $t \leqslant |\Sigma|/2$. Then G acts t-pseudo naturally on Ω with respect to M if G acts transitively on Ω and if there exists a maximal system of imprimitivity Δ for G on Ω such that $C_G(\Delta) = M$ and the action of G on Δ is isomorphic to the action of G on subsets of size t of Σ. We say that G acts pseudo naturally on Ω with respect to M if G acts 1-pseudo naturally on Ω. Also, G acts essentially on Ω with respect to M if $C_G(\Omega) \leqslant M$. We remark that if G/M is perfect (that is $|\Sigma| \geqslant 5$) and R is the minimal normal supplement to M in G (see Lemma 2.8) then Ω is essential if and only if R acts non-trivially on Ω. The reader should notice that in the case $M = 1$ a pseudo natural action does not have to be natural. In fact it is easy to see that even the right regular permutation action is pseudo natural.

If p is a prime, then $d(p) = 1$ if $p \neq 2$, and $d(p) = 2$ if $p = 2$.

For a real number x, let $[x]$ be the largest integer less than or equal to x.

For a group G let $G^{(0)} = G$ and, inductively, $G^{(i+1)} = (G^{(i)})'$. Furthermore, put $G^\infty = \bigcap_{i=1}^\infty G^{(i)}$ and note that, if G is finite, G^∞ is the largest perfect subgroup of G. If G is solvable, let $\mathrm{der}(G)$ be the smallest non-negative integer i with $G^{(i)} = 1$.

If G is a group acting on a set Ω, then a system of imprimitivity Δ for G on Ω is a set of proper subsets of Ω such that $|D| \geqslant 2$ for at least one D in Δ, $D^g \in \Delta$, for all $D \in \Delta$ and $g \in G$, and Ω is the disjoint union of the members of Δ. If G is a group acting on a vector space V, then a system of imprimitivity Δ for G on V is a set of proper subspaces of V such that $D^g \in \Delta$, for all $D \in \Delta$ and $g \in G$, and V is the direct sum of the members of Δ. We say that G acts primitively on I (or on V) if G has no system of imprimitivity on I (or on V).

Acknowledgements

I would like to thank Jon Hall and Dick Phillips for many fruitful discussions on the topics in this paper.

2. Preliminaries

Throughout this chapter K is a finite field, V a finite dimensional vector space over K and $Q = \mathrm{GL}_K(V)$.

Lemma 2.1   *Let σ be a field automorphism of K of order 1 or 2 and let s be a σ-sesquilinear form on V.*

(a) *There exists a subspace U of V such that $\dim U \geqslant \frac{1}{4}(\dim V - 4)$ and $s|_{U \times U} = 0$.*

(b) *Let $G \leqslant Q$ with $s(u, v) = s(u^g, v^g)$ for all u, v in V and all g in G. Let $Z = Z(Q) \cap G$, $m = |G/Z|$ and W a subspace of V. Then there exists a subspace X of W with*

$$\dim X \geqslant (\dim W - 4\frac{4^m - 1}{3})/4^m$$

such that $s|_{U \times U} = 0$, where $U = \langle X^G \rangle$.

Proof. (a) Define $t : V \times V \to K$ by $t(u, v) = s(u, v) + s(v, u)^\sigma$. Then t is a symmetric or unitary form on V. Therefore there exists a subspace W of V with $\dim W \geqslant (\dim V - 2)/2$ such that $t|_{W \times W} = 0$. Indeed, the worst possible case is when V has dimension $2k$ and t has Witt index $k - 1$. If $\sigma \neq 1$, pick λ in K with $\lambda \lambda^\sigma = -1$; otherwise let $\lambda = 1$. Then, restricted to W, λs is a skew symmetric or unitary form and there exists a subspace U of W with $\dim U \geqslant (\dim W - 1)/2$ such that $\lambda s|_{U \times U} = 0$. Here the worst possible case occurs if s restricted to W is unitary and W is odd dimensional, or if $\mathrm{char}\, K = 2$ and s restricted to W is symmetric but not alternating. This proves (a).

(b) Let $g \in G$ and Y a subspace of V. Define a sesquilinear form s_g on Y by $s_g(u, v) = s(u, v^g)$. By (a) there exists a subspace R of Y with $s_g|_{R \times R} = 0$

and $\dim R \geqslant (\dim Y - 4)/4$. Let $T = \{g_1, \ldots, g_m\}$ be a transversal to Z in G. Then by an easy induction proof there exists a subspace X in W with $\dim X \geqslant (\dim W - 4\frac{4^m-1}{3})/4^m$ and $s_{g_i}|_{X \times X} = 0$ for all $1 \leqslant i \leqslant m$. Put $U = \langle X^G \rangle$ and note that $U = \langle X^T \rangle$. Let $u, v \in X$. Then $s(u, v^{g_i}) = s_{g_i}(u, v) = 0$. Thus $s(u, w) = 0$ for all $u \in X$ and $w \in U$. Since s is G-invariant, $s(u, v) = 0$ for all $u, v \in U$. $\square$

Lemma 2.2 *There exists an increasing function f defined on the positive integers and independent of K and V with the following property: if $G \leqslant Q$, $Z = G \cap Z(Q)$, q is a G-invariant quadratic, symplectic or unitary form on V and X is a subspace of V of dimension at least $f(|G/Z|)$, then there exists $0 \neq x \in X$ such that $\langle x^G \rangle$ is singular with respect to q.*

Proof. Let $f(m) = 2 \cdot 4^m + 4(\frac{4^m-1}{3})$. If q is symplectic or unitary, the assertion follows directly from Lemma 2.1(b). The same is true if q is quadratic and char K is odd. So suppose q is quadratic and char $K = 2$. Let s be the symplectic form associated to q. Then Lemma 2.1(b) provides a subspace Y in X of dimension at least two such that s vanishes on $\langle Y^G \rangle$. Pick $0 \neq x \in Y$ with $q(x) = 0$. Then q vanishes on $\langle x^G \rangle$. $\square$

Lemma 2.3 *Let f, G, Z, V, and q be as in Lemma 2.2. Put $m = |G/Z|$ and let $\mathcal{S}$ be the set of G-invariant singular subspaces of V.*

(a) *Let M be a maximal element in $\mathcal{S}$. Then $\dim M > (\dim V - f(m))/2$.*

(b) *If $\dim V \geqslant 2m + f(m)$ and $x \in G$ with $\mathrm{pdeg}_V(x) > f(m)$, then there exists $M \in \mathcal{S}$ such that x does not act as a scalar on M.*

(c) *Let t be a positive integer and H a subset of G such that, for all $x \in H$, $\mathrm{pdeg}_V(x) > 2m(|H|t - 1) + f(m)$. If $\dim V > 2tm|H| + f(m)$, then there exists $M \in \mathcal{S}$ with $\dim M \leqslant t|H|m$ and $\mathrm{pdeg}_M(x) \geqslant t$ for all $x \in H$.*

(d) *Put $g(m) = 2m^2 + f(m)$. If $\mathrm{pdeg}_V(x) > g(m)$ for all $x \in G \setminus Z$, then there exists $U \in \mathcal{S}$ such that no element of $G \setminus Z$ acts as a scalar on U.*

Proof. (a) Let M be a maximal element in $\mathcal{S}$. Then G normalizes no non-trivial singular subspace of $M^\perp/M$. By Lemma 2.2, $\dim M^\perp/M < f(m)$ and so

$$\dim V = \dim V/M^\perp + \dim M^\perp/M + \dim M < \dim M + f(m) + \dim M.$$

Thus (a) holds.

(b) Suppose that x acts as a scalar on every element of $\mathcal{S}$ and let M and N be maximal elements of $\mathcal{S}$. Let $0 \neq u \in N$ and put $U = K\langle u^G \rangle$. Then clearly $\dim U \leqslant m$. By (a), and since $\dim V \geqslant 2m + f(m)$, $\dim M > m$. Hence $M \cap U^\perp \neq 0$. Since $(M \cap U^\perp) + U$ is singular, we conclude that x acts as the same scalar on M, $(M \cap U^\perp) + U$ and N. Since this is true for all such M and N, x acts as a scalar on $\langle \mathcal{S} \rangle$. By Lemma 2.2 applied with X as a complement to $\langle \mathcal{S} \rangle$ in V, we have $\dim V/\langle \mathcal{S} \rangle \leqslant f(m)$ and thus $\mathrm{pdeg}(x) \leqslant f(m)$, a contradiction.

(c) Suppose first that $|H| = 1$ and let $H = \{x\}$. By induction on t, there exists N in $\mathcal{S}$ with $\dim N \leqslant (t-1)m$ such that $\mathrm{pdeg}_N(x) \geqslant t - 1$ (choose $N = 0$ if $t = 1$). Let $W = N^\perp/N$. Then $\dim W \geqslant \dim V - 2\dim N > 2m + f(m)$ and

$\mathrm{pdeg}_W(x) \geqslant \mathrm{pdeg}_V(x) - 2\dim N > f(m)$. So by (b) there exists a G-invariant singular subspace X in W such that x does not act as scalar on X. Replacing X by $K\langle u^G \rangle$, where $u \in X$ with $u^x \notin Ku$, we may assume that $\dim X \leqslant m$. Let M be the inverse image of X in V. Then

$$\mathrm{pdeg}_M(x) \geqslant \mathrm{pdeg}_N(x) + \mathrm{pdeg}_X(x) \geqslant (t-1) + 1 = t.$$

The case $|H| > 1$ follows with a similar argument and induction on $|H|$.

(d) Let H be a set of representatives of the non-trivial cosets of Z in G. Note that $\mathrm{pdeg}_V(x) > g(m)$ implies $\dim V > g(m)$. Apply (c) with $t = 1$. □

Lemma 2.4 *Let $H = \mathrm{PGL}_K(V)$ or a finite symmetric group. Then there exist increasing functions h, k defined on the positive integers and independent of H such that each of the following two statements holds.*

(a) *If $x \in S \leqslant H$ with S solvable, then $\mathrm{der}(\langle x^S \rangle) \leqslant h(\mathrm{pdeg}_H(x))$.*

(b) *Let $x \in H'$ with $|x| = p$, p a prime, and, if $H = \mathrm{PGL}_K(V)$, $|x| = 2$. If $\mathrm{pdeg}_V(x) \geqslant k(t)$, then there exists a p-subgroup S of H' with $x \in S$ and $\mathrm{der}(\langle x^S \rangle) \geqslant t$.*

Proof. For (a) see [6, Proposition 1] and for (b) see [1, 2.1, 2.4]. □

Lemma 2.5 *Let $H = Q$ or a finite symmetric group. There exists an increasing function l defined on the positive integers and independent of H with the following property:*

Let $G \leqslant H$ and $N \lhd G$ with $G/N \in \mathcal{A} \cup \mathcal{L}$. Let $x \in G$ with $|x| = p$, p a prime, and, if $G/N \in \mathcal{L}$, $|x| = 2$. If $\mathrm{pdeg}_{G/N}(x) \geqslant l(m)$ then $\mathrm{pdeg}_H(x) > m$.

Proof. Let k, h be the functions given by Lemma 2.4 and define $l(m) = k(h(m)+1)$. If $\mathrm{pdeg}_{G/N}(x) \geqslant l(m)$, then by Lemma 2.4(b) there exists a p-group S in G with $x \in S$ and $\mathrm{der}(\langle x^S \rangle N/N) \geqslant h(m) + 1$. So $\mathrm{der}(\langle x^S \rangle) > h(m)$ and, by Lemma 2.4(a), $\mathrm{pdeg}_H(x) > m$. □

Lemma 2.6 *Let $G \leqslant Q$ and Δ a system of imprimitivity of G on V. Let $U \in \Delta$ and E a subgroup of $\mathrm{GL}_K(U)$ with $N_G(U)/C_G(U) \leqslant E$. Suppose that G acts transitively on Δ. Then there exists $H \leqslant N_Q(\Delta)$ with $G \leqslant H$, $N_H(U)/C_H(U) = E$ and $H \cong E \wr \mathrm{Sym}(\Delta)$.*

Proof. Let I be a transversal to $N_G(U)$ in G with $1 \in I$. Define $F \leqslant Q$ by F normalizes U, $F/C_F(U) = E$ and $[W, F] = 0$ for all $W \in \Delta \smallsetminus \{U\}$. Define an action of $\mathrm{Sym}(I)$ on V as follows. For $\pi \in \mathrm{Sym}(I)$ and $u_i \in U$, $i \in I$, let

$$\left(\sum_{i \in I} u_i^i \right)^\pi = \sum_{i \in I} u_i^{i^\pi}.$$

Let $H = \langle F, \mathrm{Sym}(I) \rangle$, where we view $\mathrm{Sym}(I)$ as a subgroup of Q by the above action. Then clearly H normalizes Δ, $N_H(U)/C_H(U) = E$, $\mathrm{Sym}(I) \cong \mathrm{Sym}(\Delta)$ and

$H \cong E \wr \mathrm{Sym}(\Delta)$. It remains to prove that $G \leqslant H$. Let $g \in G$ and define $\pi \in \mathrm{Sym}(I)$ and $n_i \in N_G(U)$, $i \in I$, by $ig = n_i i^\pi$. Let $h = g\pi^{-1}$. Then

$$\left(\sum_{i \in I} u_i^i\right)^h = \left(\sum_{i \in I} u_i^{ig}\right)^{\pi^{-1}} = \left(\sum_{i \in I} u_i^{n_i i^\pi}\right)^{\pi^{-1}} = \sum_{i \in I} u_i^{n_i i}.$$

Pick f_i in F such that $f_i n_i^{-1}$ centralizes U and $\pi_i \in \mathrm{Sym}(I)$ with $1^{\pi_i} = i$. Then, since $u_i^{i \pi_i^{-1}} = u_i^{i^{\pi_i^{-1}}} = u_i^1 = u_i$,

$$u_i^{i(\pi_i^{-1} f_i \pi_i)} = u_i^{f_i \pi_i} = u_i^{n_i \pi_i} = u_i^{n_i i}.$$

It follows that $h = \prod_{i \in I} \pi_i^{-1} f_i \pi_i$. So $h \in H$ and $G \leqslant H$. $\qquad\square$

Lemma 2.7 *Let $M \leqslant Q$, Δ be the set of components of M and $H = N_Q(M)$. Assume that $M = \langle \Delta \rangle$, that H acts irreducibly and primitively on V and that H acts transitively on Δ. Then there exist a cyclic group S and $m \geqslant 1$ such that $H/C_H(\Delta) \cong \mathrm{Sym}_m \wr S$, where the wreath product is built via the regular permutation representation of S.*

Proof. Let U be an irreducible KM-submodule in V. Since H is primitive on V, V is a direct sum of KM-submodules isomorphic to U. Put $D = \mathrm{Hom}_{KM}(V,V)$, $h = \dim_K V / \dim_K U$ and $E = \mathrm{Hom}_{KM}(U,U)$. Consequently E is a field and $D \cong \mathrm{Hom}_E(E^h, E^h)$ as KH-modules. Put $F = \mathrm{Hom}_{DM}(V,V)$. Then F is a finite field isomorphic to E, and $H/C_H(F)$ acts as a group of field automorphisms on F. Hence $H/C_H(F)$ is cyclic. Pick L in Δ and define $S = H/C_H(F)N_H(L)$ and $m = |\Delta|/|S|$. Since $H/C_H(F)$ is cyclic and H acts transitively on Δ, S is cyclic and independent of the choice of L. Moreover, $|S|$ is the number of orbits of $C_H(F)$ on Δ, $C_H(F)$ and $C_H(F)N_H(L)$ have the same orbits on Δ and m is the length of each of those orbits. Note that V is a vector space over F, the elements of H act semilinearly with respect to F and every irreducible FM-submodule in V is isomorphic to U as FM-module. Let Γ be an orbit for $C_H(F)$ on Δ. We will prove next that:

(*) *Let $\pi \in \mathrm{Sym}(\Delta)$ such that π fixes all elements in $\Delta \setminus \Gamma$. Then π is induced by some element $g \in C_H(F)$.*

Note that as FM-modules

$$U \cong \bigotimes_{x \in \Delta} Y_x,$$

where Y_x is an irreducible Fx-submodule in U and the tensor product is built over F. Fix $L \in \Gamma$ and for each $x \in \Gamma$ pick $h(x)$ in $C_H(F)$ with $x = L^{h(x)}$. Put $Y = Y_L$ and $Z = \bigotimes_{x \in \Gamma} Y_x$. Then Y_x and $Y^{h(x)}$ are both irreducible Fx-submodules of U and so isomorphic as Fx-modules. Put $W = \bigotimes_{x \in \Gamma} Y^{h(x)}$. Then Z and W are isomorphic as $F\langle \Gamma \rangle$-modules. Define a map α,

$$\alpha : \prod_{x \in \Gamma} Y^{h(x)} \longrightarrow W,$$

by

$$\{y_x^{h(x)}\}_{x\in\Gamma} \mapsto \otimes_{x\in\Gamma} y_{x^{\pi-1}}^{h(x)}$$

where $y_x \in Y$ for each $x \in \Gamma$. Then it is readily verified that α is F-linear in each of the coordinates and so induces an F-linear map β from W to W. Let $T = \otimes_{x\in\Delta\setminus\Gamma} Y_x$. Define $\gamma : T \otimes W \to T \otimes W$ by $(t \otimes w)^g = t \otimes w^\beta$. It is easily checked that β normalizes $\langle\Gamma\rangle$ in $\mathrm{GL}_F(W)$ and so γ normalizes M in $\mathrm{GL}_F(T \otimes W)$ and acts as π on Δ. Since U and $T \otimes W$ are isomorphic as FM-modules and V is the direct sum of FM-modules isomorphic to U, there exists g in $\mathrm{GL}_F(V)$ such that g normalizes M and acts as π on Δ. Since $N_{\mathrm{GL}_F(V)}(M) = C_H(F)$, (*) is proved. $\square$

By (*), $C_H(F)$ induces all possible permutations of Δ which normalize the orbits of $C_H(F)$ on Δ. Hence the same is true for $C_H(F)N_H(L)$ in place of $C_H(F)$, $C_H(F)N_H(L)/C_H(\Delta) \cong \mathrm{Sym}_m^{|S|}$ and $H/C_H(\Delta) \cong \mathrm{Sym}_m \wr S$. $\square\square$

Lemma 2.8 *Let G be a finite group and N a normal subgroup of G such that G/N is perfect. Then there exists a unique subnormal subgroup R of G which is minimal with respect to $G = RN$.*

Proof. Let R_1 and R_2 be minimal subnormal supplements to N in G. Let K_i be proper normal subgroups of G with $R_i \leqslant K_i$ for $i = 1, 2$. Then $G = K_1N = K_2N$. Since G/N is perfect, $G = [K_1, K_2]N$. Let R be a minimal subnormal supplement to $[K_1, K_2] \cap N$ in $[K_1, K_2]$. By induction on $|G|$, R and R_i are both the unique minimal subnormal supplement to $N \cap K_i$ in K_i for $i = 1, 2$. Thus $R_1 = R = R_2$. $\square$

Lemma 2.9 *Let G_1, G, N_1 and N be subgroups of Q such that $N_1 \lhd G_1$, $N \lhd G$, G_1/N_1 and G/N are perfect and simple, $G/N \in \mathcal{L}$, $G_1 \leqslant G$ and $G_1 \cap N \leqslant N_1$. Let R_1 be a minimal subnormal supplement to N_1 in G_1. Suppose that each of the following two statements holds:*

   (i) *There exists x in R_1 with $|xZ(Q)| = 2$ and $\mathrm{pdeg}_{G/N}(x) \geqslant l(2m+f(m))$, where $m = |G_1/G_1 \cap Z(Q)|$ and f and l are as in Lemmata 2.2 and 2.5, respectively.*

   (ii) *$C_G(N/O_p(G)) \leqslant N$, where p is the characteristic of K.*

Then there exists $G_2 \leqslant Q$ and $N_2 \lhd G_2$ such that $G_1 \leqslant G_2$, $G_1 \cap N_2 \leqslant N_1$ and all non-abelian composition factors of G_2/N_2 are alternating groups.

Proof. The proof is by induction on $|V|$. Let R be a minimal subnormal supplement to N in G. By Lemma 2.8, R is unique and so normal in G. Since G/N is simple, we conclude that R is contained in every subnormal subgroup H of G with $H \not\leqslant N$. Moreover, since G/N is perfect, $R'N = G$ and so $R = R'$. Hence the three subgroup lemma implies:

(*) *If M is a normal subgroup of G with $[M, R] \neq 1$, then $[[M, R], R] \neq 1$.*

Assume first that G acts reducibly on V. By (ii), $[R, N] \not\leqslant O_p(G)$ and so $[R, N]$ does not act unipotently on V. Hence there exists a chief factor W for G on V with $[W, [R, N]] \neq 0$. Let $\bar{G} = G/C_G(W)$. Then $O_p(\bar{G}) = 1$. Suppose that $C_G(W) \not\leqslant N$;

then $R \leqslant C_G(W)$, a contradiction. Hence $C_G(W) \leqslant N$ and $\bar{G}/\bar{N} \cong G/N$. If $\bar{G}_1 = \bar{N}_1$, then

$$G_1 \leqslant C_{G_1}(W)N_1 \leqslant (G_1 \cap N)N_1 = N_1,$$

which is a contradiction. Hence $\bar{G}_1/\bar{N}_1 \cong G_1/N_1$. By choice of W, $\bar{R} \not\leqslant C_{\bar{G}}(\bar{N})$ and so $C_{\bar{G}}(\bar{N}) \leqslant \bar{N}$. It is now easy to verify that $(\bar{G}_1, \bar{N}_1, \bar{G}, \bar{N}, \bar{R}_1, \bar{x}, W)$ fulfills the assumptions of the lemma. So by induction there exists a subgroup $\bar{G}_2$ of $\mathrm{GL}_K(W)$ and $\bar{N}_2 \lhd \bar{G}_2$ such that $\bar{G}_1 \leqslant \bar{G}_2$, $\bar{G}_1 \cap \bar{N}_2 \leqslant \bar{N}_1$ and all non-abelian composition factors of $\bar{G}_2/\bar{N}_2$ are alternating groups. Let $W = X/Y$ for some KG-submodules X, Y in V. Let G_2 and N_2 be the largest subgroups of Q which normalize X and Y and satisfy $G_2/C_{G_2}(W) = \bar{G}_2$ and $N_2/C_{N_2}(W) = \bar{N}_2$. Then (G_2, N_2) fulfills the conclusion of the lemma.

Assume next that G acts irreducibly but imprimitively on V. Let Δ be a system of imprimitivity for G on V. Suppose that $R_1 \not\leqslant C_G(\Delta)$. Then $C_{G_1}(\Delta) \leqslant N_1$. Note that $N_Q(\Delta)/C_Q(\Delta) \cong \mathrm{Sym}(\Delta)$. Put $G_2 = N_Q(\Delta)$ and $N_2 = C_Q(\Delta)$. Then (G_2, N_2) fulfills the conclusion of the lemma.

So we may assume that $R_1 \leqslant C_G(\Delta)$. Since $G_1 \cap N \leqslant N_1$, $R_1 \not\leqslant N$ and we get $C_G(\Delta) \not\leqslant N$ and $R \leqslant C_G(\Delta)$. Pick U in Δ and for any $X \subseteq G$ put $\bar{X} = N_X(U)C_G(U)/C_G(U)$. Let S be the minimal subnormal supplement to $N_N(U)$ in $N_G(U)$. Since $RN_N(U) = N_G(U)$, $S \leqslant R$. In particular, S is normal in R and so subnormal in G. Since $S \not\leqslant N$, we get $S = R$ by minimality of R.

Suppose that $C_{\bar{G}}(\bar{N}) \not\leqslant \bar{N}$. Then, since $R = S$, $\bar{R} \leqslant C_{\bar{G}}(\bar{N})$. So $[N, R, R]$ centralizes U and, since G acts irreducibly on V and $[N, R, R]$ is normal in G, $[N, R, R] = 1$, a contradiction to (*). Thus $C_{\bar{G}}(\bar{N}) \leqslant \bar{N}$. If $\bar{G} = \bar{N}$, then we have $N_G(U) \leqslant C_G(U)N_N(U)$ and so $R = S \leqslant C_G(U)$, a contradiction. Thus $\bar{G}/\bar{N} \cong G/N$. If $\bar{G}_1 = \bar{N}_1$, then

$$R_1 \leqslant N_{G_1}(U) \leqslant C_{G_1}(U)N_{N_1}(U) \leqslant C_G(U)N_N(U).$$

Hence $C_G(U) \not\leqslant N$ and $\bar{G} = \bar{N}$, contradiction. Thus $\bar{G}_1/\bar{N}_1 \cong G_1/N_1$. It is now easily verified that $(\bar{G}_1, \bar{N}_1, \bar{G}, \bar{N}, \bar{R}_1, \bar{x}, U)$ fulfills the assumptions of the lemma. So by induction there exists a subgroup $\bar{G}_2$ of $\mathrm{GL}_K(U)$ and $\bar{N}_2 \lhd \bar{G}_2$ such that $\bar{G}_1 \leqslant \bar{G}_2$, $\bar{G}_1 \cap \bar{N}_2 \leqslant \bar{N}_1$ and all non-abelian composition factors of $\bar{G}_2/\bar{N}_2$ are alternating groups. Apply Lemma 2.6 to G_1 in place of G and with $E = \bar{G}_2$. Put $G_2 = H$ and $H_2 = C_{G_2}(\Delta)$ and let N_2 be the largest normal subgroup of G_2 contained in H_2 with $N_2 C_{G_2}(U)/C_{G_2}(U) = \bar{N}_2$. Then $H_2/N_2 \cong (\bar{G}_2/\bar{N}_2)^{|\Delta|}$ and so every non-abelian composition factor of G_2/N_2 is an alternating group. Note that $\bar{G}_1 \cap \bar{N}_2 \leqslant \bar{N}_1$. Since $C_{G_1}(U) \leqslant C_G(U) \cap G_1 \leqslant N \cap G_1 \leqslant N_1$, we conclude that $G_1 \cap N_2 \leqslant N_1$. Therefore (G_2, N_2) fulfills the conclusion of the lemma.

Assume last that G acts irreducibly and primitively on V. Let X be any normal subgroup of G. If X had more than one Wedderburn component on V, these Wedderburn components would form a system of imprimitivity for G on V. Thus V is a direct sum of isomorphic irreducible KX-submodules in V. In particular, $Z(X)$ is cyclic. Let M be a normal subgroup of G in N minimal with respect to $[M, R] \neq 1$. By (*), $[M, R, R] \neq 1$ and so $M = [M, R] \leqslant R$. It follows that $C_M(R) \leqslant Z(M)$. Since $Z(M)$ is cyclic and R is perfect, $[Z(M), R] = 1$ and so $C_M(R) = Z(M)$. Put $\bar{M} = M/Z(M)$. Then $\bar{M}$ is a minimal normal subgroup of $G/Z(M)$ and so $\bar{M}$ is the direct product of simple groups.

Suppose first that $\bar{M}$ is perfect. Then $M = E(M)$. Let Δ be the set of components of M and note that G acts transitively on Δ. Assume that $R \leqslant C_G(\Delta)$. Since R is perfect and the outer automorphism group of any finite simple group is solvable we conclude that R induces inner automorphisms on $\bar{M}$. Thus $R \leqslant MC_G(\bar{M})$, $C_G(\bar{M}) \not\leqslant N$, $R \leqslant C_G(\bar{M})$ and $[M, R, R] = 1$, a contradicton to $M = [M, R]$. Therefore $R \not\leqslant C_G(\Delta)$ and so $C_G(\Delta) \leqslant N$. Put $G_2 = N_Q(M)$ and $N_2 = C_{G_2}(\Delta)$. Then $G_1 \cap N_2 \leqslant C_G(\Delta) \leqslant N$ and so $G_1 \cap N_2 \leqslant G_1 \cap N \leqslant N_1$. By Lemma 2.7, every non-abelian composition factor of G_2/N_2 is an alternating group and so (G_2, N_2) fulfills the conclusion of the lemma.

Suppose next that $\bar{M}$ is an elementary abelian q-group for some prime q. Since $M' \leqslant Z(M)$, M' is elementary abelian and cyclic. Thus $|M'| = q$. Define a symplectic form on $\bar{M}$ by $s(a, b) = [a, b]$. By Lemma 2.5 applied to $H = \mathrm{GL}_{\mathbb{F}_p}(\bar{M})$ we have that $\mathrm{pdeg}_{\bar{M}}(x) > 2m + f(m)$. Now $m \geqslant |G_1/C_{G_1}(\bar{M})|$ and so by Lemma 2.3(b) there exists a G_1-invariant subspace $\bar{A}$ in $\bar{M}$ such that $[\bar{A}, x] \neq 1$ and s vanishes on $\bar{A}$. Let A be the inverse image of $\bar{A}$ in M. Then A is abelian. Let

$$\Delta = \{C_V(B) \mid B \leqslant A, \ A/B \text{ cyclic and } C_V(B) \neq 0\}$$

and note that V is the direct sum of the elements of Δ. Let $G_2 = N_Q(\Delta)$ and $N_2 = C_Q(\Delta)$. Then G_2/N_2 is the direct product of symmetric groups and $G_1 M \leqslant G_2$.

Suppose that $G_1 \cap N_2 \not\leqslant N_1$. Then $R_1 \leqslant N_2$ and $[R_1, M] \leqslant N_2$. Pick $D \in \Delta$. Then $[R_1, M]$ normalizes $C_A(D)$ and so $[[R_1, M], C_A(D)] \leqslant C_M(D) \cap M' = 1$. Since also $[C_A(D), M, R_1] = 1$, the three subgroup lemma yields $[C_A(D), R_1, M] = 1$. Thus $[C_A(D), R_1] \leqslant C_M(D) \cap Z(M) = 1$. Furthermore, R_1 is perfect and $A/C_A(D)$ is cyclic. Thus R_1 centralizes A, a contradiction to $[\bar{A}, x] \neq 1$.

Hence $G_1 \cap N_2 \leqslant N_1$, (G_2, N_2) fulfills the conclusion of the lemma, and the proof of Lemma 2.9 is completed. $\square$

Corollary 2.10   *Lemma 2.9 remains true if Q is replaced by $\mathrm{PSL}_K(V)$.*

Proof.   Apply Lemma 2.9 to the inverse images in $\mathrm{GL}_K(V)$, intersect the resulting G_2 and N_2 with $\mathrm{SL}_K(V)$ and then look at the images in $\mathrm{PSL}_K(V)$. $\square$

Lemma 2.11   *Let I be a finite set, for $i \in I$ let L_i be a perfect simple group, and let $M \leqslant \prod_{i \in I} L_i$. For $J \subseteq I$, let $L_J = \prod_{j \in J} L_j$, $M_J = M \cap L_J$, and let M^J be the projection of M onto L_J.*

(a) *If $M^i = L_i$ for all $i \in I$, then there exists a partition Π of I such that $M = \prod_{\pi \in \Pi} M_\pi$ and, for all $\pi \in \Pi$ and $i \in \pi$, the projection of M_π onto L_i is an isomorphism.*

(b) *Put $J = \{i \in I \mid M^i = L_i\}$ and $K = I \setminus J$. If L_i is finite and $L_i \cong L_j$ for all $i, j \in I$, then $M = M_J \times M_K$.*

Proof.   (a) Let $\Pi = \{J \subseteq I \mid M_J \neq 1 \text{ and } M_K = 1 \text{ for all } K \subset J\}$. Let $i \in \pi \in \Pi$ and let ϕ be the projection map from M_π to L_i. Then $\ker \phi \leqslant M_{\pi \setminus \{i\}} = 1$ and ϕ is one to one. Moreover, M, and so L_i, normalizes the image of ϕ. Since L_i is simple, we conclude that ϕ is onto and so ϕ is an isomorphism. If $\pi' \in \Pi$ with $\pi \cap \pi' \neq \emptyset$, then $1 \neq [M_\pi, M_{\pi'}] \leqslant M_{\pi \cap \pi'}$ and so $\pi = \pi \cap \pi' = \pi'$.

It remains to show that $M = M^*$, where $M^* = \prod_{\pi \in \Pi} M_\pi$. For m in M let $S(m) = \{i \in I \mid m_i \neq 1\}$. We will prove by induction on $|S(m)|$ that $m \in M^*$. Without loss $m \neq 1$. Then $M_{S(m)} \neq 1$ and so there exists $\pi \in \Pi$ with $\pi \subseteq S(m)$. Pick $i \in \pi$ and $n \in M_\pi$ with $n_i = m_i$. Then $S(mn^{-1}) \subseteq S(m) \smallsetminus \{i\}$ and so, by induction, $mn^{-1} \in M^*$. Clearly $n \in M^*$ and so $m \in M^*$.

(b) By (a), M/M_K is a direct product of simple groups isomorphic to L_i. Note that M/M_J is a subdirect product of proper subgroups of L_i and so has no composition factor isomorphic to L_i. So no non-trivial factor group of M/M_K is isomorphic to a factor group of M/M_J. Thus $M/M_K M_J = 1$, $M = M_K M_J$ and $M = M_J \times M_K$. $\square$

Lemma 2.12   *Let L be a perfect, finite, simple group, n a positive integer, $T = L^n$, h an automorphism of T of order q, q a prime, I the set of components of T, t the number of non-trivial orbits for h on I, M an h-invariant subgroup of T and $K = \{g \in T \mid [g, h] \in M\}$. Then one of the following holds:*

(i) *M contains a component of T;*

(ii) *$|K| \leqslant (\frac{3}{4})^t |T|$.*

Proof.   We assume without loss that M does not contain a component of T and $t > 0$. We use the notation introduced in Lemma 2.11 with $L_D = D$ for all $D \in I$. We may assume:

(*) *If J is an h-invariant subset of I such that M^J does not contain a component of T and either $M^{I \smallsetminus J}$ does not contain a component of T or h acts trivially on $I \smallsetminus J$, then $J = \emptyset$ or $J = I$.*

Indeed suppose that (*) is false. Then, by induction on n, $|K^J| \leqslant (\frac{3}{4})^s |L_J|$ and $|K^{I \smallsetminus J}| \leqslant (\frac{3}{4})^{t-s} |L_{I \smallsetminus J}|$, where s is the number of non-trivial orbits for h on J. Hence (ii) holds in this case.

Using Lemma 2.11 we will prove next that one of the following holds:

(1) *The projection of M to L is not onto and h acts transitively on I.*

(2) *The projection of M to L is an isomorphism and h acts transitively on I.*

(3) *There exists an h-invariant partition Π of I such that $\Pi \neq \{I\}$, h acts transitively on Π, $M = \prod_{\pi \in \Pi} M_\pi$ and, if $D \in \pi \in \Pi$, the projection of M_π to D is an isomorphism.*

Indeed, put $J = \{D \in I \mid M^D = D\}$. Then, by Lemma 2.11(b), $M = M_J \times M_{I \smallsetminus J}$. So $M^{I \smallsetminus J} = M_{I \smallsetminus J}$ and neither M^J nor $M^{I \smallsetminus J}$ contains a component of T. Thus, by (*), $I = J$ or $J = \emptyset$. If $J = \emptyset$, let J^* be any h-orbit on I. Then, by (*), $J^* = I$ and (1) holds. So we may assume $I = J$ and thus $M^D = D$ for all $D \in I$. Let Π be the partition of I given by Lemma 2.11. Then Π is clearly h-invariant. Let Δ be an h-orbit on Π and put $J' = \bigcup \Delta$. Similarly, as above, neither $M^{J'}$ nor $M^{I \smallsetminus J'}$ contains a component of T and so, by (*), $J' = I$. If $\Pi \neq \{I\}$, (3) holds. So assume $\Pi = \{I\}$ and let J_* be any non-trivial h-orbit on I. If $|I \smallsetminus J_*| \geqslant 2$, $M^{I \smallsetminus J_*}$ does not contain a component of T and, if $|I \smallsetminus J_*| \leqslant 1$, h acts trivially on $I \smallsetminus J_*$. Thus in any case $I = J_*$ by (*), and (2) holds.

Suppose first that (1) or (3) holds. In case (1) put $\Pi = I$. Pick $\pi \in \Pi$ and let $g \in T$. Then $g = g_1 g_2^h \cdots g_q^{h^{q-1}}$ for some $g_i \in L_\pi$. So

$$[g, h] = (g_1^{-1} g_q)(g_2^{-1} g_1)^h \cdots (g_q^{-1} g_{q-1})^{h^{q-1}}.$$

Thus if $g \in K$ we conclude from $M = \prod_{\rho \in \Pi} M_\rho$ that

$$g_1 M_\pi = g_2 M_\pi = \ldots = g_q M_\pi.$$

Hence

$$|K| \leqslant |M_\pi|^q |L_\pi / M_\pi| = |T| / |L_\pi / M_\pi|^{q-1}.$$

It remains to show that $|L_\pi / M_\pi|^{q-1} \geqslant (4/3)^t$. If (1) holds, $t = 1$ and this is obvious. If (3) holds, $t = |\pi|$. Since M contains no components of T, $|\pi| > 1$ and so $t > 1$. Moreover, $|M_\pi| = |L|$ and so

$$|L_\pi / M_\pi|^{q-1} = |L|^{(t-1)(q-1)} \geqslant |L|^{t-1} \geqslant 4^{t-1} = (4/3)^t 3^t / 4 \geqslant (4/3)^t.$$

Suppose next that (2) holds. Then $M \cong L$, $t = 1$ and $C_T(h) \cong L$. Define $Y = \{[k, h] \mid k \in K\}$. Then $Y \subseteq M$ and $|Y| \leqslant |L|$. Let $k, l \in T$. Then $[k, h] = [l, h]$ if and only if $lk^{-1} \in C_T(h)$. Thus $|K| = |Y||C_T(h)| = |Y||L| \leqslant |L|^2$. If $q > 2$ we conclude that $|K| \leqslant |L|^2 \leqslant \frac{3}{4}|L||L|^2 \leqslant \frac{3}{4}|L|^q$. If $q = 2$, h inverts all elements of Y. So since M is not abelian it follows from a well-known exercise [3, 2.9 #12, p. 71] that $|Y| \leqslant \frac{3}{4}|M|$ and so $|K| \leqslant \frac{3}{4}|L|^2 = \frac{3}{4}|T|$, completing the proof of the lemma. $\square$

Lemma 2.13 *Let Ω be a finite set, $H \leqslant \mathrm{Sym}(\Omega)$, $H^* \subseteq H \smallsetminus \{1\}$ and h, k positive integers. If $\deg_\Omega(x) \geqslant hkw(H)|H||H^*|$ for all $x \in H^*$, then there exists a subset Γ of Ω and a partition Δ of Γ into subsets of size h such that H normalizes Δ and $\deg_\Delta(x) \geqslant k$ for all $x \in H^*$.*

Proof. Let $\mathcal{W}$ be a set of representatives of the isomorphism classes of transitive permutation representations of H. We note that $|O| \leqslant |H|$ for all $O \in \mathcal{W}$. For $O \in \mathcal{W}$, let $r(O)$ be the number of H-orbits on Ω isomorphic to O.

Let $x \in H^*$. We claim that there exists $O_x \in \mathcal{W}$ such that $x \notin C_H(O_x)$ and $r(O_x) \geqslant hk|H^*|$. Indeed, let $\mathcal{W}_x = \{O \in \mathcal{W} \mid x \notin C_H(O)\}$. Then

$$\deg_\Omega(x) = \sum_{O \in \mathcal{W}_x} r(O) \deg_O(x).$$

Since $\deg_O(x) \leqslant |O| \leqslant |H|$, $|\mathcal{W}_x| \leqslant |\mathcal{W}| = w(H)$ and $\deg_\Omega(x) \geqslant hkw(H)|H||H^*|$, we conclude that $r(O_x) \geqslant hk|H^*|$ for at least one O_x in $\mathcal{W}_x$.

Let Y be a subset of H^* maximal with respect to the existence of pairwise distinct H-orbits $O(y, i, j)$, $y \in Y$, $1 \leqslant i \leqslant h, 1 \leqslant j \leqslant k$, in Ω such that $O(y, i, j)$ is isomorphic to O_y. Suppose that $Y \neq H^*$ and pick u in $H^* \smallsetminus Y$. Since we have $r(O_u) \geqslant hk|H^*| \geqslant hk|Y| + hk$ there are at least hk H-orbits on Ω which are isomorphic to O_u and distinct from the $O(y, i, j)$, a contradiction to the maximal choice of Y.

Thus $Y = H^*$. Let $\phi(x, i, j)$ be an H-isomorphism from O_x to $O(x, i, j)$. For $d \in O_x$, $x \in H^*$ and $1 \leqslant j \leqslant k$ put

$$D(x, j, d) = \{\phi(x, i, j)(d) \mid 1 \leqslant i \leqslant h\}.$$

Also, put

$$\Delta = \{D(x, j, d) \mid x \in H^*, \, 1 \leqslant j \leqslant k, \, d \in O_x\}.$$

Note that $D(x, j, d)^g = D(x, j, d^g)$ for all $g \in H$ and so H normalizes Δ. For $x \in H^*$ pick d in O_x with $d^x \neq d$. Then $D(x, j, d)^x \neq D(x, j, d)$ for all $1 \leqslant j \leqslant k$, and so $\deg_\Delta(x) \geqslant k$. $\qquad \square$

Lemma 2.14 *Let Ω be a finite set, $G \leqslant \mathrm{Alt}(\Omega)$, $N \vartriangleleft G$ with $G/N \cong \mathrm{Alt}_n, n \geqslant 5$, $H \leqslant G$ with $H \cap N = 1$, u a positive integer with $u \geqslant (2 \log_{4/3} |H|) + 2$, and $k = \max(l(u), 5uw(H)|H|^2, 9|H|^2)$, where l is as in Lemma 2.5. If $\mathrm{pdeg}_{G/N}(x) \geqslant k$ for all $1 \neq x \in H$ then one of the following holds:*

(a) *H has a regular orbit on Ω.*

(b) *G has a t-pseudo natural orbit on Ω with respect to N, where t is a positive integer with $t \leqslant |H| - 2$.*

Proof. Consider a counter example with $|\Omega|$ minimal. Let R be the minimal subnormal supplement to N in G provided by Lemma 2.8. Let O be a G-orbit on Ω with $R \not\leqslant C_G(O)$. Then

$$(G/C_G(O), \, NC_G(O)/C_G(O), \, HC_G(O)/C_G(O), \, O)$$

fulfills the assumption of the lemma. We conclude that $\Omega = O$ and so G acts transitively on Ω.

Suppose that G acts imprimitively on Ω and let Δ be a maximal system of imprimitivity for G on Ω. If N is not transitive on Ω, the orbits of N on Ω form a system of imprimitivity for G on Ω and we can and do choose a system Δ such that $N \leqslant C_G(\Delta)$.

If $C_G(\Delta) \leqslant N$, then

$$(G/C_G(\Delta), \, NC_G(\Delta)/C_G(\Delta), \, HC_G(\Delta)/C_G(\Delta), \, \Delta)$$

fulfills the assumption of the lemma. But then (a) or (b) holds for Δ and so also for Ω.

Hence $C_G(\Delta) \not\leqslant N$, $R \leqslant C_G(\Delta)$, $N \not\leqslant C_G(\Delta)$, N is transitive on Ω and R is not transitive on Ω. Let O be an orbit for R on Ω. For $X \leqslant G$ put $X_0 = N_X(O)$. Let R^* be the minimal subnormal supplement to N_0 in G_0. Then $R^* \vartriangleleft R$, R^* is a subnormal supplement to N in G, and thus $R = R^*$. If $C_{G_0}(O) \not\leqslant N_0$, $R = R^* \leqslant C_{G_0}(O)$. Since R is normal in G and G is transitive we conclude that $R = 1$, a contradiction. Thus $C_{G_0}(O) \leqslant N_0$. If H_0 has a regular orbit on O, H has a regular orbit on Ω. Hence, by minimality of $|\Omega|$, there exists a maximal system of imprimitivity Γ_0 for G_0 on O such that $N_0 = C_{G_0}(\Gamma_0)$. It follows that N_0 is not transitive on O and so N is not transitive on Ω, a contradiction.

We proved that G acts primitively on Ω. Let M be the stabilizer in G of some point in Ω. Since H has no regular orbit on Ω, $H \cap M^g \neq 1$ for all $g \in G$. Let T be a minimal normal subgroup of G with $T \leqslant R$. Then $H \cap M^t \neq 1$ for all $t \in T$. By the pigeon hole principle there exists $1 \neq h \in H$ with $|S| \geqslant |T|/(|H|-1)$, where $S = \{s \in T \mid h \in M^{s^{-1}}\}$. Without loss $|h| = p$, p a prime. Pick $s_0 \in S$. Replacing M by $M^{s_0^{-1}}$ and S by $S^{s_0^{-1}}$ we may assume that $1 \in S$ and so $h \in M$. Let $s \in S$. Then $h \in M^{s^{-1}}$ and so $h^s \in M$. It follows that $[s,h] = h^{-s}h \in M \cap T$ for all $s \in S$.

Suppose that T is an elementary abelian q-group for some prime q. Then T acts regularly on Ω and so $C_G(T) \leqslant T$ and $T \cap M = 1$. In particular, $S \leqslant C_T(h)$ and, since $|S| \geqslant |T|/(|H|-1)$, $|T/C_T(h)| \leqslant |T/S| \leqslant |H|$. Thus

$$\operatorname{pdeg}_T(h) = \log_q |T/C_T(H)| \leqslant \log_q |H| \leqslant \log_{4/3} |H|.$$

On the other hand, by assumption, $\operatorname{pdeg}_{G/N}(h) \geqslant l(u)$ and so by Lemma 2.5 applied to '$H = \operatorname{GL}_{\mathbb{F}_q}(T)$' and '$x = h$', $\operatorname{pdeg}_T(x) \geqslant u > \log_{4/3}(|H|)$, a contradiction.

Hence T is the direct product of perfect simple groups. Let I be the set of components of T. Note that T acts transitively on Ω and so $G = MT$ and M acts transitively on I. In particular, M contains no component of T.

Suppose that $C_G(I) \leqslant N$. Let t be the number of non-trivial orbits of h on I. Then, by Lemma 2.12, $|S| \leqslant (3/4)^t|T|$. Since $|S| \geqslant |T|/|H|$ we conclude that

$$t \leqslant \log_{4/3}(|T|/|S|) \leqslant \log_{4/3} |H|.$$

Since $t = (p-1/p)\operatorname{pdeg}_I(h) \geqslant \frac{1}{2}\operatorname{pdeg}_I(h)$ we conclude that $\operatorname{pdeg}_I(h) \leqslant 2\log_{4/3} |H|$, a contradiction to $\operatorname{pdeg}_{G/N}(h) \geqslant l(2\log_{4/3} |H|)$ and Lemma 2.5.

Thus $C_G(I) \not\leqslant N$ and so $R \leqslant C_G(I)$, i.e. R normalizes all components of T. Since the outer automorphism group of every finite simple group is solvable and since R is perfect we conclude that R induces inner automorphisms on T. Thus $R \leqslant TC_G(T)$ and $TC_G(T) \not\leqslant N$. Recall that $T \leqslant R$. If $C_G(T) \not\leqslant N$ we get $T \leqslant R \leqslant C_G(T)$, a contradiction. Hence $C_G(T) \leqslant N$, $T \not\leqslant N$ and $R \leqslant T$. It follows that $R = T$, $R \cap N = 1$, $R \cong \operatorname{Alt}_n$ and $G = R \times N$.

Suppose that $N \neq 1$. Then, since G is primitive, N is transitive and R is regular. So $R \cap M = 1$ and $S \subseteq C_R(h)$. Since $|T|/|S| \leqslant |H|-1$, $|R/C_R(h)| \leqslant |H|-1$. Since $R \cong \operatorname{Alt}_n$ where $n \geqslant 5$, R has no subgroup of index less than n. So

$$|H| - 1 \geqslant n \geqslant \operatorname{pdeg}_{G/N}(h) \geqslant |H|,$$

which is a contradiction.

Therefore $N = 1$ and $G \cong \operatorname{Alt}_n$. Let $\Lambda = \{1, 2, \ldots, n\}$ with G acting naturally on Λ. Let $1 \neq x \in H$. Then $\deg_\Lambda(x) = \operatorname{pdeg}_{G/N}(x) \geqslant 5uw(H)|H|$ and so, by Lemma 2.13 applied to $(H, H \smallsetminus \{1\}, 5, u, \Lambda)$ in place of (H, H^*, h, k, Ω), there exists a subset Γ of Λ and a partition Δ of Γ into subsets of size five such that H normalizes Δ and $\deg_\Delta(x) \geqslant u$ for all $x \in H$, $x \neq 1$. Let $T^* = (C_{\operatorname{Alt}(\Lambda)}(\Delta) \cap C_{\operatorname{Alt}(\Lambda)}(\Lambda \smallsetminus \Gamma))'$ and $M^* = M \cap T^*$. Note that T^* is the direct product of alternating groups of degree five and the action of H on the components of T^* is isomorphic to the action on Δ. If M^* does not contain a component of T^*, then using Lemma 2.12 we get the same contradiction as in the case $C_G(I) \leqslant N$ with T replaced by T^*. Hence M contains a component of T^* and in particular an element acting as a three cycle on Λ. If M acts primitively on Λ, we conclude from [4, II 4.5(c)] that $M = G$, a contradiction.

Thus M does not act primitively on Λ. Note that $\deg_\Lambda(x) > 8|H|$ for all elements $1 \neq x \in H$. It follows that, for $x \in H \smallsetminus \{1\}$, there exist pairwise distinct elements a_x, b_x, c_x, d_x in Λ such that $a_x^x = c_x$ and $b_x^x = d_x$.

Assume that M acts transitively and imprimitively on Λ, and let Θ be a system of imprimitivity for M on Λ. Put $k = |\Lambda|/|\Theta|$. If $|\Theta| \geqslant 3|H|$, then, for $x \in H \smallsetminus \{1\}$, we can choose disjoint sets $\alpha_x, \gamma_x, \delta_x$ of size k in Λ with $a_x, b_x \in \alpha_x$, $c_x \in \gamma_x$ and $d_x \in \delta_x$. Choose $g \in G$ such that Θ^g contains α_x, γ_x and δ_x, for all $x \in H \smallsetminus \{1\}$. Since $H \cap M^g \neq 1$, there exists $1 \neq x \in H$ with $x \in M^g \leqslant N_G(\Theta^g)$. Since $a_x^x = c_x$, $\alpha_x^x = \gamma_x$, and, since $b_x^x = d_x$, $\alpha_x^x = \delta_x$, a contradiction to $\gamma_x \neq \delta_x$. Thus $|\Theta| \leqslant 3|H|$ and, since $|\Lambda| \geqslant \deg_\Lambda(x) \geqslant 9|H|^2$, $k \geqslant 3|H|$. It follows that we can choose disjoint sets α and δ of size k in Λ with $a_x, b_x, c_x \in \alpha$ and $d_x \in \delta$, for all $x \in H \smallsetminus \{1\}$. As above, choose $g \in G$ with $\alpha, \delta \in \Theta^g$, pick $1 \neq x \in H$ with $x \in M^g \leqslant N_G(\Theta^g)$ and conclude that $\alpha = \alpha^x = \delta$, a contradiction.

Thus M does not act transitively on Λ, so $M = N_G(\Theta)$ for some $\Theta \subseteq \Lambda$ with $|\Theta| \leqslant |\Lambda \smallsetminus \Theta|$. Suppose that $|\Theta| \geqslant |H| - 1$. Then there exists $g \in G$ with $a_x \in \Theta^g$ and $c_x \in \Lambda \smallsetminus \Theta^g$ for all $x \in H \smallsetminus \{1\}$. It follows that $H \cap M^g = 1$, a contradiction. $\square$

Lemma 2.15 *Let G be a LFS-group and $\{(G_i, N_i) \mid i \in I\}$ a sectional cover for G.*

(a) *There exists a Kegel cover $\{(H_j, M_j) \mid j \in J\}$ such that for all $j \in J$ there exists $i \in I$ with $N_i \leqslant M_j \trianglelefteq H_j \trianglelefteq G_i$.*

(b) *For $i \in I$ let M_i be a normal subgroup of G_i. Then at least one of the sets $\{(G_i, M_i) \mid i \in I\}$ and $\{(M_i, M_i \cap N_i) \mid i \in I\}$ is a sectional cover for G.*

(c) *$\{(G_i^\infty, G_i^\infty \cap N_i) \mid i \in I\}$ is a sectional cover for G.*

(d) *Let $\mathcal{E}$ be a class of groups such that $K \in \mathcal{E}$ for each $i \in I$ and each non-abelian composition factor K of G_i/N_i. Then there exists a Kegel cover for G all of whose factors are in $\mathcal{E}$.*

Proof. (a) Let E be a non-trivial finite subgroup of G and $1 \neq e \in E$. Since G is simple, $E \leqslant \langle e^G \rangle$ and, since G is locally finite, $E \leqslant \langle e^F \rangle$ for some finite subgroup F of G. Similarly $F \leqslant \langle e^{T_e} \rangle$ for some finite subgroup T_e of G. Then $E \leqslant \langle e^{(e^{T_e})} \rangle$. Let T be the finite subgroup of G generated by E and all the T_e, $1 \neq e \in E$. Pick $i \in I$ with $T \leqslant G_i$ and $T \cap N_i = 1$. Put $H_E = \langle E^{G_i} \rangle N_i$. Clearly EN_i does not lie in any normal subgroup of G_i properly contained in H_E and, in particular, it is not in the intersection of the maximal normal subgroup of H_E containing N_i. Thus there exists a maximal normal subgroup M_E of H_E with $E \not\leqslant M_E$ and $N_i \leqslant H_E$. Suppose that $1 \neq e \in E$ with $e \in M_E$. Then $\langle e^{G_i} \rangle \leqslant H_E$ and $E \leqslant \langle e^{(e^{T_e})} \rangle \leqslant \langle e^{(e^{G_i})} \rangle \leqslant \langle e^{H_e} \rangle \leqslant M_E$, a contradiction. Thus $E \cap M_E = 1$. It follows that $\{(H_E, M_E) \mid E$ a non-trivial finite subgroup of $G\}$ is a Kegel cover that fulfills (a).

(b) Assume that $\{(G_i, M_i) \mid i \in I\}$ is not a sectional cover for G. Then there exists a finite subgroup H of G such that $H \cap M_i \neq 1$ for all $i \in I$ with $H \leqslant G_i$. Without loss $H \leqslant G_i$ for all $i \in I$. For $1 \neq h \in H$ let $I_h = \{i \in I \mid h \in M_i\}$. Then I is the finite union of these I_h and so there exists $1 \neq h \in H$ such that

$\{(G_i, N_i) \mid i \in I_h\}$ is a sectional cover for G. We may assume that $I = I_h$. Let E be any finite subgroup of G. Since G is LFS, there exists a finite subgroup T in H with $h \in T$ and $E \leqslant \langle h^T \rangle$. Pick $i \in I$ with $T \leqslant G_i$ and $T \cap N_i = 1$. Then $E \leqslant \langle h^T \rangle \leqslant M_i$ and $E \cap (N_i \cap M_i) \leqslant T \cap N_i = 1$. Thus $\{(M_i, M_i \cap N_i) \mid i \in I\}$ is a sectional cover for G.

(c) Otherwise we conclude from (b) that $\{(G_i, G_i^\infty) \mid i \in I\}$ is a sectional cover for G. Hence, by (a), G has a Kegel cover all of whose factors are of prime order. Thus G is of prime order, a contradiction since G is infinite.

(d) By (a) there exists a Kegel cover all of whose factors are abelian or lie in $\mathcal{E}$. Since G is not abelian, (d) holds.   □

3. Kegel Covers

Throughout this chapter G is a non-finitary LFS-group and $\mathcal{K}$ is a Kegel cover for G. For $K \in \mathcal{K}$ let H_K and M_K be defined by $K = (H_K, M_K)$ and put $\bar{K} = H_K/M_K$. If $\mathcal{E}$ is a class of groups, $\mathcal{K}_\mathcal{E} = \{K \in \mathcal{K} \mid \bar{K} \in \mathcal{E}\}$. For $K \in \mathcal{K}_\mathcal{F}$ pick a finite field F_K, a finite dimensional vector space V_K over F_K and a non-degenerate symplectic, orthogonal or unitary form q_K on V_K such that $\bar{K} \cong \mathrm{P}\Omega(V_K, q_K)$. For $K \in \mathcal{K}_\mathcal{L}$, pick a finite field F_K and a finite dimensional vector space V_K over F_K such that $\bar{K} \cong \mathrm{PSL}(V_K, q_K)$. We view V_K as a projective module for H_K. For $K \in \mathcal{K}_\mathcal{A}$ pick a set Ω_K with $K \cong \mathrm{Alt}(\Omega_K)$. For a finite subset T of G let

$$\mathcal{K}(T) = \{(H, M) \in \mathcal{K} \mid T \subseteq H \text{ and } T \cap M \subseteq \{1\}\}.$$

Lemma 3.1 *Let k be a positive integer and $X \subseteq G \smallsetminus \{1\}$ with $|X|$ finite. Put*

$$\mathcal{J} = \{K \in \mathcal{K}_{\mathcal{L} \cup \mathcal{A}}(X) \mid \mathrm{pdeg}_{\bar{K}}(x) \geqslant k \text{ for all } x \in X\}.$$

Then $\mathcal{J}$ is a Kegel cover for G. In particular, at least one of $\mathcal{K}_\mathcal{A}$, $\mathcal{K}_\mathcal{L}$ and $\mathcal{K}_\mathcal{F}$ is a Kegel cover for G.

Proof. Without loss $\mathcal{K} = \mathcal{K}(X)$. By induction on $|X|$ we may assume that $|X| = 1$. Let $x \in X$ and suppose that $\mathcal{J}$ is not a Kegel cover for G. Then $\mathcal{K} \smallsetminus \mathcal{J}$ is a Kegel cover for G. By the classification of finite simple groups there exists a natural number t such that every finite simple group not contained in $\mathcal{C} \cup \mathcal{A}$ has a faithful projective representation of dimension at most t. Let $s = \max\{k, t\}$. Then $\mathrm{pdeg}_{\bar{K}}(x) \leqslant s$ for all $K \in \mathcal{K} \smallsetminus \mathcal{J}$. Thus, by [2, (3.1)], G has a faithful projective representation U with $\mathrm{pdeg}_U(x) \leqslant s$. Since $G = \langle x^G \rangle$, G is finitary, a contradiction.   □

Proposition 3.2

(a) *Suppose $\mathcal{K} = \mathcal{K}_\mathcal{F}$. Let $J = \{(K, U) \mid K \in \mathcal{K}$ and U is a singular subspace of $V_K\}$, and for $j = (K, U) \in J$ put $H_j = N_{H_K}(U)$ and*

$$M_j = \{x \in H_j \mid x \text{ acts as a scalar on } U\}.$$

Then $\mathrm{PSL}_{F_K}(U) \leqslant H_j/M_j \leqslant \mathrm{PGL}_{F_K}(U)$, $\{(H_j, M_j) \mid j \in J\}$ is a sectional cover for G and $\{(H_j^\infty, H_j^\infty \cap M_j) \mid j \in J\}$ is a Kegel cover for G with all factors in $\mathcal{L}$.

(b) *G has a Kegel cover $\mathcal{J}$ with $\mathcal{J} = \mathcal{J}_A$ or $\mathcal{J} = \mathcal{J}_\mathcal{L}$.*

Proof. (a) Let T be a finite subgroup of G and $k = g(|T|)$, where the function g is defined in Lemma 2.3(d). By Lemma 3.1 there exists $K \in \mathcal{K}$ with $T \leqslant H_K$ and $\mathrm{pdeg}_{V_K}(t) \geqslant k$ for all $1 \neq t \in T$. By Lemma 2.3(d) there exists a T-invariant singular subspace U in V_K such that no element of $T \smallsetminus \{1\}$ acts as a scalar on U. Thus $T \leqslant H_{(K,U)}$ and $H \cap M_{(K,U)} = 1$. So $\{(H_j, M_j) \mid j \in J\}$ is a sectional cover for G.

Using Witt's theorem we get $\mathrm{PSL}_{F_K}(U) \leqslant H_j/M_j \leqslant \mathrm{PGL}_{F_K}(U)$ for all $j \in J$. The last claim in (a) follows from Lemma 2.15(c) and since $\mathrm{PGL}_{F_K}(V_K)^\infty \in \mathcal{L} \cup \{1\}$.

(b) By Lemma 3.1 we may assume that $\mathcal{K}_{\mathcal{F}}$ is a Kegel cover. Then (a) provides a Kegel cover all of whose factors are in $\mathcal{L}$. $\qquad\square$

Theorem 3.3 *Let G be an LFS-group which is neither finitary nor of alternating type. Then G is of p-type for some prime p and there exists a Kegel cover $\mathcal{K}$ for G such that $\mathcal{K} = \mathcal{K}_{\mathcal{L}(p)}$ and, for all $K \in \mathcal{K}$, $H_K/O_p(H_K)$ is the central product of perfect central extensions of groups in $\mathcal{C}(p)$.*

Proof. Put

$$\mathcal{P} = \{ (H, M) \mid H \leqslant G, \; M \lhd H, \; H \text{ is finite, and } H/M \text{ is perfect and simple} \}$$

and note that $\mathcal{P}$ is a Kegel cover for G. By assumption G is not of alternating type and so $\mathcal{P}_\mathcal{A}$ is not a Kegel cover.

(1) $\{(H, M) \mid H \leqslant G, \; M \lhd H, \; H$ *is finite and every non-abelian composition factor of H/M is in $\mathcal{A}\}$ is not a sectional cover for G.*

Otherwise Lemma 2.15 provides a Kegel cover with alternating factors.

(2) *If $\mathcal{T}$ is a Kegel cover for G, then $\mathcal{T}_\mathcal{C}$ is a Kegel cover as well.*

By Lemma 3.1, $\mathcal{T}_\mathcal{A}$ or $\mathcal{T}_\mathcal{C}$ is a Kegel cover. By assumption $\mathcal{T}_\mathcal{A}$ is not a Kegel cover.

Let $\mathcal{J}$ be the set of all pairs $(H, M) \in \mathcal{P}_\mathcal{C}$ such that there exists a prime p with $C_H(M/O_p(H)) \leqslant M$ and $\mathcal{P}_{\mathcal{L}(p)}(H) \neq \emptyset$.

(3) *$\mathcal{J}$ is not a Kegel cover.*

Suppose that $\mathcal{J}$ is a Kegel cover. Using Lemma 2.9 and its Corollary 2.10 we will derive a contradiction to (1). Let T be any finite subgroup of G and pick $(G_1, N_1) \in \mathcal{J}(T)$. Let R_1 be the minimal normal supplement to N_1 in G_1. Since R_1 is of even order there exists $x \in R_1$ with $|x| = 2$. Let $k = l(2|G_1| + f(|G_1|))$, where l and f are as in Lemmata 2.5 and 2.2. By Lemma 3.1 there exists (G_0, N_0) in $\mathcal{J}(G_1)$ with $\mathrm{pdeg}_{G_0/N_0}(x) \geqslant k$. By definition of $\mathcal{J}$ there exists a prime p with $C_{G_0}(N_0/O_p(G_0)) \leqslant N_0$ and $K \in \mathcal{P}_{\mathcal{L}(p)}(G_0)$. Hence, by Corollary 2.10 applied to $\mathrm{PSL}_{F_K}(V_K) \cong \bar{K}$ and the images of G_0, N_0, G_1, N_1 in $\mathrm{PSL}_{F_K}(V_K)$ in place of G, N, G_1, N_1, there exists $G_2 \leqslant H_K$ and $N_2 \lhd G_2$ such that $G_1 \leqslant G_2$, $G_1 \cap N_2 \leqslant N_1$ and every non-abelian composition factor of G_2/N_2 is in $\mathcal{A}$. Note that $T \leqslant G_1 \leqslant G_2$ and $T \cap N_2 \leqslant T \cap N_1 = 1$. Since T was an arbitrary finite subgroup of G, we get a contradiction to (1).

(4) *There exists a prime p such that G is of p-type. In particular, $\mathcal{P}_{\mathcal{L}(p)}$ is a Kegel cover.*

Note that G is of p-type if and only if $\mathcal{P} \setminus \mathcal{P}_{\mathcal{C}(p)}$ is not a Kegel cover. Suppose that, for all primes p, $\mathcal{P} \setminus \mathcal{P}_{\mathcal{C}(p)}$ is a Kegel cover. Then, by (2) and Proposition 3.2, $\mathcal{P}_{\mathcal{L} \setminus \mathcal{L}(p)}$ is a Kegel cover for all primes p. Let T_0 be a finite subgroup of G and note that, by Lemma 2.15(c), there exists a perfect finite subgroup T of G with $T_0 \leqslant T$. By Lemma 3.1(a), there exists $K \in \mathcal{P}_{\mathcal{L}}(T)$ with $\dim_{F_K} V_K > |T|^2$. For $1 \neq t \in T$, pick $v_t \in V_K$ with $v_t^t \notin F_K v_t$. Put $U = F_K \langle v_t^T \mid t \in T \setminus \{1\} \rangle$. Then $\dim_{F_K} U \leqslant |T|^2$ and U is a proper T-invariant subspace of V_K such that no non-trivial element of T acts as a scalar on U. Let $H = N_{H_K}(U)^\infty$ and let M consist of all the elements in H which act as a scalar on U. Since T is perfect, $T \leqslant H$ and so $(H, M) \in \mathcal{P}_{\mathcal{L}}(T)$. Let $q = \mathrm{char} F_K$. Since $\mathcal{P}_{\mathcal{L} \setminus \mathcal{L}(q)}$ is a Kegel cover for G, there exists a prime p distinct from q such that $\mathcal{P}_{\mathcal{L}(p)}(H) \neq \varnothing$. Note that $O_p(H/H \cap M_K) = 1$ and $C_H(M/H \cap M_K) \leqslant M$. Thus $C_H(M/O_p(H)) \leqslant M$ and $(H, M) \in \mathcal{J}(T) \subseteq \mathcal{J}(T_0)$. It follows that $\mathcal{J}$ is a Kegel cover, a contradiction to (3).

Hence there exists a prime p such that $\mathcal{P} \setminus \mathcal{P}_{\mathcal{C}(p)}$ is a not a Kegel cover. This implies that $\mathcal{P}_{\mathcal{C}(p)}$ is a Kegel cover and so, by Proposition 3.2(a), $\mathcal{P}_{\mathcal{L}(p)}$ is a Kegel cover.

For a finite group H and a prime p let $F_p^*(H)$ be defined by $F^*(H/O_p(H)) = F_p^*(H)/O_p(H)$.

(5) $\{(H, F_p^*(H)) \mid H \leqslant G, H \text{ finite}\}$ *is not a sectional cover for G.*

Suppose $\{(H, F_p^*(H)) \mid H \leqslant G, H \text{ finite}\}$ is a sectional cover for G. Note that $C_H(H/O_p(H)) \leqslant F_p^*(H)$. Hence, by Lemma 2.15(a),

$$\mathcal{I} = \{(H, N) \in \mathcal{P} \mid C_H(N/O_p(H)) \leqslant N\}$$

is a Kegel cover for G. If $\mathcal{I}_{\mathcal{F}}$ is a Kegel cover, then the Kegel cover provided by Proposition 3.2(a) is contained in $\mathcal{I}_{\mathcal{L}}$. Thus in any case $\mathcal{I}_{\mathcal{L}}$ is a Kegel cover and so, by (4), $\mathcal{I}_{\mathcal{L}(p)}$ is a Kegel cover. In particular, $\mathcal{P}_{\mathcal{L}(p)}(H) \neq \varnothing$ for all finite subgroups H of G, and $\mathcal{I}_{\mathcal{L}(p)} \subseteq \mathcal{J}$, a contradiction to (3).

(6) $\{F_p^*(H) \mid H \leqslant G, H \text{ finite}\}$ *is a sectional cover for G.*

This follows immediately from Lemma 2.15(b) and (5).

Put $\mathcal{M} = \{H \leqslant G \mid H \text{ finite and perfect, } H = F_p^*(H)\}$. It follows by (6) and Lemma 2.15(c) that $\mathcal{M}$ is a sectional cover for G. For $H \in \mathcal{M}$, let $\mathrm{Sol}(H)$ be the largest solvable normal subgroup of H. Then $H/\mathrm{Sol}(H)$ is the direct product of non-abelian simple groups. Let $H_p/\mathrm{Sol}(H)$ be the product of the components of $H/\mathrm{Sol}(H)$ contained in $\mathcal{C}(p)$.

(7) $\{H_p \mid H \in \mathcal{M}\}$ *is a sectional cover for G and $\{(H, H_p) \mid H \in \mathcal{M}\}$ is not a sectional cover.*

Otherwise we conclude from (6) and Lemma 2.15(b) that $\{(H, H_p) \mid H \in \mathcal{M}\}$ is a sectional cover for G. Hence by Lemma 2.15(a) there exists a Kegel cover for G none of whose factors are in $\mathcal{C}(p)$, a contradiction to (4).

Applying Lemma 2.15(a) to the sectional cover $\{H_p \mid H \in \mathcal{M}\}$ we get a Kegel cover $\mathcal{K}$ for G such that $H/O_p(H)$ is the central product of perfect central extensions of elements in $\mathcal{C}(p)$ for all $(H, N) \in \mathcal{K}$. Using Proposition 3.2(a) we can choose $\mathcal{K}$ such that in addition $\mathcal{K} = \mathcal{K}_{\mathcal{L}(p)}$. Thus (b) holds. $\qquad \square$

Theorem 3.4 *Let G be of alternating type and $\mathcal{K}$ a Kegel cover for G with $\mathcal{K} = \mathcal{K}_{\mathcal{A}}$.*

(a) *One of the following holds:*

    (a1) *There exists a Kegel cover $\mathcal{J} \subseteq \mathcal{K}$ such that for all $J, K \in \mathcal{J}$ with $H_J \leqslant H_K$ every essential orbit of H_J on Ω_K is pseudo natural with respect to M_J.*

    (a2) *For all finite subgroups T in G,*

$$\mathcal{K}_R(T) = \{K \in \mathcal{K}(T) \mid T \text{ has a regular orbit on } \Omega_K\}$$

    is a Kegel cover for G.

(b) *If G is countable, there exists a Kegel sequence $\{K_n \mid n \geqslant 1\} \subseteq \mathcal{K}$ for G such that one of the following holds:*

    (b1) *For all $n < m$ all essential orbits of H_{K_n} on Ω_{K_m} are pseudo natural with respect to M_{K_n}.*

    (b2) *For all $n < m$, H_{K_n} has a regular orbit on Ω_{K_m}.*

Proof. Suppose (a) is false. Then there exists a finite subgroup H of G such that $\mathcal{K}_R(H)$ is not a Kegel cover. Hence $\mathcal{K} \setminus \mathcal{K}_R(H)$ is a Kegel cover and we may assume that $\mathcal{K}_R(H) = \emptyset$. Let k be defined as in Lemma 2.14. By Lemma 3.1 we may assume that $\mathcal{K} = \mathcal{K}(H)$ and $\deg_{\Omega_K}(h) \geqslant k$ for all $h \in H \setminus \{1\}$ and all $K \in \mathcal{K}$. Let $\mathcal{J}$ be the set of all $K \in \mathcal{K}$ such that, for all $L \in \mathcal{K}$ with $H_K \leqslant H_L$, all essential orbits of H_K on Ω_L are pseudo natural with respect to M_K.

Since (a1) does not hold, $\mathcal{J}$ is not a Kegel cover and so there exists a finite subgroup T of G with $\mathcal{J}(T) = \emptyset$. Then $\mathcal{K}(T) \cap \mathcal{J} = \emptyset$ and so for all $K \in \mathcal{K}(T)$ there exists $X(K) \in \mathcal{K}$ such that $H_K \leqslant H_{X(K)}$ and not all essential orbits of H_K on $\Omega_{X(K)}$ are pseudo natural. Since H has no regular orbit on $\Omega_{X(K)}$, we conclude from Lemma 2.14 that all essential orbits for H_K on $\Omega_{X(K)}$ are t-pseudo natural for some t. Hence H_K has a t-pseudo natural orbit on $\Omega_{X(K)}$ with $t > 1$. Pick $K_0 \in \mathcal{K}(T)$ and inductively define $K_n = X(K_{n-1})$ for $n \geqslant 1$. Put $\Omega_n = \Omega_{K_n}$ and $H_n = H_{K_n}$. Pick $n \geqslant 1$ and $\omega \in \Omega_n$ with $|C_H(\omega)|$ mimimal. Since H has no regular orbit on Ω_n, there exists $1 \neq h \in C_H(\omega)$. Let O be a t-pseudo natural orbit for H_n on Ω_{n+1} with $t > 1$ and Δ a maximal system of imprimitivity for H_n on O such that the action of H_n on Δ is isomorphic to the action of H_n on subsets of size i in Ω_n, where $2 \leqslant i \leqslant |\Omega_n|/2$. Note that $\deg_{\Omega_n}(h) \geqslant k \geqslant 2|H|$ and thus there exists $a \in \Omega_n \setminus \omega^H$ with $a^h \neq a$. Moreover, $|\Omega_n| - i \geqslant |H|$ and there exists a subset U of size i in Ω_n with $a \in U$, $a^h \notin U$ and $\omega^H \cap U = \{\omega\}$. Then $N_H(U) \leqslant C_H(\omega)$ and $h \notin N_H(U)$. Let D be the element of Δ corresponding to U and pick $\sigma \in D$. Then

$C_H(\sigma) \leqslant N_H(D) = N_H(U) < C_H(\omega)$, a contradiction to the minimality of $|C_H(\omega)|$. This completes the proof of (a).

(b) Note that every Kegel cover for a countable group contains a Kegel sequence. Hence (a1) implies (b1) and we may assume that (a2) holds. Let $G = \{g_n \mid n \geqslant 1\}$. Pick $K_1 \in \mathcal{K}_R(\langle g_1 \rangle)$, and inductively $K_n \in \mathcal{K}_R(\langle H_{K_{n-1}}, g_n \rangle)$, $n \geqslant 2$. Then it follows that $\{K_n \mid n \geqslant 1\}$ is a Kegel sequence for G which fulfills (b2). $\qquad \square$

Remark 3.5 *Let H be a group and $N \lhd H$ with $H/N \cong \mathrm{Alt}_n$. Then every regular orbit for H is pseudo natural with respect to N. It follows that, under the assumptions of Theorem 3.4(a) and (b), statements (a2) and (b2), respectively, are always true if regular is replaced by pseudo natural.*

4. Maximal Subgroups

Theorem 4.1 *Let G be a countable, non-finitary LFS-group and H a finite subgroup of G. Then H is contained in a maximal subgroup of G. In particular, G has maximal subgroups.*

Proof. The idea is to find a Kegel sequence $\{(H_n, M_n) \mid n \geqslant 1\}$ of G and maximal subgroups T_n of H_n such that $H \leqslant T_1$, $T_n \leqslant T_{n+1}$ and $H_n \not\leqslant T_{n+1}$. Given such a sequence put $T = \bigcup_{n=1}^{\infty} T_n$. Then T is a maximal subgroup of G. Indeed, since T_n is maximal in H_n and $H_n \not\leqslant T_{n+1}$, $H_n \cap T_{n+1} = T_n$. Thus $H_n \cap T = T_n$ and, in particular, $T \neq G$. Let $x \in G \setminus T$. Then for all n with $x \in H_n$, $H_n = \langle x, T_n \rangle \leqslant \langle x, T \rangle$ and so $G = \langle x, T \rangle$.

By Theorem 3.3 and Remark 3.5 we can find a Kegel sequence $\mathcal{K} = \{K_n \mid n \geqslant 1\}$ of G such that one of the following holds (with $X_n = X_{K_n}$ for $X = H, M, V, F$ and Ω).

(1) There exists a prime p such that $\mathcal{K} = \mathcal{K}_{\mathcal{L}(p)}$ and, for all $n \geqslant 1$, $H_n/O_p(H_n)$ is the central product of perfect central extensions of groups in $\mathcal{C}(p)$.

(2) For all $n < m$, H_n has a pseudo natural orbit on Ω_m with respect to M_n.

Suppose first that (1) holds. Therefore, by Lemma 3.1 we can find n such that $\dim_{F_n} V_n > |H|$. Without loss $n = 1$. Then H does not act irreducibly on V_1 and so there exists a proper H-submodule U in V_1. Put $T_1 = N_{H_1}(U)$. Then T_1 is a maximal subgroup of H_1 containing H. Moreover, by the structure of H_1, $O_p(T_1) \neq O_p(H_1)$. Inductively, we will find a maximal subgroup T_{n+1} of H_{n+1} such that $T_n \leqslant T_{n+1}$, $H_n \not\leqslant T_{n+1}$ and $O_p(T_{n+1}) \neq O_p(H_{n+1})$. Since $O_p(T_n) \neq O_p(H_n)$, there exists a chief factor for H_n on V_{n+1} not centralized by $O_p(T_n)$, so T_n is not irreducible on this chief factor and there exists a T_n-submodule U_n in V_{n+1} which is not normalized by H_n. Put $T_{n+1} = N_{H_{n+1}}(U_n)$. Then T_{n+1} has all the desired properties and $\bigcup_{n=1}^{\infty} T_n$ is a maximal subgroup of G containing H.

Suppose next that (2) holds. By Lemma 3.1 we may assume that, for all n, $|\Omega_{n+1}| > 2|H_n|$ and $|\Omega_1| > 2|H|$. Thus H normalizes a proper subset Γ_0 in Ω_1 with $|\Gamma_0| < |\Omega_1|/2$. Let $T_1 = N_{G_1}(\Gamma_0)$. Then T_1 is a maximal subgroup of G_1 which contains H and does not act transitively on Ω_1. Inductively we will find a maximal subgroup T_{n+1} of H_{n+1} such that $T_n \leqslant T_{n+1}$, $H_n \not\leqslant T_{n+1}$ and T_{n+1} does not act transitively on Ω_{n+1}. Let O be a pseudo natural orbit for H_n on Ω_{n+1} with respect to M_n. Since T_n does not act transitively on Ω_n, T_n does not act transitively on O. Hence there exists a T_n-orbit Γ_n of O such that H_n does not normalize Γ_n. Then $|\Gamma_n| \leqslant |T_n| \leqslant |H_n| < |\Omega_{n+1}|/2$. Put $T_{n+1} = N_{H_{n+1}}(\Gamma_n)$. Then T_{n+1} has all the desired properties and $\bigcup_{n=1}^{\infty} T_n$ is a maximal subgroup of G containing H. □

5. An Abstract Characterization of Finitary, Locally Finite, Simple Groups

Lemma 5.1 *Let p be a prime, G a group and S a Sylow p-subgroup of G.*

(a) *If $G = \mathrm{PSL}_n(p^k)$, then $\mathrm{der}(S) = [\log_2 n]$.*

(b) *If $G = \mathrm{Sym}(n)$, then $\mathrm{der}(S) = [\log_p n]$.*

(c) *If $G = \mathrm{Alt}_n$ and $p = 2$ then $\mathrm{der}(S) = [\log_2 n]$, provided that $n \geqslant 6$, whereas $\mathrm{der}(S) = [\log_2 n] - 1$ for $n \leqslant 5$.*

Proof. For (a) and (b) see [4, III 16.3, 15.3] while the proof for (c) is similar to the one for (b) in [4]. □

Lemma 5.2 *Let p be a prime, F a finite field with char $F = p$, V a finite dimensional vector space over F, P a p-subgroup of $\mathrm{GL}_F(V)$, $k \geqslant 1$ and $x \in P$ with $\deg_V(x) \geqslant |P|2^k$. Then there exists a p-subgroup P^* of $\mathrm{GL}_F(V)$ containing P with $\mathrm{der}(\langle x^{P^*} \rangle) \geqslant k$.*

Proof. Pick a chain $0 = V_0 \leqslant V_1 \leqslant V_2 \leqslant \ldots \leqslant V_l$ of FP-submodules in V of maximal length with respect to $\dim V_i/V_{i-1} \leqslant |P|$ and $[V_i/V_{i-1}, x] \neq 0$ for all $1 \leqslant i \leqslant l$. Suppose that $l < 2^k$. Then $\dim V_l \leqslant l|P| < 2^k|P| \leqslant \deg_V(x)$ and so $[V, x] \not\leqslant V_l$. Pick $v \in V$ with $[v, x] \notin V_l$ and put $V_{l+1} = F\langle v^P \rangle + V_l$. Then $\dim V_{l+1}/V_l \leqslant |P|$ and $[V_{l+1}/V_l, x] \neq 0$, a contradiction to the maximality of the length of the chain.

Thus $l \geqslant 2^k$. Pick $x_i \in V_i \setminus V_{i-1}$ with $[x_i, x] \notin V_{i-1}$ and put $y_i = [x_i, x]$. Then there exists a basis $v_1, v_2, \ldots$ of V such that P normalizes the corresponding flag and such that there exist indices $i_1 < i_2 < \ldots < i_{2l}$ with $v_{i_{2j-1}} = y_j$ and $v_{i_{2j}} = x_j$. Let P^* be the full stabilizer of this flag and let S be the largest subgroup of $\mathrm{GL}_F(V)$ with the following properties:

S centralizes all v_i with $i \notin \{i_1, \ldots, i_{2l}\}$,
S centralizes $x_j + y_j$ for all j, and
S stabilizes the flag $0 \leqslant Fx_1 \leqslant F\langle x_1, x_2 \rangle \leqslant \ldots \leqslant F\langle x_1, \ldots, x_l \rangle$.

Put $X = F\langle x_1, \ldots, x_l \rangle$. Then S normalizes X, S is isomorphic to a Sylow p-subgroup of $\mathrm{GL}_F(X)$ and

$$[y_j, S] = [x_j, S] \leqslant F\langle x_1, \ldots, x_{j-1} \rangle \leqslant F\langle v_1, \ldots, v_{i_{2(j-1)}} \rangle.$$

It follows that $S \leqslant P^*$. Since $x_j^x = x_j + y_j$, S centralizes X^x. Thus $\langle S, S^x \rangle$ normalizes X^x, $[S, x]C_{P^*}(X^x) = S^x C_{P^*}(X^x)$ and $[S, x]$ acts as a full Sylow p-subgroup of $\mathrm{GL}_F(X)$ on X^x. By Lemma 5.1, $\mathrm{der}([S, x]) \geqslant [\log_2 l]$. Since $l \geqslant 2^k$, we conclude that $\mathrm{der}(\langle x^{P^*} \rangle) \geqslant \mathrm{der}([S, x]) \geqslant k$. □

Lemma 5.3 *Let p be a prime, Ω a finite set, $G = \mathrm{Alt}(\Omega)$ or $G = \mathrm{Sym}(\Omega)$, P a p-subgroup of G, $k \geqslant 1$ and $x \in P$ with $\deg_\Omega(x) \geqslant 2w(P)|P|p^k$. Then there exists a p-subgroup P^* of G containing P with $\mathrm{der}(\langle x^{P^*}\rangle) \geqslant k$.*

Proof. By Lemma 2.13 there exists a subset Γ of Ω and a partition Δ of Γ into subsets of size $2p^k$ such that P normalizes Δ and x acts non-trivially on Δ. Let $D \in \Delta$ with $D^x \neq D$. Pick $S \leqslant G$ such that S centralizes $\Omega \setminus D$, S acts as a full Sylow p-subgroup of $\mathrm{Alt}(D)$ on D and $N_P(D)$ normalizes S. Put $P^* = \langle S, P\rangle$. Then P^* is a p-group and $[S, x]$ acts as a full Sylow p-subgroup of $\mathrm{Alt}(D^x)$ on D^x. Since $|D| = 2p^k$ we conclude from Lemma 5.1(b)(c) that $\mathrm{der}([S, x]) \geqslant k$. Thus $\mathrm{der}(\langle x^{P^*}\rangle) \geqslant k$. $\square$

Lemma 5.4 *Let $G = \mathrm{SL}_F(V)$, F a finite field and V a finite dimensional vector space over F. Let p be a prime with $\mathrm{char}\, F \neq p$ and put $d = d(p)$. If P is p-subgroup of G, $x \in P$, $k \geqslant 1$, and $y \in \langle x^P\rangle^{(d)}$ with $\deg_V(y) \geqslant 2w(P)|P|^2 p^{k+1}$, then there exists a p-subgroup P^* of G with $P \leqslant P^*$ and $\mathrm{der}(\langle x^{P^*}\rangle) \geqslant k$.*

Proof. Let T be a Sylow p-subgroup of $\mathrm{GL}_F(V)$ with $P \leqslant T$ and $e \geqslant 1$ minimal with pd dividing $|K|^e - 1$. Then there exists an FT-submodule U in V with $\dim V/U < e$ and a system of imprimitivity Δ for T on U such that $\dim D = e$ for all $D \in \Delta$. Note that $e \leqslant p$. Let $D \in \Delta$ with $\langle x^P\rangle \leqslant N_T(D)$. Since $N_T(D)/C_T(D)$ has derived length at most d and $y \in \langle x^P\rangle^{(d)}$ we conclude that $[D, y] = 0$. Hence

$$2w(P)|P|^2 p^{k+1} \leqslant \deg_V(y) \leqslant e|\Delta \setminus C_\Delta(\langle x^P\rangle)| \leqslant p|\Delta \setminus C_\Delta(\langle x^P\rangle)|.$$

On the other hand, $|\Delta \setminus C_\Delta(\langle x^P\rangle)| \leqslant |P| \deg_\Delta(x)$ and so $\deg_\Delta(x) \geqslant 2w(P)|P|p^k$. Let $H = N_G(\Delta)$ and $\bar H = H/C_G(\Delta)$. Then $\bar H \cong \mathrm{Sym}(\Delta)$ or $\mathrm{Alt}(\Delta)$ and by Lemma 5.3 there exists a p-subgroup $\bar P^*$ of $\bar H$ with $\bar P \leqslant \bar P^*$ and $\mathrm{der}(\langle \bar x^{\bar P^*}\rangle) \geqslant k$. Let P^* be a Sylow p-subgroup of the inverse image of $\bar P^*$ in H with $P \leqslant P^*$. Then $\mathrm{der}(\langle x^{P^*}\rangle) \geqslant k$ and Lemma 5.4 is proved. $\square$

Theorem 5.5 *Let G be a LFS-group, p a prime, and $x \in G$ with $|x|$ a power of p. If G has no Kegel cover with all of its factors in $\mathcal{A} \cup \mathcal{L}(p)$, put $d = d(p)$ and assume that $\langle x^S\rangle^{(d)} \neq 1$ for some p-subgroup S of G with $x \in S$. Then G is finitary if and only if $\langle x^T\rangle$ is solvable for all p-subgroups T of G with $x \in T$.*

Proof. If G is finitary, then, by [6, Prop 1], $\langle x^T\rangle$ is solvable for all p-subgroups T of G with $x \in T$. So suppose that G is not finitary. If G has a Kegel cover with all of its factors in $\mathcal{A} \cup \mathcal{L}(p)$, put $S = \langle x\rangle$ and $d = 0$. By Proposition 3.2 there exists a Kegel cover $\mathcal{K}$ for G with $\mathcal{K} = \mathcal{A} \cup \mathcal{L}$ and, if $d = 0$, $\mathcal{K} = \mathcal{A} \cup \mathcal{L}(p)$. We will first prove:

> (*) *If P is a finite p-subgroup of G with $S \leqslant P$, then there exists a finite p-subgroup P^* of G with $P \leqslant P^*$ and $\mathrm{der}(\langle x^P\rangle) < \mathrm{der}(\langle x^{P^*}\rangle)$.*

Let $k = \mathrm{der}(\langle x^P\rangle) + 1$ and pick $1 \neq y \in \langle x^P\rangle^{(d)}$. By Lemma 3.1 there exists $K \in \mathcal{K}(P)$ with $\mathrm{pdeg}_{\bar K}(y) \geqslant 2w(P)|P|^2 p^{k+1}$. Thus Lemmata 5.2, 5.3 and 5.4 provide a p-subgroup $\bar P^*$ of $\bar K$ with $\bar P \leqslant \bar P^*$ and $\mathrm{der}(\langle \bar x^{\bar P^*}\rangle) \geqslant k$. Let P^* be a Sylow

p-subgroup of the inverse image of $\bar{P}^*$ in H_K with $P \leqslant P^*$. Then $\mathrm{der}(\langle x^{P^*}\rangle) \geqslant k$ and (*) is proved.

Let $P_0 = S$ and inductively define $P_{i+1} = P_i^*$. Put $T = \bigcup_{i=1}^{\infty} P_i$. Then T is a p-subgroup of G such that $\langle x^T \rangle$ is not solvable. $\qquad \square$

6. Countable LFS-Groups which are not Absolutely Simple

In this chapter we will construct countable LFS-groups G which possess an ascending series

$$M_1 \trianglelefteq M_2 \trianglelefteq M_3 \trianglelefteq \dots$$

of proper subgroups of G such that $G = \bigcup_{i=1}^{\infty} M_i$. The main step in the construction is the following lemma:

Lemma 6.1 *Let H be a perfect finite group. Then there exists a perfect finite group H^* containing H and a function X which associates to each subgroup A of H a subgroup $X(A)$ of H^* such that*

(a) $H \leqslant \langle h^{H^*} \rangle$ *for all* $1 \neq h \in H$.
(b) $X(A) \cap H = A$ *for all* $A \leqslant H$.
(c) *If* $A \leqslant B \leqslant H$, *then* $A \trianglelefteq B$ *if and only if* $X(A) \trianglelefteq X(B)$.
(d) $X(H) \trianglelefteq \langle X(H)^{H^*} \rangle$.

Proof. Let S be any finite simple group such that there exists a monomorphism $\alpha : H \to S$ and let T be any non-trivial finite perfect group. Furthermore, let S and T act transitively and non-trivially on the sets I and J, respectively. We assume that $0 \in I$ and $\{0, 1\} \subseteq J$. Let $K = H \wr_I S$. For $i \in I$ let $\beta_i : H \to K$ be the canonical isomorphism between H and the i-th component of the base group of K and let β be the canonical monomorphism from S to K. Let $H^* = K \wr_J T$ and for $j \in J$ let $\gamma_j : K \to H^*$ be the canonical isomorphism between K and the j-th component of the base group of H^*. Define $\rho : H \to H^*$ by $\rho(h) = \gamma_0(\beta_0(h))\gamma_1(\beta(\alpha(h)))$. Then ρ is clearly a monomorphism. For $A \leqslant H$ let $X(A)$ be the set of elements in the base group of H^* such that the projection onto the 0-th component is contained in $\gamma_0(\prod_{i \in I} \beta_i(A))$. Identifying H with $\rho(H)$ we see immediately that (b) and (c) hold. Now $\langle X(H)^{H^*} \rangle$ is the base group of H^* and so (d) holds. Moreover, (a) is readily verified. $\qquad \square$

We are now able to construct locally finite simple groups which are not absolutely simple. Let G_1 be any non-trivial perfect finite group, and inductively let $G_{i+1} = G_i^*$ and X_i any function from the subgroups of G_i to the subgroups of G_{i+1} which fulfills Lemma 6.1. Let $G = \bigcup_{i=1}^{\infty} G_i$. Then, by (a) in Lemma 6.1, G is a locally finite simple group.

Put $M_{1,1} = 1$, $M_{1,2} = G_1$ and then, inductively, define $M_{n+1,j} = X_n(M_{n,j})$, for $1 \leqslant j \leqslant 2n$, $M_{n+1,2n+1} = \langle X(G_n)^{G_{n+1}} \rangle$ and $M_{n+1,2n+2} = G_{n+1}$. Then, by induction and Lemma 6.1, $M_{n,i} \trianglelefteq M_{n,i+1}$ for all $1 \leqslant i \leqslant 2n+1$ and $M_{n+1,i} \cap G_n = M_{n,i}$ for all $1 \leqslant i \leqslant 2n$. Put $M_i = \bigcup_{n \geqslant \frac{i}{2}} M_{n,i}$. Then $G_n \leqslant M_{2n}$, $G_n \cap M_i = M_{n,i}$ for all $i \leqslant 2n$, $M_i \trianglelefteq M_{i+1}$ and $G = \bigcup_{i=1}^{\infty} M_i$.

References

1. J. I. Hall and B. Hartley, A group theoretical characterization of simple, locally finite, finitary linear groups, *Arch. Math.* 60 (1993), 108–114.
2. J. I. Hall, Infinite alternating groups as finitary linear transformation groups, *J. Algebra* 119 (1988), 337–359.
3. I. N. Herstein, *Topics in Algebra*, John Wiley & Sons, New York, 1975.
4. B. Huppert, *Endliche Gruppen I*, Springer-Verlag, Berlin, 1983.
5. O. Kegel and B. A. F. Wehrfritz, *Locally Finite Groups*, North-Holland, 1973.
6. U. Meierfrankenfeld, R. E. Phillips and O. Puglisi, Locally solvable finitary linear groups, *J. London Math. Soc.* (2) 47 (1991), 31–40.
7. R. E. Phillips, On absolutely simple locally finite groups, *Rend. Sem. Mat. Univ. Padova* 79 (1988), 213–220.
8. C. E. Praeger and A. E. Zalesskiĭ, Orbit lengths of permutation groups and group rings of locally finite simple groups of alternating type, preprint, 1993.

INERT SUBGROUPS IN SIMPLE LOCALLY FINITE GROUPS*

V. V. BELYAEV
Krasnoyarsk Building Engineering Institute
Svobodnyi Prosp. 80
660062 Krasnoyarsk
Russia

Abstract. The paper is a brief survey of recent results about inert subgroups in locally finite groups and their use in characterizations of linear and finitary locally finite groups.

Key words: locally finite groups, simple groups, inert subgroups, signalizers, finitary groups, alternating groups.

The classification of the finite simple groups (CFSG), which was announced at the beginning of the eighties, exercised a very stimulating influence on the study of simple locally finite groups. A close connection between these classes was described by O. H. Kegel, and, thanks to this, finite simple groups can be used in investigating simple locally finite ones. This connection has a local character, and is nowadays formalized in terms of Kegel covers. But this connection is not so close as to allow us to classify infinite simple locally finite groups on the basis of CFSG. The amount of freedom in ways of constructing infinite simple locally finite groups is so great that the question of composing a complete list of such groups does not even arise. Nevertheless, many outstanding problems in the theory of locally finite groups demonstrate that it is necessary to introduce some systematic order into the class of simple locally finite groups.

The first important step in the business of introducing order into this class was taken at the beginning of the eighties, when, on the basis of CFSG, it was shown [1, 8, 11, 15] that every infinite simple locally finite group is either isomorphic to a group of Lie type of finite rank, defined over a locally finite field, or is an enormous group (this term was proposed by Kegel and Wehrfritz in [12] to denote those simple locally finite groups, not satisfying a certain collection of finiteness conditions). This division of the class of all simple locally finite groups into two subclasses, sharply distinguished by their properties, on the one hand allowed many questions about finiteness conditions to be answered, and, on the other, directed the main attention to the class of enormous groups, where chaos reigned.

Hall's universal group, and infinite finitary alternating groups, form only a small family of examples from the whole class of enormous groups. Although the local structures of universal and alternating groups are identical (each can be represented as a direct limit of finite alternating groups), their global properties are completely different and are extremes of their kind in the spectrum of properties of enormous groups. Thus, for example, every countable simple locally finite group can be embed-

---

* Translated from Russian by B. Hartley

B. Hartley et al. (eds.), Finite and Locally Finite Groups, 213–218.
© *1995 Kluwer Academic Publishers. Printed in the Netherlands.*

ded in Hall's universal group, while at the same time every infinite simple subgroup of an alternating group is isomorphic to an alternating group (possibly of a different cardinality). This local similarity, combined with an evident global dissimilarity, is not peculiar to this example. It is a typical phenomenon among enormous groups, and makes it more difficult to study them. Therefore, in order to divide the class of enormous groups into finer subclasses, we need ideas which take into account both the global and local behaviour of these groups.

A fundamental method of studying simple locally finite groups, based on the classification of Kegel covers, is described in Hartley's article [10] in these Proceedings. Important and deep results in this direction have been obtained by Meierfrankenfeld and can be found in Hartley [10] and Meierfrankenfeld [13]. Another approach, which was developed in [2, 3, 4, 5, 6, 7], is given here. This approach is based on the concept of an inert subgroup, where a subgroup H of a group G is called *inert* if $|H : H \cap H^g| < \infty$ for all $g \in G$. (We note that, initially, such subgroups were called *kernels* in [2]. The term 'inert subgroup' was suggested by Kegel.) Also, several of the following results are discussed in [14, Section 10].

Inert subgroups arise in a natural fashion in many areas of group theory. For example, $SL_n(\mathbb{Z})$ is an inert subgroup of $SL_n(\mathbb{Q})$. The stabilizer of a vertex in the automorphism group of a locally finite connected graph serves as another example. As a rule, the special nature of the embedding of the inert subgroups in these examples is used to define some topology on the group. The connection between inert subgroups and topological groups is not coincidental. In fact, any open subgroup of a locally compact group is inert. Inert subgroups of countable locally finite groups have a different nature, and their appearance is connected with the consideration of certain special infinite sequences of finite subgroups, which it is convenient to define with the help of the notion of a signalizer. A finite subgroup K of G is called an *H-signalizer* if $K \cap H = 1$ and $H \leqslant N_G(K)$. We denote the set of all H-signalizers of G by $\mho(G, H)$. We can use signalizers to construct inert subgroups in countable locally finite groups in the following way.

Let $G = \{g_0, g_1, \ldots\}$ be a countable locally finite group and $\{H_i \mid i = 0, 1, \ldots\}$ be an infinite sequence of finite subgroups of G satisfying the condition

$$H_{n+1} \in \mho(G, \langle H_0, \ldots, H_n, g_0, \ldots, g_n \rangle)$$

for all n. Let $H = \langle H_0, H_1, \ldots \rangle$. Then, clearly, $H \cap H^{g_n} \geqslant \langle H_{n+1}, H_{n+2}, \ldots \rangle$, and so H is an inert subgroup of G.

Thus, in the class of countable locally finite groups, there is a close connection between the ideas of inert subgroups and signalizers. Moreover, many results that are formulated in terms of inert subgroups can be reformulated in terms of signalizers, and vice versa. The investigations into inert subgroups and signalizers, carried out in [3, 4, 5, 6, 7], have shown that these notions are useful in studying the connection between the local and global structure of locally finite groups. With the help of the set $\mho(G, H)$, it is easy to study rather subtle properties of the embedding of H in G, and it is convenient to formulate certain asymptotic properties of the embedding of a finite subgroup in the whole group in terms of inert subgroups. All this, in its turn, allows us to characterize certain classes of simple locally finite groups in terms of the structure of inert subgroups in the countable case, or in terms of the saturation of the set $\mho(G, H)$ by signalizers of a particular type in the case of groups of arbitrary cardinality.

It was shown in [4] that the structure of a proper inert subgroup in a simple group is far from arbitrary (a *proper inert subgroup* is one which is neither G nor a finite subgroup of G). In addition to the obvious property of residual finiteness, a proper inert subgroup of a simple group has a series of special properties that enables us to give an *a priori* estimate for the possible types of inert subgroups in simple groups.

Theorem 1   *Let H be an inert subgroup of a simple locally finite group G. Then one of the following conditions holds.*

(1) $H = G$.

(2) H is finite.

(3) H is an infinite locally normal group.

(4) There exists a prime p such that $F(H) = O_p(H) \neq 1$, and $H/O_p(H)$ is locally normal; further $\mathrm{FC}(H) = \mathrm{FC}(O_p(H)) = 1$.

(5) $F(H) = \mathrm{FC}(H) = 1$, $\mathrm{Sol}(H) \neq 1$, and $H/\mathrm{Sol}(H)$ is locally normal.

(6) $F(H) = \mathrm{FC}(H) = \mathrm{Sol}(H) = 1$, and $1 \neq H \neq G$.

Here, $F(H)$ is the locally nilpotent radical of H, $\mathrm{Sol}(H)$ is the locally soluble radical of H, and $\mathrm{FC}(H) = \{g \in H \mid |H : C_H(g)| < \infty\}$.

Of course, a simple locally finite group may contain inert subgroups of different types. For example, it is not difficult to show that Hall's universal group contains inert subgroups of all six types allowed by Theorem 1. It would be interesting to know the answer to the following.

Problem 2   *Which combinations of types of inert subgroups can be realized in a simple locally finite group?*

The first step in solving this problem was taken in [2, 3], where locally finite groups were studied which contain only trivial inert subgroups, that is, inert subgroups of the first two types. The following turned out to be true.

Theorem 3   *If G is a countable simple locally finite group, then the following two conditions are equivalent.*

(1) G is a group of Lie type.

(2) G contains only inert subgroups of types (1) and (2).

Thus, every countable enormous simple locally finite group contains a non-trivial inert subgroup. This fact can be interpreted as follows.

Theorem 4   *A countable simple locally finite group can be densely embedded in a non-discrete locally compact group if and only if G is enormous.*

In [3], a characterization of groups of Lie type was also obtained without the assumption of countability.

Theorem 5 *Let G be an infinite simple locally finite group. Then the following conditions are equivalent.*

(1) *G is of Lie type.*

(2) *G contains a finite subgroup H such that $\mho(G, H)$ contains only the trivial subgroup.*

(3) *Every residually finite subgroup of G is almost nilpotent.*

In [6], a second step was taken in studying the types of inert subgroups in countable simple locally finite groups.

Theorem 6 *Let G be a countable simple locally finite group. Then the following conditions are equivalent.*

(1) *G is isomorphic to an alternating group of finitary permutations.*

(2) *G contains only inert subgroups of the first three types.*

This characterization of the alternating group was interpreted in [6] in terms of signalizers and residually finite subgroups.

Theorem 7 *Let G be an infinite simple locally finite group. Then the following conditions are equivalent.*

(1) *G is isomorphic to an alternating group of finitary permutations.*

(2) *Every residually finite subgroup of G is locally normal.*

(3) *Every finite subgroup K of G has a non-trivial K-signalizer and G contains a non-trivial finite subgroup H for which $\langle X \mid X \in \mho(G, H) \rangle \leqslant C_G(H)$.*

It is easy to see that studying countable simple locally finite groups containing no inert subgroup of type (5) or (6) amounts to considering the situation in which, for every proper inert subgroup H of G and some prime p, the factor group $H/O_p(H)$ is locally normal. In terms of signalizers, this is equivalent to the existence of a non-trivial finite subgroup K such that $[K, X] \leqslant O_p(X)$ for all $X \in \mho(G, K)$. The analysis of this situation in [7] showed that every such group has a faithful finitary representation (for the definition of finitary groups see the articles of Phillips [14] and Hall [9]). In fact the following was established in [7].

Theorem 8 *Let G be a countable simple locally finite group. Then the following conditions are equivalent.*

(1) *G is a finitary group.*

(2) *There exists a prime p such that if H is any proper inert subgroup of G then $H/O_p(H)$ is locally normal.*

(3) *G contains no inert subgroup of type (6).*

Thus, a countable simple locally finite group containing no inert subgroup of type (6) cannot contain one of type (5), and such a group is finitary. We also remark that, for the proper inert subgroups of G that are not locally normal, the prime p in (2) of Theorem 8 is uniquely defined and is the characteristic of the field over which the finitary representation is defined.

The proof of Theorem 8, as given in [7], is based on a more general characterization of periodic simple finitary groups in terms of signalizers. For simplicity, the signalizer characterization given in [7] is divided into two parts below (Theorems 9 and 10). In Theorem 9, the discussion centres around properties of signalizers of finitary groups that depend on the characteristic of the base field, and thus have an outer character with respect to the group itself. These properties of the signalizers can be weakened to such an extent that it becomes possible to formulate them in terms which do not depend on the characteristic of the field. Theorem 10 asserts that further weakened properties of signalizers characterize periodic simple finitary groups among simple locally finite groups.

Theorem 9 *Let G be a simple periodic finitary group defined over a field of characteristic p. Then G contains a non-trivial finite subgroup H such that*

(1) *if $p = 0$, then $[H, X] = 1$ for all $X \in \mho(G, H)$,*

(2) *if $p > 0$, then $[H, X] \leqslant O_p(X)$ for all $X \in \mho(G, H)$.*

Theorem 10 *The following conditions on a simple locally finite group G are equivalent.*

(1) *G is a finitary group.*

(2) *G has a finite subgroup H such that $C_H(X/\mathrm{Sol}(X)) \neq 1$ for all $X \in \mho(G, H)$.*

A further characterization of periodic simple finitary groups was obtained in [7], occupying an intermediate position between the characterizations in terms of inert subgroups and signalizers. In stating it, we use the term *semisimple group* to denote a non-trivial group with trivial locally soluble radical.

Theorem 11 *The following conditions on a simple locally finite group G are equivalent.*

(1) *G is finitary.*

(2) *Every countable semisimple residually finite subgroup of G contains a non-trivial finite normal subgroup.*

(3) *Every countable semisimple residually finite subgroup of G is locally normal.*

We should also note that simple periodic finitary groups have been completely classified by J. Hall, using CFSG; see his article [9] for more details. The infinite groups of this type are precisely the groups of Lie type, alternating groups of finitary permutations, and finitary groups of classical type.

Finally we give two characterizations of simple locally finite non-finitary groups, in the languages of inert subgroups and signalizers. To state them we require the following additional definitions.

Let H be a subgroup of a group G. If X and Y are H-signalizers in G, such that $[X, Y] \leqslant X \cap Y$ and $H \leqslant XY$, then X and Y are said to form an H-pair. The subgroup H is said to be *diagonal* if $\mho(G, H)$ contains some H-pair, and *nondiagonal* otherwise. An expression of G as the product of two proper inert subgroups A and B is called an *inert factorization* if, for all finite subsets K of G,

$$G = \bigcap_{g \in K} A^g \cdot \bigcap_{g \in K} B^g.$$

There is a close connection between the notions of diagonal subgroup and inert factorization. It was shown in [5] that a countable locally finite group in which each finite subgroup is diagonal has an inert factorization. It is not hard to show that the converse is true in locally finite groups of any cardinality; that is, in a locally finite, inertly factorizable group, every finite subgroup is diagonal. Taking this into account, we obtain the following two assertions from the results of [7].

Theorem 12   *A countable simple locally finite group G is not finitary if and only if G has an inert factorization.*

Theorem 13   *A simple locally finite group G is non-finitary if and only if every finite subgroup of G is diagonal.*

References

1. V. V. Belyaev, Locally finite Chevalley groups, in *Studies in Group Theory*, Urals Scientific Centre of the Academy of Sciences of the USSR, Sverdlovsk, 1984, pp. 39–50.
2. V. V. Belyaev, Kernels in countable locally finite groups, *Seventeenth All-Union Algebra Congress, Abstracts of Talks*, Part 1, Minsk, 1983, p. 24.
3. V. V. Belyaev, Locally finite groups with a finite inseparable subgroup, *Sibirsk. Mat. Zh.* 34 (1993), 23–41.
4. V. V. Belyaev, Inert subgroups in infinite simple groups, *Sibirsk. Mat. Zh.* 34 (1993), 17–23.
5. V. V. Belyaev, Simple locally finite groups expressible as the product of two inert subgroups, *Algebra i Logika* 31 (1992), 360–369.
6. V. V. Belyaev, Local characterizations of infinite alternating groups and groups of Lie type, *Algebra i Logika* 31 (1992), 369–391.
7. V. V. Belyaev, Local characterizations of periodic simple groups of finitary transformations, *Algebra i Logika* 32 (1993), 201–223.
8. A. V. Borovik, Embeddings of finite Chevalley groups and periodic linear groups, *Sibirsk. Mat. Zh.* 24 (1983), 26–35.
9. J. Hall, Locally finite simple groups of finitary linear transformations, *these Proceedings.*
10. B. Hartley, Simple locally finite groups, *these Proceedings.*
11. B. Hartley and G. Shute, Monomorphisms and direct limits of finite groups of Lie type, *Quart. J. Math. Oxford* (2) 35 (1984), 49–71.
12. O. H. Kegel and B. A. F. Wehrfritz, *Locally Finite Groups*, North-Holland, Amsterdam, 1973.
13. U. Meierfrankenfeld, Non-finitary locally finite simple groups, *these Proceedings.*
14. R. Phillips, Finitary linear groups: a survey, *these Proceedings.*
15. S. Thomas, The classification of the simple periodic linear groups, *Arch. Math.* 41 (1983), 103–116.

GROUP RINGS OF SIMPLE LOCALLY FINITE GROUPS

A. E. ZALESSKIĬ*
Institute of Mathematics
of the Academy of Sciences of Belaruse
Minsk, 220072
Belaruse†

Abstract. This is an expository paper which describes some new ideas and results on the group rings of simple locally finite groups. The problem of describing the two-sided ideal lattice is restated in terms of the representation theory of finite groups. This leads to various asymptotic problems for representations of finite groups. The problem is also linked with describing permutation representations, satisfying some finiteness condition, of simple locally finite groups. The ground field is mainly of characteristic 0. The modular case is described briefly in the last section.

Key words: locally finite group, locally finite Lie algebra, group ring, prime ideal, positive definite function.

The aim of these notes is to explain new ideas which link group-theoretical and ring-theoretical problems arising in the theory of group rings of locally finite groups with asymptotic problems in the representation theory of finite groups and permutation representations of finite groups. We focus our attention on simple locally finite groups, one of the richest and most interesting class of groups. The theory of this class has been developing very rapidly during the last decade. No doubt, the simplicity assumption will be weakened in future investigations of group rings. Our discussion concentrates on the study of two-sided ideal lattices of complex group rings of simple locally finite groups. Modular group rings are briefly discussed in the last section. We also outline the problem of describing certain central positive definite functions on locally finite groups which appeared in the theory of $\mathbb{C}^*$-algebras. Section 1 contains a general survey of the circle of problems, and the further sections are devoted to explaining various details which will help the reader to understand the essence of the problems. Of course, the proofs of more or less difficult results have not been included. Finally, I remark that these notes are a shortened and revised version of my lecture notes in the University of East Anglia [60].

I am happy to use this occasion to express my sincere thanks to Professor Roger Bryant who made numerous suggestions for improving the English version of this paper.

* This work was supported in part by International Science Foundation Grant No. RWXOOO.

† Current address: Department of Mathematics, Bilkent University, Bilkent 06533, Ankara, Turkey; e-mail: zales@fen.bilkent.edu.tr.

B. Hartley et al. (eds.), Finite and Locally Finite Groups, 219–246.

1. The Problems and Main Results

The problem of describing the lattice of ideals (we always mean two-sided ones) of the group rings of locally finite groups over fields of characteristic 0 can be completely translated into the language of representation theory of finite groups. In this way a number of new problems on representations of finite groups arise concerned with restrictions of representations of finite groups to small subgroups. It is not true to say that this connection between group rings of locally finite groups and representation theory has not been observed earlier. However, sufficient attention has not been given to it. It should be noted that in a more general context, namely for arbitrary locally semisimple algebras, this connection is reflected in the notion of the Bratteli diagram (see Vershik and Kerov [54], Bratteli [6], Elliott [10]). However, the group ring of a locally finite group is not given in terms of a Bratteli diagram, although the problem considered reduces to determining the Bratteli diagram or obtaining crucial information about it.

In these notes we assume that all groups in question are countable. This is in fact a purely technical assumption which allows us to make the exposition more transparent. Under this assumption any locally finite group G is a union of finite subgroups each contained in another. Thus, let $G = \bigcup_{i \in \mathbb{N}} G_i$ where $G_1 \subset G_2 \subset \ldots$ are finite groups. If F is a field and H a finite group, we denote by $\mathrm{Irr}_F H$ the collection of representatives of classes of equivalent irreducible representations of H over F (or FH-modules). If M is an FH-module (or a representation of H over F) then we write $\mathrm{Irr} M$ for the set of irreducible constituents of M (their multiplicities are not taken into account). Let FG denote the group ring (more precisely, the group algebra) of G over F. The following definition is crucial for what follows.

Definition 1.1 *Let $\Phi_i \subset \mathrm{Irr}_F G_i$ be a subset. We say that the collection $\Phi = \{\Phi_i\}$, $i = 1, 2, \ldots$, is an* inductive system *of modules (or representations) if for any $i, j \in \mathbb{N}$ with $i < j$ we have*

$$\Phi_i = \{\mathrm{Irr}(\sigma|_{G_i}) \mid \sigma \in \Phi_j\};$$

that is, Φ_i coincides with the set of all irreducible constituents of the restrictions of representations from Φ_j to G_i.

Proposition 1.2 *Let F be a field of characteristic 0. Then there exists an order reversing isomorphism of partially ordered sets between the two-sided proper ideals of FG and the inductive systems of FG-modules. (The order relations are those of inclusion.)*

Thus, the problem of determining ideals of the group ring of a locally finite group G is equivalent to that of determining inductive systems. The latter is a problem in the representation theory of finite groups. Being well developed, the representation theory provides much machinery for analyzing it. Proposition 1.2 shows that the lattice of the inductive systems does not depend on the choice of $\{G_i\}$.

Observe that every group homomorphism $G \longrightarrow H$ extends to a group ring homomorphism $FG \longrightarrow FH$ whose kernel is an ideal of FG. Hence the lattice of normal subgroups of G is a sublattice of the lattice of ideals of FG. As a first stage it is reasonable to exclude the influence of normal subgroups. Thus we come to the problem:

Problem 1.3 *Determine the ideal lattices of the group algebras of (infinite) simple locally finite groups.*

A more special question goes back to Kaplansky's lecture notes [30, p. 37] of 1965:

Problem 1.4 *Determine the groups whose group algebra ideal lattice is trivial.*

The sense of the term 'trivial' in this context is as follows.

As well as the zero ideal, every group ring contains a proper ideal which is the kernel of the homomorphism $FG \longrightarrow FH$ where $H = \{1\}$ is the unity group. (This ideal is called the augmentation ideal; it is of codimension 1 in FG.) So this ideal is trivial in the sense that it comes from a trivial homomorphism $G \longrightarrow \{1\}$ and has the same nature in any group ring. All other ideals will be called nontrivial. Also, the ideal lattice is called nontrivial if it contains a nontrivial ideal, and trivial otherwise. It is not hard to show that Problem 1.4 is equivalent to the problem of determining those groups G such that the augmentation ideal of the group algebra FG (for a given F) is a simple ring. So Problem 1.4 is very natural from a ring theoretical point of view. Problem 1.4 is important also for general group ring theory: it asks for the groups whose group rings are simplest from the ring-theoretical point of view.

Although the question is old enough, the history of studying it is not rich. In fact, until recent times the only significant result was that of Bonvallet, Hartley, Passman and Smith [4] of 1976:

Theorem 1.5 *Let p and r be distinct prime numbers and F a field of characteristic p or 0. Let G be a group. Suppose that for any finite set of elements $1 = x_0, \ldots, x_n$ in G there exist $y_0, \ldots, y_n \in G$ such that the group $\langle x_i y_j x_i^{-1} \mid i, j = 0, \ldots, n \rangle$ is elementary abelian of order $r^{(n+1)^2}$. Then the group ring FG has no nontrivial ideals.*

The groups in Theorem 1.5 may be locally finite. For them Theorem 1.5 was particularly surprising. The assumption in Theorem 1.5 is very restrictive and the main examples of groups satisfying the assumptions of [4] were the universal locally finite group of P. Hall and algebraically closed groups. Such groups are very rare; this gave the impression that group rings with no nontrivial ideals are rare as well. Perhaps this was the reason why no further attempts to understand this phenomenon better were made during the eighties.

By Proposition 1.2, for F of characteristic 0 we can restate Problem 1.4 in the language of representation theory as follows.

Problem 1.6 *Determine simple locally finite groups which have no nontrivial inductive systems.*

In order to start analyzing Problem 1.6 it is reasonable first to consider a special case that can serve as a model for the general one. I did so in [57] having chosen $G_i \cong \mathrm{Alt}_{n_i}$ where Alt_k denotes the alternating group on k symbols. The set of these symbols will be denoted by $\Omega(k)$.

To state the result the following definitions are needed.

Definition 1.7   An embedding $\mathrm{Alt}_k \longrightarrow \mathrm{Alt}_l$ is called diagonal *if every orbit of* Alt_k *on* $\Omega(l)$ *of length greater than 1 is permutation isomorphic to* $\Omega(k)$.

Definition 1.8   *We say that* G *is of* diagonal type *if* G *is a union of an ascending chain of alternating groups* Alt_{k_i} $(i = 1, 2, \ldots)$ *and all the inclusions* $\mathrm{Alt}_{k_i} \subset \mathrm{Alt}_{k_{i+1}}$ *are diagonal.*

Theorem 1.9 [57] *Let* G *be a union of a sequence of alternating groups* Alt_{k_i}, $i = 1, 2, \ldots$, *such that* $\mathrm{Alt}_{k_i} \subset \mathrm{Alt}_{k_{i+1}}$ *for every* i. *Let* F *be a field of characteristic* 0. *Then* G *has a nontrivial inductive system if and only if* G *is of diagonal type.*

Groups of diagonal type form a minor subclass of the class of all groups which are unions of ascending chains of alternating groups (i.e. of groups G as in Theorem 1.9). Groups of diagonal type were utilized by group-theorists in some contexts but were not viewed as a class of groups of intrinsic interest. In particular, in [32, 6.10 and 6.12] it is proved that among them there are $2^{\aleph_0}$ non-isomorphic groups.

Unfortunately, the notion of groups of diagonal type does not seem to extend to the general case. This forces us to look for more adequate characteristics of groups with no nontrivial inductive systems. Possibly, one of them may be given in terms of permutation representations of G.

Definition 1.10   *Let* $H \subset \mathrm{Sym}(\Omega)$ *be a permutation group. For* $h \in H$ *we set* $\Omega^h = \{w \in \Omega \mid gw = w\}$, *i.e.* Ω^h *is the subset of* h-*fixed elements of* Ω. *We say that* H *is of* finite type *if* Ω *is a finite union of subsets* Ω^h *with* $h \neq 1$.

This is a fairly interesting finiteness condition for permutation groups. For locally finite groups Definition 1.10 may be stated in the following equivalent form which is more convenient for handling:

Definition 1.11   *Let* $H \subset \mathrm{Sym}(\Omega)$ *be locally finite. Then* H *is of* finite type *if there is a finite subgroup* K *of* H *which has no orbit on* Ω *of length* $|K|$. *(Recall that an orbit of length* $|K|$ *is called* regular*).*

Definition 1.12   *Let* G *be a simple locally finite group. We say that* G *is of* finite type *if* G *has a nontrivial permutation representation of finite type. Otherwise,* G *is of* infinite type.

Conjecture 1.13   *Let* F *be a field of characteristic* 0 *and* G *a simple locally finite group. Then* G *has a nontrivial inductive system if and only if* G *is of finite type.*

The 'only if' part was recently proved in [26].

Problem 1.14   *Let* G *be a simple locally finite group of finite type and* ρ *a representation as in Definition 1.12. Is it true that the annihilator of the permutation* FG-*module associated with* ρ *is not the zero ideal of* FG? *To ask a stronger question, is the annihilator of the permutation* FK-*module associated with* $\rho|_K$ *non-zero in* FK?

How are Conjecture 1.13 and Theorem 1.9 related? It is proved in [57] that G is of diagonal type provided G is both of finite type and a union of alternating groups; see also [46].

Problem 1.15 *Classify simple locally finite groups of finite type.*

Let G_ρ be the stabilizer of a point in a permutation representation ρ of G. The condition in Definition 1.11 that K has no regular orbits on Ω is equivalent to the following one: $G_\rho \cap gKg^{-1} \neq \{1\}$ for all $g \in G$. This motivates

Definition 1.16 *Let X be a proper subgroup of G. We say that X is* confined (*or* enormous) *if there exists a finite subgroup K of G such that $X \cap gKg^{-1} \neq \{1\}$ for all $g \in G$. (Both these terms have been used in papers.)*

It is amusing that the notion of a confined subgroup generalizes that of a normal subgroup. Indeed, if $X \neq \{1\}$ is normal, then K can be chosen in X. This notion seems to be very useful for understanding groups of finite type. At the current stage of studying groups of finite type the following problems are fairly exciting (the reader should also consult Hartley's article [21]):

Problem 1.17 *Given a simple locally finite group of finite type, determine explicitly its confined subgroups.*

Moreover, even the answers to the following qualitative questions are not clear:

Problem 1.18 *Is it true that every confined subgroup has only finitely many over-groups? Do confined subgroups satisfy the ascending chain condition? Is every confined subgroup contained in a maximal subgroup of G? Is the intersection of two confined subgroups again confined? Is it true that every confined subgroup of a simple group is not solvable?*

For the group of finitary permutations of an infinite set, Problem 1.17 is completely solved; see [48]. In particular, all questions of Problem 1.18 are answered positively for these groups.

Let X be a confined subgroup of a locally finite group G. As G is simple, all subgroups of G are of infinite index. Nevertheless, the group K cannot be carried outside X by conjugation. This suggests that X must in fact be big. It would be nice to find a way of estimating the size of X in comparison with G. Probably, the only possible way of doing this is to represent G as a union of finite groups G_i and to compare G_i with $G_i \cap X$. For example, one may try to study the growth of $|G_i|/|X \cap G_i|$ as a function of i. However, this direct approach does not seem to be fruitful. In [48] another growth function is considered. For any element h of G set $\mathrm{Cl}(h) = \{g^{-1}hg\}_{g \in G}$. For $x \in X$ and $g \in G$ we define

$$b_i(x, g) = |G_i \cap \mathrm{Cl}(x)|/(1 + |G_i \cap gXg^{-1} \cap \mathrm{Cl}(x)|).$$

Definition 1.19 *Let H be a subgroup of a locally finite group G. We say that H is* cobounded *if there exists a representation $G = \bigcup G_i$ such that the set*

$$\{b_i(h,g) \mid i \in \mathbb{N},\ h \in H,\ g \in G\}$$

is bounded.

Conjecture 1.20 *Let G be a simple locally finite group and $X \subset G$ a subgroup. Then X is confined if and only if X is cobounded.*

It is easy to prove this in one direction even under weaker assumptions:

Proposition 1.21 [48] *Let G be a locally finite group without finite normal subgroups other than $\{1\}$. Let $X \subset G$ be a subgroup. If X is cobounded then X is confined.*

It seems very probable that the problem of classifying two-sided ideals of complex group rings is closely connected with that of classifying normalized central indecomposable positive definite functions. This problem arose in the theory of C^*-algebras and was discussed in many papers: see the survey of Vershik and Kerov [54]. The exact sense of these terms is as follows.

Definition 1.22 *Let f be a function $G \longrightarrow \mathbb{C}$. Then f is called*

(1) normalized *if $f(1) = 1$,*

(2) central *if $f(gh) = f(hg)$ for any $g, h \in G$,*

(3) positive definite *if for any $g_1, \ldots, g_n \in G$ and $c_1, \ldots, c_n \in \mathbb{C}$ we have*

$$\sum_{k,l} f(g_k^{-1} g_l) c_k^* c_l \geqslant 0,$$

(4) indecomposable *if, for any functions f', f'' satisfying the conditions (1), (2), and (3), the equality $f = \alpha f' + (1 - \alpha) f''$ with $0 < \alpha < 1$, $\alpha \in \mathbb{R}$, implies $f' = f'' = f$.*

Here c^* denotes the complex conjugate of $c \in \mathbb{C}$. In the theory of C^*-algebras functions satisfying (1)–(4) are called characters of $\mathbb{C}G$, but to avoid confusion we do not use this term. Functions satisfying the conditions (1)–(3) constitute a convex set, so indecomposable ones just form the set of extremal points of this convex set. We denote the set of functions satisfying (1)–(3) by $\mathcal{F}(G)$ and the ones satisfying (1)–(4) by $\mathcal{E}(G)$. The set $\mathcal{F}(G)$ always contains two 'trivial' functions f_1 and f_0 where $f_1(g) = 1$ for all $g \in G$ and $f_0(g) = 0$ for all $1 \neq g \in G$; moreover, $f_1 \in \mathcal{E}(G)$. Observe that the condition (2) just means that f is constant on every conjugacy class of G. We say that $\mathcal{F}(G)$ is trivial if every function of $\mathcal{F}(G)$ is constant on $G \smallsetminus \{1\}$.

There is a program of determining $\mathcal{E}(G)$ for the most important groups; see [33, 38, 39, 49, 50, 52]. However, it is reasonable to consider the following general problem.

Problem 1.23 *Determine locally finite groups G with trivial $\mathcal{F}(G)$.*

Conjecture 1.24 *If G is of infinite type then $\mathcal{F}(G)$ is trivial.*

Probably, $\mathcal{F}(G)$ is nontrivial for almost every group of finite type. I know no group of finite type with trivial $\mathcal{F}(G)$.

In accord with Vershik and Kerov [54, Theorem 6] each function of $\mathcal{E}(G)$ is a weak (i.e. pointwise) limit of normalized irreducible characters of the finite subgroups G_i which approximate G (i.e., $G = \bigcup G_i$ and $G_i \subset G_{i+1}$ for all i). (A *normalized character* is the ratio $\chi/\chi(1)$ where χ is a character of G_i.) It follows that $\mathcal{F}(G)$ is trivial if and only if the weak limit of any convergent sequence of normalized characters of G_i is constant on $G \smallsetminus \{1\}$.

If the ground field is of prime characteristic then the connection between ideals of the group rings of locally finite groups and the representation theory of finite groups is more complicated. We have the following result (see Section 8 for a proof; a particular case was published in [58]):

Theorem 1.25 *Let G be a locally finite group and $G = \bigcup G_i$, where $G_i \subset G_{i+1}$ for all i. Let F be a field of characteristic $p > 0$. Then for any given inductive system $\Phi = \{\Phi_i\}$ there exists a largest ideal $I(\Phi)$ and a smallest ideal $K(\Phi)$ among the two-sided ideals L of FG with the property $\mathrm{Irr}(FG/L \,|_{G_i}) = \Phi_i$. Furthermore, $I(\Phi)/K(\Phi) = \mathrm{Rad}(FG/K(\Phi))$ and the map $\Phi \mapsto I(\Phi)$ is a $1-1$ correspondence between inductive systems and two-sided ideals I of FG such that FG/I is semisimple. (The inverse map is given by $\Phi_i = \mathrm{Irr}(FG/I|_{G_i})$.)*

If $\Phi_i = \mathrm{Irr}(G_i)$ for all i, then $K(\Phi) = \{0\}$ and $I(\Phi)$ is just the Jacobson radical JFG of FG. So Theorem 1.25 prompts the question whether the modular group ring of every simple locally finite group is semisimple (i.e. $\mathrm{Rad}(FG) = \{0\}$). The general question was open until recently. By a result of Formanek [13], $\mathrm{Rad}(FG) = \{0\}$ for $G \cong \mathrm{Alt}(\Omega)$ for an infinite set Ω. For linear locally finite groups some results were obtained in the seventies in [56] and [40]. Recently, Passman and Zalesskiĭ [44] proved (modulo the classification of finite simple groups) that $\mathrm{Rad}(FG) = \{0\}$ whenever G is not linear. Finally, Passman [42] has obtained the complete result, by settling the case where G is linear.

In general, it is much more difficult to describe inductive systems of modular representations than of ordinary ones. Nevertheless Theorem 1.25 can be helpful as a first approximation for determining the ideal lattice of a group ring. For instance, the following is proved on the basis of a particular case of Theorem 1.25:

Theorem 1.26 [58] *Let F, K be fields of characteristics p, k, respectively, where $p \neq k$. Suppose that K is a union of finite fields. Let $G = \mathrm{PSL}_2(K)$. Then the group ring FG has no nontrivial ideal.*

Problem 1.27 *Extend Theorem 1.26 to other groups of Lie type.*

In [46] the following result is proved.

Theorem 1.28 *Let $G = \bigcup G_i$ be an infinite simple group where the groups G_i are the covering alternating groups $\widetilde{\mathrm{Alt}}_{n_i}$. Let F be a field of characteristic $p \neq 2$. Then the group algebra FG has no two-sided proper nonzero ideals other than the augmentation ideal.*

Problem 1.29 *Let G_1 and G_2 be simple locally finite groups. Is it true that G_1 is isomorphic to G_2 if and only if $\mathbb{C}G_1 \cong \mathbb{C}G_2$?*

According to a theorem of Elliott [10] and Handelman [20] locally semisimple algebras can be classified in terms of their Grothendick groups with some additional invariants. (An exposition of this can be found in [17, Theorem 15.26].) One can hope that for simple locally finite groups G the Grothendick group of $\mathbb{C}G$ is a complete invariant of $\mathbb{C}G$:

Problem 1.30 *Let G_1 and G_2 be simple locally finite groups. Is it true that $\mathbb{C}G_1 \cong \mathbb{C}G_2$ if and only if the Grothendick groups of $\mathbb{C}G_1$ and $\mathbb{C}G_2$ are isomorphic?*

We have considered the case where G is simple, which certainly is of primary importance at this stage of our analysis. Very little is known about more general groups. The general problem in the spirit of Problem 1.4 can be stated as follows:

Problem 1.31 *Describe locally finite groups whose complex group rings contain no nonzero faithful ideals.*

An ideal I of the group ring FG is called *faithful* if $g - 1 \in I$ ($g \in G$) implies $g = 1$. The particular case of Problem 1.31 closest to the simple group one is that of perfect center-by-simple groups. This does not seem to require new methods. More mysterious is the case of groups which are elementary abelian-by-simple.

2. Examples

We distinguish several classes of groups which will play an important part in what follows; see also Hartley's paper [21] in these Proceedings.

Class 2.1 LINEAR GROUPS. *We use this term for abstract groups which have a faithful matrix representation over some field. Otherwise, we say that the group is nonlinear.*

It is known that a linear simple locally finite group is a group of Lie type over an infinite field which is a union of finite fields; see Section 5 for more details.

Class 2.2 FINITARY GROUPS. *This term is used for abstract groups which have a faithful representation by finitary transformations of a linear space over a field (see below).*

We say 'nonlinear finitary group' to exclude groups of Class 2.1. Among them are simple classical groups of infinite matrices with only finitely many non-zero elements off the diagonal. A systematic exposition of finitary group theory can be found in Phillips' paper [45] in these Proceedings. Classification of simple locally finite finitary groups is given in Hall's paper [19].

Class 2.3 NON-FINITARY GROUPS. *Formally, these are groups which are not finitary. However, they possess some common properties which may possibly allow the development of a general theory of non-finitary simple locally finite groups.*

Class 2.4 GROUPS OF DIAGONAL TYPE. *By this we mean the groups defined in Definition 1.8.*

Observe that Definition 1.8 should be considered as a preliminary version: there are hints that Definition 1.8 should be revised in order to cover Meierfrankenfeld's groups (see [22] and Meierfrankenfeld's original paper [37]). Some extensions of Definition 1.8 are proposed in [37] and [46]; however, a further improvement seems to be necessary. Possibly, it suffices to extend Definition 1.8 only to wreath products with alternating top groups as follows. Let A, B be finite permutation groups and let $\mathrm{Alt}_k, \mathrm{Alt}_l$ be alternating groups acting naturally. An embedding of the wreath products $A \wr \mathrm{Alt}_k \longrightarrow B \wr \mathrm{Alt}_l$ is called *diagonal* if the induced projection of Alt_k into Alt_l is diagonal in the sense of Definition 1.8.

As diagonal embeddings are very specific, it is clear that groups of diagonal type constitute a very small subclass among groups which are unions of alternating groups.

One may copy Definition 1.8 with classical groups instead of alternating ones. These and related groups must be split off from non-finitary groups in order to have a smoother theory of the latter.

Our first objective is to show that the complex group ring of a group G of type 2.2 or 2.4 has nontrivial two-sided ideals. We use a common argument for all these groups: we consider a suitable permutation FG-module and show that its annihilator is not zero.

Proposition 2.5 *Let G be a group of diagonal type. Then the group ring $\mathbb{C}G$ has a nontrivial ideal.*

Proof. First let $G = \mathrm{Alt}(\Omega)$ be the finitary alternating group on an infinite set Ω. Let Π be the permutation FG-module associated with the natural action of G on Ω. Then for any finite alternating group $\mathrm{Alt}_n \subset \mathrm{Alt}(\Omega)$ the restriction $\Pi |_{\mathrm{Alt}_n}$ is a direct sum of a trivial $F\mathrm{Alt}_n$-module (of infinite dimension) and the natural permutation $F\mathrm{Alt}_n$-module Π_n of dimension n. Hence the annihilator of $\Pi |_{\mathrm{Alt}_n}$ coincides with the annihilator I_n of Π_n. Then $\dim(F\mathrm{Alt}_n/I_n) \leqslant n^2$. On the other hand, $\dim \mathrm{Alt}_n = n!$. As $n! > n^2$ for $n > 3$, $I_n \neq \{0\}$ for $n > 3$. It follows that the annihilator of Π is not zero.   □

The same argument works for arbitrary groups of diagonal type. It suffices to observe that there exists a permutation FG-module Π such that the restriction

$\Pi\,|_{\mathrm{Alt}_{n_i}}$ is a direct sum of copies of trivial modules and natural $F\mathrm{Alt}_{n_i}$-modules. Although it is intuitively clear which module is meant, for a formal construction we have to use the direct limit of permutation sets. When we embed Alt_{n_i} into $\mathrm{Alt}_{n_{i+1}}$ in a diagonal way, we have an inclusion $\Omega(n_i)$ into $\Omega(n_{i+1})$ such that $\Omega(n_i)$ turns out to be an Alt_{n_i}-subset of $\Omega(n_{i+1})|_{\mathrm{Alt}_{n_i}}$. Therefore, we can take a direct limit of these permutation sets; the group $\bigcup \mathrm{Alt}_{n_i}$ acts on the resulting set Ω so that the orbits of each Alt_{n_i} on Ω are either trivial or permutation isomorphic to $\Omega(n_i)$.

Let us consider now classical groups of infinite dimension over a finite field $\mathbb{F}_q$. Let V be a vector space of infinite dimension over $\mathbb{F}_q$. Let us fix a basis B of V.

We can view the action of $G = \mathrm{GL}(V)$ on points (or lines) of V to get a permutation representation Δ of G. It has the following remarkable property. Let $b_1, \ldots, b_n$ be elements of B and $V_n = \langle b_1, \ldots, b_n \rangle$. Let $G_n \cong \mathrm{GL}(V_n)$ be the group of all elements $g \in \mathrm{GL}(V)$ such that $gV_n = V_n$ and $gb = b$ for $b \notin V_n$. The orbits of the finite groups G_n on V are either trivial or permutationally isomorphic to the orbit of a non-zero vector of V_n. Denote by Δ_n the permutation representation of G_n on V_n. Hence we see that the annihilator of the $\mathbb{F}_q G_n$-module associated with Δ coincides with that of the $\mathbb{F}_q G_n$-module associated with Δ_n. The dimension of Δ_n is q^n while the dimension of $\mathbb{F}_q G_n$ is $q^{n(n-1)/2}(q^n - 1)(q^{n-1} - 1) \cdots (q - 1)$. It follows that J_1, the annihilator of the permutation module associated with Δ, is not the zero ideal. A very similar argument leads to the same conclusion for other classical groups.

Let G be a simple finitary locally finite group. Then, obviously, G is isomorphic to an irreducible finitary group (take a nontrivial composition factor). J. Hall [19] proves that G is either locally finite classical (i.e. consists of all nondegenerate finitary transformations preserving a nondegenerate sesquilinear or quadratic form over a locally finite field) or G is a transvection group of the Cameron-Hall type [7] which contains all the groups G_n introduced above. This implies:

Proposition 2.6 *Let G be a finitary simple group over a finite field F. Then $\mathbb{C}G$ contains a nontrivial ideal.*

We see that $\Delta_{n+1}|_{G_n}$ falls into orbits isomorphic either to Δ_n or to the trivial one-point orbit. Let us extend each nontrivial orbit to $\mathrm{Alt}(\Delta_n)$; then we get a diagonal embedding of $\mathrm{Alt}(\Delta_n)$ into $\mathrm{Alt}(\Delta_{n+1})$ for each n. It follows that G is embedded into a group of diagonal type $\bigcup \mathrm{Alt}(\Delta_n)$. This may suggest that groups of diagonal type are somewhat universal with respect to the class of groups whose complex group rings have nontrivial ideals. However, a disappointing observation is that the regular representation embeds any locally finite group into a group of diagonal type.

The lattices of two-sided ideals of $\mathbb{C}G$, for G classical, are still not determined. However, for special linear groups (as well as for the Cameron-Hall groups) these are known to be descending chains:

Theorem 2.7 [25] *Let $G = \mathrm{FSL}(V)$. The two-sided ideals of $\mathbb{C}G$ form a descending chain: $J_0 \supset J_1 \supset \ldots \supset J_m \supset \ldots$ where J_0 is the augmentation ideal and, for $i > 0$, J_i is the annihilator of the permutation module associated with the action of G on the set of i-dimensional subsets of V.*

3. Inductive Systems

In this section we make further comments on the notion of an inductive system and prove Proposition 1.2. The ground field F is assumed to be of characteristic 0.

Proposition 3.1 *A collection* $\{\Lambda_i\}$, $\Lambda_i \in \mathrm{Irr} G_i$, *is an inductive system if and only if* Λ_i *is just the set of composition factors of a restriction of a representation of* G *to* G_i.

Proof. It is convenient to use module terminology. Let M be an FG-module. It is clear that $\mathrm{Irr}(M\,|_{G_i}) = \{L\,|_{G_i}|\ L \in \mathrm{Irr}(M\,|_{G_{i+1}})\}$. This implies the 'if' part for any characteristic. Conversely, let $\{\Lambda_i\}$ be an inductive system. Let $M_i = \bigoplus L_i$ where L_i runs over Λ_i. It is clear that there exists an injective homomorphism $M_i \longrightarrow M_{i+1}\,|_{G_i}$ for any $i = 1, 2, \ldots$ (we have used here complete reducibility of $M_{i+1}\,|_{G_i}$). Next take the direct limit $M = \lim M_i$ and observe that M is an FG-module. As $\{\Lambda_i\}$ is an inductive system, we have $\mathrm{Irr}(M_{i+1}\,|_{G_i}) = \Lambda_i$ and hence $\mathrm{Irr}(M_j\,|_{G_i}) = \Lambda_i$ for $j > i$. Therefore, all composition factors of $M\,|_{G_i}$ belong to Λ_i, as desired. □

The following lemma is a crucial element of the proof of Proposition 1.2. In order to state it we introduce some notation. Let F be a field of characteristic 0, H a finite group and K a two-sided ideal of FH. We put $M(K) = \mathrm{Irr}(FH/K)$ and $L(K) = \mathrm{Irr}(K)$ (we view K and FH/K as FH-modules). Recall that FH is a direct sum of minimal two-sided ideals: $FH = I_1 \oplus \cdots \oplus I_n$ where the FH-modules $I_1, \ldots, I_n$ have no common constituents. It follows that $L(K) \cap M(K) = \emptyset$. Moreover, every irreducible FH-module is a constituent of some I_j $(1 \leqslant j \leqslant n)$. Hence for any disjoint partition $\mathrm{Irr}(H) = M_1 \cup M_2$ there is a unique ideal K of FH such that $M_1 = M(K)$ and $M_2 = L(K)$. In Lemma 3.2, ϕ^H means the induced module.

Lemma 3.2 *Let* F *be a field of characteristic* 0. *Let* $B \subset H$ *be finite groups. Let* $J \subset K$ *be ideals of* FB *and* FH, *respectively. Then the following conditions are equivalent.*

  (i) $J = K \cap FB$.
  (ii) J *is the largest ideal of* FB *such that* $hJh^{-1} \subset K$ *for any* $h \in H$.
  (iii) *The following equalities hold:*

$$L(J) = \{\phi \in \mathrm{Irr} B \mid \mathrm{Irr}(\phi^H) \subseteq L(K)\}$$
$$= \{\phi \mid M(K) \cap \mathrm{Irr}(\phi^H) = 0\}$$
$$= \{\phi \mid \phi \in \mathrm{Irr}(M(K)|_B)\}.$$

  (iv) $M(J) = \mathrm{Irr}(M(K)|_B)$.

Proof. Let $J = K \cap FB$ and let J_1 be the ideal of FB which satisfies

$$L(J_1) = \{\phi \in \mathrm{Irr} B \mid \mathrm{Irr}(\phi^H) \subseteq L(K)\}.$$

By the comment prior to the Lemma, J_1 is well-defined. Let $h_1 = 1, h_2, \ldots, h_n$ be a transversal of the left cosets of B in H. By the definition of the induced module we have

$$(FB)^H = FH \otimes_{FB} FB = h_1 FB \oplus \cdots \oplus h_n FB$$

and

$$J^H = FH \otimes_{FH} J = h_1 J \oplus \cdots \oplus h_n J \subseteq K,$$

so $\phi^H \in L(K)$ for any $\phi \in L(B)$, by additivity of induction. Hence $J \subseteq J_1$. On the other hand, $\mathrm{Irr}(J_1^H) \subset \mathrm{Irr}(K)$, so $J_1 \subseteq K$, as $J_1^H = h_1 J_1 \oplus \cdots \oplus h_n J_1 \subseteq K$.   □

Next we give a slightly more precise version of Proposition 1.2.

Proposition 3.3   *Let G be a locally finite group and $G = \bigcup G_i$ with $G_i \subset G_{i+1}$ $(i = 1, 2, \ldots)$. For an ideal I of FG we put $M_i(I) = \mathrm{Irr}(FG_i/(I \cap FG_i))$. Then $M_i(I)$ is an inductive system of FG_i-modules and the map $I \mapsto \{M_i(I)\}$ is a $1-1$ correspondence between the proper two-sided ideals of FG and the inductive systems of FG_i-modules.*

Proof.   The first assertion follows from Lemma 3.2. Let $\{M_i\}$ be an inductive system. Let I_j be the ideal of FG_j such that $\mathrm{Irr}(FG_j/I_j) = M_j$. By Lemma 3.2, for $i < j$ we have $I_i = I_j \cap FG_i$. Put $I = \bigcup I_j$. It is clear that I is an ideal of FG. By the construction $I_j = I \cap FG_j$ for all j. This proves the Proposition.   □

Observe that the union of inductive systems is an inductive system associated with the intersection of the ideals.

The following provides a fairly efficient sufficient condition for the absence of inductive systems.

Proposition 3.4   *Let $G = \bigcup G_i$ with $G_i \subset G_{i+1}$ $(i = 1, 2, \ldots)$. Suppose that for every i there exists $j > i$ such that each nontrivial representation ϕ of G_j has the property $\mathrm{Irr}(\phi|_{G_i}) = \mathrm{Irr}(G_j)$. Then G has no nontrivial inductive system.*

One may give an equivalent condition in terms of group characters:

Proposition 3.5   *Let G be as in Proposition 3.4. Suppose that for every $k \in \mathbb{N}$ there exists $i \in \mathbb{N}$ with $i > k$ such that $(\chi|_{G_k}, \xi) > 0$ for any $\xi \in \mathrm{Irr}(G_k)$ and any $\chi \in \mathrm{Irr}(G_i)$ with $\chi(1) \neq 1$. Then G has no nontrivial inductive system.*

Of course, $(\chi|_{G_k}, \xi)$ stands for the usual scalar product of characters. The equivalence of Propositions 3.4 and 3.5 is a routine fact of character theory. However, we may give another sufficient condition for the absence of inductive systems which uses properties of characters in a more crucial way. We shall see that it is very helpful in practice.

Proposition 3.6   *Suppose that there exists a monotone function $f : \mathbb{N} \longrightarrow \mathbb{R}$ such that (a) $f(n) \to 0$ as $n \to \infty$ and (b) for any irreducible character χ of G_i with $\chi(1) > 1$ we have $|\chi(g)| < f(i)\chi(1)$ for all $1 \neq g \in G_i$. Then G has no nontrivial inductive system.*

Proof. We note that $|\ |$ denotes the modulus of a complex number. Let ξ be an irreducible character of G_k, where $k < i$. We have

$$
\begin{aligned}
(\chi|_{G_k}, \xi) &= |G_k|^{-1}\left(\chi(1)\xi(1) + \sum_{1 \neq g \in G_k} \chi(g)\xi(g^{-1})\right) \\
&\geqslant |G_k|^{-1}\chi(1)\xi(1) - |G_k|^{-1} \sum_{1 \neq g \in G_k} |\chi(g)||\xi(g^{-1})| \\
&> |G_k|^{-1}\chi(1)\xi(1) - f(i)\chi(1)\xi(1) \\
&\geqslant \chi(1)\xi(1)\left(|G_k|^{-1} - f(i)\right).
\end{aligned}
$$

By (a) the number in the last line is not zero for i large enough, and we are done by Proposition 3.6. □

REMARKS. (i) Let G be simple. Then $\chi(1)$ in Proposition 3.6 grows to infinity as i does. (If not, by Jordan's theorem on finite complex linear groups, for infinitely many i, G_i has a proper normal subgroup of bounded index. It follows that G is of bounded exponent. However, a locally finite simple group of bounded exponent is finite.) This implies that the multiplicities of the irreducible constituents $(\chi|_{G_k}, \xi)$ in the proof of Proposition 3.6 grow with i. Furthermore, the multiplicity of the regular representation of G_k in $\phi|_{G_k}$ grows with i.

(ii) Let us fix k and consider the restriction $\phi|_{G_k}$ where $\phi \in \mathrm{Irr}(G_i)$ is as in (i). So $\dim(\phi)$ is arbitrarily large for i large enough. In general, we have to expect that the multiplicity of every irreducible constituent of G_k in $\phi|_{G_k}$ grows to infinity with $\dim(\phi)$. If this is the case, Proposition 3.4 shows that G has no nontrivial inductive system! We must conclude that the existence of a nontrivial inductive system is an exception while the absence of them is no surprise. This intuitive observation helps us to understand things more correctly. Moreover, this shows that changing the language from group rings to representation theory should give a considerable advantage.

We may introduce a natural notion of tensor product of inductive systems.

Definition 3.7 *Let $\{\Phi_i\}$ and $\{\Psi_i\}$ be inductive systems of representations of the groups $\{G_i\}$. Denote by $\{\Phi_i \otimes \Psi_i\}$ the collection of representations $\phi \otimes \psi$ where ϕ and ψ run over Φ_i and Ψ_i, respectively. It is trivial that $\{\Phi_i \otimes \Psi_i\}$ is an inductive system. We call it the* tensor product *of $\{\Phi_i\}$ and $\{\Psi_i\}$.*

The operation of tensor product of inductive systems yields a binary operation on the ideal lattice of the group ring of every locally finite group over a field of characteristic 0. It differs from the operations of union and intersection.

Definition 3.8 *Let I and J be ideals of the group ring of a locally finite group G over a field F. Denote by $I\hat{\otimes}J$ the annihilator of the FG-module $(FG/I)\otimes(FG/J)$. We call the map $\{I, J\} \longrightarrow I\hat{\otimes}J$ the* T-operation.

Let $\{\Phi_i\}$ and $\{\Psi_i\}$ be the inductive systems corresponding to I and J, respectively, in accordance with Proposition 3.3. Then $\{\Phi_i \otimes \Psi_i\}$ corresponds to the ideal $I\hat{\otimes}J$.

Theorem 3.9 [25] *Let $G = \mathrm{FSL}(V)$ and let J be the annihilator of the permutation module associated with the action of G on the lines of the underlying space. Then every proper ideal of $\mathbb{C}G$ other than $\{0\}$ and the augmentation ideal is of the form $J\hat{\otimes}J\hat{\otimes}\cdots\hat{\otimes}J$.*

Definition 3.10 *Let Φ be an inductive system; Φ is called* indecomposable *if whenever $\Phi = \Phi' \cup \Phi''$ where Φ' and Φ'' are inductive systems we have $\Phi = \Phi'$ or $\Phi = \Phi''$.*

Recall that an ideal I of a ring R is called *prime* if for any two ideals I_1, I_2 of R such that $I_1 I_2 \subseteq I$ we have $I_1 \subseteq I$ or $I_2 \subseteq I$.

Proposition 3.11 *Let I be an ideal of the group ring FG and Φ be the inductive system afforded by I. Then I is prime if and only if Φ is indecomposable.*

Proof. Observe that for any finite subgroup H of G the group ring FH is semisimple. It follows that for any two ideals L_1, L_2 of PH we have $L_1 L_2 = L_1 \cap L_2$. Therefore the same is true for FG. Suppose that $I = I' \cap I''$ for some ideals I', I'' such that $I \neq I'$, $I \neq I''$. Let Φ' and Ψ'' be the inductive systems afforded by I' and I'', respectively. By the Jordan-Hölder theorem, the set of irreducible constituents of the restriction of the module FG/I to G_i is the union of the sets of irreducible constituents of the restrictions of the modules FG/I' and FG/I'' to G_i. So $\Phi = \Phi' \cup \Phi''$ and, by Proposition 3.3, $\Phi \neq \Phi'$ and $\Phi \neq \Phi''$. Hence Φ is decomposable. Conversely, let $\Phi = \Phi' \cup \Phi''$ with $\Phi \neq \Phi'$ and $\Phi \neq \Phi''$. Let I' and I'' be the ideals of FG afforded by Φ' and Φ'', respectively. By the Jordan-Hölder theorem, the inductive system afforded by $I' \cap I''$ is just $\Phi' \cup \Phi''$. By Proposition 3.3, $I' \cap I'' = I$.   □

Proposition 3.12 *Let $G = \bigcup_{\lambda \in \Lambda} G_\lambda$. Let $\Phi = \{\Phi_\lambda\}$ be an inductive system. Then the following assertions are equivalent:*

(1) *Φ is indecomposable, and*

(2) *for every $\lambda \in \Lambda$ there exists $\mu \in \Lambda$ and $\tau \in \Phi_\mu$ such that $\Phi_\lambda = \mathrm{Irr}(\tau|_{G_\lambda})$.*

Proof. $(2) \Rightarrow (1)$ is obvious. For $(1) \Rightarrow (2)$, put

$$d_\lambda = \max(\max_{\substack{\mu > \lambda \\ \phi \in \Phi_\mu}} |\{\mathrm{Irr}(\phi|_{G_\lambda})\}|).$$

Let $\rho \in \Phi_\mu$, $\mu > \lambda$, be such that $|\{\mathrm{Irr}(\rho|_{G_\lambda})\}| = d_\lambda$, and fix μ and ρ. For $\nu > \mu$ we put $\Phi'_\nu = \{\phi \in \Phi_\nu \mid \rho \in \mathrm{Irr}(\phi|_{G_\mu})\}$. Define Θ_ν to be

$$\Theta_\nu = \bigcup_{\substack{\nu_1 > \nu \\ \nu_1 > \lambda}} \{\mathrm{Irr}(\Phi'_{\nu_1}|_{G_\nu})\}.$$

Then $\Theta = \{\Theta_\nu\}$ is an inductive system. Set $\Phi_\lambda^\rho = \mathrm{Irr}(\rho|_{G_\lambda})$. It is clear that $\Theta_\lambda = \Phi_\lambda^\rho$. We show that $\Theta = \Phi$. If not, put $\Phi'_\sigma = \Phi_\sigma \smallsetminus \Theta_\sigma$ and $\Psi_\sigma = \bigcup_{\delta > \sigma} \{\mathrm{Irr}(\Phi'_\sigma)\}$. Then $\Psi = \{\Psi_\sigma\}_{\sigma \in \Lambda}$ is an inductive system. It is clear that $\rho \notin \Psi$. Hence $\Psi \neq \Phi$ and

$\Phi = \Theta \cup \Psi$ by the construction. This means that Φ is decomposable, a contradiction. Thus, $\Theta = \Phi$ and $\Phi_\lambda = \Theta = \Phi_\lambda^\rho$ and we are done. $\square$

As we mentioned in Section 1, the definition of an inductive system is appropriate for Lie algebras as well. However, it is more convenient to add the condition that every set Ψ_i is of finite cardinality. To be precise, we take the following:

Definition 3.13 *Let L be a locally finite Lie algebra. Let L be the union of a tower of finite dimensional subalgebras L_i, $i \in \mathbb{N}$. Let Φ_i be a finite subset of finite dimensional representations of L_i. We say that $\Phi = \{\Phi_i\}$ is an* inductive system *if for every $i, j \in \mathbb{N}$ with $j > i$ we have $\mathrm{Irr}(\Phi_j|_{L_i}) = \Phi_i$.*

A. Jilinskiĭ [28] described inductive systems for Lie algebras which are the unions of classical Lie algebras with the natural embeddings into each other (such as the embeddings $A_n \longrightarrow A_{n+1}$, $C_n \longrightarrow C_{n+1}$ and so on).

Definition 3.14 *Let L be a locally finite Lie algebra. We say that L is* locally semisimple *if every finite subset of L is contained in a semisimple finite dimensional subalgebra of L.*

Bahturin and Strade [2] have showed that there exists a simple locally finite Lie algebra which is not locally semisimple.

Proposition 3.15 *Let L be a locally finite, locally semisimple, Lie algebra over a field of characteristic 0. Then there exists a $1-1$ correspondence between the inductive systems and the two-sided ideals I of the universal enveloping algebra UL such that UL/I is a locally finite algebra.*

Proof. Let L be a union of a tower of finite dimensional subalgebras L_i ($i \in \mathbb{N}$) and let $\Phi = \{\Phi_i\}$ be an inductive system. Observe that every two-sided ideal X of UL_i of finite codimension is semi-primitive (i.e. $\mathrm{Rad}(UL_i/X) = 0$). This follows from Weyl's theorem on complete reducibility of representations of semisimple Lie algebras (indeed, if $\mathrm{Rad}(UL_i/X)$ is not zero then the regular representation of UL_i/X and hence L_i is not completely reducible).

Let I_i be the intersection of the kernels of the representations $\phi \in \Phi_i$. Since $|\Phi_i| < \infty$, the ideal I_i is of finite codimension in UL_i. Let ρ_i denote the regular UL_i/I_i-module. As the algebra UL_i/I_i is semisimple, $\mathrm{Irr}(UL_i/I_i) = \Phi_i$. Let $j \in \Lambda$, $j > i$, and $I_i' = I_j \cap UL_i$. Then UL_i/I_i' is a submodule of UL_j/I_j so $\mathrm{Irr}(UL_i/I_i') = \Phi_i$ by the definition of an inductive system. As UL_i/I_i' is semisimple as well, we have $UL_i/I_i' \cong UL_i/I_i$. It follows that $I_i' = I_i$. Therefore $I_i \subseteq I_j$ so we can take the direct limit I of the ideals I_i ($i \in \Lambda$). Then I is an ideal of UL and $I \cap UL_i = I_i$. It follows that UL/I is the direct limit of the UL_i/I_i. Hence UL/I is a locally finite locally semisimple algebra.

Conversely, let I be a two-sided ideal of UL such that UL/I is a locally finite algebra. Set $I_i = I \cap UL_i$. Then UL_i/I_i is a finitely generated subalgebra of a locally finite algebra UL/I. Hence UL_i/I_i is of finite dimension. By the observation above UL_i/I_i is semisimple. Therefore UL_i/I_i is the regular UL_i/I_i-module. It follows that the UL_i/I_i-modules UL/I and UL_i/I_i contain the same collection of irreducible constituents. Set $\Phi_i' = \operatorname{Irr}(UL/I \,|_{UL_i/I_i})$ and $\Phi_i = \operatorname{Irr}(UL/I \,|_{L_i})$. As UL_i/I_i is a semisimple finite dimensional algebra, Φ_i' is essentially the same as Φ_i. It is clear that Φ_i is an inductive system. On the other hand, $\operatorname{Irr}(UL/I \,|_{L_i}) = \operatorname{Irr}(UL_i/I_i \,|_{L_i})$ so $\operatorname{Irr}(UL_j/I_j \,|_{L_i}) = \operatorname{Irr}(UL_i/I_i \,|_{L_i})$. Hence different ideals lead to different inductive systems and conversely.   □

Question 3.16   *Does there exist an analog of Proposition* 3.15 *for the case where L is locally finite but not locally semisimple?*[1]

Definition 3.17

(1) L *is called* locally simple *if every finite subset of L is contained in a finite dimensional simple subalgebra of L.*

(2) L *is called* of diagonal type *if L is a union of a tower of finite dimensional Lie algebras L_i of classical type $(i = 1, 2, \ldots)$ such that the dimension of every irreducible constituent of L_i in L_{i+1} is either 1 or is equal to the dimension of the natural L_i-module.*

For example, if L_i is a Lie algebra of type A_n then the highest weight of the irreducible constituents may be ω_1, ω_{n-1} or 0.

Theorem 3.18   (Jilinskiĭ [29])   *Let L be a locally simple Lie algebra. L has a nontrivial inductive system if and only if L is of diagonal type.*

One can observe an analogy between Theorems 3.18 and 1.9. Also, A. Jilinskiĭ (unpublished) has described inductive systems for L to be of diagonal type.

4. Linear Groups

Modulo the classification of finite simple groups, every simple locally finite linear group is a group of Lie type over a field which is a union of finite fields (see [23, 51, 3, 5]). It follows that every simple locally finite linear group is a union of finite simple groups of the same Lie type. For such groups there is a well developed theory of characters (the theory of complex representations is much less developed). For some small groups such as $\operatorname{PSL}_2(q)$, $\operatorname{PSL}_3(q)$, $\operatorname{PSU}_3(q)$, $\operatorname{PSp}_4(q)$ there are explicit character tables. Therefore, one may ask whether the assumptions of Proposition 3.6 are satisfied for simple locally finite linear groups. We may write $G = \bigcup G_i$ where G_i is a simple finite group of Lie type. If p is the characteristic of the ground field then the G_i are of the same type and are parametrized by some powers of p (for instance, $G_i = \operatorname{PSL}_n(p^{r(i)})$ where $r(i)$ must divide $r(i+1)$). So, if an appropriate function f in Proposition 3.6 exists, it may be expressed as a function of the field parameter. Put $p^{r(i)} = q_i$. Thus, the question is whether the nontrivial characters χ

---

[1]   Very recently A. Baranov obtained an analog of Proposition 3.15 in the spirit of Theorem 1.25 for the case when L is locally perfect (i.e. L is a union of finite dimensional Lie subalgebras L_i coinciding with their derived subalgebras).

of groups $G_i(q_i)$ satisfy the condition $|\chi(g)| < f(q_i)\chi(1)$ where $g \in G_i \setminus Z(G_i)$ and f depends on G. However, examples showed that f could be chosen independently of G, and in fact the function f for PSL$_2$ was suitable for other groups of Lie type as well. Now, D. Gluck has proved the following:

Theorem 4.1 [15, 16] *Let $H = H(q)$ be a finite group of Lie type with the field parameter q. Suppose that every proper normal subgroup of H is central (that is, H is quasi-simple). Then there exists a constant C independent of H such that $|\chi(h)| < (C/q)^{1/2}\chi(1)$ for every $\chi \in \mathrm{Irr}H$ with $\chi(1) > 1$ and for every $h \in H$ with $h \in Z(H)$, where $Z(H)$ denotes the center of H.*

We emphasize that Theorem 4.1 is a result of very great generality. The estimate is valid for *every* quasi-simple group, *every* nontrivial character, *every* nonzero element! One can see that no direct analog of Theorem 4.1 is valid for the alternating groups (examine 3-cycles in the characters of Alt$_n$ of degree $n - 1$). However, some analog might be valid, for example, under an appropriate restriction which specifies a class of characters and a class of elements. Anyway, we state the 'problem':

Problem 4.2 *Obtain versions of Theorem 4.1 which apply to other groups. Extend Theorem 4.1 as far as possible.*

Theorem 4.3 *Let G be a simple locally finite linear group and F a field of characteristic 0. Then G has no nontrivial inductive system and the group ring FG has no nontrivial two-sided ideal.*

It seems to be very probable that Theorem 4.1 is true for the Brauer characters associated with primes r not dividing q and for r'-elements $h \in H$. However, this would not be helpful for applications to group rings. As we have seen in Section 1, the following conjecture is closely related to the modular version of Problem 1.4.

Conjecture 4.4 *Let F be a field of characteristic r prime to q and let H be as in Theorem 4.1. Then there exists $n \in \mathbb{N}$ such that for $m > n$ and for every nontrivial irreducible representation ϕ of $H(q^n)$ we have $\mathrm{Irr}_F(\phi|_H) = \mathrm{Irr}_F H$.*

Is Conjecture 1.13 true for linear groups, i.e. is G in Theorem 4.3 of infinite type? The answer is positive (see [24]). The following is standard:

Lemma 4.5 *Let F be an infinite field. Let $G \subseteq \mathrm{GL}_n(F)$ be a connected linear group and X a Zariski closed subgroup. Suppose that X is confined. Then X contains a nontrivial normal subgroup of G.*

Proof. Let $K \subset G$ be a finite set. Suppose that $gKg^{-1} \cap X \neq \{1\}$ for all $g \in G$. It follows that for every $g \in G$ there exists an element $k \in K$, $k \neq 1$, such that gkg^{-1} satisfies all the equations determining X. For $k \in K$ let $G_k = \{g \in G \mid gkg^{-1} \in X\}$. Then G_k is an algebraic set and we have $G = \bigcup_{1 \neq k \in K} G_k$. As G is connected, $G = G_k$ for some $k \neq 1$. So X contains the conjugacy class $\{gkg^{-1}\}_{g \in G}$. It follows that X contains a nontrivial normal subgroup of G. $\square$

The argument in [24] shows that if X is a proper subgroup of G then there exists a faithful representation $\rho : G \longrightarrow \mathrm{GL}(m, F)$ such that $\rho(X)$ is contained in a proper Zariski closed subgroup Y. In general, this requires rather careful analysis and uses the result of the classification of simple locally finite linear groups mentioned above.

We make some comments on the notion of the Zariski topology. Let F be an infinite field. A set $X \subset \mathrm{GL}(n, F)$ is called *Zariski closed* if $X = Y \cap \mathrm{GL}(n, F)$ where Y is the set of all $n \times n$ matrices over F whose entries satisfy some collection of algebraic equations (in n^2 indeterminates). Let H be a subgroup of $\mathrm{GL}(n, F)$ which is Zariski closed. A standard fact in the theory of algebraic linear groups is that H contains a subgroup H_0 such that $|H : H_0| < \infty$ and H_0 is not a union of finitely many proper Zariski closed subsets.

5. Positive Definite Functions

For the main definition and notation see Definition 1.22. Observe, however, that the condition (3) of Definition 1.22 may be described in terms of the involution $*$ of a complex group ring which extends the complex conjugacy map of $\mathbb{C}$ and maps every $g \in G$ to g^{-1}. Thus, $*$ is an anti-automorphism of $\mathbb{C}G$. We extend a function f on G to a linear function on $\mathbb{C}G$ by linearity. Then (3) is equivalent to the condition $(3')$ $f(x^*x) \geqslant 0$ for all $x \in \mathbb{C}G$.

The material of this section concerning finite groups is taken partially from [38].

Lemma 5.1   *Let $f \in \mathcal{F}(G)$. Then $f(g^{-1}) = f(g)^*$ for every $g \in G$. Hence $f(x^*) = f(x)^*$ for every $x \in \mathbb{C}G$.*

Indeed, in condition (3) of Definition 1.22 we put $n = 2$, $g_1 = g$, $g_2 = 1$, $c_1 = 1$, $c_2 = c$. Then $1 + c^*c + f(g)c^* + f(g^{-1})c \geqslant 0$. Let $i \in \mathbb{C}$, $i^2 = -1$. If we set $c = 1$ then $\mathrm{Im}(f(g)) + \mathrm{Im}(f(g^{-1})) = 0$; if we set $c = i$ then $\mathrm{Re}(f(g)) - \mathrm{Re}(f(g^{-1})) = 0$. It follows that $f(g^{-1}) = f(g)^*$.

Proposition 5.2   *Let H be a finite subgroup of G and $f \in \mathcal{F}(G)$. Then $f|_H$ is a linear combination of irreducible characters of H with non-negative coefficients (that is, $f|_H$ belongs to the convex envelope of the irreducible characters of H).*

Proof.   Let $\{\chi_1, \ldots, \chi_n\}$ be the set of all irreducible characters of H. As this set is a basis of the space of central functions on H (that is, the space of conjugacy class functions) then we can write $f = c_1\chi_1 + \cdots + c_n\chi_n$ with $c_1, \ldots, c_n \in \mathbb{C}$. Let us fix $j \in \{1, \ldots, n\}$ and for $h \in H$ put $c_j = \chi_j(h)^*$. By (3) we have

$$0 \leqslant \sum_{h,g \in H} \sum_{i=1}^{n} c_i\chi_i(h^{-1}g)\chi_j(h)\chi_j(g)^*$$
$$= \sum_i c_i \sum_h \chi_j(h) \left(\sum_g \chi_i(h^{-1}g)\chi_j(g)^*\right).$$

By the orthogonality relations (see [8, 31.16]), the expression in the brackets is equal to $\chi_j(h^{-1})|H|\delta_{ij}/\chi_j(1)$ where δ_{ij} is the Kronecker symbol. So we have

$$0 \leqslant (c_j|H|/\chi_j(1)) \cdot \sum_h \chi_j(h)\chi_j(h^{-1}).$$

Again by the orthogonality relations (see [8, 31.17]), the sum is equal to $|H|$, so the result is $0 \leqslant c_j |H|^2/\chi_j(1)$. It follows that $c_j \geqslant 0$. $\qquad\square$

Corollary 5.3 *Let G be a locally finite group and $f \in \mathcal{F}(G)$. Then $|f(g)| \leqslant 1$ for all $g \in G$.*

Proof. Let $g \in G$ and let H be a finite subgroup of G with $g \in H$. Put $\xi = \chi/\chi(1)$ for any irreducible character χ of H. By Proposition 5.2 we have $f = d_1\xi_1 + \cdots + d_n\xi_n$ with $d_i \geqslant 0$. It is well known that $|\chi(g)| \leqslant \chi(1)$ so $|\xi_i| \leqslant 1$, $i = 1, \ldots, n$. As $1 = f(1) = d_1 + \cdots + d_n$ then $d_i \leqslant 1$ and

$$\begin{aligned} |f(g)| &\leqslant d_1|\xi_1(g)| + \cdots + d_n|\xi_n(g)| \\ &\leqslant d_1 + \cdots + d_n = 1. \end{aligned}$$

This completes the proof of the corollary. $\qquad\square$

Lemma 5.4 *Let G be a finite group and $\phi \in \mathrm{Irr}_{\mathbb{C}}G$. Let χ be the character of ϕ and $\xi = \chi/\chi(1)$. Then $\xi \in \mathcal{F}(G)$.*

Proof. For a matrix a let a^* denote the complex conjugate of a and ${}^t a$ the transpose. Then the trace of $({}^t a^*)a$ is a real non-negative number. As every representation of a finite group is equivalent to a unitary one, we can assume that ϕ is unitary. Then ${}^t\phi(g)^* = \phi(g)^{-1}$. It is clear that $\xi(1) = 1$ and $\xi(gh) = \xi(hg)$. Let

$$x = \sum_{g \in G} c(g)g \in \mathbb{C}G.$$

Then

$$\begin{aligned} 0 &\leqslant \mathrm{Trace}({}^t\phi(x)^*\phi(x)) \\ &= \mathrm{Trace}\left(\sum_{g,h \in G} c(g)^* c(h)\phi(g)^{-1}\phi(h) \right) \\ &= \sum_{g,h \in G} c(g)^* c(h)\chi(g^{-1}h). \end{aligned}$$

This proves that χ, and hence ξ, is a positive definite function. $\qquad\square$

Lemma 5.5 *Let G be a finite group and $\xi \in \mathcal{E}(G)$. Then $\xi = \chi/\chi(1)$ for some irreducible character χ of G.*

Proof. Put $\xi_i = \chi_i/\chi_i(1)$. By Proposition 5.2 we can write $\xi = c_1\chi_1 + \cdots + c_n\chi_n = d_1\xi_1 + \cdots + d_n\xi_n$ where $d_1, \ldots, d_n \in \mathbb{R}$ are non-negative. By Lemma 5.4, $\xi_i \in \mathcal{E}(G)$. Therefore, as ξ is indecomposable, we have $\xi = d_i\xi_i$ for some i. Since $\xi(1) = 1 = \xi_i(1)$ then $d_i = 1$. $\qquad\square$

Proposition 5.6 *Let G be a finite group and χ an irreducible character of G. Then $\chi/\chi(1) \in \mathcal{E}(G)$.*

Proof. By Lemma 5.4, $\chi/\chi(1) \in \mathcal{F}(G)$. By Proposition 5.2, $\chi/\chi(1) = \sum \alpha_i \zeta_i$ with $\zeta_i \in \mathcal{E}(G)$. By Lemma 5.4 we have $\chi/\chi(1) = \sum \alpha_i \chi_i/\chi_i(1)$; hence

$$\chi = \sum (\alpha_i \chi(1)/\chi_i(1)) \chi_i.$$

As the χ_i are linearly independent, all the α_i but one are 0, so $\chi/\chi(1) = \alpha_i \zeta_i$ for some i. By evaluating both sides at $1 \in G$, we get $1 = \alpha_i$ as desired. □

Let $G = \bigcup G_i$ and let $\xi_i = \chi_i(1)$ be a normalized character of G_i, $i = 1, 2, \ldots$. Suppose that $\lim_{i \to \infty} \xi_i(g)$ exists for every $g \in G$ (we formally put $\xi_i(g) = 0$ if $g \in G_i$). Let $\xi(g) = \lim_{i \to \infty} \xi_i(g)$. It follows that $\xi \in \mathcal{F}(G)$. If ξ is not constant on $G \smallsetminus \{1\}$ then ξ is nontrivial. One may construct a representation of G related to ξ. Let ϕ_i $(i = 1, 2, \ldots)$ be a representation of G_i such that ϕ_i is a subrepresentation of $\phi_{i+1}|_{G_i}$. Let M_i be the module affording ϕ_i. Let us fix an embedding $M_i \longrightarrow M_{i+1}|_{G_i}$ and take the direct limit $M = \lim M_i$. It is clear that M is a G-module. Let χ_i be the character of ϕ_i. Consider the sequence $\{\chi_i(g)/\chi_i(1)\}$ for each $g \in G$. If for every $g \in G$ the limit of this sequence exists we can tie the function $\xi = \lim \chi_i(g)/\chi_i(1)$ to the representation of G afforded by M. It should be noted that the connection between M and ξ is much weaker than for finite G.

Examples. (i) Let $G = \bigcup \mathrm{Alt}_n$ such that all the embeddings $\mathrm{Alt}_n \longrightarrow \mathrm{Alt}_{n+1}$ are natural. Let π_n be the permutation module associated with the action of Alt_n on $\Omega(n)$. Let χ_n be the character of π_n and $\xi_n = \chi_n/\chi_n(1)$. Then for $a \in \mathrm{Alt}_n$ we have $\xi_n(a) = t_n(a)/n$ where $t_n(a)$ is the number of fixed points of a on $\Omega(n)$. It is clear that $\lim \xi_n(a) = 1$ for any $a \in G$. It follows that the limit function is constant so it is trivial in the sense of Definition 1.22. However, there exist nontrivial functions in $\mathcal{F}(G)$. Moreover, for this case all functions of $\mathcal{E}(G)$ were calculated by Thoma [50]; see also [52].

(ii) Let $G = \bigcup G_i$ where $G_i = \mathrm{Alt}_{2^i}$ and $\mathrm{Alt}_{2^i} \longrightarrow \mathrm{Alt}_{2^{i+1}}$ is a two-fold diagonal embedding for all i. For $g \in G$ let $i(g)$ be the least i such that $g \in G_i$. To simplify the notation let χ_i denote the character of π_{2^i} and $\xi_i = \chi_i/\chi_i(1)$. Let $t_i(g)$ be the number of fixed points of g on Ω_{2^i}. Then for $j \geqslant i(g)$ we have $\chi_j(g) = 2^{j-i(g)} t_{i(g)}(g)$ so $\xi_j(g) = t_{i(g)}(g)/2^{i(g)}$. So $\xi_j(g)$ does not depend on j and

$$\lim_{j \to \infty} \xi_j(g) = \xi(g) = t_{i(g)}(g)/2^{i(g)}.$$

A similar argument shows that for $G_i = \mathrm{Alt}_{n_i}$ with the purely diagonal embeddings $\mathrm{Alt}_{n_i} \longrightarrow \mathrm{Alt}_{n_{i+1}}$ (i.e. Alt_{n_i} has no fixed points on Ω_{n_i}) the set $\mathcal{F}(G)$ contains a nontrivial function. However, this argument does not work for all groups of diagonal type. Indeed, let $G_i = \mathrm{Alt}_{4^i}$ and let $G_i \longrightarrow G_{i+1}$ be a diagonal embedding such that G_i has $2 \cdot 4^i$ fixed points on Ω_{i+1}. Let $i(g)$ be defined as above. Then for $j = i(g) + 2k$ we have $\chi_j(g) = 4^{j-k}(4^k - 1) + 2^{2k-1} t_{i(g)}(g)$, so

$$\begin{aligned}
\xi_j(g) &= 1 - 4^{-k} + 2^{2k-1} t_{i(g)}(g)/4^j \\
&= 1 - 4^{-k} + 4^{k-j} t_{i(g)}(g)/2.
\end{aligned}$$

As $k - j = -i(g) - k$, we have $\lim_{j \to \infty} \xi_j(g) = 1$ for any $g \in G$. It follows that the function $\xi(g)$ is trivial. Of course, this fact does not imply that $\mathcal{F}(G)$ consists of trivial functions only.

Problem 5.7 *Describe $\mathcal{E}(G)$ for groups G of diagonal type.*

Definition 5.8 *Let $f \in \mathcal{F}(G)$. We put $I_f = \{x \in \mathbb{C}G \mid f(gx) = 0 \text{ for all } g \in G\}$.*

Lemma 5.9 *I_f is a two-sided ideal of $\mathbb{C}G$.*

Proof. It is clear that I_f is a left ideal. As $f(gx) = f(xg)$, it is two-sided. □

The following fact is well known (see Dixmier [9, Section 2.1]):

Lemma 5.10 *Let $x, y \in \mathbb{C}G$ and let $f \in \mathcal{F}(G)$. Then $|f(y^*x)|^2 \leqslant f(x^*x)f(y^*y)$.*

Proof. Let $a \in \mathbb{R}$. Suppose first that $f(y^*x) \in \mathbb{R}$. Then

$$0 \leqslant f(x^* + ay^*, x + ay) = f(x^*x) + a^2 f(y^*y) + af(y^*x) + af(x^*y).$$

By Lemma 5.1, $f(y^*x) = f(x^*y)^* = f(x^*y)$ as $f(y^*x) \in \mathbb{R}$. Hence

$$0 \leqslant f(x^*x) + a^2 f(y^*y) + 2af(y^*x)$$

for any $a \in \mathbb{R}$, so the discriminant $-f(x^*x)f(y^*y) + f(y^*x)^2$ is not negative. Thus, the Lemma is proved for $f(y^*x) \in \mathbb{R}$. In general, if $f(y^*x) = c$ then let $d \in \mathbb{C}$ be such that $|d| = 1$ and $cd \in \mathbb{R}$. Set $x_1 = d^{-1}x$. Then $f(y^*x_1) \in \mathbb{R}$, so

$$|f(y^*x)|^2 = |f(y^*x_1)|^2 \leqslant f(x_1^*x_1)f(y^*y) = f(x^*x)f(y^*y).$$

□

Corollary 5.11 *Let $J_f = \{x \in \mathbb{C}G \mid f(x^*x) = 0\}$. Then $J_f = I_f$.*

Indeed, by Lemma 5.10, $f(y^*x) = 0$ for any $y \in \mathbb{C}G$ and y^* runs over $\mathbb{C}G$ when y runs over $\mathbb{C}G$. □

Problem 5.12 *Let G be a simple locally finite group. Is it true that $I_f \neq \{0\}$ for every $f \in \mathcal{E}(G)$ with $f \neq f_0$?*

Problem 5.13 *Let G be a locally finite group. Is it true that $f_0 \in \mathcal{E}(G)$ if and only if G has no finite normal subgroup other than $\{1\}$?*

Proposition 5.14 *Let G be a locally finite group. If G contains a finite normal subgroup other than $\{1\}$ then $f_0 \notin \mathcal{E}(G)$.*

Proof. Let G_0 be a finite normal subgroup other than $\{1\}$. As $f_0|_{G_0} \in \mathcal{F}(G_0)$, by Proposition 5.2 and Lemma 5.4, we can write $f_0|_{G_0} = \sum \alpha_i \chi_i / \chi_i(1)$ where $\alpha_i > 0$ and $\chi_i \in \operatorname{Irr}(G_0)$. For $g \in G$ we set $\xi_i(g) = \chi_i(g)$ for $g \in G_0$ and $\xi_i(g) = 0$ otherwise. Then $\xi_i \in \mathcal{F}(G)$ and $f_0 = \sum \alpha_i \xi_i / \xi_i(1)$. Hence $f_0 \notin \mathcal{E}(G)$. $\qquad\square$

Problem 5.15 *Suppose that $\mathcal{F}(G)$ is nontrivial. Is it true that there exists a nontrivial function $f \in \mathcal{F}(G)$ such that $f(g) \geqslant 0$ for all $g \in G$?*

Proposition 5.16 *Let $f \in \mathcal{F}(G)$. For every $i \in \mathbb{N}$ let $f|_{G_i} = \sum \alpha_\chi \chi$ where $\alpha_\chi > 0$ and $\chi \in \operatorname{Irr}(G_i)$. Set $\Phi_i = \{\chi : \alpha_\chi \neq 0\}$. Then $\Phi = \{\Phi_i\}$ is an inductive system.*

Proof. This follows easily from Proposition 5.2 and Lemma 5.5. $\qquad\square$

It is not hard to show that the inductive system in Proposition 5.16 corresponds to the ideal I_f of $\mathbb{C}G$.

6. Cobounded Subgroups

Proof of Proposition 1.21. Suppose that X is not confined. Then for each G_i there exists $g_i \in G$ such that $G_i \cap g_i X g_i^{-1} = \{1\}$. Hence $|G_i \cap g_i X g_i^{-1} \cap \operatorname{Cl}(x)| \leqslant 1$ for all $x \in G$. On the other hand, as $\operatorname{Cl}(x)$ is infinite for $1 \neq x \in G$ then $|G_i \cap \operatorname{Cl}(x)|$ is not bounded. So $b_i(x, g_i)$ is not bounded as desired. $\qquad\square$

Proposition 6.1 [48] *Let $G = \operatorname{Alt}(\Omega)$ be the finitary alternating group. Let X be a confined subgroup. Then X contains as a normal subgroup of finite index a group $\operatorname{Alt}(\Omega_1) \times \cdots \times \operatorname{Alt}(\Omega_n)$ where $\Omega_1, \ldots, \Omega_n \subset \Omega$ are disjoint infinite subsets such that the set $\Omega \setminus (\Omega_1 \cup \cdots \cup \Omega_n)$ is finite.*

This follows mainly from the Jordan-Wielandt theorem that for an infinite set Ω every primitive subgroup of $\operatorname{Alt}(\Omega)$ is $\operatorname{Alt}(\Omega)$ itself.

Proposition 6.2 [48] *Let $G = \operatorname{Alt}(\Omega)$ be the finitary alternating group. Then every confined subgroup of G is cobounded.*

Proof. Let $G = \bigcup_{i \in \mathbb{N}} G_i$ where $G_i \cong \operatorname{Alt}_i$. Let X be a confined subgroup of G. By Proposition 6.1, $|X : Y| < \infty$ where $Y = \operatorname{Alt}(\Omega_1) \times \cdots \times \operatorname{Alt}(\Omega_n)$ for some $\Omega_1, \ldots, \Omega_n$. Set $R = \Omega \setminus (\Omega_1 \cup \cdots \cup \Omega_n)$ and $r = |R|$. As

$$|G_i \cap gYg^{-1} \cap \operatorname{Cl}(x)| \leqslant |G_i \cap gXg^{-1} \cap \operatorname{Cl}(x)|$$

it suffices to bound $\{b_i(x, g)\}$ for $X = Y$ and all $x \in G$. Let $x \in G$ and $k = |\sup(x)|$. Choose i such that $(1/n) - (r/in) > (1/n + 1)$, $x \in G_i$ and $(i - r)/n > k$. Then $|\sup(G_i) \cap \Omega_{j_0}| > k$ for any $g \in G$ and a suitable $j_0 \in \{1, \ldots, n\}$. (Observe that only finitely many values of i do not satisfy these conditions.) Put $e = [(i - r)/r]$; here $[\ ]$ denotes the greatest integer function. Then $|\sup(G_i) \cap \Omega_{j_0}| > e$ for any $g \in G$ and a suitable $j_0 \in \{1, \ldots, n\}$. Denote by $C_G(x)$ the centralizer of x in G. We have $|G_i \cap \operatorname{Cl}(x)| \leqslant 2|G_i|/|C_{G_i}(x)|$ because $G_i = \operatorname{Alt}_i$ and elements of Alt_i conjugate in $\mathrm{Alt}(\Omega)$ are conjugate in Sym_i. Furthermore, $C_{G_i}(x)$ is a subgroup of index 2 of $C_{\mathrm{Sym}_k}(x) = \mathrm{Sym}_{i-k}$. If follows that $|G_i \cap \mathrm{Cl}(x)| \leqslant 2i!/(i-k)!|C_{\mathrm{Sym}_k}(x)|$. Put $D_j = \Omega_j \cap \sup(G_i)$ and $d_j = |D_j|$, $j = 1, \ldots, n$. We have

$$G_i \cap gYg^{-1} \cap \mathrm{Cl}(x) \supseteq \mathrm{Cl}(x) \cap \mathrm{Alt}(D_{j_0}).$$

Therefore,

$$\begin{aligned} |G_i \cap gYg^{-1} \cap \mathrm{Cl}(x)| &\geqslant |\mathrm{Alt}(D_{j_0})|/|C_{\mathrm{Sym}_k}(x)||\mathrm{Sym}_{d_{j_0}-k}| \\ &\geqslant |C_{\mathrm{Sym}_k}(x)|^{-1} d_{j_0}!/(d_{j_0}-k)!. \end{aligned}$$

Since $d_{j_0} \geqslant e$, then $d_{j_0}!/(d_{j_0}-k)! \geqslant e!/(e-k+1)!$. Thus

$$\begin{aligned} |G_i \cap \mathrm{Cl}(x)|.|G_i \cap gYg^{-1} \cap \mathrm{Cl}(x)| &\leqslant (2i!/(i-k)!)/(e!/(e-k)!) \\ &= 2i(i-1)(i-2)\cdots(i-k+1)/(e!/(e-k)!). \end{aligned}$$

Hence for any $g \in G$ we have

$$\begin{aligned} b_i(x,g) &= |G_i \cap \mathrm{Cl}(x)|/(i + |G_i \cap gYg^{-1} \cap \mathrm{Cl}(x)|) \\ &\leqslant |G_i \cap \mathrm{Cl}(x)|/|G_i \cap gYg^{-1} \cap \mathrm{Cl}(x)| \\ &\leqslant 2i(i-1)(i-2)\cdots(i-k+1)/e(e-1)\cdots(e-k+2) \\ &\leqslant 2i^k/(e-k+2)^k \\ &= 2((e-k+2)/i)^{-k} \\ &\leqslant 2((1/n+1)-((k-2)/i))^{-k} \end{aligned}$$

as $(1/n)-(r/in) > (1/n+1)$. Since there are only finitely many values of i which do not satisfy the conditions of our choice of i, the theorem follows. $\quad\square$

7. Annihilators of Induced Modules

Let G be a locally finite simple group. According to Proposition 1.2 and Conjecture 1.13 we conjecture that the group ring $\mathbb{C}G$ has a nontrivial ideal if and only if G contains a proper confined subgroup. There is an intermediate conjecture that would serve as a bridge between these two properties. For an FG-module M we set $\mathrm{Ann}_{FG}(M) = \{x \in FG \mid xM = \{0\}\}$. It is clear that $\mathrm{Ann}_{FG}(M)$ is a two-sided ideal of FG.

Conjecture 7.1 *Let G be a simple locally finite group and let X be a proper subgroup of G. The following conditions are equivalent:*

(a) *X is confined;*
(b) *$\mathrm{Ann}_{FG}(1_X^G) \neq \{0\}$.*

In view of Problem 1.4 we conjecture also that if FG has a nontrivial ideal then there exists a proper subgroup $X \subset G$ such that $\mathrm{Ann}_{FG}(1_X^G) \neq \{0\}$. We show that (b) implies (a).

Proposition 7.2 [48] *Let G be an arbitrary group. Let F be an arbitrary field and X a subgroup of G. Let M be an FX-module. If the annihilator of the module M^G is not zero then the permutation representation of G on the set of cosets $\Omega = G/X$ is of finite type. If G is locally finite then X is confined.*

Proof. Let $0 \leqslant a \in I_X$; write $a = \sum_{i=1}^r a_i g_i$ with $g_i \in G$ and $0 \neq a_i \in F$. We may write $M^G = \bigoplus_k h_k M$ where $\{h_k\}$ is a set of representatives of the cosets G/X. Let $0 \neq m \in h_l M$. As g_i permutes the $h_k M$ then $am = 0$ implies that for every pair l, i there is $j = j(i,l) \neq i$ such that $g_i h_l X = g_j h_l X$. Hence the coset $h_l X$ is fixed by $g_j^{-1} g_i$. Therefore, $\{g_j^{-1} g_i\}_{i \leqslant i \leqslant r}$ can be taken for the finite set in Definition 1.16. $\square$

The proof of the following result in [26] is much more difficult than that of Proposition 7.2:

Theorem 7.3 *Let F be a field of characteristic 0, G a simple locally finite group and M an FG-module. Suppose that $\mathrm{Ann}_{FG}(M) \neq \{0\}$. Then G contains a confined subgroup X.*

In view of Conjecture 1.13 one can expect that $\mathrm{Ann}_{FG}(1_X^G) \neq \{0\}$. It is not clear whether the argument of [26] can be adapted to prove this.

8. Inductive Systems and Ideals of Modular Group Rings

Let the ground field F be of characteristic $p > 0$. We say that the group ring FG is modular if G contains an element of order p. We retain the definition of an inductive system given in Definition 1.1.

Recall that if A is a locally finite algebra over a field F then the Jacobson radical $\mathrm{Rad}(A)$ of A is the largest locally nilpotent ideal of A (i.e. it contains every locally nilpotent ideal of A) and $A/\mathrm{Rad}(A)$ is semisimple (i.e. $\mathrm{Rad}(A/\mathrm{Rad}(A)) = \{0\}$).

We need the following simple lemma.

Lemma 8.1 *Let $\phi : FG \longrightarrow \mathrm{End}(FG/I)$ be the natural homomorphism (i.e. FG/I is a quotient of the regular module FG) and let $\phi_i = \phi|_{FG_i}$. Then $\mathrm{Ker}(\phi_i) = I \cap FG_i$ and $\mathrm{Im}(\phi_i) = FG_i/(I \cap FG_i)$. Furthermore, $\mathrm{Irr}(FG/I|_{G_i}) = \mathrm{Irr}(FG_i/(I \cap FG_i))$.*

Proof. It is clear that $I = \mathrm{Ker}(\phi)$ and that $\mathrm{Ker}(\phi_i) = FG_i \cap \mathrm{Ker}(\phi) = FG_i \cap I$. So $\mathrm{Im}(\phi_i) = FG_i/(I \cap FG_i)$. Set $L = FG_i/(I \cap FG_i)$. As ϕ_i is in fact a representation of L, $\mathrm{Irr}(\phi_i) \subseteq \mathrm{Irr}(L)$. On the other hand, L acts regularly (hence, faithfully) on the L-submodule $FG_i/(I \cap FG_i)$ of FG/I considered as L-module. Hence we have $\mathrm{Irr}(L) \subseteq \mathrm{Irr}(\phi_i)$. Thus, $\mathrm{Irr}(FG/I|_{FG}) = \mathrm{Irr}(FG_i/(I \cap FG_i))$. It remains to observe that $\mathrm{Irr}(FG_i)$ and $\mathrm{Irr}(G_i)$ are essentially the same. $\square$

Lemma 8.2 *Let I be an ideal of FG. Put $\Phi_i = \mathrm{Irr}(FG/I|_{G_i})$. Then $\{\Phi_i\}$ is an inductive system.*

The proof is obvious.

Lemma 8.3 *Let A be an algebra of finite dimension over F and $M \supseteq N$ two-sided ideals. Then $\mathrm{Irr}(A/M) = \mathrm{Irr}(A/N)$ if and only if M/N is nilpotent.*

Proof. Put $B = A/N$, $L = M/N$. Let J be the Jacobson radical of B. Then J is nilpotent. It is well known that $\mathrm{Irr}(B/J) = \mathrm{Irr}(B)$ and J is the largest ideal with this property. This implies the Lemma. $\square$

Proof of Theorem 1.25. Given an inductive system $\Phi = \{\Phi_i\}, \Phi_i \subseteq \mathrm{Irr}(G_i)$, we define K_i to be the smallest two-sided ideal of FG_i such that $\mathrm{Irr}(FG_i/K_i) = \Phi_i$. By the Jordan-Hölder theorem the ideal is well-defined. By the definition of an inductive system, $\mathrm{Irr}(FG_j/K_j|_{G_i}) = \Phi_i$ for $j > i$. Hence $\mathrm{Irr}(FG_i/(K_j \cap FG_i)) \subseteq \Phi_i$, so $K_j \cap FG_i \supseteq K_i$. Hence $K = \bigcup K_i$ is an ideal of FG and $\mathrm{Irr}(FG/K|_{G_i}) = \Phi_i$. Let J/K be the Jacobson radical of FG/K; so FG/J is semisimple. As FG/K is a locally finite algebra, J/K is locally nilpotent. So $(J \cap FG_i)/(K \cap FG_i)$ is nilpotent for all i, and $\mathrm{Irr}(FG_i/(J \cap FG_i)) = \mathrm{Irr}(FG_i/(K \cap FG_i))$ by Lemma 8.3. Hence $\mathrm{Irr}(FG/J|_{G_i}) = \Phi_i$. Let L be an ideal of FG such that $\mathrm{Irr}(FG/L|_{G_i}) = \Phi_i$. Then $L \cap FG_i \supseteq K_i$ by Lemma 8.3, so $L \supseteq K$. As L_i/K_i is nilpotent for every i, L/K is locally nilpotent. So $L/K \subseteq \mathrm{Rad}(FG/K) = J$. Thus, the first two assertions of the Theorem are obtained if we set $K(\Phi) = K$, $I(\Phi) = J$.

If Φ and Φ' are distinct inductive systems then $\Phi_i \neq \Phi_i'$ for some i. Then $K_i \neq K_i'$, hence $K(\Phi) \neq K(\Phi')$. It follows that $I(\Phi) \neq I(\Phi')$; otherwise $I(\Phi)/(K(\Phi) \cap K(\Phi'))$ is locally nilpotent, which is impossible. Thus, the maps $\Phi \mapsto K(\Phi)$ and $\Phi \mapsto I(\Phi)$ (where Φ runs over inductive systems) are injective.

Suppose now that I is an ideal of FG such that FG/I is semisimple. Denote $\Phi_i = \mathrm{Irr}(FG/I|_{G_i})$. By Lemma 8.3, $\{\Phi_i\} = \Phi$ is an inductive system. Set $J = I(\Phi)$ and $K = K(\Phi)$. By the part of Theorem 1.25 already proved, we have $K \supseteq L \supseteq J$, so J/I is locally nilpotent. As FG/I is semisimple, $I = J = I(\Phi)$ and the theorem follows. $\square$

Lemma 8.4 *Let Φ be an inductive system. Then $K(\Phi)$ is the smallest ideal K of FG such that $K \subseteq I(\Phi)$ and $I(\Phi)/K$ is locally nilpotent.*

Proof. Let $K \subseteq I(\Phi)$ where $I(\Phi)/K$ is locally nilpotent. Then it follows that $I(\Phi) \cap FG_i \supseteq K \cap FG_i$ and $(I(\Phi) \cap FG_i)/(K \cap FG_i)$ is nilpotent. By Lemma 8.3, $\mathrm{Irr}(FG_i/(I(\Phi) \cap FG_i)) = \mathrm{Irr}(FG_i/(K \cap FG_i))$, so $\Phi(K) = \Phi$. By Theorem 1.25, $K = K(\Phi)$. $\square$

Recall Definition 3.10: an inductive system Φ is called indecomposable if for any inductive systems Φ' and Φ'' such that $\Phi = \Phi' \cup \Phi''$ we have $\Phi = \Phi'$ or $\Phi = \Phi''$.

Proposition 8.5 *Let Φ be an inductive system of G and $I = I(\Phi)$. Then I is prime if and only if Φ is indecomposable.*

Proof. Let I be a prime ideal. Let I_j $(j = 1, 2)$ be ideals of FG properly containing I such that $I_1 I_2 \subseteq I$. Then $\Phi(I_j) \subseteq \Phi$, $j = 1, 2$. As FG/I has no nil-ideals, $\Phi(I_j) \neq \Phi = \Phi(I)$ by Lemma 8.1. Furthermore, $(I_1 \cap I_2)^2 \subseteq I_1 I_2 \subseteq I$. As FG/I has no nil-ideals, $I_1 \cap I_2 = I$ and $\Phi = \Phi(I_1 \cap I_2)$. By the Jordan-Hölder theorem, $\Phi(I_1 \cap I_2) = \Phi(I_1) \cup \Phi(I_2)$. Hence $\Phi = \Phi(I_1) \cup \Phi(I_2)$, so Φ is decomposable.

Conversely, let $\Phi = \Phi_1 \cup \Phi_2$, $\Phi_j \subset \Phi$, $\Phi_j \neq \Phi$, $j = 1, 2$. Then $I(\Phi_j) \supset I$ and, by Theorem 1.25, $I(\Phi_j) \neq I$. Also, $I(\Phi) = I(\Phi_1 \cup \Phi_2) \supseteq I(\Phi_1) \cap I(\Phi_2) \supseteq K(\Phi)$ by Theorem 1.25 and the Jordan-Hölder theorem. As $I(\Phi_1) I(\Phi_2) \subseteq I(\Phi_1) \cap I(\Phi_2)$, we have $I(\Phi_1) I(\Phi_2) \subseteq I$, so I is not prime. $\square$

Proposition 8.6 *Let $G = \bigcup_i G_i$ where the G_i are finite and $G_i \subset G_{i+1}$. Let $\Phi = \{\Phi_i\}$ be an indecomposable inductive system. Then for every i there exists $j > i$ and $\tau \in \Phi_j$ such that $\Phi_i = \mathrm{Irr}(\tau|_{G_i})$.*

The proof does not differ from that of Proposition 3.12 for characteristic 0.

Problem 8.7   *Under what circumstances is it true that $I(\Phi) = K(\Phi)$? For which Φ is $I(\Phi)/K(\Phi)$ nilpotent?*

Problem 8.8   *Describe the inductive systems Φ such that for every $i \in \mathbb{N}$ and every $\phi \in \Phi_{i+1}$ the representation $\phi|_{G_i}$ is completely reducible.*

If for every $i \in \mathbb{N}$ and every $\phi \in \Phi_{i+1}$ the representation $\phi|_{G_i}$ is completely reducible then $I(\Phi) = K(\Phi)$. However, the converse is not true even in the special case $\Phi = \{\Phi_i\} = \{\mathrm{Irr}(G_i)\}$; see the comments after Theorem 1.25 in the Introduction. Of course, there are 'very few' systems Φ in Problem 8.8.

Problem 8.9   *Determine the inductive systems for the infinite alternating group for every prime p.*

Definition 8.10   *Let I be a two-sided ideal of a group ring FG. We say that I is a block ideal if there exist subgroups G_i of G with $G_i \subset G_{i+1}$ and $G = \bigcup G_i$ such that $I \cap FG_i$ is a block ideal in FG_i (i.e. FG_i is the direct sum of $I \cap FG_i$ and another two-sided ideal J_i).*

It is clear that every ideal of FG is a block ideal provided $\mathrm{char}(F) = 0$.

Problem 8.11   *For which G does the modular group ring FG contain a nonzero block ideal?*

We say that an associative algebra is locally semisimple if every finite subset is contained in a semisimple subalgebra of finite dimension (cf. Definition 3.14). There are a lot of semisimple modular group algebras. However, can it happen that a modular group algebra is locally semisimple? The following result shows that the answer is 'no'.

Theorem 8.12   *Let G be a locally finite group and F a field of characteristic $p > 0$. The group algebra FG is locally semisimple if and only if G has no element of order p.*

This is a precise version of the classical Maschke theorem for finite group algebras. The proof of Theorem 8.12 follows immediately from the following lemma about finite group algebras which seems to be new.

Lemma 8.13   *Let $H \subset K$ be finite groups and suppose that $p = \mathrm{char}(F)$ divides $|H|$. Then FH is not contained in a semisimple subalgebra of FK.*

Proof.   Suppose otherwise, and let S be a semisimple subalgebra of FK containing FH. Let R be the regular FK-module and R_S its restriction to S. As S is semisimple, R_S is a direct sum of irreducible S-modules. Obviously, the composition series of R contains an FK-module of dimension 1. It follows from the Jordan-Hölder theorem that R_S contains a submodule L of dimension 1. Let $R_S = L \oplus M$ be a direct sum of S-modules. Then $R_S|_{FH} = R|_{FH} = L|_{FH} \oplus M|_{FH}$. It is clear that $R|_{FH}$ is a free FH-module and $L|_{FH}$ is a direct summand of $R|_{FH}$. Hence $L|_{FH}$ is projective. By [12, Ch. III, Corollary 2.10] we have $1 = \dim(L) \equiv 0 \pmod{p}$, a contradiction.   □

References

1. Yu. Bahturin and H. Strade, Locally finite-dimensional simple Lie algebras, *Mat. Sb.* 185 (1994), 3–32 (in Russian).
2. Yu. Bahturin and H. Strade, Some examples of locally finite simple Lie algebras, *Arch. Math.*, to appear.
3. V. V. Belyaev, Locally finite Chevalley groups, in *Publications in Group Theory*, Sverdlovsk, 1984, pp. 39–50 (in Russian).
4. K. Bonvallet, B. Hartley, D. S. Passman and M. K. Smith, Group rings with simple augmentation ideals, *Proc. Amer. Math. Soc.* 56 (1976), 79–82.
5. A. V. Borovik, Classification of the periodic linear groups over fields of odd characteristic, *Siberian Math. J.* 25 (1984), 221–235.
6. O. Bratteli, Inductive limits of finite dimensional C^*-algebras, *Trans. Amer. Math. Soc.* 171 (1972), 195–234.
7. P. J. Cameron and J. I. Hall, Some groups generated by transvection subgroups, *J. Algebra* 140 (1991), 184–209.
8. C. W. Curtis and I. Reiner, *Representation Theory of Finite Groups and Associative Algebras*, Interscience (Wiley), New York-London, 1962.
9. J. Dixmier, C^*-algebras et leurs Representations, Gautier-Villars, Paris, 1969.
10. G. Elliott, On classification of inductive limits of sequences of semisimple finite dimensional algebras, *J. Algebra* 38 (1976), 29–44.
11. D. R. Farkas and R. L. Snider, Simple augmentation modules, *Quart. J. Math. Oxford* (2) 45 (1994), 29–42.
12. W. Feit, *The Representation Theory of Finite Groups*, North-Holland, Amsterdam, 1982.
13. E. Formanek, A problem of Passman on semisimplicity, *Bull. London Math. Soc.* 4 (1972), 375–376.
14. E. Formanek and J. Lawrence, The group algebra of the infinite symmetric group, *Isr. J. Math.* 23 (1976), 325–331.
15. D. Gluck, Character value estimates for groups of Lie type, *Pacific J. Math* 150 (1991), 279–307.
16. D. Gluck, Character value estimates for non-semisimple elements, *J. Algebra* 155 (1993), 221–237.
17. K. R. Goodearl, *Von Neumann Regular Rings*, Pitman, London, 1979.
18. J. Hall, Finitary linear groups and elements of finite degree, *Arch. Math.* 50 (1988), 315–318.
19. J. Hall, Locally finite simple groups of finitary linear transformations, *these Proceedings*.
20. D. Handelman, K_0 of von Neumann and AF C^*-algebras, *Quart. J. Math.* 29 (1978), 427–441.
21. B. Hartley, Simple locally finite groups, *these Proceedings*.
22. B. Hartley, On simple locally finite groups constructed by Meierfrankenfeld, preprint no. 1994/01, Manchester Centre for Pure Mathematics.
23. B. Hartley and G. Shute, Monomorphisms and direct limits of finite groups of Lie type, *Quart. J. Math.* 35 (1984), 49–71.
24. B. Hartley and A. E. Zalesskiĭ, On periodic linear groups: dense subgroups, permutation representations and induced modules, *Israel J. Math.* 82 (1993), 299–327.
25. B. Hartley and A. E. Zalesskiĭ, The ideal lattice of the complex group rings of finitary special and general linear groups over finite fields, *Math. Proc. Cambr. Phil. Soc.* 116 (1994), 7–25.
26. B. Hartley and A. E. Zalesskiĭ, Confined subgroups of simple locally finite groups and ideals of their group rings (in preparation).
27. K. K. Hickin, Universal locally finite central extensions of groups, *Proc. London Math. Soc.* (3) 52 (1986), 53–72.
28. A. G. Jilinskiĭ, Coherent systems of representations of inductive families of simple complex Lie algebras, Inst. Math. Acad. Sci. BSSR, preprint no. 38 (438), 1990 (in Russian).
29. A. G. Jilinskiĭ, Coherent systems of finite type of inductive systems of non-diagonal embeddings of simple Lie algebras, *Doklady AN Belorus. SSR* 36 (1992), 9–13 (in Russian).
30. I. Kaplansky, *Notes on Ring Theory*, mimeographed notes, University of Chicago, 1965.
31. O. Kegel, Über einfach, lokal endliche Gruppen, *Math. Z.* 95 (1967), 169–195.
32. O. H. Kegel and B. A. F. Wehrfritz, *Locally Finite Groups*, North-Holland, Amsterdam, 1973.
33. A. Kirillov, Positive definite functions on a group of matrices with elements from a discrete field, *Doklady AN SSSR* 162 (1965), 503–505 (in Russian).

34. P. Kleidman and M. Liebeck, On a theorem of Feit and Tits, *Proc. Amer. Math. Soc.* 107 (1987), 315–322.

35. F. Leinen, Lokale Systeme in universellen Gruppen, *Arch. Math.* 41 (1983), 401–403.

36. B. Maier, Existentiell abgeschlossene lokal endliche p-Gruppen, *Arch. Math.* 37 (1981), 113–128.

37. U. Meierfrankenfeld, Non-finitary simple locally finite groups, *these Proceedings.*

38. M. L. Nazarov, Factor-representations of the infinite spin-symmetric group, in *Differential Geometry, Lie Groups and Mechanics, Zap. Nauchn. Semin. Leningr. Otdel. Math. Inst. AN SSSR,* 181 (1990), 132–145 (in Russian).

39. S. Ovchinnikov, Positive definite functions on Chevalley groups, *Izv. VysŮch. Zav. Ser. Matem.* 1971, no. 8, 77–87 (in Russian).

40. D. S. Passman, On the semisimplicity of group rings of linear groups, II, *Pacific J. Math.* 48 (1977), 215–234.

41. D. S. Passman, *The Algebraic Structure of Group Rings,* Wiley, New York, 1977.

42. D. S. Passman, Semiprimitivity of group algebras of infinite simple groups of Lie type, *Proc. Amer. Math. Soc.* 121 (1994), 399-403.

43. D. S. Passman, Semiprimitivity of group algebras: a survey, preprint, Univ. of Wisconsin-Madison.

44. D. S. Passman and A. E. Zalesskiĭ, On the semiprimitivity of modular group algebras of locally finite simple groups, *Proc. London Math. Soc.* 67 (1993), 243–276.

45. R. Phillips, Finitary linear groups: a survey, *these Proceedings.*

46. C. Praeger and A. E. Zalesskiĭ, Orbit lengths of permutation groups, and group rings of locally finite simple groups of alternating type, *Proc. London Math. Soc.* 70 (1995), 313–335.

47. U. P. Razmyslov, Trace identities of full matrix algebras over a field of characteristic zero, *Math. USSR Isv.* 8 (1974), 727–760.

48. S. K. Sehgal and A. E. Zalesskiĭ, Induced modules and arithmetic invariants of the finitary symmetric groups, *Nova J. Algebra and Geometry* 2 (1993), 89–105.

49. H. Skudlarek, Die unzerlegbaren Charactere einiger diskreter Gruppen, *Math. Ann.* 223 (1976), 213–231.

50. E. Thoma, Die unzerlegbaren, positiv-definiten Klassenfunktionen der abzalhlbar unendlichen, symmetrischen Gruppe, *Math. Z.* 85 (1964), 40–61.

51. S. Thomas, The classification of the simple periodic linear groups, *Arch. Math.* 41 (1983), 103–116.

52. A. M. Vershik and S. V. Kerov, Asymptotic character theory of the symmetric group, *Functional Anal. i Prilozhen.* 15 (1981), 1–14 (in Russian). English translation: *Funkts. Anal.* 16 (1981), 15–27.

53. A. M. Vershik and S. V. Kerov, K_0-functor (the Grothendick group) of the infinite symmetric group, *J. Soviet Math.* 28 (1985), 549–568.

54. A. M. Vershik and S. V. Kerov, Locally semisimple algebras. Combinatorial theory and K-functor, *J. Soviet Math.* 38 (1987), 1701–1733.

55. A. M. Vershik and D. P. Kokhas, Calculation of the Grothendick group of the algebra $\mathbb{C}[PSL(2,k)]$ where k is a countable algebraically closed field, *Algebra i Analiz (Sanct-Peterburg)* 2 (1990), 98–106 (in Russian).

56. A. E. Zalesskiĭ, On group rings of linear groups, *Siberian Math. J.* 12 (1971), 246–250.

57. A. E. Zalesskiĭ, Group rings of inductive limits of alternating groups, *Leningrad Math. J.* 2 (1990), 1287–1303.

58. A. E. Zalesskiĭ, A simplicity condition of the augmentation ideal of the modular group algebra of a locally finite group, *Ukrainskiĭ Mat. J.* (1991), 1088–1091 (in Russian). English translation: *Ukrainian Math. J.* 43 (1991), 1021–1024.

59. A. E. Zalesskiĭ, Group rings of locally finite groups and representation theory, in *Proc. Internat. Conf. on Algebra Dedicated to the Memory of A. I. Mal'cev, Contemporary Math.* 131 (1992), 453–472.

60. A. E. Zalesskiĭ, *Simple Locally Finite Groups and their Group Rings,* mimeographed notes, Univ. of East Anglia, 1992.

61. A. E. Zalesskiĭ and V. N. Serezhkin, Linear groups generated by transvections, *Math. USSR Izv.* 10 (1976), 25–46.

SIMPLE LOCALLY FINITE GROUPS
OF FINITE MORLEY RANK
AND ODD TYPE

A. V. BOROVIK
Department of Mathematics
UMIST
PO Box 88
Manchester M60 1QD
United Kingdom
E-mail: borovik@lanczos.ma.umist.ac.uk

Abstract. The paper is devoted to a discussion of possible approaches to the classification of simple infinite groups of finite Morley rank. As an application of methods borrowed from finite group theory we classify locally finite simple groups of finite Morley rank with Černikov Sylow 2-subgroups (without using the classification of finite simple groups).

Key words: Morley rank, locally finite groups, simple groups, finite simple groups.

Introduction

ω-STABLE GROUPS OF FINITE MORLEY RANK

The theory of ω-stable groups and, in particular, of ω-stable groups of finite Morley rank lies on the border between group theory and model theory. The concept of ω-stability arose in model theory in the late 1960s. It is a very subtle notion and we are not in a position to discuss it here. However a modicum of model-theoretical terminology is unavoidable in any discussion of the subject.

Model theory of groups deals mostly with structural properties of groups, their subgroups and subsets which can be expressed by formulas of the first-order logic.

Let M be an algebraic structure (for example, a group or a field). A subset $X \subseteq M^n$ of a finite direct power M^n of M is called *definable* if there is a formula $\phi(x_1, \ldots, x_n, a_1, \ldots, a_m)$ in the language of M (in the case of groups, for example, this means that the only functional symbols involved in ϕ are $\cdot$ and $^{-1}$) with free variables $x_1, \ldots, x_n$ and parameters $a_1, \ldots, a_m \in M$ such that

$$X = \{(b_1, \ldots, b_n) \in M^n \mid \phi(b_1, \ldots, b_n, a_1, \ldots, a_m)\}.$$

For example, if G is a group and $a \in G$, then the centralizer $C_G(a)$ and the conjugacy class a^G are definable in G (using the parameter a):

$$C_G(a) = \{g \in G \mid ga = ag\},$$

$$a^G = \{g \in G \mid \exists y\, (y^{-1}ay = g)\}.$$

The corresponding formulas here are $xa = ax$ and $\exists y\, (y^{-1}ay = x)$.

B. Hartley et al. (eds.), Finite and Locally Finite Groups, 247–284.
© *1995 Kluwer Academic Publishers. Printed in the Netherlands.*

In an ω-stable structure M every definable set can be assigned an ordinal which is called its *Morley rank*, and we say that M has *finite Morley rank* if the ranks of all definable sets are finite. In the case of algebraic groups over algebraically closed fields this rank coincides with the algebraic dimension. Thus groups of finite Morley rank include algebraic groups over algebraically closed fields, and in general the properties of this rank function strongly resemble properties of dimension in algebraic geometry. A result of Angus Macintyre states that an infinite field has finite Morley rank if and only if it is algebraically closed. There are also interesting examples of structures of infinite Morley rank occurring in algebra, notably the *differentially closed fields* of characteristic zero.

One of the main problems in the theory of ω-stable groups is the classification of simple groups of finite Morley rank. About fifteen years ago Gregory Cherlin and Boris Zil'ber conjectured a possible solution to this problem.

Main Conjecture (G. Cherlin, B. Zil'ber) Simple infinite groups of finite Morley rank are algebraic groups over algebraically closed fields.

This is certainly the central and the most important problem of the theory of ω-stable groups. The present paper is devoted to a search for possible approaches to its resolution.

$\aleph_1$-CATEGORICAL GROUPS

If we restrict our attention to *simple* groups of finite Morley rank, then this class of groups can be given a relatively simple characterization in terms of another model-theoretical concept, categoricity.

Let $\mathrm{Th}(G)$ denote the set of all sentences of the first-order language which are true in G. Two groups G and H are called *elementarily equivalent* if $\mathrm{Th}(G) = \mathrm{Th}(H)$. For example, since the property of *commutativity* is expressed by the formula

$$\forall x \, \forall y \, (xy = yx),$$

divisibility by the infinite sequence of formulas

$$\forall x \, \exists y \, (y^2 = x), \ \forall x \, \exists y \, (y^3 = x), \ \ldots, \ \forall x \, \exists y \, (y^n = x), \ \ldots,$$

and *torsion-freeness* by the sequence of formulas

$$\forall x \, (x^2 = 1 \Rightarrow x = 1), \ \forall x \, (x^3 = 1 \Rightarrow x = 1), \ \ldots, \ \forall x \, (x^n = 1 \Rightarrow x = 1), \ \ldots,$$

it follows that any group elementarily equivalent to a torsion-free divisible abelian group is itself torsion-free, divisible, and abelian; and, conversely, one can show that any two nontrivial torsion-free divisible abelian groups are indeed elementarily equivalent.

It is known from model theory that for every infinite group G there is a group $\tilde{G}$ elementarily equivalent to G and of cardinality $\aleph_1$. A group G is called $\aleph_1$-*categorical* if $\tilde{G}$ is unique (up to isomorphism).

Since torsion-free divisible abelian groups of cardinality $\aleph_1$ are vector spaces over $\mathbb{Q}$ of dimension $\aleph_1$ and so are isomorphic, nontrivial torsion-free divisible abelian groups are $\aleph_1$-categorical. In particular, the additive group $\mathbb{Q}^+$ of rational numbers is $\aleph_1$-categorical.

Algebraically closed fields are also $\aleph_1$-categorical: in each characteristic there is exactly one algebraically closed field of cardinality $\aleph_1$, up to isomorphism. It can be deduced from this that simple algebraic groups over algebraically closed fields are $\aleph_1$-categorical.

It follows from deep model-theoretic results by John Baldwin and Boris Zil'ber that the classes of simple groups of finite Morley rank and simple $\aleph_1$-categorical groups coincide. So the Cherlin-Zil'ber conjecture can be understood as a conjecture about simple $\aleph_1$-categorical groups (indeed Zil'ber originally stated his conjecture in this form). Unfortunately the class of $\aleph_1$-categorical groups is not closed under passage to definable subgroups. On the other hand a definable subgroup of a group of finite Morley rank is again a group of finite Morley rank. For this reason it is much more convenient to work in the class of groups of finite Morley rank than in the narrower class of $\aleph_1$-categorical groups.

Instead of developing the theory of ω-stability we prefer to use an axiomatic description of groups of finite Morley rank in purely algebraic terms; the reader can find it in the next section. Meanwhile an analogy with simple algebraic groups is sufficient for the understanding of most of the present article, and one may refer to the book [24] for all necessary technical details. The reader who wishes to learn more about the model theoretical background of the theory may consult Bruno Poizat's book [55].

SIMPLE LOCALLY FINITE GROUPS OF FINITE MORLEY RANK

Our main source of inspiration for dealing with the Zil'ber-Cherlin Conjecture is the classification of simple locally finite groups of finite Morley rank. This was achieved by Simon Thomas [59] modulo the heavy use of the classification of finite simple groups. But, what is much more promising, in the narrower class of locally finite simple groups of finite Morley rank having Černikov Sylow 2-subgroups the same result can be proved *without* the use of the classification of finite simple groups. The proof of this fact can be obtained by an easy modification of the arguments from [18, 21] where simple periodic linear groups of odd characteristic were classified with the help of ideas and constructions borrowed from the classification theory of finite simple groups (but without relying on the statement of the classification theorem for finite simple groups!). It happens, quite surprisingly, that the methods of finite group theory work successfully in the domain of simple locally finite groups of finite Morley rank. One may hope that the same methods will also be fruitful in the theory of groups of finite Morley rank in general.

The aim of the present paper is to prove the Cherlin-Zil'ber conjecture for locally finite groups of finite Morley rank with Černikov Sylow 2-subgroups (Theorem 7.1, Section 7) and to show some similarities with, as well as differences from, the general case of arbitrary simple groups of finite Morley rank (Section 6). The main result of the paper, Theorem 6.19, is stated and proven in Section 6. For intermediate results we quite often give only sketches of the proofs or no proofs at all. The complete proofs will be published elsewhere. At the same time we give the complete proof of the most important technical result, Theorem 6.15. This theorem is due to John

Walter [60]. Its proof, originally given by Walter for the case of finite groups, is actually valid for groups of finite Morley rank as well, with slight modifications. This proof is given in Section 8, and it provides confirmation of one of our main theses: large portions of the theory of groups of finite Morley rank are available in work on finite group theory and need only be extracted from the appropriate papers on the classification of finite simple groups.

Acknowledgements

I would like to thank Tuna Altınel for many valuable comments. My special thanks are due to Gregory Cherlin for many stimulating conversations during my visit to Rutgers University in September 1993 and at Oberwolfach in January 1994. He also carefully read the manuscript and suggested many important corrections.

Many of the results in this paper originated in my work of 1980–82 on periodic linear groups. My warmest thanks are long due to Vladimir Nikanorovich Remeslennikov who in 1982 drew my attention to Gregory Cherlin's paper [30] on groups of finite Morley rank and conjectured that some ideas from my work could be used in this then new area of algebra. A year later Simon Thomas sent to me the manuscripts of his work on locally finite groups of finite Morley rank. Besides many interesting results and observations his manuscripts contained also an exposition of Boris Zil'ber's fundamental results on $\aleph_1$-categorical structures which were made known to many western model theorists in Wilfrid Hodges' translation of Zil'ber's paper [63] but which, because of the regrettably restricted form of publication of the Russian original, remained unknown to me. Later I had many happy opportunities to discuss model theory with Thomas, Hodges and Zil'ber; I am very grateful to them all.

1. Axioms of Morley Rank

Since we discuss only purely group-theoretical aspects of the theory of groups of finite Morley rank and ignore all model-theoretical connections and background, we will find it convenient to rely on an axiomatic approach to the theory, as developed in [24].

1.1. UNIVERSE

A *universe* is a collection of sets $\mathcal{U}$ that satisfies certain properties that we will soon list. We will refer to the elements of $\mathcal{U}$ as the $\mathcal{U}$-*definable*, or just *definable*, sets. A function will be called *definable* (in $\mathcal{U}$) if and only if its graph is. A universe is required to satisfy the following axioms:

AXIOM 1 (*Closure under Boolean operations*). If A and B are definable sets, then the sets $A \cap B$, $A \cup B$ and $A \smallsetminus B$ are also definable.

AXIOM 2 (*Closure under products*). If A and B are definable sets, then their Cartesian product $A \times B$ and the canonical projections

$$\pi_1 : A \times B \longrightarrow A, \quad \pi_2 : A \times B \longrightarrow B$$

are also definable. If $A = B$, then the diagonal

$$\Delta = \{(a, a) \,|\, a \in A\} \subset A \times A$$

is also definable. We assume also that if C is a definable subset in $A \times B$ then the images $\pi_1(C)$, $\pi_2(C)$ of C under the canonical projections are definable.

AXIOM 3 (*Finite subsets*). If A is definable and $a \in A$, then the singleton set $\{a\}$ is definable.

AXIOM 4 (*Factorization*). If $E(x, y)$ is a definable equivalence relation on a definable set A (thus in particular E is a definable subset of A^2), then the quotient A/E is encoded in the universe as follows: there is a set $\bar{A} \in \mathcal{U}$ and a surjective definable function $f : A \longrightarrow \bar{A}$ such that, for $x, y \in A$, $f(x) = f(y)$ iff $E(x, y)$ holds.

It follows from Axiom 1 that the union and the intersection of finitely many definable sets are also definable. Axioms 1 and 3 imply that the empty set $\emptyset$ and all the finite and cofinite subsets of a definable set are definable. By Axiom 2 and the definition, the image and the preimage of a definable function are definable sets.

The universe that model theorists are mainly concerned with arises in the following way. If $\mathcal{M}$ is an algebraic structure (say, a group or a field), then we say that a set X is *interpretable* in $\mathcal{M}$ if X is the quotient of a definable set modulo a definable equivalence relation. One can show that the set consisting of sets interpretable in $\mathcal{M}$ is a universe in the sense of our definition. We denote this universe by $\mathcal{U}(\mathcal{M})$.

A special case of this construction is the *universe of constructible sets* $\mathcal{U}(K)$ over an algebraically closed field K. Indeed, Alfred Tarski proved in [58] that sets definable in an algebraically closed field K by formulas of a first-order language are *constructible*, i.e. they are boolean combinations of Zariski closed subsets of K^n, $n = 1, 2, \ldots$. The validity of Axioms 1–3 for them is obvious. Bruno Poizat proved in [54] that the quotients of a constructible set with respect to constructible equivalence relations (as in Axiom 4) yield constructible sets.

Given a set of sets, there is a minimal universe that contains these sets. If $\mathcal{M}$ is an algebraic structure, the minimal universe that contains the set $\mathcal{M}$ and the graphs of the functions and the relations from the signature of $\mathcal{M}$ is just $\mathcal{U}(\mathcal{M})$.

We also say that an algebraic structure is *definable* or (as is the same) *interpretable* in the universe $\mathcal{U}$ if its underlying set, operations and predicates belong to $\mathcal{U}$. If a structure $\mathcal{N}$ is interpretable in $\mathcal{U}(\mathcal{M})$, then we say that $\mathcal{N}$ is *interpretable* in $\mathcal{M}$ and $\mathcal{M}$ *interprets* $\mathcal{N}$.

1.2. RANK

Our next concept is that of a *ranked universe*. In a ranked universe, each definable set is assigned a natural number and these natural numbers will behave like the 'dimension' of the sets they are attached to. Our axioms are such that if we attach to the constructible sets of the universe $\mathcal{U}(K)$ of an algebraically closed field K their Zariski dimensions then $\mathcal{U}(K)$ will become a ranked universe.

Let $\mathcal{U}$ be a universe. A function

$$\mathrm{rk} : \mathcal{U} \smallsetminus \{\emptyset\} \longrightarrow \mathbb{N}$$

is called a *rank* if the following axioms A–D are satisfied for all $A, B \in \mathcal{U}$.

AXIOM A (*Monotonicity of rank*). $\mathrm{rk}(A) \geqslant n + 1$ if and only if there are infinitely many pairwise disjoint, non-empty, definable subsets of A each of which has rank at least n.

AXIOM B (*Definability of rank*). If f is a definable function from A into B, then, for each integer n, the set $\{b \in B \mid \mathrm{rk}(f^{-1}(b)) = n\}$ is definable.

AXIOM C (*Additivity of rank*). If f is a definable function from A onto B and if, for all $b \in B$, $\mathrm{rk}(f^{-1}(b)) = n$ then $\mathrm{rk}(A) = \mathrm{rk}(B) + n$.

AXIOM D (*Model theorists call it 'elimination of infinite quantifiers'*). For any definable function f from A into B there is an integer m such that for any b in B the preimage $f^{-1}(b)$ is infinite whenever it contains at least m elements.

We say that a universe $\mathcal{U}$ is *ranked* if there is a rank function with the above properties.

If $\mathcal{M}$ is an algebraic structure, we say that $\mathcal{M}$ is a *ranked structure* if $\mathcal{U}(\mathcal{M})$ is a ranked universe. In this case, the rank of the definable set $\mathcal{M}$ is called the *rank* of the structure $\mathcal{M}$. As already mentioned, if K is an algebraically closed field, then the constructible universe $\mathcal{U}(\mathcal{K})$ is ranked.

Now we can give our main definition: a group G is said to be *of finite Morley rank* if it is ranked.

1.3. EXAMPLES OF GROUPS OF FINITE MORLEY RANK

We give six types of examples of groups of finite Morley rank.

Finite groups. They obviously have Morley rank 0.

Algebraic groups. An algebraic group G over an algebraically closed field K is a group of finite Morley rank (and the rank of G itself coincides with the dimension of G over K).

Abelian groups of bounded exponent. It can be proven that they have finite Morley rank.

Divisible abelian groups. A divisible abelian group A is of finite Morley rank if and only if for every prime number p the group contains only finitely many quasicyclic direct factors $\mathbb{Z}(p^\infty)$. In particular, torsion-free divisible abelian groups are of finite Morley rank, as well as the quasicyclic groups $\mathbb{Z}(p^\infty)$. Moreover, Mike Prest has shown (private communication) that a finite extension of a divisible abelian group of finite Morley rank is itself a group of finite Morley rank.

Baudisch's Monster. Andreas Baudisch [10] has constructed a nilpotent group of class 2 and exponent p, p an odd prime, which is of finite Morley rank and interprets neither an infinite field nor an infinite simple group. Baudisch's group has very interesting model-theoretic properties and most probably is non-algebraic.

Direct products. It can be shown that a direct product of two groups of finite Morley rank is a group of finite Morley rank.

1.4. GROUPS WHICH *are not* OF FINITE MORLEY RANK

In order to answer some frequently asked questions, I have prepared a list of groups which *are not* of finite Morley rank:

The *infinite cyclic group* $\mathbb{Z}$.

Free groups F_n.

Residually finite groups are, in general, not of finite Morley rank. It is known [24, Exercise 7, p. 88] that if a group G of finite Morley rank is residually finite then G is abelian-by-finite and has bounded exponent.

Real orthogonal groups $\mathrm{SO}_n(\mathbb{R})$, $n \geqslant 3$, and, more generally, simple compact Lie groups.

The special linear group $\mathrm{SL}_n(K)$, $n \geqslant 2$, where K is an infinite and *not* algebraically closed field. More generally, if K is *not* algebraically closed and infinite, the group of points $G(K)$ of a simple algebraic group defined and split over K is *not* a group of finite Morley rank. It is not known whether there is a nonsplit algebraic group with finite Morley rank.

2. Basic Properties of Groups of Finite Morley Rank

As already mentioned, when working with ranked groups it is useful to think of them as algebraic groups from which the structure of an algebraic variety has been deleted, while retaining a notion of dimension for the constructible sets. Many group-theoretical constructions in a ranked group yield definable sets: intersections and products, as well as the centralizers and normalizers of definable subgroups, are definable. Also definable are the centralizers of elements. This analogy will suffice for an understanding of the subsequent text without turning to [24], which contains proofs of all the necessary auxiliary results. The following facts illustrate the analogy with algebraic groups.

From now on G stands for a group of finite Morley rank.

2.1. GENERAL STRUCTURAL PROPERTIES

Groups of finite Morley rank satisfy many group-theoretical minimality properties. The most important of these is the descending chain condition for centralizers, which was crucial for S. Thomas' classification of simple locally finite groups of finite Morley rank [59].

Many of the minimality conditions for groups of finite Morley rank follow from the following general result.

Theorem 2.1 (Macintyre [48]; see also [24, Theorem 5.2]) *A group G of finite Morley rank satisfies the descending chain condition for chains of definable subgroups.*

In particular, we have

Corollary 2.2 [24, Lemma 5.7] *A group G of finite Morley rank contains a unique minimal definable subgroup of finite index, which is called the connected component of G and is denoted by G°.*

A group G is called *connected* if $G = G^\circ$.

It follows easily from Theorem 2.1 that, if $X \subset G$ is an arbitrary subset of a group G of finite Morley rank, then the intersection of all definable subgroups containing X is definable. We call this intersection the *definable closure of X* and denote it by $d(X)$.

Theorem 2.3 (Baldwin and Saxl [9]; see also [24, Corollaries 5.17 and 5.18]) *The centralizer $C_G(X)$ of any subset X in a group of finite Morley rank is a definable subgroup. Moreover, G satisfies the ascending and descending chain conditions for centralizers.*

The following two facts are consequences of a deep result by Boris Zil'ber, the so called Zil'ber Indecomposability Theorem [64].

Theorem 2.4 (Zil'ber [64]; see also [24, Corollary 5.28]) *If $\{H_i \mid i \in I\}$ is a family of connected definable subgroups in a group G of finite Morley rank, then the subgroup $\langle H_i \mid i \in I \rangle$ generated by them is definable and connected.*

Theorem 2.5 (Zil'ber [64]; see also [24, Corollary 5.29]) *Let $H \leqslant G$ be a definable connected subgroup. Let $X \subseteq G$ be any set. Then the subgroup $[H, X]$ is definable and connected. In particular, the commutator $[M, N]$ of two definable connected subgroups is a definable subgroup.*

Corollary 2.6 (Zil'ber [64]; see also [24, Corollary 5.30]) *Let G be a connected group of finite Morley rank. Then the members G^n of the lower central series and the members $G^{(n)}$ of the derived series of a connected group G are definable and connected.*

2.2. Special Classes of Groups

We list here some known results on abelian, nilpotent and solvable groups of finite Morley rank. In the abelian case we have a complete understanding of the situation.

Theorem 2.7 (Macintyre [48]; see also [24, Theorem 6.7]) *Let G be an abelian group of finite Morley rank. Then the following hold:*

(1) $G = D \oplus B$ *where D is a divisible group and B is a subgroup of bounded exponent.*

(2) $D \cong \bigoplus_{p \text{ prime}} (\bigoplus_{I_p} \mathbb{Z}(p^\infty)) \oplus \bigoplus_I \mathbb{Q}$ *where the index sets I_p are finite.*

(3) $G = DC$, *where D and C are definable characteristic subgroups of G, D is divisible, C has bounded exponent, and their intersection is finite. The subgroup D is connected. If G is connected, C can also be taken to be connected.*

A great deal is also known about nilpotent groups of finite Morley rank.

Theorem 2.8 (Nesin [52]; see also [24, Corollary 6.12]) *Let G be a nilpotent group of finite Morley rank. Then G is a central product $D * C$ where $D = T \times N$ and*

D is definable, connected, characteristic in G and divisible,

C is definable and of bounded exponent,

T is the torsion part of D and is abelian and divisible, and

N is a torsion-free subgroup.

Furthermore T is central in G. If G is connected, then C can be chosen to be connected and characteristic.

In the case of solvable groups we also have useful structural information.

Theorem 2.9 (Zil'ber [62], Nesin [51]; see also [24, Corollary 9.9]) *Let G be a solvable connected group of finite Morley rank. Then G' is nilpotent.*

For solvable groups of finite Morley rank we also have analogues of the theorems concerning Hall and Sylow subgroups from finite group theory (Altınel, Cherlin, Corredor and Nesin [3], Borovik and Nesin [23, 25]; see also [24, Chapter 9]). In the present paper we need very little of this theory.

Theorem 2.10 (Borovik and Nesin [25]; see also [24, Theorem 9.22]) *Let G be a connected solvable group of finite Morley rank. Denote by B the product of all definable connected subgroups of G of bounded exponent. Then B is a definable normal nilpotent connected group of bounded exponent. If, in addition, G is locally finite, then G/B is an abelian divisible group.*

We cite one more result of Boris Zil'ber; it is a corollary of his Indecomposability Theorem (cf. Theorem 2.5) and shows that simplicity of a group of finite Morley rank is, in a sense, a definable property. An analogous fact for algebraic groups is well-known.

Theorem 2.11 (Zil'ber [64]; see also [24, Exercise 1, p. 92]) *Let G be a group of finite Morley rank. If G has no nontrivial proper definable normal subgroups, then G is either a simple group or a torsion-free divisible abelian group.*

2.3. FIELDS OF FINITE MORLEY RANK

Finally we cite one of the most important results of the theory. It explains why in the Cherlin-Zil'ber Conjecture we hope to identify simple groups of finite Morley rank with simple algebraic groups over *algebraically closed fields*.

Theorem 2.12 (Macintyre [49]; see also [24, Theorem 8.1]) *Infinite fields of finite Morley rank are algebraically closed.*

3. The Cherlin-Zil'ber Conjecture

3.1. MAIN CONJECTURE

As promised, in this section we begin our discussion of a possible approach to the solution of the main problem in the theory of groups of finite Morley rank: the classification of the simple groups. For the convenience of the reader we again state the Cherlin-Zil'ber Conjecture.

Main Conjecture (G. Cherlin, B. Zil'ber) *Simple infinite groups of finite Morley rank are algebraic groups over algebraically closed fields.*

We begin with three problems (Problems A–C below) which seem to provide a natural partitioning of the conjecture. In what follows G is a connected group of finite Morley rank.

3.2. LOOKING FOR A FIELD

In any attempt to identify a simple group G of finite Morley rank with an algebraic group over an algebraically closed field we need, first of all, to construct the underlying field. This can be done in *solvable non-nilpotent subgroups* of G (if G contains any) by means of the following result, due to Boris Zil'ber, which is probably one of the two most important and most frequently applied results of the subject, the other one being Macintyre's classification of fields of finite Morley rank, Theorem 2.12.

Theorem 3.1 (Zil'ber [64]; see also [24, Theorem 9.1]) *Let $A \lhd G$ be a connected group of finite Morley rank where A and $H = G/C_G(A)$ are infinite definable abelian groups and A is a minimal normal subgroup in G. Let $R = \mathbb{Z}[H]$ be the integral group ring of H. Then the following statements are true:*

(1) *The subring $K = R/\mathrm{ann}_R(A)$ of $\mathrm{End}(A)$ is a definable algebraically closed field; in fact, there is an integer l such that every element of K can be represented as the endomorphism $\sum_{i=1}^{l} h_i$ $(h_i \in H)$.*

(2) *$A \cong K^+$, H is isomorphic to a subgroup T of K^* and H acts on A by multiplication; in other words*

$$A \rtimes H \cong \left\{ \begin{pmatrix} t & a \\ 0 & 1 \end{pmatrix} \middle| t \in T, a \in K \right\}.$$

(3) *The structure $\langle K; 0, 1, +, \cdot, ^{-1}, T \rangle$, where T is (a unary predicate for) a multiplicative subgroup of K (denoted by the same letter T), is definable in G and has finite Morley rank. Also it follows from (1) that $K = T + \cdots + T$ (l times).*

It would be nice to have in the above theorem $T = K^*$. If this is not the case, the structure of finite Morley rank $\langle K; 0, 1, +, \cdot, ^{-1}, T \rangle$, where T is an infinite proper subgroup of K^*, is called a *bad field*. No bad fields are known; we will return to the discussion of bad fields later.

Zil'ber's theorem gives no information if all connected solvable subgroups of G are *nilpotent*. This misfortune cannot occur in simple algebraic groups G where Borel subgroups (i.e. maximal connected solvable closed subgroups) are known to be non-nilpotent [42, §21.4]. Also it is well-known that the Borel subgroups of an algebraic group G over an algebraically closed field are conjugate to each other (see e.g. [42, §21.3]), and the latter fact is crucial for the structure theory of simple algebraic groups.

The situation in groups of finite Morley rank is not so well understood. We can introduce the notion of a Borel subgroup of a group G of finite Morley rank in almost the same terms: $B \leqslant G$ is Borel if B is maximal among the connected definable solvable subgroups of G. However we do not know whether Borel subgroups are non-nilpotent. The proof of the conjugacy of Borels in algebraic groups does not generalize to the context of groups of finite Morley rank because it is based on the notion of a complete variety from algebraic geometry, and this does not have a model theoretic analogue.

By definition (due to Gregory Cherlin and Bruno Poizat), a *bad group* is a non-solvable group of finite Morley rank whose proper, definable and connected subgroups are nilpotent. These groups lie at the opposite extreme from the realm of algebraic groups.

The following simple result sums up our discussion and shows the place of bad groups in the theory.

Theorem 3.2 [24, Proposition 13.2] *Let G be a connected non-nilpotent group of finite Morley rank. Then either an algebraically closed field is interpretable in G or G interprets a simple bad group.*

The possible existence of bad groups is one of the main obstacles to the classification of simple groups of finite Morley rank.

Problem A (G. Cherlin) *Are there any bad groups?*

Now we need to explain the role of bad groups in the theory.

As the next result shows, the structural properties of simple bad groups are bizarre. But even more surprising is the fact that there exist groups (not of finite Morley rank) with similar algebraic properties. Indeed, Sergei V. Ivanov constructed a 2-generated group G whose maximal subgroups are infinite cyclic and conjugate to each other [46, Problem 10.49]; see also Ivanov and Ol'shanski [43].

If one calls the maximal subgroups of Ivanov's group Borel subgroups, then G shares all the properties of a simple bad group listed in the theorem below. However, Ivanov's example is not a group of finite Morley rank, because the maximal subgroups of G are the centralizers of their non-identity elements, hence are definable, while an infinite cyclic group is not a group of finite Morley rank.

The following list of properties of bad groups is due to A. Borovik and B. Poizat (see [24, Theorem 13.3]) and L.-J. Corredor [31].

Theorem 3.3 *Let G be a simple bad group. Then the following statements hold:*

(1) *The Borel subgroups of G are conjugate to each other.*

(2) *Distinct Borel subgroups of G are disjoint.*

(3) *Every element of G lies in some Borel subgroup.*

(4) *Any finite subgroup of G is nilpotent and has odd order. In particular, G has no involutions.*

(5) *$N_G(B) = B$ for any Borel subgroup $B < G$.*

(6) *If B is any Borel subgroup of G and $a \in G \smallsetminus B$, then*

$$G = (BaB)(BaB) \cdots (BaB),$$

where the number of double cosets BaB in the product is bounded by a constant which does not depend on the choice of the element a.

We would like to stress that, in the above theorem, the most important result is the one that states that G does not have an involution.

3.3. MULTIPLICATIVE SUBGROUPS OF BAD FIELDS

By definition, a *bad field* is a ranked structure of the form $\langle K, +, \cdot, 0, A \rangle$ where $\langle K, +, \cdot \rangle$ is an algebraically closed field and A is a (predicate for a) proper infinite multiplicative subgroup of K^*.

Problem B (B. Poizat) *Are there bad fields?*

If a bad field $\langle K, +, \cdot, 0, A \rangle$ exists, then the group S of matrices over K of the form

$$\begin{pmatrix} a & b \\ 0 & a^{-1} \end{pmatrix}, \ a \in A, \ b \in K,$$

has finite Morley rank and is solvable, but not algebraic. One of the worst things that may happen is that S does not necessarily contain an involution. But, if S is algebraic (i.e. $A = K^*$), S does contain an involution.

The following simple result follows immediately from Theorem 3.1.

Theorem 3.4 *If a group of finite Morley rank G does not interpret a bad field, then every connected solvable non-nilpotent subgroup of G contains an involution.*

3.4. CHARACTERIZATION OF SIMPLE ALGEBRAIC GROUPS

If Problems A and B have negative solutions, then the rest of the classification of simple groups of finite Morley rank can be subdivided into several problems which appear to be manageable.

Let us call a group G of finite Morley rank *tame* if it does not interpret a bad group or a bad field.

Problem C *Is it true that every tame infinite simple group of finite Morley rank is an algebraic group over an algebraically closed field?*

Obviously, negative answers to Problems A and B, and a positive answer to Problem C, would provide a complete proof of the Cherlin-Zil'ber Conjecture.

We will concentrate in this paper on the discussion of possible approaches to the problem of classifying tame groups. For tame groups there is an easy but fundamental analogue of the Feit-Thompson theorem on solvability of groups of odd order: a connected tame group without involutions is nilpotent!

More precisely, the following useful result is true. Recall that a *section* of a group G is a group of the form $S = H/N$ for some subgroups $N \triangleleft H \leqslant G$. We say that S is a *definable section* if H and N are definable subgroups.

If p is a prime, a $p^\perp$-*group* is a group without elements of order p.

Theorem 3.5 [24, Theorem B.1] *Any connected definable $2^\perp$-section H of a tame group G is nilpotent.*

Proof. Since definable sections of G cannot be simple bad groups, if H is not nilpotent then one can find in H a connected solvable non-nilpotent subgroup, and the result follows from Theorem 3.4. □

This simple result shows that Problems A and B play the same role in the theory of groups of finite Morley rank as Burnside's question on the solvability of finite groups of odd order in the theory of finite groups. A systematic study of finite simple groups became possible only after W. Feit and J. Thompson proved the solvability of groups of odd order in their famous 'Odd order paper' [33]. It is noteworthy that the solution to this problem came more than half a century after the problem was first posed.

The purpose of this paper is to convince the reader that Problem C is ripe enough for an attempt at a solution. We are confident that methods of finite group theory will successfully generalize to the context of groups of finite Morley rank.

The very first idea borrowed from the theory of finite groups immediately yields remarkable simplifications in all considerations concerned with Problem C. This is the idea of a *minimal counterexample*. Indeed, we can restrict ourselves to dealing with a group G which has minimal Morley rank subject to being a counterexample to Problem C. Then we can assume without loss that every proper, simple, definable and connected section of G is a simple algebraic group over some algebraically closed field. It will be convenient to use the following two technical definitions describing this situation.

A group G of finite Morley rank is called a K-*group* if every infinite, simple, definable and connected section of G is an algebraic group over an algebraically closed field. We shall also call a group G of finite Morley rank a K*-*group* if every *proper* definable subgroup of G is a K-group. So in order to solve Problem C positively it suffices to show that

> every simple tame K*-group is a simple algebraic group over an algebraically closed field.

In the language of finite group theory this means that we are taking a resolutely revisionist approach to our subject, from the beginning. We do not expect to encounter any sporadic groups, though of course we would be delighted to do so.

3.5. LOCALLY FINITE GROUPS OF FINITE MORLEY RANK

Now let us consider what becomes of Problems A, B, and C in the context of locally finite groups of finite Morley rank. First of all, Problem A has the desired negative solution: bad locally finite groups do not exist since they are destined to be locally nilpotent by Theorem 3.3(4). (This does not require the Feit-Thompson theorem on solvability of finite groups of odd order [33], although this, certainly, yields the same result.)

On the other hand, we can say nothing about bad fields interpretable in a locally finite group of finite Morley rank, or even about locally finite bad fields. Thus Problem B remains open even in the locally finite case. But we can circumvent bad fields by considering a group G which is a counterexample of minimal possible Morley rank to the statement

'*A simple locally finite group of finite Morley rank is an algebraic group over a locally finite algebraically closed field*'.

Then G is a K*-group, and we come to the following problem.

Problem D *Let G be a simple locally finite K*-group of finite Morley rank. Prove that G is a simple algebraic group over a locally finite algebraically closed field.*

Section 7 contains a solution of Problem D under the additional assumption that Sylow 2-subgroups of G are Černikov. The success of this classification is due to the three major results from finite group theory which we can apply in the context of locally finite groups:

- classification of finite simple groups of sectional rank at most 4 (Gorenstein and Harada [38]);
- classification of finite groups with a proper 2-generated core (Aschbacher [5]);
- Aschbacher's characterization of finite Chevalley groups of odd characteristic as groups with a 'classical involution' [7].

The rest of the proof is done in the framework of the theory of groups of finite Morley rank. The reader will find in Section 6 how the above-mentioned three results from finite group theory can be adapted to the theory of groups of finite Morley rank.

4. Sylow Theory

4.1. 2-SYLOW THEOREM

A *Sylow p-subgroup* of a group G is a maximal p-subgroup in G. Obviously every p-subgroup of G lies in some Sylow p-subgroup.

Sylow 2-subgroups in groups of finite Morley rank are not necessarily definable. An easy example is provided by the group $\mathrm{PSL}_2(\mathbb{C})$: here any Sylow 2-subgroup is conjugate to the group

$$\left\{ \begin{pmatrix} \lambda & 0 \\ 0 & \lambda^{-1} \end{pmatrix} \right\} \rtimes \left\langle \begin{pmatrix} 0 & -1 \\ 1 & 0 \end{pmatrix} \right\rangle,$$

consisting of monomial matrices where $\lambda^{2^n} = 1$ for some $n \in \mathbb{N}$. Moreover, an analogous fact is valid for any algebraic group G over an algebraically closed field of characteristic not 2: a Sylow 2-subgroup P in G is not definable and contains a Prüfer 2-subgroup T of finite index. (A *Prüfer p-group* is a product of a finite number of copies of the quasicyclic group $\mathbb{Z}(p^\infty)$; the number of copies is called its *Prüfer rank*.) Recall that a p-group is called a *Černikov group* if it is a finite extension of a Prüfer p-group. So P is a Černikov 2-group. Furthermore, $d(T) = H$ is a maximal torus in G and $d(P) = HP$ is a Sylow 2-subgroup of $N = N_G(H)$. Moreover, it can be shown that P is not nilpotent, though it is solvable and abelian-by-finite. The exponent of P is infinite and all elementary abelian subgroups of P are finite and generated by no more than (Prüfer rank of T) + (the order of P/T) elements.

On the other hand, in the case of characteristic 2, the Sylow 2-subgroups in semisimple algebraic groups are precisely the maximal unipotent subgroups. Therefore they are definable, nilpotent, connected, have bounded exponent, and contain infinite elementary abelian 2-subgroups.

In order to state the Sylow theory for groups of finite Morley rank we need one more technical definition. If X is an arbitrary subgroup of a group G of finite Morley rank we set $X° = X \cap d(X)°$, which we shall call the *connected component* of X. Note that $X°$ is a subgroup of finite index in X.

Theorem 4.1 (Borovik and Poizat [26]; see also [24, Theorem 6.21, Corollary 6.22 and Theorem 10.11]) *All Sylow 2-subgroups in a group G of finite Morley rank are conjugate. If S is one of them, then $S° = B * D$ is a central product of a definable connected nilpotent subgroup B of bounded exponent and a divisible abelian group D. The subgroups B and D are uniquely determined and characteristic in S. In particular, S is nilpotent-by-finite.*

We shall call B the *bounded* or *unipotent part* of S, and D the *maximal 2-torus* of S. More generally, any divisible abelian 2-subgroup of G is called a *2-torus*. It can be shown that 2-tori in groups of finite Morley rank are Prüfer groups.

The following result is an easy consequence of Theorem 3.3 on the structure of bad groups.

Theorem 4.2 [24, Theorem B.3] *Let G be a simple tame group. Then the Sylow 2-subgroups in G are infinite.*

4.2. Groups of Odd and Even Type

First of all, we have to find some *a priori* way of separating algebraic groups of characteristic 2 from those of characteristic not 2, since the applicable methods will be different in these two cases.

The largest and hardest part of the proof of the following result was done by Tuna Altınel.

Theorem 4.3 (Altınel, Borovik and Cherlin [2]) *Let S be a Sylow 2-subgroup of a simple tame K*-group G and $S° = T * B$, where T is the maximal 2-torus and B is the bounded part of S. Then either $T = 1$ or $B = 1$.*

In the two cases of Theorem 4.3, we say that G is of *even type* if $T = 1$ and of *odd type* if $B = 1$. So we arrive at the most important bifurcation of our theory:

every simple tame group of finite Morley rank is either of odd or of even type.

An analogy with finite group theory suggests that these two classes of groups should be studied by radically different methods. In the present paper we restrict our attention (in part because of limitations of space) to the better understood class of groups of odd type.

4.3. Versions of 2-Rank

If E is a finite elementary abelian 2-group, its *2-rank* $m(E)$ is the minimal number of generators. If H is a not necessarily definable subgroup of a group G of finite Morley rank and odd type, its *2-rank* $m(H)$ is defined as the maximum of the 2-ranks of elementary abelian subgroups in H. We also introduce the *sectional 2-rank* of H

as the maximum of the 2-ranks of sections of 2-subgroups S in H, which we shall denote by $r(H)$. Also, the *Prüfer 2-rank* $pr(H)$ of H is the maximum of the Prüfer ranks of Prüfer 2-subgroups of H. If H is a definable subgroup of G and S is a Sylow 2-subgroup of H, then the *normal 2-rank* $n(H)$ is the maximum of the 2-ranks of normal elementary abelian subgroups in S.

Obviously

$$pr(G) \leqslant n(G) \leqslant m(G) \leqslant r(G).$$

Notice also that, if G is a simple algebraic group over an algebraically closed field of characteristic not 2, then, in fact, $pr(G) = n(G)$.

The following result will be very useful in the study of simple locally finite groups of finite Morley rank. It was proven by Anne MacWilliams for finite 2-groups, and the result for Černikov 2-groups follows easily from the finite case.

Theorem 4.4 (MacWilliams [50]) *Let S be a Černikov 2-group. If $n(S) \leqslant 2$, then $r(S) \leqslant 4$.*

Černikov 2-groups of normal 2-rank at least 3 have the following very important (but easily proved) property. Again it is an immediate generalization of a well-known result from finite group theory.

Lemma 4.5 *If S is a Černikov 2-group with $n(S) \geqslant 3$, then for any two four-subgroups (i.e. elementary abelian subgroups of order 4) $U, V \leqslant S$ there is a sequence of four-subgroups*

$$U = V_1, V_2, \ldots, V_n = V,$$

such that $[V_i, V_{i+1}] = 1$ for all $i = 1, 2, \ldots, n - 1$.

Finite 2-groups with this property are called, in finite group theory, *connected*. Since in our theory the term 'connected' is already reserved, we shall call Černikov 2-groups with the same property 2-*connected*.

4.4. ACTION OF INVOLUTIONS

We shall frequently use the following observation.

Theorem 4.6 *Let H be a solvable group of finite Morley rank with Černikov Sylow 2-subgroups. If V is a four-subgroup in H, then*

$$H^\circ = \langle C_{H^\circ}(v) \mid v \in V^\# \rangle.$$

Here $V^\#$ denotes the set of all non-identity elements of the group V.

Notice the following two special cases of the theorem.

(*) *If T is a 2-divisible abelian group and V a four-group acting on T, then*

$$T = \langle C_T(v) \mid v \in V^\# \rangle.$$

(**) *If Q is a solvable $2^{\perp}$-group of finite Morley rank and V is a four-group acting definably on Q, then*

$$Q = \langle C_Q(v) \mid v \in V^{\#} \rangle.$$

The latter has a well-known analogue in finite group theory (see also [24, Exercise 4, p. 175]), and the former shows that many important properties of finite groups of odd order are shared by solvable 2-divisible groups. Theorem 4.6 is an easy combination of these two special cases. The reader can find a discussion of the relevant properties of solvable groups in [24, Chapter 9], especially in Sections 9.7 and 9.8 of the book.

4.5. 2-GENERATED CORE

Let S be a Sylow 2-subgroup in a group G of finite Morley rank. We define the 2-*generated core* $\Gamma_{S,2}(G)$ as the definable closure of the group generated by all normalizers $N_G(U)$ of all subgroups $U \leqslant S$ with $\mathrm{m}(U) \geqslant 2$. Finite simple groups with a proper 2-generated core $\Gamma_{S,2}(G) < G$ are known, by Aschbacher [5]. As the reader will see from the sequel, in the finite Morley rank context an analogous result plays an even more important role. Fortunately, we do not expect the existence of proper 2-generated cores in simple infinite groups of finite Morley rank.

Problem 1 *Prove that if S is a Sylow 2-subgroup in a simple tame K*-group G then*

$$\Gamma_{S,2}(G) = G.$$

5. Structure of Subgroups in a Minimal Counterexample

5.1. GENERALIZED FITTING SUBGROUP

A group G is called *semisimple* if $G = G'$ and $G/Z(G)$ is completely reducible (i.e. $G/Z(G)$ is a direct sum of finitely many simple non-abelian subgroups). Also, G is called *quasi-simple* if $G = G'$ and $G/Z(G)$ is non-abelian simple. Thus quasi-simple groups are semisimple. Note that the only proper normal subgroups of a quasi-simple group are the central ones. Notice that quasi-simple *algebraic groups* are traditionally called *simple algebraic groups*.

The next result introduces the notion of a *component* which will be crucial in our analysis of the structure of the centralizers of involutions in groups of finite Morley rank.

Theorem 5.1 (Belegradek [11]; see also [24, Lemmas 7.6 and 7.10]) *In a group G of finite Morley rank, every quasi-simple subnormal subgroup is definable and there are only finitely many of them. We shall call them the* components *of G. The product $L(G)$ of all the components of G is a definable normal subgroup in G and every component of G is a normal subgroup of $L(G)$.*

The subgroup $L(G)$ is called the *layer* of G. We also write $E(G) = L(G)^{\circ}$. We list for convenience some further properties of the layer:

Theorem 5.2 [24, Lemma 7.10]

(1) *If $H \lhd G$ is a definable subgroup, then $L(H) \lhd G$ and $L(H)$ is definable.*

(2) *$L(G)$ is the maximal, semi-simple, normal (also subnormal) subgroup of G.*

(3) *If G is connected, then any quasi-simple subnormal subgroup of G is normal in G and connected.*

(4) *$E(G) = L(G^\circ)$ and all the components of $E(G)$ are connected.*

Now let G be any group. Let $R(G)$ be the subgroup generated by all the normal solvable subgroups of G and let $F(G)$ be the subgroup generated by all the normal nilpotent subgroups. Clearly the subgroups $R(G)$ and $F(G)$ are characteristic subgroups. They are called the *radical* and the *Fitting subgroup* of G.

Theorem 5.3 (Belegradek [11] and Nesin [53]; see also [24, Theorem 7.3]) *Let G be a group of finite Morley rank. Then $R(G)$ and $F(G)$ are definable subgroups and they are solvable and nilpotent respectively.*

We will now define the *generalized Fitting subgroup* $F^*(G)$ of a group G of finite Morley rank as

$$F^*(G) = F(G)L(G).$$

Clearly $F^*(G)$ is a definable characteristic subgroup of G. This notion exists already in finite group theory and was introduced by Bender [14]. Our definition is exactly the same as in the finite group case.

The following statement is the most important property of the generalized Fitting subgroup.

Theorem 5.4 (Nesin [24, Theorem 7.13]) *For any group G of finite Morley rank,*

$$C_G(F^*(G)) \leqslant F^*(G).$$

5.2. ALMOST SIMPLE AND SEMISIMPLE K-GROUPS

K-groups of finite Morley rank behave nicely (much better than the groups known as K-groups in finite group theory). One of the reasons for this more orderly behavior is the simple but very important observation that we do not have in our theory definable groups of field automorphisms. More precisely, the following is true.

Theorem 5.5 [24, Theorem 8.3] *Let $\mathcal{A} = \langle K, A \rangle$ be a structure consisting of an algebraically closed field K and a group A of automorphisms of K. If $\mathcal{A}$ is definable in a ranked universe, then $A = 1$.*

A group G of finite Morley rank is called *almost simple* if $F^*(G)$ is simple.

It is well-known that every automorphism of a simple algebraic group G over an algebraically closed field K is the product of an inner automorphism, a graph automorphism and a field automorphism. The product of the first two is a rational automorphism, hence it is definable in G. So by Theorem 5.5 one has the following description of definable groups of automorphisms of simple algebraic groups.

Theorem 5.6 [24, Theorem 8.4] *Let H be an almost simple group of finite Morley rank with $L = F^*(H)$ a simple algebraic group over an algebraically closed field K. Then $L = H^\circ$ and H is interpretably isomorphic to an extension of H by a finite group of rational automorphisms of H induced by symmetries of the Dynkin diagram for H.*

Corollary 5.7 *Assume that G is a K-group. If $F^*(G)^\circ$ is semisimple, then $E(G) = G^\circ$.*

Let H be a group of finite Morley rank. We shall denote by $O_\infty(H)$ the maximal definable normal torsion-free subgroup of H and by $O_p(H)$, p a prime, the maximal normal p-subgroup in H. Notice that $O_p(H)$, p a prime number, is not necessarily definable. Also we denote by $O(H)$ the maximal normal definable connected $2^\perp$-subgroup of H. This exists because it may be shown that the product of two normal definable $p^\perp$-subgroups is a $p^\perp$-group itself.

Theorem 5.8 *If H is a K-group, then the subgroups $O_p(H)$, p a prime, are nilpotent-by-finite. If, in addition, H does not interpret a bad field, then the subgroups $O_\infty(H)$ and $O(H)$ are nilpotent.*

Theorem 5.9 *Let H be a connected K-group. Assume, in addition, that either H is locally finite or does not interpret a bad field. Then $H/F(H) = T * L$ is the central product of a divisible abelian group T and a semisimple group L whose simple components are simple algebraic groups over algebraically closed fields.*

The proof of the above result involves the following remark by Tuna Altınel on central extensions of simple algebraic groups.

Theorem 5.10 (Altınel [1]; see also [2]) *Let G be a quasisimple group of finite Morley rank with $G/Z(G)$ isomorphic to a simple algebraic group over an algebraically closed field. Assume that G does not interpret a bad field. Then G is algebraic.*

This result uses the theory of perfect central extensions of simple algebraic groups. The 'no bad fields' assumption is used to eliminate the need to deal with the K-theory. It would be of interest to redo this portion without using the 'no bad fields' hypothesis, though this is irrelevant in the tame groups context. An analogue of Theorem 5.10 for locally finite groups of finite Morley rank follows easily from the theory of the Schur multiplier for finite Chevalley groups (Steinberg [57]).

Theorem 5.11 *Let G be a connected K-group of odd type and write $\overline{G} = G/O(G)$. Then $\overline{G} = F^*(\overline{G})$ and $F(\overline{G})$ is an abelian group.*

Theorem 5.12 *Let G be a connected locally finite K-group of odd type. Denote by $B(G)$ the product of all connected definable nilpotent $2'$-subgroups of bounded exponent in $R(G)$. Then $B(G)$ is a normal connected definable nilpotent $2'$-subgroup. If we write $\overline{G} = G/B(G)$ then $\overline{G} = F^*(\overline{G})$ and $F(\overline{G})$ is an abelian group.*

5.3. BALANCE

In finite group theory 'balance' is a very important concept used in the theory of so-called groups of component type (cf. Walter [60]). I do not want to discuss it here in any detail because in the context of groups of finite Morley rank it boils down to a simple statement about K-groups (which is an easy corollary of the reductivity of involution centralizers in simple algebraic groups of characteristic not 2).

Theorem 5.13 *Let G be a K-group and let $t \in G$ be an involution. Then we have $O(C_G(t)) \leqslant O(G)$ and $B(O(C_G(t))) \leqslant B(O(G))$.*

5.4. GENERATION OF K-GROUPS

The following fact is one of the observations which dramatically simplify proofs in the theory of groups of finite Morley rank in comparison with proofs in finite group theory. Recall that a *four-group* is an elementary abelian group of order 4.

Theorem 5.14 *Let H be a K-group of finite Morley rank of odd type. If V is a four-subgroup of H, then*

$$H^\circ = \langle C_{H^\circ}(v)^\circ \mid v \in V^\# \rangle.$$

We shall use in the sequel some useful notation: if G is a group of finite Morley rank, V is a four-subgroup and H is a definable subgroup of G, we write

$$\Gamma_V(H) = \langle C_{H^\circ}(v)^\circ \mid v \in V^\# \rangle.$$

In this notation the conclusion of Theorem 5.14 may be stated as

$$H^\circ = \Gamma_V(H).$$

Proof. We can assume without loss that H is a counterexample to the theorem of least possible Morley rank. Thus, if $F < H^\circ$ is a proper definable connected subgroup and W is a four-subgroup of H normalizing F, then $\Gamma_W(F) = F$.

First consider the case when $L = F^*(H)$ is a direct product of simple algebraic groups over algebraically closed fields and $Z(L) = 1$. (Recall that 'a simple algebraic group' means actually 'a quasisimple algebraic group'; it is well-known that simple algebraic groups have finite centers (cf. [42] or [56]).) If the Prüfer 2-rank of L is at least 3, then $n(H) \geqslant 3$ and Sylow 2-subgroups of H are 2-connected by virtue of Lemma 4.5. If V and U are two commuting four-subgroups, then for $u \in U^\#$ we have

$$C_L(u)^\circ = \Gamma_V(C_L(u)^\circ) \leqslant \Gamma_V(L)$$

and therefore

$$\Gamma_U(L) = \langle C_L(u)^\circ \mid u \in U^\# \rangle \leqslant \Gamma_V(L).$$

Analogously $\Gamma_V(L) \leqslant \Gamma_U(L)$ and thus $\Gamma_V(L) = \Gamma_U(L)$. Therefore in order to prove the theorem in this case it is enough to find a four-subgroup $U \leqslant L$ such that $\Gamma_U(L) = L$. If $L = L_1 \times \cdots \times L_n$ is the decomposition of L into the product of simple algebraic groups, take maximal tori $T_i < L_i$, $i = 1, 2, \ldots, n$, and form

$T = T_1 \times \cdots \times T_n$. Then T contains the maximal divisible subgroup of a Sylow 2-subgroup in H. Now we can take any four-subgroup $U < T$. Then if B is any Borel subgroup of L (i.e. the product of Borel subgroups $B_i < L_i$) containing T, we have $\Gamma_U(B) = B$ by Theorem 4.6. Since each L_i is generated by the Borel subgroups containing T_i, we have $L = \Gamma_U(L)$.

So we can assume without loss that L has Prüfer 2-rank at most 2 and thus H is one of the groups $\mathrm{PSL}_2(K)$, $\mathrm{PSL}_2(K_1) \times \mathrm{PSL}_2(K_2)$, $\mathrm{PSL}_3(K)$, $\mathrm{PSp}_4(K)$, $\mathrm{G}_2(K)$ for algebraically closed fields K, K_1, K_2 of characteristics not 2. The groups $\mathrm{PSL}_2(K)$, $\mathrm{PSp}_4(K)$ and $\mathrm{G}_2(K)$ have no rational outer automorphisms [57]. Also, any involutory rational outer automorphism of the group $L \cong \mathrm{PSL}_3(K)$ is conjugate by an element of L to the inverse-transpose automorphism. Finally, it is easy to see that, if $L \cong \mathrm{PSL}_2(K_1) \times \mathrm{PSL}_2(K_2)$ and $V \not< L$, then $K_1 \cong K_2$ and an element of V swaps the two copies of $\mathrm{PSL}_2(K_1)$ in the direct product. In all cases $V \cap L \neq 1$.

At this point it would be convenient for us to exclude from consideration the case when $L \cong \mathrm{PSL}_2(K_1) \times \mathrm{PSL}_2(K_2)$ and $V < L$. Then V normalizes the components $L_1 \cong \mathrm{PSL}_2(K_1)$ and $L_2 \cong \mathrm{PSL}_2(K_2)$, whereupon the problem reduces to proving that $\Gamma_V(L_i) = L_i$, $i = 1, 2$. Thus we can exclude this case without loss of generality.

Now in all the remaining cases LV is an algebraic group. Every involution of L lies in (and hence centralizes) a maximal torus, and therefore we can assume without loss of generality that $\Gamma_V(L)$ contains a maximal torus from L. Moreover, $\Gamma_V(L)$, being generated by the connected Zariski closed subgroups $C_L(v)^\circ$, is itself a Zariski closed subgroup of L. Thus $\Gamma_V(L)$ is a 'group of maximal rank' in the sense of algebraic group theory. These groups are known to be 'subsystem subgroups of G' [56, Proposition 3.1], and, combining the description of the subsystem subgroups in L with the information on the centralizers of involutions in groups of type PSL_2, PSL_3, PSp_4, G_2 (see, for example, Wong [61] for a description of the involution centralizers in $\mathrm{PSp}_4(K)$, Gorenstein and Harada [37] for $\mathrm{G}_2(K)$, Iwahori [44] for a general discussion of the structure of involution centralizers in simple algebraic groups of characteristic not 2), one can easily show that $L = \Gamma_V(L)$.

Now we want to lift one restriction in the above consideration: instead of assuming that $Z(L) = 1$ we now suppose that $Z = Z(L)$ is a finite group. Write $\overline{H} = H/Z$ and use bars for images. It is easy to see that for an involution $v \in V^\#$ one has

$$\overline{C_H(v)} = C_{\overline{H}}(\overline{v}),$$

and therefore

$$\overline{\Gamma_V(H)} = \Gamma_{\overline{V}}(\overline{H}) = \overline{L} = \overline{H}^\circ,$$

thus proving the theorem in this, slightly more general, case as well.

Before considering the general case we prove the following intermediate result.

Let H be a group of finite Morley rank and suppose that $L = F^*(H)$ is a central product of simple algebraic groups over algebraically closed fields of characteristic not 2. If V is a four-subgroup of L, then L is generated by connected solvable V-invariant subgroups.

Indeed, since we know that

$$L = \langle C_L(v)^\circ \mid v \in V^\# \rangle,$$

it is enough to prove that for any two involutions $v, w \in V^{\#}$ the group $C_L(v)$ is generated by w-invariant connected solvable subgroups.

Let $L = L_1 * \cdots * L_n$ where the L_i are simple algebraic groups, $i = 1, 2, \ldots, n$. If the involution v normalizes L_i, then v induces on L_i a rational automorphism (Theorem 5.6); moreover, v is semisimple, so $C_{L_i}(v)$ is a reductive algebraic group. If v does not normalize L_i, then $L_i^v L_i$ is v-invariant and $C_{L_i^v L_i}(v)$ is a homomorphic image of L_i. It follows from these observations that $C_L(v)$ is a central product of reductive algebraic groups. Now the desired statement follows easily from the following property of algebraic groups.

Let M be a reductive algebraic group over an algebraically closed field of characteristic not 2 and w an involutory rational automorphism of M. Then M is generated by w-invariant connected solvable subgroups.

This statement is easy to prove. All we need is to show that M is generated by proper w-invariant Zariski closed subgroups, because after reaching this conclusion we can use induction on the dimension of the group M. We may also assume without loss that $Z(M) = 1$. But it can be easily checked that in this case w centralizes a four-subgroup U in M. By the previous discussion $M = \langle C_M(u)^\circ \,|\, u \in U^{\#} \rangle$ is the required generation by proper w-invariant Zariski closed subgroups.

Now we can return to the proof of the theorem and consider the general case. Let $Q = S(H)^\circ$ be the connected component of the solvable radical of H and $\overline{H} = H/Q$. As usual, we use bars to denote images in $\overline{H}$.

Notice that, if $Q = H^\circ$, then the result follows from Theorem 4.6. So we can assume without loss of generality that $\overline{L} = F^*(\overline{H})$ is the central product of simple algebraic groups over algebraically closed fields. We already know that $\overline{L}$ is generated by connected solvable $\overline{V}$-invariant definable subgroups. But then H° (which is the full preimage of $\overline{L}$ in H) is generated by V-invariant connected solvable definable V-invariant subgroups. If S is one of them, then $S = \Gamma_V(S) \leqslant \Gamma_V(H)$. This proves $H^\circ = \Gamma_V(H)$. $\square$

6. Groups of Odd Type

In this section we start a systematic study of groups of odd type. First of all we restate Problem C for the class of groups of odd type.

Problem E *Is it true that all simple tame K^*-groups of odd type are algebraic groups over algebraically closed fields of characteristic not 2?*

However, keeping in mind applications to locally finite groups of finite Morley rank, we shall prove some results for a certain subclass of the class of K^*-groups which is less restrictive than the class of tame K^*-groups. We shall call a group of finite Morley rank a K^{***}-*group* if proper definable quasisimple sections of G are simple algebraic groups over algebraically closed fields. Notice that, by virtue of Theorem 5.10 and remarks after it, the two classes of groups we are specially interested in, tame K^*-groups and locally finite K^*-groups, are K^{***}-groups.

Notice also that we do not include in the concept of a K***-group the assumption that G is a tame group (otherwise our results on K***-groups would not cover the case of locally finite groups).

It has already been mentioned that in a group G of odd type elementary abelian 2-subgroups of G are finite and have uniformly bounded orders. So the ranks $\mathrm{pr}(G)$, $\mathrm{n}(G)$, $\mathrm{m}(G)$, and $\mathrm{r}(G)$ are well defined.

6.1. CENTRALIZERS OF INVOLUTIONS IN GROUPS OF ODD TYPE

Since a Černikov 2-group has only finitely many classes of conjugate involutions, and since Sylow 2-subgroups in a group of odd type are Černikov and conjugate, we immediately have the following useful observation.

Theorem 6.1 *In a group G of odd type every definable subgroup possesses only finitely many conjugate classes of involutions.*

If t is an involution we write $C_t = C_G(t)$.

Notice, first of all, that in a simple group of finite Morley rank the centralizer of any involution is infinite; this follows from the following simple result.

Theorem 6.2 (Borovik [19]; see also [24, Theorem 10.5]) *If the centralizer C_t of an involution t in a group G of finite Morley rank is finite, then G° is abelian, $t \in G \smallsetminus G^\circ$ and t inverts every element in G°, i.e. $g^t = g^{-1}$ for all $g \in G^\circ$.*

The first step in the classification of groups of odd type is to show that the structure of centralizers of involutions resembles the structure found in one of the simple algebraic groups of characteristic not 2. We cite the following well-known fact, which follows easily from the reductivity of the centralizers of semisimple rational automorphisms of reductive groups [17]; see also [29, Theorem 3.5.4].

Theorem 6.3 *If H is a simple algebraic group over an algebraically closed field K of characteristic different from 2 and t is an involutory definable automorphism of H, then $F(C_t) = Z(C_t^\circ)$ is a finite extension of an algebraic torus over K and $E(C_t) = L_1 \cdots L_n$ is the central product of quasi-simple subgroups L_i which are algebraic groups over K. Also $O(C_t) = 1$ and $C_t^\circ = F(C_t)^\circ E(C_t)$.*

We shall say that G satisfies the B-conjecture if $C_t^\circ = F(C_t)^\circ E(C_t)$ for any involution $t \in G$. This notion generalizes to the context of finite Morley rank the B-conjecture of J. Thompson in finite group theory. When we investigate whether the centralizers of involutions in simple groups of odd type resemble the centralizers of involutions in simple algebraic groups of characteristic not 2, we investigate, first of all, the validity of the B-conjecture.

Let us compare Theorem 6.3 with what we have in groups of finite Morley rank. The next simple result is immediate from the properties of K-groups (Theorem 5.11).

Theorem 6.4 *Let G be a simple K***-group of finite Morley rank and odd type. If $t \in G$ is an involution, then $\overline{C_t^\circ} = C_t^\circ/O(C_t)$ is a central product of a semisimple*

*group all of whose components are simple algebraic groups over algebraically closed
fields of characteristic different from 2 and an abelian divisible group. In particular,*
$\overline{C_t^o} = F(\overline{C_t^o})E(\overline{C_t^o})$.

If G is tame then, in addition, $O(C_t)$ is a nilpotent group.

We see that one of the ways to prove the B-conjecture for groups of finite Morley
rank and odd type is to show that, in a tame simple K*-group G, $O(C_t) = 1$ for all
involutions $t \in G$. So we have the following problem.

Problem 2   *Prove that, in a simple tame K*-group G of odd type, $O(C_t) = 1$ for
any involution $t \in G$.*

When G has normal 2-rank $n(G) \geqslant 3$, we can develop an approach to Problem 2
based on the well-known idea of the signalizer functor, which will be discussed later.
But, if the normal 2-rank of G is 1 or 2, we do not have in our possession this
powerful method. We would like to emphasize the challenge posed by groups of
small normal rank. Their classification will be a substitute in our theory for the
theorem of Daniel Gorenstein and Koichiro Harada [38] on finite groups of sectional
2-rank at most 4.

Problem 3   *Prove that if $n(G) \leqslant 2$ then a simple tame K*-group G is isomorphic
to one of the groups $\mathrm{PSL}_2(K)$, $\mathrm{PSL}_3(K)$, $\mathrm{PSp}_4(K)$, $\mathrm{G}_2(K)$ over some algebraically
closed field K of characteristic not 2.*

6.2. SIGNALIZER FUNCTORS

From now on we assume that the Prüfer 2-rank of G is bigger than 2.

A possible approach to Problem 2 is well-known to finite group theorists and
based on the following observation which follows easily from Theorem 5.13 (or from
Theorem 6.3).

Theorem 6.5 [24, Theorem B.29] *For any two commuting involutions t and s in
a tame K*-group G we have*

$$O(C_t) \cap C_s = O(C_s) \cap C_t.$$

In this situation, finite group theorists say that $\theta(t) = O(C_t)$ is a *signalizer functor*. More precisely, for any involution $s \in G$, let $\theta(s) \leqslant O(C_s)$ be some connected
definable normal subgroup of C_s. We say that θ is a signalizer functor if, for any
commuting involutions $t, s \in G$,

$$\theta(t) \cap C_s = \theta(s) \cap C_t.$$

The signalizer functor θ is *complete* if for any elementary abelian subgroup $E \leqslant G$
of order at least 8 the subgroup

$$\theta(E) = \langle \theta(t) \mid t \in E^* \rangle$$

is a connected $2^{\perp}$-subgroup and

$$C_{\theta(E)}(s) = \theta(s)$$

for any $s \in E^*$. Finally, a signalizer functor θ is *non-trivial* if $\theta(s) \neq 1$ for some involution $s \in G$, and *nilpotent* if all the subgroups $\theta(t)$ are nilpotent.

The following theorem is a generalization of results of David Goldschmidt [34] and Helmut Bender [16] on signalizers in finite groups.

Theorem 6.6 (Borovik [22]; see also [24, Theorem B.30]) *Any nilpotent signalizer functor θ on a group G of finite Morley rank is complete.*

Now let G be a tame K*-group of odd type and S a Sylow 2-subgroup in G. Since we assume that the normal 2-rank of G is at least 3, we can find in S a normal elementary abelian subgroup E of order at least 8. If $s \in S$ is any involution and $\theta(s) \neq 1$, then let us take $D = C_E(s)$; then $|D| \geqslant 4$. Obviously D normalizes $\theta(s)$ and

$$\theta(s) = \langle \theta(s) \cap C_t \,|\, t \in D^\# \rangle \leqslant \theta(E)$$

by Theorem 4.6. So, if we assume that $\theta(s) \neq 1$ for some involution $s \in P$, then we have $\theta(E) \neq 1$.

Now we can state our main result about signalizer 2-functors.

Theorem 6.7 *In the notation above, if a tame K*-group G of finite Morley rank and odd type has normal 2-rank at least 3 and admits a non-trivial nilpotent signalizer functor θ, then*

$$\Gamma_{S,2}(G) \leqslant N_G(\theta(E)).$$

In particular, G has a proper 2-generated core.

Corollary 6.8 *If G is a tame simple K*-group of odd type such that $\mathrm{n}(G) \geqslant 3$ and $O(C_t) \neq 1$ for some involution $t \in G$, then G has a proper 2-generated core.*

Proof of Theorem 6.7. For an arbitrary four-subgroup $W < S$ write

$$\theta(W) = \langle \theta(w) \,|\, w \in W^\# \rangle.$$

Notice that, if V and W are two commuting four-subgroups in S, $[V, W] = 1$, and $v \in V^\#$, then $\theta(v)$ is W-invariant and, by Theorem 4.6,

$$
\begin{aligned}
\theta(v) &= \langle \theta(v) \cap C_w \,|\, w \in W^\# \rangle \\
&\leqslant \langle \theta(w) \,|\, w \in W^\# \rangle \\
&= \theta(W).
\end{aligned}
$$

Therefore $\theta(V) \leqslant \theta(W)$. Analogously $\theta(W) \leqslant \theta(V)$ and $\theta(V) = \theta(W)$.

Recall that the Sylow subgroup S is 2-connected by Lemma 4.5. So, if now U is a four-subgroup from E and V is an arbitrary four-subgroup in S, there is a sequence of four-subgroups

$$U = V_0, V_1, \ldots, V_n = V$$

such that $[V_i, V_{i+1}] = 1$ for $i = 0, 1, \ldots, n-1$. Therefore, by the previous paragraph, $\theta(V) = \theta(U)$. But

$$
\begin{aligned}
\theta(U) &= \langle \theta(u) \mid u \in U^\# \rangle \\
&= \langle C_{\theta(E)}(u) \mid u \in U^\# \rangle \\
&= \theta(E),
\end{aligned}
$$

by Theorem 4.6. Therefore $\theta(V) = \theta(E)$ for any four-subgroup $V < S$.

If now $P \leqslant S$ is a subgroup of 2-rank at least 2, $g \in N_G(P)$ and $V \leqslant P$ is a four-subgroup, we have

$$
\theta(E)^g = \theta(V)^g = \theta(V^g) = \theta(E)
$$

and $N_G(P) \leqslant N_G(\theta(E))$. Therefore $\Gamma_{S,2}(G) \leqslant N_G(\theta(E))$.   □

Finally, we have one more version of a nilpotent signalizer functor on G, which will be used in the context of locally finite groups of finite Morley rank and odd type. (See Theorem 2.10 for the definition of the subgroup $B(H)$ for a solvable subgroup H.)

Theorem 6.9   *In a* K***-*group* G *of odd type,* $\theta(t) = B(O(C_t))$ *is a nilpotent signalizer functor on* G.

Proof. This follows at once from Theorem 5.13 and Theorem 2.10.   □

The role of the signalizer functor $B(O(C_t))$ is shown by the following simple result.

Theorem 6.10   *If* G *is a locally finite* K*-*group and* $B(O(C_t)) = 1$ *for any involution* $t \in G$, *then* G *satisfies the* B-*conjecture.*

One more application of signalizer functors gives us a result on Sylow 2-subgroups in the locally finite case.

Theorem 6.11   *If* G *is a locally finite* K*-*group and* $n(G) \geqslant 3$, *then either* G *has a proper 2-generated core or the Sylow 2-subgroups in* G *are infinite.*

Proof.   Assume that Sylow 2-subgroups in G are finite. By virtue of Theorem 6.7 it is sufficient to construct a nilpotent nontrivial signalizer functor on G. Since $\theta(t) = B(O(C_t))$ is a nilpotent signalizer functor by Theorem 6.9, we can assume without loss that $B(O(C_t)) = 1$ for every involution $t \in G$. Since Sylow 2-subgroups in the K-subgroup C_t are finite, C_t is solvable and $C_t^\circ = O(C_t)$. But, since $B(O(C_t)) = 1$, the subgroups $O(C_t)$ are abelian and divisible by Theorem 2.10. Now it is not difficult to show that $\theta(t) = O(C_t)$ is a nilpotent signalizer functor on G. So θ is complete by Theorem 6.6, and, by Theorem 6.7, G has a proper 2-generated core.   □

6.3. ASCHBACHER'S COMPONENT ANALYSIS

In this subsection we consider groups which satisfy the B-conjecture.

For a definable subgroup $H \leqslant G$ denote by $\mathcal{E}(H)$ the set of all infinite components of the centralizers of involutions in H. Set $\mathcal{E} = \mathcal{E}(G)$.

First of all we have to characterize the groups $\mathrm{PSL}_2(K)$ as the only groups of odd type with $\mathcal{E} = \emptyset$. The next theorem is proved under the assumption that Sylow 2-subgroups in G are infinite. This assumption is satisfied by tame groups (Theorem 4.2). In the case of locally finite groups it is not very restrictive for our purposes in view of Theorem 6.11.

From now on we assume that G is a simple K***-group of odd type.

Theorem 6.12 *Assume that a simple K***-group G of odd type satisfies the B-conjecture and* $\mathrm{n}(G) \geqslant 3$. *Assume also that Sylow 2-subgroups in G are infinite. Then either* $\mathcal{E} \neq \emptyset$ *or G has a proper 2-generated core.*

Theorem 6.12 follows almost immediately from the following two more general results. The first of them is an analogue of a theorem of Ali Asar [4] for locally finite groups (not necessarily of finite Morley rank).

Theorem 6.13 *Let S be a Sylow 2-subgroup in a simple K***-group G of odd type and $T = S^\circ$ the maximal 2-torus in S. Assume that $T \neq 1$. If S contains a four-subgroup V with the property that $T \lhd C_v$ for all $v \in V^\#$, then G has a proper 2-generated core.*

Theorem 6.14 *Assume that G is a simple K***-group of odd type and $\mathrm{n}(G) \geqslant 3$. Let V be a four-subgroup in G and S a Sylow 2-subgroup of G containing S. Then*

$$\Gamma_{S,2}(G) \leqslant N_G(\Gamma_V(G)).$$

In particular, if $\Gamma_V(G) \neq G$, then G has a proper 2-generated core.

Proof of Theorem 6.14. First we will prove the following claim.

If U is a four-subgroup in S and $[U, V] = 1$, then $\Gamma_U(G) = \Gamma_V(G)$.

Indeed, let u be an arbitrary involution in U. Then $V \leqslant C_u$ and, by Theorem 5.14,

$$C_u^\circ = \Gamma_V(C_u^\circ) \leqslant \Gamma_V(G).$$

Therefore

$$\Gamma_U(G) = \langle C_u^\circ \mid u \in U^\# \rangle \leqslant \Gamma_V(G).$$

Analogously $\Gamma_V(G) \leqslant \Gamma_U(G)$ and $\Gamma_U(G) = \Gamma_V(G)$.

Now, by Lemma 4.5, S is 2-connected and, therefore, $\Gamma_U(G) = \Gamma_V(G)$ for any four-subgroups $U, V \leqslant S$. If P is an arbitrary subgroup of S with $\mathrm{m}(P) \geqslant 2$, then P contains some four-subgroup U. Take an arbitrary element $g \in N_G(P)$. We have

$$\Gamma_V(G)^g = \Gamma_U(G)^g = \Gamma_{U^g}(G) = \Gamma_V(G),$$

so $N_G(P) \leqslant N_G(\Gamma_V(G))$. By definition of a proper 2-generated core this means that $\Gamma_{S,2}(G) \leqslant N_G(\Gamma_V(G))$. □

Proofs of Theorems 6.13 *and* 6.12.   Under the assumptions of Theorem 6.13, we have $T \unlhd \Gamma_V(G)$, and, since G is simple, $\Gamma_V(G) < G$. Therefore Theorem 6.13 follows immediately from Theorem 6.14. Now we turn to Theorem 6.12. Since $n(G) \geqslant 3$, we can find in a Sylow 2-subgroup S of G a normal elementary abelian subgroup E. Let $T = S^\circ$ be the maximal 2-torus of S. We assume that S is infinite, so $T \neq 1$. The 2-torus T acts on E by conjugation and induces a finite group of automorphisms of E. But T, being a divisible abelian group, does not have proper subgroups of finite index; therefore $[T, E] = 1$.

Assume now that $\mathcal{E} = \emptyset$ and consider an arbitrary involution $u \in E$. By the B-conjecture, $C_u^\circ = F(C_u^\circ)E(C_u)$. But $E(C_u) = 1$, so C_u° is nilpotent. Since $T = S^\circ$ is connected, $T \leqslant C_u^\circ$ and, by Theorem 2.8, $T \unlhd C_u$. We find ourselves in the conditions of Theorem 6.13, and by virtue of this theorem G has a proper 2-generated core. $\square$

Let t be an involution in G. A component $A \unlhd E(C_t)$ is called *intrinsic* if $t \in Z(A)$. Following Aschbacher [7], we call an involution *classical* if its centralizer contains an intrinsic component isomorphic to $\mathrm{SL}_2(K)$ for an algebraically closed field K.

The following problem is possibly the most important for the identification of simple tame groups of odd type as simple algebraic groups. It is an exact analogue of Aschbacher's famous 'Classical Involution Theorem' [7].

Problem 4   *Assume that a simple tame* K*-*group (or* K***-*group)* G *satisfies the B-conjecture. Prove that if* G *contains a classical involution then* G *is a simple algebraic group over an algebraically closed field.*

The role of Problem 4 is shown by the following result. Its proof is adapted, with considerable simplifications, but without introducing a single new idea, from the proof of Theorem III in Walter [60]. This proof is independent of the rest of the paper and is postponed to Section 8.

Theorem 6.15   (Walter [60, Theorem III])   *Assume that a simple* K***-*group* G *of odd type satisfies the B-conjecture. If* $\mathcal{E}$ *contains a simple algebraic group of Lie rank at least 2, then* G *possesses a classical involution.*

The next lemma (which is a special case of Lemma 8.1 below) describes possible interactions between components in $E(C_t)$ and $E(C_s)$ for commuting involutions t and s.

Lemma 6.16   *Let* t *and* s *be involutions in* G, $C = C_s$, $t \in C$ *and* L *a component in* $E(C_C(t))$. *Then there exists a component* $K \unlhd E(C)$ *such that either* $K = K^t$ *and* $L \unlhd E(C_K(t))$ *or* $K \neq K^t$ *and* $L = C_{K^t K}(t)$.

Proof.   This is analogous to the proof of Lemma 2.7(2) in Aschbacher [6]. $\square$

It follows easily from Lemma 6.16 that, if Problem 4 has a positive solution, we can assume, without loss, that all components in $\mathcal{E}$ are isomorphic to $\mathrm{PSL}_2(K)$ for some (possibly different) algebraically closed fields K.

A quasisimple subgroup $A \in \mathcal{E}$ is called a *prestandard subgroup* in G if $A \lhd E(C_t)$ for any involution $t \in C_G(A)$. If, in addition, $[A, A^h] \neq 1$ for any $h \in G$, then A is *standard* in G. This is an analogue of the well-known notion of a standard subgroup in a finite group, due to Michael Aschbacher [6].

The next theorem is an analogue of a result by Michael Aschbacher and Gary Seitz on standard subgroups in finite simple groups [8].

Theorem 6.17 *Let A be a prestandard subgroup in a simple K***-group G of odd type. Then either $\mathrm{m}(C_G(A)) = 1$ or G has a proper 2-generated core.*

Proof. This is immediate from Theorem 6.14: if V is a four-subgroup in $C_G(A)$, then $A \lhd C_v^\circ$ for all $v \in V^\#$; consequently $A \lhd \Gamma_V(G)$ and therefore $\Gamma_V(G) < G$. $\square$

6.4. COMPONENTS OF PSL$_2$-TYPE

Now we have arrived at the final stage of our analysis.

Theorem 6.18 *Assume that a simple K***-group G of odd type satisfies the B-conjecture, $\mathcal{E} \neq \emptyset$ and all components in $\mathcal{E}$ are isomorphic to groups $\mathrm{PSL}_2(K)$ for some (possibly different) algebraically closed fields K. Then either G has a proper 2-generated core or $\mathrm{n}(G) \leqslant 2$.*

Proof. Notice that, since $\mathcal{E}$ consists only of groups $\mathrm{PSL}_2(K)$, we can further improve Lemma 6.16:

If t and s are commuting involutions in G, $C = C_s$ and L is a component in $E(C_C(t))$, then there exists a component $K \lhd E(C)$ such that $K \neq K^t$ and $L = C_{K^t K}(t)$.

Now we can repeat the proofs of Lemmas 3.5 and 3.6 in [21] and obtain the following result.

Choose $A \in \mathcal{E}$ with the maximal possible value of $\mathrm{m}(C_G(A))$. Then A is prestandard.

Now, by Theorem 6.17, we have $\mathrm{m}(C_G(A)) = 1$. By virtue of the maximality of $\mathrm{m}(C_G(A))$ we conclude that $\mathrm{m}(C_G(A)) = 1$ for any component $A \in \mathcal{E}$.

Assume by way of contradiction that $\mathrm{n}(G) \geqslant 3$. Let S be a Sylow 2-subgroup in G, $T = S^\circ$ the maximal 2-torus in S and $E \lhd S$ a normal elementary abelian subgroup of order at least 8. We already noticed in the proof of Theorem 6.13 that $[T, E] = 1$. If for all $t \in E^\#$ the centralizers C_t are solvable-by-finite, then by Theorem 6.13 either G has a proper 2-generated core or $\mathrm{n}(G) \leqslant 1$. So we can assume without loss of generality that for at least one involution $t \in E$ there is a component $A \lhd E(C_t)$ such that $A \cong \mathrm{PSL}_2(K)$. We also know that $\mathrm{m}(C_G(A)) = 1$ or G has a proper 2-generated core (Theorem 6.17). Therefore we can assume without loss of generality that A is the only component in $E(C_t)$ and $A = E(C_t) \lhd C_t$. In particular, the subgroup $E \leqslant C_t$ normalizes A. Consider the subgroup EA. Since E can induce on A only rational automorphisms (Theorem 5.6), and $\mathrm{PSL}_2(K)$ does not admit outer rational automorphisms, $EL = D \times A$ for some subgroup $D \leqslant E$. Moreover, $\mathrm{PSL}_2(K)$ obviously has normal 2-rank 1; so $\mathrm{m}(E \cap L) \leqslant 1$ and $\mathrm{m}(D) \geqslant 2$, which contradicts the fact that $\mathrm{m}(C_G(A)) \leqslant 1$. $\qquad\square$

6.5. MAIN THEOREM

Now, summarizing our arguments, we can state the main result of the paper.

Theorem 6.19 *Let G be a simple K*-group of finite Morley rank and odd type. Assume also that G is tame or locally finite. Then one of the following statements is true.*

(1) $\mathrm{n}(G) \leqslant 2$.
(2) G *has a proper 2-generated core.*
(3) G *satisfies the B-conjecture and contains a classical involution.*

Proof. First of all, G contains an involution by Theorem 4.2 or by Feit and Thompson [33]. Moreover, the centralizers of involutions in G are infinite by Theorem 6.2.

By Theorems 6.5 and 6.9, $\theta(t)$ defined by

$$\theta(t) = O(C_t) \quad \text{or} \quad \theta(t) = B(O(C_t))$$

(depending on whether G is tame or locally finite, respectively) is a nilpotent signalizer functor. If θ is non-trivial and $\mathrm{n}(G) \geqslant 3$, then θ is complete by Theorem 6.6 and, by Theorem 6.7, G has a proper 2-generated core. Thus we can assume that $\theta(t) = 1$ for all involutions $t \in G$. But then it follows from Theorems 6.4 and 6.10 that G satisfies the B-conjecture.

If G is tame, Sylow 2-subgroups in G are infinite by Theorem 4.2. If G is locally finite, then either G has a proper 2-generated core or Sylow 2-subgroups of G are infinite (Theorem 6.11).

Therefore we can assume without loss that in both cases Sylow 2-subgroups of G are infinite. Now, by Theorem 6.12, $\mathcal{E} \neq \emptyset$ and we can apply Aschbacher's Component Analysis (Subsection 6.3). If $\mathcal{E}$ contains a component of Prüfer 2-rank at least 2, then G possesses a classical involution (Theorem 6.15). So we can assume that all components in $\mathcal{E}$ are isomorphic to $\mathrm{SL}_2(K)$ or $\mathrm{PSL}_2(K)$. If $\mathcal{E}$ contains a component $\mathrm{SL}_2(K)$, then this component is obviously intrinsic and thus G has a classical involution. So we can assume that all components in $\mathcal{E}$ are isomorphic to groups $\mathrm{PSL}_2(K)$. But then by Theorem 6.18 we have $\mathrm{n}(G) \leqslant 2$. $\qquad\square$

7. Locally Finite Groups of Odd Type

As promised, we can now easily obtain a classification of locally finite groups of finite Morley rank and odd type.

The following result is due to Simon Thomas [59], who classified locally finite groups G of finite Morley rank without any assumptions on Sylow 2-subgroups in G. His proof used the classification theorem for finite simple groups.

Theorem 7.1 *A simple locally finite group G of finite Morley rank and odd type is isomorphic to a simple algebraic group over an algebraically closed locally finite field of odd characteristic.*

Proof. Let G be a counterexample to the theorem of least possible Morley rank. Then every proper definable subgroup of G is a K-group and by Theorem 6.19 we have to consider one of the following situations:

(1) *the normal 2-rank of G is at most 2;*
(2) *G has a proper 2-generated core;*
(3) *G has a classical involution.*

We know that G has Černikov Sylow 2-subgroups. This allows us to use results of Kegel [45] and prove that

> G *is the union of a Kegel cover (G_i, N_i), $i \in I$, in which the Kegel kernels $N_i \lhd G_i$ are subgroups of odd order.*

(For the definition of Kegel covers and a discussion of relevant results see Hartley [40].)

Write $\overline{G}_j = G_j/N_j$ and use bars for images. It is easy to see that, if $n(G) \leqslant 2$, then, since G has sectional 2-rank $r(G) \leqslant 4$ by Theorem 4.4, all the Kegel factors $\overline{G}_i$, $i = 1, 2, \ldots$, have sectional 2-rank at most 4 and therefore are known, by Gorenstein and Harada [38]. Analogously, if G has a proper 2-generated core, then, starting with some i, all the Kegel factors $\overline{G}_j$, $j > i$, have proper 2-generated cores and are known, by Aschbacher [5]. Finally, if $L \cong \mathrm{SL}_2(K)$ is an intrinsic component in the centralizer of a classical involution $t \in G$, then L is a subnormal subgroup of C_t and thus $L \cap G_i$ is a subnormal subgroup in $C_{G_i}(t)$, as soon as $t \in G_i$. Again, starting with some i, for all $j > i$ the commutator $L_j = (L \cap G_j)'$ is isomorphic to $\mathrm{SL}_2(\mathbb{F}_{q_j})$ for some q_j, is subnormal in $C_{G_j}(t)$ and contains t. So, for almost all j, the Kegel factor $\overline{G}_j$ has an intrinsic component $\overline{L}_j \cong \mathrm{SL}_2(\mathbb{F}_{q_j})$ of the centralizer of an involution $C_{\overline{G}_j}(\bar{t})$. By Aschbacher [7], $\overline{G}_j$ is a known finite simple group.

So we can assume without loss that all $\overline{G}_i$ are known finite simple groups. Since the 2-ranks of the G_i are bounded, it is not difficult to check that almost all the factors $\overline{G}_i$ are groups of Lie type of the same odd characteristic p. Their Lie ranks are bounded, so we can assume without loss that all $\overline{G}_i$ are groups of the same Lie type. Now it is not difficult to prove that, in fact, $N_i = 1$. After this it follows easily from any of the four papers Belyaev [12, 13], Borovik [18, 20], Hartley and Shute [41] or Thomas [59] that G is a Chevalley group over a locally finite field K of the same characteristic $p > 2$ (see also Liebeck and Seitz [47] for another proof, and Hartley [40, §1.2] for a general discussion). We need to notice now that K is interpretable in G (see, for example, Thomas [59]) and thus is algebraically closed by Macintyre's theorem (Theorem 2.12). Therefore G is a simple algebraic group over an algebraically closed field K. □

8. The Proof of Walter's Theorem

This, the last section of the paper, contains a proof of Theorem 6.15. We have adapted it, with some minor changes, from J. Walter's memoir [60], where an analogue of the theorem for finite simple groups was proven. Walter's original theorem is much more technical and difficult than our simplified and more straightforward version for groups of finite Morley rank. Nevertheless most arguments can be copied almost word for word.

For the convenience of the reader we repeat the statement of the theorem.

Theorem 6.15 *Let G be a simple infinite* K***-*group of finite Morley rank and odd type. Assume also that G satisfies the B-conjecture and $\mathcal{E} = \mathcal{E}(G)$ contains a component of Lie rank at least 2. Then G contains a classical involution.*

8.1. Notation

We shall say that a group $L \leqslant G$ is of SL_2-type (analogously of SO_n-type, Spin_n-type, etc.) if L is isomorphic to $SL_2(K)$ (or $SO_n(K)$, $\mathrm{Spin}_n(K)$, etc., respectively) for some algebraically closed field K.

8.2. Properties of Components of the Centralizers of Involutions

We shall frequently use the following general (and easy to prove) property of the involution centralizers.

Lemma 8.1 (Walter [60, Proposition I.1.1]) *Let X be a K-group of finite Morley rank and assume that $X^\circ = F(X^\circ)E(X)$. Let t be an involution in X and L a component in $E(C_X(t))$. Then there exists a component $K \trianglelefteq E(X)$ such that either*

$$K^t = K \text{ and } L \trianglelefteq E(C_K(t))$$

or

$$K^t \neq K \text{ and } L \trianglelefteq E(C_{K^t K}(t)).$$

We shall denote the two clauses of the lemma by $L \hookrightarrow K$ and $L \hookrightarrow K^t K$.

Lemma 8.2 (Walter [60, Lemma I.4.7]) *Assume that G satisfies the B-conjecture. Let L_1 be a component in $E(C_{t_0})$ for some involution $t_0 \in G$. Suppose that there exists an involution $t_1 \in Z(L_1)$ such that*

$$L_1 \hookrightarrow L_2^{t_0} L_2$$

where L_2 is a component of $E(C_{t_1})$. Then there exists an involution $t_2 \in Z(L_2)$ such that $L_2 \hookrightarrow M_2$, where M_2 is a component of $E(C_{t_2})$. Furthermore, L_1 is a homomorphic image of L_2, and either $L_2 = M_2$ or $L_1 \trianglelefteq E(C_{M_2}(t_1))$.

Proof. Because $L_1 \hookrightarrow L_2^{t_0} L_2$, L_1 is a homomorphic image of L_2. Now, take an involution $t_2 \in Z(L_2)$ so that $t_2 t_1 \in Z(L^{T_0})$.

By Lemma 8.1, either $L_2 \hookrightarrow M_2$ or $L_2 \hookrightarrow M_2^{t_1} M_2$ where M_2 is a component in $E(C_{t_2})$. Now, since $[L_2^{t_0}, t_2] = 1$, it follows that $L_2^{t_0} \leqslant E(C_{t_2})$ and $t_2 t_1 \in E(C_{t_2})$. But $t_2 \in Z(L_2) \leqslant E(C_{t_2})$, so $t_1 \in E(C_{t_2})$. This forces $M_2^{t_1} = M_2$. So $L_2 \hookrightarrow M_2$.

Suppose that $L_2 \neq M_2$. Then t_1 does not centralize M_2. So $t_2 t_1$ also does not centralize M_2, $L_2^{t_0} \leqslant E(C_{t_2})$ does not centralize M_2, and thus $L_2^{t_0} \hookrightarrow M_2$ or $L_2^{t_0} \hookrightarrow M_2^{t_1} M_2$. But then $t_2 t_1 \in M_2^{t_1} M_2$ normalizes M_2. Since t_2 also normalizes $M_2 \trianglelefteq E(C_{t_2})$, so does t_1. This shows that $L_2^{t_0} \hookrightarrow M_2$. Thus $L_1 = C_{L_2^{t_0} L_2}(t_0) \leqslant M_2$ and so L_1 is a component in $E(C_{M_2}(t_1))$. $\square$

Lemma 8.3 (Walter [60, Corollary I.4.9]) *Let t_0 be an involution in G and L_1 a component of type SL_2 in $E(C_{t_0})$. Let t_1 be an involution from $Z(L_1)$. Then one of the following holds.*

(1) $L_1 \hookrightarrow L_2$, where L_2 is a t_0-invariant component in $E(C_{t_1})$.

(2) $L_1 \hookrightarrow L_2^{t_0} L_2$, where L_2 is a component in $E(C_{t_1})$ of type SL_2 and $L_2 \hookrightarrow M_2$ where M_2 is a t_1-invariant component in $E(C_{t_2})$ and $t_2 \in Z(L_2)^\#$.

Proof. This is immediate from Lemma 8.1 and Lemma 8.3. $\qquad\square$

8.3. THE THREE COMPONENTS LEMMA

Assume that G satisfies the conditions of the theorem and assume, by way of contradiction, that G does not contain a classical involution.

The following lemma considers a special configuration of three SL_2-components in the centralizer of a four-subgroup and is of crucial importance for the proof of the theorem.

Lemma 8.4 (Walter [60, Proposition I.4.3]) *Let $W \leqslant G$ be a four-group. Set $W^\# = \{z_1, z_2, z_3\}$, and suppose that there exist components $L_i \trianglelefteq E(C_G(W))$ of SL_2-type with $z_i \in L_i$, $i = 1, 2, 3$. Then, for some i, $L_i \trianglelefteq E(C_{z_i})$. In particular, G has a classical involution.*

Proof. Assume the contrary and set $H_1 = C_{z_1}$. By the choice of G, L_1 is not a component of H_1.

Now, by virtue of Lemma 8.1, $L_1 \leqslant K_1$ for some component $K_1 \trianglelefteq E(H_1)$. Then $[K_1, z_j] \neq 1$ for $j = 2, 3$, since $L_1 \trianglelefteq E(C_{K_1}(z_j))$. As $L_j \leqslant H_1$, it follows that $L_j \leqslant K_1$ since L_j is intrinsic (cf. Lemma 8.1).

A survey of the structure of the centralizers of involutions in Chevalley groups shows that K_1 is the group $\mathrm{Spin}_7(F)$ for some (algebraically closed) field F. This may be verified by inspecting [7, Table 14.5] or [8]. (Also, using [27] and [39], it follows that K_1 has a Dynkin diagram of type B_3 since what is required is that a Dynkin diagram or a Dynkin diagram extended by adjoining a highest root should break up into three isolated nodes after the deletion of a single node.)

Let $L = L_1 L_2 L_3$. Then $L \trianglelefteq C_{K_1}(z_2) = C_{K_1}(W)$. We wish to discuss the structure of $N_{K_1}(W)$. For this purpose, set $\overline{K}_1 = K_1/Z(K_1)$ and denote images by bars. Then $\overline{K}_1 \cong SO_7(F)$ for an algebraically closed field F of characteristic not 2, and $\overline{K}_1$ may be presented as acting on an orthogonal geometry (V, f) where V is a 7-dimensional vector space over F and f is a nondegenerate symmetric bilinear form. Now each involution of $SO(V, f)$ is a product of an even number of mutually commuting symmetries, and among these only those which are products of four mutually commuting symmetries lift to involutions in K_1 (cf. Dieudonne [32, §11.7]). Thus $SO_7(F)$ has one class of involutions which are images of involutions in $\mathrm{Spin}_7(F)$. Since the number of mutually commuting symmetries in the factorization of an involution is the dimension of its (negative) eigenspace, $\overline{z}_2$ is represented as an involution with a 4-dimensional eigenspace V_2. The orthogonal complement $V_1 = V_2^\perp$ is its other eigenspace. The subgroup $\overline{L}_2 \overline{L}_3$ acts on V_2 as $SO_4(F)$ and the group $\overline{L}_1$ acts on V_1 as $SO_3(F) \cong PSL_2(F)$. Now take V_3 to be a 3-dimensional nonisotropic subspace of V_2 with a nonisotropic orthogonal basis x_1, x_2, x_3. Take a nonisotropic vector $x_4 \in V_1$ and take $\overline{z}_4 = \overline{t}_1 \overline{t}_2 \overline{t}_3 \overline{t}_4$ where $\overline{t}_i$ is the symmetry of V relative to the hyperplane $x_i^{\perp}$. So $\overline{z}_4 \in \mathrm{SO}_7(F) = \overline{K}_1$ and there exists an involution $z_4 \in K_1$ which maps onto $\overline{z}_4$. Then $\overline{z}_4$ is conjugate to $\overline{z}_2$ in $\overline{K}_1$ and hence z_4 is conjugate to either z_2 or to $z_3 = z_2 z_1$ in K_1.

Now the eigenspace $V_4 = \langle x_1, x_2, x_3, x_4 \rangle$ intersects V_1 in a 1-dimensional space $\langle x_4 \rangle$ and intersects V_2 in a 3-dimensional space $\langle x_1, x_2, x_3 \rangle$. Thus

$$E(C_{\overline{K}_1}(\overline{z}_2, \overline{z}_4)) \cong \mathrm{SO}_3(F) \cong \mathrm{PSL}_2(F).$$

It follows that $\overline{L}_2^{\overline{z}_4} = \overline{L}_3$, so $L_2^{z_4} = L_3$ and $z_2^{z_4} = z_3$. Consequently $[z_1, z_4] = 1$ and $z_4 \in N_G(W)$. Using z_2 and $K_2 = [E(C_{z_2}), L_2]$ in place of z_1 and K_1, we may also obtain that z_1 and z_2 are conjugate in $N_G(W)$. In particular, as z_4 is conjugate to either z_2 or z_3 in K_1, it is conjugate to z_1 in G.

Let $L_1' = E(C_{L_2 L_3}(z_4))$. Then L_1' is a component in $E(C_G(z_1, z_4))$ and $z_1 = z_2 z_3$ belongs to L_1'. On the other hand, set $K_4 = K_1^g$ where $g \in G$ is chosen so that $z_4 = z_1^g$. Now if $[K_4, L_1'] = 1$, we have $[K_4, z_1] = 1$ since $z_1 \in L_1'$. Then $K_4 \leqslant E(C_{z_1})$. Because $z_4 \in K_4$ and $[K_1, z_4] = K_1$, we have $K_4 = K_1$, which is not possible. Hence $[K_4, L_1'] = K_4$. But, as L_1' is intrinsic in $E(C_G(z_4, z_1))$, $L_1' \leqslant K_4$. Then, as z_1 is a noncentral involution of K_4, $E(C_{K_4}(z_1)) = L_4' L_1' L_5'$ where L_4' and L_5' are elements of $E(C_{K_4}(z_1))$ containing z_4 and $z_5 = z_1 z_4$ respectively. This, in turn, means that $L_4' \leqslant E(C_{K_1}(z_4))$ since $z_4 \in K_1$. Then $L_5' \leqslant K_1$ as well. Hence $C_{K_1}(z_1, z_4) \trianglelefteq L_1' L_4' L_5'$. But $z_2 \in N_{K_1}(\langle z_1, z_4 \rangle)$ and $[z_2, L_1'] = 1$. Since $z_4^{z_2} = z_5$, $L_1'^{z_2} = L_5'$.

Set $L_6' = E(C_{L_4' L_5'}(z_2))$. Then $z_1 = z_4 z_5 \in L_6'$. Now $z_1 \in \langle z_2, z_4 \rangle$. Hence L_1' and L_6' are two distinct components of $E(C_G(\langle z_2, z_4 \rangle))$ whose centers contain z_1. On the other hand, by Lemma 8.1, either a component $L_\alpha \trianglelefteq E(C_G(\langle z_2, z_4 \rangle))$ whose center contains z_1 is contained in K_1 or $[K_1, L_\alpha] = 1$ since $z_1 \in K_1 \cap L_\alpha$. In the latter case, L_α is a component of $E(C_{z_1})$; however, this contradicts the choice of G inasmuch as $\langle z_1, z_4 \rangle$ is conjugate to $W = \langle z_1, z_2 \rangle$ in C_{z_1}. Hence we conclude that the components in $E(C_G(z_1, z_4))$ of SL$_2$-type which contain z_1 are components in $E(C_{K_1}(z_4))$. Thus the components of $E(C_G(z_2, z_4))$ of type SL$_2$ which contain z_1 are components in $E(C_{K_1}(z_2, z_4)) = E(C_{K_1}(W\langle z_4 \rangle)) = E(C_L(z_4))$. But, since $\overline{z}_4$ does not have an eigenspace containing V_1, $[L_1, \overline{z}_4] = \overline{L}_1$. Therefore, $L_1' = E(C_{K_1}(z_4))$. Thus L_6' is not a component in $E(C_{K_1}(z_2, z_4))$. This contradiction proves the lemma. $\square$

8.4. Strongly Intrinsic Components

Let t_0 be an involution in G and let $L \trianglelefteq E(C_{t_0})$ be an intrinsic component. Recursively, we define L to be *strongly intrinsic* in C_{t_0} if either

(*) L has type SL$_2$, or

(**) given an involution $t_1 \in L$ such that $C_L(t_1)$ has an intrinsic component L_1 of type SL$_2$, there exists a strongly intrinsic component L_2 in $C_L(t_0 t_1)$.

Set $t_2 = t_0 t_1$ and note that $C_L(t_1) = C_L(t_2)$. We note that (**) has the consequence that $L_1 L_2 \trianglelefteq E(C_L(t_2))$ with $t_1 \in L_1$ and $t_2 \in L_2$.

Let L be a simple algebraic group over an algebraically closed field of characteristic not 2 and $\tilde{L}$ its simply connected covering group. We say that L is 2-*universal* if the kernel of the canonical homomorphism $\pi : \tilde{L} \longrightarrow L$ has odd order. Except for 2-universal simple algebraic groups of type Spin_{4m}, all 2-universal simple algebraic groups have cyclic centers. Denote by $\mathrm{Spin}^*_{4m}(K)$ any factor group of $\mathrm{Spin}_{4m}(K)$ by a subgroup of order 2. This type includes both the orthogonal groups $\mathrm{SO}_{4m}(K)$ and the half-spin groups; the latter are not isomorphic to $\mathrm{SO}_{4m}(K)$ except when $m = 2$.

Lemma 8.5 (Walter [60, Lemma I.4.4]) *Let t_0 be an involution in G. Let L be an intrinsic component of C_{t_0}. Then, if L has 2-universal type or has type Spin^*_{4m}, L is strongly intrinsic in C_{t_0}.*

Proof. We may assume without loss that $G = L$ and so we work in the context of simple algebraic groups. Now the result follows immediately from [60, Lemma 4.4] where it is proved for finite groups of Lie type over fields of odd order.   □

Let L be a definable connected quasisimple subgroup of the group G. Let t be an involution in $N_G(L)$. A component L_1 of $E(C_L(t))$ is said to be an *anchored component* of $C_L(t)$ provided $Z(L_1) \leqslant Z(L)$.

Lemma 8.6 (Walter [60, Proposition I.4.5]) *Let t_0 be an involution in G and $L \trianglelefteq E(C_{t_0})$ an intrinsic component, and let t be an involution in $N_G(L)$. Suppose that $C_L(t)$ contains an anchored SL_2-component L_1. Then L is strongly intrinsic in C_{t_0}.*

Proof. We may take $G = L$. By virtue of Lemma 8.4, we may assume that L is neither 2-universal nor of type Spin^*_{4m}. Let $\tilde{L}$ be a 2-universal group covering L. We may assume that $Z(\tilde{L})$ is a 2-group. Let Z_L be the kernel of the covering mapping. Then $1 < Z_L < Z(\tilde{L})$. Because $\tilde{L}$ does not have type Spin_{4m}, $Z(\tilde{L})$ is cyclic. But, because $Z(\tilde{L}_1) \leqslant Z(\tilde{L})$, L_1 lifts to a subgroup $\tilde{L}_1$ of $\tilde{L}$ containing $Z(\tilde{L})$. This gives a nontrivial central extension of SL_2, which is not possible by our assumption.   □

Lemma 8.7 (Walter [60, Proposition I.4.10]) *Assume that the centralizer of some involution in G contains a strongly intrinsic component. Then G has a classical involution.*

Proof. Take L to be a strongly intrinsic component of $E(C_{t_0})$ for some involution $t_0 \in G$. Then either (*) or (**) in the definition of a strongly intrinsic component holds for L. We can assume without loss that G has no classical involutions and (**) holds in L for some involution $t \in L$.

Among all strongly intrinsic components of the centralizers of involutions in G choose L to maximize $\mathrm{rk}(L)$, and let L_1, L_2, t_0, t_1, and $t_2 = t_0 t_1$ be given as in (**). Then L_1 has type SL_2, and L_2 is strongly intrinsic. Set $\overline{L} = L/Z(L)$ and denote images by bars. Here are the possibilities for the pair $(\overline{L}, \overline{L}_2)$ as taken from [7, Table 14.5]:

$$(\mathrm{PSL}_{2n}, \mathrm{SL}_{2n-2}), \ n \geqslant 2;$$

$$(\mathrm{PSp}_{2n}, \mathrm{Sp}_{2n-2}), \ n \geqslant 3;$$

$$(\mathrm{PSO}_n, \mathrm{SL}_2 * \mathrm{SO}_{n-4}), \ n \geqslant 7;$$

$$(\mathrm{E}_7, \mathrm{SO}_{12}).$$

Notice that $L_1 \leqslant E(C_{t_1})$ and $L_2 \leqslant E(C_{t_1})$. Hence $t_0 = t_1 t_2 \in E(C_{t_1})$. Thus $L_1 \leqslant L_3$ where L_3 is a component of $E(C_{t_1})$. But also $L_2 \leqslant L_4$, where L_4 is a component of $E(C_{t_1})$. Then if $L_3 \neq L_4$ it follows that $[L_3, t_2] = 1$ and $[L_3, t_0] = 1$. But then $L_3 = L_2$ and t_1 is a classical involution. Therefore, $L_4 = L_3$.

Thus L_1 and L_2 are components of $E(C_{L_3}(t_2))$. Let $\overline{L}_3 = L_3/Z(L_3)$ and denote images by bars. Then $\overline{L}_1$ is of type $\mathrm{PSL}_2 \cong \mathrm{SO}_3$ and $\overline{L}_2 \neq 1$. This implies that $\overline{L}_3$ has type PSO_{m_3}. Then, because $\overline{L}_2$ is intrinsic, it has type SO_{2m_2} or $E(C_{\overline{L}_3}(\overline{t}_2))$ has a third component $\overline{M}_2$ such that $\overline{L}_2 \cong \overline{M}_2$ where these are of type SL_2 and $\overline{L}_2 \overline{M}_2 \cong \mathrm{SO}_4(F)$. But L_2 has the same type as $\overline{L}_2$, and it appears as a component in $E(C_L(t_1))$. This allows us to identify the type of L. We have from the table that when $\overline{L}_2$ is not of type SL_2 (which means that $\overline{L}_2$ is of orthogonal type), L must have orthogonal type. But when L_2 has type SL_2, $E(C_G(t_1, t_2))$ has a component M_2 of SL_2-type mapping onto $\overline{M}_2$, and either t_2 or $t_0 = t_1 t_2$ belongs to $Z(M_2)$. In this case $E(C_{L/Z(L)}(t_1))$ has at least one component of type SL_2 and one of type SO_{2m_2} or SL_2. This implies that L has orthogonal type.

When L_2 has type SL_2, it follows that $\overline{L}_3$ has type SO_7 and L_3 has type Spin_7. We argue that this is also true when L_2 has type SO_{2m_2}. Indeed, now $E(C_L(t_1))$ must have a factor of type $\mathrm{SO}_4 \cong \mathrm{SL}_2 * \mathrm{SL}_2$. Then there exists a second component $M_1 \trianglelefteq E(C_L(t_1))$ of SL_2-type, and clearly either $t_1 \in M_1$ or $t_2 = t_0 t_1 \in M_1$. But in the former case we may replace L_1 by M_1 in the preceding argument to obtain that $M_1 \hookrightarrow M_3$ where M_3 is a component of $E(C_{t_1})$ such that $[M_3, L_4] = M_3 = L_4$. Then $M_3 = L_3$ and $L_4 = L_3$. Hence M_1 is a component in $E(C_L(t_1))$. When $t_2 \in M_1$, then $[L_3, M_1] = L_3$ inasmuch as $t_2 \notin Z(L_3)$. This implies that $M_1 \leqslant L_3$ again. This means that $E(C_{\overline{L}_3}(L_2))$ now contains a component of SL_2-type. So $E(C_{\overline{L}_3}(t_2))$ has a factor of type SO_4, as well as a factor of type SO_3. So again L_3 has type Spin_7.

Thus take $W = \langle t_1, t_2 \rangle$ in order to apply Lemma 8.4. The structure of Spin_7 shows that $E(C_G(W))$ has the structure asserted in the hypothesis of Lemma 8.4. But then $E(C_{t_1})$ has an intrinsic component of type SL_2. $\square$

8.5. End of the Proof

Now we are in a position to complete the proof of the theorem.

Let L be a component of Lie rank at least 2 in $E(C_{t_0})$ for some involution $t_0 \in G$. Then L contains a classical involution t_1. Let L_1 be a component of type SL_2 in $E(C_L(t_1))$. By Lemma 8.3 we may suppose that $L_1 \hookrightarrow L_2$ where L_2 is a component of $E(C_{t_1})$ with $t_1 \in L_1$. If $L_1 = L_2$, then t_1 is a classical involution in G. So we can assume without loss that $L_2 > L_1$. Then L_1 is an anchored SL_2-component of $C_{L_2}(t_1)$, and thus L_2 is a strongly intrinsic component by virtue of Lemma 8.6. But now G has a classical involution by Lemma 8.7. $\square$ $\square$

References

1. T. Altınel, *Groups of Finite Morley Rank with Strongly Embedded Subgroups*, Ph. D. thesis, Rutgers University, 1994.
2. T. Altınel, A. Borovik and G. Cherlin, Groups of mixed type, in preparation.

3. T. Altınel, G. Cherlin, L.-J. Corredor and A. Nesin, A Hall theorem for ω-stable groups, to appear.
4. A. O. Asar, On a problem of Kegel and Wehrfritz, J. Algebra 59 (1979), 47–55.
5. M. Aschbacher, Finite groups with a proper 2-generated core, Trans. Amer. Math. Soc. 197 (1974), 87–112.
6. M. Aschbacher, On finite groups of component type, Illinois J. Math. 19 (1975), 87–115.
7. M. Aschbacher, A characterization of Chevalley groups over fields of odd order. I, II, Ann. Math. 106 (1977), 353–398, 399–468. Corrections: Ann. Math. 111 (1980), 411–414.
8. M. Aschbacher and G. M. Seitz, On groups with a standard component of known type, Osaka J. Math. 13 (1976), 439–482.
9. J. T. Baldwin and J. Saxl, Logical stability in group theory, J. Austral. Math. Soc. (Ser. A) 21 (1976), 267–276.
10. A. Baudisch, A new uncountably categorical pure group, I, II, preprints A93-40 and A94-23, Freie Universität Berlin, 1993, 1994.
11. O. V. Belegradek, On groups of finite Morley rank, in Abstracts of the Eighth International Congress of Logic, Methodology and Philosophy of Science, LMPS '87, Moscow, USSR, 17-22 August 1987, Moscow, 1987, pp. 100–102.
12. V. V. Belyaev, On locally finite Chevalley groups, in 17-th All-Union Algebraic Conference, Leningrad, 1981, Part 2, p. 17 (in Russian).
13. V. V. Belyaev, Locally finite Chevalley groups, in Investigations in Group Theory, Urals Scientific Centre, Sverdlovsk, 1984, pp. 39–50 (in Russian).
14. H. Bender, On groups with abelian Sylow 2-subgroups, Math. Z. 117 (1970), 164–176.
15. H. Bender, Transitive Gruppe gerader ordnung, in denen jede Involution genau einen Punkt festlässt, J. Algebra 17 (1971), 175–204.
16. H. Bender, Goldschmidt's 2-signalizer functor theorem, Israel J. Math. 22 (1975), 208–213.
17. A. Borel and J. Tits, Groupes réductifs, Publ. IHES 27 (1965), 55–151.
18. A. V. Borovik, Periodic linear groups of odd characteristic, Soviet Math. Dokl. 26 (1982), 484–486.
19. A. V. Borovik, Involutions in groups with dimension, preprint (Acad. Nauk SSSR, Sibirsk. Otdel. Vychisl. Tsentr), No. 512, Novosibirsk, 1982 (in Russian).
20. A. V. Borovik, Embeddings of finite Chevalley groups and periodic linear groups, Siberian Math. J. 24 (1983), 843–851.
21. A. V. Borovik, Classification of periodic linear groups over fields of odd characteristic, Siberian Math. J. 25 (1984), 217–235.
22. A. V. Borovik, On signalizer functors for groups of finite Morley rank, in Soviet-French Colloquium on Model Theory, Karaganda, 1990, p. 11 (in Russian).
23. A. Borovik and A. Nesin, On the Schur-Zassenhaus theorem for groups of finite Morley rank, J. Symbolic Logic 57 (1992), 1469–1477.
24. A. V. Borovik and A. Nesin, Groups of Finite Morley Rank, Oxford University Press, 1994.
25. A. V. Borovik and A. Nesin, Schur-Zassenhaus theorem revisited, J. Symbolic Logic 59 (1994), 283–391.
26. A. V. Borovik and B. Poizat, Tores et p-groupes, J. Symbolic Logic 55 (1990), 478–491.
27. N. Burgoyne and C. Williamson, Centralizers of semisimple groups, Pacific. J. Math. 72 (1977), 341–350.
28. R. Carter, Simple Groups of Lie Type, Wiley-Interscience, 1972 .
29. R. Carter, Finite groups of Lie Type. Conjugacy Classes and Complex Characters, Wiley-Interscience, 1985.
30. G. Cherlin, Groups of small Morley rank, Ann. Math. Logic 17 (1979), 53–74.
31. L.-J. Corredor, Bad groups of finite Morley rank, J. Symbolic Logic, 54 (1989), 768–773.
32. J. Dieudonné, La Géométrie des Groupes Classiques, Springer-Verlag, 1955.
33. W. Feit and J. Thompson, Solvability of groups of odd order, Pacific J. Math. 13 (1963), 775–1029.
34. D. M. Goldschmidt, 2-signalizer functors on finite groups, J. Algebra 21 (1972), 321–340.
35. D. Gorenstein, Finite Groups, Chelsea Publishing Company, New York, 1980.
36. D. Gorenstein, Finite Simple Groups. An Introduction to Their Classification, Plenum Press, 1982.
37. D. Gorenstein and K. Harada, Finite simple groups of low 2-rank and the families $G_2(q)$, $D_4^2(q)$, q odd, Bull. Amer. Math. Soc. 77 (1971), 829–862.

38. D. Gorenstein and K. Harada, Finite groups whose 2-subgroups are generated by at most 4 elements, *Mem. Amer. Math. Soc.* 147 (1974).

39. M. E. Harris, Finite groups containing an intrinsic 2-component of Chevalley type over a field of odd order, *Trans. Amer. Math. Soc.* 272 (1982), 1–65.

40. B. Hartley, Simple locally finite groups, *these Proceedings*.

41. B. Hartley and G. Shute, Monomorphisms and direct limits of finite groups of Lie type, *Quart. J. Math.* 35 (1984), 49–71.

42. J. E. Humphreys, *Linear Algebraic Groups*, Springer-Verlag, Berlin-New York, 2nd edition, 1981.

43. S. V. Ivanov and A. Yu. Ol'shanski, Some applications of graded diagrams in combinatorial group theory, to appear.

44. N. Iwahori, Centralizers of involutions in finite Chevalley groups, in *Seminar on Algebraic Groups and Related Finite Groups*, Lect. Notes in Math. 131, Springer-Verlag, Berlin, 1970.

45. O. H. Kegel, Über einfache local endliche Gruppen, *Math. Z.* 95 (1967), 169–195.

46. *Kourovka Notebook (Unsolved Problems in Group Theory)*, 10th edition, Institute of Mathematics SO AN SSSR, Novosibirsk, 1986.

47. M. W. Liebeck and G. M. Seitz, Subgroups generated by root elements in groups of Lie type, *Ann. Math.* 139 (1994), 293–361.

48. A. Macintyre, On ω_1-categorical theories of abelian groups, *Fund. Math.* 70 (1971), 253–270.

49. A. Macintyre, On ω_1-categorical theories of fields, *Fund. Math.* 71 (1971), 1–25.

50. A. MacWilliams, On 2-groups with no normal abelian subgroups of rank 3, and their occurrence as Sylow 2-subgroups of finite simple groups, *Trans. Amer. Math. Soc.* 150 (1970), 345–408.

51. A. Nesin, Solvable groups of finite Morley rank, *J. Algebra* 121 (1989), 26–39.

52. A. Nesin, Poly-separated and ω-stable nilpotent groups, *J. Symbolic Logic* 56 (1991), 915–931.

53. A. Nesin, Generalized Fitting subgroup of a group of finite Morley rank, *J. Symbolic Logic* 56 (1991), 1391–1399.

54. B. Poizat, Une théorie de Galois imaginaire, *J. Symbolic Logic* 48 (1983), 1151–1170.

55. B. Poizat, *Groupes Stables*, Nur Al-Mantiq Wal-Ma'rifah, Villeurbanne, France, 1987.

56. G. M. Seitz, Algebraic groups, *these Proceedings*.

57. R. Steinberg, *Lectures on Chevalley groups*, Yale University, 1967.

58. A. Tarski, *A Decision Method for Elementary Algebra and Geometry*, 2nd ed., Berkeley, University of California Press, 1951.

59. S. Thomas, The classification of simple periodic linear groups, *Arch. Math.* 41 (1983), 103–116.

60. J. H. Walter, The B-Conjecture; characterizations of Chevalley groups, *Memoirs Amer. Math. Soc.* 61 (1986), 1–196.

61. W. J. Wong, A characterization of the finite projective symplectic groups $PSp_4(q)$, *Trans. Amer. Math. Soc.* 139 (1969), 1–35.

62. B. Zil'ber, Groups with categorical theories, in *Fourth All-Union Symposium on Group Theory*, Math. Inst. Sibirsk. Otdel. Akad. Nauk SSSR, 1973, pp. 63–68 (in Russian).

63. B. Zil'ber, The structure of models of categorical theories and the problem of axiomatizability, manuscript deposited with VINITI, Dep. No. 2800–77 (in Russian).

64. B. Zil'ber, Groups and rings whose theory is categorical, *Fund. Math.* 55 (1977), 173–188.

EXISTENTIALLY CLOSED GROUPS IN SPECIFIC CLASSES

F. LEINEN
Fachbereich Mathematik
Universität Mainz
D-55099 Mainz
Germany
e-mail: leinen@mzdmza.zdv.uni-mainz.de
leinen@mat.mathematik.uni-mainz.de

Abstract. This survey article is intended to make the reader familiar with the algebraic structure of existentially closed groups in specific group classes, and with the ideas and methods involved in this area of group theory. We shall try to give a fairly complete account of the theory, but there will be a certain emphasis on classes of nilpotent groups, locally finite groups, and extensions.

Key words: algebraically closed groups, existentially closed groups, amalgamation, back and forth, completions, constructions, equations and inequalities, forcing, homogeneity, injectivity, universality.

1. Introduction

In order to give a precise definition of the notions *existentially closed* (*e. c.*) and *algebraically closed* (*a. c.*) we shall need to make use of the language of model theory. Algebraists who are not familiar with the beginnings of model theory may consult [43] or [44]. However, we shall translate the definition immediately into algebraic language. And, for most of the present article, this latter version will be fully sufficient.

A structure G in a class $\mathfrak{X}$ of similar structures is said to be *existentially closed* (resp. *algebraically closed*) *in* $\mathfrak{X}$ if, for every (positive) existential formula $\phi(\bar{x})$ and for every tuple $\bar{g}$ of elements from G, we have $G \models \phi(\bar{g})$ whenever $H \models \phi(\bar{g})$ for some $H \in \mathfrak{X}$ containing G as a substructure. In the case when the underlying first-order language contains no relation symbols, this is equivalent to saying that every finite set of equations and inequalities (resp. of equations only) with coefficients from G, which is solvable in some $H \in \mathfrak{X}$ containing G, already has a solution in G itself. Especially in the class of all fields, this latter statement has a natural meaning. And, in fact, it follows from Hilbert's Nullstellensatz that the algebraically closed fields are precisely the e. c. (and the a. c.) structures in the class of all fields. In the usual language of groups, an equation (resp. inequality) always takes the form $w = 1$ (resp. $w \neq 1$) for some word w. The language of lattice-ordered groups however involves a relation symbol. So at least in this case one has to appeal to the original model theoretic definition of e. c. and a. c. structures. The same applies when considering (finitary) linear groups.

If the class $\mathfrak{X}$ under consideration is *inductive*, i. e. closed with respect to the formation of unions of chains, then a straightforward recursive completion procedure

B. Hartley et al. (eds.), Finite and Locally Finite Groups, 285–326.
© *1995 Kluwer Academic Publishers. Printed in the Netherlands.*

(see [40, 1.1.3]) shows that every $\mathfrak{X}$-structure G is contained in an e. c. $\mathfrak{X}$-structure $\overline{G}$. And if $\mathfrak{X}$ is also closed with respect to the formation of substructures, then one can ensure that $|\overline{G}| \leqslant \max\{\aleph_0, |G|\}$. In this situation the e. c. $\mathfrak{X}$-structures reflect those properties of $\mathfrak{X}$-structures which can be formulated in terms of existential formulas and which are realizable relative to finitely generated substructures. Therefore the e. c. $\mathfrak{X}$-structures are often very rich and homogeneous structures which combine a variety of especially striking properties. On the other hand the study of their properties can only be based on a deep knowledge or analysis of constructions within the given class $\mathfrak{X}$, which may well be of independent interest.

In the present survey article, $\mathfrak{X}$ will always be a class of groups. Already around 1950, the algebraic version of the notion 'a. c.' was introduced by W. R. Scott for the class of all groups [93], and by T. Szele for the class of all abelian groups [94]. But then it took until the early seventies for a broader interest in e. c. structures to arise. At that time model theorists like A. Macintyre [64], P. Eklof and G. Sabbagh [12], *et al.* produced so-called generic structures via forcing methods due to A. Robinson; and generic structures are in particular e. c. However, algebraic properties of e. c. structures were studied only for a very few specific classes. Much progress was made then with e. c. structures in the class of all groups, one milestone being the results obtained independently by O. Belegradek in his Ph. D. thesis [5] (see [6, 7]) and by M. Ziegler in his Habilitation dissertation (see [101]). Large parts of this theory, with fascinating results reflecting many relationships between infinite group theory, model theory and recursion theory, are now well documented in the book of G. Higman and E. Scott [39].

In the present survey article, we shall therefore not dwell on this subject. Instead, it is our intention to make the reader familiar with results concerning the algebraic structure of e. c. and a. c. groups *in specific group classes*, and with the ideas and techniques employed to prove them. All of this goes back to an increasingly fruitful period of research over the past 15 years. Although we shall try to give a fairly complete account of the theory, our emphasis will be on classes of nilpotent groups, locally finite groups, and extensions. The organization of the article may be read off from the following list of contents.

1. Introduction
2. Group class notation
3. The philosophy of constructions
4. Algebraically closed groups versus existentially closed groups
5. Soluble groups
6. Nilpotent groups
7. Locally finite groups
8. Periodic locally soluble groups and locally profinite groups
9. Torsion-free locally nilpotent groups
10. Locally FC-groups
11. Extensions
12. Finitary linear groups

At the end we shall supply an extensive bibliography for the reader who would like to see further details.

We apologise for omitting lattice-ordered groups completely. The main reason is that large parts of the theory of e. c. lattice-ordered groups are of a model theoretic nature, and therefore do not match the intention of this article. A second reason is that we assume that most readers are not familiar with features of lattice-ordered groups—and this makes it hard to appreciate the structural results. The interested reader is therefore referred directly to the book of A. Glass [19, Section 12], and to the articles by A. Glass and K. Pierce [20]–[22], D. Saracino and C. Wood [89, 90], and V. Weispfennig [98].

2. Group Class Notation

We shall denote specific group classes by the following symbols.

$\mathfrak{A}, \mathfrak{N}, \mathfrak{N}_c$	:	abelian groups, nilpotent groups (of class at most c);
$\mathfrak{S}, \mathfrak{S}_d$	:	soluble groups (of derived length at most d);
$\mathfrak{F}, \mathfrak{F}_p$, FC	:	finite groups, finite p-groups, groups with finite conjugacy classes;
$\mathfrak{X}_\pi, \mathfrak{X}^+, \mathfrak{X}^e$	:	$\mathfrak{X}$-groups whose torsion-elements are π-elements, torsion-free $\mathfrak{X}$-groups, $\mathfrak{X}$-groups of exponent e (note that $\mathfrak{X}_\pi = \mathfrak{X}$ if π is the set of all primes, while $\mathfrak{X}_\pi = \mathfrak{X}^+$ if π is empty);
$\mathrm{L}\mathfrak{X}$	:	groups whose finite subsets are contained in an $\mathfrak{X}$-subgroup (locally $\mathfrak{X}$-groups);
$\mathfrak{X}(A)$	:	$\mathfrak{X}$-groups whose center contains the distinguished abelian group A;
$\mathfrak{X}(N, A)$	:	groups G with normal subgroup N such that $C_G(N)N/N \in \mathfrak{X}$ and $G/C_G(N)N \leqslant A \leqslant \mathrm{Out}(N)$;
$\mathfrak{L}(n, c), \mathfrak{L}(c)$	:	groups with a faithful linear representation (of degree n) over a field of characteristic c.

Note that classes of the form $\mathrm{L}\mathfrak{X}$ are always inductive from their definition, while classes like $\mathfrak{N}_c, \mathfrak{S}_d$ or $\mathfrak{L}(n, c)$ are inductive since they are closed under the formation of ultraproducts (see [11, Corollary 6.2]).

3. The Philosophy of Constructions

Let G be an a. c. or e. c. $\mathfrak{X}$-group. Starting from the definition there is just one way to obtain information about the algebraic structure of G: find a construction which produces an $\mathfrak{X}$-group $H \geqslant G$, and try to get hold of finite systems of equations (and inequalities) with coefficients from G which admit a solution in H. Provided that these systems are sufficiently complicated, this will entail some interesting consequences for our group G. Sometimes, constructions have already been carried out. For example, by a well-known theorem of A. I. Mal'cev [76], every $\mathfrak{N}_c^+$-group resp. $\mathrm{L}\mathfrak{N}^+$-group is contained in a divisible $\mathfrak{N}_c^+$-group resp. $\mathrm{L}\mathfrak{N}^+$-group.

Proposition 3.1 *A. c. $\mathfrak{N}_c^+$-groups and a. c. $\mathrm{L}\mathfrak{N}^+$-groups are divisible.*

For abelian groups we even have a converse, which may be seen as a sharp analogue of Hilbert's Nullstellensatz.

Theorem 3.2 [94, Lemma 7] and [12] *Let* $\mathfrak{X} \in \{\mathfrak{A} \cap \mathrm{L}\mathfrak{F}_\pi, \mathfrak{A}_\pi\}$. *Then the group* G *is a.c. in* $\mathfrak{X}$ *if and only if* G *is a divisible* $\mathfrak{X}$-*group.*

However, a.c. $\mathfrak{X}$-groups cannot in general be characterized as those $\mathfrak{X}$-groups which solve single equations: see [2, Example 7.1] for $\mathfrak{X} = \mathfrak{N}_2^+$.

Another technique employed very frequently is the embedding of G into a wreath product. The *unrestricted wreath product* $W = A \,\mathrm{Wr}\, B$ of two groups A and B is the split extension $W = \{f \cdot b \mid f : B \to A,\, b \in B\}$ of the cartesian power A^B by B, where conjugation with elements from the *top group* B permutes the factors of the *base group* A^B via right multiplication, i. e.

$$(f_2 b_2) \cdot (f_1 b_1) = (f_2 f_1^{b_2^{-1}}) \cdot (b_2 b_1),$$

where

$$(b)(f_2 f_1^{b_2^{-1}}) = (b)f_2 \cdot (bb_2)f_1 \quad \text{for all } b \in B.$$

For any normal subgroup N of G there always exists the so-called KKF-embedding $\sigma : G \to N \,\mathrm{Wr}\, G/N$, which goes back to Krasner, Kaloujnine and Frobenius: for a fixed transversal $T = \{t_{Ng} \mid g \in G\}$ of N in G with $N \cdot t_{Ng} = Ng$, the embedding σ is given by $g\sigma = f_g \cdot (Ng)$ where $(Nh)f_g = t_{Nh} \cdot g \cdot t_{Nhg}^{-1}$ for all $h \in G$. In our setup, it is the interesting and usually very difficult task to find $\mathfrak{X}$-subgroups of $N \,\mathrm{Wr}\, G/N$ which contain $G\sigma$.

Let us illustrate the wreath product technique on an elementary level. Suppose that G is an a.c. $\mathfrak{X}$-group, let $C_n = \langle c \rangle$ be the cyclic group of order n, and consider $W_n = G \,\mathrm{Wr}\, C_n$. Identify G with the constant functions in $G^{C_n} \leqslant W_n$ (*diagonal subgroup*), and define $\tau : G \to G^{C_n} \leqslant W_n$ via $(1)(g\tau) = g$ and $(d)(g\tau) = 1$ for all $d \in C_n \smallsetminus 1$ (*1-component*). Then every $g \in G$ has the n-th root $(g\tau) \cdot c$ in W_n. Thus G is divisible provided that $W_n \in \mathfrak{X}$ for all n. Replacing C_n by more complicated groups leads to the following result.

Proposition 3.3 [51, Theorem 2.7] *Let* G *be an a.c.* $\mathrm{L}\mathfrak{X}$-*group where* $\mathfrak{X}$ *is closed under the formation of subgroups, extensions, and cartesian powers of finitely generated* $\mathfrak{X}$-*groups. Then* G *is verbally complete if and only if every element of infinite order in* G *has infinite height in* G. *In particular, periodic a.c.* $\mathrm{L}\mathfrak{X}$-*groups are verbally complete.*

Here G is said to be *verbally complete* if for every $g \in G$ and for every non-trivial reduced word $w(x_1, \ldots, x_r)$ there exist $g_1, \ldots, g_r \in G$ such that $g = w(g_1, \ldots, g_r)$. Note that the conditions of the proposition are satisfied for example for the classes $\mathfrak{X} \in \{\mathfrak{S}_\pi, \mathrm{L}\mathfrak{F}_\pi, \mathrm{L}(\mathfrak{F}_\pi \cap \mathfrak{S})\}$.

A third general embedding technique, due to B. Wehrfritz [47, p. 180], applies to $\mathrm{L}\mathfrak{F}$-groups G. Consider the right regular representation $\rho : G \to \mathrm{Sym}(G)$. Then $G\rho$ is contained in the so-called *constricted symmetric group* $\mathrm{CSym}(G)$, consisting of all $\pi \in \mathrm{Sym}(G)$ for which there exists a finite subgroup $F \leqslant G$ such that $(gF)\pi = gF$ for all $g \in G$. The point here is that $\mathrm{CSym}(G)$ is locally finite again. For example, the fact that every isomorphism between two finite subgroups of $G\rho$ is induced by conjugation with some $\pi \in \mathrm{CSym}(G)$ [47, Lemma 6.3] has the following consequence.

Theorem 3.4   *Every isomorphism between two finite subgroups of an a. c. $L\mathfrak{F}$-group G is induced by conjugation with some element from G. In particular, all the automorphisms of G are locally inner.*

Homogeneity of this kind is a quite common feature of countable e. c. $\mathfrak{X}$-groups.

Up to now we have just tried to gather some *ad hoc* constructions. A more systematic approach is the *method of amalgamation*. Consider an $\mathfrak{X}$-group U which solves some interesting finite system $\mathcal{S}$ of equations (and inequalities). Let V be the subgroup of U generated by the coefficients of $\mathcal{S}$, and suppose that V is contained in the e. c. $\mathfrak{X}$-group G. We call $G \cup U | V$ an *amalgam* of groups. The big question now is: under which conditions are we able to produce a *completion* of the amalgam within the class $\mathfrak{X}$, that is, a group $H \in \mathfrak{X}$ and embeddings $\alpha : G \to H$ and $\beta : U \to H$ such that $\alpha|_V = \beta|_V$? If such a completion exists, then the system $\mathcal{S}\alpha$ can be solved in $U\beta \leqslant H$, and hence $\mathcal{S}$ also has a solution in the e. c. $\mathfrak{X}$-group G. Note that, in the class of all groups, the amalgamated free product $G *_V U$ always completes the amalgam $G \cup U | V$. For a variety $\mathfrak{X}$, one may try to consider $G *_V U$ modulo the corresponding verbal subgroup. In other situations, it has been possible to extend wreath product embeddings of V to G and U.

A third amalgamation method is the *permutational product*, due to B. H. Neumann [80]. Choose left transversals S of V in G, and T of V in U. Then the amalgam $G \cup U | V$ acts faithfully on $\Omega = S \times T \times V$ via $(s,t,v)^g = (\tilde{s},t,\tilde{v})$ and $(s,t,v)^u = (s,\hat{t},\hat{v})$, where $\tilde{s}\tilde{v} = svg$ and $\hat{t}\hat{v} = tvu$. The subgroup of $\mathrm{Sym}(\Omega)$ generated by G and U completes the amalgam and is called the permutational product $P = P(G \cup U | V; S, T)$. Again the problem lies in finding criteria as to when P is contained in the given class $\mathfrak{X}$. This usually requires a slick choice of transversals S and T. However, if G and U are locally finite, and if V is finite, then P is always locally finite again [81, Theorem 5.2]. Since the group table of a finite group can be expressed as a finite system of equations and inequalities, a repeated application of the permutational product yields the following injectivity property of e. c. $L\mathfrak{F}$-groups.

Theorem 3.5   *Let G be an e. c. $L\mathfrak{F}$-group, and let V be a finite subgroup of a countable $L\mathfrak{F}$-group U. Then every embedding of V into G can be extended to an embedding of U into G. In particular, G contains a copy of every countable $L\mathfrak{F}$-group.*

Injectivity of this kind is a second typical feature of e. c. $\mathfrak{X}$-groups. In fact, the e. c. structures in a subgroup closed class $\mathfrak{X} \subseteq L\mathfrak{F}$ are precisely the $\mathfrak{X}$-groups which admit all possible injections of finitely generated $\mathfrak{X}$-groups. This observation certainly motivated B. Maier to introduce and study so-called *closed* $L\mathfrak{X}$-groups [72]. Let $\mathfrak{X}$ be a class of finitely generated groups. The $L\mathfrak{X}$-group G is said to be *closed* in $L\mathfrak{X}$ if for every amalgam $G \cup U | V$ with $U, V \in \mathfrak{X}$, which can be completed in $L\mathfrak{X}$, the inclusion map $\mathrm{id} : V \to G$ lifts to an embedding $U \to G$. Obviously, every closed $L\mathfrak{X}$-group is e. c. in $L\mathfrak{X}$, and the converse holds in classes of $L\mathfrak{F}$-groups. If $\mathfrak{X}$ contains only countably many isomorphism types, and if the underlying language is countable, then every countable $L\mathfrak{X}$-group is contained in a countable closed $L\mathfrak{X}$-group [72, Proposition 1.4]. However, the notion 'closed' is in general much stronger than 'e.c.': closed groups are usually those e.c. groups with the highest degree of homogeneity.

Let $\mathfrak{X}$ be as above. On the basis of a good amalgamation criterion for $\mathfrak{X}$-groups it is then possible to show by means of a *back and forth* argument that there exists a unique countable closed L$\mathfrak{X}$-group. In order to formulate this precisely, we denote by $\mathcal{O}_{\mathfrak{X}}(U)$ the (possibly empty) class of all $\mathfrak{X}$-groups U^* which contain the $\mathfrak{X}$-group U and satisfy the following condition: whenever $V \cup W | U^*$ is an amalgam of $\mathfrak{X}$-groups, then the amalgam $V \cup W | U$ can be completed within the class $\mathfrak{X}$. Vividly stated, the groups in $\mathcal{O}_{\mathfrak{X}}(U)$ *control* amalgamation over U within $\mathfrak{X}$.

Theorem 3.6 [72, Theorem 5.7] *Let $\mathfrak{X}$ be as above, and let G and H be countable closed L$\mathfrak{X}$-groups with the property that $\mathfrak{X}$-subgroups U of G resp. H are always contained in $\mathcal{O}_{\mathfrak{X}}(U)$-subgroups of G resp. H. Then any embedding $\alpha : U^* \to H$, where $G \geqslant U^* \in \mathcal{O}_{\mathfrak{X}}(U)$, gives rise to an isomorphism $G \to H$ which induces α on U.*

The idea here is to extend isomorphisms between $\mathfrak{X}$-subgroups of G and H not only to larger $\mathfrak{X}$-subgroups, but rather to their *controllers*. This guarantees that the process of amalgamation can be carried on step by step (cf. proof of Theorem 6.6). As an example, it follows that there is a unique countable e.c. L$\mathfrak{F}$-group. This group was discovered by P. Hall [25]. On the other hand, there exists a criterion as to when there are many countable closed L$\mathfrak{X}$-structures.

Theorem 3.7 [72, Theorem 4.1] *Let $\mathfrak{X}$ be as above. If there exists an $\mathfrak{X}$-group U such that $\mathcal{O}_{\mathfrak{X}}(U)$ is empty, then there are $2^{\aleph_0}$ countable closed L$\mathfrak{X}$-groups.*

Note that 2^{κ} is the largest possible number of non-isomorphic groups of the infinite cardinality κ. Theorem 3.7 is proved by using the model theoretic method of *forcing*. In many cases, this method allows us to construct e.c. $\mathfrak{X}$-groups with certain properties, for example e.c. $\mathfrak{N}_{c,\pi}$-groups $(\pi \neq \emptyset)$ which are periodic [69, Theorem 2]. Roughly speaking, the philosophy here is to build up an e.c. group recursively as the union of an ascending chain of finitely generated subgroups, which control first-order properties of those members of the chain which could follow at later stages of the construction. Since we are mainly interested in the algebraic properties of e.c. groups, and since this technique requires a deeper model theoretic background, we shall not pursue it in this article. The interested reader is referred to [40] and [43].

Formally, the definition of existential closure can of course be modified by allowing systems of equations and inequalities of larger cardinalities. However, in this setup it is usually very difficult to find useful constructions: most results concerning the completion of amalgams $G \cup U | V$ can only be proved for finitely generated groups V.

4. Algebraically Closed versus Existentially Closed Groups

Before entering into the theory of e.c. groups in specific classes, let us consider the relationships between a.c. and e.c. $\mathfrak{X}$-groups. It has already been mentioned that a. c. fields are e. c. too. A corresponding result holds in the class of all groups [79]. The idea is of course to replace any inequality by a finite set of equations such that solutions of the inequality lead to solutions of the equations, and *vice versa*. This is always possible if every $\mathfrak{X}$-group embeds into a simple $\mathfrak{X}$-group. For suppose that G is a non-trivial a. c. $\mathfrak{X}$-group and that the inequality $w(\bar{c}, \bar{x}) \neq 1$ with coefficients $\bar{c}$ from G has the solution $\bar{h}$ in some $\mathfrak{X}$-group $H \geqslant G$. We may assume now that H is simple. Hence a fixed $g \in G \smallsetminus 1$ lies in the normal closure of $w(\bar{c}, \bar{h})$. Therefore H solves the equation $v(\bar{c}, \bar{x}, \bar{y}) = g$, where the word v takes the form $w(\bar{c}, \bar{x})^{\epsilon_1 y_1} \cdots w(\bar{c}, \bar{x})^{\epsilon_r y_r}$ for suitable $\epsilon_i \in \{\pm 1\}$. Now every solution $\bar{a}, \bar{b}$ in G to the equation $v(\bar{c}, \bar{x}, \bar{y}) = g$ satisfies $w(\bar{c}, \bar{a}) \neq 1$. This method works for example with $\mathfrak{X} = \mathrm{L}\mathfrak{F}$. Although $\mathrm{L}(\mathfrak{F}_\pi \cap \mathfrak{S})$-groups are highly non-simple, the argument can even be extended to give a characterization of a. c. $\mathrm{L}(\mathfrak{F}_\pi \cap \mathfrak{S})$-groups as certain quotients of e. c. $\mathrm{L}(\mathfrak{F}_\pi \cap \mathfrak{S})$-groups [62], since the lattice of their normal subgroups is totally ordered via inclusion.

A general formulation has been proved by P. Bacsich.

Theorem 4.1 [1, Theorem 2.1] *Let $\mathfrak{X}$ be an inductive class of structures which can be axiomatized in a countable first-order language. If every $\mathfrak{X}$-structure embeds into a simple one, then the non-trivial a. c. $\mathfrak{X}$-structures are simple and e. c. in $\mathfrak{X}$.*

Here $G \in \mathfrak{X}$ is said to be *simple* if every non-constant morphism $G \to H$, with $H \in \mathfrak{X}$, is injective. A more transparent proof of Theorem 4.1 can be found in [11, Theorem 6.5]. The theorem applies for example to lattice-ordered groups and to totally ordered groups [19, Theorem 12B], [20, 22].

However, a. c. $\mathfrak{X}$-groups are in general not e. c. in $\mathfrak{X}$. From Theorem 3.2 the a. c. $\mathfrak{A}$-groups are the direct sums of Prüfer p-groups and copies of the rationals. There is no way to express, by a first-order formula, that an element of an $\mathfrak{A}$-group has infinite order. But, as soon as inequalities enter the picture, we can at least blow up the torsion subgroup.

Theorem 4.2 [12] *Let $\mathfrak{X} \in \{\mathfrak{A}_\pi, \mathfrak{A} \cap \mathrm{L}\mathfrak{F}_\pi\}$. Then the e. c. $\mathfrak{X}$-groups are precisely the divisible $\mathfrak{X}$-groups with infinite p-rank for all primes $p \in \pi$.*

Note that the only e. c. $\mathfrak{A}$-groups which are closed are those with infinite torsion-free rank. D. Saracino and C. Wood have obtained a corresponding result for countable $\mathfrak{N}_2^e$-groups.

Theorem 4.3

(a) [88, Section 6] *Any given finite direct power of the cyclic group C_e (resp. $C_{e/2}$ when $2 | e$) occurs as the center of some countable a. c. $\mathfrak{N}_2^e$-group.*

(b) [88, Theorem 3.5] *There is a unique countable e. c. $\mathfrak{N}_2^e$-group, and its center has infinite p-rank for all prime divisors p of e.*

(c) *The assertions (a) and (b) hold, correspondingly, if we remove the bound on the exponent and replace 'countable' by 'countable periodic' everywhere, and C_e by $\mathbb{Q}/\mathbb{Z}$.*

On the other hand, the non-trivial a. c. $\mathfrak{A}^+$-groups are e. c. in $\mathfrak{A}^+$ by Theorems 3.2 and 4.2, and this can be generalized to the classes $\mathfrak{N}_c^+$ (Theorem 6.3).

The class LFC can be viewed as a well-behaved join of the classes $\mathfrak{A}$ and L$\mathfrak{F}$. Since LFC-groups are extensions of L$\mathfrak{F}$-groups by $\mathfrak{A}^+$-groups, the following theorem of F. Haug is in line with the above results.

Theorem 4.4 [26, Satz 10.2] *The a. c.* LFC-*groups are precisely the groups which are e. c. in one of the classes* $\mathfrak{A}^+$ *and* LFC.

In the remainder of this article we shall not pursue the relationships between a. c. and e. c. $\mathfrak{X}$-groups any more. We shall state structural results for a. c. groups whenever possible, but our main interest will be in e. c. $\mathfrak{X}$-groups, since they are more homogeneous.

5. Soluble Groups

A natural construction within the class $\mathfrak{S}_d$ seems to be the wreath product, since its derived length can be well controlled. For example, we may consider the KKF-embedding $G \to G^{(i)}$ Wr $G/G^{(i)}$, where $G^{(i)}$ denotes the i-th term of the derived series of the e. c. $\mathfrak{S}_d$-group G. Define iterated commutators recursively via $w_1(x_1, x_2) = [x_1, x_2]$ and

$$w_{i+1}(x_1, \ldots, x_{2^{i+1}}) = [w_i(x_1, \ldots, x_{2^i}), w_i(x_{2^i+1}, \ldots, x_{2^{i+1}})]$$

for all $i \geqslant 1$. Using the above KKF-embedding, B. Maier has shown that every g in $G \setminus G^{(i)}$ has certain conjugates $g_1, \ldots, g_{2^{d-i}} \in G$ such that $w_{d-i}(g_1, \ldots, g_{2^{d-i}}) \neq 1$. In particular, $\langle g^G \rangle \notin \mathfrak{S}_{d-i}$ and $G^{(i)} = \{g \in G \mid \langle g^G \rangle \in \mathfrak{S}_{d-i}\}$. Therefore the derived series of G is very rigid.

Theorem 5.1 [67, Satz 6] *Let G be an e. c. group in one of the classes $\mathfrak{S}_{d,\pi}$ or* L$\mathfrak{F}_\pi \cap \mathfrak{S}_d$ *where $d \geqslant 2$. Then $G^{(i)}$ $(1 \leqslant i \leqslant d-1)$ is the unique maximal normal $\mathfrak{S}_{d-i}$-subgroup of G, and in particular the unique normal $\mathfrak{S}_{d-i}$-subgroup of G with $\mathfrak{S}_i$-factor group. Moreover, every $g \in G^{(i)}$ is a word of the form $g = w_i(g_1, \ldots, g_{2^i})$ for some $g_1, \ldots, g_{2^i} \in G$.*

E. c. $\mathfrak{S}_2$-groups have been studied by D. Saracino.

Theorem 5.2

(a) [85, Proposition 11] *If G is e. c. in $\mathfrak{S}_2$, then $G/G^{(1)}$ and $G^{(1)}$ are e. c. in* $\mathfrak{A}$.

(b) [85, Proposition 9] *There exists a periodic e. c. $\mathfrak{S}_2$-group.*

The group in part (b) is constructed by forcing methods. In this context it must be pointed out that—in contrast to the situation for $\mathfrak{A}$-groups—the element g of the e. c. $\mathfrak{S}_2$-group G has infinite order modulo $G^{(1)}$ if and only if the first-order sentence $\forall x_1, x_2 \, \exists y_1, y_2 \, [x_1, x_2] = [g, [y_1, y_2]]$ is satisfied in G [85, Proposition 7]. In view of the fact that finitely generated $\mathfrak{S}_2$-groups are residually finite, the assertion in part (b) is related to the result that every e. c. L($\mathfrak{F}_\pi \cap \mathfrak{N}$)-group is e. c. in L$\mathfrak{N}_\pi$ ([68, Satz 6] and [52, Theorem 3.7]). See also Section 6 for the class L$\mathfrak{F} \cap \mathfrak{N}_2$. In fact, Theorem 5.2(b) is a consequence of the following general argument.

Theorem 5.3 *Let $\mathfrak{X}$ be a class of groups which is closed under the formation of subgroups and quotients. If every finitely generated $\mathfrak{X}$-group is residually finite, then every finite $\mathfrak{X}$-group is contained in a countable locally finite e. c. L$\mathfrak{X}$-group.*

Proof. A bijection $f : \omega \times \omega \to \omega$ is given by $(i,j)f = \frac{1}{2}(i+j)(i+j+1) + j$. Note that $i \leqslant (i,j)f$ for all $i,j \in \omega$. Now, let G_0 be any given finite $\mathfrak{X}$-group. The desired e. c. L$\mathfrak{X}$-group G will be the union of an ascending chain of finite $\mathfrak{X}$-groups G_n ($n \in \omega$) which can be constructed recursively from G_0 by the following procedure. Let $\{S_{n,k}\}_{k \in \omega}$ be a list of all finite systems of equations and inequalities with coefficients from G_n. If S_{nf-1} is not solvable in any L$\mathfrak{X}$-supergroup of G_n, then let $G_{n+1} = G_n$. Otherwise, there exists a solution $\overline{h}$ to S_{nf-1} in some finitely generated $\mathfrak{X}$-group $H = \langle G_n, \overline{h} \rangle$. Choose a normal subgroup N of finite index in H such that $G_n \cap N = 1$ and such that $w(\overline{c}, \overline{h}) \notin N$ for every inequality $w(\overline{c}, \overline{x}) \neq 1$ in S_{nf-1} with coefficients $\overline{c} \in G_n$. Then G_n can be identified canonically with $G_n N / N$, and $\overline{hN}$ is a solution to S_{nf-1} in the finite $\mathfrak{X}$-group $G_{n+1} = H/N \geqslant G_n$. This completes the recursion. Since every finite system of equations and inequalities with coefficients from $G = \bigcup_{n \in \omega} G_n$ is of the form $S_{n,k}$ for some $n,k \in \omega$, the countable L$\mathfrak{F}$-group G is e. c. in L$\mathfrak{X}$. $\square$

Recently, O. Belegradek has shown that there are no periodic e. c. $\mathfrak{S}_d$-groups for any $d \geqslant 3$ (see [8]). This follows from the construction of a center-by-$\mathfrak{S}_2$ group containing elements x, y, z which satisfy $x^{-1}[x,y] = [x^2, z] = 1 \neq [x, z]$; these relations imply that x has infinite order.

Questions 5.4

(a) *What can be said about the structure of e. c. L$\mathfrak{F} \cap \mathfrak{S}_d$-groups ?*

(b) *If G is an e. c. $\mathfrak{S}_d$-group, is it true that $G/G^{(i)}$ is e. c. in $\mathfrak{S}_i$ and that $G^{(i)}$ is e. c. in $\mathfrak{S}_{d-i}$ for $1 \leqslant i \leqslant d-1$?*

(c) *Is every a. c. $\mathfrak{S}_d$-group divisible ?*

(d) *If G is an e. c. $\mathfrak{S}_d^+$-group, does there exist for every $N \trianglelefteq G$ an $i \leqslant d-1$ such that $G^{(i+1)} \leqslant N \leqslant G^{(i)}$?*

(e) *Does a (countable) e. c. $\mathfrak{S}_d$-group necessarily split over each term of its derived series ?*

The last two questions are motivated by corresponding results about e. c. $\mathfrak{N}_c^+$-groups (cf. Section 6) resp. e. c. L($\mathfrak{F}_\pi \cap \mathfrak{S}$)-groups (Theorem 8.7).

In [54] the author has tried to develop an amalgamation technique for soluble groups via wreath product embeddings. In analogy with amalgamation theorems for $\mathfrak{F}_p$-groups [38] and for L($\mathfrak{F}_\pi \cap \mathfrak{S}$)-groups [50, Theorem 2.1], the basic idea here is that an amalgam $G \cup H|U$ of $\mathfrak{S}$-groups should be completable in $\mathfrak{S}$, provided that there exist normal series with abelian factors in G and H which induce the same series in U via intersection, and provided that the actions of G and H on the factors of these series are somehow compatible. The unsettled question is as follows.

Question 5.5 *Suppose that $G \cup H|U$ is an amalgam of soluble groups, and that $M \cup N|V$ is an amalgam of abelian groups, such that M is a $\mathbb{Z}G$-module, N*

is a $\mathbb{Z}H$*-module, and* V *is* U*-invariant. Under which conditions can we complete* $M \cup N | V$ *within an abelian group* Z *and extend the actions of* G *and* H *on* M *and* N *(resp.) to actions on* Z *such that* G *and* H *generate a soluble subgroup of* $\mathrm{Aut}(Z)$ *?*

Any half-way satisfactory answer would open the possibility of making further progress in the area of e. c. $\mathfrak{S}_d$-groups via the method of amalgamation.

6. Nilpotent Groups

Let $\zeta_i(G)$ resp. $\gamma_i(G)$ be the i-th term of the upper resp. lower central series of the group G. Moreover, let $[g,_i h]$ denote the iterated commutator $[g, h, \ldots, h]$ of length $i+1$. The first major observation about $\mathfrak{N}_c$-groups G is that a wreath product construction allows us to solve the equation $g = [x,_i y]$ with coefficient $g \in \zeta_{c-i}(G)$ in an $\mathfrak{N}_c$-supergroup of G. Note that wreath products of nilpotent groups are in general far from being nilpotent; so one really has to consider proper subgroups of wreath products here. As a consequence, a. c. nilpotent groups only have a single central series of length c (cf. Theorem 5.1).

Theorem 6.1 [69, Theorem 1] *The upper and the lower central series of an a. c. group* G *in any of the classes* $\mathfrak{N}_{c,\pi}$ *and* $\mathrm{L}\mathfrak{F}_\pi \cap \mathfrak{N}_c$ *coincide, and every* $g \in \gamma_i(G)$ *is of the form* $g = [h,_{i-1} k]$ *for some* $h, k \in G$.

Central product constructions show that the center of every e. c. group in any of the classes $\mathfrak{N}_{c,\pi}$ and $\mathrm{L}\mathfrak{F}_\pi \cap \mathfrak{N}_c$ is e. c. in $\mathfrak{A}_\pi$ resp. $\mathrm{L}\mathfrak{F}_\pi \cap \mathfrak{A}$.

6.1. TORSION-FREE NILPOTENT GROUPS

Among e. c. nilpotent groups, the e. c. $\mathfrak{N}_c^+$-groups are now the ones best understood. The main reason seems to be the uniqueness of roots in $\mathfrak{N}_c^+$-groups. We will give a detailed exposition of these groups, since their theory shows typical features which also occur in other group classes, and since the general results presented here are only indicated roughly in the literature. E. c. $\mathfrak{N}_c^+$-groups have been studied by B. Baumslag and F. Levin [2], D. Saracino [86, 87], R. Hessenkamp [32] and B. Maier [67, 69, 70], [72, Section 8.8]. The core of the theory is the following amalgamation theorem due to B. Maier.

Theorem 6.2 [71, Satz 1] *An amalgam* $G \cup H | U$ *of* $\mathfrak{N}_c^+$*-groups can be completed in the class* $\mathfrak{N}_c^+$ *if and only if there exist central series* $G = G_1 \geqslant \ldots \geqslant G_{c+1} = 1$ *in* G *and* $H = H_1 \geqslant \ldots \geqslant H_{c+1} = 1$ *in* H *such that*

(1) $G_i \cap U = H_i \cap U$ *for all* i, *and*

(2) $[G_i, G_j] \leqslant G_{i+j}$ *and* $[H_i, H_j] \leqslant H_{i+j}$ *for all* $i, j \leqslant c$.

The completion can in fact be chosen so that one of its central series of length c *induces the given series in* G *and in* H.

Idea of Proof. The conditions are obviously necessary: intersect G and H with the terms of the lower central series of a completion in order to obtain series satisfying the conditions (1) and (2). Conversely, we may restrict ourselves to subamalgams consisting of finitely generated $\mathfrak{N}_c^+$-groups, since an ultraproduct argument then allows us to complete the original amalgam $G \cup H | U$ in $\mathfrak{N}_c^+$ [11, Theorem 6.1]. Next, we replace all the finitely generated groups G, H, U, G_i and H_i by their divisible hulls. Then the assumptions of the theorem are still valid, and the factors G_i/G_{i+1} and H_i/H_{i+1} are finite-dimensional $\mathbb{Q}$-vector spaces. By recursively choosing copies of $\mathbb{Q}$ from the lowest possible layers of the given series, we see that G and H are iterated semidirect products which have the form $G = (\cdots(U \rtimes A_1) \rtimes \cdots) \rtimes A_r$ and $H = (\cdots(U \rtimes B_1) \rtimes \cdots) \rtimes B_s$, where $A_i \cong \mathbb{Q} \cong B_j$. Proceeding by induction we may assume that $s = 1$ and that the amalgam $(U A_1 \cdots A_{r-1}) \cup (U B_1) | U$ has a completion V with certain properties regarding central series. It is then possible to construct a supergroup W of V, to which the actions of A_r and B_1 on U extend in such a way that A_r and B_1 stabilize the same central series in W. Note that A_r and B_1 then generate a nilpotent group of automorphisms of W. The assumptions of the theorem even allow us to design W in such a way that the semidirect product $W \rtimes \langle A_r, B_1 \rangle$ has the right class of nilpotency. $\qquad\square$

Theorem 6.3   *Let $G \in \mathfrak{N}_c^+$ where $c \geqslant 2$. Then the following are equivalent.*

(1) *G is an e. c. $\mathfrak{N}_c^+$-group.*

(2) *G is a non-trivial a. c. $\mathfrak{N}_c^+$-group.*

(3) *G is non-trivial, divisible, and has a unique central series of length c; and whenever $G \cup U | V$ is an amalgam of $\mathfrak{N}_c^+$-groups with $\zeta_1(U) \leqslant V$, which can be completed in $\mathfrak{N}_c^+$, and where $U/\zeta_1(U)$ is finitely generated, then $\mathrm{id} : V \to G$ can be extended to an embedding $U \to G$.*

Proof.   (2) $\Rightarrow$ (3): Let $Z = \zeta_1(U)$. Clearly $U = (\cdots(Z \rtimes \langle b_1 \rangle) \rtimes \cdots) \rtimes \langle b_s \rangle$ and $V = \langle Z, a_1, \ldots, a_r \rangle$. Now only finitely many equations are required in order to express the coefficients a_ν and the commutators $[b_i, b_j]$ as words in the b_k ($k < i < j$) and coefficients from Z. Any solution to these equations in the a. c. $\mathfrak{N}_c^+$-group G gives rise to a homomorphism $\alpha : U \to G$ with $\alpha|_V = \mathrm{id}_V$. Because of $Z \cap \mathrm{Ker}\,\alpha = 1$, the homomorphism α is injective.

The implication (3) $\Rightarrow$ (1) will be shown after the proof of Proposition 6.5. $\square$

The amalgamation criterion in Theorem 6.2 can be controlled quite easily in finitely generated $\mathfrak{N}_c^+$-groups.

Proposition 6.4   *Let U be a finitely generated subgroup of the a. c. $\mathfrak{N}_c^+$-group G. Then there exists a finitely generated subgroup U^* of G containing U such that every central series of length c in U^* induces the series $\{U \cap \gamma_i(G) \mid 1 \leqslant i \leqslant c+1\}$ in U via intersection. The same holds with $U \cdot \zeta_1(G)$ and $U^* \cdot \zeta_1(G)$ in place of U resp. U^*.*

Proof. By Theorem 6.1, each of the finitely many generators of $U \cap \gamma_i(G)$ ($2 \leqslant i \leqslant c$) has the form $[h, _{i-1} k]$ for suitable elements $h, k \in G$. Let $\mathcal{S}$ be the set of all finitely generated subgroups of G which contain U and all these elements h, k. Then $U \cap \gamma_i(G) = U \cap \gamma_i(V)$ for all $V \in \mathcal{S}$ and all i. Suppose now that there exists i

such that the $\mathbb{Q}$-dimension of the divisible hull Z_V of

$$(U \cap \zeta_{c+1-i}(V))\gamma_i(G)/\gamma_i(G) \cong (U \cap \zeta_{c+1-i}(V))/(U \cap \gamma_i(V))$$

is positive for all $V \in \mathcal{S}$. Then $1 \neq \bigcap \{Z_V \mid V \in \mathcal{S}\} \leqslant \zeta_{c+1-i}(G)/\gamma_i(G)$, in contradiction to Theorem 6.1. This shows that some $U^* \in \mathcal{S}$ has the property that $U \cap \gamma_i(U^*) = U \cap \zeta_{c+1-i}(U^*)$ for all i. $\qquad\square$

Note that *every* finitely generated $\mathfrak{N}_c^+$-group U is contained in some a. c. $\mathfrak{N}_c^+$-group and hence in a finitely generated $\mathfrak{N}_c^+$-group which controls amalgamation over U. In fact, the series of isolators of the $\gamma_i(U)$ can be controlled by some finitely generated $\mathfrak{N}_c^+$-supergroup of U [71, Satz 3]. It follows from Proposition 6.4 and Theorem 3.6 that there exists a unique countable closed $\mathfrak{N}_c^+$-group. We shall see now that e. c. $\mathfrak{N}_c^+$-groups are almost closed but do not necessarily have a large center. In the following, $\mathcal{S}(G)$ will denote the set of all subgroups U of the group G such that $\zeta_1(G) \leqslant U$ and such that $U/\zeta_1(G)$ is finitely generated.

Proposition 6.5

    (a) *For each* $1 \leqslant n \leqslant \omega$ *there exists a countable a. c. $\mathfrak{N}_c^+$-group G_n such that* $\zeta_1(G_n) = \mathbb{Q}^n$.

    (b) *Let H be a. c. in $\mathfrak{N}_c^+$. Then every group $U \in \mathcal{S}(H)$ is contained in some $W \in \mathcal{S}(H)$ such that $\zeta_1(W) = \zeta_1(H)$.*

Proof. (a) As in the proof of Theorem 5.3, the group G_n can be constructed as the union of an ascending chain of subgroups $G_{n,\nu}$ $(\nu < \omega)$ such that $\mathbb{Q}^n = \zeta_1(G_{n,\nu})$, such that $G_{n,\nu}/\mathbb{Q}^n$ is finitely generated, and such that for each finite set of equations $\mathcal{S}$ with coefficients from $G_{n,\nu}$ there exists $\mu \geqslant \nu$ such that $\mathcal{S}$ has a solution in $G_{n,\mu+1}$, provided that it can be solved in some $\mathfrak{N}_c^+$-supergroup of $G_{n,\mu}$. Here the center can be kept small by a repeated application of the following procedure: as long as $\mathbb{Q}^n$ has a non-trivial direct complement K in $\zeta_1(G_{n,\mu+1})$, replace $G_{n,\mu+1}$ by $G_{n,\mu+1}/K$ and identify $G_{n,\mu}$ canonically with $G_{n,\mu}K/K$.

    (b) Let $U \leqslant U^* \in \mathcal{S}(H)$ as in Proposition 6.4. Then $U^* = U_0 \cdot \zeta_1(H)$ for some finitely generated group U_0, where the rank n of $U_0 \cap \zeta_1(H)$ is positive. Let $\overline{U}_0$ be the subgroup of H generated by U_0 and the divisible hull $\mathbb{Q}^n$ of $U_0 \cap \zeta_1(H)$. Let $A = \zeta_1(U_0) \times \langle z_1 \rangle \times \cdots \times \langle z_c \rangle$ with $\langle z_i \rangle \cong \mathbb{Z}$, and let T be the stability group of the series $\{(\zeta_1(U_0) \cap \gamma_i(H)) \times \langle z_i \rangle \times \cdots \times \langle z_c \rangle \mid 1 \leqslant i \leqslant c+1\}$ in A. Then every central series of length c in the semidirect product $S = (A \rtimes T) \in \mathfrak{N}_c^+$ induces the above series in A, and $\zeta_1(S) = U_0 \cap \zeta_1(H)$. Let $\overline{S}$ be the central product of S and $\mathbb{Q}^n$ over $\zeta_1(S)$. An application of Theorem 6.3 to the amalgam $G_n \cup \overline{S}|\mathbb{Q}^n$ allows us to assume that $\overline{S} \leqslant G_n$. The presence of T implies $A \cap \gamma_i(G_n) = A \cap \gamma_i(H)$ for all i. Therefore the amalgam $G_n \cup \overline{U}_0|A\mathbb{Q}^n$ has a completion in $\mathfrak{N}_c^+$, and Theorem 6.3 allows us to assume that $\overline{U}_0 \leqslant G_n$. From the construction in (a) we see that $\overline{U}_0$ is contained in some $W_0 \in \mathcal{S}(G_n)$ with $\zeta_1(W_0) = \mathbb{Q}^n$. Let W be the central product of W_0 and $\zeta_1(H)$ over $\mathbb{Q}^n$. Clearly $\zeta_1(W) = \zeta_1(H)$. Now $U^* \leqslant W$ ensures that the amalgam $H \cup W|U$ can be completed in $\mathfrak{N}_c^+$, and Theorem 6.3 yields an embedding $\alpha : W \to H$ with $\alpha|_U = \mathrm{id}|_U$. $\qquad\square$

Note that we have only used the implication $(2) \Rightarrow (3)$ of Theorem 6.3 in the proof of Proposition 6.5(b) We shall now complete the proof of Theorem 6.3.

Proof of Theorem 6.3. $(3) \Rightarrow (1)$: Let $Z = \zeta_1(G)$. Let $\mathcal{S}$ be a finite set of equations and inequalities with coefficients $g_1, \ldots, g_r \in G$ and with a solution $h_1, \ldots, h_s$ in the divisible $\mathfrak{N}_c^+$-group $H \geqslant G$. Then $Z \leqslant \zeta_1(H)$ by assumption. Also, since Z is divisible, there exists a homomorphism $\varphi : \zeta_1(H) \to Z$ with $\varphi|_Z = \mathrm{id}_Z$ and such that $(v(h_1, \ldots, h_s))\varphi \neq 1$ for every inequality $v(x_1, \ldots, x_s) \neq 1$ from $\mathcal{S}$ with $v(h_1, \ldots, h_s) \in \zeta_1(H)$. Now $N = \mathrm{Ker}\,\varphi$ satisfies $N \cap Z = 1$; hence $N \cap G = 1$, and so we may identify G canonically with the subgroup GN/N of H/N. Note that $\mathcal{S}$ still has the solution $h_1 N, \ldots, h_s N$ in $H/N \in \mathfrak{N}_c^+$. And again, $\zeta_1(GN/N) \leqslant \zeta_1(H/N)$. By repeating this argument we can produce an ascending chain of normal subgroups $N_\alpha \trianglelefteq H$ ($\alpha \leqslant \kappa$) (where we take unions at limit stages), such that $G \cong GN_\alpha/N_\alpha$ and such that $\mathcal{S}$ has the solution $h_1 N_\alpha, \ldots, h_s N_\alpha$ in H/N_α. Finally, H/N_κ will satisfy $\zeta_1(H/N_\kappa) = \zeta_1(GN_\kappa/N_\kappa)$. We may therefore assume that $\zeta_1(H) = Z$. Proceeding as in the proof of Proposition 6.5(a) we can produce an a. c. $\mathfrak{N}_c^+$-group $L \geqslant H$ with $\zeta_1(L) = Z$.

Let $V = \langle Z, g_1, \ldots, g_r \rangle$ and $U = \langle V, h_1, \ldots, h_s \rangle$. By Proposition 6.5(b), the group U is contained in some $W \in \mathcal{S}(L)$ such that $\zeta_1(W) = Z$. Since L is a completion of the amalgam $G \cup W|V$, condition (3) ensures that there is an embedding $W \to G$ which maps $h_1, \ldots, h_s$ to a solution of $\mathcal{S}$ in G. $\qquad \square$

In analogy with $\mathfrak{A}^+$-groups, the countable e. c. $\mathfrak{N}_c^+$-groups can be parametrized by the $\mathbb{Q}$-dimensions of their centers. Of course, the countable closed $\mathfrak{N}_c^+$-group is the one whose center has infinite $\mathbb{Q}$-dimension.

Theorem 6.6

(a) For each $1 \leqslant n \leqslant \omega$ the group G_n is the unique countable e. c. $\mathfrak{N}_c^+$-group with $\zeta_1(G_n) = \mathbb{Q}^n$.

(b) An isomorphism $\alpha : U \to V$ between finitely generated subgroups of G_n can be extended to an automorphism of G if and only if $(U \cap \gamma_i(G_n))\alpha = V \cap \gamma_i(G_n)$ for all i.

(c) $\mathrm{Aut}(G_n)$ acts transitively on the non-trivial elements of each layer of the lower central series of G_n.

(d) $|\mathrm{Aut}(G_n)| = 2^{\aleph_0}$.

Proof. (a) Let H be any countable e. c. $\mathfrak{N}_c^+$-group with $\zeta_1(H) = \mathbb{Q}^n$. Applying Propositions 6.4 and 6.5(b) we find that every $U \in \mathcal{S}(H)$ is contained in some $U^* \in \mathcal{S}(H)$ such that every central series of length c in U^* induces the series $\{U \cap \gamma_i(H) \mid 1 \leqslant i \leqslant c + 1\}$ in U, and such that $\zeta_1(U^*) = \zeta_1(H)$. Consider an embedding $\alpha : U^* \to G_n$. If $U \leqslant W \in \mathcal{S}(H)$, then $\alpha|_U$ can be extended to an embedding $\beta : W^* \to G_n$ as follows. The presence of U^* allows us to complete the amalgam $W^* \cup G_n|U \equiv U\alpha$ in $\mathfrak{N}_c^+$, where U is identified canonically with $U\alpha$. Because of $\zeta_1(W^*) = \mathbb{Q}^n \leqslant U$, the existence of β follows from Theorem 6.3.

The above method can now be combined with a *back and forth* argument in order to build up an isomorphism $H \to G_n$ via successive extensions along countable chains and their controllers in $\mathcal{S}(H)$ resp. $\mathcal{S}(G_n)$ with union H resp. G_n.

(b) This follows in the same manner as the uniqueness of G_n.

(c) Apply (b) with $U = \langle g \rangle$ and $V = \langle h \rangle$, where $g, h \in \gamma_i(G_n) \smallsetminus \gamma_{i+1}(G_n)$.

(d) Application of the construction in Proposition 6.5(a), with the central product C of $\aleph_0$ copies of G_n in place of $G_{n,0}$, shows that C is contained in G_n. In particular, G_n contains a copy D of an infinite direct power of $\mathbb{Q}$, such that $D \cap \gamma_1(G_n) = 1$. Now (b) allows us to construct $2^{\aleph_0}$ automorphisms of G_n from isomorphisms between finitely generated subgroups of G_n which act distinctly on D.   □

It should also be possible to show that normal subgroups of G_n are always enclosed by two subsequent terms of the lower central series in G_n (cf. [70, Theorem 4.2]). In connection with Theorem 6.6(c) this implies that the $\gamma_i(G_n)$ are the only characteristic subgroups of G_n. However, the central quotients of e. c. $\mathfrak{N}_c^+$-groups are not e. c. in $\mathfrak{N}_{c-1}^+$ for $c \geqslant 3$ [32, Satz 5.7]. This corresponds to results about e. c. central extensions (cf. Theorems 7.7 and 8.13).

Questions 6.7

(a) Can $\mathrm{Aut}(G_n)$ be described in a way similar to that in Theorem 8.12 ?

(b) Is every automorphism of G_n which stabilizes the lower central series of G_n a locally inner automorphism ?

(c) Is it possible to calculate the Schur multipliers of e. c. $\mathfrak{N}_c^+$-groups ?

For positive results concerning the last question in other classes of groups see Theorem 7.7, and also Sections 8.2 and 10.1.

6.2. PERIODIC NILPOTENT GROUPS

E. c. periodic nilpotent groups have been studied by D. Saracino and C. Wood [88], and by B. Maier [75]. Since every periodic nilpotent group is the direct product of its maximal p-subgroups, it is a straightforward fact that the e. c. $\mathfrak{N}_c^e$-groups (resp. $\mathrm{L}\mathfrak{F} \cap \mathfrak{N}_c$-groups) are precisely the direct products of e. c. $\mathrm{L}\mathfrak{F}_p \cap \mathfrak{N}_c$-groups (resp. $\mathrm{L}\mathfrak{F}_p \cap \mathfrak{N}_c$-groups). The analysis of e. c. $\mathfrak{N}_c^p$-groups can be based on an analogue of Theorem 6.2.

Theorem 6.8 [75, Theorem 1.2] *Let p be a prime. If $p > c$, then Theorem 6.2 holds word for word with the class $\mathfrak{N}_c^p$ in place of $\mathfrak{N}_c^+$.*

Idea of Proof. Forget divisibility, and follow the proof of Theorem 6.2. Because $c < p$, results of P. Hall [24, Section 4] yield that the $\mathfrak{N}_c$-group $W \rtimes \langle A_r, B_1 \rangle$ (where $A_r \cong C_p \cong B_1$) is regular, generated by elements of order p, and hence of exponent p.   □

Theorem 6.8 allows us to proceed as for the class $\mathfrak{N}_c^+$ provided that we can show that e. c. $\mathfrak{N}_c^p$-groups have a unique central series of length c. This is accomplished as follows.

Proposition 6.9 [75, Theorem 2.1] *Let $U \in \mathfrak{N}_c^p$, where p is a prime and $p > c$. Then the following are equivalent.*

(1) U *has a unique central series of length c.*

(2) *Every amalgam $G \cup H | U$ of $\mathfrak{N}_c^p$-groups over the given group U can be completed in $\mathfrak{N}_c^p$.*

Proof. $(1) \Rightarrow (2)$: This follows from Theorem 6.8.

$(2) \Rightarrow (1)$: Suppose that $\gamma_i(U) \neq \zeta_{c+1-i}(U)$ for some i. Choose i maximal, and let $u \in \zeta_{c+1-i}(U) \smallsetminus \gamma_i(U)$. Let T_1, T_2 be two copies of the group of all upper unitriangular $(c+1) \times (c+1)$-matrices over $\mathbb{F}_p$. Then $p > c$ ensures that $T_1 \in \mathfrak{N}_c^p$. Choose $a \in \gamma_{i-1}(T_1) \smallsetminus \gamma_i(T_1)$ and $b \in \gamma_i(T_2) \smallsetminus \gamma_{i+1}(T_2)$. Using Theorem 6.8 it is then possible to complete the amalgams $U \cup T_1 | \langle u \rangle \equiv \langle a \rangle$ and $U \cup T_2 | \langle u \rangle \equiv \langle b \rangle$ within some $G \in \mathfrak{N}_c^p$ resp. $H \in \mathfrak{N}_c^p$. But then it is impossible to complete the amalgam $G \cup H | U$ within some $W \in \mathfrak{N}_c^p$, since this would imply $u = a \notin \gamma_i(T_1) = \zeta_{c+1-i}(T_1) \geqslant \zeta_{c+1-i}(W) \cap T_1 \geqslant \zeta_{c+1-i}(W) \cap \langle u \rangle$ while $u = b \in \gamma_i(T_2) \cap \langle u \rangle \leqslant \gamma_i(W) \cap \langle u \rangle$. $\square$

A corresponding result holds for $\mathfrak{N}_c^+$-groups [71, Satz 2]. Groups with the above property (2) are called *amalgamation bases* in $\mathfrak{N}_c^p$. Since the class $\mathfrak{N}_c^p$ is first-order axiomatizable by $\forall \exists$-sentences, every e.c. $\mathfrak{N}_c^p$-group is an amalgamation base in $\mathfrak{N}_c^p$ [43, 3.2.7].

Corollary 6.10 *Every e. c. $\mathfrak{N}_c^p$-group (where p is a prime and $p > c$) has a unique central series of length c.*

We can now prove the analogue of Proposition 6.4 and obtain a unique countable closed $\mathfrak{N}_c^p$-group. However, since $\mathfrak{N}_c^p$ is a class of $L\mathfrak{F}$-groups, every closed $\mathfrak{N}_c^p$-group is e. c. in $\mathfrak{N}_c^p$.

Theorem 6.11 [75, Theorem 3.5] *There exists a unique countable e. c. $\mathfrak{N}_c^p$-group for every prime p with $p > c$.*

Questions 6.12

(a) *Which further properties does the above countable e. c. $\mathfrak{N}_c^p$-group have ?*

(b) *What happens for $p \leqslant c$?*

(c) *Can the above argument for $\mathfrak{N}_c^p$-groups be carried over to other classes of $L\mathfrak{F}_p \cap \mathfrak{N}_c$-groups, e. g. to regular groups or to powerful groups ?*

Results about e. c. $\mathfrak{N}_c^e$-groups for proper prime-power exponents e exist only for $c = 2$.

Proposition 6.13 [88, Theorem 2.7] *Let G be an e. c. $\mathfrak{N}_2^e$-group, where $3 \leqslant e = p^s$ for some prime p. Then $\gamma_2(G) = \zeta_1(G)$ is an infinite direct power of copies of the cyclic group C_{p^s} (for $p > 2$) resp. $C_{2^{s-1}}$ (for $p = 2$).*

This follows from central amalgamations with the free $\mathfrak{N}_2^e$-group. As mentioned in Theorem 4.3(b), there is a unique countable e. c. $\mathfrak{N}_2^e$-group. This group and its automorphisms can be characterized with the help of the following notion. A subset

$\{a_n \mid n < \omega\}$ of the countable $\mathfrak{N}_2^{p'}$- group G is said to be a *tidy basis* if

(1) $G/\gamma_2(G)$ is the direct sum of the $\langle a_n \gamma_2(G)\rangle$;

(2) for every $n < \omega$ there is a unique $j < n$ such that $[a_j, a_n] \neq 1$; and

(3) whenever $n, i, k, r < \omega$ and $x, y \in \gamma_2(G)$ satisfy $p^r \leqslant \exp(G/\gamma_2(G))$, $r + k \leqslant s$, $o(y) = p^k$, and $1 < o(x) \leqslant \max\{o(a_1), \ldots, o(a_n), p^{r+k}\}$, then there are infinitely many $m < \omega$ such that $o(a_m \gamma_2(G)) = p^r$, $a_m^{p^r} = y$, and $[a_i, a_m] = x$ while $[a_j, a_m] = 1$ for all $i \neq j \leqslant n$.

Theorem 6.14 [88, Section 5] *Let e be a prime power, $e \geqslant 3$.*

(a) *Every countable e. c. $\mathfrak{N}_2^e$-group has a tidy basis.*

(b) *Let G and H be countable e. c. $\mathfrak{N}_2^e$-groups. If A and B are tidy bases of G resp. H, then any isomorphism $\alpha : \gamma_2(G) \to \gamma_2(H)$ can be extended to an isomorphism $\beta : G \to H$ which maps A onto B.*

Corresponding results apply to periodic e. c. $\mathfrak{N}_2$-groups. In fact these groups can be axiomatized by a set of $\forall\exists$-sentences expressing the following properties [88, 3.10]:

(1) the group is nilpotent of class two;

(2) its center consists of commutators and is divisible of infinite p-rank for every prime p; and

(3) there exist solutions to sets of equations similar to those in condition (3) of the definition of a tiny basis (for all p and all s).

Since these $\forall\exists$-sentences can always be realized in a $\mathrm{L}\mathfrak{F} \cap \mathfrak{N}_2$-supergroup of any given $\mathrm{L}\mathfrak{F} \cap \mathfrak{N}_2$-group [88, Theorem 4.1], it follows that the periodic e. c. $\mathfrak{N}_2$-groups are precisely the e. c. $\mathrm{L}\mathfrak{F}\cap\mathfrak{N}_2$-groups. Moreover there exist 2^κ e. c. $\mathrm{L}\mathfrak{F}\cap\mathfrak{N}_2$-groups of each infinite cardinality κ, since there is a $\mathrm{L}\mathfrak{F} \cap \mathfrak{N}_2$-group which realizes more than κ quantifier-free types over some subset of cardinality κ [88, Corollary 4.2].

Question 6.15 *Can the above results be extended in some way to the classes $\mathfrak{N}_3^e$ resp. $\mathrm{L}\mathfrak{F} \cap \mathfrak{N}_3$?*

6.3. Mixed Nilpotent Groups

In certain classes of nilpotent groups the e. c. objects decompose nicely. We shall denote the torsion subgroup of the $\mathfrak{N}$-group G by $T(G)$, and let $G^e = \langle g^e \mid g \in G\rangle$.

Theorem 6.16

(a) [74, Proposition 3] *Suppose that G is e. c. in the class of all $\mathfrak{N}_c$-groups whose torsion subgroup has exponent e. If $c < p$ for each prime divisor p of e, then $G = T(G) \times G^e$, and $T(G)$ is e. c. in $\mathfrak{N}_c^e$, while G^e is a. c. in $\mathfrak{N}_c^+$.*

(b) [75, Proposition 3.7 and Theorem 3.8] *Suppose that G is e. c. in the class $\mathfrak{N}_c(p)$ of all $\mathfrak{N}_c$-groups H such that $H/\zeta_1(H)$ has exponent p. Then we have $G = T(G) \times A$, where $T(G)$ is e. c. in $\mathrm{L}\mathfrak{F} \cap \mathfrak{N}_c(p)$, while A is a. c. in $\mathfrak{A}^+$. Also, if $c \leqslant p$, then there is a unique countable e. c. group in $\mathrm{L}\mathfrak{F} \cap \mathfrak{N}_c(p)$.*

Idea of Proof. Once it is established that $T(G)$ is a direct factor of G, it is straightforward that the factors are e. c. resp. a. c. in the corresponding classes. In case (a) it follows from the hypotheses on c and e that every element of G^e is an e-th power in G [3, Corollary 2.31]. Hence $T(G) \cap G^e = 1$, and G embeds into $G/G^e \times G/T(G)$. This embedding is surjective, since the e. c. group G solves the equations $g = xy^e$ and $x^e = 1$ for every $g \in G$ (note that $G/T(G) \in \mathfrak{N}_c^+$ has a divisible hull).

In case (b), central product constructions show that G^p is divisible. Therefore $G = T(G) \cdot G^p = T(G) \times A$ for a torsion-free direct complement A to $T(G) \cap G^p$ in $\zeta_1(G)$. The uniqueness of the countable e. c. group in $\mathrm{L}\mathfrak{F} \cap \mathfrak{N}_c(p)$ for $c \leqslant p$ follows by the same procedure as in the class $\mathfrak{N}_c^p$ for $c < p$ (Theorem 6.11) from a characterization of the amalgamation bases in $\mathfrak{N}_c(p)$ [75, Theorem 2.2]. □

E. c. objects G in the class of all $\mathfrak{N}_c$-groups H with $T(H) \in \mathfrak{N}_c(p)$ need not decompose directly over $T(G)$: if G contains any non-trivial split extension of the form $(C_{p^\infty} \times C_{p^\infty}) \rtimes C_\infty$, then $C_{p^\infty} \times C_{p^\infty} \leqslant \zeta_1(T(G))$. There also exists a result about the structure of the torsion subgroup of e. c. $\mathfrak{N}_{c,\pi}$-groups.

Theorem 6.17 [69, Section 4] *Let G be an e. c. $\mathfrak{N}_{c,\pi}$-group, and let $T = T(G)$.*

(a) *The set D of all elements of infinite height in T forms a divisible subgroup of $\zeta_1(T)$, and $D = \gamma_c(T) \times A$ where $\gamma_c(T) = D \cap \zeta_1(G)$ and $A \cap \zeta_1(G) = 1$.*
(b) *If $c = 2$, then $D = \zeta_1(T)$ and $T = E \times A$, where E is periodic e. c. in $\mathfrak{N}_{c,\pi}$.*

There is not much more to be said in general about e. c. $\mathfrak{N}_{c,\pi}$-groups ($c \geqslant 2$), since there exist too many countable such objects. In [41] W. Hodges constructs for each couple $\pi_0 \subseteq \pi$ of sets of primes a countable $\mathfrak{N}_{c,\pi}$-group G_{π_0} which contains a π_0-divisible element admitting no π_0-roots modulo $\gamma_2(H)$ in any $\mathfrak{N}_{c,\pi}$-group $H \geqslant G_{\pi_0}$. Suppose now that π is infinite. Then there are $2^{\aleph_0}$ possible choices of π_0, while every countable $\mathfrak{N}_{c,\pi}$-group can contain at most countably many of the groups G_{π_0}. This implies that there are $2^{\aleph_0}$ countable e. c. $\mathfrak{N}_{c,\pi}$-groups. The result does in fact also hold for finite sets π.

Theorem 6.18 [69, Theorem 5] *For every non-empty set π of primes and every $c \geqslant 2$ there exist 2^κ e. c. $\mathfrak{N}_{c,\pi}$-groups of each infinite cardinality κ.*

This is proved by controlling p-heights ($p \in \pi$) modulo the derived subgroup.

7. Locally Finite Groups

7.1. UNCOUNTABLE GROUPS, AUTOMORPHISMS, GROUP RINGS, AND PERMUTATION REPRESENTATIONS

From Section 3 it is already known that there exists a unique countable e. c. $\mathrm{L}\mathfrak{F}$-group ULF, and that every countable $\mathrm{L}\mathfrak{F}$-group embeds into ULF. Some authors have tried to check out how far such properties extend to e. c. $\mathrm{L}\mathfrak{F}$-groups of larger cardinalities. We shall call an $\mathrm{L}\mathfrak{F}$-group κ-*universal* if it is e. c. in $\mathrm{L}\mathfrak{F}$ of cardinality κ and contains a copy of *every* $\mathrm{L}\mathfrak{F}$-group of cardinality κ.

Theorem 7.1

(a) [66] *There are 2^κ e. c. $L\mathfrak{F}$-groups of each uncountable cardinality κ.*

(b) [33] *There exist $2^{\aleph_1}$ e. c. $L\mathfrak{F}$-groups of cardinality $\aleph_1$ whose uncountable subgroups do not belong to any proper variety.*

(c) [23] *If the uncountable cardinal κ satisfies $\kappa = \kappa^{\aleph_0}$, then there is no κ-universal $L\mathfrak{F}$-group.*

Part (a) follows from stability theory. Theorems 3.4 and 3.5 allow us to axiomatize the e. c. $L\mathfrak{F}$-groups by a $L_{\omega_1,\omega}$-sentence; and the cartesian power C_κ of κ copies of $\text{Sym}(3)$ realizes more than κ types over the corresponding direct power of $\text{Sym}(3)$. In this way one even obtains 2^κ pairwise non-embeddable e. c. $L\mathfrak{F}$-groups of each regular uncountable cardinality κ. Part (b) is a considerable improvement of the above observation that the metabelian group $C_{\aleph_1}$ contains a subgroup of cardinality $\aleph_1$ which does not embed into every uncountable e. c. $L\mathfrak{F}$-group (see also [65]). Part (c) is just one of several criteria under which there exist no κ-universal $L\mathfrak{F}$-groups. The condition $\kappa = \kappa^{\aleph_0}$ is satisfied for example when the generalized continuum hypothesis GCH holds and κ has uncountable cofinality.

Question 7.2 *Is the existence of a $\aleph_1$-universal $L\mathfrak{F}$-group consistent with* ZFC *?*

It follows from Theorem 3.4 that every isomorphism between finite subgroups of ULF extends to various isomorphisms between larger finite subgroups, and hence to $2^{\aleph_0}$ automorphisms of ULF. In fact, a copy of $\text{Sym}(\omega)$ is contained in the outer automorphism group of ULF [34, Theorem 4(d)]. In contrast to this, there exist e. c. $L\mathfrak{F}$-groups which are *complete*, i. e. each of whose automorphisms is inner.

Theorem 7.3 [10, 33, 96] *Let κ^+ denote the successor cardinal of an infinite cardinal κ. Assume $\kappa = \aleph_0$, or* GCH, *or $\kappa = \kappa^{\aleph_0}$. Then there exist 2^{κ^+} complete e. c. $L\mathfrak{F}$-groups of cardinality κ^+.*

Such groups G are always constructed as the union of an ascending chain of subgroups G_α $(\alpha < \kappa^+)$ of cardinalities κ (where unions are to be taken at limit stages), using the fact that, whenever φ is an outer automorphism of G, then there exists a closed unbounded subset $A \subset \kappa^+$ such that φ restricts to an outer automorphism of G_α for all $\alpha \in A$ [96, Lemma 12]. The groups G_α have therefore to be designed in such a way that outer automorphisms of previous members of the chain cannot be extended to G_α.

In [33] K. Hickin performed such a construction for $\kappa = \aleph_0$, with each $G_\alpha \cong$ ULF. The embeddings $G_\alpha \to G_{\alpha+1}$ are built up via amalgamations along an ascending chain of finite subgroups in G_α in such way that, for any given infinite set π of primes, the Sylow p-subgroups $(p \in \pi)$ of certain 2-generated subgroups are larger than their index. Different choices of π then yield $2^{\aleph_0}$ non-isomorphic complete e. c. $L\mathfrak{F}$-groups G of cardinality $\aleph_1$. Further properties can be built in here: see for example Theorem 7.1(b). In this context it should also be mentioned that the embedding technique of [33] allows us to show that a copy of every countably infinite $L\mathfrak{F}_\pi$-group occurs as a maximal π-subgroup in ULF, provided that the set π does not contain every prime [34, Theorem 4(c)].

In [96] S. Thomas proves Theorem 7.2 assuming GCH. This allows him to restrict himself to the consideration of a single outer automorphism of G_α at each stage $\alpha < \kappa^+$. His groups also enjoy the property that their soluble subgroups have cardinality at most κ. In [10] M. Dugas and R. Goebel prove the theorem for $\kappa = \kappa^{\aleph_0}$. Here the technique is to sandwich $G_{\alpha+1}$ between the restricted and the unrestricted wreath product of a very carefully chosen $\mathfrak{A}$-group A_α by G_α.

Let us now consider permutation representations. A permutation group G which acts on a set Ω is said to be *homogeneous* if every permutation group isomorphism $\alpha : (\Omega_U, U) \rightarrow (\Omega_V, V)$ between finitely generated subgroups U, V of G and unions Ω_U, Ω_V of finitely many orbits under the action of U resp. V is induced by conjugation with some $g \in G$. As in Section 3 it is straightforward to show that a L$\mathfrak{F}$-group $G \leqslant \mathrm{Sym}(\Omega)$ is e. c. in the class of all locally finite permutation groups if and only if it is homogeneous and contains a copy of $(n, \mathrm{Sym}(n))$ for every $n < \omega$. Obviously these groups are *highly transitive* (i. e. n-transitive for every $n < \omega$), and there exists a unique countable object, which in turn must be ULF.

Theorem 7.4 [45] ULF *has a highly transitive and homogeneous permutation representation.*

As a further feature of this representation, the stabilizer $S \leqslant$ ULF of any finite subset $\Omega_0 \subset \Omega$ acts again as an e. c. locally finite permutation group on $\Omega \setminus \Omega_0$, whence $S \cong$ ULF. Thus, every point stabilizer $M \leqslant$ ULF is a maximal subgroup in ULF such that $M^{g_1} \cap \cdots \cap M^{g_r} \cong$ ULF for all $g_1, \ldots, g_r \in$ ULF. K. Hickin has used the method of *tree-limits*, due to S. Shelah, in order to stretch O. H. Kegel's representation of ULF in Theorem 7.4 to cardinality $2^{\aleph_0}$.

Theorem 7.5 [35] $\mathrm{Sym}(\omega)$ *contains* $2^{\aleph_0}$ *e. c.* L$\mathfrak{F}$-*groups* G_α $(\alpha < 2^{\aleph_0})$ *which have cardinality* $2^{\aleph_0}$, *act highly transitively and homogeneously on* ω, *and have the property that no uncountable subgroup of* G_α *is isomorphic to a subgroup of* G_β *for all* $\beta \neq \alpha$.

These groups have a local system of permutation subgroups isomorphic to ULF in Kegel's representation.

Finally we would like to mention an astonishing observation concerning group rings of e. c. L$\mathfrak{F}$-groups, which is due to K. Bonvallet, B. Hartley, D. Passman and M. Smith, and which holds equally well for e. c. groups.

Theorem 7.6 [9] *Let G be an e. c.* L$\mathfrak{F}$-*group, and let K be any field. Then the group ring KG is primitive, and its augmentation ideal is the only proper ideal in KG.*

Aside from the simplicity of G, the basic tool in the proof of this theorem is the fact that every finite subgroup F of G is contained as the top group in a finite subgroup of G of the form $A \mathrm{Wr} F$, where A is an elementary abelian q-group for some prime $q \neq \mathrm{char}\, K$. Theorem 7.6 does not hold in general for simple L$\mathfrak{F}$-groups, a counterexample being the alternating group on an infinite set. Quite recently, an intense study of the ideal lattices of group rings of infinite simple L$\mathfrak{F}$-groups by means of representation theoretic methods was initiated by A. Zalesskiĭ. In particular, the circumstances under which the augmentation ideal is the only proper ideal have been investigated [99], and there exist quite a number of infinite simple L$\mathfrak{F}$-groups with this property. For a detailed account of this theory, the reader is referred to A. Zalesskiĭ's article in these Proceedings, or to [100].

7.2. Central Extensions and Approximating Sequences

The group ULF is an interesting example of a countable simple L$\mathfrak{F}$-group. A slick method of producing more groups of this nature has been found by K. Hickin [34]. Consider central quotients of e. c. objects in the class L$\mathfrak{F}(A)$ of all L$\mathfrak{F}$-groups whose center contains the distinguished L$\mathfrak{F} \cap \mathfrak{A}$-group A. Formally, first-order properties of L$\mathfrak{F}(A)$-groups can be formulated in a language which contains constants for the elements in A. In this sense, the finitely generated L$\mathfrak{F}(A)$-groups are the L$\mathfrak{F}(A)$-groups U with $|U : A| < \infty$.

At least for finite A, the e. c. L$\mathfrak{F}(A)$-groups occur naturally as the centralizers of A in e. c. L$\mathfrak{F}$-supergroups of A. Proceeding as in Section 3, one can characterize the e. c. L$\mathfrak{F}(A)$-groups G in terms of homogeneity and injectivity with respect to finitely generated L$\mathfrak{F}(A)$-groups (see [82]). In particular, if A is a π-group, then every isomorphism between finite π'-subgroups of $H = G/A$ is induced by conjugation in H. In the sequel, we shall call groups H with the latter property π'-homogeneous. The usual wreath product methods also show that G is verbally complete with $\zeta_1(G) = A$. In particular, $|G/A| \geqslant |A|$. We now come to a key observation.

Theorem 7.7 [34, Theorem 1(a)] and [82, (3.4) and Theorem A] *For every e. c.* L$\mathfrak{F}(A)$-*group* G, *the central quotient* G/A *is a simple group with Schur multiplier* $M(G/A) = A$.

Since G/A is perfect, the second assertion can be deduced from the following proposition.

Proposition 7.8 *If* G *is e. c. in* L$\mathfrak{F}(A)$, *then the homology groups* $H_n(G, \mathbb{Z})$ *are trivial for all* $n < \omega$.

Proof. (cf. [60, Theorem 6.6]) By [97, Theorem 1.7] it suffices to show that G is *pseudo-mitotic*, i. e. that for every finite subgroup $U \leqslant G$ there exists an embedding $\alpha : U \to G$ and an element $g \in G$ such that $[U, U\alpha] = 1$ and $(h\alpha)^g = h \cdot (h\alpha)$ for all $h \in U$. To this end, identify G with the diagonal subgroup of $G \operatorname{Wr} U \in$ L$\mathfrak{F}(A)$. Consideration of the top group leads to an embedding $\alpha : U \to G$ satisfying $[U, U\alpha] = 1$. Define $\beta : U \to G$ via $h\beta = h \cdot (h\alpha)$ for all $h \in U$. If A is non-trivial, the verbal completeness of G ensures, furthermore, that $a \in \langle x^G \rangle^{(1)}$ for some fixed $a \in A \smallsetminus 1$ and all $x \in (U\alpha \cup U\beta) \smallsetminus 1$ (see [61, 3.2.3]). The latter implies that $A \cap U\alpha = 1 = A \cap U\beta$, whence the isomorphism $\alpha^{-1}\beta : U\alpha \to U\beta$ is induced by conjugation in G.   □

Note that pseudo-mitoticity is a specialized form of mitoticity (see [4]). In [82] R. Phillips proves Theorem 7.7 for countable groups A and G via his theory of approximating sequences, which will be described below.

For finite $\mathfrak{A}$-groups A, the arguments of [33] yield $2^{\aleph_1}$ e.c. $\text{L}\mathfrak{F}(A)$-groups of cardinality $\aleph_1$ [34, Theorem 3(d)]. On the other hand, Theorem 3.6 leads to a unique countable e.c. $\text{L}\mathfrak{F}(A)$-group $\text{ULF}(A)$ for every countable $\text{L}\mathfrak{F} \cap \mathfrak{A}$-group A. As the Schur multipliers A of the central quotients $H_A = \text{ULF}(A)/A$ vary, the groups H_A must be pairwise non-isomorphic.

Theorem 7.9

(a) [34, Theorem 5] *The number of conjugacy classes in H_A of elements of prime power order q equals the cardinality of A/A^q.*

(b) *For each set π of primes there exist $2^{\aleph_0}$ countable π'-homogeneous simple $\text{L}\mathfrak{F}$-groups which are not p-homogeneous for any prime $p \in \pi$.*

In (a) the bijection between A/A^q and the conjugacy classes of elements of prime power order q is defined via $C \mapsto (gA)^q$ for any $gA \in C$.

The fact that $A = M(H_A)$ also affects the question of the ways the group H_A can be built up from finite simple groups (see also B. Hartley's article in these Proceedings). Recall that every countable simple $\text{L}\mathfrak{F}$-group H is the union of an ascending chain of finite subgroups U_n ($n < \omega$) such that each U_{n+1} has a maximal normal subgroup M_{n+1} satisfying $U_n \cap M_{n+1} = 1$ [47, Section 4.A]. For example, ULF is such a union, where all the U_n are alternating groups or alternatively belong to any given classical family of finite simple groups [48]. In [82] R. Phillips studies *approximating sequences* of the groups H_A, i.e. ascending chains of finite subgroups of H_A with union H_A.

Theorem 7.10 [82, Theorem B] *For every countable $\text{L}\mathfrak{F} \cap \mathfrak{A}$-group A the group H_A has an approximating sequence of direct products of groups of type PSL (resp. PSU).*

Idea of Proof. Let $g \in \text{ULF}(A)$ be fixed. Then $\langle g \rangle$ can be identified with a subgroup of the center of a suitable finite group V of type SL (resp. SU). By injectivity, we may therefore assume that $g \in V \leqslant \text{ULF}(A)$. This shows that every element, and hence every finite subgroup, of $\text{ULF}(A)$ is contained in a finite perfect subgroup U of $\text{ULF}(A)$. Next, let $A_1 \times \cdots \times A_r$ be a decomposition of $A \cap U$ as a direct product of cyclic groups A_i. Since U is perfect, there exist representations φ_i of U into $S_i = \text{SL}_{m_i}(p_i)$ (resp. $S_i = \text{SU}_{m_i}(p_i)$) with $A \cap U \cap \text{Ker}\,\varphi_i = \langle A_j \mid j \neq i \rangle$ and $\zeta_1(S_i) = (A \cap U)\varphi_i \times B_i$. We can therefore identify U with a subgroup of the direct product W of the groups S_i/B_i. By injectivity we may then assume that $U \leqslant W \leqslant \text{ULF}(A)$ and $A \cap W = \zeta_1(W)$.   □

Question 7.11 *Which other families of finite simple groups can be chosen in Theorem 7.10 in place of PSL and PSU ?*

Consider an approximating sequence $\{U_n\}_{n<\omega}$ of H_A, where each U_n is a direct product of finite simple groups. Let $c(U_n)$ be the number of composition factors of U_n, and let $r(U_n)$ be the rank of $M(U_n)$. Using well-known properties of Schur multipliers it can be shown that there are severe restrictions on $c(U_n)$ and $r(U_n)$.

Theorem 7.12   [82, Theorem C] *In the above notation, the following hold.*

(a) *If A has infinite rank, then $lim_{n\to\infty} r(U_n) = \infty$ and $lim_{n\to\infty} c(U_n) = \infty$ for any approximating sequence as above.*

(b) *If A has finite rank r, then $r(U_n) \geqslant r$ and $c(U_n) \geqslant r/2$ for almost all terms U_n of any approximating sequence of H_A as above; however there exists an approximating sequence as above such that $c(U_n) \leqslant r$ for all $n < \omega$.*

In particular, H_A has an approximating sequence of finite simple groups whenever the rank r of A is 1, while there is no such sequence for $r \geqslant 3$.

Questions 7.13

(a) *Which families of finite simple groups occur in the rank 1 case ?*

(b) *For which $\mathfrak{A}$-groups A of rank 2 does H_A have an approximating sequence of finite simple groups ?*

Inspired by the above results, R. Phillips and S. Schuur have constructed for every set π of primes and every set $\{k_p \mid p \in \pi\} \subseteq \omega$ a countable π'-homogeneous L$\mathfrak{F}$-group with the properties that it contains a copy of every finite group, can be approximated by alternating groups or by any classical family of finite simple groups, and has exactly $\min\{n, k_p\} + 1$ conjugacy classes of elements of order p^n for every $p \in \pi$ [83, Theorem 3]. S. Schuur has then shown in [91] that every L$\mathfrak{F}$-group G of cardinality κ can be embedded into a π'-homogeneous L$\mathfrak{F}$-group of cardinality $\max\{\aleph_0, \kappa\}$ which contains a copy of every finite group and has the above numbers of conjugacy classes. The proofs are based on an iteration of the right regular embedding into a modified constricted symmetric group.

In [92] S. Schuur presents two constructions of $\mathrm{ULF}(A)$ as a direct limit of finite groups U_n $(n < \omega)$. In the first construction these groups are direct products of general linear groups, where the embeddings $U_n \to U_{n+1}$ are similar to the embedding $U \to W$ in the proof of Theorem 7.10. The second construction resembles P. Hall's construction [25] of ULF. The basic step is to embed U_n right regularly into the centralizer $C_{\mathrm{Sym}(U_n)}(U_n \cap A)$.

8. Periodic Locally Soluble Groups and Locally Profinite Groups

8.1. General Results

The theory of e. c. $\mathrm{L}(\mathfrak{F}_\pi \cap \mathfrak{S})$-groups has almost exclusively been developed by the author. Parts of it can be proved in greater generality. In this section we shall therefore work under the following hypothesis.

Hypothesis 8.1 $\mathfrak{X}$ *will denote a class of groups which is closed with respect to the formation of subgroups, quotients, extensions, and of cartesian powers of finitely generated $\mathfrak{X}$-groups.*

The most natural examples for $\mathfrak{X}$ are the classes $\mathfrak{S}$, $\mathrm{L}\mathfrak{F}_\pi$ and $\mathrm{L}(\mathfrak{F}_\pi \cap \mathfrak{S})$. All of our results can be proved by using the following construction, which combines features of KKF-embeddings, permutational products, and the right regular representation into the constricted symmetric group.

Construction 8.2 [58, Section 2] *Let* $^{-}: G \to H$ *be a homomorphism of* $L\mathfrak{X}$-*groups with kernel* N. *Let* $U \leqslant V$ *be finitely generated subgroups of* G. *Choose left transversals* R *of* $U \cap N$ *in* U, *and* S *of* UN *in* G, *and* T *of* $\overline{G}$ *in* H. *Consider* $W = G \operatorname{Wr} H$ *as a permutation group on* $\Omega = G \times H$ *in the usual way:*

$$(g_0, h_0)^{fh} = (g_0 \cdot (h_0)f, h_0 h) \text{ for all } f : H \to G, \ h \in H.$$

Define embeddings $\sigma : H \to W$ *and* $\tau : G \to W$ *via* $(g_0, t_0 \overline{s}_0 \overline{r}_0)^{h\sigma} = (g_1, t_1 \overline{s}_1 \overline{r}_1)$ *where* $t_1 \overline{s}_1 \overline{r}_1 = t_0 \overline{s}_0 \overline{r}_0 h$ *and* $g_1 = g_0 s_0 s_1^{-1}$, *and via* $(g_0, t_0 \overline{s}_0 \overline{r}_0)^{g\tau} = (g_2, t_2 \overline{s}_2 \overline{r}_2)$ *where* $t_2 \overline{s}_2 \overline{r}_2 = t_0 \overline{s}_0 \overline{r}_0 \overline{g}$ *and* $g_2 = g_0 s_0 r_0 g r_2^{-1} s_2^{-1}$. *Moreover, a third embedding* $\mu : V \operatorname{Wr} \overline{U} \to W$ *is given via* $(f\overline{u})\mu = f^* \overline{u}$ *where* $(t\overline{sr})f^* = s \cdot (\overline{r})f \cdot s^{-1}$. *Then* $H\sigma$, $G\tau$ *and* $\operatorname{Im} \mu$ *are contained in the union* W_0 *of all split extensions* $W_X = \Delta_X \rtimes H\sigma$, *where* X *ranges over the finitely generated subgroups of* G, *and where*

$$\Delta_X = \{ f : H \to G \,|\, (t\overline{sr})f \in sXs^{-1} \} \leqslant W.$$

Therefore Hypothesis 8.1 ensures that $W_0 \in L\mathfrak{X}$. *Moreover,* $\tau|_U = \kappa\mu$ *for the usual KKF-embedding* $\kappa : U \to (U \cap N) \operatorname{Wr} \overline{U} \leqslant V \operatorname{Wr} \overline{U}$.

The construction yields a completion of the amalgam $G \cup (V \operatorname{Wr} \overline{U}) | U \equiv U\kappa$ for any given $U \leqslant V \leqslant G$. Hence it makes KKF-embeddings accessible for the study of e. c. $L\mathfrak{X}$-groups.

Theorem 8.3 [51, Theorem 4.7] *and* [58] *Let* G *be an a. c.* $L\mathfrak{X}$-*group. If* M/N *is a chief factor in* G, *then* $M = \langle g^G \rangle$ *for every* $g \in M \setminus N$, *and every* $h \in N$ *occurs as any given non-trivial reduced word in elements from* M. *In particular, the normal subgroup lattice of* G *is totally ordered.*

Proof. For $g \in M \setminus N$ and $h \in N$, let $U = \langle g, h \rangle$. From the verbal completeness of G (Proposition 3.3) we can choose $U \leqslant V \leqslant G$ such that h is any given non-trivial reduced word $w(x_1, \ldots, x_r)$ in elements from $V^{(1)}$. Using the above notation we now have $h\kappa = w(z_1, \ldots, z_r)$ for suitable $z_i \in \langle g\kappa^{(V \operatorname{Wr} \overline{U})} \rangle$ (see [61, 3.2.3]). $\square$

The above theorem is analogous to Theorems 5.1 and 6.1 about e. c. soluble and nilpotent groups. Note that e. c. $L\mathfrak{F}_\pi$-groups are only simple if π is the set of all primes [95, Proposition 1]. It can also be shown that every countable e. c. $L\mathfrak{X}$-group has $2^{\aleph_0}$ automorphisms, and that automorphisms of finite order of any e. c. $L\mathfrak{X}$-group G which stabilize the unique chief series in G are locally inner [51, Section 5].

More information can be obtained if we assume that the chief factors of $L\mathfrak{X}$-groups belong to a proper variety. In this situation, Theorem 8.3 allows us to express elements from N via certain words in M. This yields particularly nice results for $L\mathfrak{G}$-groups.

Theorem 8.4 [51, Section 4] *and* [58] *Let* $\mathfrak{X} \in \{\mathfrak{S}, L(\mathfrak{F}_\pi \cap \mathfrak{S})\}$ *and let* G *be an a. c.* $L\mathfrak{X}$-*group. If* $\mathfrak{X} = \mathfrak{S}$, *then let* π *be the set of all primes.*

(a) *Every fixed non-trivial coset* xN *in a p-chief factor* M/N *of* G *contains elements of order* p, *and these are all conjugate in* G. *This applies correspondingly to elements in non-trivial cosets of torsion-free chief factors.*

(b) *If $\mathfrak{X} \neq L\mathfrak{F}_p$, then every chief factor of G is non-central and infinite.*

(c) *For each $p \in \pi$ there is a p-chief factor between any two given chief factors of G. Moreover, G does not have a maximal normal subgroup; and, if G is e. c. in $L\mathfrak{X}$, then there is also no minimal normal subgroup in G.*

(d) *On every finitely generated subgroup of a chief factor in G, π-power automorphisms are induced via conjugation in G.*

(e) *Every subnormal subgroup of G is sandwiched between the terms of a chief factor in G.*

Part (e) follows from the observation that, whenever W is a wreath product with base group Ω, then $\Omega^{(2)} \leqslant [z, \langle z^W \rangle^{(1)}, \langle z^W \rangle^{(1)}]$ for every $z \in W \smallsetminus \Omega$. It would be very interesting to determine the action of G on its chief factors more precisely.

Question 8.5 *Does every e. c. $L(\mathfrak{F}_\pi \cap \mathfrak{S})$-group act transitively on the non-trivial elements of each of its chief factors via conjugation ?*

If G is countable e. c. in $L(\mathfrak{F}_\pi \cap \mathfrak{S})$ or $L\mathfrak{S}$, then its chief factors are ordered like the rationals. However, there exist additional normal subgroups K_r, corresponding to the irrational real numbers r. Let s_r be an ascending sequence of rationals converging to r; then K_r is the union of the terms of the chief factors corresponding to the elements in s_r. This argument applies to any Dedekind cut of the order type of the chief series in any e. c. $L\mathfrak{X}$-group.

Questions 8.6

(a) *Does every linear order without endpoints occur as the order type of the unique chief series of an e. c. $L(\mathfrak{F}_\pi \cap \mathfrak{S})$-group ?*

(b) *Do torsion-free chief factors occur necessarily in e. c. $L\mathfrak{S}$-groups, and do they occur densely ?*

(c) *Is every normal subgroup K of an e. c. group in $L\mathfrak{X} \in \{L(\mathfrak{F}_\pi \cap \mathfrak{S}), L\mathfrak{S}\}$, which satisfies $K \neq \langle g^G \rangle$ for all $g \in G$, itself e. c. in $L\mathfrak{X}$?*

(d) *How many countable e. c. $L(\mathfrak{F}_\pi \cap \mathfrak{S})$-groups exist for $|\pi| \geqslant 2$?*

Question 8.6(c) is motivated by a partial answer for e. c. $L\mathfrak{X}$-groups [58, Theorem 3.1], and by a positive answer for e. c. $L\mathfrak{F}_p$-groups [52, Theorem 4.1(c)]. Moreover, there exists a unique countable e. c. $L\mathfrak{F}_p$-group for every prime p (see Section 8.2). Both results follow from a nice amalgamation theorem for $L\mathfrak{F}_p$-groups (Theorem 8.10). The difficulties in finding an amalgamation theorem for $L(\mathfrak{F}_\pi \cap \mathfrak{S})$-groups are similar to those mentioned at the end of Section 5. At present it is only possible to complete amalgams of $L(\mathfrak{F}_\pi \cap \mathfrak{S})$-groups over a common supersoluble finite subgroup [58, Section 5], or amalgams of a $L(\mathfrak{F}_\pi \cap \mathfrak{S})$-group with a $L(\mathfrak{F}_\pi \cap \mathfrak{N})$-group over a common finite subgroup [50, Theorem 2.1], when there exist compatible chief series. As a consequence, every countable locally supersoluble $L\mathfrak{F}_\pi$-group H can be embedded into any e. c. $L(\mathfrak{F}_\pi \cap \mathfrak{S})$-group G in such a way that prescribed chief factors of G induce prescribed factors of a given chief series in H via intersection [50, Section 3]. A corresponding version can certainly be proved for $L\mathfrak{S}$-groups. An extension to embeddings of countable $L(\mathfrak{F}_\pi \cap \mathfrak{S})$-groups would yield a unique countable

e. c. $L(\mathfrak{F}_\pi \cap \mathfrak{S})$-group via *back and forth*. Note, however, that there exist $2^{\aleph_0}$ count-able e. c. $L\mathfrak{S}$-groups [51, Section 6], since there are $2^{\aleph_0}$ finitely generated $\mathfrak{S}$-groups. We finally mention an interesting splitting property of e. c. $L(\mathfrak{F}_\pi \cap \mathfrak{S})$-groups.

Theorem 8.7 [58, Section 4] *If G is e. c. in $L(\mathfrak{F}_\pi \cap \mathfrak{S})$, and if $K \trianglelefteq G$ satisfies $K \neq \langle g^G \rangle$ for all $g \in G$, then the preimage in G of every countable subgroup of G/K splits over K.*

In particular, countable e. c. $L(\mathfrak{F}_\pi \cap \mathfrak{S})$-groups G split over such normal subgroups K. Note that any complement H to K is a. c. with respect to sets of equations *of any size* which are solvable in G [16, Theorem 1].

H. Ensel has obtained corresponding results for Sylow tower groups. For any totally ordered set $(\pi, \preceq)$ of primes we let $L(\mathfrak{F} \cap \mathfrak{S})_{(\pi, \preceq)}$ be the class of all $L(\mathfrak{F}_\pi \cap \mathfrak{S})$-groups which have a normal series with p-factors S_p/T_p $(p \in \pi)$ ordered according to $\preceq$.

Theorem 8.8 [15] *Let G be an e. c. $L(\mathfrak{F} \cap \mathfrak{S})_{(\pi, \preceq)}$-group with Sylow factors S_p/T_p $(p \in \pi)$. If G splits over each S_p and each T_p, then the quotients G/S_p and G/T_p are e. c. Sylow tower groups in the corresponding classes, and G has a unique chief series of dense order type without endpoints. Moreover the chief factors M/N of G with $S_p \leqslant N \leqslant M \leqslant T_p$ are infinite if and only if p is not maximal in π.*

It follows from the well-known theorem of Schur and Zassenhaus that every *count-able* Sylow tower group splits as in the above theorem.

Questions 8.9

   (a) *Does every e. c. Sylow tower group split as in Theorem 8.8?*

   (b) *Does every e. c. $L(\mathfrak{F}_\pi \cap \mathfrak{S})$-group G split over each of its normal subgroups K satisfying $K \neq \langle g^G \rangle$ for all $g \in G$?*

By applying Jensen's principle $\diamondsuit$, S. Thomas has constructed $2^{\aleph_1}$ complete e. c. groups of cardinality $\aleph_1$ in each of the classes $L\mathfrak{X} \in \{L\mathfrak{F}_\pi, L(\mathfrak{F}_\pi \cap \mathfrak{S})\}$ [95, Theorem 1] (cf. Theorem 7.3) as direct limits of a fixed countable e. c. $L\mathfrak{X}$-group G. One of the crucial observations here is that, whenever $\alpha \in \mathrm{Aut}(G)$, and whenever U is a finite subgroup of G, then α leaves every subgroup of G containing U invariant if and only if α is conjugation with some element from U [95, Theorem 2]. Each of Thomas' complete groups has only countably many conjugacy classes, and hence only countably many chief factors [95, Section 5]. Complete groups of cardinality κ^+ in the classes $L\mathfrak{X}$ have also been constructed by M. Dugas and R. Goebel [10] for every infinite cardinal κ satisfying $\kappa = \kappa^{\aleph_0}$. However these groups need not be e. c. in $L\mathfrak{X}$. The above results are especially striking for $L\mathfrak{F}_p$-groups, since every countably infinite $L\mathfrak{F}_p$-group has $2^{\aleph_0}$ automorphisms [84].

8.2. Locally Finite p-Groups

The theory of e. c. $L\mathfrak{F}_p$-groups parallels the theory of e. c. $L\mathfrak{F}$-groups nicely, because there exists a handy amalgamation theorem which generalizes [38].

Theorem 8.10 [52, Theorem 3.1] *An amalgam $G \cup H | U$ of $\text{L}\mathfrak{F}_p$-groups over the finite common subgroup U can be completed in $\text{L}\mathfrak{F}_p$ if and only if there exist chief series in G and H which induce the same chief series in U.*

This can be proved by considering a permutational product with respect to transversals which respect the given chief series in G and H; such a permutational product is a $\text{L}\mathfrak{F}$-subgroup of an iterated wreath product of $\text{L}\mathfrak{F}_p$-groups. Note that chief factors of $\text{L}\mathfrak{F}_p$-groups are always cyclic of order p. Since every e. c. $\text{L}\mathfrak{F}_p$-group G has a unique chief series S (Theorem 8.3), it follows easily that every finite subgroup $U \leqslant G$ is contained in a finite subgroup $U^* \leqslant G$ which controls amalgamation over U, because every chief series in U^* induces the series $U \cap S$ in U [68, Hilfssatz 1]. In fact it can be shown that *any* given chief series in U is controlled by a $\mathfrak{F}_p$-supergroup of U [55]. Consider the right regular representation $\rho : U \to \text{Sym}(U)$. Then there is a one-to-one correspondence between the chief series in $U\rho$ and the Sylow p-subgroups of $\text{Sym}(U)$ which contain $U\rho$, such that the particular chief series in $U\rho$ is induced from every chief series in the particular Sylow p-subgroup of $\text{Sym}(U)$.

It now follows from Theorem 3.6 that there is a unique countable e. c. $\text{L}\mathfrak{F}_p$-group E_p [68]. Because of the above observation, E_p can be constructed as a direct limit of $\mathfrak{F}_p$-groups U with respect to right regular representations $U \to P \in \text{Syl}_p(\text{Sym}(U))$ [53, 55]. Moreover, the e. c. $\text{L}\mathfrak{F}_p$-groups can be characterized as in Section 3 in terms of injectivity.

Theorem 8.11 [52, Theorem 3.4] *The $\text{L}\mathfrak{F}_p$-group G is e. c. in $\text{L}\mathfrak{F}_p$ if and only if*
  (1) *every finite subgroup U of G is contained right regularly in a copy $U^* \leqslant G$ of a Sylow p-subgroup of $\text{Sym}(U)$, and*
  (2) *whenever $U^* \leqslant V \in \mathfrak{F}_p$, then $\text{id} : U \to G$ can be extended to an embedding $V \to G$.*

The above theorem immediately yields an axiomatization of the e. c. $\text{L}\mathfrak{F}_p$-groups by a sentence in $L_{\omega_1, \omega}$. So it follows as in Theorem 7.1(a) that there are 2^κ e. c. $\text{L}\mathfrak{F}_p$-groups of each uncountable cardinality κ (see [68, Satz 8]). An explicit construction of $2^{\aleph_1}$ e. c. $\text{L}\mathfrak{F}_p$-groups of cardinality $\aleph_1$ is given in [52, Section 7] by iterating embeddings of E_p onto proper normal subgroups of itself, where we can choose between the lower terms of chief factors and the 'irrational' normal subgroups at each step. The resulting groups have chief series with pairwise distinct order-types, and all their proper normal subgroups are countable.

With the help of the amalgamation theorem and *back and forth* we can also describe the automorphism group of E_p in a quite detailed way.

Theorem 8.12 [52, Theorem 6.2] *Let S be the unique chief series in E_p, and let Σ be the group of all automorphisms of E_p which stabilize S.*

  (a) *An isomorphism $\alpha : U \to V$ between finite subgroups of E_p extends to an automorphism of E_p if and only if it maps the series $S \cap U$ onto $S \cap V$.*
  (b) *Σ is the group of all locally inner automorphisms of E_p.*

(c) $\mathrm{Aut}(E_p)/\Sigma$ *is isomorphic to the wreath product* $C_{p-1}\,\mathrm{Wr}_{\mathbb{Q}}\,A(\mathbb{Q})$ *of the cyclic group* C_{p-1} *of order* $p-1$ *by the group* $A(\mathbb{Q})$ *of all order-preserving automorphisms of* $\mathbb{Q}$. *Here the elements from the top group permute the chief factors of* E_p *naturally, while the base group leaves every normal subgroup of* E_p *invariant and acts on the chief factors in any way we please.*

In particular, any two elements of equal order in E_p are mapped onto each other under $\mathrm{Aut}(E_p)$, and so E_p is characteristically simple. This allows us to proceed as in Theorem 7.6 in order to show that, whenever the characteristic of the field K is different from p, then the augmentation ideal is the only candidate for a proper characteristic ideal in the group ring KE_p. Moreover, KE_p is primitive [68, pp. 124–125]. And the first assertion still holds in modular characteristic [56].

Finally, as in Section 7.2, we can consider the classes $\mathrm{L}\mathfrak{F}_p(A)$ of central extensions of $\mathrm{L}(\mathfrak{F}_p \cap \mathfrak{A})$-groups A. For countable A there exists a unique countable e.c. $\mathrm{L}\mathfrak{F}_p(A)$-group E_A [61, Theorem A]. It follows, as in Proposition 7.8, that the integer homology groups of every e.c. $\mathrm{L}\mathfrak{F}_p(A)$-group G are trivial, whence A is the Schur multiplier of the perfect quotient G/A. Another way to prove $A = M(E_A/A)$ is described in [61, Section 5]: here E_A is constructed as a direct limit of certain wreath products, whose Schur multipliers can be sufficiently controlled to ensure that their direct limit $M(E_A)$ must be trivial.

The central quotient G/A of an e.c. $\mathrm{L}\mathfrak{F}_p(A)$-group G still satisfies all the properties discovered in Section 8.1 for e.c. $\mathrm{L}\mathfrak{F}_p$-groups. As A varies, we obtain the following result.

Theorem 8.13 [61, Theorem B] *There exist* $2^{\aleph_0}$ *countable* $\mathrm{L}\mathfrak{F}_p$-*groups which are verbally complete, characteristically simple, have a unique chief series (whose order-type is dense without endpoints), and in which normality is transitive.*

Note that the central quotients of e.c. $\mathrm{L}\mathfrak{F}_p(A)$-groups (and of the groups T_A in Theorem 10.7) are the only known examples of non-soluble, locally nilpotent groups in which normality is transitive. Since Schur multipliers are compatible with direct limits, it is apparent that there are restrictions on the local systems of G/A [61, Corollary 2]. For example, if $\exp(A) \geqslant p^2$, then G/A cannot have a local system of Sylow p-subgroups of finite symmetric groups.

Question 8.14 *Which kinds of subgroups can form a local system of* E_A/A *?*

Almost certainly E_A can be constructed by iterating right regular embeddings of $\mathfrak{F}_p(A)$-groups of the form $U \to P \in \mathrm{Syl}_p(C_{\mathrm{Sym}(U)}(A))$. The automorphisms of E_A have been described in [14] along lines similar to Theorem 8.12. Here the quotient modulo the inner automorphisms is isomorphic to $\mathrm{Aut}(A) \times (C_{p-1}\,\mathrm{Wr}_{\mathbb{Q}}\,A(\mathbb{Q}))$.

Question 8.15 *Is every automorphism of* E_A/A *induced from an automorphism of* E_A *which centralizes* A *?*

8.3. Locally Profinite Groups

Hypothesis 8.16 *In this section $\mathfrak{X}$ will always be a class of finite groups which is closed with respect to the formation of subgroups, quotients, and extensions.*

Applying the argument of Theorem 5.3 it is easy to construct a countable L$\mathfrak{X}$-group which is e. c. in the class LR$\mathfrak{X}$ of all groups whose finitely generated subgroups are residually an $\mathfrak{X}$-group. In fact, if $\mathfrak{X} \in \{\mathfrak{F}, \mathfrak{F}_p\}$, then *every* e. c. L$\mathfrak{X}$-group is e. c. in LR$\mathfrak{X}$ (copy the proofs of [68, Satz 6] and [52, Theorem 3.7]). It is therefore a natural question to ask which properties of e. c. L$\mathfrak{X}$-groups extend to e. c. LR$\mathfrak{X}$-groups. Since E. S. Golod and R. I. Grigorchuk have produced $2^{\aleph_0}$ finitely generated R$\mathfrak{F}_p$-groups, there also exist $2^{\aleph_0}$ countable e. c. LR$\mathfrak{X}$-groups, and so we should not expect quite as nice results here as for the class L$\mathfrak{X}$.

Because it is not so easy to find valuable constructions within the class LR$\mathfrak{X}$, we rather consider the class LC$\mathfrak{X}$ of all *locally co-$\mathfrak{X}$ groups*, i. e. of all groups G whose finitely generated subgroups are equipped with a fixed pro-$\mathfrak{X}$ topology in such a way that, for any pair $U \leqslant V$ of finitely generated subgroups of G, the given pro-$\mathfrak{X}$ topology on V induces the given pro-$\mathfrak{X}$ topology on U. Since every LC$\mathfrak{X}$-group G is contained in the direct limit of the pro-$\mathfrak{X}$ completions of its finitely generated subgroups, it is clear that every e. c. locally pro-$\mathfrak{X}$ group is, in particular, e. c. in LC$\mathfrak{X}$.

By taking direct limits of certain quotients of free products of finitely generated subgroups it can be shown that an amalgam $G \cup H|U$ of LC$\mathfrak{X}$-groups can be completed in LC$\mathfrak{X}$ if and only if the local topologies fit together in such a way that amalgams of certain $\mathfrak{X}$-quotients of finitely generated subgroups of G and H can be completed in $\mathfrak{X}$ [59, Theorem 2.1]. It follows that a finite set of equations and inequalities over the e. c. LC$\mathfrak{X}$-group G has a solution in G if and only if it has a solution in $\mathfrak{X}$-supergroups of almost all $\mathfrak{X}$-quotients of almost all finitely generated subgroups of G [59, Theorem 2.3]. A useful corollary is as follows.

Theorem 8.17 [59, Corollary 2.5] *The $\forall\exists$-sentences which hold in every e. c. LC$\mathfrak{X}$-group are precisely those which hold in every countable e. c. L$\mathfrak{X}$-group which is e. c. in LC$\mathfrak{X}$.*

We note a particularly striking consequence for LC$\mathfrak{F}_p$-groups.

Theorem 8.18 [59, Section 5] *Every e. c. LC$\mathfrak{F}_p$-group G has a unique chief series, the chief factors are central and cyclic of order p, and the order type is dense without endpoints. Moreover, normality is transitive in G.*

This implies the following highly non-obvious fact: *every* LC$\mathfrak{F}_p$-group has a chief series with central and cyclic factors of order p. The proof of Theorem 8.18 depends on the fact that, whenever M/N is a chief factor in E_p, then every $h \in M$ is the product of a bounded number of conjugates of any given $g \in M \setminus N$. Therefore the theorem could probably be extended to LC$(\mathfrak{F}_\pi \cap \mathfrak{S})$-groups if the action of countable e. c. L$(\mathfrak{F}_\pi \cap \mathfrak{S})$-groups on their chief factors was better understood (e. g. transitive).

Of course, the characterization of conjugacy of *finite* subgroups can be transferred from E_p to e. c. LC$\mathfrak{F}_p$-groups via $\forall\exists$-sentences. But apart from this, conjugacy in

e. c. $\mathrm{LC\mathfrak{F}}_p$-groups G is not really well understood. At least it is possible to show that an element $g \in G$ of infinite order is conjugate to g^n ($n \in \mathbb{Z}$) if and only if $n \equiv 1 \pmod{p}$ [59, Theorem 5.5].

Since every element of **ULF** is a product of two conjugates of any other given element $g \in$ **ULF** (see [47, Theorem 6.1(d)]), it follows from Theorem 8.17 that the same holds for every e. c. $\mathrm{LC\mathfrak{F}}$-group G, whence G is simple. Moreover, no element of infinite order in G is conjugate to all of its non-trivial powers [59, Section 6].

9. Torsion-Free Locally Nilpotent Groups

The analogy between the amalgamation theorems 6.2 and 8.10 allows us to develop a theory of e. c. $\mathrm{L\mathfrak{N}}^+$-groups which parallels that of e. c. $\mathrm{L\mathfrak{F}}_p$-groups. In fact, $\mathrm{L\mathfrak{N}}^+$-groups seem to be related to $\mathrm{L\mathfrak{F}}_p$-groups in very much the same way that unipotent (finitary) linear groups in characteristic 0 are related to those in characteristic $p > 0$. In [72, Section 8.7] B. Maier has indicated in outline some basic results about e. c. $\mathrm{L\mathfrak{N}}^+$-groups, and we shall try to provide proofs here. A normal series of a $\mathrm{L\mathfrak{N}}^+$-group G will be called an *isolated rank 1 series* (I_1-*series*) if its factors M/N are isolated and central in G/N of rank 1. The I_1-series in $\mathrm{L\mathfrak{N}}^+$-groups correspond to the chief series in $\mathrm{L\mathfrak{F}}_p$-groups. Note that every finitely generated $\mathfrak{N}^+$-group has an I_1-series.

Proposition 9.1 *Every* $\mathrm{L\mathfrak{N}}^+$*-group has an* I_1*-series.*

Proof. Let $\mathcal{S}$ be a non-refinable series of isolated normal subgroups of the $\mathrm{L\mathfrak{N}}^+$-group G. Consider a factor M/N in $\mathcal{S}$. Assume that M/N is not central. Without loss we may assume $N = 1$. Denote isolators of subgroups by bars. Choose $g \in M \smallsetminus 1$. Then $g \in M = \overline{[\langle g^G \rangle, G]}$. Hence there exists a finitely generated subgroup U of G such that $g \in U$ and $g \in \overline{[\langle g^U \rangle, U]}$. But this contradicts the existence of an I_1-series in U. $\quad\square$

Proposition 9.2 *An amalgam* $G \cup H | U$ *of countable* $\mathrm{L\mathfrak{N}}^+$*-groups over the finitely generated common subgroup* U *can be completed in* $\mathrm{L\mathfrak{N}}^+$ *if and only if there exist* I_1*-series* $\mathcal{S}_G$ *in* G *and* $\mathcal{S}_H$ *in* H *such that* $U \cap \mathcal{S}_G = U \cap \mathcal{S}_H$. *The completion can in fact be so chosen that one of its* I_1*-series induces* $\mathcal{S}_G$ *in* G *and* $\mathcal{S}_H$ *in* H.

Proof. For amalgams of finitely generated groups this follows immediately from Theorem 6.2, since the condition (2) there can be met by repeating terms in the given series [71, Korollar 5]. Now let G and H be unions of ascending chains of finitely generated subgroups G_n resp. H_n ($n < \omega$) with $G_0 = U = H_0$. Put $W_0 = U$, and successively complete $G_n \cup W_{n-1} | G_{n-1}$ in some finitely generated $V_n \in \mathfrak{N}^+$ and $V_n \cup H_n | H_{n-1}$ in some finitely generated $W_n \in \mathfrak{N}^+$, such that some I_1-series in W_n induces $G_n \cap \mathcal{S}_G$ in G_n and $H_n \cap \mathcal{S}_G$ in H_n. The direct limit of the groups W_n is a completion of $G \cup H | U$. $\quad\square$

We shall now construct controllers for the amalgamation of finitely generated $\mathfrak{N}^+$-groups.

Proposition 9.3 [71, Satz 3] *For every given I_1-series S in a finitely generated $\mathfrak{N}^+$-group U there exists a finitely generated $\mathfrak{N}^+$-group $U^* \geqslant U$ such that every I_1-series in U^* induces S in U.*

Proof. Let S be the series $U = U_1 \geqslant \ldots \geqslant U_{c+1} = 1$, and choose $h_i \in U_i \smallsetminus U_{i+1}$ such that $U_i = \langle h_i, U_{i+1} \rangle$ for all i. It suffices to produce a $\mathfrak{N}^+$-group $U^* \geqslant U$ such that $h_{i+1} \in [h_i, U^*, U^*]$ for all i. By induction we may assume that there is a finitely generated $\mathfrak{N}^+$-group $V \geqslant U$ such that $h_{i+1} \in [h_i, V, V]$ for all $i \geqslant 2$, and such that some I_1-series S^* in V induces S in U. Let $T = \langle a, b \rangle$ be the free $\mathfrak{N}_2^+$-group of rank 2. Let $W \in \mathfrak{N}^+$ be a completion of the amalgam $T \cup (V \times \langle c \rangle)|\langle a, [a, b] \rangle$, where a and $[a, b]$ are identified with h_1 and c, respectively, such that some I_1-series in W induces S^* in V in such a way that the factor containing c lies above the factor containing h_2. Now any completion U^* of the amalgam $T \cup W|\langle a, [a, b] \rangle$, where a and $[a, b]$ are identified with c and h_2, respectively, has the desired property $h_2 \in [h_1, U^*, U^*]$.   □

An application of Theorem 3.6 now yields the following $\mathfrak{L}\mathfrak{N}^+$-counterpart to E_p.

Theorem 9.4

   (a) *There exists a unique countable closed $\mathfrak{L}\mathfrak{N}^+$-group E_+.*

   (b) *The lattice of all divisible normal subgroups of E_+ is totally ordered. There exists a unique I_1-series in E_+. Its factors are ordered like the rationals.*

   (c) *An isomorphism $\alpha : U \to V$ between finitely generated subgroups of E_+ can be extended to an automorphism of E_+ if and only if $(U \cap K)\alpha = V \cap K$ for every divisible normal subgroup K of E_+.*

   (d) *$\mathrm{Aut}(E_+)$ acts transitively on the non-trivial elements of E_+. In particular, E_+ is characteristically simple.*

Proof. (a) Let G be a countable closed $\mathfrak{L}\mathfrak{N}^+$-group, and consider an I_1-series S in G. Let U be any finitely generated subgroup of G. By Proposition 9.3 there is a finitely generated $\mathfrak{N}^+$-group $U^* \geqslant U$ which controls $U \cap S$. By Proposition 9.2 we can complete the amalgam $G \cup U^*|U$ in $\mathfrak{L}\mathfrak{N}^+$, whence the closed $\mathfrak{L}\mathfrak{N}^+$-group G contains a copy of U^* above U. Now (a) follows from Theorem 3.6.

   (b) E_+ is divisible, by Proposition 3.1. Let S be a non-refinable series of divisible normal subgroups in E_+. Then S is an I_1-series, and the argument in (a) yields that S induces a fixed I_1-series in every finitely generated subgroup of E_+. Therefore S must be unique.

   Assume now that there exist factors M_1/N_1 and M_2/N_2 in S with $N_1 = M_2$. The divisibility of E_+ implies $\zeta_2(E_+/N_2) = \zeta_1(E_+/N_2)$, whence M_1/N_2 is central of rank 2, in contradiction to the uniqueness of S. This shows that the factors of S are ordered densely. Assume finally that E_+ has a minimal divisible normal subgroup M. Fix $h \in M \smallsetminus 1$, and amalgamate E_+ with the free $\mathfrak{N}_2^+$-group $T = \langle a, b \rangle$ of rank 2 over the common subgroup $\langle h \rangle \equiv \langle a \rangle$. Since E_+ is closed, we may assume that $M \leqslant T \leqslant E_+$. But then $[M, E_+] \neq 1$, a contradiction.

   (c), (d) Apply *back and forth* as in Theorem 6.6.   □

Questions 9.5

(a) *Which other properties does E_+ share with E_p?*

(b) *What can be said about uncountable closed $L\mathfrak{N}^+$-groups?*

(c) *Does Proposition 9.2 extend to $L\mathfrak{N}^+$-groups G and H of arbitrary cardinality?*

Concerning the second question, we can at least produce $2^{\aleph_1}$ closed $L\mathfrak{N}^+$-groups of cardinality $\aleph_1$ with unique I_1-series of pairwise distinct order types, in the same way as was indicated in Section 8.2 for $L\mathfrak{F}_p$-groups.

10. Locally FC-Groups

10.1. ARBITRARY TORSION

The class LFC of all groups whose finitely generated subgroups have finite conjugacy classes comprises all $\mathfrak{A}$-groups and all $L\mathfrak{F}$-groups. In every LFC-group G the elements of finite order form a $L\mathfrak{F}$-subgroup $\tau(G)$ with $\mathfrak{A}^+$-quotient. The rank of $G/\tau(G)$ is called the *rank* of G. Inspired by the results about e. c. groups in the classes $L\mathfrak{F}$ and $\mathfrak{A}$, F. Haug studied e. c. LFC-groups in [26]–[29]. In this context a puzzling pecularity arises: the *inevitable center* $\eta(G)$ (i. e. the set of all $g \in G$ which lie in the center of *any* LFC-group $H \geqslant G$) can be non-trivial [27, Section 2]. It is shown in [27] that $\eta(G)$ can be described internally as

$$\eta(G) \;=\; \bigcap_{f:G\to\omega} \langle g^{f(g)} \mid g \in G \rangle \;=\; \bigcap_{f:B\to\omega} \langle b^{f(b)} \mid b \in B \rangle^{(1)}$$

for any set $B \subseteq G$ which is maximal independent modulo $\tau(G)$. It follows that $\eta(G) \leqslant \tau(G)$ and that $\eta(G)$ is trivial whenever G is countable. Some other conditions which ensure that $\eta(G)$ is trivial can be found in [27, Section 3]. However, the inevitable center still seems to be rather mysterious. The study of e. c. LFC-groups can be based on the following amalgamation theorem, which can be proved by forming a permutational product with respect to a clever choice of transversals.

Theorem 10.1 [27, Corollary 1.3] *An amalgam $G \cup H|U$ of LFC-groups over the finitely generated common subgroup U has a completion in LFC if and only if $U \cap \eta(G) \leqslant \zeta_1(H)$ and $U \cap \eta(H) \leqslant \zeta_1(G)$.*

This allows us to give the usual characterization of e. c. LFC-groups in terms of homogeneity and injectivity: the LFC-group G is e. c. in LFC if and only if $G/\tau(G)$ is divisible and G is homogeneous and injective with respect to finitely generated $\mathrm{FC}(\eta(G))$-groups [28, Theorem 1.1]. In particular, $\tau(G)$ is an e. c. $L\mathfrak{F}(\eta(G))$-group [29, Corollary 1.4], and $\zeta_1(G) = \eta(G)$. Also, if G is countable, then the triviality of $\eta(G)$ implies that $\tau(G) \cong \mathrm{ULF}$. Another observation is that sets of equations and inequalities can be solved in LFC-groups without increasing the rank. Hence we obtain the following classification of countable e. c. LFC-groups.

Theorem 10.2 [28, Section 2] *For each $\rho \leqslant \omega$ there exists a unique countable e. c. LFC-group G_ρ of rank ρ, and G_ρ is a split extension of ULF by $\mathbb{Q}^\rho$.*

Clearly G_ω is the unique countable closed LFC-group. In [28, Section 3] a direct limit construction for the groups G_ρ is provided, which is similar to the constructions of ULF and E_p via right regular representations.

The study of $\mathrm{Aut}(G_\rho)$ is of interest, since $\mathrm{Aut}(G_\rho)$ embeds into $\mathrm{Aut}(\tau(G_\rho))$ via restriction to the subgroup $\tau(G_\rho)$. Correspondingly, the canonical homomorphism $\mathrm{Aut}(G_\rho) \to \mathrm{Aut}(G_\rho/\tau(G_\rho))$ is surjective, and its kernel consists of the locally inner automorphisms of G_ρ [28, Theorem 4.6]. Let U be a finitely generated subgroup of the group G. A locally inner automorphism of G is said to be U-*locally inner* if it is locally induced by conjugation with elements from U.

Theorem 10.3 [28, Section 4] *For each finitely generated subgroup U of G_ρ the U-locally inner automorphisms of G_ρ form a group A_U which is isomorphic to the profinite completion of U. Moreover, the groups A_U generate a divisible LFC-subgroup $\mathrm{FInn}(G_\rho)$ of $\mathrm{Aut}(G_\rho)$ with torsion subgroup isomorphic to ULF. Also, if $\rho > 0$, then $\mathrm{FInn}(G_\rho)/\mathrm{Inn}(G_\rho) \cong \mathbb{R}^+$.*

Uncountable e. c. LFC-groups have been studied in [29].

Theorem 10.4 [29, Theorem 3.1] *For each $\mathrm{L}\mathfrak{F} \cap \mathfrak{A}$-group A and all cardinals $\kappa \geqslant \rho \geqslant \max\{\aleph_0, |A|\}$ there exist 2^κ e. c. LFC-groups of cardinality κ, with rank ρ and inevitable center A.*

A slick direct limit construction, which uses the density of $\mathbb{Q}$ in $\mathbb{R}$, allows us to produce an e. c. LFC-group which is the split extension of ULF by a direct power Q of $2^{\aleph_0}$ copies of $\mathbb{Q}$ [29, Theorem 1.5]. In particular, $\mathrm{Aut}(\mathrm{ULF})$ contains a copy of Q. Further constructions of e. c. LFC-groups with prescribed properties can be found in [26, Section 9] and [29, Section 3]. In analogy to Theorem 7.3 Jensen's principle $\diamondsuit$ ensures the existence of $2^{\aleph_1}$ e. c. LFC-groups G of cardinality $\aleph_1$ with $\mathrm{Aut}(G) = \mathrm{FInn}(G)$ [26, Section 7].

Every e. c. LFC-group G has a torsion-free divisible Schur multiplier of rank $\binom{\rho}{2}$, where ρ is the rank of G. Moreover, $M(G/\eta(G)) \cong M(G) \oplus \eta(G)$ [29, Lemma 2.1]. In fact, the e. c. LFC-groups can be characterized as being the stem extension of their inevitable center by a quotient with torsion-free divisible Schur multiplier [29, Theorem 2.3]. Here the central quotient determines the e. c. LFC-group if and only if the inevitable center has finite exponent [29, Theorem 2.5].

10.2. p-Torsion

The close analogy between e. c. groups in $\mathrm{L}\mathfrak{F}(A)$ and $\mathrm{L}\mathfrak{F}_p(A)$ on the one hand, and between e. c. groups in $\mathrm{L}\mathfrak{F}(A)$ and LFC on the other hand, suggests the study of e. c. groups in the class LFC_p of all LFC-groups G with $\tau(G) \in \mathrm{L}\mathfrak{F}_p$. In [60] the author has considered countable e. c. $\mathrm{LFC}_p(A)$-groups G, where $A \in \mathrm{L}\mathfrak{F}_p \cap \mathfrak{A}$. Since only elements of infinite order in G can induce p'-automorphisms on G-factors of $\tau(G)$, we can express by a finite set of equations and inequalities the statement that an element in G has infinite order. This has the following consequence.

Theorem 10.5 [60, Theorem 2.5] *Every countable e. c. $\mathrm{LFC}_p(A)$-group is closed in $\mathrm{LFC}_p(A)$. In particular, countable e. c. $\mathrm{LFC}_p(A)$-groups have infinite rank.*

The same holds for e. c. $\mathrm{LFC}_p(A)$-groups of arbitrary cardinality, if A has finite exponent.

Question 10.6 *Is every e. c. $\mathrm{LFC}_p(A)$-group closed in $\mathrm{LFC}_p(A)$?*

Since there is no control on the torsion in permutational products, the basic tool in studying countable e. c. $\mathrm{LFC}_p(A)$-groups is a wreath product construction, where the countability assumption allows us to keep the conjugacy classes finite locally. The main results are as follows.

Theorem 10.7

(a) *For every countable $\mathrm{L}\mathfrak{F}_p \cap \mathfrak{A}$-group A there exist $2^{\aleph_0}$ countable e. c. $\mathrm{LFC}_p(A)$-groups.*

(b) *These groups G have a unique isomorphism type T_A of torsion subgroup, and G/T_A is divisible.*

(c) *The group T_A/A is verbally complete, characteristically simple, and normality is transitive in T_A/A.*

(d) *For every countable e. c. $\mathrm{LFC}_p(A)$-group G, the G-composition factors of T_A/A are infinite elementary abelian, and G acts on them in the same way as $K^\times$ acts on K^+ right regularly, where K denotes the algebraic closure of $\mathbb{F}_p$.*

(e) *T_A/A has a unique G-composition series, and its factors are ordered like the rationals.*

(f) *$H_n(T_A, \mathbb{Z}) = 0$ for all $n < \omega$, and $A = M(T_A/A)$.*

Part (a) follows from Theorem 3.7, since amalgamation of finitely generated $\mathrm{LFC}_p(A)$-groups cannot be controlled. The philosophy of Theorem 10.7 is that T_A can only be distinguished from the countable e. c. $\mathrm{LFC}_p(A)$-group E_A by the fact that its 'unique series' has infinite elementary-abelian factors.

11. Extensions

The results about e. c. central extensions of $\mathrm{L}\mathfrak{F}$- and $\mathrm{L}\mathfrak{F}_p$-groups, as presented in Sections 7.2 and 8.2, suggest that we should consider the more general situation of arbitrary extensions of a fixed normal subgroup. A first approach in this direction was undertaken by K. Hickin and R. Phillips [37], and later extended by K. Hickin [36] and F. Haug [30]. The major difficulty here is, of course, to find suitable constructions. It turns out that the following classes of extensions can be handled.

Let $\mathfrak{X}$ be a class of groups, let N be a fixed group, and consider a distinguished subgroup A of its outer automorphism group $\mathrm{Out}(N)$. Then $\mathfrak{X}(N, A)$ will denote the class of all groups G with $N \trianglelefteq G$, such that $G^0/N \in \mathfrak{X}$ and $G^* \leqslant A$. Here, $G^0 = C_G(N) \cdot N$ is the largest subgroup of G inducing only inner automorphisms on N via conjugation, while $G^* = G/G^0$ consists of the outer automorphisms of N induced from conjugation with elements in G. Again, first-order properties of $\mathfrak{X}(N, A)$-groups may be formulated in a language containing constants for the elements in N and in A.

Most of the theory of e. c. extensions deals with the class $\mathfrak{X} = \mathrm{L}\mathfrak{F}$. In some cases corresponding results can be proved for $\mathfrak{X} = \mathrm{L}\mathfrak{F}_p$, where one has to assume, in addition, that certain chief series match (cf. Section 8.2). In order to simplify our presentation we shall not pursue this variation here. By studying permutational products resp. quotients of free products, the following amalgamation results have been obtained.

Theorem 11.1 [37, Lemma 2] and [30, Section 2] *Suppose that* $A \in \mathrm{L}\mathfrak{F}$. *The amalgam* $G \cup H | U$ *of* $\mathrm{L}\mathfrak{F}(N, A)$-*groups can be completed within* $\mathrm{L}\mathfrak{F}(N, A)$ *provided that one of the following conditions holds.*

(1) $\exp(\zeta_1(N))$ *and* U/N *are finite.*

(2) G/N *and* H/N *are countable, and* U/N *is finite.*

(3) A *is trivial, and* H/N *is finite.*

In the above cases, all the quotients modulo N are locally finite. Note, also, that every $\mathfrak{X}(N, 1)$-group G is the central product of N and $C_G(N)$. Condition (2) in Theorem 11.1 cannot be weakened to uncountable groups G/N: [31] contains an example of a non-completable amalgam $G \cup H | U$ of $\mathrm{L}\mathfrak{F}(N, A)$-groups, where A is elementary abelian, N is free abelian, H/N is finite, and G/N is uncountable.

Question 11.2 *Can condition* (2) *in Theorem* 11.1 *be relaxed to uncountable groups* G/N *and* H/N, *provided that* $\zeta_1(N)$ *is periodic* (*of infinite exponent*) ?

[30, Section 2] also contains results about completions in classes $\mathfrak{X}(N, A)$, where $\mathfrak{X}$ is closed with respect to the formation of subgroups and extensions, and where the existence of a completion within $\mathfrak{X}$ of the corresponding factor amalgam modulo N is assumed. We say that $\mathfrak{X}(N, A)$ has the *finite amalgamation property*, denoted by $\mathfrak{F}$-AP, if $\mathfrak{X}(N, A)$ admits the completion of amalgams $G \cup H | U$ where $H/N \in \mathfrak{F}$. Classes $\mathfrak{X} \subseteq \mathrm{L}\mathfrak{F}$ such that $\mathfrak{X}(N, A)$ has $\mathfrak{F}$-AP are supplied by Theorem 11.1. In this situation, the e. c. $\mathfrak{X}(N, A)$-groups can often be characterized in the usual way in terms of *local injectivity* and *local homogeneity*, i. e., injectivity and homogeneity with respect to finitely generated subgroups of $\mathfrak{X}(N, A)$-groups. We call such groups *locally universal*.

Theorem 11.3 [30, Theorems 3.1 and 3.3] *Let* $\mathfrak{X}$ *be a class of* $\mathrm{L}\mathfrak{F}$-*groups, let* $A \leqslant \mathrm{Out}(N)$ *be locally finite, and suppose that* $\mathfrak{X}(N, A)$ *has* $\mathfrak{F}$-AP. *The e. c.* $\mathfrak{X}(N, A)$-*groups are precisely the locally universal* $\mathfrak{X}(N, A)$-*groups, provided that one of the following conditions holds.*

(1) $A = 1$.

(2) $N/\zeta_1(N) \in \mathrm{L}\mathfrak{F}$, *and* A *centralizes* $N/\tau(N)$ (*where* $\tau(N)$ *denotes the torsion subgroup of the* LFC-*group* N).

Sometimes, however, there exist e. c. extensions which are even injective and homogeneous with respect to finitely generated $\mathfrak{X}(N, A)$-groups, i. e., with respect to $\mathfrak{X}(N, A)$-groups U such that U/N is finitely generated. We call such extensions *universal*.

Theorem 11.4 [37, Theorems 1 and 4], [30, Theorem 3.5 and Proposition 3.6] *Let* $\mathfrak{X} \subseteq \mathrm{L}\mathfrak{F}$, *and suppose that* $\mathfrak{X}(N, A)$ *has* $\mathfrak{F}$-*AP. Then every* $\mathfrak{X}(N, A)$-*group* G *is contained in a universal* $\mathfrak{X}(N, A)$-*group* H *with* $|H/N| = \max\{\aleph_0, |A|, |\zeta_1(N)|, |G/N|\}$, *and there exists at most one such object* H *with* $|H/N| = \aleph_0$.

In the case of central extensions (N abelian, $A = 1$), local injectivity resp. local homogeneity is equivalent to injectivity resp. homogeneity ([30, Lemma 3.8]). The uniqueness statement in Theorem 11.4 allows the construction of universal $\mathrm{L}\mathfrak{F}(N, A)$-groups as the union of an ascending chain of countable groups as in Theorem 11.4 also in the following situation.

Theorem 11.5 [37, Theorem 6] *Suppose that* $\zeta_1(N)$ *is countable periodic, and that* A *is an* $\mathrm{L}\mathfrak{F}$-*subgroup of* $\mathrm{Out}(N)$ *with* $|A| \leqslant \aleph_1$. *Then there exist universal* $\mathrm{L}\mathfrak{F}(N, A)$-*groups* G *with* $|C_G(N)| = \aleph_0$.

Theorem 11.5 gains its importance from the following observation.

Proposition 11.6 [37, (1.7), Theorems 3 and 5], [30, Theorem 4.1] *Let* G *be a universal* $\mathrm{L}\mathfrak{F}(N, A)$-*group.*

(a) $C_G(N)$ *is an e. c.* $\mathrm{L}\mathfrak{F}(\zeta_1(N), 1)$-*group.*

(b) *If* N *and* A *are locally finite, then* $G^* = A$.

(c) *If* $A \in \mathrm{L}\mathfrak{F}$, *then* $N = C_G(C_G(N))$. *In particular,* $G^* \leqslant \mathrm{Out}(C_G(N))$.

Suppose now that Z is a countable $\mathrm{L}\mathfrak{F} \cap \mathfrak{A}$-group. Then a combination of Theorem 11.5 and Proposition 11.6 allows us to embed *every* $\mathrm{L}\mathfrak{F}$-group of cardinality $\aleph_1$ into the outer automorphism group of the unique countable e. c. $\mathrm{L}\mathfrak{F}(Z)$-group $\mathrm{ULF}(Z)$ (cf. Section 7.2). A corresponding result holds for $\mathrm{L}\mathfrak{F}_p$-groups.

Question 11.7 *Does every group of cardinality* $\aleph_1$ *embed into* $\mathrm{Out}(\mathrm{ULF})$ *?*

In fact, if N is countable, with periodic center, and with no normal subgroup isomorphic to $\mathrm{ULF}(\zeta_1(N))$, and if A is a $\mathrm{L}\mathfrak{F}$-subgroup of $\mathrm{Out}(N)$ of cardinality $\aleph_1$, then the continuum hypothesis permits us to construct $\aleph_2$ non-isomorphic universal $\mathrm{L}\mathfrak{F}(N, A)$-groups G_α ($\alpha < \omega_2$) with $|C_{G_\alpha}(N)| = \aleph_0$ for all α [36, Theorem 3].

In order to describe the conjugation action of e. c. $\mathrm{L}\mathfrak{F}(N, A)$-groups G on the centralizer $C_G(N)$ in more detail, the notion of a locally homogeneous action has been created in [37]. Let Z be an abelian group, let $C \in \mathrm{L}\mathfrak{F}(Z, 1)$, and suppose that $A \in \mathrm{L}\mathfrak{F}$. A homomorphism $A \to \mathrm{Aut}(C)$ is said to be a *locally homogeneous action* of A on C if the following holds for every finite subgroup A_0 of A: whenever $U \cup V | W$ is an amalgam of $\mathfrak{F}(Z, 1)$-groups such that $W \leqslant U \leqslant C$ are A_0-invariant, and whenever the action of A_0 on W extends to an action of A_0 on V, then there exists an A_0-embedding $V \to C$.

Theorem 11.8 [37, Theorem 7] and [30, Corollary 4.8] *Let* G *be an e. c.* $\mathrm{L}\mathfrak{F}(N, A)$-*group, where* $A \in \mathrm{L}\mathfrak{F}$. *Let* $C = C_G(N)$ *and* $Z = \zeta_1(N)$. *If* C *is e. c. in* $\mathrm{L}\mathfrak{F}(Z, 1)$ *and* $C_K(N) = Z$ *for some* $N \leqslant K \leqslant G$, *then conjugation brings about a locally homogeneous action of* K/N *on* C.

There also exists a converse: certain locally homogeneous actions of countable
L$\mathfrak{F}$-groups A on groups C with $|C/N| = \aleph_0$ allow the construction of universal ex-
tensions as a quotient of the split extension $C \rtimes A$ [30, Theorems 4.9 and 5.15]. The
existence of a group K as in Theorem 11.8 is guaranteed for countable A, provided
that $\zeta_1(N)$ is trivial or N is abelian [30, Proposition 4.10]. Using the method of
tree-limits, K. Hickin has constructed $2^{\aleph_0}$ inequivalent locally homogeneous actions
of any given countably infinite L$\mathfrak{F}$-group on the unique countable e. c. L$\mathfrak{F}(A)$- resp.
L$\mathfrak{F}_p(A)$-groups ULF(A) resp. E_A [36, Section 2]. This result has been obtained
independently with a different method by F. Haug [30, Section 6].

The following questions are motivated by the fact that every epimorphic image
of an e. c. L$\mathfrak{F}_p$-group is e. c. in L$\mathfrak{F}_p$ or in L$\mathfrak{F}_p(C_p)$, and that every epimorphic image
of a countable e. c. L($\mathfrak{F}_\pi \cap \mathfrak{S}$)-group is e. c. in L($\mathfrak{F}_\pi \cap \mathfrak{S}$) or in the class of all those
L($\mathfrak{F}_\pi \cap \mathfrak{S}$)-groups which have a minimal normal subgroup containing C_p for some
prime $p \in \pi$ [62, Theorems C and D].

Questions 11.9   Let $\mathfrak{X} = $ L($\mathfrak{F}_\pi \cap \mathfrak{S}$).

(a) Let M/N be a chief factor of the e. c. $\mathfrak{X}$-group G, and denote epimorphic
    images modulo N by bars. Is $\overline{G}$ e. c. in $\mathfrak{X}(\overline{M}, \overline{G}^*)$?

(b) Does there exist an e. c. $\mathfrak{X}$-group G such that, in the above notation, $\overline{G}$ is
    universal in $\mathfrak{X}(\overline{M}, \overline{G}^*)$ for every chief factor M/N of G ?

Finally, in [30, Section 5], F. Haug studies e. c. L$\mathfrak{F}(Z, 1)$-groups, where Z is abelian
but not necessarily periodic. Such groups are LFC-groups, and hence the investigations here may be seen as a continuation of both the theory of e. c. L$\mathfrak{F}(A)$-groups
(cf. Section 7.2) and the theory of e. c. LFC-groups (cf. Section 10). The major results are displayed in the following theorem. The proof of parts (c) and (d) requires
knowledge of the cohomological extension theory.

Theorem 11.10 [30, Section 5] Let G be e. c. in L$\mathfrak{F}(Z, 1)$, where $Z \in \mathfrak{A}$. Let
$Q = G/Z$, and denote epimorphic images modulo torsion subgroups by bars.

(a) $\zeta_1(G) = Z$, and $G^{(1)} = \tau(G)$ is e. c. in L$\mathfrak{F}(\tau(Z))$, and $\overline{G} \cong \overline{Z}$ is divisible.

(b) $Q/Q^{(1)}$ is divisible of p-rank $\beta_p(Z) = dim_{\mathbb{F}_p}(\overline{Z}/\overline{Z}^p)$, and $\tau(G)/\tau(Z) \cong Q^{(1)}$.

(c) $G/\tau(Z)$ is isomorphic to the direct product of $\overline{G}$ and Q with amalgamated
    factor group $G/(Z \cdot \tau(G))$. In particular, the isomorphism types of $\overline{Z}$ and Q
    determine $G/\tau(Z)$.

(d) If Z is the direct product of a π-group and a torsion-free π-divisible group,
    then Q and Z determine G.

(e) If $\overline{Z}$ has rank $\rho \leqslant \omega$, then G splits over $\tau(G)$ and Q splits over $Q^{(1)}$.

In view of Theorems 7.7 and 11.10(c),(d) the question arises as to how far the
structure of an e. c. L$\mathfrak{F}(Z, 1)$-group G or the structure of Z is determined by the
central quotient G/Z. In general, neither G/Z nor $G/\tau(Z)$ determine Z [30,
Lemma 5.8 and Corollary 5.9]. However it is possible to relate the class of all central
quotients of e. c. L$\mathfrak{F}(Z, 1)$-groups with the structure of Z.

Theorem 11.11 [30, Theorem 5.13] *Let Y and Z be non-periodic $\mathfrak{A}$-groups. Then the following are equivalent.*

(1) $\{G/Y \mid G$ *is e. c. in* $L\mathfrak{F}(Y,1)\} = \{H/Z \mid H$ *is e. c. in* $L\mathfrak{F}(Z,1)\}$.

(2) $\tau(Y) \cong \tau(Z)$, *and* $\beta_p(Y) \cong \beta_p(Z)$ *for all primes p.*

(3) $\tau(Y) \cong \tau(Z)$, *and the completions of* $\overline{Y} = Y/\tau(Y)$ *and* $\overline{Z} = Z/\tau(Z)$, *with respect to the topology where the* $\overline{Y}^n$ *resp.* $\overline{Z}^n$ *($n \in \omega$) form a basis of open neighborhoods of the identity, are isomorphic.*

12. Finitary Linear Groups

12.1. LINEAR GROUPS

H. Mez has determined the e. c. linear groups of fixed degree in [78]. It is pretty obvious that the e. c. $\mathfrak{L}(1,c)$-groups are precisely the divisible $\mathfrak{A}_{c'}$-groups of p-rank 1 for every $p \in c'$ [12]. The general classification of Mez is as follows.

Theorem 12.1 [78, Theorem 2]

(a) *A group G is e. c. in $\mathfrak{L}(n,c)$ if and only if there exists an e. c. $\mathfrak{L}(1,c)$-group A and an a. c. field K of characteristic c such that G is the central product of A and $\mathrm{SL}_n(K)$ over $\{a \in A \mid a^n = 1\}$.*

(b) *The e. c. $\mathfrak{L}(n)$-groups with a faithful representation in characteristic c are precisely the e. c. $\mathfrak{L}(n,c)$-groups.*

Idea of Proof. (a) Clearly every e. c. $\mathfrak{L}(n,c)$-group G may be considered as a subgroup of some $\mathrm{GL}_n(K)$, where K is an a. c. field of characteristic c. In particular, G contains a copy of every finite subgroup of $\mathrm{GL}_n(K)$. If $c > 0$, then a conjugate of G inside $\mathrm{GL}_n(K)$ contains $\mathrm{SL}_n(c^k)$ and hence also contains a non-scalar diagonal matrix and elementary matrices which differ from the identity in just one non-diagonal entry. From this information, the desired structure of G can be built up. The case $c = 0$ is more difficult.

Conversely, it has to be shown that every group G of the form $A \cdot \mathrm{SL}_n(K)$ is e. c. in $\mathfrak{L}(n,c)$. The problem is the transition from abstract groups to matrix groups. Here one uses the following fact: whenever $U \leqslant G \in \mathfrak{L}(n,c)$ are such that G is e. c. in $\mathfrak{L}(n,c)$, and U is e. c. with respect to finite sets of equations and inequalities which can be solved in G, then U is also e. c. in $\mathfrak{L}(n,c)$.

(b) Non-trivial representations of $\mathrm{SL}_n(K)$, where K is a. c. of characteristic c, are only possible in characteristic c. $\qquad\square$

The above considerations can be refined to give a characterization of e. c. $\mathfrak{L}(n) \cap \mathfrak{S}$-groups as central products of e. c. $\mathfrak{L}(1,c)$-groups with maximal solvable subgroups of groups of type $\mathrm{GL}_n(K)$; see [77] and [78, Proposition 4].

12.2. FINITARY LINEAR GROUPS

A group G of automorphisms of a vector space V is said to be *finitary linear* if the endomorphisms $g - 1$ ($g \in G$) have finite ranks. In particular, every finitely generated subgroup $U \leqslant G$ has finite *degree*

$$d(U) \; = \; \min \{ \dim V_0 \mid [V, U] \leqslant V_0 \leqslant V \text{ and } V = V_0 + C_V(U) \}$$

where $[V, U] = \langle V(g - 1) \mid g \in U \rangle$. Examples of finitary linear groups are the *stable linear groups* $\lim_{n \to \infty} \mathrm{GL}_n(K)$ (where $\mathrm{GL}_n(K)$ embeds canonically into the left upper corner of $\mathrm{GL}_{n+1}(K)$), or the locally nilpotent groups of D. H. McLain. A detailed exposition of the present state of knowledge about finitary linear groups is given in the article by R. Phillips in these Proceedings. In order to make the class of all finitary linear groups inductive, we consider pairs (G, d), where G is finitary linear with respect to the *degree function* d which assigns to each finitely generated subgroup U of G its degree $d(U)$. Embeddings of such pairs always have to respect the degree functions. In abuse of notation we shall usually write G instead of (G, d). E. c. finitary linear groups G have been considered by O. Kegel and D. Schmidt [46].

Theorem 12.2 [46] *If G is e. c. finitary linear, then G has trivial center, and its commutator subgroup has a local system of stable special linear groups over an a. c. field. In particular, $G^{(1)}$ is simple.*

Idea of Proof. The center can obviously be killed by embedding G onto the 1-component of $G \, \mathrm{Wr} \, C_p$. Now every finitely generated subgroup of G is contained in a finitely generated subgroup U of G such that U acts irreducibly on $[V, U]$. Consider an ascending chain of such subgroups U_i $(i < \omega)$ of G, and let $n_i = d(U_i)$. It turns out that the subgroups $\overline{U}_i = \langle g \in G \mid d(\langle U_i, g \rangle) = n_i \rangle$ are e. c. in $\mathfrak{L}(n_i, c)$, where c is the underlying characteristic. Therefore $\overline{U}_i^{\,(1)} \cong \mathrm{SL}_{n_i}(K)$ for some a. c. field K of characteristic c (Theorem 12.1), and $W = \bigcup_{i < \omega} \overline{U}_i^{\,(1)}$ is a stable special linear group on the subspace $\bigcup_{i < \omega} [V, U_i]$ of V. Moreover it can be shown that W contains a conjugate of every finitely generated subgroup of G. It follows that the subgroups of type W form a local system in G. $\square$

A countable e. c. unipotent finitary linear group has been constructed in a paper by O. Puglisi and the author [63]. Let K be a fixed a. c. field, and let T_n be the group of all upper unitriangular $(3^n \times 3^n)$-matrices over K, with the natural degree function corresponding to its canonical representation on the K-space V_n with basis $B_n = \{ v_i \mid i \in I_n \}$ where $I_n = \{ \frac{a}{3^n} \mid a \in \mathbb{Z}, |a| < \frac{3^n}{2} \}$. Then an embedding $\alpha_n : T_n \to T_{n+1}$, which preserves the degree function on T_n, is given via

$$v_{(3a-1)/3^{n+1}} - v_{a/3^n}, \; v_{(3a+1)/3^{n+1}} \in C_{V_{n+1}}(T_n \alpha_n) \quad \text{for } |a| < \frac{3^n}{2}.$$

If one considers an element $g \in T_n$ as a matrix with respect to the basis B_n, then the embedding α_n replaces each column of $g - 1$ by two neighbouring identical columns immediately followed by a zero column. The direct limit L_K of the groups T_n $(n < \omega)$ with respect to the embeddings α_n acts as a unipotent finitary linear group on the K-space $V = \bigcup_{n < \omega} V_n$.

Theorem 12.3 [63, Theorem 5.1] *For each a. c. field K, the above group L_K is e. c. in the class of all unipotent finitary linear groups.*

This result holds because the embeddings α_n are constructed in such a way that L_K is injective with respect to unipotent finitary linear groups of finite degrees over the field K. Although the McLain group of order type $\mathbb{Q}$ embeds every unipotent finitary linear group which is the union of an ascending chain of subgroups of finite degrees [63, Theorem 2.3], it can be shown that McLain groups fail to be e. c. unipotent finitary linear, since the duplication of columns is missing there. Even as an abstract group, L_K is not isomorphic to a McLain group, since the intersections of maximal abelian normal subgroups behave slightly differently [63, Theorem 4.7].

References

1. P. D. Bacsich, Cofinal simplicity and algebraic closedness, *Algebra Universalis* 2 (1972), 354–360.

2. B. Baumslag and F. Levin, Algebraically closed torsionfree nilpotent groups of class 2, *Comm. Algebra* 4 (1976), 533–560.

3. G. Baumslag, *Lectures on Nilpotent Groups*, Conf. Board Math. Sci., Amer. Math. Soc., Providence, RI, 1971.

4. G. Baumslag, E. Dyer and A. Heller, The topology of discrete groups, *J. Pure Appl. Algebra* 16 (1980), 1–47.

5. O. V. Belegradek, *Algebraically Closed Groups*, Cand. Sc. thesis, Novosibirsk, 1974.

6. O. V. Belegradek, Algebraically closed groups, *Algebra and Logic* 13 (1974), 135–143.

7. O. V. Belegradek, Elementary properties of algebraically closed groups, *Fund. Math.* 98 (1978), 83–101.

8. O. V. Belegradek, On inevitable submodels of existentially closed models (in preparation).

9. K. Bonvallet, B. Hartley, D. S. Passman and M. K. Smith, Group rings with simple augmentation ideals, *Proc. Amer. Math. Soc.* 56 (1976), 79–82.

10. M. Dugas and R. Göbel, On locally finite p-groups and a problem of P. Hall's, *J. Algebra* 159 (1993), 115–138.

11. P. C. Eklof, Ultraproducts for algebraists, in J. Barwise (ed.), *Handbook of Mathematical Logic*, North-Holland, Amsterdam, 1977, pp. 105–137.

12. P. C. Eklof and G. Sabbagh, Model-completions and modules, *Ann. Math. Logic* 2 (1970/71), 251–295.

13. H. Ensel, *Existentiell abgeschlossene Gruppen in lokal endlich–auflösbaren Gruppenklassen*, Diplomarbeit, Mainz, 1987.

14. H. Ensel, Die Automorphismengruppe der abzählbaren existentiell abgeschlossenen $\mathcal{P}_A$-Gruppe E_A, *Arch. Math.* 51 (1988), 198–203.

15. H. Ensel, Existentiell abgeschlossene Sylowturmgruppen, *Arch. Math.* 51 (1988), 385–392.

16. M. Erdélyi, On n-algebraically closed groups, *Publ. Math. Debrecen* 7 (1960), 310–315.

17. S. Gacsalyi, On algebraically closed abelian groups, *Publ. Math. Debrecen* 2 (1952), 292–296.

18. D. Giorgetta and S. Shelah, Existentially closed structures in the power of the continuum, *Ann. Pure Appl. Logic* 26 (1984), 123–148.

19. A. M. W. Glass, *Ordered Permutation Groups*, University Press, Cambridge, 1981.

20. A. M. W. Glass and K. R. Pierce, Equations and inequations in lattice-ordered groups, in J. E. Smith, G. O. Kenny and R. N. Ball (eds.), *Ordered Groups: Proc. Boise State Conf. 1978*, Marcel Dekker, New York, 1980, pp. 141–171.

21. A. M. W. Glass and K. R. Pierce, Existentially complete abelian lattice-ordered groups, *Trans. Amer. Math. Soc.* 261 (1980), 255–270.

22. A. M. W. Glass and K. R. Pierce, Existentially complete lattice-ordered groups, *Israel J. Math.* 36 (1980), 257–272.

23. R. Grossberg and S. Shelah, On universal locally finite groups, *Israel J. Math.* 44 (1983), 289–302.

24. P. Hall, A contribution to the theory of groups of prime-power order, *Proc. London Math. Soc.* (2) 36 (1932), 29–95.

25. P. Hall, Some constructions for locally finite groups, *J. London Math. Soc.* 34 (1959), 305–319.

26. F. Haug, *Existenziell abgeschlossene LFC-Gruppen*, Dissertation, Tübingen, 1987.

27. F. Haug, An amalgamation theorem for locally FC-groups, *J. London Math. Soc.* 43 (1991), 421–430.

28. F. Haug, Countable existentially closed locally FC-groups, *J. Algebra* 143 (1991), 1–24.

29. F. Haug, Existentially closed locally FC-groups, *Comm. Algebra* 21 (1993), 4513–4539.

30. F. Haug, *Existentially Closed Locally Finite Extensions*, Habilitationsschrift, Tübingen, 1994.

31. F. Haug, U. Meierfrankenfeld and R. E. Phillips, An amalgamation theorem for finite group extensions, *Arch. Math.* 57 (1991), 325–331.

32. R. Hessenkamp, *Algebraisch abgeschlossene torsionsfreie nilpotente Gruppen*, Diplomarbeit, Freiburg i. Br., 1981.

33. K. K. Hickin, Complete universal locally finite groups, *Trans. Amer. Math. Soc.* 239 (1978), 213–227.

34. K. K. Hickin, Universal locally finite central extensions, *Proc. London Math. Soc.* (3) 52 (1986), 53–72.

35. K. K. Hickin, Some applications of tree-limits to groups. Part I, *Trans. Amer. Math. Soc.* 305 (1988), 797–839.

36. K. K. Hickin, Locally homogeneous actions and universal extensions of groups, *J. London Math. Soc.* (2) 47 (1993), 269–284.

37. K. K. Hickin and R. E. Phillips, Universal locally finite extensions of groups, in *Group Theory, Conf. Proc. Bressanone/Brixen 1989, Rend. Circ. Mat. Palermo* (II), Suppl. no. 23 (1990), pp. 143–171.

38. G. Higman, Amalgams of *p*-groups, *J. Algebra* 1 (1964), 301–305.

39. G. Higman and E. Scott, *Existentially Closed Groups*, Clarendon Press, Oxford, 1988.

40. J. Hirschfeld and W. H. Wheeler, *Forcing, Arithmetic, Division Rings*, Springer-Verlag, Berlin-Heidelberg-New York, 1975.

41. W. Hodges, Interpreting number theory in nilpotent groups, *Arch. Math. Logik Grundlag.* 20 (1980), 103–111.

42. W. Hodges, Groupes nilpotents existentiellement clos de classe fixée, *Mém. Soc. Math. de France*, Nouv. Sér. 16 (1984), 1–10.

43. W. Hodges, *Building Models by Games*, University Press, Cambridge, 1985.

44. W. Hodges, *Model Theory*, University Press, Cambridge, 1993.

45. O. H. Kegel, Examples of highly transitive permutation groups, *Rend. Sem. Mat. Univ. Padova* 63 (1980), 295–300.

46. O. H. Kegel and D. Schmidt, Existentially closed finitary linear groups, in C. M. Campbell and E. F. Robertson (eds.), *Groups—St. Andrews 1988*, vol. 2, University Press, Cambridge, 1991, pp. 353–360.

47. O. H. Kegel and B. A. F. Wehrfritz, *Locally Finite Groups*, North-Holland, Amsterdam, 1973.

48. F. Leinen, Lokale Systeme in universellen Gruppen, *Arch. Math.* 41 (1983), 401–403.

49. F. Leinen, *Existenziell abgeschlossene LX-Gruppen*, Dissertation, Freiburg i. Br., 1984.

50. F. Leinen, Existentially closed groups in locally finite group classes, *Comm. Algebra* 13 (1985), 1991–2024.

51. F. Leinen, Existentially closed LX-groups, *Rend. Sem. Mat. Univ. Padova* 75 (1986), 191–226.

52. F. Leinen, Existentially closed locally finite *p*-groups, *J. Algebra* 103 (1986), 160–183.

53. F. Leinen, A uniform way to control chief series in finite *p*-groups and to construct the countable algebraically closed locally finite *p*-groups, *J. London Math. Soc.* (2) 33 (1986), 260–270.

54. F. Leinen, An amalgamation theorem for soluble groups, *Canad. Math. Bull.* 30 (1987), 9–18.

55. F. Leinen, Chief series and right regular representations of finite *p*-groups, *J. Austral. Math. Soc.* (A) 44 (1988), 225–232.

56. F. Leinen, Group rings of existentially closed locally finite *p*-groups, *Publ. Math. Debrecen* 35 (1988), 289–294.

57. F. Leinen, *Existentiell abgeschlossene Gruppen*, Habilitationsschrift, Mainz, 1990.

58. F. Leinen, Uncountable existentially closed groups in locally finite group classes, *Glasgow Math. J.* 32 (1990), 153–163.

59. F. Leinen, Existentially closed locally cofinite groups, *Proc. Edinburgh Math. Soc.* 35 (1992),

233–253.

60. F. Leinen, Countable closed LFC-groups with p-torsion, *Trans. Amer. Math. Soc.* 336 (1993), 193–217.

61. F. Leinen and R. E. Phillips, Existentially closed central extensions of locally finite p-groups, *Math. Proc. Camb. Phil. Soc.* 100 (1986), 281–301.

62. F. Leinen and R. E. Phillips, Algebraically closed groups in locally finite group classes, in O. H. Kegel, F. Menegazzo and G. Zacher (eds.), *Group Theory, Conf. Proc. Brixen/Bressanone 1986*, Springer-Verlag, Berlin-Heidelberg-New York, 1987, pp. 85–102.

63. F. Leinen and O. Puglisi, Unipotent finitary linear groups, *J. London Math. Soc.* (2) 48 (1993), 59–76.

64. A. Macintyre, On algebraically closed groups, *Ann. Math.* 96 (1972), 53–97.

65. A. Macintyre, Existentially closed structures and Jensen's principle $\Diamond$, *Israel J. Math.* 25 (1976), 202–210.

66. A. Macintyre and S. Shelah, Uncountable universal locally finite groups, *J. Algebra* 43 (1976), 168–175.

67. B. J. Maier, *Existenziell abgeschlossene Gruppen in nilpotenten Gruppenklassen*, Dissertation, Freiburg i. Br., 1981.

68. B. J. Maier, Existenziell abgeschlossene lokal endliche p-Gruppen, *Arch. Math.* 37 (1981), 113–128.

69. B. J. Maier, On existentially closed and generic nilpotent groups, *Israel J. Math.* 46 (1983), 170–188.

70. B. J. Maier, Existentially closed torsion-free nilpotent groups of class three, *J. Symbolic Logic* 49 (1984), 220–230.

71. B. J. Maier, Amalgame torsionsfreier nilpotenter Gruppen, *J. Algebra* 99 (1986), 520–547.

72. B. J. Maier, On countable locally described structures, *Ann. Pure Appl. Logic* 35 (1987), 205–246.

73. B. J. Maier, On universal nilpotent groups of class two, *Comm. Algebra* 16 (1988), 1453–1456.

74. B. J. Maier, On universal nilpotent groups, *Algebra Universalis* 26 (1989), 202–207.

75. B. J. Maier, On nilpotent groups of exponent p, *J. Algebra* 127 (1989), 279–289.

76. A. I. Mal'cev, Nilpotent torsion-free groups, *Izv. Akad. Nauk SSSR, Ser. Mat.* 13 (1949), 201–212 (in Russian).

77. H.-C. Mez, *Existentiell abgeschlossene lineare Gruppen*, Diplomarbeit, Freiburg i. Br., 1979.

78. H.-C. Mez, Existentially closed linear groups, *J. Algebra* 76 (1982), 84–98.

79. B. H. Neumann, A note on algebraically closed groups, *J. London Math. Soc.* 27 (1952), 247–249.

80. B. H. Neumann, Permutational products of groups, *J. Austral. Math. Soc.* 1 (1960), 299–310.

81. B. H. Neumann, On amalgams of periodic groups, *Proc. Royal Soc. London* (A) 255 (1960), 477–489.

82. R. E. Phillips, Existentially closed locally finite central extensions; multipliers and local systems, *Math. Z.* 187 (1984), 383–392.

83. R. E. Phillips and S. E. Schuur, Partial homogeneity in locally finite groups, *Math. Z.* 194 (1987), 293–308.

84. O. Puglisi, A note on the automorphism group of a locally finite p-group, *Bull. London Math. Soc.* 24 (1992), 437–441.

85. D. Saracino, Wreath products and existentially complete solvable groups, *Trans. Amer. Math. Soc.* 197 (1974), 327–339.

86. D. Saracino, Existentially complete nilpotent groups, *Israel J. Math.* 25 (1976), 241–248.

87. D. Saracino, Existentially complete torsion-free nilpotent groups, *J. Symb. Logic* 43 (1978), 126–134.

88. D. Saracino and C. Wood, Periodic existentially closed nilpotent groups, *J. Algebra* 58 (1979), 189-207.

89. D. Saracino and C. Wood, Finitely generic abelian lattice-ordered groups, *Trans. Amer. Math. Soc.* 277 (1983), 113–123.

90. D. Saracino and C. Wood, An example in the model theory of abelian lattice-ordered groups, *Algebra Universalis* 19 (1984), 34–37.

91. S. E. Schuur, Controlling conjugacy classes in embeddings of locally finite groups, *Rocky Mountain J. Math.* 20 (1990), 215–221.

92. S. E. Schuur, Direct limit constructions of the countable ULF(A) group, *Bull. London Math. Soc.*, to appear in 1994.

93. W. R. Scott, Algebraically closed groups, *Proc. Amer. Math. Soc.* 2 (1951), 118–121.

94. T. Szele, Ein Analogon der Körpertheorie für abelsche Gruppen, *Publ. Math. Debrecen* 188 (1950), 167–192.

95. S. Thomas, Complete existentially closed locally finite groups, *Arch. Math.* 44 (1985), 97–109.

96. S. Thomas, Complete universal locally finite groups of large cardinality, in J. B. Paris, A. J. Wilkie and G. M. Wilmers (eds.), *Logic Colloquium '84, Manchester*, North-Holland, Amsterdam, 1986, pp. 277–301.

97. K. Varadarajan, Pseudo-mitotic groups, *J. Pure Appl. Algebra* 37 (1985), 205–213.

98. V. Weispfennig, Model theory of abelian ℓ-groups, in A. M. W. Glass and W. C. Holland (eds.), *Lattice-Ordered Groups*, Kluwer Acad. Publishers, Dordrecht, 1989, pp. 41–79.

99. A. E. Zalesskiĭ, Group rings of locally finite groups and representation theory, in *Proc. Intern. Algebra Conf. in Honour of A. I. Mal'cev, Novosibirsk 1989*, Part I, *Contemp. Math.* 131, Amer. Math. Soc., Providence, RI, 1992, pp. 453–472.

100. A. E. Zalesskiĭ, *Simple Locally Finite Groups and their Group Rings*, TEMPUS Lecture Notes 2, Norwich.

101. M. Ziegler, Algebraisch abgeschlossene Gruppen, in S. I. Adian, W. W. Boone and G. Higman (eds.), *Word Problems II, The Oxford Book*, North-Holland, Amsterdam, 1980, pp. 449–576.

GROUPS ACTING ON POLYNOMIAL ALGEBRAS

R. M. BRYANT
Department of Mathematics
UMIST
PO Box 88
Manchester M60 1QD
United Kingdom
e-mail: bryant@umist.ac.uk

Abstract. Let K be an infinite field of non-zero characteristic. Then, for any positive integer r, the general linear group $\mathrm{GL}_r(K)$ has a natural action on the polynomial algebra $K[x_1,\ldots,x_r]$ so that this algebra becomes a $\mathrm{GL}_r(K)$-module. The submodule structure of $K[x_1,\ldots,x_r]$ was determined by S. R. Doty and L. Krop. The first part of this paper gives a full account of the main results of this theory. In the second part, the theory is applied to give a new proof of the result that if G is any finite subgroup of $\mathrm{GL}_r(K)$ and $K[x_1,\ldots,x_r]$ is regarded as a KG-module then $K[x_1,\ldots,x_r]$ is asymptotically close to being a free KG-module.

Key words: polynomial algebras, general linear group, representation theory, finite group actions.

Introduction

Let K be a field and let R be the polynomial algebra $K[x_1,\ldots,x_r]$, where r is a positive integer. Then the general linear group $\mathrm{GL}_r(K)$ acts on R in a natural way (by homogeneous linear substitutions), so that R can be regarded as a $K\mathrm{GL}_r(K)$-module. In the case where K has characteristic 0 the submodule structure of R is easy to determine (see Section 1 below). When K is finite the corresponding question is much harder and only partial results are available (see, for example, L. Krop [12,13] and L. G. Kovács [11]). In the case where K is infinite of non-zero characteristic the submodule structure of R was completely determined by S. R. Doty [5,6] and (independently) by Krop (pp. 383–388 in [12]). The main results in this case are surprisingly simple and form a very elegant theory. I give a self-contained description of the rudiments of this theory in Sections 1 and 2 below together with complete proofs of the main results. The approach used is based on the paper by Kovács [11]. Another account of this material has been given by Doty [7]. Professor I. D. Suprunenko has kindly informed me that some of the theory was also discovered independently by A. M. Adamovich [1] (but I have not seen this paper).

Let K be an infinite field of non-zero characteristic p and, for each non-negative integer n, let R_n be the subspace of R spanned by all monomials of degree n. Clearly R_n is a $K\mathrm{GL}_r(K)$-submodule of R. It is easily seen that the determination of the submodules of R can be reduced to the determination of the submodules of each R_n. The submodules of R_n form a lattice under the operations of addition (join) and intersection, and it turns out that this lattice is finite and distributive. Consequently every submodule of R_n can be written (essentially uniquely) as a sum

B. Hartley et al. (eds.), Finite and Locally Finite Groups, 327–346.
© *1995 Kluwer Academic Publishers. Printed in the Netherlands.*

(join) of submodules which are join-irreducible in the sense of lattice theory. The submodules of R_n are fully determined from information which describes the join-irreducible elements of the lattice and describes whether or not $U \subseteq V$ holds for given join-irreducible elements U and V. This information is provided by the first main result of the theory. Before stating it we need some notation.

Let T_n be the set of all $(n+1)$-tuples $(t_0, t_1, \ldots, t_n)$ where each t_j is an integer satisfying $0 \leqslant t_j \leqslant r(p-1)$ and $n = t_0 + t_1 p + \cdots + t_n p^n$. For $s, t \in T_n$ with $s = (s_0, s_1, \ldots, s_n)$, $t = (t_0, t_1, \ldots, t_n)$, write $s \preccurlyeq t$ if

$$s_0 + s_1 p + s_2 p^2 + \cdots + s_{k-1} p^{k-1} \leqslant t_0 + t_1 p + t_2 p^2 + \cdots + t_{k-1} p^{k-1}$$

for $k = 1, \ldots, n$. Clearly $\preccurlyeq$ is a partial order on T_n. For non-negative integers m and k let $R_m^{p^k}$ be the K-span of the monomials u^{p^k} where u is a monomial of degree m. For $t \in T_n$ with $t = (t_0, t_1, \ldots, t_n)$ write

$$W_t = R_{t_0} R_{t_1}^p \cdots R_{t_n}^{p^n}.$$

In other words, W_t is the submodule of R_n spanned by all monomials $u_0 u_1^p \cdots u_n^{p^n}$ where, for $j = 0, \ldots, n$, u_j is a monomial of degree t_j. The first main result is the following theorem (see Doty [6] and pp. 383–387 in Krop [12]).

Theorem 1.4 *The submodules of R_n form a finite distributive lattice under $+$ and $\cap$ whose non-zero join-irreducible elements are the modules W_t with $t \in T_n$. For $s, t \in T_n$, $W_s \subseteq W_t$ if and only if $s \preccurlyeq t$.*

It should be noted that Doty states the result in a slightly different way using the idea of 'carry patterns' in base p arithmetic. This approach provides an attractive alternative to the one used here.

Let us illustrate Theorem 1.4 by examining the special case where $r = 2$, $p = 3$ and $n = 12$.

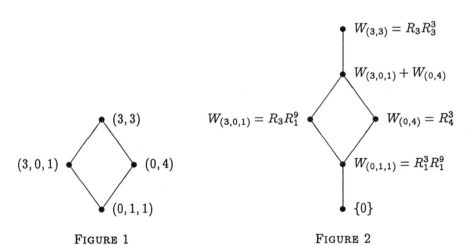

FIGURE 1   FIGURE 2

Here, by the definition of T_n, we see that T_{12} consists of the 13-tuples $(0, 1, 1, 0, \ldots, 0)$, $(3, 0, 1, 0, \ldots, 0)$, $(0, 4, 0, \ldots, 0)$ and $(3, 3, 0, \ldots, 0)$, which may be written in the abbreviated form $(0, 1, 1)$, $(3, 0, 1)$, $(0, 4)$ and $(3, 3)$. By the definition of $\preccurlyeq$ we see that this partial order on T_{12} is as shown in Figure 1.

Thus, by virtue of Theorem 1.4, the only non-zero submodules of R_{12} are $W_{(0,1,1)}$, $W_{(3,0,1)}$, $W_{(0,4)}$, $W_{(3,0,1)} + W_{(0,4)}$ and $W_{(3,3)}$. Furthermore the lattice of submodules is as shown in Figure 2. In particular, $R_{12} = R_3 R_3^3$ (which is easy to verify directly).

We return to the general theory. Since each W_t is non-zero and join-irreducible it has a unique maximal proper submodule, denoted by $\mathrm{rad}(W_t)$, and $W_t/\mathrm{rad}(W_t)$ is an irreducible $K\mathrm{GL}_r(K)$-module, denoted by $\sigma(W_t)$. It is an easy corollary of Theorem 1.4 that the factors in every composition series of R_n are isomorphic to the modules $\sigma(W_t)$, $t \in T_n$. These modules are described up to isomorphism by the other main result of the theory, due to Krop [12].

Theorem 2.2 Let $t \in T_n$ where $t = (t_0, t_1, \ldots, t_n)$. Then

$$\sigma(W_t) \cong \sigma(R_{t_0}) \otimes \sigma(R_{t_1}^p) \otimes \cdots \otimes \sigma(R_{t_n}^{p^n}).$$

Since $0 \leqslant t_j \leqslant r(p-1)$ for each j and $\sigma(R_{t_j}^{p^j})$ can be regarded as coming from $\sigma(R_{t_j})$ by a 'Frobenius twist', Theorem 2.2 describes the composition factors of R_n in terms of the modules $\sigma(R_0)$, $\sigma(R_1)$, $\ldots$, $\sigma(R_{r(p-1)})$.

Following a description of this basic general theory in Sections 1 and 2, a few more details are worked out in Section 3 and then the theory is used in Section 4 to obtain a new and much more satisfactory proof of a theorem about symmetric powers of representations of finite groups which first appeared in [2].

Let K be any field and let G be a finite subgroup of $\mathrm{GL}_r(K)$. Thus R can be regarded as a KG-module. The theorem asserts that, in a certain asymptotic sense, R is close to being a free KG-module. To be precise, let c be the number of scalar matrices in G and for each non-negative integer n let $f(R_n \oplus R_{n+1} \oplus \cdots \oplus R_{n+c-1})$ be defined in such a way that

$$f(R_n \oplus \cdots \oplus R_{n+c-1}) \dim(R_n \oplus \cdots \oplus R_{n+c-1})$$

is the largest dimension of a free KG-submodule of $R_n \oplus R_{n+1} \oplus \cdots \oplus R_{n+c-1}$. (It is easy to see that no direct sum of fewer than c homogeneous components of R can have a non-zero free KG-submodule.) Then the theorem may be stated as follows.

Theorem 4.3 $\displaystyle\lim_{n \to \infty} f(R_n \oplus \cdots \oplus R_{n+c-1}) = 1.$

Finally, at the end of Section 4, Theorem 4.3 is compared with the work of D. J. Glover [8]. It is shown that Glover's work yields the following result.

Theorem 4.6 Let K be the field of p elements, p a prime, and let $G = \mathrm{GL}_2(K)$, $R = K[x_1, x_2]$. Then, for all n,

$$R_n \oplus R_{n+1} \oplus \cdots \oplus R_{n+p-2}$$

has a free KG-submodule of bounded codimension.

In the case under consideration, and with c as defined above, $c = p - 1$. Thus Theorem 4.6 is a strengthening of Theorem 4.3 in this special case.

It is a pleasure for me to acknowledge my indebtedness to Dr L. G. Kovács for guidance on many of the topics treated here. Dr Kovács also very kindly sent me a copy of an early draft of his paper [11] containing additional details not to be found in [11] and allowed me to draw on these details in the composition of my paper. I am also grateful to Dr Grant Walker for the suggestion that I consider Glover's work. Finally, I thank both Dr Kovács and Dr Walker for several other comments on a preliminary version of this paper.

1. The Submodule Structure

As explained in the Introduction, the results in this section and the next all originate in the theory developed by Doty and Krop. The main result of this section is Theorem 1.4 (stated in the Introduction). Let K be an infinite field and write $R = K[x_1, \ldots, x_r]$. (We shall soon assume that K has non-zero characteristic, but we do not do so yet.) Write R as the direct sum of its homogeneous components, namely $R = R_0 \oplus R_1 \oplus \cdots$ where, for each n, R_n is the K-span of the set of all monomials of degree n. There is one monomial $x_1^{a_1} \cdots x_r^{a_r}$ of degree n corresponding to each r-tuple $(a_1, \ldots, a_r)$ of non-negative integers satisfying $a_1 + \cdots + a_r = n$, and these monomials form a basis for R_n. Thus R_n has dimension equal to the binomial coefficient

$$\binom{n + r - 1}{r - 1}.$$

In particular, R_1 has basis $\{x_1, \ldots, x_r\}$ and R_0 may be identified with K.

Let M denote the set $M_r(K)$ of all $r \times r$ matrices over K and regard M as a monoid (semigroup with identity) under the operation of matrix multiplication. (We shall never use the operation of addition in M.) For $u \in R$ we write u as $u(x_1, \ldots, x_r)$ to emphasise its expression as a 'word' in $x_1, \ldots, x_r$ and to allow us to write $u(v_1, \ldots, v_r)$, where $v_1, \ldots, v_r \in R$, for the image of u under the K-algebra endomorphism of R which maps x_i to v_i $(1 \leqslant i \leqslant r)$. Less formally $u(v_1, \ldots, v_r)$ comes from u by the substitution which replaces x_i by v_i $(1 \leqslant i \leqslant r)$.

For $\phi \in M$, let us write $\phi = (\alpha_{ij})$ in the usual matrix notation, where $\alpha_{ij} \in K$ $(1 \leqslant i, j \leqslant r)$, and define

$$u\phi = u\left(\sum_{j=1}^{r} \alpha_{1j} x_j, \ldots, \sum_{j=1}^{r} \alpha_{rj} x_j\right).$$

Thus $u\phi$ is the image of u under the 'homogeneous linear substitution' which replaces x_i by $\sum_j \alpha_{ij} x_j$ for each i. It is then easy to verify that R becomes a right (unitary) module for the monoid M over the field K. We call such modules KM-modules though we do not need to consider the 'monoid-ring' KM. In fact, not only is R a KM-module, but each element of M acts as a K-algebra endomorphism, so we have $(uv)\phi = (u\phi)(v\phi)$ for all $u, v \in R$ and all $\phi \in M$. Clearly each component R_n is a submodule of R.

Let Γ denote the general linear group $\mathrm{GL}_r(K)$. Then Γ is a submonoid of M, and so R can be regarded as a $K\Gamma$-module. Note that the concept of a $K\Gamma$-module is the same whether we regard Γ as a group or as a monoid. Clearly each KM-submodule of R is a $K\Gamma$-submodule. The first result shows that the converse is also true.

Lemma 1.1 *Let U be a $K\Gamma$-submodule of R.*
  (i) *U is spanned over K by the monomials it contains.*
  (ii) *U is a KM-submodule of R.*

Proof. (i) R is a vector space direct sum $R = \bigoplus Ku$, where the sum is over all monomials u. Let D be the submonoid of M consisting of all diagonal matrices. Then each Ku is D-invariant: in other words, each monomial u is an eigenvector for each element of D. It is easy to verify (using the fact that K is infinite) that if u and v are distinct monomials then there exists $\phi \in \Gamma \cap D$ such that ϕ has distinct eigenvalues on u and v. Thus $R = \bigoplus Ku$ expresses R as a direct sum of pairwise non-isomorphic $K(\Gamma \cap D)$-modules. Since U is a $K(\Gamma \cap D)$-submodule of R it follows that U is a direct sum of some subset of the set of all Ku. This gives (i).

(ii) By (i), U is D-invariant. Also, by assumption, U is Γ-invariant. But it is easy to verify that M is generated as a monoid by D and Γ. The result follows. $\square$

By virtue of Lemma 1.1, R has the same submodule structure whether regarded as $K\Gamma$-module or KM-module. It is convenient to work mostly with M. Also, by Lemma 1.1(i), every submodule U of R can be written as a module direct sum

$$U = (R_0 \cap U) \oplus (R_1 \cap U) \oplus \cdots.$$

Thus to determine the submodule structure of R it suffices to determine the submodule structure of each R_n. (A small variation in the proof of Lemma 1.1 shows that the KM-submodules of R_n are the same as the submodules of R_n under the action of the special linear group $\mathrm{SL}_r(K)$.)

The submodules of R_n form a lattice under the operations $+$ and $\cap$. By Lemma 1.1(i) this lattice is isomorphic to a sublattice of the lattice formed by the subsets of the set of all monomials of degree n. Thus the lattice of submodules of R_n is finite and distributive.

If K has characteristic 0 then it is well known and easy to prove that each module R_n is irreducible. So from now on in Sections 1, 2 and 3 we assume that K *has prime characteristic p.*

If U is a submodule of R and k is a non-negative integer we define U^{p^k} to be the K-span of $\{u^{p^k} \mid u \in U\}$. This is a submodule of R because $(u^{p^k})\phi = (u\phi)^{p^k}$ for all $u \in U$, $\phi \in M$. By Lemma 1.1, U is spanned by a set of monomials $\{u_i \mid i \in I\}$. Since $(u+v)^{p^k} = u^{p^k} + v^{p^k}$ for all $u, v \in U$ it follows that U^{p^k} is spanned by the set of monomials $\{u_i^{p^k} \mid i \in I\}$. Thus, for example, if $r \geq 2$ and $m, k \geq 1$, then $R_m^{p^k}$ is a proper submodule of R_{mp^k}.

Similarly, if U and V are submodules of R we define UV to be the K-span of $\{uv \mid u \in U, v \in V\}$. This is again a submodule of R and it is spanned by all monomials uv where u is a monomial of U and v is a monomial of V. Addition and intersection of submodules have already been mentioned. These operations allow many new submodules to be formed. For example, we obtain $R_p R_1^{p^2} + R_{p+1}^p$ as a submodule of R_{p^2+p}. In fact, the main result (Theorem 1.4) shows that every submodule of R can be obtained as a sum of products of modules all of which have the form $R_m^{p^k}$.

We shall now begin the study of the submodules of R_n for a fixed non-negative integer n. The case $n = 0$ is of course trivial, so when necessary for notational or other reasons we assume $n > 0$. Let T_n^* be the set of all infinite sequences $(t_0, t_1, t_2, \ldots)$ consisting of non-negative integers t_j, all but finitely many of which are 0, satisfying

$$n = t_0 + t_1 p + t_2 p^2 + \cdots.$$

Note that $t_j = 0$ for all $j > n$ and T_n^* is finite. For $s, t \in T_n^*$ with $s = (s_0, s_1, \ldots)$, $t = (t_0, t_1, \ldots)$, write $s \preccurlyeq t$ if

$$s_0 + s_1 p + s_2 p^2 + \cdots + s_{k-1} p^{k-1} \leqslant t_0 + t_1 p + t_2 p^2 + \cdots + t_{k-1} p^{k-1}$$

for every non-negative integer k. (Both sums are interpreted as 0 when $k = 0$.) It is clear that $\preccurlyeq$ gives a partial order on T_n^*.

For each $t \in T_n^*$ where $t = (t_0, t_1, \ldots)$ write

$$W_t = R_{t_0} R_{t_1}^p R_{t_2}^{p^2} \cdots = R_{t_0} R_{t_1}^p \cdots R_{t_n}^{p^n}.$$

Thus each W_t is a submodule of R_n.

For a short while it will be convenient to work with a polynomial algebra $\hat{R}$ which is larger than R. (The reader who does not wish to see complete proofs can ignore $\hat{R}$, Lemma 1.2, and the proof of Lemma 1.3.) We take

$$\hat{R} = K[x_1, \ldots, x_r, y_{1,0}, \ldots, y_{n,n}]$$

with new indeterminates $y_{m,j}$ for $1 \leqslant m \leqslant n$, $0 \leqslant j \leqslant n$. Let $\hat{R}_0$, $\hat{R}_1$, ... be the homogeneous components of $\hat{R}$ and let $\hat{M}$ be the monoid of matrices corresponding to homogeneous linear substitutions of $\hat{R}$. For each $t \in T_n^*$ where $t = (t_0, t_1, \ldots)$ let $\hat{W}_t = \hat{R}_{t_0} \hat{R}_{t_1}^p \hat{R}_{t_2}^{p^2} \cdots$ and let $\hat{u}_t$ be the element of $\hat{R}$ defined by

$$\hat{u}_t = (y_{1,0} \cdots y_{t_0,0})(y_{1,1} \cdots y_{t_1,1})^p (y_{1,2} \cdots y_{t_2,2})^{p^2} \cdots.$$

(When $t_j = 0$ the 'empty' product $y_{1,j} \cdots y_{t_j,j}$ is interpreted as 1.)

Lemma 1.2

(i) *For all $t \in T_n^*$, $\hat{W}_t = \hat{u}_t K \hat{M}$ (the submodule generated by $\hat{u}_t$).*

(ii) *For $s, t \in T_n^*$, $\hat{W}_s \subseteq \hat{W}_t$ if and only if $s \preccurlyeq t$.*

Proof. (i) Clearly $\hat{u}_t \in \hat{W}_t$. But if v is any monomial of $\hat{W}_t$ then v has the form

$$v = (v_{1,0} \cdots v_{t_0,0})(v_{1,1} \cdots v_{t_1,1})^p (v_{1,2} \cdots v_{t_2,2})^{p^2} \cdots,$$

where each $v_{m,j}$ belongs to $\{x_1, \ldots, x_r, y_{1,0}, \ldots, y_{n,n}\}$. Thus $v = \hat{u}_t \phi$ for some $\phi \in \hat{M}$. The result follows.

(ii) Let $s = (s_0, s_1, \ldots)$ and $t = (t_0, t_1, \ldots)$ and suppose first that $\hat{W}_s \subseteq \hat{W}_t$. Then $\hat{u}_s \in \hat{W}_t$ and so we can write $\hat{u}_s = u_0 u_1^p u_2^{p^2} \cdots$ where, for each j, u_j is a monomial of degree t_j. Thus, for each k, the factor $u_k^{p^k} u_{k+1}^{p^{k+1}} \cdots$ of $\hat{u}_s$ is a p^k-th power and has degree $t_k p^k + t_{k+1} p^{k+1} + \cdots$. But it is clear from the definition of $\hat{u}_s$ that every factor of $\hat{u}_s$ which is a p^k-th power must divide

$$(y_{1,k} \cdots y_{s_k,k})^{p^k} (y_{1,k+1} \cdots y_{s_{k+1},k+1})^{p^{k+1}} \cdots.$$

Thus

$$t_k p^k + t_{k+1} p^{k+1} + \cdots \leqslant s_k p^k + s_{k+1} p^{k+1} + \cdots.$$

Since $\sum_j s_j p^j = \sum_j t_j p^j = n$ we obtain $\sum_{j<k} s_j p^j \leqslant \sum_{j<k} t_j p^j$. Thus $s \preccurlyeq t$.

For the converse we must prove that $\hat{W}_s \subseteq \hat{W}_t$ when $s \preccurlyeq t$. It is clearly enough to prove this in the case where s is maximal subject to $s \preccurlyeq t$ and $s \neq t$. Let k be the least integer such that $s_k \neq t_k$. Then $s_k < t_k$. Since $\sum_j s_j p^j = \sum_j t_j p^j$, s_k and t_k are congruent modulo p. It follows by maximality that $s_k = t_k - p$, $s_{k+1} = t_{k+1} + 1$ and $s_j = t_j$ for all $j \notin \{k, k+1\}$. Clearly every monomial in $\hat{R}_{t_k-p} \hat{R}^p_{t_{k+1}+1}$ belongs to $\hat{R}_{t_k} \hat{R}^p_{t_{k+1}}$. Hence $\hat{R}^{p^k}_{t_k-p} \hat{R}^{p^{k+1}}_{t_{k+1}+1} \subseteq \hat{R}^{p^k}_{t_k} \hat{R}^{p^{k+1}}_{t_{k+1}}$ and so $\hat{W}_s \subseteq \hat{W}_t$. $\square$

Now let T_n be the subset of T_n^* consisting of those elements $t = (t_0, t_1, \ldots)$ such that $0 \leqslant t_j \leqslant r(p-1)$ for all j. Clearly T_n is finite and $\preccurlyeq$ restricts to a partial order on T_n. Since $t_j = 0$ for $j > n$ this definition of T_n is essentially the same as that given in the Introduction.

Suppose u is a monomial of R_n. Then u has the form $u = x_1^{a_1} \cdots x_r^{a_r}$ where $a_1, \ldots, a_r$ are non-negative integers satisfying $a_1 + \cdots + a_r = n$. Write the numbers $a_1, \ldots, a_r$ to base p; namely, for each i,

$$a_i = a_{i0} + a_{i1} p + a_{i2} p^2 + \cdots,$$

where $0 \leqslant a_{ij} \leqslant p-1$ for all j. For each $j \geqslant 0$, let t_j be the sum of the digits in $a_1, \ldots, a_r$ corresponding to p^j; namely, $t_j = a_{1j} + \cdots + a_{rj}$. Thus $0 \leqslant t_j \leqslant r(p-1)$ for all j and $t_0 + t_1 p + t_2 p^2 + \cdots = n$. We define $t(u) = (t_0, t_1, \ldots)$. Thus $t(u) \in T_n$. Note also that we can write u in the form

$$u = (x_1^{a_{10}} \cdots x_r^{a_{r0}})(x_1^{a_{11}} \cdots x_r^{a_{r1}})^p (x_1^{a_{12}} \cdots x_r^{a_{r2}})^{p^2} \cdots.$$

Thus $u \in R_{t_0} R^p_{t_1} R^{p^2}_{t_2} \cdots = W_{t(u)}$.

Lemma 1.3

(i) *For each $s \in T_n^*$ there exists $s' \in T_n$ such that $s' \preccurlyeq s$ and $W_{s'} = W_s$.*

(ii) *For each $t \in T_n$ there exists a monomial u of R_n such that $t = t(u)$.*

(iii) *For every monomial u of R_n, $W_{t(u)} = uKM$.*

(iv) *For $s, t \in T_n$, $W_s \subseteq W_t$ if and only if $s \preccurlyeq t$.*

Proof. (i) Let $s = (s_0, s_1, \ldots)$. If $s \in T_n$ we can take $s' = s$. So assume $s \notin T_n$. Thus there exists k such that $s_k > r(p-1)$. Let

$$s^* = (s_0, \ldots, s_{k-1}, s_k - p, s_{k+1} + 1, s_{k+2}, \ldots).$$

Consequently $s^* \in T_n^*$ and $s^* \preccurlyeq s$. Since $s_k > r(p-1)$, every monomial in R_{s_k} must have a factor x_i^p for some i. It follows that $R_{s_k} R_{s_{k+1}}^p = R_{s_k-p} R_{s_{k+1}+1}^p$ and so $R_{s_k}^{p^k} R_{s_{k+1}}^{p^{k+1}} = R_{s_k-p}^{p^k} R_{s_{k+1}+1}^{p^{k+1}}$. Therefore $W_s = W_{s^*}$. But s^* is smaller than s in the obvious lexicographic ordering on T_n^*, so the result now follows by induction.

(ii) This is easy.

(iii) Let $u = x_1^{a_1} \cdots x_r^{a_r}$ and use the notation associated with u introduced immediately before the statement of the lemma. It is convenient to rename those indeterminates of $\hat{R}$ which occur in $\hat{u}_t$. For each j, the indeterminates $y_{1,j}, \ldots, y_{t_j,j}$ are renamed as

$$z_{1,j}^{(1)}, \ldots, z_{1,j}^{(a_{1j})}, z_{2,j}^{(1)}, \ldots, z_{2,j}^{(a_{2j})}, \ldots, z_{r,j}^{(1)}, \ldots, z_{r,j}^{(a_{rj})},$$

respectively. (Here there are no elements $z_{i,j}^{(k)}$ listed when $a_{ij} = 0$.) Write $t = t(u)$. Then

$$\hat{u}_t = (z_{1,0}^{(1)} \cdots z_{1,0}^{(a_{10})} \cdots z_{r,0}^{(1)} \cdots z_{r,0}^{(a_{r0})})(z_{1,1}^{(1)} \cdots z_{1,1}^{(a_{11})} \cdots z_{r,1}^{(1)} \cdots z_{r,1}^{(a_{r1})})^p \cdots.$$

Let ϕ be the element of $\hat{M}$ such that

$$x_1 \phi = z_{1,0}^{(1)} + \cdots + z_{1,0}^{(a_{10})} + z_{1,1}^{(1)} + \cdots + z_{1,1}^{(a_{11})} + \cdots,$$

$$\vdots$$

$$x_r \phi = z_{r,0}^{(1)} + \cdots + z_{r,0}^{(a_{r0})} + z_{r,1}^{(1)} + \cdots + z_{r,1}^{(a_{r1})} + \cdots,$$

and such that ϕ fixes all other indeterminates of $\hat{R}$.

For each non-negative integer m, let $\langle m \rangle_p$ denote the multinomial coefficient defined by

$$\langle m \rangle_p = \frac{m!}{(1!)^{m_0}(p!)^{m_1}(p^2!)^{m_2} \cdots},$$

where $m = m_0 + m_1 p + m_2 p^2 + \cdots$ is the base p expansion of m. It is straightforward to verify that the integer $\langle m \rangle_p$ can be written as a product of binomial coefficients each of which has the form

$$\binom{ap^i}{p^i}$$

where i is a non-negative integer and a is a positive integer not divisible by p. None of these binomial coefficients is divisible by p, so $\langle m \rangle_p$ is not divisible by p.

We regard u as an element of $\hat{R}$ and apply ϕ. It is straightforward to verify that, when $u\phi$ is expressed as a linear combination of monomials, the coefficient of $\hat{u}_t$ is the element of K given by $\langle a_1 \rangle_p \cdots \langle a_r \rangle_p$. Thus, since K has characteristic p, $\hat{u}_t$ occurs with non-zero coefficient in $u\phi$. By Lemma 1.1(i) applied to $\hat{R}$ it follows that $\hat{u}_t \in uK\hat{M}$.

Let v be any element of W_t. Then $v \in \hat{W}_t$ and so, by Lemma 1.2, $v \in \hat{u}_t K\hat{M}$. Thus $v \in uK\hat{M}$. Write u as $u(x_1, \ldots, x_r)$. Since $v \in uK\hat{M}$ it can be written as a linear combination of elements each of which has the form $u(w_1, \ldots, w_r)$ where each w_i is a linear combination of $x_1, \ldots, x_r$ and the extra indeterminates $y_{m,j}$.

(Here we have reverted to the original names of the indeterminates of $\hat{R}$.) We apply the K-algebra homomorphism $\pi : \hat{R} \longrightarrow R$ satisfying $x_1\pi = x_1, \ldots, x_r\pi = x_r$ and $y_{m,j}\pi = 0$ for all m, j. Then $v\pi = v$ and each $w_i\pi$ is a linear combination of $x_1, \ldots, x_r$. It follows that $v \in uKM$. Hence $W_t \subseteq uKM$. Since $u \in W_t$ the proof is complete.

(iv) Suppose that $s, t \in T_n$. If $s \preccurlyeq t$ then $\hat{W}_s \subseteq \hat{W}_t$, by Lemma 1.2, and so, using the homomorphism $\pi : \hat{R} \longrightarrow R$ defined in (iii), $W_s = \hat{W}_s\pi \subseteq \hat{W}_t\pi = W_t$. Conversely, suppose that $W_s \subseteq W_t$. By (ii) there exists a monomial u such that $s = t(u)$. Thus $u \in W_s \subseteq W_t$. Therefore $uK\hat{M} \subseteq \hat{W}_t$. But, by the proof of (iii), $\hat{u}_s \in uK\hat{M}$. Thus $\hat{u}_sK\hat{M} \subseteq \hat{W}_t$ and so, by Lemma 1.2(i), $\hat{W}_s \subseteq \hat{W}_t$. Lemma 1.2(ii) now gives $s \preccurlyeq t$ as required. $\qquad\square$

We shall say that a module W is *join-irreducible* if it cannot be expressed as the sum of two proper submodules. (In the lattice of submodules of R_n under the operations $+$ and $\cap$, the join-irreducible elements in this sense are the same as the join-irreducible elements in the sense of lattice theory.)

Let $t \in T_n$ and choose a monomial u such that $t = t(u)$. If W_t is the sum of two submodules U and V then, by Lemma 1.1, $u \in U$ or $u \in V$. Since $W_t = uKM$ by Lemma 1.3, it follows that $W_t = U$ or $W_t = V$. Thus W_t is join-irreducible. It is also non-zero. Conversely, suppose that W is a non-zero join-irreducible submodule of R_n. By Lemma 1.1, W is generated by monomials. Hence it is the sum of a finite number of modules of the form W_t. Thus $W = W_t$ for some t.

We have now proved the following theorem (see Doty [6] and pp. 383–387 in Krop [12]).

Theorem 1.4 *The submodules of R_n form a finite distributive lattice under $+$ and $\cap$ whose non-zero join-irreducible elements are the modules W_t with $t \in T_n$. For $s, t \in T_n$, $W_s \subseteq W_t$ if and only if $s \preccurlyeq t$.* $\qquad\square$

Since every submodule of R_n is a finite sum of join-irreducible modules, Theorem 1.4 gives very strong information about the submodule lattice of R_n. The information is complete except for the combinatorial problems associated with dealing with the partially ordered set $(T_n, \preccurlyeq)$.

For each $t \in T_n$, W_t is non-zero and join-irreducible, so it has a unique maximal proper submodule which we shall denote by $\mathrm{rad}(W_t)$. The module $W_t/\mathrm{rad}(W_t)$ is irreducible (in the ordinary sense of module theory) and we denote this irreducible module by $\sigma(W_t)$.

Corollary 1.5 *If $\{0\} = U_0 \subsetneq U_1 \subsetneq \cdots \subsetneq U_k = R_n$ is a composition series of the module R_n then there is a bijection $\theta : \{1, \ldots, k\} \longrightarrow T_n$ such that $U_i/U_{i-1} \cong \sigma(W_{i\theta})$ for all i. In particular, $k = |T_n|$.*

Proof. For each $i \in \{1, \ldots, k\}$, U_i is a sum of join-irreducibles and so there exists $t \in T_n$ such that $U_i = U_{i-1} + W_t$. If $s, t \in T_n$ and $U_{i-1} + W_s = U_{i-1} + W_t = U_i$

then distributivity gives

$$W_s = (W_s \cap U_{i-1}) + (W_s \cap W_t).$$

Hence, since W_s is join-irreducible, $W_s = W_s \cap W_t$. Similarly, $W_t = W_s \cap W_t$ and so $W_s = W_t$. Thus we may define $\theta : \{1, \ldots, k\} \longrightarrow T_n$ so that, for each i, $i\theta$ is the unique element of T_n satisfying $U_i = U_{i-1} + W_{i\theta}$. Therefore, for each i,

$$U_i/U_{i-1} \cong W_{i\theta}/(U_{i-1} \cap W_{i\theta}).$$

Consequently $U_{i-1} \cap W_{i\theta} = \mathrm{rad}(W_{i\theta})$ and $U_i/U_{i-1} \cong \sigma(W_{i\theta})$. Clearly θ is one-one. Finally, for all $t \in T_n$ there exists i such that $U_i = U_{i-1} + W_t$ (by taking i minimal subject to $W_t \subseteq U_i$). Hence θ is onto. $\qquad \square$

2. The Irreducible Constituents

In Corollary 1.5 we saw that the modules $\sigma(W_t)$ with $t \in T_n$ are the 'irreducible constituents' of R_n in the sense that they occur, up to isomorphism, as the factors in every composition series of R_n. In this section we shall investigate the structure of these modules. The main result is Theorem 2.2.

For $k \geqslant 0$ and $0 \leqslant m \leqslant r(p-1)$, we have $R_m^{p^k} = W_t$ where t is the element of T_{mp^k} which has entry m in the position corresponding to p^k and all other entries 0. In particular, $R_m^{p^k}$ is join-irreducible.

Lemma 2.1 Let $k \geqslant 0$ and $0 \leqslant m \leqslant r(p-1)$. Then $\mathrm{rad}(R_m^{p^k}) = \{0\}$ if $0 \leqslant m \leqslant p-1$ and $\mathrm{rad}(R_m^{p^k}) = R_{m-p}^{p^k} R_1^{p^{k+1}}$ if $p \leqslant m \leqslant r(p-1)$.

Proof. We first calculate $\mathrm{rad}(R_m)$. For $0 \leqslant m \leqslant p-1$, $(m, 0, 0, \ldots)$ is the only element of T_m and so $\mathrm{rad}(R_m) = \{0\}$. Suppose that $p \leqslant m \leqslant r(p-1)$. (This can only occur if $r \geqslant 2$.) Then $(m-p, 1, 0, \ldots) \in T_m$ and it is easily verified that if $t \in T_m$ and $t \neq (m, 0, 0, \ldots)$ then $t \preccurlyeq (m-p, 1, 0, \ldots)$: hence

$$\mathrm{rad}(R_m) = W_{(m-p,1,0,\ldots)} = R_{m-p} R_1^p.$$

For each submodule X of R_m, X^{p^k} is a submodule of $R_m^{p^k}$, and it is straightforward to verify that the map $X \longmapsto X^{p^k}$ gives an isomorphism between the lattice of submodules of R_m and the lattice of submodules of $R_m^{p^k}$. Hence $\mathrm{rad}(R_m^{p^k}) = (\mathrm{rad}(R_m))^{p^k}$, which gives the required result. $\qquad \square$

Note that, because of Lemma 2.1, we have $\sigma(R_m^{p^k}) \cong R_m^{p^k}$ if $0 \leqslant m \leqslant p-1$ and $\sigma(R_m^{p^k}) = R_m^{p^k}/R_{m-p}^{p^k} R_1^{p^{k+1}}$ if $p \leqslant m \leqslant r(p-1)$.

From KM-modules U and V we may form the tensor product $U \otimes V$ by tensoring as vector spaces over the field K. (All tensor products in this paper will be over K.) Furthermore, $U \otimes V$ may be regarded as a KM-module by taking the usual 'diagonal' action of M: $(u \otimes v)\phi = u\phi \otimes v\phi$ for all $u \in U$, $v \in V$, $\phi \in M$.

The following result is Theorem 2.3.2 of Krop [12]. The proof given here is based on that in the draft paper of L. G. Kovács referred to at the end of the Introduction.

Theorem 2.2 *Let $t \in T_n$ where $t = (t_0, t_1, \ldots)$. Then*

$$\sigma(W_t) \cong \sigma(R_{t_0}) \otimes \sigma(R_{t_1}^p) \otimes \sigma(R_{t_2}^{p^2}) \otimes \cdots.$$

(Since $t_j = 0$ for $j > n$ and $\sigma(R_0^{p^j}) \cong K$, the formally infinite tensor product should be interpreted as the finite one $\sigma(R_{t_0}) \otimes \cdots \otimes \sigma(R_{t_n}^{p^n})$.)

Proof. Recall that $W_t = R_{t_0} R_{t_1}^p \cdots R_{t_n}^{p^n}$. Thus there is a K-space epimorphism

$$\mu : R_{t_0} \otimes R_{t_1}^p \otimes \cdots \otimes R_{t_n}^{p^n} \longrightarrow W_t$$

which maps each tensor $u_0 \otimes u_1 \otimes \cdots \otimes u_n$ to the product $u_0 u_1 \cdots u_n$. This is easily seen to be a KM-module epimorphism because each element of M has diagonal action on the tensor product and acts on R as a K-algebra endomorphism. Also, the natural maps $\nu_j : R_{t_j}^{p^j} \longrightarrow \sigma(R_{t_j}^{p^j})$ give rise to a KM-module epimorphism

$$\nu : R_{t_0} \otimes R_{t_1}^p \otimes \cdots \otimes R_{t_n}^{p^n} \longrightarrow \sigma(R_{t_0}) \otimes \sigma(R_{t_1}^p) \otimes \cdots \otimes \sigma(R_{t_n}^{p^n}).$$

Let $U = \ker \nu$ (the kernel of ν) and, for $k = 0, \ldots, n$, let

$$U_k = (\ker \nu_k) \otimes \bigotimes_{j \neq k} R_{t_j}^{p^j} = \mathrm{rad}(R_{t_k}^{p^k}) \otimes \bigotimes_{j \neq k} R_{t_j}^{p^j}.$$

Clearly each U_k is contained in U, and it is easily checked that $\sum_k U_k$ has the same dimension as U. Hence $U = \sum_k U_k$.

We shall show that, for each k, $U_k \mu \subseteq \mathrm{rad}(W_t)$. If $0 \leqslant t_k \leqslant p - 1$ then we have $\mathrm{rad}(R_{t_k}^{p^k}) = \{0\}$, by Lemma 2.1, and so $U_k = \{0\}$. Thus we need only consider the case where $p \leqslant t_k \leqslant r(p-1)$. Then, by Lemma 2.1, $\mathrm{rad}(R_{t_k}^{p^k}) = R_{t_k-p}^{p^k} R_1^{p^{k+1}}$ and so

$$\begin{aligned}
U_k \mu &= R_{t_k-p}^{p^k} R_1^{p^{k+1}} \prod_{j \neq k} R_{t_j}^{p^j} \\
&= R_{t_0} \cdots R_{t_{k-1}}^{p^{k-1}} R_{t_k-p}^{p^k} R_{t_{k+1}+1}^{p^{k+1}} R_{t_{k+2}}^{p^{k+2}} \cdots.
\end{aligned}$$

Thus $U_k \mu = W_s$ where s is the element of T_n^* given by

$$s = (t_0, \ldots, t_{k-1}, t_k - p, t_{k+1} + 1, t_{k+2}, \ldots).$$

(Note that it may not be true that $s \in T_n$.) Clearly $s \preccurlyeq t$ and $s \neq t$. By Lemma 1.3(i), there exists $s' \in T_n$ such that $s' \preccurlyeq s$ and $W_{s'} = W_s = U_k \mu$. Then we have $s' \preccurlyeq t$ and $s' \neq t$. Therefore, by Lemma 1.3(iv), $U_k \mu \subsetneqq W_t$. Thus $U_k \mu \subseteq \mathrm{rad}(W_t)$ as required.

Since $U = \sum_k U_k$ it follows that $U\mu \subseteq \mathrm{rad}(W_t)$. Now $\mathrm{rad}(W_t)$ is the kernel of the natural map $\delta : W_t \longrightarrow \sigma(W_t)$, so we have shown that $(\ker \nu)\mu \subseteq \ker \delta$ or, in other words, $\ker \nu \subseteq \ker(\mu\delta)$. Thus the image of $\mu\delta$ is isomorphic to a factor module of the image of ν : that is, $\sigma(W_t)$ is isomorphic to a factor module of $\sigma(R_{t_0}) \otimes \cdots \otimes \sigma(R_{t_n}^{p^n})$. To complete the proof we show that the modules have the same dimension.

If u is a monomial which belongs to W_t but not to $\mathrm{rad}(W_t)$ then it follows that $W_t = W_{t(u)} + \mathrm{rad}(W_t)$, so $W_{t(u)} = W_t$ and consequently $t(u) = t$. Conversely, if u is a monomial such that $t(u) = t$ then $u \in W_t$ but $u \notin \mathrm{rad}(W_t)$. Hence $\dim(\sigma(W_t))$ is the number of monomials u such that $t(u) = t$. This is the number of choices of integers

$$a_{10}, a_{11}, \ldots, a_{20}, a_{21}, \ldots, a_{r0}, a_{r1}, \ldots$$

satisfying $0 \leqslant a_{ij} \leqslant p-1$ for all i, j and $a_{1j} + \cdots + a_{rj} = t_j$ for all j. It follows that $\dim(\sigma(W_t)) = d_0 d_1 d_2 \cdots$ where, for each j, d_j is the number of choices of integers $a_{1j}, a_{2j}, \ldots, a_{rj}$ satisfying $0 \leqslant a_{ij} \leqslant p-1$ for all i and $a_{1j} + \cdots + a_{rj} = t_j$. Note that d_j is equal to the number of monomials u such that

$$t(u) = (0, \ldots, 0, t_j, 0, \ldots).$$

Hence, for all j, $d_j = \dim(\sigma(R_{t_j}^{p^j}))$, which gives the required result. □

If an element ϕ of M is represented on $\sigma(R_{t_k})$ by a matrix A then it is easy to verify that ϕ is represented on $\sigma(R_{t_k}^{p^k})$ by the matrix $A^{(k)}$ where each entry of $A^{(k)}$ is the p^k-th power of the corresponding entry of A. Thus the module $\sigma(R_{t_k}^{p^k})$ may be thought of as coming from $\sigma(R_{t_k})$ by a 'Frobenius twist' (using the k-th power of the Frobenius map $\alpha \mapsto \alpha^p$ from K to K). Thus Theorem 2.2 may be phrased as saying that every $\sigma(W_t)$ is a 'twisted' tensor product of modules belonging to the finite set

$$\{\sigma(R_0), \sigma(R_1), \ldots, \sigma(R_{r(p-1)})\}.$$

This can be viewed as a special case of Steinberg's tensor product theorem [14].

3. Two Further Lemmas

In this section we shall derive two additional results which will be useful in Section 4. As before, K is an infinite field of prime characteristic p and $R = K[x_1, \ldots, x_r]$. We shall continue to use the notation developed so far, but in this section we shall assume that $r \geqslant 2$.

Recall that, by Corollary 1.5, every composition series of the KM-module R_n has length $|T_n|$. We shall first explore, to a very limited extent, the way that this number varies with n.

Let $t \in T_n$ where $t = (t_0, t_1, \ldots)$ and write

$$m_1 = t_0 p^0 + t_r p^r + t_{2r} p^{2r} + \cdots,$$
$$m_2 = t_1 p^1 + t_{r+1} p^{r+1} + t_{2r+1} p^{2r+1} + \cdots,$$

$$\vdots$$

$$m_r = t_{r-1} p^{r-1} + t_{2r-1} p^{2r-1} + t_{3r-1} p^{3r-1} + \cdots.$$

Thus $m_1 + m_2 + \cdots + m_r = n$. Since $0 \leqslant t_j \leqslant r(p-1) \leqslant p^r - 1$ for all j, it follows that $t_0 p^0 + t_r p^r + \cdots$ is the expansion of m_1 to base p^r, $t_1 p^0 + t_{r+1} p^r + \cdots$ is the expansion of m_2/p to base p^r, and so on. Hence the r-tuple $(m_1, m_2, \ldots, m_r)$

determines t uniquely. But the number of r-tuples of non-negative integers satisfying $m_1 + \cdots + m_r = n$ is

$$\binom{n+r-1}{r-1},$$

which is the dimension of R_n. Hence $|T_n| \leqslant \dim(R_n)$. This, of course, we knew already because $|T_n|$ is the length of a composition series of R_n.

But we can do rather better using our assumption that $r \geqslant 2$. Note that we have $r(p-1) \leqslant rp - 2 \leqslant p^r - 2$. Thus none of the digits t_j occurring in the expressions to base p^r takes the value $p^r - 1$; hence there are restrictions on the numbers $m_1, \ldots, m_r$. We shall only exploit the restriction on m_1.

Write $q = p^r$ and, for each non-negative integer n, let $D(n)$ denote the set of all non-negative integers m such that $m \leqslant n$ and the digit $q - 1$ does not occur in the base q expression for m. Let $E(n)$ be the set of all r-tuples $(m_1, \ldots, m_r)$ of non-negative integers such that $m_1 + \cdots + m_r = n$ and $m_1 \in D(n)$. From the discussion above we know that $|T_n| \leqslant |E(n)|$.

Suppose that $n \geqslant 1$ and let k be the positive integer satisfying $q^{k-1} \leqslant n \leqslant q^k - 1$. Then

$$|D(n)| \leqslant |D(q^k - 1)| = (q-1)^k.$$

Hence

$$\frac{|D(n)|}{n} \leqslant \frac{(q-1)^k}{q^{k-1}} = q\left(\frac{q-1}{q}\right)^k,$$

and so

$$\lim_{n \to \infty} \left(|D(n)|/n\right) = 0.$$

Lemma 3.1 *Suppose $r \geqslant 2$. Then*

$$\lim_{n \to \infty} \frac{|T_n|}{\dim(R_n)} = 0.$$

Proof. For each $x \in D(n)$, the number of elements of $E(n)$ which have the form $(x, m_2, \ldots, m_r)$ is

$$\binom{n-x+r-2}{r-2}.$$

Hence

$$|T_n| \leqslant |E(n)| = \sum_{x \in D(n)} \binom{n-x+r-2}{r-2} \leqslant |D(n)|\binom{n+r-2}{r-2}.$$

This gives

$$\frac{|T_n|}{\dim(R_n)} \leqslant \frac{|D(n)|\binom{n+r-2}{r-2}}{\binom{n+r-1}{r-1}} = \frac{|D(n)|(r-1)}{n+r-1}.$$

Thus

$$\lim_{n \to \infty} \frac{|T_n|}{\dim(R_n)} = 0,$$

as required.   □

The other result we need is a characterisation of the action of the scalar elements of $\mathrm{GL}_r(K)$.

Lemma 3.2 *Let $r \geqslant 2$ and $0 < m < r(p-1)$. Then the only elements of $\mathrm{GL}_r(K)$ which have scalar action on $\sigma(R_m)$ are the scalar matrices of $\mathrm{GL}_r(K)$.*

Proof. Recall that $\sigma(R_m) = R_m/\mathrm{rad}(R_m)$. Also, by Lemma 2.1, $\mathrm{rad}(R_m) = \{0\}$ if $0 < m \leqslant p-1$ and $\mathrm{rad}(R_m) = R_{m-p}R_1^p$ if $p \leqslant m < r(p-1)$. Thus $\sigma(R_m)$ has a basis consisting of the cosets represented by the monomials of R_m of the form $x_1^{a_1} \cdots x_r^{a_r}$ where $0 \leqslant a_i \leqslant p-1$ for all i.

Let $\Gamma = \mathrm{GL}_r(K)$, let Z be the normal subgroup of Γ consisting of all scalar matrices and let N be the normal subgroup of Γ consisting of all elements which have scalar action on $\sigma(R_m)$. Clearly $Z \leqslant N$. Suppose, to get a contradiction, that $Z < N$. It is well known and easy to prove that Γ/Z has trivial centre. Hence the mutual commutator subgroup $[\Gamma, N]$ is not contained in Z.

Let S be the normal subgroup of Γ consisting of all matrices of determinant 1. Thus, in standard notation, $S = \mathrm{SL}_r(K)$ and the group $S/S \cap Z$ is the simple group $\mathrm{PSL}_r(K)$. Since Γ/S is abelian, $[\Gamma, N] \leqslant S \cap N$. Thus $S \cap Z < S \cap N \leqslant S$ and we obtain $S \leqslant N$.

Let λ be a non-zero element of K such that $\lambda^2 \neq 1$ and let g be the diagonal matrix of $\mathrm{GL}_r(K)$ with diagonal entries $\lambda, \lambda^{-1}, 1, \ldots, 1$. Then $g \in N$ because $g \in S$. Since $0 < m < r(p-1)$ we can find integers $a_1, \ldots, a_r$ such that $a_1 + \cdots + a_r = m$, $0 \leqslant a_1 < p-1$, $0 < a_2 \leqslant p-1$, and $0 \leqslant a_i \leqslant p-1$ for $i = 3, \ldots, r$. Then

$$(x_1^{a_1} x_2^{a_2} x_3^{a_3} \cdots x_r^{a_r})g = \lambda^{a_1 - a_2} x_1^{a_1} x_2^{a_2} x_3^{a_3} \cdots x_r^{a_r},$$

whereas

$$(x_1^{a_1+1} x_2^{a_2-1} x_3^{a_3} \cdots x_r^{a_r})g = \lambda^{a_1 - a_2 + 2} x_1^{a_1+1} x_2^{a_2-1} x_3^{a_3} \cdots x_r^{a_r}.$$

Since g acts as a scalar on $\sigma(R_m)$ we obtain $\lambda^{a_1 - a_2 + 2} = \lambda^{a_1 - a_2}$ and so $\lambda^2 = 1$. This is the required contradiction.   □

4. Finite Groups

In this section we shall apply the preceding theory to obtain a new and improved proof of the main result of [2]: this is Theorem 4.3 below.

Let K be any field (not necessarily infinite of non-zero characteristic). Let G be a finite group and let V be a (finite-dimensional right) KG-module which is faithful for G (that is, every non-trivial element of G has non-trivial action on V). A classical theorem of Burnside (Section 226 of [4]) states that if K is the field of complex numbers then every irreducible KG-module is isomorphic to a constituent of some tensor power of V. The same result holds for arbitrary fields and has been strengthened and generalised in various ways by various authors. The version that will be useful here is the following result, taken from [3].

Proposition 4.1 *Suppose $G \smallsetminus \{1\}$ is the union of subsets $S_1, \ldots, S_m$ and suppose that, for $i = 1, \ldots, m$, V_i is a KG-module containing an element v_i such that v_i does not belong to $K v_i S_i$ (the K-span in V_i of the set $\{v_i g \mid g \in S_i\}$). Then $V_1 \otimes \cdots \otimes V_m$ has a submodule isomorphic to the regular KG-module.*

Proof. We first show by induction on i that, for $i = 1, \ldots, m$, the element $v_1 \otimes \cdots \otimes v_i$ of $V_1 \otimes \cdots \otimes V_i$ does not belong to $K(v_1 \otimes \cdots \otimes v_i)(S_1 \cup \ldots \cup S_i)$. The case $i = 1$ is trivial. Suppose the result holds for a given value of i, where $i < m$. Then $V_1 \otimes \cdots \otimes V_i$ has a vector space decomposition $U \oplus U'$ where $v_1 \otimes \cdots \otimes v_i \in U$ and $(v_1 \otimes \cdots \otimes v_i)g \in U'$ for all $g \in S_1 \cup \ldots \cup S_i$. Similarly $V_{i+1} = V \oplus V'$ where $v_{i+1} \in V$ and $v_{i+1} g \in V'$ for all $g \in S_{i+1}$. Hence

$$V_1 \otimes \cdots \otimes V_{i+1} \cong (U \otimes V) \oplus (U \otimes V') \oplus (U' \otimes V) \oplus (U' \otimes V').$$

Identifying the two sides in the obvious way we find that $v_1 \otimes \cdots \otimes v_{i+1} \in U \otimes V$, whereas we have $(v_1 \otimes \cdots \otimes v_{i+1})g \in U' \otimes (V \oplus V')$ for all $g \in S_1 \cup \ldots \cup S_i$ and $(v_1 \otimes \cdots \otimes v_{i+1})g \in (U \oplus U') \otimes V'$ for all $g \in S_{i+1}$. Thus

$$(v_1 \otimes \cdots \otimes v_{i+1})g \in (U \otimes V') \oplus (U' \otimes V) \oplus (U' \otimes V')$$

for all $g \in S_1 \cup \ldots \cup S_{i+1}$, which yields the inductive step.

The case $i = m$ shows that

$$v_1 \otimes \cdots \otimes v_m \notin K(v_1 \otimes \cdots \otimes v_m)(G \smallsetminus \{1\}).$$

It follows easily that the elements $(v_1 \otimes \cdots \otimes v_m)g$, with $g \in G$, are linearly independent. Hence they span a regular KG-module. □

For any (finite-dimensional) KG-module U we define $f(U)$ to be the rational number satisfying $0 \leqslant f(U) \leqslant 1$ such that $f(U)\dim(U)$ is the largest dimension of a free KG-submodule of U. Thus $f(U)$ measures the extent to which U is free. Let H be a subgroup of G. A KG-module which is isomorphic to a module induced from a KH-module will be called H-*induced*. For every KG-module U we define $f_H(U)$ to be the rational number satisfying $0 \leqslant f_H(U) \leqslant 1$ such that $f_H(U)\dim(U)$ is the largest dimension of a direct summand of U which is H-induced. We also write $f'_H(U) = 1 - f_H(U)$. It follows from Theorem VII.7.23 of [10] that if K' is an extension field of K and we regard $K' \otimes U$ as a $K'G$-module, then $f(K' \otimes U) = f(U)$. We shall also need the following result.

Lemma 4.2 *Let U and V be KG-modules and let H be a subgroup of G. Then*

$$f'_H(U \otimes V) \leqslant f'_H(U) f'_H(V).$$

Proof. We can write U and V as direct sums of KG-modules, $U = U_0 \oplus U_1$ and $V = V_0 \oplus V_1$, where U_0 and V_0 are H-induced while $\dim(U_1) = f'_H(U)\dim(U)$ and $\dim(V_1) = f'_H(V)\dim(V)$. Thus

$$U \otimes V \cong (U_0 \otimes V_0) \oplus (U_0 \otimes V_1) \oplus (U_1 \otimes V_0) \oplus (U_1 \otimes V_1).$$

By Lemma VII.4.15 of [10], the first three summands on the right-hand side are H-induced. Hence

$$f'_H(U \otimes V) \dim(U \otimes V) \leqslant \dim(U_1 \otimes V_1),$$

which gives the required result.   □

Let $R = K[x_1, \ldots, x_r]$, where r is a positive integer, and let R_0, R_1, ... be the homogeneous components of R. Regard R as a $K\mathrm{GL}_r(K)$-module as before, and let G be any finite subgroup of $\mathrm{GL}_r(K)$. Then we may regard R as a KG-module by restricting from $\mathrm{GL}_r(K)$ to G. Let H be the subgroup of G consisting of all scalar matrices belonging to G, and write $c = |H|$. Clearly H is a cyclic central subgroup of G. Let h be a generator of H, and suppose that ξ is the scalar corresponding to h. Then ξ is a primitive c-th root of unity in K and the characteristic of K does not divide c.

Let Q be a regular KH-module. Then we can write $Q = Q_0 \oplus Q_1 \oplus \cdots \oplus Q_{c-1}$ where, for $i = 0, 1, \ldots, c-1$, Q_i is a one-dimensional module on which h acts as the scalar ξ^i. Write P and P_i for the KG-modules induced from Q and Q_i, respectively. Thus P is a regular KG-module and $P = P_0 \oplus P_1 \oplus \cdots \oplus P_{c-1}$. Note that h acts on P_i as the scalar ξ^i, and each P_i is both projective and injective.

Since every KH-module is isomorphic to a direct sum of modules from the set $\{Q_0, \ldots, Q_{c-1}\}$, every H-induced KG-module is isomorphic to a direct sum of modules from the set $\{P_0, \ldots, P_{c-1}\}$. In particular, every H-induced module is both projective and injective. It also follows that if U is H-induced and h acts on U as the scalar ξ^i, where $0 \leqslant i \leqslant c-1$, then U is isomorphic to a direct sum of modules all isomorphic to P_i.

Now h acts on R_n as the scalar ξ^i where $0 \leqslant i \leqslant c-1$ and $n \equiv i$ (modulo c). Thus a submodule of R_n is H-induced if and only if it is isomorphic to a direct sum of copies of P_i.

We shall be interested in locating free KG-submodules of R. As we shall see, when n is sufficiently large, $R_n \oplus \cdots \oplus R_{n+c-1}$ contains a regular KG-module. Note that it follows from the previous discussion that no sum of fewer than c components R_j can contain a regular KG-module. The main result of this section is the following theorem.

Theorem 4.3 Let $R = K[x_1, \ldots, x_r]$ and let G be a finite subgroup of $\mathrm{GL}_r(K)$ acting on R by homogeneous linear substitutions. Let c be the number of scalar elements of G. Then

$$\lim_{n \to \infty} f(R_n \oplus \cdots \oplus R_{n+c-1}) = 1.$$

Theorem 4.3 is the main result of [2]. It asserts that, in an asymptotic sense, R is close to being a free KG-module.

Let K' be an extension field of K and write $R' = K'[x_1, \ldots, x_r]$ with homogeneous components R'_0, R'_1, ... Then we can regard G as a subgroup of $\mathrm{GL}_r(K')$ acting on R', and there is a $K'G$-module isomorphism

$$R'_n \oplus \cdots \oplus R'_{n+c-1} \cong K' \otimes (R_n \oplus \cdots \oplus R_{n+c-1}).$$

Hence, by the property of f already mentioned,

$$f(R'_n \oplus \cdots \oplus R'_{n+c-1}) = f(R_n \oplus \cdots \oplus R_{n+c-1}).$$

Thus it suffices to prove Theorem 4.3 with K replaced by K'. Consequently we may assume that K is infinite. The main step in the proof of Theorem 4.3 is the following result.

Theorem 4.4 *Let $R = K[x_1, \ldots, x_r]$ and let G be a finite subgroup of $\mathrm{GL}_r(K)$ acting on R by homogeneous linear substitutions. Let H be the subgroup consisting of all scalar elements of G. Then*

$$\lim_{n \to \infty} f_H(R_n) = 1.$$

Theorem 4.3 follows from Theorem 4.4 quite easily. The details are given in Section 2 of [2], and, for brevity, we omit them here. In the case where K has characteristic 0, Theorem 4.4 is a consequence of the Corollary of [9], and we shall not give a proof in this case. Thus we only prove Theorem 4.4 in the case where K is an infinite field of prime characteristic p. (In fact, the characteristic 0 case can be obtained from the characteristic p case by considering a prime p which does not divide the order of G.) If $r = 1$ then $H = G$ and $f_H(R_n) = 1$ for all n, so Theorem 4.4 is trivial in this case. Thus we further assume that $r \geqslant 2$. Consequently, all the results of Sections 1, 2 and 3 are now available for our use. We shall prove Theorem 4.4 in the form

$$\lim_{n \to \infty} f'_H(R_n) = 0.$$

Recall that R_n, regarded as a $K\mathrm{GL}_r(K)$-module, has a composition series with factors isomorphic to the modules $\sigma(W_t)$, $t \in T_n$. We regard these factors as KG-modules, although as KG-modules they are not irreducible in general (and we shall exploit this very fact).

The proof of Theorem 4.4 goes roughly as follows. By Theorem 2.2, each $\sigma(W_t)$ is a tensor product. We prove that when n is 'large' the number of factors in the tensor product is 'large' for 'most' values of $t \in T_n$. Proposition 4.1 and Lemma 4.2 are used to show that when the number of factors of $\sigma(W_t)$ is large then $f'_H(\sigma(W_t))$ is small. It follows that, when n is large, $f'_H(R_n)$ is small.

Lemma 4.5 *Write $b = |G|/c$, where c is the number of scalar elements of G. Let $m_1, \ldots, m_b$ be positive integers satisfying $0 < m_i < r(p-1)$ for $i = 1, \ldots, b$, and let $k(1), \ldots, k(b)$ be non-negative integers. Then*

$$\sigma(R_{m_1}^{p^{k(1)}}) \otimes \cdots \otimes \sigma(R_{m_b}^{p^{k(b)}}),$$

regarded as a KG-module, has a non-zero H-induced submodule.

Proof. Let $m_1 p^{k(1)} \equiv l$ (modulo c) where $0 \leqslant l < c$. Thus h acts as the scalar ξ^l on $\sigma(R_{m_1}^{p^{k(1)}})$. Write

$$U = \sigma(R_{m_1}^{p^{k(1)}}) \oplus R_0 \oplus \cdots \oplus R_{l-1} \oplus R_{l+1} \oplus \cdots \oplus R_{c-1}.$$

It is clear that U, regarded as a KH-module, contains a regular KH-module. Hence there exists an element v_1 of U such that $v_1 \notin Kv_1(H \smallsetminus \{1\})$.

Let $\{g_1, \ldots, g_b\}$ be a complete set of coset representatives for H in G, with $g_1 = 1$. For $i = 2, \ldots, b$, the element g_i does not have scalar action on $\sigma(R_{m_i}^{p^{k(i)}})$, by Lemma 3.2. It follows easily that there is a non-zero element v_i of $\sigma(R_{m_i}^{p^{k(i)}})$ which is not an eigenvector for g_i. Thus $v_i \notin Kv_ig_i = Kv_ig_iH$. But

$$G \smallsetminus \{1\} = (H \smallsetminus \{1\}) \cup g_2H \cup \ldots \cup g_bH.$$

Therefore, by Proposition 4.1, the module $U \otimes \sigma(R_{m_2}^{p^{k(2)}}) \otimes \cdots \otimes \sigma(R_{m_b}^{p^{k(b)}})$ contains a submodule isomorphic to P, the regular KG-module. Let j be the integer satisfying $0 \leqslant j < c$ and

$$j \equiv m_1 p^{k(1)} + \cdots + m_b p^{k(b)} \pmod{c}.$$

Then $\sigma(R_{m_1}^{p^{k(1)}}) \otimes \cdots \otimes \sigma(R_{m_b}^{p^{k(b)}})$ is the summand of $U \otimes \sigma(R_{m_2}^{p^{k(2)}}) \otimes \cdots \otimes \sigma(R_{m_b}^{p^{k(b)}})$ consisting of those elements on which h acts as the scalar ξ^j. Hence the module $\sigma(R_{m_1}^{p^{k(1)}}) \otimes \cdots \otimes \sigma(R_{m_b}^{p^{k(b)}})$ has a submodule isomorphic to P_j. This is a non-zero H-induced submodule, so the proof is complete. □

Let a be the maximum of the dimensions of the modules $\sigma(R_0), \ldots, \sigma(R_{r(p-1)})$. Thus $\dim(\sigma(R_m^{p^k})) \leqslant a$ for all integers m satisfying $0 \leqslant m \leqslant r(p-1)$ and all non-negative integers k. Note that $a > 1$ since $\dim(\sigma(R_1)) = r > 1$. On the other hand, $\sigma(R_0)$ and $\sigma(R_{r(p-1)})$ have dimension 1. Thus $\sigma(R_m^{p^k})$ has dimension 1 if $m = 0$ or $m = r(p-1)$. Suppose $t \in T_n$ for some n and let $t = (t_0, t_1, \ldots)$. Then, by Theorem 2.2,

$$\sigma(W_t) \cong \sigma(R_{t_0}) \otimes \sigma(R_{t_1}^p) \otimes \sigma(R_{t_2}^{p^2}) \otimes \cdots,$$

where $0 \leqslant t_j \leqslant r(p-1)$ for all j. It follows that $\dim(\sigma(W_t)) \leqslant a^{l(t)}$, where $l(t)$ is the number of values of j for which $0 < t_j < r(p-1)$.

Let $\mathcal{C}$ be the set of all modules of the form $\sigma(R_{m_1}^{p^{k(1)}}) \otimes \cdots \otimes \sigma(R_{m_b}^{p^{k(b)}})$, where $b = |G|/c$ and $0 < m_i < r(p-1)$ for $i = 1, \ldots, b$, as in Lemma 4.5. Since H-induced modules are injective, Lemma 4.5 shows that $f_H(W)\dim(W) \geqslant 1$ for all $W \in \mathcal{C}$. But $\dim(W) \leqslant a^b$ for all $W \in \mathcal{C}$. Hence $f_H(W) \geqslant 1/a^b$ for all $W \in \mathcal{C}$. It follows that there exists a number γ, with $0 < \gamma < 1$, such that $f'_H(W) \leqslant \gamma$ for all $W \in \mathcal{C}$. Since $a > 1$ we can choose γ so that, in addition, $\gamma > 1/a$.

For each n and each $t \in T_n$ write $d(t) = \dim(\sigma(W_t))$, $f_H(t) = f_H(\sigma(W_t))$ and $f'_H(t) = 1 - f_H(t)$. Then $d(t) \leqslant a^{l(t)}$ and so $l(t) \geqslant \log_a d(t)$. Therefore $l(t) \geqslant e(t)b$, where $e(t)$ is the integer part of $\frac{1}{b}\log_a d(t)$. It follows that we can write W_t as a tensor product of modules in such a way that at least $e(t)$ of these modules belong to $\mathcal{C}$. Hence, by Lemma 4.2, $f'_H(t) \leqslant \gamma^{e(t)}$. But $0 < \gamma < 1$ and $e(t) > \frac{1}{b}\log_a d(t) - 1$. Therefore

$$f'_H(t) \leqslant \gamma^{\frac{1}{b}\log_a d(t)-1} = \gamma^{-1}d(t)^{\frac{1}{b}\log_a \gamma}.$$

Write $\kappa = 1 + \frac{1}{b}\log_a \gamma$. Then we have

$$f'_H(t) \leqslant \gamma^{-1}d(t)^{\kappa-1}.$$

Now we use the fact that R_n has a series with factors isomorphic to the modules $\sigma(W_t)$, $t \in T_n$. In particular,

$$\dim(R_n) = \sum_{t \in T_n} d(t).$$

Since H-induced modules are injective and projective,

$$f_H(R_n)\dim(R_n) \geqslant \sum_{t \in T_n} f_H(t)d(t)$$

and so

$$f_H'(R_n)\dim(R_n) \leqslant \sum_{t \in T_n} f_H'(t)d(t).$$

It follows that

$$f_H'(R_n) \leqslant \gamma^{-1} \frac{\sum_t d(t)^\kappa}{\dim(R_n)}.$$

Since $1/a < \gamma < 1$ we have $0 < \kappa < 1$. By elementary calculus, if $x(t)$, $t \in T_n$, are non-negative real numbers satisfying the constraint $\sum_t x(t) = \dim(R_n)$ then the maximum value of $\sum_t x(t)^\kappa$ occurs when all $x(t)$ take the same value d. Clearly, $d = \dim(R_n)/|T_n|$. Hence,

$$f_H'(R_n) \leqslant \gamma^{-1} \frac{|T_n|d^\kappa}{\dim(R_n)} = \gamma^{-1} \left(\frac{|T_n|}{\dim(R_n)} \right)^{1-\kappa}.$$

But $|T_n|/\dim(R_n) \to 0$ as $n \to \infty$, by Lemma 3.1. Hence $f_H'(R_n) \to 0$. This completes the proof of Theorem 4.4. $\square$

Finally we show that the work of Glover [8] yields the following result.

Theorem 4.6 *Let K be the field of p elements, p a prime, and let $G = \mathrm{GL}_2(K)$, $R = K[x_1, x_2]$. Then, for all n,*

$$R_n \oplus R_{n+1} \oplus \cdots \oplus R_{n+p-2}$$

has a free KG-submodule of bounded codimension.

In the case under consideration $c = p - 1$ and so Theorem 4.6 is a strengthening of Theorem 4.3 in this case.

Sketch of Proof. The proof is heavily dependent on [8] and we omit some details which are straightforward but laborious.

In Section 6 of [8], Glover defines certain KG-modules Q_k for $k = 0, 1, \ldots, p$ and considers the modules $Q_k \otimes D^j$ where D^j denotes the j-th tensor power of the one-dimensional module D which affords the determinant representation of G. Let X be the direct sum of the modules $Q_k \otimes D^j$ for $k = 0, 1, \ldots, p$, and $j = 0, 1, \ldots, p-2$. Then it may be verified from the results of [8] that X is isomorphic to the regular KG-module.

For each non-negative integer n let

$$A_n = R_n \oplus R_{n+1} \oplus \cdots \oplus R_{n+p-2}.$$

By (6.7) of [8], for every non-negative integer m,

$$R_{m+p(p-1)} \cong R_m \oplus (Q_k \otimes D^j),$$

where $m + 1 = j(p+1) + k$ and $0 \leqslant k \leqslant p$. Using this result it may be verified that

$$A_{n+(p+1)p(p-1)} \cong A_n \oplus X,$$

a result which was also obtained in some unpublished work of L. G. Kovács: I thank Dr Kovács for informing me of this work. It follows that each module A_n is isomorphic to the direct sum of a free module and a module A_m with $m < p(p^2 - 1)$. Hence A_n has a free submodule of bounded codimension. $\square$

References

1. A. M. Adamovich, The structure of some indecomposable representations of GL_n over a field of finite characteristic, in *Algebra, Logic and Number Theory, The 7th Special Conference of the Mechanics and Mathematics Department of the Moscow State University, February–March, 1985*, Moscow, 1986, pp. 8–12.
2. R. M. Bryant, Symmetric powers of representations of finite groups, *J. Algebra* 154 (1993), 416–436.
3. R. M. Bryant and L. G. Kovács, Tensor products of representations of finite groups, *Bull. London Math. Soc.* 4 (1972), 133–135.
4. W. Burnside, *Theory of Groups of Finite Order*, 2nd ed., Cambridge University Press, 1911, Dover Publ., 1955.
5. S. R. Doty, The submodule structure of certain Weyl modules for groups of type A_n, *J. Algebra* 95 (1985), 373–383.
6. S. Doty, Submodules of symmetric powers of the natural module for GL_n, in *Invariant Theory, Proceedings of an AMS Special Session held October 31–November 1, 1986: Contemp. Math.* 88 (1989), 185–191.
7. S. Doty, The symmetric algebra and representations of general linear groups, in *Proceedings of the Hyderabad Conference on Algebraic Groups, Hyderabad, India, December 1989*, Manoj Prakashan, pp. 123–150.
8. D. J. Glover, A study of certain modular representations, *J. Algebra* 51 (1978), 425–475.
9. R. Howe, Asymptotics of dimensions of invariants for finite groups, *J. Algebra* 122 (1989), 374–399.
10. B. Huppert and N. Blackburn, *Finite Groups II*, Springer-Verlag, Berlin, Heidelberg, New York, 1982.
11. L. G. Kovács, Some representations of special linear groups, in *The Arcata Conference on Representations of Finite Groups, Arcata, California, 1986: Proc. Sympos. Pure Math.* 47 (1987), 207–218.
12. L. Krop, On the representations of the full matrix semigroup on homogeneous polynomials, *J. Algebra* 99 (1986), 370–421.
13. L. Krop, On the representations of the full matrix semigroup on homogeneous polynomials, II, *J. Algebra* 102 (1986), 284–300.
14. R. Steinberg, *Lectures on Chevalley Groups*, Yale University, New Haven, 1967.

CHARACTERS AND SETS OF PRIMES FOR SOLVABLE GROUPS

I. M. ISAACS
University of Wisconsin
Mathematics Department
Madison WI 53706
USA
E-mail: isaacs@math.wisc.edu

Abstract. Given a set π of prime numbers, the π-partial characters of a finite group G are defined as the restrictions of the ordinary characters of G to the π-elements of G. In the case where G is solvable, or even just π-separable, these π-partial characters are under good control and they can be used to prove new results about the ordinary characters of such groups. In these lectures, several of the key results that form the foundation of the theory are presented without proof. These are then used to develop the theory and derive a number of applications.

Key words: partial characters, Hall subgroups, π-separable groups, monomial characters, Fong characters.

1. Introduction and Motivation

There is a surprisingly deep connection between the arithmetic structure of a finite solvable group G and the properties of its set $\mathrm{Irr}(G)$ of ordinary, complex-valued irreducible characters. It is our purpose in these lectures to describe some of this still developing 'π-theory' to an audience that contains people whose primary interest is in not-quite-finite groups. We want to show some of what is going on in finite group theory today, especially in that part of finite group theory not concerned with the classification and properties of finite simple groups.

Since we have given expository lectures elsewhere [16] on the foundations of π-theory, our emphasis here is on more recent developments and on applications. Some of this material is new and is presented here with complete proofs. We omit proofs, however, for most of the fundamental facts of π-theory and we even refuse to give one of the principal definitions. Nevertheless, we have organized the presentation so that it should be intelligible even to students familiar with only the basics of character theory. (The first six chapters of [15], for example, are more than sufficient.)

The motivation for the material we are about to discuss is the observation that in the situation of solvable groups, there is a more intimate than usual connection between R. Brauer's 'modular' character theory and the classical 'ordinary' character theory of Frobenius and Schur. To explain this, let us review a bit of the Brauer theory. (Chapter 15 of [15] provides the necessary background, but none of the material there should actually be necessary in order to follow the present discussion.)

Suppose that F is an algebraically closed field of prime characteristic p and let G be an arbitrary finite group. Given any finite dimensional FG-module M, Brauer

B. Hartley et al. (eds.), Finite and Locally Finite Groups, 347–376.

described a process (which we need not discuss here) for constructing the associated 'Brauer character' φ of M. This is a *complex*-valued class function defined on the set of p-regular elements of G, the elements, that is, having order not divisible by p. It is clear from its construction that the Brauer character φ associated with M depends only on the isomorphism type of M as an FG-module.

We introduce some notation that will facilitate the discussion. If π is an arbitrary (but fixed) set of prime numbers, we write G^0 to denote the set of π-elements of G, the elements having orders divisible only by primes in π. We denote by $\mathrm{cf}(G^0)$ the complex vector space of class functions defined on G^0. (These are, of course, the functions having constant values on each of the conjugacy classes of π-elements of G.) In Brauer's situation, we take $\pi = p'$, the complement in the set of all primes of the singleton set $\{p\}$, and we see that the Brauer characters of G are members of the space $\mathrm{cf}(G^0)$.

Brauer's construction shows that the set $\mathrm{Br}(G)$ of all Brauer characters of G is closed under addition. In fact, the Brauer character φ associated with a nonzero module M is exactly the sum of the Brauer characters corresponding to the composition factors of M. Furthermore, $\varphi(1) = \dim_F(M)$ is a positive integer, and it follows that every Brauer character can be written as a finite sum of *irreducible* Brauer characters, which we define to be those nonzero Brauer characters that cannot be decomposed as a sum of two others. (We write $\mathrm{IBr}(G)$ to denote the set of irreducible Brauer characters of G.) It follows from all of this that the modules that give rise to irreducible Brauer characters are necessarily simple. Furthermore, it is a theorem that the Brauer characters associated with the different (nonisomorphic) simple modules are linearly independent members of $\mathrm{cf}(G^0)$, and it follows that every simple FG-module gives rise to an irreducible Brauer character. Furthermore, Brauer proved that the number of isomorphism types of simple FG-modules is exactly equal to the number of p-regular classes in G, and this, of course, is the dimension (over $\mathbb{C}$) of the vector space $\mathrm{cf}(G^0)$. Putting all of this together, we have the following.

Theorem 1.1 (Brauer) *The set* $\mathrm{IBr}(G)$ *is a basis for the complex vector space* $\mathrm{cf}(G^0)$.

Brauer suggested that the real importance of his modular character theory is that it gives information about the ordinary characters of G that does not appear to be otherwise accessible. (While it is clear that Brauer's theory is interesting and important in its own right, it seems to this author that the value of any mathematical theory is enhanced if the theory has the power to answer questions that can be stated outside of itself.) One reason that Brauer's modular theory has implications for ordinary characters is apparent from the following result, which at first may seem somewhat surprising. To state it, we introduce a bit more notation. As usual, we write $\mathrm{cf}(G)$ to denote the space of all complex-valued class functions on G and we observe that restriction to the subset $G^0 \subseteq G$ of π-elements defines a linear transformation $\chi \mapsto \chi^0$ of $\mathrm{cf}(G)$ onto $\mathrm{cf}(G^0)$. If we start with a character $\chi \in \mathrm{Char}(G) \subseteq \mathrm{cf}(G)$, we refer to its restriction χ^0 as a π-*partial character* of G.

Theorem 1.2 (Brauer) *If* $\pi = p'$, *then every* π-*partial character of* G *is a Brauer character of* G *for the prime* p. *In other words,* $\mathrm{Char}(G)^0 \subseteq \mathrm{Br}(G)$.

Theorems 1.1 and 1.2 provide the starting point for much of Brauer's theory. For example, if $\chi \in \text{Irr}(G)$, we can uniquely write the Brauer character χ^0 as a sum of irreducible Brauer characters. This enables us to define a bipartite graph having vertex set $\text{Irr}(G) \cup \text{IBr}(G)$, where we join $\chi \in \text{Irr}(G)$ and $\varphi \in \text{IBr}(G)$ if φ occurs as one of the summands of χ^0. The connected components of this graph are the *blocks* (or *p-blocks*) of G. Many of the most important of Brauer's results, including his three 'main theorems', are statements about the p-blocks of a group and of certain of its subgroups.

For our purposes, however, we carry the story in a different direction and we consider what additional information is available when G is solvable. In general, the set $\text{Char}(G)^0$ of π-partial characters of G can be a proper subset of the set $\text{Br}(G)$ of Brauer characters. (We are continuing, of course, to take $\pi = p'$.) When G is solvable, however, equality always holds.

Theorem 1.3 (Fong-Swan) *Let G be solvable. Then every Brauer character φ of G has the form $\varphi = \chi^0$, where χ is some ordinary character of G.*

It should be clear that if $\varphi \in \text{IBr}(G)$ and $\varphi = \chi^0$ with $\chi \in \text{Char}(G)$, then χ is necessarily irreducible. Also, in order to prove Theorem 1.3, it suffices to consider only irreducible Brauer characters of G and to show that each one can be 'lifted' to an ordinary irreducible character of G. (In other words, given $\varphi \in \text{IBr}(G)$, it suffices to find $\chi \in \text{Irr}(G)$ such that $\varphi = \chi^0$.) The Fong-Swan theorem was first proved, or at least essentially proved, by P. Fong and then, later, it was stated by R. Swan. A proof more in the spirit of these lectures, however, can be found in [6]. To explain the significance of the Fong-Swan theorem for our purposes, we digress briefly to discuss the general case.

For an arbitrary finite group G and an arbitrary set π of primes, we say that a nonzero π-partial character of G is *irreducible* if it cannot be written as a sum of two other nonzero π-partial characters and we write $I_\pi(G)$ to denote the set of irreducible π-partial characters of G. Since every nonzero π-partial character has a positive integer value at the identity of G, it follows (just as in the case of Brauer characters) that every π-partial character of G can be written as a sum of members of $I_\pi(G)$. Unlike the situation for Brauer characters, however, it is easy to read off $I_\pi(G)$ from the character table of G. First, delete all columns of the table that correspond to classes that are not π-classes. (Although most published character tables label the columns with information giving the orders of the corresponding elements, this extra information is not really necessary. By a result of G. Higman that appears as Theorem 8.21 of [15], the set of primes dividing the order of an element can be determined from the character table itself, even if the columns are unlabeled.) From the resulting rectangular matrix, delete all rows that can be written as sums of other rows. What remains is the table of irreducible π-partial characters of G.

In the case where G is solvable and $\pi = p'$ is the complement of a single prime, we know by the Fong-Swan theorem that $\text{Br}(G) = \text{Char}(G)^0$. It follows that the irreducible Brauer characters of G are exactly the irreducible π-partial characters of G; in other words, $\text{IBr}(G) = I_\pi(G)$ in this situation. It follows that, for solvable groups, the irreducible Brauer characters can easily be determined from the character table. (Actually, we cheated a little before. In general, the sets $\text{Br}(G)$ and $\text{IBr}(G)$

are *not* unambiguously and canonically determined by G; their construction depends on a certain choice. In the case where G is solvable and $\mathrm{IBr}(G)$ can be determined from the character table of G, however, the ambiguity disappears and the Brauer characters turn out to be independent of any choice made in their construction.)

In complete generality, we see that, since $\mathrm{Char}(G)$ spans $\mathrm{cf}(G)$, it follows that $\mathrm{Char}(G)^0$ spans $\mathrm{cf}(G^0)$ and thus $I_\pi(G)$ spans $\mathrm{cf}(G^0)$. In the case where G is solvable and π is the complement of a prime, Brauer's theory, combined with the Fong-Swan theorem, tells us that $I_\pi(G) = \mathrm{IBr}(G)$ is also linearly independent, and so forms a basis for $\mathrm{cf}(G^0)$. For nonsolvable groups, a bit of experimentation with character tables quickly shows that $I_\pi(G)$ usually has cardinality greater than the number of π-classes of G, and so it is not generally linearly independent, even in the case where π is the complement of a prime. What is perhaps the first surprise of this theory is that, when G is solvable, $I_\pi(G)$ is always linearly independent, even when π is not the complement of a prime and Brauer's theory does not apply.

Theorem 1.4 *Let G be solvable and suppose that π is an arbitrary set of prime numbers. Then $I_\pi(G)$ is a basis for $\mathrm{cf}(G^0)$.*

The first proof of Theorem 1.4 appeared in [9]. An alternative and simpler argument based on ideas in the paper [20] can be found in [13], but in order to obtain the deeper properties of $I_\pi(G)$ that we shall discuss here the original approach appears to be necessary.

We close this introduction with the remark that for Theorem 1.4 (and for the Fong-Swan theorem) the assumption that G is solvable is somewhat too strong; the 'correct' hypothesis is that G should be π-separable. Recall that this means that each composition factor (or, equivalently, each chief factor) of G is either a π-group or a π'-group. When $\pi = p'$ as in the Brauer situation, π-separability is exactly equivalent to p-solvability, and this is the right hypothesis for the Fong-Swan theorem. In fact, π-separability is the natural hypothesis for virtually everything we discuss in these lectures, and we shall state our results accordingly. (Little will be lost, however, if 'solvable' is read wherever 'π-separable' is written.)

The modest increase in generality from solvable to π-separable groups is, unfortunately, not completely without cost. We shall need, for instance, the facts that Hall π-subgroups exist in G, that all of them are conjugate and that every π-subgroup of G is contained in one of them. It is well known that, for solvable groups, these results of P. Hall are fairly elementary. These facts about Hall π-subgroups also hold more generally for π-separable groups, but in that setting some of the assertions rely on the conjugacy part of the Schur-Zassenhaus theorem. The proof of that seems to depend on the decidedly not elementary odd-order theorem of Feit and Thompson.

2. Fong Characters and Applications

If φ is a π-partial character of G and $H \subseteq G$ is an arbitrary subgroup, then the restriction φ_H is obviously a π-partial character of H; it is just $(\chi_H)^0$, where $\chi^0 = \varphi$. If H is a π-subgroup, then clearly φ_H is an ordinary character of H. In the special case where H is a Hall π-subgroup of G and φ is irreducible (in other words, $\varphi \in I_\pi(G)$), we say that an (ordinary) irreducible character $\alpha \in \mathrm{Irr}(H)$ is a *Fong character* belonging to φ if α is one of the irreducible constituents of φ_H having smallest degree. (An easy exercise that might help to solidify the various definitions is to show that if $H \subseteq G$ is a Hall π-subgroup then the Fong characters of H are exactly those irreducible characters that occur as least degree irreducible constituents of χ_H as χ runs over $\mathrm{Irr}(G)$. In particular, every linear character of H is a Fong character belonging to some member of $\mathrm{I}_\pi(G)$.)

If G is π-separable, so that a Hall π-subgroup $H \subseteq G$ is guaranteed to exist, it is obvious that associated to every member $\varphi \in \mathrm{I}_\pi(G)$ there is at least one Fong character $\alpha \in \mathrm{Irr}(H)$. Also, if β is any other irreducible constituent of φ_H having the same degree as α, then β is also a Fong character belonging to φ. In particular, it follows that the action of the normalizer $N_G(H)$ on $\mathrm{Irr}(H)$ permutes the set of Fong characters of H belonging to each irreducible π-partial character φ of G. (It is common for there to be more than one Fong character of H belonging to a given irreducible π-partial character. Indeed, there may even be more than orbit under the action of $N_G(H)$.) Since the Hall π-subgroups of G are all conjugate, it is immediate that the number of Fong characters of H that belong to φ and their common degree are independent of the choice of the particular Hall π-subgroup H. Our fundamental result about Fong characters is the following.

Theorem 2.1 *Suppose G is π-separable and let $\varphi \in \mathrm{I}_\pi(G)$. Choose a Hall π-subgroup $H \subseteq G$ and let $\alpha \in \mathrm{Irr}(H)$ be a Fong character belonging to φ. The following then hold.*

(a) *α occurs with multiplicity 1 as a constituent of φ_H.*

(b) *α is not a constituent of θ_H if $\varphi \neq \theta \in \mathrm{I}_\pi(G)$.*

(c) *$\alpha(1) = \varphi(1)_\pi$, the π-part of the degree of φ.*

Note that part (b) of the above result justifies our saying that the Fong character α 'belongs' to φ. It asserts that an irreducible character of the Hall subgroup H can belong (as a Fong character) to at most one irreducible π-partial character of G. We stress, however, that in general not every irreducible character of H is a Fong character.

We should mention that we have named these objects 'Fong characters' because in the classical case, where $\pi = p'$, they appear in the paper [5] of P. Fong. Fong shows that the 'projective indecomposable' character Φ_φ corresponding to an irreducible Brauer character φ of a p-solvable group G can be written in the form α^G, where α is a certain irreducible character of a p-complement of G and $\alpha(1) = \varphi(1)_{p'}$. It is not hard to see that Fong's character α is, according to our definition, a Fong character belonging to φ.

Our assertion in Theorem 1.4 that $\mathrm{I}_\pi(G)$ is linearly independent when G is solvable (and, in fact, when G is π-separable) is an immediate consequence of Theorem 2.1.

Corollary 2.2 *If G is π-separable, then $\mathrm{I}_\pi(G)$ is a basis for $\mathrm{cf}(G^0)$. Furthermore, if $\theta \in \mathrm{cf}(G^0)$ and*

$$\theta = \sum_{\varphi \in \mathrm{I}_\pi(G)} a_\varphi \varphi,$$

then the coefficients a_φ are determined by the formula $a_\varphi = [\theta_H, \alpha]$, where H is a Hall π-subgroup of G and $\alpha \in \mathrm{Irr}(H)$ is a Fong character belonging to φ.

Proof. Given $\theta \in \mathrm{cf}(G^0)$, we have seen that it is possible to write $\theta = \sum a_\varphi \varphi$ with complex coefficients a_φ, where φ runs over $\mathrm{I}_\pi(G)$. Now fix $\mu \in \mathrm{I}_\pi(G)$, let $H \subseteq G$ be a Hall π-subgroup and choose a Fong character $\alpha \in \mathrm{Irr}(H)$ belonging to μ. By Theorem 2.1(b), we have $[\varphi_H, \alpha] = 0$ if $\varphi \neq \mu$, and thus

$$[\theta_H, \alpha] = \sum_\varphi a_\varphi [\varphi_H, \alpha] = a_\mu [\mu_H, \alpha] = a_\mu \,,$$

where the last equality is a consequence of Theorem 2.1(a). The stated formula for the coefficients is now proved and in particular, if $\theta = 0$, we see that all coefficients $a_\varphi = 0$. This proves the linear independence of $\mathrm{I}_\pi(G)$. $\square$

By definition, we know that an arbitrary π-partial character θ of a group G is a sum of irreducible π-partial characters. It can be written, therefore, as a linear combination $\theta = \sum a_\varphi \varphi$, where φ runs over $\mathrm{I}_\pi(G)$ and the coefficients a_φ are nonnegative integers, at least one of which is nonzero. If G is π-separable, then Corollary 2.2 guarantees that these coefficients are uniquely determined. We shall often refer to the coefficient a_φ as the *multiplicity* with which φ occurs in θ and if this multiplicity is nonzero we say that φ is an *irreducible constituent* of θ.

We mention that we could proceed from this point to construct (for π-separable groups) an analog of Brauer's theory of p-blocks. The starting point for this, of course, is to consider the bipartite graph with vertex set $\mathrm{Irr}(G) \cup \mathrm{I}_\pi(G)$, where $\varphi \in \mathrm{I}_\pi(G)$ is joined to $\chi \in \mathrm{Irr}(G)$ if φ is an irreducible constituent of χ^0. Such a theory was developed by M. Slattery [21], and it does lead to interesting results, but we choose to go in a different direction here.

In these lectures, we shall occasionally offer 'applications' of our theory: corollaries whose statements do not mention any of our technical definitions. We believe that such easily stated results, which do not appear to follow via the usual techniques, demonstrate the existence of a deeper structure to the representation theory of solvable and related groups than had been previously recognized. (We cannot actually prove, of course, that these applications truly require the theory we are developing here and it would be interesting to see direct proofs if they could be found.) As our first and very modest example of an application, we offer the following consequence of Theorem 2.1. (Note that this result is false in general for nonsolvable groups, even if we limit our attention to Sylow subgroups in place of Hall subgroups.)

Corollary 2.3   *Let H be a Hall subgroup of a solvable group G and suppose that λ is a linear character of H. Then every irreducible constituent of least degree of λ^G occurs with multiplicity 1 and has degree dividing $|G : H|$.*

Proof. Let $\chi \in \mathrm{Irr}(G)$ be a constituent of least degree of λ^G and let π denote the set of prime divisors of $|H|$. We claim that the π-partial character χ^0 is irreducible. Otherwise we could write $\chi^0 = \xi^0 + \eta^0$ for some characters $\xi, \eta \in \mathrm{Char}(G)$, each having degree less than $\chi(1)$. Since λ is a constituent of $\chi_H = \xi_H + \eta_H$, it would follow that some irreducible constituent of ξ or η occurs as a constituent of λ^G, and this would contradict the choice of χ. We see now that λ is a Fong character belonging to χ^0 and the result follows from parts (a) and (c) of Theorem 2.1.   □

What we believe is a striking application is the following.

Theorem 2.4   *Let H be a Hall π-subgroup of a π-separable group G and suppose that $\psi \in \mathrm{Char}(H)$ has the property that $\psi(x) = \psi(y)$ whenever x and y are G-conjugate elements of H. Then there exists $\chi \in \mathrm{Char}(G)$ such that $\chi_H = \psi$.*

The condition that ψ is constant on intersections of H with classes of G is clearly necessary for the extendibility of ψ, but that this condition should also be sufficient seems rather surprising. (In fact, it is easy to see that it is *not* sufficient without the π-separability assumption.) In the situation where H is normal in G, this condition is equivalent to the invariance of ψ in G, and in that case (at least when ψ is irreducible) the sufficiency is a well known result of P. X. Gallagher. (See, for example, Theorem 8.16 of [15].) In fact, our Theorem 2.4 is a proper generalization of Gallagher's result since we have replaced his requirement that the Hall π-subgroup is normal with the much weaker assumption that G is π-separable.

Proof of Theorem 2.4.   Define $\theta \in \mathrm{cf}(G^0)$ as follows. If $g \in G$ is a π-element, choose $x \in H$ conjugate to g in G and set $\theta(g) = \psi(x)$. By the hypothesis on ψ, the function θ is well defined and since $\mathrm{I}_\pi(G)$ spans $\mathrm{cf}(G^0)$, we can write

$$\theta = \sum_{\varphi \in \mathrm{I}_\pi(G)} a_\varphi \varphi$$

with some complex coefficients a_φ. By Corollary 2.2, we have $a_\varphi = [\theta_H, \alpha]$, where $\alpha \in \mathrm{Irr}(H)$ is any Fong character belonging to φ. By construction, however, we know that $\theta_H = \psi$ is a character, and it follows that $a_\varphi = [\psi, \alpha]$ is a nonnegative integer.

For each π-partial character $\varphi \in \mathrm{I}_\pi(G)$, choose $\xi_\varphi \in \mathrm{Char}(G)$ such that $(\xi_\varphi)^0 = \varphi$ and set $\chi = \sum a_\varphi \xi_\varphi$. Since we have seen that the coefficients a_φ are nonnegative integers (and at least one of them is clearly positive), it follows that χ is a character. Furthermore,

$$\chi_H = (\chi^0)_H = \theta_H = \psi,$$

as required.   □

We will not give, or even sketch, either of the two known methods of proof of Theorem 2.1 here. The easier of these, which yields little beyond the desired result, appears in the mostly expository paper [13]. The alternative (and original) approach is that of [9], which lays the foundation for the deeper parts of the theory that we are about to discuss. For an outline and discussion of the proof in [9], the student may also wish to consult the survey paper [16], which overlaps the present lectures somewhat.

3. Canonical Lifts

Of course, it is obvious that, given an irreducible π-partial character φ of G, there exists at least one (necessarily irreducible) ordinary character χ of G that 'lifts' φ.

(In other words, χ is chosen so that $\chi^0 = \varphi$.) In fact, for each π-separable group G, it is possible to pick out in advance a certain subset of $\mathrm{Irr}(G)$, which we denote $\mathrm{B}_\pi(G)$, such that the map $\chi \mapsto \chi^0$ defines a bijection from $\mathrm{B}_\pi(G)$ onto $\mathrm{I}_\pi(G)$.

We stress that this set $\mathrm{B}_\pi(G)$ is 'canonical'; its construction is completely determined by the structure of G and does not depend on any arbitrary choices. Of course, if we actually gave the definition of $\mathrm{B}_\pi(G)$ this point would be obvious and would require no special mention, but, nevertheless, we choose (for at least two reasons) not to present the construction of this set of irreducible characters here. The definition is complicated and requires several technical lemmas even to describe it. (In particular, we would need to present π-special characters and character factorization theory, and that would carry us far from our goal in these lectures.) Equally significant in our decision not to present the definition of $\mathrm{B}_\pi(G)$ is the fact that it is seldom of practical use for deciding whether or not a particular irreducible character is or is not actually in the set. The reason for this is that the definition is inductive, and so one would have to construct the sets $\mathrm{B}_\pi(H)$ for certain subgroups $H \subseteq G$ before determining $\mathrm{B}_\pi(G)$. In fact, one of the open problems in this theory is to find a good characterization of the members of $\mathrm{B}_\pi(G)$.

The precise definition of the set $\mathrm{B}_\pi(G)$ for π-separable groups G is given in the paper [9] and it is discussed in the survey paper [16]. Since we recognize that the audience for these lectures may feel uncomfortable without at least one concrete example in hand of a character that lies in $\mathrm{B}_\pi(G)$, we mention that the principal character of any π-separable group G always lies in this set. More generally, for those familiar with π-special characters of π-separable groups as defined by D. Gajendragadkar, we mention that the π-special characters of G all lie in $\mathrm{B}_\pi(G)$. In fact, the π-special characters of G are exactly those members of $\mathrm{B}_\pi(G)$ whose degrees are π-numbers.

In order to help explain what we mean by the assertion that $\mathrm{B}_\pi(G)$ is canonical, we present the following result. The main benefit to us of the construction of this set of characters, however, will not become apparent until we discuss the interaction between $\mathrm{B}_\pi(G)$ and the the normal structure of G in the next two sections.

Corollary 3.1 *Let G be π-separable. Then the values of the characters in $\mathrm{B}_\pi(G)$ all lie in the cyclotomic field $\mathbb{Q}_m$, where $m = |G|_\pi$, the π-part of the group order.*

Proof. Let $\chi \in \mathrm{B}_\pi(G)$ and write $|G| = n$. We know that $\mathbb{Q}(\chi)$ is contained in the cyclotomic field $\mathbb{Q}_n$, and our task is to show that it is contained in the subfield $\mathbb{Q}_m$. To see this, let σ lie in the Galois group $\mathrm{Gal}(\mathbb{Q}_n/\mathbb{Q}_m)$ and note that $\chi^\sigma \in \mathrm{Irr}(G)$. In fact, because $\chi \in \mathrm{B}_\pi(G)$ and the construction of this set is canonical, it follows that $\chi^\sigma \in \mathrm{B}_\pi(G)$.

Clearly, $\chi(x) \in \mathbb{Q}_m$ for π-elements $x \in G$, and thus $(\chi^\sigma)^0 = (\chi^0)^\sigma = \chi^0$. Since the map $\xi \mapsto \xi^0$ is a bijection from $\mathrm{B}_\pi(G)$ onto $\mathrm{I}_\pi(G)$, it is in particular injective, and we conclude that $\chi = \chi^\sigma$. Since $\sigma \in \mathrm{Gal}(\mathbb{Q}_n/\mathbb{Q}_m)$ was arbitrary, the result follows. $\square$

As an application, we mention a strong form of the Fong-Swan theorem. This result first appeared in [6] and, in fact, it preceded and to some extent motivated the construction of $\mathrm{B}_\pi(G)$. Recall that a character is said to be *p-rational* if its values lie in a field obtained by adjoining to $\mathbb{Q}$ a root of unity having p'-order.

Corollary 3.2 *Let G be p-solvable and suppose that $\varphi \in \mathrm{IBr}(G)$. Then there exists a p-rational character $\chi \in \mathrm{Irr}(G)$ such that $\chi^0 = \varphi$.*

4. Normal Subgroups

Since we are dealing with solvable (or at least π-separable) groups, we generally have an abundance of normal subgroups available, and so it is reasonable to ask how the theory of π-partial characters relates to the normal structure of our group. We know, for example, that if $H \subseteq G$ is an arbitrary subgroup and φ is an irreducible ordinary or π-partial character of G then φ_H can be uniquely decomposed as a sum of irreducible ordinary or π-partial characters of H (its irreducible constituents). If H is normal in G, then, in the case of ordinary characters, Clifford's theorem applies and we can say much more: the irreducible constituents of φ_H constitute a single orbit of the action of G on the characters of H and they all occur with equal multiplicity. Since Clifford's theorem also works for irreducible Brauer characters, it is perhaps not too surprising that there is also an analog for π-partial characters of π-separable groups. The key to the proof of this is a strong interaction between the set $\mathrm{B}_\pi(G)$ and the normal structure of the group. (A full proof of the following result can be found in [9].)

Theorem 4.1 *Let G be π-separable and suppose $N \triangleleft G$. Assume $\chi \in \mathrm{B}_\pi(G)$ and $\psi \in \mathrm{B}_\pi(N)$. The following then hold.*

(a) *Every irreducible constituent of χ_N lies in $\mathrm{B}_\pi(N)$.*

(b) *If G/N is a π-group, then every irreducible constituent of ψ^G lies in $\mathrm{B}_\pi(G)$.*

(c) *If G/N is a π'-group, then exactly one irreducible constituent ξ of ψ^G lies in $\mathrm{B}_\pi(G)$. Furthermore, ξ occurs with multiplicity 1 in ψ^G.*

Of course, parts (b) and (c) of the above result do not describe ψ^G in all possible cases, but, since G is π-separable, any chief series running from N up to G has factors that are either π-groups or π'-groups, and so either (b) or (c) can be applied to each factor, as appropriate. This yields, for example, the following easy corollary.

Corollary 4.2 *Let $N \triangleleft G$, where G is π-separable, and let $\psi \in \mathrm{Irr}(N)$. Then $\psi \in \mathrm{B}_\pi(N)$ iff some irreducible constituent of ψ^G lies in $\mathrm{B}_\pi(G)$.*

Given Theorem 4.1(a) and what we already know, it is not hard to prove 4.1(c), and we give that argument here. We begin with an easy, general lemma.

Lemma 4.3 *Let $H \subseteq G$ and suppose that $\theta \in \mathrm{I}_\pi(H)$. Then there exists $\varphi \in \mathrm{I}_\pi(G)$ such that θ is a constituent of φ_H.*

Proof. Choose $\psi \in \mathrm{Irr}(H)$ with $\psi^0 = \theta$ and choose an (ordinary) irreducible constituent χ of ψ^G. By Frobenius reciprocity, ψ is a constituent of χ_H, and thus $\theta = \psi^0$ is a constituent of $(\chi^0)_H$. It follows that θ is a constituent of φ_H for some irreducible constituent φ of χ^0.   □

In the situation of Lemma 4.3, we shall often say that 'φ lies over θ' or that 'θ lies under φ'. More formally, we write $\mathrm{I}_\pi(G|\theta)$ to denote the (nonempty) set of π-partial characters $\varphi \in \mathrm{I}_\pi(G)$ such that θ is a constituent of φ_H.

Proof of Theorem 4.1(c). Write $\theta = \psi^0$ and observe that $\theta \in \mathrm{I}_\pi(N)$ since $\psi \in \mathrm{B}_\pi(N)$. By Lemma 4.3, choose $\varphi \in \mathrm{I}_\pi(G|\theta)$ and let ξ be the unique member of $\mathrm{B}_\pi(G)$ such that $\xi^0 = \varphi$. Then θ is a constituent of $(\xi_N)^0$ and hence θ is a constituent of η^0 for some ordinary irreducible constituent η of ξ_N. By Theorem 4.1(a), however, we know that η lies in $\mathrm{B}_\pi(N)$, and thus η^0 is irreducible and we have $\eta^0 = \theta$. But also $\psi \in \mathrm{B}_\pi(N)$ and $\psi^0 = \theta$, and it follows that $\eta = \psi$ and ξ is a constituent of $\eta^G = \psi^G$, as desired.

We need to show that ξ is the unique irreducible constituent of ψ^G that lies in $\mathrm{B}_\pi(G)$ and that it occurs with multiplicity 1 as a constituent of ψ^G. For this purpose, choose a Hall π-subgroup $H \subseteq N$ and note that H is actually a Hall π-subgroup of G since we are assuming that G/N is a π'-group. Let $\alpha \in \mathrm{Irr}(H)$ be a Fong character belonging to θ and observe that

$$\alpha(1) = \theta(1)_\pi = \psi(1)_\pi = \xi(1)_\pi = \varphi(1)_\pi \,,$$

where the penultimate equality holds because G/N is a π'-group, and so $\xi(1)/\psi(1)$ is a π'-number. (See Corollary 11.29 of [15], for example.) We see now that α is an irreducible constituent of φ_H that has the right degree to be a Fong character belonging to φ, and thus α is a Fong character belonging to φ. Since α uniquely determines φ and φ uniquely determines ξ, it follows that ξ is unique. (Note that, in general, α is *not* uniquely determined, but this does not matter because any choice of α determines the same character $\xi \in \mathrm{B}_\pi(G)$.) Finally, we know that α occurs with multiplicity 1 as a constituent of φ_H, and thus θ occurs with multiplicity 1 as a constituent of φ_N. It follows that $1 = [\xi_N, \psi] = [\xi, \psi^G]$, as required.   □

Sometimes, Theorem 4.1 is powerful enough to determine completely the set $\mathrm{B}_\pi(G)$. Consider, for example, the case where $G = \mathrm{Sym}_4$, the symmetric group on four letters. The only interesting cases for this group are $\pi = \{2\}$ and $\pi = \{3\}$, and we shall consider these separately. First, we 'construct' $\mathrm{Irr}(G)$.

The unique chief series for G is $1 < K < A < G$, where K has order 4 and A is the alternating group of order 12. The trivial character of 1 lies under four linear characters of K. One of these is the trivial character of K, which lies under exactly three linear characters of A, one of which is trivial. The other three linear characters of K all lie under a unique irreducible character θ of A of degree 3. At this point we have mentioned all irreducible characters of A and we work up to G. The trivial character of A lies under two linear characters of G; the other two linear characters of A lie under a single degree 2 irreducible character ψ of G and θ lies under exactly two irreducible characters α and β of G, each having degree 3.

Suppose now that $\pi = \{2\}$. The trivial character of the trivial subgroup lies in $\mathrm{B}_\pi(1)$, and so, by Theorem 4.1(b), all four irreducible characters of K lie in $\mathrm{B}_\pi(K)$.

By Theorem 4.1(c) it follows that $\theta \in \mathrm{B}_\pi(A)$ and that exactly one of the three linear characters of A lies in $\mathrm{B}_\pi(A)$, and this must be the trivial character, which always lies in $\mathrm{B}_\pi(\ )$. Finally, since G/A is a π-group, we deduce that the two linear characters of G lie in $\mathrm{B}_\pi(G)$, as do both α and β. As a check, we count that $|\mathrm{B}_\pi(G)| = 4$, and this is correct since, in general, $|\mathrm{B}_\pi(G)| = |\mathrm{I}_\pi(G)|$ is the number of π-classes of G.

Next, we work through the case where $\pi = \{3\}$. Again, the trivial character lies in $\mathrm{B}_\pi(1)$, and so, by Theorem 4.1(c), exactly one of the four linear characters of K lies in $\mathrm{B}_\pi(K)$, and this must be the principal character. It follows that $\mathrm{B}_\pi(A)$ consists of the three linear characters of A and hence $\mathrm{B}_\pi(G)$ has exactly two members: ψ and one of the two linear characters, which is necessarily the principal character. As a check, we observe that G has exactly two π-classes.

Our calculations have shown that only the principal character of $G = \mathrm{Sym}_4$ lies in both $\mathrm{B}_2(G)$ and $\mathrm{B}_3(G)$. In fact, it is generally true that $\mathrm{B}_\pi(G) \cap \mathrm{B}_{\pi'}(G) = \{1_G\}$, and we leave this as an exercise. (This assertion can be proved using only the results we have already stated, without knowing the actual definition of $\mathrm{B}_\pi(G)$, which we have suppressed.)

5. Clifford Theory

As was promised in the previous section, we present an analog of Clifford's theorem for π-partial characters.

Corollary 5.1 *Let $N \triangleleft G$, where G is π-separable, and suppose $\varphi \in \mathrm{I}_\pi(G)$. Then, for some positive integer e and distinct π-partial characters $\theta_i \in \mathrm{I}_\pi(N)$, we have $\varphi_N = e \sum \theta_i$, where the constituents θ_i constitute a single orbit of the action of G on $\mathrm{I}_\pi(N)$. Furthermore, e divides $|G : N|$ and $e = 1$ if G/N is a π'-group.*

Proof. Let $\chi \in \mathrm{B}_\pi(G)$ with $\chi^0 = \varphi$. By Clifford's theorem for ordinary characters, we can write $\chi_N = e \sum \psi_i$, where the irreducible constituents ψ_i are distinct and constitute a single orbit of the action of G on $\mathrm{Irr}(G)$. Also, the theory of projective representations (see Corollary 11.29 of [15], for example) guarantees that the integer e divides $|G : N|$.

By Theorem 4.1, we have $\psi_i \in \mathrm{B}_\pi(N)$ for each subscript i, and thus the π-partial characters $\theta_i = (\psi_i)^0$ are irreducible and distinct. Since G acts transitively on the characters ψ_i, it is clear that it also acts transitively on the θ_i. Finally, if G/N is a π'-group, Theorem 4.1(c) gives $e = 1$, and the proof is complete. $\square$

But there is more to Clifford theory than this. The key observation for ordinary characters is the following 'Clifford correspondence', as given in Theorem 6.11 of [15]. Suppose $N \triangleleft G$ and $\psi \in \mathrm{Irr}(N)$ and let T be the stabilizer in G of ψ. Then character induction defines a bijection from the set $\mathrm{Irr}(T|\psi)$ onto the set $\mathrm{Irr}(G|\psi)$. Furthermore, if $\eta \in \mathrm{Irr}(T|\psi)$ and $\eta^G = \chi$, then $[\chi_N, \psi] = [\eta_N, \psi]$. We shall see that it is not too hard to prove an analog of the Clifford correspondence for π-partial characters. (This appeared first in [11].)

We need to define induction of π-partial characters first. If $H \subseteq G$ and $\gamma \in \mathrm{cf}(H)$, we know that the induced class function $\gamma^G \in \mathrm{cf}(G)$ is defined by the simple formula

$$\gamma^G(x) = \frac{1}{|H|} \sum_{g \in G} \gamma^*(gxg^{-1}),$$

where γ^* is defined so that it agrees with γ on H and vanishes on $G - H$. We can use exactly the same formula to define the induction map $\mathrm{cf}(H^0) \to \mathrm{cf}(G^0)$, and it is clear that, with this definition, we have $(\gamma^G)^0 = (\gamma^0)^G$ for all class functions γ of H. It is then immediate that θ^G is a π-partial character of G whenever θ is a π-partial character of H. (We mention that in the Brauer case, where $\pi = p'$, it is true, but not a triviality, that the Brauer character corresponding to an induced module is the induced Brauer character.)

It is useful to note that if $\theta^G \in \mathrm{I}_\pi(G)$ then necessarily $\theta \in \mathrm{I}_\pi(H)$. As a consequence, we see that if $\chi \in \mathrm{Irr}(G)$ and it happens that χ^0 is irreducible (for instance, if $\chi \in \mathrm{B}_\pi(G)$) then, whenever we can write $\chi = \psi^G$ for some character ψ of a subgroup $H \subseteq G$, we can conclude that ψ^0 is irreducible. Even if $\chi \in \mathrm{B}_\pi(G)$, however, it is not necessarily true in this situation that $\psi \in \mathrm{B}_\pi(H)$.

Some caution should be exercised because Frobenius reciprocity does not usually hold for π-partial characters, even in solvable groups. If $H \subseteq G$ and θ and φ are respectively irreducible π-partial characters of H and of G, it is not necessarily true that θ is a constituent of φ_H iff φ is a constituent of θ^G. In fact, neither of these two implications is true in general. The situation is not completely out of control, however, because of the following simple observation.

Lemma 5.2 *Let $H \subseteq G$ and suppose $\theta \in \mathrm{I}_\pi(H)$ and $\varphi \in \mathrm{I}_\pi(G)$.*

(a) *If $\theta^G = \varphi$, then θ is a constituent of φ_H.*

(b) *If $\varphi_H = \theta$, then φ is a constituent of θ^G.*

Proof. We can make compatible choices of ordinary characters $\psi \in \mathrm{Irr}(H)$ and $\chi \in \mathrm{Irr}(G)$ such that $\psi^0 = \theta$ and $\chi^0 = \varphi$. In the situation of (a), choose ψ first and take $\chi = \psi^G$ and, for (b), choose χ first and take $\psi = \chi_H$. Since ψ is a constituent of χ_H, assertion (a) follows and, similarly, (b) is a consequence of the fact that χ is a constituent of ψ^G. $\square$

Theorem 5.3 *Let $\theta \in \mathrm{I}_\pi(N)$, where $N \triangleleft G$ and G is π-separable, and let T be the stabilizer of θ in G. Then induction defines a bijection from $\mathrm{I}_\pi(T|\theta)$ onto $\mathrm{I}_\pi(G|\theta)$. Furthermore, if $\mu \in \mathrm{I}_\pi(T|\theta)$ and $\mu^G = \varphi$, then*

(a) *μ is the unique irreducible constituent of φ_T that lies over θ, and*

(b) *the multiplicities of θ in μ_N and in φ_N are equal.*

Proof. Given $\varphi \in \mathrm{I}_\pi(G|\theta)$, we first show that we can write $\varphi = \mu^G$, where μ satisfies (a) and (b). (Note that the uniqueness assertion in (a) is an immediate consequence of (b), and so it need not be proved separately.)

Choose $\chi \in \mathrm{B}_\pi(G)$ such that $\chi^0 = \varphi$. By Theorem 4.1, all irreducible constituents of χ_N lie in $\mathrm{B}_\pi(N)$, and it follows that, for some unique irreducible constituent ψ

of χ_N, we have $\psi^0 = \theta$ and the multiplicity e of ψ as a constituent of χ_N is exactly equal to the multiplicity of θ as a constituent of φ_N. Also, by the uniqueness of ψ, we see that T is exactly the stabilizer in G of ψ.

By the Clifford correspondence (for ordinary characters), there exists $\eta \in \mathrm{Irr}(T)$ such that $\eta^G = \chi$ and $\eta_N = e\psi$. Writing $\eta^0 = \mu$, we see that $\mu^G = (\eta^G)^0 = \varphi$, and so $\mu \in \mathrm{I}_\pi(T)$. Since η is a constituent of χ_T, it follows that μ is a constituent of φ_T, as required in (a). Also, $\eta_N = e\psi$, and thus $\mu_N = e\theta$ and it follows that $\mu \in \mathrm{I}_\pi(T|\theta)$ and that the multiplicity of θ as a constituent of μ_N is e, which is also the multiplicity of θ as a constituent of φ_N. This proves (b) and the rest of (a) and the assertion of the first paragraph is proved.

We can now construct our bijection. Given $\mu \in \mathrm{I}_\pi(T|\theta)$, let $\varphi \in \mathrm{I}_\pi(G|\mu)$. (This is possible by Lemma 4.3.) We have just shown that φ is induced from the unique irreducible constituent of φ_T that lies over θ. But μ lies under φ and over θ and we conclude that $\varphi = \mu^G$. Induction does, therefore, define a map $\mathrm{I}_\pi(T|\theta) \to \mathrm{I}_\pi(G|\theta)$ and we know by the first paragraph that this map is surjective.

We continue to assume that μ and φ are as above, and, in particular, that $\mu^G = \varphi$. If also $\nu \in \mathrm{I}_\pi(T|\theta)$ and $\nu^G = \varphi$, then, by Lemma 5.2, we know that ν lies below φ. Since ν also lies above θ, the uniqueness in (a) guarantees that $\mu = \nu$, and hence the induction map is injective. $\qquad\square$

When we try to prove a theorem by induction on the group order, it would be convenient if we were able to establish that a character $\chi \in \mathrm{Irr}(G)$ can be obtained by induction from a proper subgroup. The Clifford correspondence guarantees that this will occur if there exists a normal subgroup N of G such that χ_N is not homogeneous (a multiple of a single irreducible character of N). Unfortunately, there is no easily checked general criterion sufficient to establish the existence of such a normal subgroup. The situation is better, however, for π-partial characters of π-separable groups, where we have the following useful tool.

Lemma 5.4   Let $\varphi \in \mathrm{I}_\pi(G)$, where G is π-separable. Assume that $\varphi(1)$ is not a π-number and choose $N \triangleleft G$ maximal with the property that φ_N has an irreducible constituent θ such that $\theta(1)$ is a π-number. (Note that N necessarily exists.) Let T be the stabilizer of θ in G. Then $T < G$ and, in fact, $|G : T|$ is not a π-number.

Proof.   Assume that $|G : T|$ is a π-number. Since $\varphi(1)$ is not a π-number, we know that $N < G$ and we can choose a chief factor M/N of G. Let γ be an irreducible constituent of φ_M that lies over θ and observe that, by the maximality of N, the degree $\gamma(1)$ cannot be a π-number.

By Corollary 5.1, however, $\gamma(1) = et\theta(1)$, where $e \mid |M : N|$ and $t = |M : M \cap T|$ is the size of the orbit of θ under the action of M. Since $\theta(1)$ is a π-number and $t = |MT : T|$ divides the π-number $|G : T|$, we deduce that e is not a π-number, and hence M/N is not a π-group. It follows that M/N is a π'-group and $e = 1$ is a π-number by Corollary 5.1. This is the desired contradiction. $\qquad\square$

Corollary 5.5   Let $\varphi \in \mathrm{I}_\pi(G)$, where G is π-separable. Then there exist a subgroup $H \subseteq G$ and a π-partial character $\nu \in \mathrm{I}_\pi(H)$ such that $\nu(1)$ is a π-number and $\nu^G = \varphi$.

Proof. We can assume that $\varphi(1)$ is not a π-number, or else we can simply take $H = G$ and $\nu = \varphi$. It follows by Lemma 5.4 and the Clifford correspondence (Theorem 5.3) that we can write $\varphi = \mu^G$, where $\mu \in I_\pi(T)$ and $T < G$. Working by induction on $|G|$, we see that the inductive hypothesis applied to T completes the proof. $\square$

An immediate (known) application is the following.

Corollary 5.6   *Let G be p-solvable and suppose $\varphi \in \mathrm{IBr}(G)$. Then φ is induced from some Brauer character with p'-degree.*

6. Vertices

In Corollary 5.5, we showed that if G is π-separable and $\varphi \in I_\pi(G)$ then it is always possible to find a subgroup $H \subseteq G$ and a π-partial character $\nu \in I_\pi(H)$ such that $\varphi = \nu^G$ and $\nu(1)$ is a π-number. We shall say in this situation that the ordered pair (H, ν) is a π-*inducing pair* in G, and that it *belongs* to φ. In general, the pair (H, ν) is not uniquely determined by φ and it is not even unique up to conjugacy. (Note that G really does act by conjugation on the set of all π-inducing pairs belonging to φ, but this action is not, in general, transitive.) We ask if there is anything at all that we can say in the direction of uniqueness.

Suppose (H, ν) is a π-inducing pair belonging to φ. Since $\varphi(1) = \nu(1)|G : H|$ and $\nu(1)$ is a π-number, we see that $\varphi(1)_{\pi'} = |G : H|_{\pi'}$, and we conclude that the π'-part of the order of $|H|$ is uniquely determined by φ. If Q is a Hall π'-subgroup of H, therefore, it follows that $|Q|$ is uniquely determined.

In order to see what more we might hope to say, we return for motivation to the 'classical' Brauer case where $\pi = p'$ and Q is a Sylow p-group of H. In this situation, $\varphi \in \mathrm{IBr}(G)$, and so φ corresponds to some simple FG-module M, where F is an algebraically closed field of characteristic p. We know, furthermore, that the module M is induced from some (necessarily simple) FH-module V and that $\dim_F(V) = \nu(1)$ is not divisible by p. We can now apply the beautiful theory of vertices developed by J. A. Green.

If X is an indecomposable FG-module (where G is an arbitrary finite group), Green showed that X determines a unique conjugacy class of p-subgroups of G, referred to as the 'vertices' for X. Also, if X is induced from some FH-module Y, where $H \subseteq G$, then every vertex for Y is automatically a vertex for X. Green also proved that if P is a vertex for X then $(\dim_F(X))_p \geqslant |G : P|_p$.

Applying all of this to our simple (and therefore indecomposable) FG-module $M = V^G$, where G is p-solvable, we see that a vertex $P \subseteq H$ for V is a vertex for M and that $|H : P|_p \leqslant (\dim_F(V))_p = \nu(1)_p = 1$. In other words, $P \in \mathrm{Syl}_p(H)$. We have shown, therefore, that in the case where $\pi = p'$ the mysterious π-complement Q of H turns out to be a vertex for M, and thus it is uniquely determined by φ up to conjugacy in G.

By analogy with the case where $\pi = p'$, we shall say that a π'-subgroup $Q \subseteq G$ is a *vertex* for $\varphi \in I_\pi(G)$ provided that Q is a Hall π'-subgroup of H for some π-inducing pair (H, ν) belonging to φ. Given $\varphi \in I_\pi(G)$, where G is π-separable, we know by Corollary 5.5 that vertices for φ always exist, but for general prime sets π

we certainly cannot hope to use Green's theory to prove that they are all conjugate. Nevertheless, it is true that these vertices are conjugate. A proof of this appears in [18], and we shall present a slight variation on that argument here.

Theorem 6.1   Let $\varphi \in I_\pi(G)$, where G is π-separable. Then all vertices for φ are conjugate in G.

To prove Theorem 6.1, we construct an undirected graph $\mathcal{G} = \mathcal{G}(\varphi)$ whose vertex set is the collection of all π-inducing pairs belonging to φ. We join two such pairs in $\mathcal{G}$ if either is properly 'contained' in the other. More precisely, we join the distinct π-inducing pairs (U, μ) and (V, ν) if either $U \subseteq V$ and $\mu^V = \nu$ or if $V \subseteq U$ and $\nu^U = \mu$. In the first case, where $\mu^V = \nu$, we note that the index $|V : U|$ divides $\nu(1)$ and is therefore a π-number. Similarly, when $\nu^U = \mu$, we conclude that $|U : V|$ is a π-number.

Assume, as usual, that G is π-separable. Suppose (V, ν) and (U, μ) are adjacent nodes in $\mathcal{G}$ and choose Hall π'-subgroups Q of U and R of V. In the case where $U \subseteq V$ and $|V : U|$ is a π-number, we see that Q is a Hall π'-subgroup of V and it is thus conjugate to R in V. Otherwise, $V \subseteq U$ and R is a Hall π'-subgroup of U, and thus R is conjugate to Q in U. In either case, therefore, R and Q are conjugate in G.

It follows from this that if π-inducing pairs (U, μ) and (V, ν) are in the same connected component of the graph $\mathcal{G}(\varphi)$ then the Hall π'-subgroups of U and of V are all conjugate in G. The same conclusion obviously continues to hold under the weaker assumption that the connected components of (U, μ) and (V, ν) in $\mathcal{G}$ are conjugate under the action of G. (Note that the conjugation action of G on the set of π-inducing pairs belonging to $\varphi \in I_\pi(G)$ induces graph automorphisms of $\mathcal{G}(\varphi)$, and it follows that G permutes the connected components of this graph.) It should now be clear that to prove Theorem 6.1 it suffices to establish the following. As we shall see, this result also has other applications.

Theorem 6.2   Let G be π-separable and suppose $\varphi \in I_\pi(G)$. Then the connected components of the graph $\mathcal{G}(\varphi)$ are all conjugate in G.

We need the following somewhat technical lemma. We shall assume it temporarily and proceed with the proof of Theorem 6.2.

Lemma 6.3   Let $U \subseteq G$, where G is π-separable, and suppose $\mu \in I_\pi(U)$ and $\mu^G = \varphi \in I_\pi(G)$. If $N \triangleleft G$ and the irreducible constituents of φ_N have π-degree, then $|NU : U|$ is a π-number.

Proof of Theorem 6.2.   If $\varphi(1)$ is a π-number, then (G, φ) is a π-inducing pair belonging to φ and it is joined to every node in the graph $\mathcal{G}(\varphi)$. In this case there is nothing further to prove since the graph has just one connected component.

We can now assume that $\varphi(1)$ is not a π-number. By Lemma 5.4 there exists a subgroup $N \triangleleft G$ and an irreducible constituent θ of φ_N such that $\theta(1)$ is a π-number and the stabilizer T of θ is a proper subgroup of G. Also, by Theorem 5.3 (the

Clifford correspondence) there is a unique π-partial character $\xi \in I_\pi(T|\theta)$ such that $\xi^G = \varphi$.

We shall show by induction on $|G|$ that the connected components of any two given π-inducing pairs (U, μ) and (V, ν) belonging to φ are conjugate. By Lemma 6.3 we know that $|NU : U|$ is a π-number, and thus $\mu^{NU}(1)$ is a π-number and we have $(\mu^{NU})^G = \varphi$. It follows that (NU, μ^{NU}) is a π-inducing pair belonging to φ, and, furthermore, it is joined to (U, μ) in the graph $\mathcal{G}(\varphi)$. It does not change the connected component, therefore, if we replace the given pair (U, μ) by the pair (NU, μ^{NU}). We can thus assume that $N \subseteq U$ and similarly that $N \subseteq V$.

By Corollary 5.1, all irreducible constituents of φ_N are conjugate to θ in G. Since $\mu^G = \varphi$, it follows from Lemma 5.2 that φ lies over μ, and thus μ lies over some G-conjugate of θ. If we replace the pair (U, μ) by a suitable conjugate, therefore, we can assume that μ lies over θ and, similarly, we can assume that ν lies over θ.

By the Clifford correspondence we can write $\mu = \eta^U$ for some π-partial character $\eta \in I_\pi(U \cap T)$, where η lies over θ. It follows that $(U \cap T, \eta)$ is also a π-inducing pair belonging to φ and that it is joined to (U, μ) in the graph $\mathcal{G}$. It is no loss, therefore, to replace (U, μ) by $(U \cap T, \eta)$, and thus we can assume that $U \subseteq T$ and, similarly, that $V \subseteq T$.

Next, we observe that $(\mu^T)^G = \varphi$, and so μ^T is irreducible and it lies over μ by Lemma 5.2. It follows that μ^T lies over θ and we deduce that $\mu^T = \xi$, the Clifford correspondent of φ with respect to θ. Similarly $\nu^T = \xi$, and so each of the pairs (U, μ) and (V, ν) is a π-inducing pair belonging to ξ in T.

Because $T < G$, the inductive hypothesis in T guarantees that we can replace (U, μ) by a suitable T-conjugate and assume that (U, μ) and (V, ν) lie in the same connected component of the graph $\mathcal{G}(\xi)$. It is clear, however, that $\mathcal{G}(\xi)$ is a subgraph of $\mathcal{G}(\varphi)$, and the result follows. $\square$

We work now toward a proof of Lemma 6.3.

Lemma 6.4   *Let G be π-separable and suppose that $G = NV$, where $N \lhd G$ and $|G : V|$ is a π'-number. Suppose that ν is a π-partial character of V and that ν^G is irreducible. If $(\nu^G)_N$ has an irreducible constituent having π-degree, then $V = G$.*

Proof.   Write $\nu^G = \varphi$ and observe that it is no loss to assume that N is maximal among normal subgroups of G such that φ_N has an irreducible constituent with π-degree. Since $|G : V|$ is a π'-number dividing $\varphi(1)$, we can assume that $\varphi(1)$ is not a π-number, and thus, by Lemma 5.4, the stabilizer in G of each irreducible constituent of φ_N is proper.

Write $U = N \cap V$ and let μ be an irreducible constituent of ν_U. Since φ lies over ν by Lemma 5.2, it also lies over μ, and thus some irreducible constituent θ of φ_N must also lie over μ.

Let H be a Hall π-subgroup of U and observe that $|N : U| = |G : V|$ is a π'-number, so that H is actually a Hall π-subgroup of N. Let $\alpha \in \mathrm{Irr}(H)$ be a Fong character belonging to θ. Since $\theta(1)$ is a π-number, Theorem 2.1(c) tells us that $\theta(1) = \alpha(1)$, and thus $\theta_H = \alpha$. It follows that μ lies over α and, since θ is the unique irreducible π-partial character of N that lies over α, it follows that θ is also unique over μ. We deduce that the stabilizer S of μ in V also stabilizes θ. Since we know that the stabilizer of θ is proper in G, we conclude that $NS < G$.

Let $\sigma \in I_\pi(S)$ be the Clifford correspondent for ν with respect to μ. Write $\tau = \sigma^{NS}$ and observe that $\tau^G = \sigma^G = \nu^G = \varphi$ is irreducible. We conclude that τ is irreducible and that φ lies over τ. The irreducible constituents of τ_N, therefore, lie among the irreducible constituents of φ_N and, in particular, they have π-degree. Since $NS < G$, we can work by induction on the group order and conclude via the inductive hypothesis applied to NS (with S and σ in the roles of V and ν) that $S = NS$. Thus $N \subseteq S \subseteq V$, and it follows that $V = NV = G$, as required. $\square$

Finally, we can prove Lemma 6.3 and thereby complete the proofs of Theorems 6.1 and 6.2.

Proof of Lemma 6.3. Since the result is trivial if $N = 1$, we can work by induction on $|N|$ and assume that $N > 1$. Choose $M \triangleleft G$ such that N/M is a chief factor of G and observe that an irreducible constituent of φ_M has degree dividing that of an irreducible constituent of φ_N, and so this degree is a π-number. By the inductive hypothesis, we know that $|MU : U|$ is a π-number, and so it suffices to show that $|NU : MU|$ is also a π-number. But $|NU : MU|$ divides the order of the chief factor N/M and thus it is either a π-number or a π'-number. We can thus assume that $|NU : MU|$ is a π'-number and we work to show that $NU = MU$.

Write $H = NU$ and $V = MU$ and observe that $NV = H$ and $|H : V|$ is a π'-number. Let $\nu = \mu^V$ and $\eta = \nu^H$ and observe that since $\nu^G = \varphi$ is irreducible, both ν and η are irreducible. Also, since $\eta^G = \varphi$, we know that φ lies over η, and thus the irreducible constituents of η_N lie among those of φ_N. These irreducible constituents have π-degree, and so Lemma 6.4 applies to H. We conclude that $V = H$, as required. $\square$

There is more that can be said about these π-vertices in π-separable groups, and we state a few results without proof. The first of these can be found in [18].

Theorem 6.5 *Let G be π-separable and suppose $Q \subseteq G$ is a vertex for $\varphi \in I_\pi(G)$. Then there exist Hall π'-subgroups X and Y of G such that $X \cap Y = Q$.*

We mentioned earlier that it is possible (for π-separable groups G) to construct a theory of π-blocks parallel to Brauer's development of p-blocks. (Actually, we should call these π'-blocks, since the Brauer case occurs when $\pi = p'$.) These blocks, studied by M. Slattery in [21] and in later papers, are sets consisting of irreducible ordinary and π-partial characters of G. Slattery associated with each block B a certain (unique up to conjugacy) π'-subgroup $D \subseteq G$ called a 'defect group' for B, and he showed that all of the characters and π-partial characters in B have degrees divisible by $|G : D|_{\pi'}$. Furthermore, B contains a 'height zero' π-partial character φ and it is easy to see from Slattery's work that the vertices of φ are exactly the defect groups of B. In particular, it follows from Theorem 6.5 that each defect group can be written as an intersection of two Hall π'-subgroups.

The classical fact that a defect group for a p-block is always an intersection of two Sylow p-subgroups is a well known result of Green and holds for all finite groups. Its proof is highly module theoretic. Our discussion here, applied to the case where

$\pi = p'$, yields a character theoretic, module-free proof of Green's result for p-solvable groups.

Finally, we mention another result from [18].

Theorem 6.6   *Let G be π-separable and assume that the Hall π'-subgroups of G are nilpotent. Let $Q \subseteq G$ be any π'-subgroup and write $N = N_G(Q)$. Then the numbers of π-partial characters in $\mathrm{I}_\pi(G)$ and in $\mathrm{I}_\pi(N)$ that have vertex Q are equal.*

Of course, in the classical case where $\pi = p'$, the nilpotence assumption in Theorem 6.6 holds automatically, and we deduce that the following strong form of Alperin's weight conjecture holds for p-solvable groups.

Corollary 6.7   *Let $P \subseteq G$, where P is a p-group and G is p-solvable, and write $N = N_G(P)$. Suppose that F is an algebraically closed field of characteristic p. Then the numbers of isomorphism types of simple FG-modules having vertex P and simple FN-modules having vertex P are equal.*

Corollary 6.7 definitely would not continue to hold if the hypothesis that G is p-solvable were dropped. Alperin's conjecture, however, asserts that if we let P run over a set of representatives of all conjugacy classes of p-subgroups of an arbitrary finite group G and we sum the numbers of simple FG-modules having vertex P and the numbers of simple FN-modules having vertex P then these two sums should be equal. To date, this conjecture remains unproved.

7. M-groups

It seems that some of the most intractable problems in solvable group character theory concern M-groups. It has been possible to make a little progress in this area using π-theory, however, and it is the purpose of this section to describe one of the outstanding problems and to show in some detail how π-theory is relevant.

We begin with a brief review. An irreducible character χ of a finite group G is said to be *monomial* if $\chi = \lambda^G$ for some linear character λ of some subgroup of G and G is an *M-group* or a *monomial group* if every character $\chi \in \mathrm{Irr}(G)$ is monomial. The fundamental result in the theory of these groups is due to K. Taketa who showed that M-groups are necessarily solvable. (See Corollary 5.13 of [15].) What seems to make the subject so difficult is that, except for the obvious consequence of Taketa's theorem that every subgroup of an M-group must be solvable, there are absolutely no other conclusions possible about the structure of an arbitrary subgroup of an M-group. It was proved by E. C. Dade that *every* finite solvable group is a subgroup of some M-group. (See [1].)

Given a solvable group H, Dade's construction (which proceeds via wreath products) results in a monomial supergroup $G \supseteq H$ such that all prime divisors of the index $|G : H|$ are already divisors of $|H|$. This suggests the question of whether or not the opposite extreme can occur: if H is not itself an M-group, is it possible to embed it in an M-group G so that no prime divisor of $|G : H|$ divides $|H|$? Equivalently, we are asking whether or not Hall subgroups of M-groups must be M-groups. This question remains unresolved.

Recall that an irreducible character $\chi \in \mathrm{Irr}(G)$ is said to be *primitive* if there is no proper subgroup $H < G$ and character $\psi \in \mathrm{Irr}(H)$ such that $\psi^G = \chi$. In some sense, primitivity is the opposite of monomiality and it should be clear that the only way that an irreducible character can be both primitive and monomial is that it be linear. If the conjecture that Hall subgroups of M-groups must be M-groups is really true, then the following would be a trivial corollary. The fact that it is possible to prove this theorem directly might be viewed as a bit of evidence in favor of the conjecture.

Theorem 7.1 *Let H be a Hall subgroup of an M-group G. Then every primitive character of H is linear.*

The first proof of this result appeared in [11], but the argument we give here is a variation on this that was inspired by some work of G. Navarro in [19]. The first part of the argument seems interesting in its own right, and we state it in a form somewhat more general than we need. In particular, we shall assume 'quasiprimitivity' in place of 'primitivity', and so we digress briefly to discuss this.

A character $\chi \in \mathrm{Irr}(G)$ is said to be *quasiprimitive* if for every normal subgroup $N \triangleleft G$, the restriction χ_N is a multiple of a single irreducible character of N. It should be clear from the Clifford correspondence that a primitive character is necessarily quasiprimitive, since otherwise it would be induced from a proper stabilizer of some irreducible character of a normal subgroup. Generally, quasiprimitivity is a proper generalization of primitivity, but a result of T. R. Berger (see Theorem 11.33 of [15]) asserts that if G is solvable then every quasiprimitive character of G is primitive.

Theorem 7.2 *Let H be a Hall π-subgroup of a π-separable group G and suppose $\alpha \in \mathrm{Irr}(H)$ is quasiprimitive. Then there exist a subgroup U with $H \subseteq U \subseteq G$ and a π-partial character $\mu \in \mathrm{I}_\pi(U)$ such that $\mu_H = \alpha$ and μ^G is irreducible. In this situation, α is a Fong character belonging to μ^G.*

As will be apparent from the proof of Theorem 7.2, the assumption that α is quasiprimitive actually is more natural (as well as slightly more general) than a primitivity assumption would have been. When we apply Theorem 7.2, however, this becomes a distinction without a difference. This is because in that case G will be an M-group, and so its subgroup H will be solvable and Berger's theorem applies.

Proof of Theorem 7.2. Choose $\varphi \in \mathrm{I}_\pi(G)$ lying over α and observe that if $\varphi(1)$ is a π-number then a Fong constituent of φ_H has degree equal to $\varphi(1)_\pi = \varphi(1)$, and so φ_H is irreducible. In this case we can take $U = G$ and $\mu = \varphi$ and there is nothing further to prove.

Assume now that $\varphi(1)$ is not a π-number. Then Lemma 5.4 applies and we can choose a subgroup $N \triangleleft G$ such that the irreducible constituents of φ_N have π-degree and have stabilizers that are proper in G.

Write $K = N \cap H$ and let β be an irreducible constituent of α_K. Since φ lies over α it also lies over β, and so some irreducible constituent θ of φ_N lies over β. Since K is a Hall π-subgroup of N and $\theta(1)$ is a π-number, we can reason as we did in the first paragraph of the proof and deduce that θ_K is irreducible, and hence

$\theta_K = \beta$. It follows that β is a Fong character belonging to θ and thus θ is the unique irreducible π-partial character of N that lies over β. This uniqueness guarantees that the stabilizer of β in H stabilizes θ.

Since α is quasiprimitive and $K \lhd H$, we know that $\alpha_K = e\beta$ for some integer e. It follows that β is invariant in H, and thus H is contained in the stabilizer T of θ in G.

We know that $T < G$, and so we can work by induction on the order of G and apply the inductive hypothesis to T. We conclude, therefore, that there is a subgroup U with $H \subseteq U \subseteq T$ and a π-partial character $\mu \in I_\pi(U)$ such that $\mu_H = \alpha$ and μ^T is irreducible. Writing $\eta = \mu^T$, we see by Lemma 5.2 that η lies over μ, and thus it lies over both α and β and it follows that some irreducible constituent of η_N lies over β. But we know that θ is the unique member of $I_\pi(N)$ that lies over β, and we deduce that $\eta \in I_\pi(T|\theta)$. By the Clifford correspondence (Theorem 5.3), we conclude that $\mu^G = \eta^G$ is irreducible, as required.

Finally, writing $\varphi = \mu^G$, we see that φ lies over μ, and hence it lies over α. Also, $\varphi(1) = |G : U|\mu(1)$. But $|G : U|$ is a π'-number and $\mu(1) = \alpha(1)$ is a π-number, and so $\varphi(1)_\pi = \alpha(1)$. It follows that α is a Fong character belonging to φ, as asserted. $\square$

We shall need the following lemma, which is an immediate consequence of the Mackey decomposition theorem.

Lemma 7.3 Let $H \subseteq G$ and suppose $\alpha \in \mathrm{Irr}(H)$ is primitive. If $\chi \in \mathrm{Irr}(G)$ and $\chi_H = \alpha$, then χ is primitive.

Proof of Theorem 7.1. Let π be the set of prime divisors of $|H|$ and apply Theorem 7.2 to G to produce a pair (U, μ), where $H \subseteq U \subseteq G$ and $\mu \in I_\pi(U)$ satisfies $\mu_H = \alpha$ and $\mu^G = \varphi$ for some irreducible π-partial character φ of G. In particular, in the language of the previous section, (U, μ) is a π-inducing pair belonging to φ.

Suppose that some pair (V, ν) is joined to (U, μ) in the graph $\mathcal{G}(\varphi)$. By definition, there are two possibilities: either $U \subseteq V$ and $\mu^V = \nu$ or else $V \subseteq U$ and $\nu^U = \mu$. Also, since the pairs (U, μ) and (V, ν) must be distinct, it follows that $U \neq V$.

In the first case, where $U < V$, we know that $|V : U|$ divides $\nu(1)$ and so this index is a nontrivial π-number. This is impossible, however, because U contains the Hall π-subgroup H of G. In the other case, we have $V < U$ and $\nu^U = \mu$. Choose $\psi \in \mathrm{Irr}(V)$ with $\psi^0 = \nu$ and observe that $(\psi^U)_H = \alpha$. Since α is primitive, it follows from Lemma 7.3 that $V = U$, and this is a contradiction. There is no pair, therefore, joined to (U, μ).

We conclude that the connected component of $\mathcal{G}(\varphi)$ containing (U, μ) consists of just that one node. By Theorem 6.2, therefore, every π-inducing pair (V, ν) belonging to φ is G-conjugate to (U, μ). In particular, it follows that $\mu(1) = \nu(1)$ whenever $\nu^G = \varphi$ and $\nu(1)$ is a π-number.

Now choose $\chi \in \mathrm{Irr}(G)$ such that $\chi^0 = \varphi$. Since G is an M-group, χ is monomial and we can write $\chi = \lambda^G$ for some linear character λ of some subgroup $V \subseteq G$. Let $\nu = \lambda^0$ and observe that $\nu^G = \varphi$ and that $\nu(1) = 1$ is a π-number. It follows that $\alpha(1) = \mu(1) = \nu(1) = 1$, as required. $\square$

We close this section by mentioning a similar result related to one of the other major open problems concerning M-groups. It was asked whether or not normal subgroups of M-groups had to be monomial and this was answered in the negative by Dade, who constructed an M-group of order $2^9 \cdot 7$ having a non-M-subgroup of index 2. (See [2] for this clever example, but note that there is an error in this paper that Dade corrected in [3]. See also [12] for information about this type of construction.) The prime 2 plays a crucial role in Dade's example, and this suggests the question of whether or not a normal subgroup of an odd-order M-group must be monomial. This remains open, but we note that, if it were true, the following theorem (proved in [10]) would be an immediate consequence.

Theorem 7.4 *Let G be an M-group of odd order and suppose that H is a subnormal subgroup of G. Then every primitive irreducible character of H is linear.*

8. Odd Order Groups

As we suggested at the end of the previous section, there exist properties enjoyed by groups of odd order that are not shared by solvable groups in general. Consider the following result, for example.

Theorem 8.1 *Suppose G is solvable of odd order and let $H \subseteq G$ be a π-complement in G. Write $\mathcal{O} \subseteq \mathrm{Irr}(G)$ to denote the set consisting of those irreducible constituents of the permutation character $(1_H)^G$ that occur with odd multiplicity. Then $\mathcal{O} = \mathrm{B}_\pi(G)$ and, in particular, $|\mathcal{O}|$ is equal to the number of classes of π-elements in G.*

Note that the last assertion of Theorem 8.1 can be viewed as an 'application' of π-theory since it can be understood without reference to the material we have been discussing. It seems unlikely, however, that a direct method for computing $|\mathcal{O}|$ could be found. Theorem 8.1 appears in [14] and we will not give the complete proof here. We will prove, however, the easier containment: that $\mathrm{B}_\pi(G) \subseteq \mathcal{O}$.

We said earlier that in practice it is generally difficult to determine whether or not a character $\chi \in \mathrm{Irr}(G)$ lies in the set $\mathrm{B}_\pi(G)$ (where G is π-separable, of course). A key ingredient in the proof of Theorem 8.1 is an easy criterion for membership in $\mathrm{B}_\pi(G)$ that works whenever $|G|$ is odd. We shall present and prove the validity of that criterion here and discuss some other consequences of it.

As usual, let π be a set of primes. If δ is a complex root of unity, we shall say that δ is a π-root if its multiplicative order is a π-number. Consider now a field E with $\mathbb{Q} \subseteq E \subseteq \mathbb{C}$. We shall say that an automorphism $\sigma \in \mathrm{Gal}(E/\mathbb{Q})$ is *magic* (with respect to π) if $\delta^\sigma = \delta$ for all π-roots $\delta \in E$ and $\delta^\sigma = \bar{\delta}$, the complex conjugate, for all π'-roots $\delta \in E$. Note that if σ is a magic automorphism, then σ^2 fixes all roots of unity in E.

Lemma 8.2 *Fix the prime set π and let E be a finite degree Galois field extension of $\mathbb{Q}$ in $\mathbb{C}$. Then E has a magic field automorphism.*

Proof. Let n be the maximum integer such that the cyclotomic extension $\mathbb{Q}_n$ of $\mathbb{Q}$ is contained in E. Factor $n = uv$, where $u = n_\pi$ and $v = n_{\pi'}$, and observe that every π-root in E is a u-th root of unity and every π'-root in E is a v-th root of unity.

Since every automorphism of $\mathbb{Q}_n$ extends to E, it suffices to find $\sigma \in \mathrm{Gal}(\mathbb{Q}_n/\mathbb{Q}_u)$ such that σ behaves like complex conjugation on $\mathbb{Q}_v$.

It is well known that $\mathbb{Q}_n = \langle \mathbb{Q}_u, \mathbb{Q}_v \rangle$ and that $\mathbb{Q}_u \cap \mathbb{Q}_v = \mathbb{Q}$. We conclude via Galois theory that restriction to $\mathbb{Q}_v$ defines an isomorphism from $\mathrm{Gal}(\mathbb{Q}_n/\mathbb{Q}_u)$ onto $\mathrm{Gal}(\mathbb{Q}_v/\mathbb{Q})$. The result now follows. $\square$

Given any finite group G, suppose $E \subseteq \mathbb{C}$ is a Galois field extension of $\mathbb{Q}$ that contains a primitive $|G|$-th root of unity. Of course, E exists, and by Lemma 8.2 it has a magic automorphism σ. In this situation, we shall say that σ is *magic for G* and we observe that σ is then also magic for every subgroup of G. We can now give the promised characterization of $\mathbf{B}_\pi(G)$ when G has odd order.

Theorem 8.3 *Suppose G is solvable of odd order. Fix a set π of primes and let σ be a magic field automorphism for G with respect to π. Then a character $\chi \in \mathrm{Irr}(G)$ lies in $\mathbf{B}_\pi(G)$ iff $\chi^\sigma = \chi$.*

Proof. If $\chi \in \mathbf{B}_\pi(G)$, then, by Corollary 3.1, we know that the values of χ lie in $\mathbb{Q}_m$, where $m = |G|_\pi$. Since σ acts trivially on this subfield, it follows that $\chi^\sigma = \chi$, as required.

Assume now that $\chi^\sigma = \chi$. To show that $\chi \in \mathbf{B}_\pi(G)$, we can assume that $G > 1$ and we work by induction on the group order. Let N be a maximal normal subgroup of G and observe that $\langle \sigma \rangle$ permutes $\mathrm{Irr}(N)$ into orbits of size at most 2. Since χ is invariant under σ and $\chi(1)$ is odd, it follows that σ must fix some irreducible constituent ψ of χ_N, and so, by the inductive hypothesis, $\psi \in \mathbf{B}_\pi(N)$. If G/N is a π-group, it follows from Theorem 4.1(b) that $\chi \in \mathbf{B}_\pi(G)$, as required. We can thus assume that G/N is a π'-group.

By Theorem 4.1(c), we know that some character $\xi \in \mathrm{Irr}(G|\psi)$ lies in $\mathbf{B}_\pi(G)$. But G is solvable, and so G/N has prime order and hence either ψ^G is irreducible and $|\mathrm{Irr}(G|\psi)| = 1$ or else $\chi_N = \psi = \xi_N$. In the first case, $\chi = \xi$ and there is nothing further to prove. We can assume, therefore, that χ and ξ are extensions of ψ and it follows that $\xi = \lambda\chi$ for some uniquely determined linear character $\lambda \in \mathrm{Irr}(G)$ with $N \subseteq \ker(\lambda)$. Since $\xi \in \mathbf{B}_\pi(G)$, we know (by the first part of the proof) that $\xi^\sigma = \xi$, and thus

$$\xi = \xi^\sigma = \lambda^\sigma \chi^\sigma = \lambda^\sigma \chi$$

and we deduce from the uniqueness of λ that $\lambda^\sigma = \lambda$. But G/N is a π'-group, and it follows that $\lambda^\sigma = \overline{\lambda}$. Since $|G/N|$ is odd and $\lambda = \overline{\lambda}$, we conclude that λ is the trivial character and $\chi = \xi \in \mathbf{B}_\pi(G)$, as required. $\square$

Suppose G is solvable of odd order and let $\chi \in \mathbf{B}_\pi(G)$. It is now easy to see that if K is a Hall π'-subgroup of G then the principal character 1_K occurs with odd multiplicity in the restriction χ_K, and thus χ lies in the set $\mathcal{O}$ of Theorem 8.1. In this situation, we can prove a much more general result concerning the restriction of χ to an arbitrary subgroup.

Corollary 8.4 *Let $U \subseteq G$, where G is solvable of odd order, and let $\chi \in \mathbf{B}_\pi(G)$. Then χ_U has an odd-multiplicity irreducible constituent $\psi \in \mathbf{B}_\pi(U)$. In particular, if χ_U is irreducible, then $\chi_U \in \mathbf{B}_\pi(U)$.*

Proof. Let σ be a magic field automorphism for G and observe that $\langle \sigma \rangle$ permutes $\mathrm{Irr}(U)$ and that all orbits have size at most 2. Since $\chi^\sigma = \chi$ and $\chi(1)$ is odd, we can reason as we did in the proof of Theorem 8.3 and deduce that some irreducible constituent ψ of χ_U must be fixed by σ and, in fact, ψ can be chosen to have odd multiplicity in χ_U. By Theorem 8.3, we have $\psi \in \mathrm{B}_\pi(U)$, as required. $\qquad\square$

Of course, if U is a π'-group in Corollary 8.4, then $|\mathrm{B}_\pi(U)| = 1$, and hence $\psi = 1_U$. This shows that $[\chi_U, 1_U]$ is odd whenever U is a π'-subgroup of an odd-order-group G and $\chi \in \mathrm{B}_\pi(G)$.

It should be clear that a result similar to Corollary 8.4 holds with induction in place of restriction.

Corollary 8.5 *Let $U \subseteq G$, where G is solvable of odd order, and let $\psi \in \mathrm{B}_\pi(U)$. Then ψ^G has an odd-multiplicity irreducible constituent $\chi \in \mathrm{B}_\pi(G)$. In particular, if ψ^G is irreducible, then $\psi^G \in \mathrm{B}_\pi(G)$.*

Much less obvious is the following converse to the last assertion in Corollary 8.5.

Theorem 8.6 *Let $U \subseteq G$, where G is solvable of odd order, and let $\psi \in \mathrm{Irr}(U)$. If $\psi^G \in \mathrm{B}_\pi(G)$, then $\psi \in \mathrm{B}_\pi(U)$.*

To prove this, we ask 'why' ψ induces irreducibly to G. One possibility, for example, is that there exists $N \subseteq U$ with $N \lhd G$ such that U is exactly the stabilizer in G of some irreducible constituent θ of ψ_N. In this situation, the character $\chi = \psi^G$ is guaranteed to be irreducible by the Clifford correspondence and we say that the equation $\chi = \psi^G$ is a *Clifford induction* in G. It is easy to prove the case of Theorem 8.6 for Clifford induction.

Lemma 8.7 *Let $U \subseteq G$, where G is solvable of odd order, and let $\psi \in \mathrm{Irr}(U)$. Suppose that $\psi^G = \chi$ is a Clifford induction and that $\chi \in \mathrm{B}_\pi(G)$. Then $\psi \in \mathrm{B}_\pi(U)$.*

Proof. There exists a subgroup $N \subseteq U$ such that $N \lhd G$ and $\psi \in \mathrm{Irr}(U|\theta)$ for some character $\theta \in \mathrm{Irr}(N)$. Since $\chi \in \mathrm{B}_\pi(G)$ and χ lies over θ, we know by Theorem 4.1 that $\theta \in \mathrm{B}_\pi(N)$, and thus both χ and θ are fixed by an appropriate magic field automorphism σ, as in Theorem 8.3. Since χ and θ uniquely determine ψ via the Clifford correspondence, it follows that ψ is fixed by σ, and we deduce that $\psi \in \mathrm{B}_\pi(U)$, as required. $\qquad\square$

It is not true, even in groups of odd order, that every irreducible induction of characters is a Clifford induction. In some sense, however, Clifford inductions determine all irreducible inductions in these groups, and this will enable us to deduce Theorem 8.6 from Lemma 8.7. To make this precise, we need to consider the Clifford-induction graph $\mathcal{C} = \mathcal{C}(\chi)$, which we are about to define. This is somewhat analogous to the graph we considered in Section 6, but here we start with an ordinary character $\chi \in \mathrm{Irr}(G)$, and not a π-partial character. If $U \subseteq G$ and $\psi \in \mathrm{Irr}(U)$, we say that (U, ψ) is an *inducing pair* belonging to χ if $\psi^G = \chi$. We take the collection of all inducing pairs belonging to χ as the vertex set for the graph $\mathcal{C}(\chi)$ and we join two such pairs (X, ξ) and (Y, η) if they are distinct and either $X \subseteq Y$ and $\xi^Y = \eta$ is a Clifford induction in Y or $Y \subseteq X$ and $\eta^X = \xi$ is a Clifford induction in X.

Theorem 8.8   *Let $\chi \in \mathrm{Irr}(G)$ have odd degree and assume that G is p-solvable for all primes p dividing $\chi(1)$. Then the Clifford-induction graph $\mathcal{C}(\chi)$ is connected.*

This theorem is sufficiently useful in solvable-group character theory that we have decided to offer a proof here, despite the fact that the result is not really π-theory. (Unfortunately, we cannot give a completely self-contained argument and at one point we shall need to appeal to an external reference.) Observe that, although to prove Theorem 8.6 we only need the case where G has odd order, we have stated a somewhat more general result by weakening this condition. It is not possible, however, to drop the oddness hypothesis entirely. We defer the proof of 8.8 until after we present its application.

Proof of Theorem 8.6.   Given $\chi \in \mathrm{B}_\pi(G)$, we wish to show that if (U, ψ) is any inducing pair belonging to χ then $\psi \in \mathrm{B}_\pi(U)$. Since (G, χ) is one of the nodes of the connected graph $\mathcal{C} = \mathcal{C}(\chi)$, it suffices to show that if two pairs (X, ξ) and (Y, η) are adjacent in $\mathcal{C}$ and $\xi \in \mathrm{B}_\pi(X)$ then $\eta \in \mathrm{B}_\pi(Y)$.

There are two possibilities. If $X \subseteq Y$ and $\xi^Y = \eta$, then $\eta \in \mathrm{B}_\pi(Y)$ by Corollary 8.5. Otherwise, $Y \subseteq X$ and $\eta^X = \xi$ is a Clifford induction in X. In this case, Lemma 8.7 guarantees that $\eta \in \mathrm{B}_\pi(Y)$.   □

The following somewhat technical result is needed in the proof of Theorem 8.8. It can be found as the case of Theorem 3.1 of [7] where the given normal subgroup is taken to be the whole group. (See also Theorem 6.2 of [8] for a more elementary approach.)

Theorem 8.9   *Let G be p-solvable, where p is an odd prime. Suppose $U \subseteq G$ is a maximal subgroup and that its index is divisible by p. Write $L = \mathrm{core}_G(U)$, the largest normal subgroup of G contained in U, and suppose that $\chi \in \mathrm{Irr}(G)$ is induced from a character of U. Then χ_L is not homogeneous.*

Proof of Theorem 8.8.   Let (U, ψ) be an inducing pair belonging to χ. We can assume that $U < G$ and we prove by induction on $|G|$ that there is a path connecting (U, ψ) to (G, χ) in the Clifford-induction graph $\mathcal{C}(\chi)$.

Since $U < G$, we can choose a maximal subgroup M of G with $U \subseteq M$ and we write $\mu^M = \zeta \in \mathrm{Irr}(M)$. Working in M, we see that (U, ψ) belongs to ζ, and so, by the inductive hypothesis applied in M, there is a path in $\mathcal{C}(\zeta)$ connecting (U, ψ) to (M, ζ). It is clear, however, that $\mathcal{C}(\zeta)$ is a subgraph of $\mathcal{C}(\chi)$, and thus it suffices to find a path connecting (M, ζ) to (G, χ) in $\mathcal{C}(\chi)$. It is no loss, therefore, to assume that U is a maximal subgroup of G.

Let $L = \mathrm{core}_G(U)$ and choose an irreducible constituent θ of ψ_L. Let T be the stabilizer of θ in G and write $S = U \cap T$, so that S is the stabilizer of θ in U. Let $\xi \in \mathrm{Irr}(S|\theta)$ be the Clifford correspondent of ψ with respect to θ and write $\eta = \xi^T$. Since η lies over θ and $\eta^G = (\xi^T)^G = (\xi^U)^G = \psi^G = \chi$, it follows that η is irreducible and that it is the Clifford correspondent of χ with respect to θ.

By Theorem 8.9, we know that $T < G$. The pair (T, η) is joined to (G, χ) in the graph $\mathcal{C}(\chi)$ and the pair (S, ξ) is either equal to or joined to (U, ψ), depending on whether $S = U$ or $S < U$. Also, since $\eta^G = \chi$, the graph $\mathcal{C}(\eta)$ is a subgraph of $\mathcal{C}(\chi)$

and the inductive hypothesis applied in the proper subgroup T implies that there is a path connecting (S, ξ) to (T, η) in $\mathcal{C}(\eta)$. It follows in this case that there is a path in $\mathcal{C}(\chi)$ that connects (U, ψ) to (G, χ), as required. $\qquad\square$

We close this section with the following unexpected theorem concerning characters of groups of odd order. Although this result is not tightly connected to π-theory, we mention it as another striking example of the fact that odd-order groups behave differently from groups in general and even from other solvable groups. A proof of this result can be found in [17], which also contains an example of a solvable group (of even order) where the conclusion fails.

Theorem 8.10 Let $H \subseteq G$, where G is solvable of odd order. If $\chi \in \mathrm{Irr}(G)$, then χ_H has an irreducible constituent having degree dividing $\chi(1)$.

9. More Hall Extendibility

Suppose that G is π-separable and let $H \subseteq G$ be a Hall π-subgroup. We return to the question that was settled in Theorem 2.4: given a character of H, when is it the restriction of some character of G? We showed in Theorem 2.4 that the obvious necessary condition is also sufficient: the given character must have equal values on G-conjugate elements of H.

In the case where we start with an irreducible character of the Hall subgroup H, there is another necessary and sufficient condition available. As far as we can see, this condition has no practical significance, but it does seem somewhat surprising, and perhaps, therefore, it is interesting. In some sense, this result appears to be deeper than Theorem 2.4 since its proof relies not only on the set $\mathrm{I}_\pi(G)$ and the corresponding Fong characters but also on facts derived from the existence of the set $\mathrm{B}_\pi(G)$ and its properties. (A special case of this result appears in [14].)

Theorem 9.1 Let $H < G$ be a proper Hall π-subgroup, where G is π-separable, and let $\alpha \in \mathrm{Irr}(H)$. Then α extends to a character of G iff there exist a prime p not in π and a character χ of G such that α is the unique irreducible constituent of χ_H that has multiplicity not divisible by p.

Note that the necessity of the condition in Theorem 9.1 is obvious since if α does extend to G we can take χ to be any extension of α. Then α is the unique irreducible constituent of χ_H and it has multiplicity 1. We can thus take p to be any prime not in π. (The assumption that the Hall π-subgroup H is proper in G guarantees that π is not the set of *all* prime numbers.) What we find surprising about this result is the sufficiency: that it is possible to prove that α extends to G without *first* assuming that $\alpha(x) = \alpha(y)$ for all G-conjugate elements $x, y \in H$. Also, it seems amusing that it is irrelevant whether or not p divides the order of the group.

We begin with a lemma that lies at the heart of the proof.

Lemma 9.2 Let H be a Hall π-subgroup of a π-separable group G and fix a prime p not in π. Let $\varphi \in \mathrm{I}_\pi(G)$ and suppose $\alpha \in \mathrm{Irr}(H)$ is the unique irreducible constituent of φ_H that occurs with multiplicity not divisible by p. Then $\varphi_H = \alpha$.

Proof. We can certainly assume that $H < G$. Let $N = O^\pi(G)$ and $M = O^{\pi'}(N)$ and observe that $M < N$ since G is π-separable and $N = O^\pi(N) > 1$. Write $K = MH$ so that $NK = G$ and $N \cap K = M$, and thus $K < G$.

By hypothesis, no irreducible constituent of φ_H other than α can have multiplicity equal to 1 and it follows that α is the unique Fong character belonging to φ. In particular, α has minimum degree among the irreducible constituents of φ_H and it occurs with multiplicity 1. Let $\nu \in I_\pi(K)$ be a constituent of φ_K that lies over α and observe that α is then an irreducible constituent of ν_H of least degree, and thus α is a Fong character belonging to ν. It also follows that ν has multiplicity 1 as a constituent of φ_K.

If δ is any irreducible constituent of φ_K other than ν, then a Fong constituent β of δ_H is an irreducible constituent of φ_H different from α, and hence its multiplicity in φ_H is divisible by p. Since β lies under no member of $I_\pi(K)$ other than δ, it follows that the multiplicity of δ as a constituent of φ_K is also divisible by p. We conclude from this that if $\mu \in I_\pi(M)$ then the multiplicities with which μ appears in φ_M and in ν_M are congruent mod p.

Since $M \lhd G$, the irreducible constituents of φ_M all have equal multiplicity, say e. Choose one of these constituents μ lying under ν and let μ' be an arbitrary irreducible constituent of φ_M. Writing m and m' to denote the multiplicities with which μ and μ' appear in ν_M, we have $m \equiv e \equiv m'$ mod p and, of course, $m \neq 0$. In fact, by Corollary 5.1, the multiplicity m of μ as a constituent of ν_M must divide $|K : M| = |G : N|$, which is a π-number and thus is not divisible by p. It follows that $m' \neq 0$ and this shows that every irreducible constituent of φ_M actually lies under ν. We conclude that these constituents are all conjugate in K.

Let T be the stabilizer of μ in G. Then $TK = G$ and hence $|G : T|$ is a π-number. Since N/M is a normal π'-subgroup of G/M, it follows that $N \subseteq T$ and μ is invariant in N. Let θ be an irreducible constituent of φ_N lying over μ and observe that θ_M is a multiple of μ. Since N/M is a π'-group, we conclude from Corollary 5.1 that $\theta_M = \mu$.

Now $\varphi(1)/\theta(1)$ and $\nu(1)/\mu(1)$ divide $|G : N| = |K : M|$, and so they are π-numbers. Since $\theta(1) = \mu(1)$, we deduce that $\varphi(1)_{\pi'} = \nu(1)_{\pi'}$. Also, α is a Fong character belonging both to φ and to ν, and hence $\varphi(1)_\pi = \alpha(1) = \nu(1)_\pi$. We deduce that $\varphi(1) = \nu(1)$, and thus $\varphi_K = \nu$.

We have $\nu_H = \varphi_H$ and hence α is the unique irreducible constituent of ν_H that has multiplicity not divisible by p. The inductive hypothesis applied in $K < G$ thus yields that $\nu_H = \alpha$ and the result follows. $\square$

Proof of Theorem 9.1. As we remarked earlier, the necessity of the condition is clear. Assume then that there exist $\chi \in \mathrm{Char}(G)$ and a prime number p not in π such that α is the unique irreducible constituent of χ_H that occurs with multiplicity not divisible by p. We must show that α extends to G.

Write $\chi^0 = \sum a_\mu \mu$, where the sum runs over $\mu \in I_\pi(G)$ and the coefficients a_μ are nonnegative integers. If $\beta \in \mathrm{Irr}(H)$ is a Fong character belonging to $\mu \in I_\pi(G)$, then, clearly, the multiplicity of β as a constituent of χ_H is exactly equal to a_μ. If $\beta \neq \alpha$, therefore, we deduce that $p | a_\mu$. But p cannot divide all of the coefficients a_μ since otherwise χ_H would be p times a character and so it would have no irreducible constituent with multiplicity not divisible by p. We deduce, therefore, that α is a Fong character belonging (say) to $\varphi \in I_\pi(G)$ and that $p \nmid a_\varphi$, but that $p \mid a_\mu$ for all $\mu \in I_\pi(G)$ different from φ.

If $\gamma \in \mathrm{Irr}(H)$ with $\gamma \neq \alpha$, we know that the multiplicity of γ in χ_H is divisible by p. Thus

$$0 \equiv [\chi_H, \gamma] = \sum_{\mu \in I_\pi(G)} a_\mu [\mu_H, \gamma] \equiv a_\varphi [\varphi_H, \gamma] \mod p$$

and, since $a_\varphi \not\equiv 0 \mod p$, we deduce that γ occurs in φ_H with multiplicity divisible by p. Since α occurs with multiplicity 1 in φ_H and every other irreducible character γ of H occurs with multiplicity divisible by p, we are in the situation of Lemma 9.2. We conclude, therefore, that $\varphi_H = \alpha$ and it follows that any ordinary character that lifts φ extends α. This completes the proof. $\square$

10. Supermonomiality

In this final section, we consider another application of π-theory to the study of monomiality. By definition, if $\chi \in \mathrm{Irr}(G)$ is a monomial character, then χ is induced from a linear character of some subgroup. There is no reason to believe, however, that *every* primitive character that induces χ must be linear. Even in the case where G is an M-group, if $\chi = \psi^G$, where ψ is primitive character of some subgroup, it is not generally true that ψ must be linear. When G is an M-group of odd order, however, there is no known example of a nonlinear primitive character of a subgroup that induces irreducibly to the whole group. It is tempting to conjecture that this never happens.

We shall say that a character $\chi \in \mathrm{Irr}(G)$ is *supermonomial* if every primitive character of a subgroup of G that induces χ is linear. Since every irreducible character is certainly induced from some primitive character, it should be clear that an irreducible character is supermonomial iff every character that induces it is monomial. In particular, supermonomial characters are necessarily monomial.

It is not hard to see that if it were true (as we suggested) that every irreducible character of an odd-order M-group is supermonomial it would follow easily that a normal subgroup of an odd-order M-group must be an M-group. (This would resolve, therefore, one of the principal open questions concerning M-groups.)

Unfortunately, there is little that we are able to say in the direction of proving that all irreducible characters of odd-order M-groups are supermonomial. Since it is easy to produce examples of odd-order groups that have monomial but not supermonomial irreducible characters, it follows that, to establish the 'conjecture', one could not hope to prove supermonomiality one character at a time; somehow, the global hypothesis that the group is an M-group would have to be used.

In certain situations, however, it is possible to prove that a particular character is supermonomial, and we would like to present two closely related sufficient conditions here. Their proofs are based on a deep theorem of E. C. Dade and they also involve results we have been discussing here.

Theorem 10.1 *Let G be p-solvable, where p is an odd prime. If $\chi \in \mathrm{Irr}(G)$ is monomial and $\chi(1)$ is a power of p, then χ is supermonomial.*

Theorem 10.2 *Let G be solvable of odd order and fix a prime p. If $\chi \in B_p(G)$ is monomial, then it is supermonomial.*

The work of Dade that underlies these results can be found in [4]. The main theorem of Dade's paper is that if $\chi \in \mathrm{Irr}(G)$ is monomial and has p-power degree, where G is p-solvable and p is an odd prime, then the irreducible constituents of χ_N are monomial for all normal subgroups N of G. What we need here, however, is not exactly this theorem, but instead the following slightly more powerful result that can be proved by a minor modification of Dade's argument and from which Dade's theorem easily follows. (At our suggestion, Dade included the statement of this result and a sketch of its proof at the end of [4].)

Theorem 10.3 (Essentially Dade) *Let G be p-solvable, where p is an odd prime, and suppose $\chi \in \mathrm{Irr}(G)$ has p-power degree. Given $N \lhd G$, choose an irreducible constituent θ of χ_N and let T be the stabilizer of θ in G. Then χ is monomial iff its Clifford correspondent $\eta \in \mathrm{Irr}(T|\theta)$ is monomial.*

Proof of Theorem 10.1. Consider the Clifford-induction graph $C = C(\chi)$ of Theorem 8.8. Recall that the vertices of C are the inducing pairs (U, ψ) with $\psi^G = \chi$ and observe that it suffices to show that, for every such pair, the character ψ is monomial. Note that $\psi(1)$ divides $\chi(1)$, and so $\psi(1)$ is a p-power in this situation.

If (X, ξ) and (Y, η) are adjacent vertices in C, then since each of ξ and η has p-power degree, Theorem 10.3 tells us that ξ is monomial iff η is monomial. By Theorem 8.8, the graph is connected and we know that at least one vertex, namely (G, χ), has a monomial second component. It follows that the second component of every vertex is monomial, as required. $\square$

We shall need the following easy lemma.

Lemma 10.4 *Let $U \subseteq G$, where G is solvable of odd order. Suppose that $\mu \in I_\pi(U)$ and $\mu^G = \varphi \in I_\pi(G)$. If $\xi \in B_\pi(U)$ with $\xi^0 = \mu$ and $\chi \in B_\pi(G)$ with $\chi^0 = \varphi$, then $\xi^G = \chi$.*

Proof. Since $(\xi^G)^0 = \varphi$ is irreducible, we deduce that $\xi^G \in \mathrm{Irr}(G)$, and it follows by Corollary 8.5 that $\xi^G \in B_\pi(G)$. But $(\xi^G)^0 = \varphi = \chi^0$, and thus $\xi^G = \chi$. $\square$

We can now use Theorem 10.1 and a number of our results in π-theory (applied in the case where $\pi = \{p\}$) to prove Theorem 10.2. For notational simplicity, we shall write p in place of $\{p\}$ in what follows.

Proof of Theorem 10.2. Let $\varphi = \chi^0 \in I_p(G)$ and consider the graph $G = G(\varphi)$ of Theorem 6.2. Recall that the vertices of G are the p-inducing pairs belonging to φ. Given any such pair (U, μ), we know that μ is a p-partial character having p-power degree and we shall show that μ is monomial as a p-partial character. (In other words, μ is induced from a linear p-partial character of some subgroup of U.)

By hypothesis, χ is monomial, and therefore φ is monomial as a p-partial character. Let $W \subseteq G$ and $\gamma \in I_p(W)$ be such that $\gamma(1) = 1$ and $\gamma^G = \varphi$. By Theorem 6.2, if we replace the pair (W, γ) by an appropriate conjugate, we can assume that (W, γ)

and the given pair (U, μ) lie in the same connected component of $\mathcal{G}$. To prove that μ is monomial, it thus suffices to show that if (U, μ) and (V, ν) are adjacent vertices of the graph and ν is monomial, then μ is monomial too. This is obvious if $V \subseteq U$ and $\nu^U = \mu$, and so we can assume that $U \subseteq V$ and $\mu^V = \nu$. Also, since ν is monomial, we can assume that $W \subseteq V$ and $\gamma^V = \nu$.

Let η be the unique character in $B_p(V)$ such that $\eta^0 = \nu$. By Lemma 10.4, we have $\eta = \lambda^V$, where $\lambda \in B_p(W)$ and $\lambda^0 = \gamma$. In particular, η is monomial, since $\lambda(1) = \gamma(1) = 1$. But also $\eta(1) = \nu(1)$ is a p-power, and thus η is actually supermonomial by Theorem 10.1. By Lemma 10.4 again, we have $\eta = \xi^V$, where $\xi \in B_p(U)$ and $\xi^0 = \mu$. Since $\xi^V = \eta$ is supermonomial, we deduce that ξ is monomial, and thus μ is monomial, as claimed.

Now let $U \subseteq G$ and suppose $\psi \in \mathrm{Irr}(U)$ is primitive and $\psi^G = \chi$. Write $\mu = \psi^0$ and note that $\mu^G = \varphi$ and hence $\mu \in I_p(U)$. We claim that μ is primitive as a p-partial character.

Suppose $W \subseteq U$ and $\delta \in I_p(W)$ is such that $\delta^U = \mu$. To show that $W = U$, we observe first that since $\psi^G = \chi \in B_p(G)$, it follows by Theorem 8.6 that $\psi \in B_p(U)$. Let $\xi \in B_p(W)$ with $\xi^0 = \delta$ and apply Lemma 10.4 to conclude that $\xi^U = \psi$. Since ψ is primitive, however, it follows that $W = U$, and thus μ is primitive, as claimed.

By Corollary 5.5, we know that μ is induced from some p-partial character having p-power degree, but since μ is primitive, we deduce that μ has p-power degree, Thus (U, μ) is a p-inducing pair belonging to φ and by the first part of the proof, μ is monomial. But since μ is also primitive, we have $\mu(1) = 1$ and thus $\psi(1) = 1$, as required. □

References

1. E. C. Dade, see page 585 of B. Huppert, *Endliche Gruppen I*, Springer-Verlag, Berlin, 1967.
2. E. C. Dade, Normal subgroups of M-groups need not be M-groups, *Math. Zeit.* 133 (1973), 313–317.
3. E. C. Dade, Characters of groups with normal extra-special subgroups, *Math. Zeit.* 152 (1976), 1–31.
4. E. C. Dade, Monomial characters and normal subgroups, *Math. Zeit.* 178 (1981), 401–420.
5. P. Fong, Solvable groups and modular representation theory, *Trans. Amer. Math. Soc.* 103 (1962), 484–494.
6. I. M. Isaacs, Lifting Brauer characters of p-solvable groups, *Pacific J. Math.* 53 (1974), 171–188.
7. I. M. Isaacs, Primitive characters, normal subgroups and M-groups, *Math. Zeit.* 177 (1981), 267–284.
8. I. M. Isaacs, On the character theory of fully ramified sections, *Rocky Mountains J. Math.* 13 (1983), 689–698.
9. I. M. Isaacs, Characters of π-separable groups, *J. Algebra* 86 (1984), 98–128.
10. I. M. Isaacs, Characters of subnormal subgroups of M-groups, *Arch. Math.* 42 (1984), 509–515.
11. I. M. Isaacs, Fong characters in π-separable groups, *J. Algebra* 99 (1986), 89–107.
12. I. M. Isaacs, Dual modules and group actions on extra-special groups, *Rocky Mountains J. Math.* 18 (1988), 505–517.
13. I. M. Isaacs, Partial characters of π-separable groups, *Progress in Math.* 95 (1991), 273–287.
14. I. M. Isaacs, Characters and Hall subgroups of groups of odd order, *J. Algebra* 157 (1993), 548–561.
15. I. M. Isaacs, *Character Theory of Finite Groups*, Dover, New York, 1994.
16. I. M. Isaacs, The π-character theory of solvable groups, *J. Austral. Math. Soc.* (Ser. A) 57 (1994), 81–102.

17. I. M. Isaacs, Constituents of restricted and induced characters in odd order groups, *J. Algebra*, to appear.

18. I. M. Isaacs and G. Navarro, Weights and vertices for characters of π-separable groups, *J. Algebra*, to appear.

19. G. Navarro, Primitive characters of subgroups of M-groups, submitted to *Math. Zeit.*

20. G. Robinson and R. Staszewski, On the representation theory of π-separable groups, *J. Algebra* 119 (1988), 226–232.

21. M. Slattery, π-blocks of π-separable groups, I, *J. Algebra* 102 (1986), 60–77.

CHARACTER THEORY AND LENGTH PROBLEMS

A. TURULL*
*Department of Mathematics
University of Florida
Gainesville, Florida 32611
USA
e-mail: turull@math.ufl.edu*

Abstract. We discuss some length questions in the theory of finite solvable groups for which character theory has proven useful. We discuss some open problems, some techniques of proof for various character and module theoretic results arising in this context, as well as the applications of these results to the length problems themselves. As this is an expository paper, most of our results appeared elsewhere. An exception is a linear bound on the Fitting height of groups accepting a fixed point free abelian group of automorphisms of square-free odd exponent, which, although it follows easily from known results, does not seem to have been noticed before.

Key words: solvable group, finite group, character, length, regular orbit.

1. Introduction

Many problems in the theory of finite solvable groups, or, more generally, in the theory of finite groups with many normal subgroups, involve length functions. If some length of a finite group is known, how large can some other related length of the same group possibly be? The known information is sometimes character theoretical. For example, we may want to bound the derived length of a finite solvable group in terms of the number of distinct degrees of irreducible characters of the group, see for example [28, 29, 31, 43, 44, 50, 68]. It should be noted, however, that even the full character table does not determine the derived length of a p-group [56, 57, 58]. Or one can find bounds on length functions from the number of prime divisors in the character degrees, or related invariants; see for example [5, 6, 13, 26, 27, 32, 33, 34, 47, 55, 59, 60, 61, 78]. At other times, the information may not be character theoretical. For example, one may want to give bounds on the p-length of a finite p-solvable group from information about its Sylow p-subgroups, or bound the nilpotent length of the group from some information about its automorphism group.

In either case, certain aspects of character theory are often (but not always[1]) useful. In the study of solvable groups often the same issues play an important role: the permutation structure of finite modules; the existence or non-existence of regular orbits in these modules; the exceptions that one has to deal with; the representation

---

* Partially supported by N.S.F.
[1] Most notably, the study of the connections between p-groups and Lie algebras can be more appropriate for certain problems. For further details on these, we refer the reader to the article by Shalev elsewhere in these Proceedings.

B. Hartley et al. (eds.), Finite and Locally Finite Groups, 377–400.

theory of groups with a normal extra-special subgroup whose center is central in the whole group. These issues, of course, are never exactly the same for different problems, but similarities do exist.

In this paper, we focus on certain questions related to p-lengths and Fitting heights. We describe some of the results and conjectures relating to these. We then explain some of the character theoretical techniques that are useful in these problems. These character theoretical techniques, or some modifications of them, are also useful in many other length problems, but these other connections would take us too far afield. Many of the results mentioned here are influenced by results of Berger, who wrote a survey article on a similar topic [3] in 1980. We refer the reader to [3] for a different perspective on some of the issues discussed below.

During and after the conference some participants found typographical errors and suggested improvements on an earlier version of the paper. J. Hall pointed out a slip in the proof of Theorem 3.5 included in it. The present paper has been corrected in the light of this input. My thanks to all.

2. Some p-Length Inequalities

In the study of finite solvable groups the inequalities between length functions feature prominently.[2] Probably the most well known of them originates in the celebrated paper [38] of Hall and Higman:

Theorem 2.1 *Let p be a prime and let G be a finite p-solvable group. Then*

$$d_p(G) \geqslant l_p(G),$$

$$e_p(G) \geqslant l_p(G) \quad \text{if } p \text{ is not a Fermat prime, and}$$

$$e_p(G) \geqslant \lfloor \frac{1}{2}(l_p(G)+1) \rfloor \quad \text{if } p \text{ is a Fermat prime.}$$

Furthermore these inequalities are best-possible.

Hall and Higman proved the theorem for p odd in 1956, and Bryukhanova [7, 8] for $p = 2$ in 1979 and 1981. The crucial fact that Hall and Higman proved is the following.

Lemma 2.2 (Hall and Higman) *Let AG be a finite group, which is the semi-direct product of a normal solvable p'-group G and a cyclic p-group A. Suppose that A acts faithfully on G and let M be a module for AG in characteristic p, on which G acts faithfully. Then the restriction of M to A contains a regular[3] submodule, except possibly when some section of G is isomorphic to an extra-special q-group (for some prime q) and $p^\alpha = q^\beta + 1$.*

---

[2] For convenience, we describe our notation for the various length functions in an appendix. Other notation is standard and can be found in books such as [35, 45, 46, 49, 62]. We refer the reader to these books for background.

[3] A regular submodule is just a free submodule of rank one.

The exception is genuine, in the sense that, whenever we have any two primes p and q satisfying $p^\alpha = q^\beta + 1$, then there is an extra-special q-group Q of order $q^{2\beta+1}$, a faithful action of a cyclic group C of order p^α on Q centralizing Q', and a faithful module M for CQ of dimension q^β, so that no regular submodule for C exists in M. A little elementary number theory shows that there are only three ways for primes p and q to satisfy the equation $p^\alpha = q^\beta + 1$. Namely, $p = 2$ and $q = 2^\alpha - 1$ is a Mersenne prime (with $\beta = 1$); or $q = 2$ and $p = 2^{2^n} + 1$ is a Fermat prime (with $\alpha = 1$ and $\beta = 2^n$); or $p = 3$ and $q = 2$ and $3^2 = 2^3 + 1$. In any case, either $p = 2$ and q is a Mersenne prime or $q = 2$ and p is a Fermat prime.

Lemma 2.2 provides much useful information for the proof of Theorem 2.1. If p is neither 2 nor a Fermat prime, then we have no exception to the existence of the regular submodule. This fact, together with induction, is the basis of the proof of Hall and Higman. When p is a Fermat prime, the failure of the regular submodule in Lemma 2.2 to exist forces Hall and Higman to a more careful analysis. They prove that there will, in any case, be a submodule which will be indecomposable of dimension at least $|A| - 1$, and obtain information about the cases where there is no regular submodule, especially in the case $p = 3$. This allows them to show the inequality involving $d_p(G)$. The inequality involving $e_p(G)$ simply changes because of the exceptions in Lemma 2.2, and they prove a variation of Lemma 2.2 which proves the weaker inequality. The case $p = 2$ proved to be much more difficult. More than 20 years passed until the corresponding inequalities were finally fully established by Bryukhanova. The exceptions to Lemma 2.2 are again responsible for much of the difficulty. The way to overcome them is again to look at a more complicated representation theoretical situation which will be simple enough to handle, but rich enough not to have exceptions in the conclusion. She can no longer assume that A is cyclic because of them, and has to work with more complicated p-groups ($p = 2$).

The exceptions to Lemma 2.2 play a role in many problems relating to p-lengths and Fitting heights. For this reason they are investigated in many papers dealing with these subjects. In addition, they are investigated in articles such as [22, 40, 53]. Other length problems depend on generalizations of Lemma 2.2 to cases where A is no longer cyclic. For example, T. Berger and his collaborators have investigated extensively the case A nilpotent in a long series of papers; see [4] for the latest in this series.

We are far from knowing all we want to know about the p-length of p-solvable groups. The following conjecture is mentioned in [3].

Conjecture 2.3 *Let p be a prime. There exists a linear function f_p such that if G is a p-solvable group and P is a subgroup of G of order p^k contained in precisely one Sylow p-subgroup of G then*

$$l_p(G) \leqslant f_p(k).$$

We know that, under the hypotheses of the conjecture, an exponential bound on the p-length does exist for p odd; see [64]. However, we do not yet know whether any bound exists for $p = 2$. By [42], if $G = PH$, where H is a normal subgroup of G and

P acts fixed point freely on every P-invariant p'-section of H, then G has a unique Sylow p-subgroup containing P. In this latter context, the bound $f_p(k) = k+1$ for P cyclic of odd order was established by Espuelas [21]. He also showed that the same inequality holds more generally [23], although this falls short of a full proof of the conjecture. In the case where P is cyclic of order p^k, a further strengthening of the hypotheses on G in Conjecture 2.3 is to assume that the Hughes subgroup generated by all the elements of G whose order does not divide p^k is a proper subgroup of G. Under these hypotheses with $p = 2$, Turau [70] has given the bound $k + 1$ on the 2-length of G. A number of papers also study the Fitting length of the Hughes subgroup of a solvable group G in some cases where it is a proper subgroup of G [1, 17, 18, 19, 20, 36, 63].

3. Some Fitting Height Inequalities

There are connections between the Fitting height of a finite solvable group and its exponent [52]. The Fitting height can also be bounded by linear functions on the number of generators needed to generate its nilpotent injectors or its Sylow subgroups [51]. The relation between the Fitting height of a finite solvable group and its Carter subgroups seems more difficult to understand. From Dade's fundamental paper [15], we know that the Fitting height of a finite solvable group is bounded by an exponential function of the number of primes (counting multiplicity) that divide the order of its Carter subgroups. Since the method of proof that Dade follows concentrates only on very special aspects of the finite solvable group, he conjectured in his paper that perhaps one could show that there is a linear bound. Even though Dade's paper [15] was published in 1969, we still do not know whether the conjecture is true.

Conjecture 3.1 (Dade) *There is a linear function g such that whenever G is a finite solvable group and C is a Carter subgroup of G then*

$$\mathrm{h}(G) \leqslant g(\Omega(|C|)). \tag{1}$$

We do know, by [15], that there is an exponential bound for $\mathrm{h}(G)$ in terms of $\Omega(|C|)$, but no linear, or even polynomial, bound has been found. No bound lower than exponential is known even in the case where the Carter subgroup is assumed to be cyclic. A Carter subgroup is any nilpotent self-normalizing subgroup of a finite solvable group. Every finite solvable group has a unique conjugacy class of Carter subgroups. It follows from the last two sentences that, if G is a finite solvable group and C is one of its Carter subgroups, then C is a Carter subgroup of every subgroup of G that contains C and, for each normal subgroup N of G, CN/N is a Carter subgroup of G/N. The following is a special case of Conjecture 3.1.

Conjecture 3.2 *There is a linear function f such that, whenever C is a nilpotent finite group which acts on the solvable finite group G in such a way that $\mathrm{C}_G(C) = 1$, then*

$$\mathrm{h}(G) \leqslant f(\Omega(|C|)). \tag{2}$$

Indeed, under these hypotheses, C is a self-normalizing nilpotent subgroup of CG, that is, a Carter subgroup of CG. Hence, by setting $f = g$, any function f that satisfies Equation 1 will also satisfy Equation 2. A simple trick allows us to go from Equation 2 to Equation 1, but at the price of going from linear to quadratic functions.

Remark 3.3   *If f satisfies Equation 2, then we may set*

$$g(n) = n + \sum_{i=1}^{n} f(i),$$

and then g satisfies Equation 1. Hence, if f is linear, at least a quadratic g can be found.

Proof.   Let $G \neq 1$ be a solvable group and let C be its Carter subgroup, and argue that f satisfies Equation 1 by induction on $\Omega(|C|) = n$. Let h be the largest integer such that $C \cap \mathrm{F}_h(G) = 1$. Since C is self-normalizing, we have $C_{\mathrm{F}_h(G)}(C) = 1$. By Equation 2, since $h = \mathrm{h}(\mathrm{F}_h(G))$, we have

$$h \leqslant f(\Omega(|C|)).$$

Now $C\mathrm{F}_{h+1}(G)/\mathrm{F}_{h+1}(G)$ is a Carter subgroup of $G/\mathrm{F}_{h+1}(G)$. Hence, by induction,

$$\mathrm{h}(G) - (h+1) \leqslant (n-1) + \sum_{i=1}^{n-1} f(i).$$

The result follows.   □

Hence the two conjectures are closely related. Are any special cases known for these conjectures? A proof of a *full special case* of any of these two conjectures would be, in my mind, the following kind of result. Some conditions are imposed on the structure of C but no special conditions are imposed on the structure of G. The conditions on C should be such that, under them, $\Omega(|C|)$ is unbounded. Then, with these extra assumptions, one of the conjectures is proved. In contrast to this, a *partial special case* would be one where further conditions on the structure of G are imposed as well. There are partial special cases of Conjecture 3.2 known, especially under the assumption that $(|C|, |G|) = 1$, as we will see in the next section. Furthermore, the assumption that C be a p-group for some prime p automatically implies that $(|C|, |G|) = 1$, and so we will not discuss it separately. However, leaving these aside, I have not seen in print a proof of either of these two conjectures in any full special case (in the above sense). We offer here the following full special case of Conjecture 3.2, which follows easily from the results of Dade [15].

Theorem 3.4   *Let C be a finite abelian group with square-free odd exponent. Suppose it acts fixed point freely on the finite solvable group G. Then*

$$\mathrm{h}(G) \leqslant 5\Omega(|C|).$$

Proof. We offer a quick sketch. All references and unexplained terms in this proof relate to [15]. Set $h = \mathrm{h}(G)$. We may assume $h \geqslant 6$. By Lemmas 8.1 and 8.2 we have a C-invariant *Fitting chain* (in Dade's sense) $A_1, \ldots, A_h$ of sections of G. This even yields a C-invariant *augmented Fitting chain*. Since C is nilpotent, it is a Carter subgroup of any semidirect product of it with a section of G. Hence, C centralizes no non-trivial section of G. In particular, C acts fixed point freely on any section of the $A_1, \ldots, A_h$. We claim that this implies that $h \leqslant 5\Omega(|C|)$. We prove this assertion, which refers only to *augmented Fitting chains* and no longer to the group G, by induction on $\Omega(|C|)$. Once the assertion is proved, the theorem follows immediately. Since C acts fixed point freely on A_1, and the exponent of C is square-free, there is a subgroup P of C of prime order, such that P does not centralize A_1. Since C is abelian, P is a normal subgroup of C and, since the exponent of C is odd, $|P|$ is an odd prime. By Theorem 2.7 and Theorem 2.13, there is a C-invariant *augmented Fitting chain* $D_6, \ldots, D_h$, where each section is centralized by P. Now C/P acts on the new *chain* $D_6, \ldots, D_h$, fixed point freely on each of its sections. By induction, it follows that $h - 5 \leqslant 5\Omega(|C/P|)$. Hence the result.   □

The bound obtained in the above theorem has the right order of magnitude. The constant 5 might be too large, however. In the case where C is cyclic and $\Omega(|C|) = 2$, the bound 2 is obtained in [14], instead of the 10 given by the above theorem.

It is tempting to think that the condition that C is nilpotent in Conjecture 3.2 is unnecessary. This is, however, not the case at all. The following elegant result of S. D. Bell and B. Hartley [2] shows that the condition cannot simply be dropped.

Theorem 3.5 (Bell and Hartley) *Let A be any finite non-nilpotent group and H be any finite group. Then there exists a finite group G, on which A acts fixed-point-freely, such that H is a homomorphic image of G. Further, if H is solvable, so is G.*

Proof. We offer a slight simplification of [2]. Assume the theorem is false, and consider a counterexample A with minimum order $|A|$. Then A is not nilpotent, but each proper factor group of A will be nilpotent, since otherwise A would have a non-faithful action contradicting our assumption. Since A itself is not nilpotent, $A/\mathrm{Z}(A)$ is not nilpotent either, and this implies that $\mathrm{Z}(A) = 1$. Suppose B is a non-nilpotent proper subgroup of A. Then, by the minimality, B acts fixed point freely on a group G satisfying the conclusion of the theorem. A standard construction, analogous to the induction of modules, shows that A acts fixed point freely on the cartesian product $G \times \cdots \times G$ of $[A : B]$ copies of G. But this contradicts the fact that A is a counterexample to the theorem. Hence, every proper subgroup of A is nilpotent. This implies that A is solvable.

I claim we can find a finite abelian group L on which A acts in such a way that there is a complement A^* to L in AL, which is not conjugate to A, and such that $\mathrm{C}_{AL}(A)$ and $\mathrm{C}_{AL}(A^*)$ are both trivial. Let us first see how this claim implies that the group G does exist. Let G be the wreath product of AL with H. Hence G has a normal subgroup B isomorphic to the direct product of $|H|$ copies of AL and B has a complement H in G, which simply permutes these copies regularly. Clearly, H is a homomorphic image of G. Because $A \cong AL/L \cong A^*$, there is a natural isomorphism

$a \mapsto a^*$ from A to A^*. There is an injective homomorphism from A to B,

$$a \mapsto (a^*, a, a, \ldots, a),$$

where the first coordinate is a^* and all the others are a. Let $\tilde{A}$ be the image of A under this homomorphism. This injective homomorphism followed by conjugation yields an action of A on G. Suppose $g \in C_G(A)$. Then $g = hb$ where $h \in H$ and $b \in B$. Now $\tilde{A}^{hb} = \tilde{A}$. Since A and A^* are not conjugate in AL, it follows that h must fix the first coordinate of B, and since H acts regularly this implies $h = 1$. Now $g = b \in C_B(A)$. But B is a direct product of copies of AL and both A and A^* act fixed point freely on AL. It follows that $g = 1$. Hence, assuming the claim, the theorem is established.

We now prove that the claim does hold. There is some prime p such that A is not p-nilpotent. Let P be the projective cover of the trivial FA module, where F is some finite splitting field for A in characteristic p. Suppose, for a moment, that all composition factors of P are isomorphic to the trivial module. Then the kernel of P consists of exactly all p'-elements of A. This is because certainly every p-singular element acts non-trivially on P, and the action of any p'-element decomposes P into a direct sum of trivial modules for that element. But then it would follow that A is p-nilpotent, since $A/\ker(P)$ is a p-group and $\ker(P)$ is a p'-group. Hence not all composition factors of P are trivial.

Let V be a submodule of P maximal subject to the condition that P/V has a non-trivial composition factor. Then the socle S/V of P/V is a non-trivial simple module L and we take T/S to be any simple submodule of P/S. We set $M = T/V$. The module M has a non-trivial simple submodule L, M/L is a trivial module, and M is indecomposable.

Consider the semidirect product AM. In it, there is the normal subgroup AL. Let $m \in M$ be such that $m \notin L$, and set $A^* = mAm^{-1}$. Notice that $N_{AM}(A) = AC_M(A) = A$. From this it follows that A^* is not conjugate to A in AL. Furthermore both $C_{AL}(A)$ and $C_{AL}(A^*)$ are trivial. This concludes the proof of the claim and the theorem.   □

4. Coprime Action

The above construction always yields group actions such that $|A|$ and $|G|$ are not relatively prime. If instead we insist that $|A|$ and $|G|$ should be relatively prime, the situation changes significantly. We no longer have to necessarily work with fixed point free action, or with A nilpotent.

Hypothesis 4.1   *Let A be a finite group that acts on the finite solvable group G, and assume $(|A|, |G|) = 1$.*

Under coprime action, we have some useful properties which are not valid in general. For example, whenever S/R is a section of G which is stabilized by a subgroup B of A, then $C_{S/R}(B) = RC_S(B)/R$. We say that the centralizer of B in S covers the centralizer of B in S/R. Another useful property which holds under coprime action, but not necessarily in general, is that, for each prime p, there will be an A-invariant Sylow p-subgroup of G. These properties are the basis of many of the proofs that we will discuss now. A way to investigate the Fitting height of G is via its A-irreducible towers [72].

Definition 4.2 *We say that a sequence $(\hat{P}_i)$, $i = 1, \ldots, h$, of A-invariant subgroups of G is an* irreducible A-*tower of height h if the following are satisfied.*

(1) $\hat{P}_i$ *is a p_i-group (p_i some prime) for $i = 1, \ldots, h$.*

(2) $\hat{P}_i$ *normalizes $\hat{P}_j$, for $i < j$.*

(3) *We set $P_h = \hat{P}_h$ and $P_i = \hat{P}_i / C_{\hat{P}_i}(P_{i+1})$ for $i = 1, \ldots, h-1$, and $P_1 \neq 1$.*

(4) $p_i \neq p_{i+1}$ *for $i = 1, \ldots, h-1$.*

(5) $\phi(\phi(P_i)) = 1$, $\phi(P_i) \subseteq Z(P_i)$ *and, if $p_i \neq 2$, $\exp(P_i) = p_i$ for $i = 1, \ldots, h$ and $\hat{P}_{i-1}$ centralizes $\phi(P_i)$ for $i = 2, \ldots, h$.*

(6) P_1 *is elementary abelian.*

(7) *There exists H_i an elementary abelian subgroup of P_{i-1} normalized by A such that $[H_i, P_i] = P_i$ for $i = 2, \ldots, h$.*

(8) *If $Q \subseteq \hat{P}_i$ for some i, Q is normalized by $A\hat{P}_1 \cdots \hat{P}_{i-1}$ and its image in P_i is not contained in $\phi(P_i)$, then $Q = \hat{P}_i$.*

The point of this definition is that it incapsulates in terms of its subgroups a minimal A-invariant subgroup of G of Fitting height h and it tells something about its Fitting factors. The following lemma (Lemma 1.4 in [72]) is proved by considering the minimal case and using the existence of A-invariant Sylow subgroups in invariant subgroups of G.

Lemma 4.3 *Assume Hypothesis 4.1. Then there is an A-irreducible tower of height $h = \mathrm{h}(G)$ consisting of subgroups of G.*

If $(\hat{P}_i)$, $i = 1, \ldots, h$, is an A-irreducible tower and B is a normal subgroup of A then $(C_{\hat{P}_i}(B))$, $i = 1, \ldots, h$, is a sequence of A-invariant subgroups of $C_G(B)$ with the property that $C_{\hat{P}_i}(B)$ normalizes $C_{\hat{P}_j}(B)$ for $i < j$. If these actions are sufficiently non-trivial then the Fitting height of $C_G(B)$ will not be much smaller than that of G.

Theorem 4.4 *Assume Hypothesis 4.1 and let $(\hat{P}_i)$, $i = 1, \ldots, h$, be an irreducible A-tower of height h consisting of A-invariant subgroups of G. Suppose A is of prime order and centralizes $\hat{P}_k$ (possibly with $k = 0$ and $\hat{P}_k = 1$). Then there exists some $j \geqslant k$ such that $(C_{\hat{P}_i}(A))$, $i = 1, \ldots, j-1, j+1, \ldots, h$, satisfies conditions (1), (2) and (3) of Definition 4.2.*

For the full proof see [72]. The proof proceeds by examining very carefully the first P_i on which A acts non-trivially. For convenience, say that this first group is simply P_{k+1}. Since centralizers of A on sections of G are covered by centralizers on G itself, one can focus attention on how $\hat{P}_k P_{k+1}$ acts on $P_{k+2}/\phi(P_{k+2})$, and on trying to show that the action of $\hat{P}_k$ on the centralizer of A on $P_{k+2}/\phi(P_{k+2})$ is sufficiently non-trivial. This becomes a representation theoretical problem, involving some relatively uncomplicated groups. For simplicity, we take a cyclic subgroup T of $\hat{P}_k$, and we concentrate our attention on the action of ATP_{k+1} on the module M arising from $P_{k+2}/\phi(P_{k+2})$. We need to show, among other things, that T acts non-trivially on $C_M(A)$. The proof is complicated however by the fact that the naive representation theoretical statement describing only ATP_{k+1} and its module is sometimes false. Here is an example for which there is this difficulty. Let p be an odd prime and assume that $p^n + 1 = 2|A|$. Let P be an extraspecial group of order p^{2n+1} and exponent p. Set T to be a cyclic group of order 2. Then the direct product AT acts as automorphisms of P centralizing P'. There is an ATP-module of dimension p^n, such that T acts trivially on $C_M(A)$.

This difficulty can happen, however, only in certain cases when $p_k = 2$ and the exponent of P_k is 2. In this case, we need to examine the action of $\hat{P}_{k-1}$. If $\hat{P}_{k-1} = 1$, it is enough to notice that then P' is centralized by A and acts non-trivially on $C_M(A)$; we do not need to use P_k itself in this case. If $\hat{P}_{k-1} \neq 1$, and it acts non-trivially on P', then again we may drop $\hat{P}_k$. Finally, we have the case where $\hat{P}_{k-1} \neq 1$ and $\hat{P}_{k-1}$ acts trivially on P'. In this case, we have more than just a cyclic group T of order 2 acting on P/P' and centralizing P', and the counterexample does not occur. Indeed in this case we show again that the action of $\hat{P}_k$ on $C_M(A)$ is non-trivial.

Theorem 4.5 *Assume that Hypothesis 4.1 holds with A solvable. Then*

$$\mathrm{h}(G) \leqslant 2\,\mathrm{l}(A) + \mathrm{h}(C_G(A)).$$

Proof. (See Corollary 3.2 in [72].) Use induction on $|A|$. Suppose first that $|A|$ is not a prime number. Since A is solvable, there is a normal subgroup B of A of prime index. By induction

$$\mathrm{h}(G) \leqslant 2\,\mathrm{l}(B) + \mathrm{h}(C_G(B)).$$

Since A/B acts on $C_G(B)$, we have, again by induction,

$$\mathrm{h}(C_G(B)) \leqslant 2 + \mathrm{h}(C_G(A)).$$

Hence we have the result in this case. Hence, we must next assume that A is of prime order. In this case, the previous theorem implies that $C_G(A)$ contains $(C_{\hat{P}_i}(B))$, $i = 1, \ldots, j-1, j+1, \ldots, h$, satisfying conditions (1), (2) and (3) of Definition 4.2. Since $p_i \neq p_{i+1}$, in the new sequence of groups no two consecutive ones are p-groups for the same prime except possibly the two on either side of $\hat{P}_j$. This implies that $\mathrm{h}(C_G(A)) \geqslant \mathrm{h}(G) - 2$. This concludes the proof of the theorem. $\qquad\square$

Less is known about the case where A is not solvable. Results that are discussed in the next section suggest that the invariant $\mathrm{l}(A)$, rather than $\Omega(|A|)$ (which is equal to $\mathrm{l}(A)$ if A is solvable), is the appropriate one for generalizing Theorem 4.5 to non-solvable groups A.

Question 4.6 *Can the hypothesis that A is solvable be dropped in Theorem 4.5?*

The previous proof reduces the question to the case where A is a simple group. Work of Kurzweil [54] shows that, at least for some A, h(G) can be bounded by some function of h($C_G(A)$) and A.

The inequality in Theorem 4.5 is the best possible in the following sense. For each finite solvable group A and each *positive* h, there exists a finite solvable group G and an action of A on G in such a way that the hypotheses of the theorem are satisfied with $h = $ h($C_G(A)$) and h(G) $= 2\,$l(A)$+$h($C_G(A)$), see [73]. If $h = 0$ (the fixed point free case), however, the inequality does not seem to be the best possible. We discuss this case in the following section. Some results can also be obtained with the intermediate assumption that the order of $C_G(A)$ is not necessarily 1 but is bounded. For these see [41, 76] and references in these papers. Other related results appear in [66, 67].

5. Fixed Point Free Coprime Action

Theorem 4.5 implies that, when a finite solvable group A acts fixed point freely and coprimely on the finite solvable group G, then necessarily h(G) $\leqslant 2\,$l(A). Even though the inequality in Theorem 4.5 is the best possible for each h($C_G(A)$) > 0, it seems to be twice what it should be in the fixed point free case.

Conjecture 5.1   *Let A be a finite group and let G be a finite solvable group on which A acts such that* $(|A|, |G|) = 1$ *and* $C_G(A) = 1$. *Then*

$$\text{h}(G) \leqslant \text{l}(A).$$

This conjecture has encouraged work on the systematic calculation of the invariant l(A). The following easy lemma reduces the calculation to simple groups.

Lemma 5.2   *Let A be any finite group, and let $C_1, \ldots, C_\alpha$ be its composition factors. Then*

$$\text{l}(A) = \sum_{i=1}^{\alpha} \text{l}(C_i).$$

Hence, l(A) is simply the sum of the corresponding lengths of its composition factors. In particular, for each finite solvable group A, we have l(A) $= \Omega(|A|)$.

Proof.   Use induction on $|A|$. If A is a simple group, there is nothing to prove. So suppose that A has a normal subgroup N, with $A \neq N \neq 1$. By induction, it is enough to show that l(A) = l(A/N) + l(N). By selecting a chain of length l(A/N) in A/N and a chain of length l(N) in N and putting one after the other, we see that

$$\text{l}(A) \geqslant \text{l}(A/N) + \text{l}(N).$$

Now let $1 = A_0 < A_1 < \ldots < A_l = A$ be a chain of subgroups of A of length $l = $ l(A). We obtain two chains (with possible repeats); one is a chain of subgroups of A/N and the other a chain of subgroups of N:

$$A_0 N/N \leqslant A_1 N/N \leqslant A_2 N/N \leqslant \ldots \leqslant A_l N/N,$$
$$A_0 \cap N \leqslant A_1 \cap N \leqslant A_2 \cap N \leqslant \ldots \leqslant A_l \cap N.$$

If we had $A_i N = A_{i+1} N$ and $A_i \cap N = A_{i+1} \cap N$, then because $A_i < A_{i+1}$ we would have $A_i = A_{i+1}$, a contradiction. Hence, l is less than or equal to the sum of the lengths of the two strictly increasing sequences generated by removing the repeats. It follows that $l(A) \leqslant l(A/N) + l(N)$. This concludes the proof of the lemma.   □

This lemma then reduces the calculation of $l(A)$ to the case where A is a simple group. The value for A an alternating group is given by the following theorem [11].

Theorem 5.3   *Let n be a positive integer. Then*

$$l(\mathrm{Sym}_n) = \lfloor \frac{3n - 1}{2} \rfloor - b_n$$

where b_n is the number of ones in the base two expansion of n.

The proof does use the classification of the finite simple groups. Work on the calculation of $l(A)$ for other finite simple groups can be found, for example, in [65, 69, 77], as well as in more recent work of D. Brozovic. This work makes use of the classification of the finite simple groups.

We do know that the inequality in Conjecture 5.1, if correct, is the best possible in a very strong sense. We now present a construction that proves this. We will later describe how, in the presence of regular orbits, the general situation is quite similar to our construction. The following theorem appears in [74]. It provides the required example in the case $B = 1$.

Theorem 5.4   *Let A be any finite group and let B be a subgroup of A. Then there exists some solvable group G such that A acts fixed point freely and coprimely on G,*

$$\bigcap_{a \in A} [B, G]^a = 1,$$

and $\mathrm{h}(G) = l(A : B)$.

Proof.   Work by induction on $l(A : B)$. If $l(A : B) = 0$, then we may simply take $G = 1$. Hence, we assume that $l(A : B) > 0$. Let C be a subgroup of A strictly containing B and such that $l(A : C) = l(A : B) - 1$. Apply induction to A, C, and obtain a solvable group G_0, such that A acts fixed point freely and coprimely on G_0,

$$\bigcap_{a \in A} [C, G_0]^a = 1,$$

and $\mathrm{h}(G_0) = l(A : C)$. Since $[C, G_0]$ is a normal subgroup of G_0, and A acts on G_0 by automorphisms, for each α, if the α-th descending Fitting subgroup of G_0 is contained in $[C, G_0]$, then it is also contained in all A-conjugates of $[C, G_0]$, and so it must then be 1. Hence

$$\mathrm{h}(G_0) = \mathrm{h}(G_0 / \bigcap_{a \in A} [C, G_0]^a) \leqslant \mathrm{h}(G_0/[C, G_0]) \leqslant \mathrm{h}(G_0).$$

It follows that

$$\mathrm{h}(G_0/[C, G_0]) = \mathrm{h}(G_0) = l(A : C).$$

Since B is a proper subgroup of C, the induced character $1_B{}^C$ contains some irreducible character χ of C such that $\chi \neq 1_C$. By Frobenius reciprocity, we have $1_B \subseteq \chi|_B$. By Dirichlet's Theorem, we may take p to be a prime not dividing the order of AG_0, for which $\mathrm{GF}(p)$ is a splitting field for all subgroups of AG_0. Let N_1 be an irreducible $\mathrm{GF}(p)C$-module affording χ as its Brauer character. We have $C_{N_1}(C) = 0$, and $C_{N_1}(B) \neq 0$. Let N_0 be some irreducible $\mathrm{GF}(p)G_0/[C, G_0]$-module such that $\mathrm{h}(G_0/\ker(N_0)) = \mathrm{h}(G_0)$. Notice that C centralizes $G_0/[C, G_0]$. Then we may obtain the irreducible CG_0-module $N = N_1 \otimes N_0$ where C acts on N_1 and G_0 acts on N_0. We have $C_N(C) = 0$ and $C_N(B) \neq 0$. Furthermore, $[C, G_0] \subseteq \ker(N)$ and $\mathrm{h}(G_0 N) = \mathrm{l}(A : B)$.

Set M to be N^{AG_0}, the module N induced from CG_0 to AG_0. Further set $H = G_0 M$. Again by Frobenius reciprocity, $C_M(A) = 0$. It follows that $C_H(A) = 1$. Write $M = N \oplus R$ where R is the sum of all the A-conjugates of N which are distinct from N. The sum is direct by the definition of the induced module M. Notice that, since the actions of B and of G_0 on N commute, $\ker(N)([B, N] \oplus R)$ is a normal subgroup of BH. It follows that $[B, H] \subseteq \ker(N)([B, N] \oplus R)$. Since $C_N(B) \neq 0$, it follows that

$$M \nsubseteq [B, H].$$

Set $G = H/\bigcap_{a \in A}[B, H]^a$. Because $(|A|, |H|) = 1$, A acts fixed point freely on G. Furthermore, it follows immediately from our construction that $\bigcap_{a \in A}[B, G]^a = 1$. Finally, by our construction, M is the $\mathrm{l}(A : C)$-th descending Fitting subgroup of H. Since $M \nsubseteq \bigcap_{a \in A}[B, H]^a$, it follows that $\mathrm{h}(G) > \mathrm{l}(A : C)$. This immediately implies that $\mathrm{h}(G) = \mathrm{l}(A : B)$, and completes the proof of the theorem. $\square$

Conjecture 5.1 is known to hold in a fairly large class of groups. We now discuss these results and how they are obtained. One cannot proceed as in Theorem 4.5 and use induction on normal subgroups of A, because for them the action is no longer fixed point free. One needs, instead, to work with A and try to show that part of A is acting trivially on G or some appropriate factor group of G. It is easy to give an equivalent form of the conjecture in terms of representation theory. The argument is a straightforward application of the Fong-Swan Theorem.

Conjecture 5.5   *Let A be a finite group and let G be a finite solvable group on which A acts such that $(|A|, |G|) = 1$, $C_G(A) = 1$ and $\mathrm{h}(G) = \mathrm{l}(A)$. Let M be any AG-module over the complex numbers which is faithful for G. Then $C_M(A) \neq 0$.*

Routine reductions get us to the case where M is irreducible. Then we let N be a homogeneous component of $M|_G$, and let $BG = \mathrm{N}_{AG}(N)$ be the inertia group of N, where B is a subgroup of A. By Clifford theory, the module M is induced from N viewed as a BG-module. By Frobenius reciprocity, $C_N(B) = 0$. Hence, representation theory leads us to the study of the BG-module N, which is homogeneous for G. The fixed point free action of A on G seems difficult to use directly. In order to use the full action of A on G we will need to prove a more precise theorem, as we will see below. For the time being let us investigate BG and N in a more general setting. First, of course, N may be induced from some module. The following lemma applies to this case.

Lemma 5.6   *Let B be a finite group and let G be a finite solvable group on which B acts such that $(|B|, |G|) = 1$. Let N be an irreducible complex BG-module such that $N|_G$ is homogeneous. Let H be some B-invariant subgroup of G and let L be some complex BH-module such that the induced module L^{BG} is isomorphic to N. Then there exist elementary abelian B-invariant sections $S_1, \ldots, S_n$ of G with the following property. Let $\alpha_1, \ldots, \alpha_m$ be any set of representatives for the orbits of B on the direct product*

$$S_1 \times \cdots \times S_n.$$

Then,

$$N|_B \cong \bigoplus_{i=1}^{m} [L|_{C_B(\alpha_i)}]^B.$$

Proof.   For a proof see Lemma 1.7 in [75] and apply the Mackey formula. Here is a quick sketch in case BH is maximal in BG (the general case follows fairly easily from this one). Let V a subgroup of G which is normal in BG and minimal with respect to not being contained in H. By the maximality of BH, $HV = G$. Further, V' must be contained in H, so that $V' \subseteq H \cap V$. It follows that $H \cap V$ is normalized by BH and by V, so that $H \cap V$ is a normal subgroup of BG contained in V. The minimality of V then implies that $V/(H \cap V)$ is a BG-chief factor. We set $S_1 = V/(H \cap V)$ and $n = 1$. The cosets of BH in BG are then in one to one correspondence with the elements of S_1. The action of B by left multiplication on these cosets is permutation isomorphic to the action of B on the chief factor S_1, where the B-BH-double cosets in BG correspond naturally to the B-orbits of S_1. If $\alpha \in S_1$, then there is some representative $r \in V$ of α in V which in centralized by every element of B that centralizes α (this is because $(|B|, |G|) = 1$). Obviously $C_B(\alpha) = C_B(r) \subseteq B \cap (BH)^r$. But, in fact, we even have $C_B(\alpha) = B \cap (BH)^r$. Indeed, suppose $b \in B \cap (BH)^r$. Then $[b, r^{-1}] \in (BH) \cap V = H \cap V$, which means that b also fixes the coset α, and $b \in C_B(\alpha)$. Picking the various r as representatives for the double cosets, we can apply the Mackey formula to complete the proof. $\square$

This lemma explains why an assumption on the existence of regular orbits on finite B-modules would be useful.

Definition 5.7   *Let B be a finite group and let it act on a solvable finite group G with $(|B|, |G|) = 1$. We say that B acts on G with regular orbits if for each elementary abelian B-invariant section S of G there is some $v \in S$ such that $C_B(v) = C_G(S)$. The B-orbit that v generates is called a regular orbit.*

Not all actions are with regular orbits, of course, but many of them are. We will discuss some results on regular orbits in the next section. We are now ready to state the following theorem [75] which gives the most general answer known to Conjecture 5.1. To obtain a special case of Conjecture 5.1 simply set $A_0 = 1$ in the theorem. The conjecture is also known to hold for A quaternion, dihedral or generalized quaternion, by other methods; see [72].

Theorem 5.8   *Let A be a finite group acting on the solvable group G such that $(|A|, |G|) = 1$ and $C_G(A) = 1$. Assume that every proper subgroup of A acts on G with regular orbits. Suppose A_0 is a subgroup of A such that $\bigcap_{a \in A}[A_0, G]^a = 1$. Then $\mathrm{h}(G) \leqslant \mathrm{l}(A : A_0)$.*

Proof.   For a complete proof see [75]. We offer here a short outline. Assume the theorem is false and consider a counterexample: a finite group A acting on a finite solvable group G_0 satisfying the conditions of the theorem, but not the conclusion, with minimum $|AG_0| + [A : A_0]$. We set $h = \mathrm{h}(G_0) - 1$. By Lemma 4.3, there exists an irreducible A-tower $(\hat{P}_i)$, $i = 1, \ldots, h+1$, of height $h+1$, consisting of subgroups of G_0. The minimality of the counterexample implies that for every subgroup C of A which strictly contains A_0, and every $h + 1 \geqslant h_0 \geqslant h + 1 - \mathrm{l}(C : A_0)$, we have

$$\hat{P}_{h_0} = \langle [C, \hat{P}_{h_0}]^{\hat{P}_1 \cdots \hat{P}_{h_0}} \rangle. \tag{3}$$

This is because, if (3) does not hold, A acts fixed point freely on $G_0/\bigcap_{a \in A}[C, G_0]^a$ and

$$\mathrm{h}(G_0 / \bigcap_{a \in A}[C, G_0]^a) \geqslant h_0.$$

The minimality of our counterexample then yields

$$\mathrm{h}(G_0 / \bigcap_{a \in A}[C, G_0]^a) \leqslant \mathrm{l}(A : C).$$

These two inequalities imply $h_0 \leqslant \mathrm{l}(A : C)$. On the other hand, we assume $h_0 \geqslant h + 1 - \mathrm{l}(C : A_0)$ and $h + 1 > \mathrm{l}(A : A_0)$, so that $h_0 > \mathrm{l}(A : A_0) - \mathrm{l}(C : A_0) \geqslant \mathrm{l}(A : C)$. This contradiction shows (3).

Since $\bigcap_{a \in A}[A_0, G_0]^a = 1$, $\hat{P}_{h+1} \not\subseteq [A_0, G_0]$, which implies that

$$\hat{P}_{h+1}/\phi(\hat{P}_{h+1})(\hat{P}_{h+1} \cap [A_0, G_0])$$

is a non-trivial A_0G_0-module (over a finite field) where A_0 acts trivially. We set $G = \hat{P}_1 \cdots \hat{P}_h$. By considering the action of A_0G on this module and using theorems of Clifford, Maschke and Fong-Swan, one obtains a subgroup B of A containing A_0 and a complex irreducible BG-module N with the properties $N|_G$ is homogeneous, $\hat{P}_h \not\subseteq \ker(N)$, $C_N(B) = 0$ and $A_0 \subseteq \ker(N)$. The group B will centralize at least some non-trivial abelian normal subgroup of $G/(G \cap \ker(N))$, and therefore B is a proper subgroup of A. Hence, by hypothesis, B acts with regular orbits on G. By applying Clifford theory to the module N with respect to $\hat{P}_h$, we see that N is induced from a BH-module L such that $L|_{\hat{P}_h}$ is homogeneous and $\hat{P}_h \subseteq H$, and, if $\hat{P}_h$ is not abelian, we even have $\hat{P}_{h-1} \subseteq H$. We apply Lemma 5.6. It is easy to see that B has a regular orbit on $S_1 \times \cdots \times S_n$ generated by some α_i. By Frobenius reciprocity, we then have that $L|_{C_B(\alpha_i)}$ does not contain the trivial $C_B(\alpha_i)$-module. Hence, if we set $C = C_B(\alpha_i)$, we have $C_L(C) = 0$. In particular, C strictly contains A_0.

Suppose first that $\hat{P}_h$ is abelian. Then it projects into the center of $BH/\ker(L)$, so that $[C, \hat{P}_h] \subseteq \ker(L)$. Since C centralizes $S_1 \times \cdots \times S_n$, and $\ker(N) = \bigcap[\ker(L)]^r$, where the r range over certain elements of G centralized by C, it follows that $[C, \hat{P}_h] \subseteq \ker(N)$. Since $\hat{P}_h \not\subseteq \ker(N)$, this implies that

$$\hat{P}_h \neq \langle [C, \hat{P}_h]^{\hat{P}_1 \cdots \hat{P}_h} \rangle.$$

This contradicts (3) with $h_0 = h$, which is applicable because $l(C : A_0) \geqslant 1$. This contradiction completes the proof in this case.

Suppose now that $\hat{P}_h$ is not abelian. Considerations similar to the above yield that the hypotheses of the following theorem are satisfied with $CH/\ker(L)$ taking the place of AG, the image of $\hat{P}_h$ in place of P, the image of $\hat{P}_{h-1}$ in place of Q, and the character afforded by L in place of χ. Hence, the conclusion holds, and $\chi|_C$ must contain in particular the trivial character. Since we know that $C_L(C) = 0$, this is a contradiction and completes the proof of the theorem. □

Theorem 5.9 *Let AG be a finite group with G a normal solvable subgroup of AG and $(|A|, |G|) = 1$. Let P be an extraspecial p-subgroup of G, for some prime p, such that P is normal in AG and $Z(P) \subseteq Z(AG)$. Let Q be an A-invariant q-subgroup of G, with q a prime, $q \neq p$, such that $(Q/C_Q(P))/Z(Q/C_Q(P))$ is elementary abelian, $[Q, P] = P$ and, in addition, if $q \neq 2$, then $\exp(Q/C_Q(P)) = q$. Assume that $QC_G(P/P')$ is normal in AG, that $Q \subseteq \langle [B, Q]^G \rangle$ for every subgroup B of A such that $l(B) \geqslant 2$, and that $P \subseteq \langle [C, P]^G \rangle$ for every subgroup C of A such that $l(C) \geqslant 1$. Assume furthermore that every subgroup D of A acts with regular orbits. Let χ be a character of AG such that $\chi|_P$ is faithful. Then $\chi|_A$ contains the regular A-character.*

Proof. For a full proof see Theorem 1.8 in [75]. It is, of course, not true that if A is any finite group acting faithfully on an extraspecial group P, and χ is some faithful character of the semidirect product, then $\chi|_A$ contains a regular module. The exceptions to this are the reason why one needs to have such precise hypotheses in this theorem. With the exception of the existence of regular orbits, these hypotheses do hold, however, in the cases we are interested in. Because of these exceptions the proof of the theorem has to make full use of its hypotheses. After the usual reduction to an irreducible χ and even to a χ such that $\chi|_P$ is already irreducible, the proof proceeds by investigating the action of AG on P/P'. It reduces to the case where AG acts irreducibly on P/P' by considering certain tensor products of characters. Then it considers the action of Q on P/P' and reduces further to the case where this action is homogeneous. This crucial reduction is achieved because, whenever P/P' is induced as an AG-module, then χ is *tensor induced* from a smaller module. We can not go into a discussion of tensor induction here. It is however an essential tool in the understanding of the representations of groups with a normal extraspecial subgroup.

The tensor induction allows us to concentrate on the case where Q acts homogeneously (and faithfully) on P/P'. This makes Q either extraspecial or cyclic. On the other hand, our hypotheses imply that $l(C_A(Q)) \leqslant 1$, so that $C_A(Q)$ is either 1 or cyclic of prime order. The representation theory of $C_A(Q)QP$ can be studied in some detail. One can show that, in the cases of interest, for every irreducible character λ of $C_A(Q)$, there is a linear character μ of Q such that $A/C_A(Q)$ generates a regular orbit in its action on μ and $\lambda \otimes \mu \subseteq \chi|_{C_A(Q)Q}$. This is easily seen to imply that $\lambda^A \subseteq \chi|_A$. As this holds for each $\lambda \in \mathrm{Irr}(A)$, the result then follows. This overall scheme is complicated by the many exceptions that have to be dealt with: not all characters of the form $\lambda \otimes \mu$ need be contained in $\chi|_{C_A(Q)Q}$; tensor induction may not always give the desired result; etc. The proof comes down to the explicit calculation of the values of $\chi|_{C_A(Q)Q}$ in the cases where either P or Q is an extraspecial 2-group.   □

6. Regular Orbits

We have seen how useful it is to know that a group A acting on a finite module M has a regular orbit, i.e. there is some vector $v \in M$ such that $C_A(v) = C_A(M)$. In this section we investigate some results that guarantee the existence of such orbits in certain cases. Obviously not all groups have regular orbits in their action on finite modules. For example, if M is a finite vector space over some finite field, the group $\mathrm{GL}(M)$ acts naturally on it, but will not have regular orbits unless the dimension of M is less than 2. For this reason, all theorems that assert the existence of regular orbits have some extra conditions on the relationship between A and M. For our purposes, the condition $(|A|, |M|) = 1$ is a natural one to assume. We will see that this simple condition takes us very far in many cases, especially when A is nilpotent. Other conditions have been proposed in the literature. For example, Espuelas [25] uses the assumptions A of odd order, complete reducibility of M and $M = P/P'$ where P is some extraspecial group of odd order normalized by A. Carlip [12] assumes that A is nilpotent and contained in some solvable group G which acts faithfully on M (a vector space over a field in characteristic p) with $O_p(G) = 1$ and both $|G|$ and p odd; see also [27].

When $(|A|, |M|) = 1$, the structure of the module M is determined by the field F and the Brauer character χ of the module. Hence from this information we expect to be able to decide whether or not the module has a regular orbit. Indeed there is a simple formula that gives the exact number of regular orbits (see Lemma 1.4 in [75]):

Lemma 6.1   *Let A be a finite group and let F be a finite field with q elements, $(|A|, q) = 1$. Let M be an FA-module with Brauer character χ. Assume that A acts faithfully on M. Then the number of regular orbits of A on M (as a permutation action) is*

$$\frac{1}{|A|} \sum_{S \subseteq S_0} (-1)^{|S|} q^{f(S)}$$

where S_0 is the set of subgroups of A of prime order and, for S a subset of S_0,

$$f(S) = \frac{1}{|\langle S \rangle|} \sum_{x \in \langle S \rangle} \chi(x),$$

where $\langle S \rangle$ is the subgroup of A generated by the union of the elements of S.

Proof.   Let n be the number of regular orbits of A on M and let Ω be the union of the regular orbits of A on M. The elements of Ω are exactly those of M whose centralizer in A is trivial, that is does not contain any subgroup of prime order. Hence

$$\Omega = M \setminus \bigcup_{\alpha \in S_0} \mathrm{C}_A(\alpha).$$

The inclusion exclusion principle then implies

$$|\Omega| = q^{\chi(1)} + \sum_{\substack{S \subseteq S_0 \\ S \neq \emptyset}} (-1)^{|S|} |\bigcap_{\alpha \in S} \mathrm{C}_M(\alpha)|.$$

Since $\bigcap_{\alpha \in S} \mathrm{C}_M(\alpha) = \mathrm{C}_M(\langle S \rangle)$ and, since $(|A|, q) = 1$, the dimension of the centralizer is simply the multiplicity of the trivial character in χ, so

$$\dim_F(\mathrm{C}_M(\langle S \rangle)) = \frac{1}{|\langle S \rangle|} \sum_{x \in \langle S \rangle} \chi(x) = f(S).$$

It follows that

$$|\Omega| = \sum_{S \subseteq S_0} (-1)^{|S|} q^{f(S)}.$$

Since $|\Omega| = n|A|$, the formula follows immediately. $\square$

This lemma implies that, in a sense, most of the irreducible modules of a given group A do contain regular orbits. For each finite group and each of its irreducible characters, there are infinitely many finite modules (in different finite characteristics) that have the given character as their Brauer character. Almost all of them do have a regular orbit.[4]

Corollary 6.2   *Let A be a finite group. Then there is only a finite number of isomorphism types of finite irreducible A-modules M for which $(|A|, |M|) = 1$ and A does not have a regular orbit on M.*

Proof.   Consider a finite irreducible module M for A over a finite field K, with $(|K|, |A|) = 1$. By considering M over a larger field $F = \mathrm{End}_{KA}(M)$, if necessary, we may assume that M is absolutely irreducible. Let χ be the Brauer character of M (considered over F). By the previous lemma, there is a polynomial $P_\chi(x)$ of degree $\chi(1)$ such that A does not have a regular orbit on M if and only if $P_\chi(|F|) = 0$. Hence, for each χ, the number of F for which M does not have a regular orbit is finite (in fact less than or equal to $\chi(1)$). Hence, the number of possible K is likewise finite. Since there are only finitely many $\chi \in \mathrm{Irr}(A)$, the result follows. $\square$

The actual calculation of this polynomial is often difficult, because it involves the study of the restriction of χ to many of the subgroups of A. For this reason, other methods need to be used to prove general statements about the existence of

---

[4] One can also obtain, with more work, similar finiteness results allowing A to vary among certain classes of groups. For example, in [37] it is shown that there is, up to isomorphism, only a finite number of pairs (A, M), where A is a finite quasi-simple group, M is a finite faithful module for A with $[A, M] = M$ and $(|A|, |M|) = 1$, and A does not have a regular orbit on M.

regular orbits. Either the module M can be induced from some smaller module N, or the module M is primitive. A second method of showing that certain modules do have regular orbits is to understand to what extent regular orbits for N give rise to regular orbits for M (when M is induced from N), and to understand the orbits on primitive modules. We start by examining the induced case.

Let M be a faithful module for a finite group A and let S be some subgroup of A, N some S-module such that $N^A = M$. From the action of $S/\ker(N)$ on N, we need to understand the action of A on M. The best way to do this is via the Frobenius map. This very important map gives us a clear understanding of what the group $A/\ker(M)$ can be, but is not emphasized in the standard group theory texts. For this reason we now treat this in some detail.

The Frobenius map is a homomorphism of A into the wreath product of Sym_n and $S/\ker(N)$. We want to make this completely explicit. For this we need to establish our conventions about composition of functions. Since nobody writes characters on the right of their argument, I see no advantage, only confusion, in writing any other functions that way. Hence, I always use functions on the left of their argument, and compose them from right to left. With these conventions it is easy to define the symmetric group Sym_n explicitly. Sym_n is the set of bijections $\pi : \{1, \ldots, n\} \to \{1, \ldots, n\}$ under composition. If $\pi, \pi' \in \mathrm{Sym}_n$ then, for all $i = 1, \ldots, n$, $\pi \circ \pi'(i) = \pi(\pi'(i))$. One can also use composition of functions to define the wreath product in an efficient way. I use the notation $\mathrm{Sym}_n \wr B$ (where B is a group) for the wreath product, since I feel that the acting group should be on the left.

Definition 6.3 *Let B be a finite group and $n \geqslant 1$ an integer. We define the* wreath product $\mathrm{Sym}_n \wr B$ *as the set of pairs (π, f), where $\pi \in \mathrm{Sym}_n$ and $f : \{1, \ldots, n\} \to B$, with the product operation defined, for $(\pi, f), (\pi', f') \in \mathrm{Sym}_n \wr B$, by*

$$(\pi, f)(\pi', f') = (\pi \circ \pi', (f \circ \pi')f'),$$

where the juxtaposition of the two functions $f \circ \pi$ and f' simply means pointwise multiplication.

Definition 6.4 *Let A be a finite group and let S be a subgroup of A of index n. Let K be a normal subgroup of S and set $B = S/K$. Let $x_1, \ldots, x_n \in A$ be representatives for the left cosets $x_1 S, \ldots, x_n S$ of S in A. Then the* Frobenius map $\varphi : A \to \mathrm{Sym}_n \wr B$ *is defined as follows. For each $a \in A$, $\varphi(a) = (\pi, f) \in \mathrm{Sym}_n \wr B$ where $\pi : \{1, \ldots, n\} \to \{1, \ldots, n\}$ is the map such that*

$$a x_i S = x_{\pi(i)} S \qquad \text{for all} \quad i = 1, \ldots, n,$$

and $f : \{1, \ldots, n\} \to B$ is the map defined by

$$f(i) = x_{\pi(i)}^{-1} a x_i K \in S/K \qquad \text{for all} \quad i = 1, \ldots, n.$$

Theorem 6.5 *With the hypotheses and notation of Definition 6.4, the Frobenius map $\varphi : A \to \mathrm{Sym}_n \wr B$ is a group homomorphism with kernel $\bigcap_{a \in A} K^a$.*

Proof. This consists of straightforward computations which we omit; see, for example, p. 53 in [71].   □

If B (and S) act on the module N, then $\mathrm{Sym}_n \wr B$ acts on $M = N \oplus \cdots \oplus N$ in a natural way. The action of A on this module via the Frobenius homomorphism makes M into the induced module $M = N^A$. Likewise, $\mathrm{Sym}_n \wr B$ acts on $M' = N \otimes \cdots \otimes N$ in a natural way. The action of A on this module via the Frobenius homomorphism makes M' into the tensor induced module $M' = N^{\otimes A}$. Hence, the Frobenius map gives a convenient framework to understand both induction and tensor induction. Using this, it is easy to give some conditions on the existence of regular orbits for $A/\mathrm{ker}(M)$ on M from the existence of regular orbits of B on N. As an example we give one such result that treats the case where S is a normal subgroup of A. The proposed point of view is, however, even more useful when S is no longer normal in G.

Lemma 6.6   *Let A be a finite group and let S be a normal subgroup of A. Let N be a finite S-module and let $B = S/\mathrm{ker}(N) \neq 1$. Let $M = N^A$ be the induced module and assume that A acts faithfully on M. Suppose that B has a regular orbit on N. Then the following hold.*

(1) *If A does not have any section isomorphic to a wreath product $\mathbb{Z}_\alpha \wr \mathbb{Z}_\beta$ (for integers α, β greater than 1), then A has a regular orbit in its action on M.*

(2) *If $A = \mathbb{Z}_p \wr \mathbb{Z}_p$, where $p = 2^m - 1$ is a Mersenne prime, and N is an irreducible module for $\mathbb{Z}_p$ with $|N| = 2^m$, then A does not have a regular orbit on M.*

Proof.   Let $n = |A/S|$. Since the Frobenius map is injective in this case, we may view A itself as a subgroup of $\mathrm{Sym}_n \wr B$. Since $B \neq 1$, the action of B on N has at least two orbits: a regular orbit containing the element $\omega \in N$ say, and $0 \in N$.

Suppose that A does not have a regular orbit. Consider the element

$$(\omega, \ldots, \omega) \in N \oplus \cdots \oplus N.$$

Its centralizer in $\mathrm{Sym}_n \wr B$ is Sym_n, so that, since A does not have a regular orbit on M, $A \cap \mathrm{Sym}_n \neq 1$. Next consider the element

$$(0, \omega, \ldots, \omega) \in N \oplus \cdots \oplus N.$$

Since S is a normal subgroup of A, if an element of A fixes any coset of S in A by left multiplication, it must be in S. This means that it fixes all cosets, i.e. that it is in $B \times \cdots \times B$. Hence, $C_A((0, \omega, \ldots, \omega)) \subseteq B \times \cdots \times B$. It then follows that $C_A((0, \omega, \ldots, \omega)) \subseteq B \times 1 \times \cdots \times 1$. We now have some $\sigma \in A \cap \mathrm{Sym}_n$ which does not fix any point, and some $(\lambda, 1, \ldots, 1) \in A \cap (B \times 1 \times \cdots \times 1)$ which is not 1. Hence a section of A is isomorphic to some wreath product, a contradiction.

Now assume that the conditions of (2) are satisfied[5]. Then there are exactly two orbits of B on N. Any element of $N \oplus \cdots \oplus N$ with at least one component equal to zero has a non-trivial centralizer in $B \times \cdots \times B$. If all components are not zero,

---

[5] A counting argument can be used to prove that A does not have any regular orbit. I think, however, that the argument that follows is more instructive.

by multiplying by an appropriate element of $B \times \cdots \times B$ we may assume that all components are equal, in which case its centralizer in A is not trivial. Hence, A does not have a regular orbit on M.   □

When A is acting primitively (and faithfully) on M, then every normal abelian subgroup of A needs to be cyclic (although not necessarily central). This is a powerful condition for a solvable group and substantially restricts its structure. A description of A in this case appears, for example, in Corollary 1.10 in [62]. Using this restriction one can show the existence of regular orbits in certain cases. If A is supersolvable then A is isomorphic to a subgroup of $\Gamma L(1, GF(p^m))$, the semilinear group of dimension 1 over a field $GF(p^m)$ in characteristic p. Certain subgroups of $\Gamma L(1, GF(p^m))$ do not have a regular orbit.

Definition 6.7 *Let p and q be primes, and let e be a positive integer. Let S be the subgroup of order q of the cyclic group $\mathrm{Gal}(GF(p^{q^e})/GF(p))$. Let*

$$N(p^{q^e}, q) = \{\lambda \in GF(p^{q^e})^{\times} : \prod_{\sigma \in S} \sigma(\lambda) = 1\},$$

and

$$GN(p^{q^e}, q) = S \cdot N(p^{q^e}, q)$$

be the semidirect product of S and $N(p^{q^e}, q)$. This semidirect product is a subgroup of $\Gamma L(1, GF(p^{q^e}))$.

The group $GN(p^{q^e}, q)$, just defined, never has a regular orbit on its natural module $GF(p^{q^e})$ of dimension q^e over $GF(p)$. However, if these groups are all absent from a supersolvable group with no wreath product sections, then A does have regular orbits.

Theorem 6.8 *Let A be a supersolvable group with no section isomorphic to $\mathbb{Z}_\alpha \wr \mathbb{Z}_\beta$ and no section isomorphic to $GN(p^{q^e}, q)$, for any prime q. Let A act faithfully on a finite module M over the field $GF(p)$, where p does not divide $|A|$. Then A has a regular orbit on M.*

Proof. See [71].   □

If we further assume A to be nilpotent then the number of sections that need to be omitted is decreased. The following result is due to B. Hargraves, who improved on earlier results of T. Berger.

Theorem 6.9 *Let A be a nilpotent group and suppose that no section of it is isomorphic to $\mathbb{Z}_2 \wr \mathbb{Z}_2$ or to $\mathbb{Z}_p \wr \mathbb{Z}_p$, for p any Mersenne prime. Let M be any finite faithful module such that $(|A|, |M|) = 1$. Then A has a regular orbit on M.*

Proof. See [39]. Other treatments and related results appear in [9, 10, 16, 24, 30, 48, 54, 62].   □

7. Concluding Remarks

When combining Theorem 5.8 with some regular orbit results one can obtain explicit cases of Conjecture 5.1. Here are some examples that appear in [75].

Theorem 7.1 *Let $A \cong SL_2(2^n)$, $Sz(2^{n+1})$ or M_{11}. Suppose that A acts on the finite (solvable) group G with $C_G(A) = 1$ and $(3|A|, |G|) = 1$. Then*

$$h(G) \leqslant l(A).$$

Theorem 7.2 *Let A be a finite group of operators on the finite (solvable) group G such that $(|A|, |G|) = 1$ and $C_G(A) = 1$. Assume further that every proper subgroup of A is nilpotent, $\mathbb{Z}_2 \wr \mathbb{Z}_2$-free and $\mathbb{Z}_p \wr \mathbb{Z}_p$-free for all Mersenne primes p. Then*

$$h(G) \leqslant l(A).$$

Theorem 7.3 *Let A be a finite supersolvable group of operators on the (solvable) finite group G such that $(|A|, |G|) = 1$ and $C_G(A) = 1$. Assume further that no section of A is isomorphic to $\mathbb{Z}_\alpha \wr \mathbb{Z}_\beta$ or to $GN(p^{q^e}, q)$, for any primes p and q, where p divides $|G|$. Then*

$$h(G) \leqslant l(A).$$

Although these results are fairly general, they do not cover all cases. We have seen how the existence of regular orbits helps prove Conjecture 5.1 in the cases where they exist. We have also seen that we have techniques that allow us to study when it is that modules fail to have regular orbits. Perhaps what we need for further progress in the fixed point free question is to find some property of finite modules, related to the existence of regular orbits, but weaker, which we can use in the proof of the conjecture, but which does not have exceptions. If this condition is simple enough, perhaps our methods could prove that it holds always.

Appendix: Notation

We describe some of the notation that we use. We let G be a finite group and p be a prime. Then we use, as usual, G' for the commutator subgroup of G; $Z(G)$ for the center of the group G; $\phi(G)$ for the Frattini subgroup of G; $O_p(G)$ for the largest normal p-subgroup of G, and $O_{p'}(G)$ for the largest normal p'-subgroup of G. In addition, we use the following notation.

$O_{p'p}(G)$: The preimage in G of $O_p(G/O_{p'}(G))$.

$F(G)$: The Fitting subgroup of G, i.e. the largest normal nilpotent subgroup of G.

$F_m(G)$: The m-th Fitting subgroup of G, defined inductively by $F_0(G) = 1$ and $F_m(G)/F_{m-1}(G) = F(G/F_{m-1}(G))$.

$\exp(G)$: The exponent of G, the smallest positive integer n such that $g^n = 1$ for all $g \in G$.

$e_p(G)$: The non-negative integer such that $p^{e_p(G)}$ is the p-part of $\exp(G)$.

$\mathrm{dl}(G)$: The derived length of G, the smallest non-negative integer n, such that $G^{(n)} = 1$.

$d_p(G)$: The derived length of a Sylow p-subgroup of G.

$l_p(G)$: The p-length of G. For example, $l_p(G) \leqslant 1$ if and only if $O_{p'pp'}(G) = G$.

$h(G)$: The Fitting height of G, also called the nilpotent length of G, the smallest integer h such that $\mathbf{F}_h(G) = G$.

$l(G)$: The length of a longest chain of subgroups of G. For example, when G is solvable $l(G) = \Omega(|G|)$.

$l(G : H)$: The length of a longest chain of subgroups of G which starts with the subgroup H of G. Of course, $l(G) = l(G : 1)$.

$\omega(n)$: The number of prime divisors of the number n, not counting multiplicities. For example $\omega(12) = 2$.

$\Omega(n)$: The number of prime divisors of the number n, counting multiplicities. For example $\Omega(12) = 3$.

$\lfloor a \rfloor$: The largest integer less than or equal to the real number a.

References

1. J. Abu-Joukha, The Fitting length of finite groups admitting an automorphism of prime order, *Doga Mat.* 15 (1991), 121–128.
2. S. D. Bell and B. Hartley, A note on fixed-point-free actions of finite groups, *Quart. J. Math. Oxford* (2) 41 (1990), 127–130.
3. T. Berger, Representation theory and solvable groups: length type problems, in *The Santa Cruz Conference on Finite Groups*, Proc. Symposia Pure Math. 37 (1980), 431–441.
4. T. Berger, B. Hargraves and C. Shelton, The regular module problem III, *J. Algebra* 131 (1990), 74–91.
5. Y. G. Berkovich, Relations between some invariants of finite solvable groups, *Soobshch. Akad. Nauk Gruzin. SSR* 123 (1986), 469–472.
6. Y. G. Berkovich, The existence of normal subgroups of a finite group, *Publ. Math. Debrecen* 37 (1990), 1–13.
7. E. G. Bryukhanova, The 2-length and 2-period of a finite solvable group, *Algebra i Logika* 18 (1979), 9–31.
8. E. G. Bryukhanova, Connection between the 2-length and the derived length of a Sylow 2-subgroup of a finite solvable group, *Mat. Zametki* 29 (1981), 161–170.
9. P. J. Cameron, Regular orbits of permutation groups on the power set, *Discrete Math.* 62 (1986), 307–309.
10. P. J. Cameron, P. M. Neumann, J. Saxl, On groups with no regular orbits on the set of subsets, *Arch. Math.* 43 (1984), 295–296.
11. P. J. Cameron, R. Solomon and A. Turull, Chains of subgroups in symmetric groups, *J. Algebra* 127 (1989), 340–352.
12. W. Carlip, Regular orbits of nilpotent subgroups of solvable groups, *Illinois J. Math.* 38 (1994), 198–222.
13. C. Casolo, Finite groups with small conjugacy classes, *Manuscripta Math.* 82 (1994), 171–189.

14. Cheng, Kai-Nah, Finite groups admitting fixed point free automorphisms of order pq, *Proc. Edinburgh Math. Soc.* 30 (1987), 51–56.

15. E. C. Dade, Carter subgroups and Fitting heights of finite solvable groups, *Illinois J. Math.* 13 (1969), 449–513.

16. F. Dalla Volta, Regular orbits for projective orthogonal groups over finite fields of odd characteristic, *Geom. Dedicata* 32 (1989), 229–245.

17. G. Ercan and I. Ş. Güloğlu, On the Fitting length of generalized Hughes subgroup, *Arch. Math.* 55 (1990), 5–9.

18. G. Ercan and I. Ş. Güloğlu, On the Fitting length of $H_{pq}(G)$, *Arch. Math.* 56 (1991), 214–217.

19. G. Ercan and I. Ş. Güloğlu, On the Fitting length of $H_n(G)$, *Rend. Sem. Mat. Univ. Padova* 89 (1993), 171–175.

20. A. Espuelas, Generalized fixed point free automorphisms, *Arch. Math.* 47 (1986), 211–214.

21. A. Espuelas, A noncoprime Shult type theorem, *Math. Z.* 196 (1987), 323–329.

22. A. Espuelas, On fixed-point-free elements, *Math. Proc. Cambridge Philos. Soc.* 103 (1988), 207–211.

23. A. Espuelas, On a conjecture of Thompson, *J. Algebra* 145 (1989), 94–112.

24. A. Espuelas, On the Fitting length conjecture, *Arch. Math.* 53 (1989), 524–527.

25. A. Espuelas, A theorem of Hall-Higman type for groups of odd order, *Arch. Math.* 55 (1990), 218–223.

26. A. Espuelas, Large character degrees for groups of odd order, *Illinois J. Math.* 35 (1991), 499–505.

27. A. Espuelas and G. Navarro, Regular orbits of Hall π-subgroups, *Manuscripta Math.* 70 (1991), 255–260.

28. M.-J. Felipe, *Correspondencias de Caracteres en Acción Coprima*, Ph. D. thesis, Universitat de València, 1994.

29. A. K. Feyzioğlu, Charaktergrade und die Kommutatorlange in auflösbaren Gruppen, *J. Algebra* 126 (1989), 225–251.

30. P. Fleischmann, Finite groups with regular orbits on vector spaces, *J. Algebra* 103 (1986), 211–215.

31. D. Gluck, Bounding the number of character degrees of a solvable group, *J. London Math. Soc.* (2) 31 (1985), 457–462.

32. D. Gluck, A conjecture about character degrees of solvable groups, *J. Algebra* 140 (1991), 26–35.

33. D. Gluck, Primes dividing character degrees and orbit sizes, *Proc. Amer. Math. Soc.* 101 (1987), 219–225.

34. D. Gluck and O. Manz, Prime factors of character degrees of solvable groups, *Bull. London Math. Soc.* 19 (1987), 431–437.

35. D. Gorenstein, *Finite Groups*, Harper and Row, 1968.

36. I. Ş. Güloğlu, On the Fitting length of generalized Hughes subgroup, *Doga Mat.* 14 (1990), 96–105.

37. J. I. Hall, M. W. Liebeck and G. M. Seitz, Generators for finite simple groups, with applications to linear groups, *Quart. J. Math. Oxford* (2) 43 (1992), 441–458.

38. P. Hall and G. Higman, On the p-length of p-soluble groups and reduction theorems for Burnside's Problem, *Proc. London Math. Soc.* (3) 6 (1956), 1–24.

39. B. Hargraves, The existence of regular orbits for nilpotent groups, *J. Algebra* 72 (1981), 54–100.

40. B. Hartley, Some theorems of Hall-Higman type for small primes, *Proc. London Math. Soc.* (3) 41 (1980), 340–362.

41. B. Hartley and I. M. Isaacs, On characters and fixed points of coprime operator groups, *J. Algebra* 131 (1990), 342–358.

42. B. Hartley and A. Rae, Finite p-groups acting on p-soluble groups, *Bull. London Math. Soc.* 5 (1973), 197–198.

43. E. Horvath, On certain properties of characters determined by centralizers, *Publ. Math. Debrecen* 36 (1989), 115–118.

44. E. Horvath, Bounding functions concerning character degrees of solvable groups, *Publ. Math. Debrecen* 38 (1991), 231–235.

45. B. Huppert, *Endliche Gruppen I*, Springer-Verlag, 1983.

46. B. Huppert and N. Blackburn, *Finite Groups II*, Springer-Verlag, 1982.

47. B. Huppert,   Research in representation theory at Mainz (1984–1990), in *Representation Theory of Finite Groups and Finite Dimensional Algebras*, Bielefeld, 1991, pp. 17–36.

48. B. Huppert and O. Manz, Orbit sizes of p-groups, *Arch. Math.* 54 (1990), 105–110.

49. I. M. Isaacs, *Character Theory of Finite Groups*, Academic Press, 1976.

50. I. M. Isaacs, Number of modular character degrees and lengths for solvable groups, *J. Algebra* 148 (1992), 264–273.

51. G. Jones, The influence of nilpotent subgroups on the nilpotent length and derived length of a finite group, *Proc. London Math. Soc.* (3) 49 (1984), 343–360.

52. E. I. Khukhro, Finite groups of period $p^\alpha q^\beta$, *Algebra i Logika* 17 (1978), 727–740.

53. W. Knapp and P. Schmid, Theorem B of Hall-Higman revisited, *J. Algebra* 73 (1981), 376–385.

54. H. Kurzweil, Auflösbare Gruppen auf denen nicht auflösbare Gruppen operieren, *Manuscripta Math.* 41 (1983), 233–305.

55. U. Leisering, Ordinary and p-modular character degrees of solvable groups, *J. Algebra* 136 (1991), 401–431.

56. S. Mattarei, Character tables and metabelian groups, *J. London Math. Soc.* (2) 46 (1992), 92–100.

57. S. Mattarei, An example of p-groups with identical character tables and different derived lengths, *Arch. Math.* 62 (1994), 12-20.

58. S. Mattarei, On character tables of wreath products, *J. Algebra*, to appear.

59. O. Manz, Arithmetical conditions on character degrees and group structure, in *The Arcata Conference on Representations of Finite Groups (Arcata, Calif., 1986)*, pp. 65–69.

60. O. Manz, On the modular version of Ito's theorem on character degrees for groups of odd order, *Nagoya Math. J.* 105 (1987), 121–128.

61. O. Manz and T. Wolf, Brauer characters of q'-degree in p-solvable groups, *J. Algebra* 115 (1988), 75–91.

62. O. Manz and T. Wolf, *Representations of Solvable Groups*, Cambridge University Press, Cambridge, 1993.

63. T. Meixner, The Fitting length of solvable H_{p^n}-groups, *Israel J. Math.* 51 (1985), 68–78.

64. A. Rae, Sylow p-subgroups of finite p-soluble groups, *J. London Math. Soc.* (2) 7 (1973) 117–123. Corrigendum, *J. London Math. Soc.* (2) 11 (1975), 11.

65. G. Seitz, R. Solomon and A. Turull, Chains of subgroups in groups of Lie type II, *J. London Math. Soc.* (2) 42 (1990), 93–100.

66. P. Shumyatskii, A four-group of automorphisms with small number of fixed points, *Algebra i Logika* 30 (1991), 735–746.

67. P. Shumyatsky, Elementary abelian operator groups, *Manuscripta Math.* 82 (1994), 105–111.

68. M. C. Slattery, Character degrees and derived length in p-groups, *Glasgow Math. J.* 30 (1988), 221–230.

69. R. Solomon and A. Turull, Chains of subgroups in groups of Lie type I, *J. Algebra* 132 (1990), 174–184.

70. V. Turau, The 2-length of the Hughes subgroup, *Israel J. Math.* 62 (1988), 206–212.

71. A. Turull, Supersolvable automorphism groups of solvable groups, *Math. Z.* 183 (1983), 47–73.

72. A. Turull, Fitting height of groups and of fixed points, *J. Algebra* 86 (1984), 555–566.

73. A. Turull, Examples of centralizers of automorphism groups, *Proc. Amer. Math. Soc.* 91 (1984), 537–539.

74. A. Turull, Generic fixed point free action of arbitrary finite groups, *Math. Z.* 187 (1984), 491–503.

75. A. Turull, Fixed point free action with regular orbits, *J. reine angew. Math.* 371 (1986), 67–91.

76. A. Turull, Groups of automorphisms and centralizers, *Math. Proc. Cambridge Philos. Soc.* 107 (1990), 227–238.

77. A. Turull and A. Zame, Number of prime divisors and subgroup chains, *Arch. Math.* 55 (1990), 333–341.

78. Y. Q. Wang, The p-parts of Brauer character degrees in p-solvable groups, *Pacific J. Math.* 148 (1991), 351–367.

FINITE p-GROUPS

A. SHALEV
Institute of Mathematics
The Hebrew University
Jerusalem 91904
Israel
E-mail: shalev@math.huji.ac.il

Abstract. This paper is devoted to the study of finite p-groups and their inverse limits, with special emphasis on recent developments and applications of Lie theory (in a broad sense, which includes Lie algebras and Lie groups). The main topics covered are: fixed point theorems, powerful groups and p-adic analytic groups, the Restricted Burnside Problem, the proof of the Coclass Conjectures, and the structure of some just-infinite pro-p groups.

Key words: finite p-groups, Lie algebras, automorphisms, fixed points, p-adic analytic groups, the restricted Burnside problem, coclass.

1. Synopsis

The theory of finite p-groups has undergone significant changes in the past decade. New methods have been introduced and some difficult problems solved. Moreover, progress in the field of finite p-groups has been reflected gradually into several other areas, such as the theory of arbitrary finite groups and the study of certain classes of infinite groups. One reason is that a number of group-theoretic problems can be reduced to questions about p-groups. These include the Restricted Burnside Problem and enumeration problems, as well as certain problems regarding automorphisms and fixed points.

Now is perhaps the right time to record the developments in p-group theory in a concise manner, and to list some of the remaining open problems. Since our main object of study here is finite p-groups, we shall not elaborate on applications to other areas, though a few examples will be mentioned. Also, within p-group theory itself there are many important aspects upon which we shall not touch; these include representations, cohomology, and algorithmic aspects. In fact, a Leitmotiv in many of the results presented here is having a Lie-theoretic proof. So the emphasis of this survey will be on those parts of p-group theory which are influenced by Lie theory. The reason for this emphasis (except for personal taste) is two-fold: many of the recent results—as well as many of the profound ones—in the theory of finite p-groups seem to stem from Lie theory. However, we do include a long introductory chapter which points to some other directions. Excellent chapters on p-groups can be found in some group theory textbooks by Huppert, Gorenstein, Suzuki, and others. See also Hartley's essay on nilpotent groups [23]. It is noteworthy that a book on finite p-groups is now being written by A. Mann [66].

Here are some words on the structure of this paper. In Chapter 2 we introduce

B. Hartley et al. (eds.), Finite and Locally Finite Groups, 401–450.
© 1995 *Kluwer Academic Publishers. Printed in the Netherlands.*

some basic notions and outline several aspects of p-group theory. Chapter 3 is devoted to classical theorems on the structure of p-groups which admit automorphisms with few fixed points. Results by Higman, Alperin, Khukhro and Medvedev are discussed; they all deal with automorphisms of prime order. In Chapter 4 we focus on powerful p-groups, stating several results of Lubotzky and Mann. We include Lazard's fundamental characterization of p-adic analytic pro-p groups in terms of powerful groups, as well as more recent characterizations of those analytic groups. Chapter 5 is devoted to the famous Restricted Burnside Problem, solved by Zelmanov in 1989. We outline the work of Kostrikin on the case of prime exponents, and Zelmanov's work on the general case. We also consider a number of applications of Zelmanov's theorem (to p-adic analytic groups, to Engel groups, etc.); these include results by Shalev, Wilson and Martinez. In Chapter 6 we return, somewhat better equipped, to the study of almost fixed-point-free automorphisms. We show how to apply p-adic ideas and powerful p-groups in particular to the study of certain automorphisms of composite order. This chapter includes recent results of Shalev, Khukhro and Medvedev. Chapter 7 describes the coclass theory for p-groups and pro-p groups. The various approaches to the coclass conjectures of Leedham-Green and Newman are outlined; these include works by Leedham-Green, Newman, McKay, Plesken, Donkin, Mann, Zelmanov, and Shalev. We also discuss related questions in Lie algebras, and some other 'narrowness conditions'. Finally, in Chapter 8 we focus on infinite pro-p groups all of whose proper images are finite. These pro-p groups, which are termed *just-infinite*, may be thought of as the 'simple objects' in the universe of infinite pro-p groups; thus a natural classification problem may be raised. We present the known types of just-infinite pro-p groups in a certain hierarchy. This chapter is based on works by Leedham-Green, York, Lubotzky, Shalev, and others.

I am grateful to many people for keeping me informed of their recent work on finite p-groups. These include S. Blackburn, N. Boston, A. Caranti, B. Kahn, C. R. Leedham-Green, A. Mann, C. Martinez, Yu. Medvedev, J. Neubüser, M. F. Newman, E. A. O'Brien, L. Pyber, C. M. Scoppola, A. Vera-López, T. S. Weigel and E. I. Zelmanov. Some of their results are quoted here with their kind permission.

I have also included some results (or better, observations) which occurred to me during the preparation of these notes; for example, we show that p-groups of bounded rank have many conjugacy classes (see §4.5) and study conjugacy classes in groups of given coclass (§7.6); we also show (in §2.7) that a 'random' finite group has a 2-subgroup of very small index.

This survey contains lists of open problems which I hope will stimulate further work on the subject.

2. Getting Started

2.1. FILTRATIONS

Throughout these notes p denotes a fixed prime.

A p-group is a torsion group in which the order of every element is a power of p. Thus p-groups can be finite or infinite. The infinite p-groups can be further divided into locally finite and non locally finite ones. In this paper p-groups are assumed to be finite, unless otherwise stated. The most basic—yet significant—fact about

p-groups is that non-trivial p-groups G have non-trivial centre $Z(G)$. This can easily be seen by looking at the conjugation action of G on itself and analyzing the sizes of the orbits.

Let us say that a *filtration* of a p-group G is a series of normal subgroups $G = G_1 \supseteq G_2 \supseteq G_3 \supseteq \ldots$ such that $G_i = 1$ for all large i. Such a filtration is called a *central series* if $G_i/G_{i+1} \subseteq Z(G/G_{i+1})$ for all i. This amounts to saying that $[G_i, G] \subseteq G_{i+1}$, where, for subgroups $H, K \subseteq G$, $[H, K]$ denotes the subgroup generated by all commutators $[x, y] = x^{-1}y^{-1}xy$ ($x \in H, y \in K$).

The existence of a non-trivial centre implies that p-groups are nilpotent, namely, that they admit a central series. Two natural central series are *the upper central series* $Z_i = Z_i(G)$ defined by $Z_1 = Z(G)$ and $Z_{i+1}/Z_i = Z(G/Z_i)$; and the *lower central series* $\gamma_i = \gamma_i(G)$ defined by $\gamma_1 = G$ and $\gamma_{i+1} = [\gamma_i, G]$. Note that the subgroup $\gamma_2(G) = [G, G]$ is also denoted by G' (the *commutator subgroup*, or the *derived subgroup* of G). The smallest index i such that $Z_i(G) = G$ is called the nilpotency class of G, or simply the class of G. Thus abelian groups have class one, and non-abelian groups with central commutator subgroups are of class two. In general G has class c if and only if $\gamma_c(G) \neq 1$ but $\gamma_{c+1}(G) = 1$. We denote the class of G by $c(G)$. By refining a central series of G we see that a group of order p^n has a descending series $G = G_0 \supset G_1 \supset \ldots \supset G_n = 1$, where each G_i is a normal subgroup of G and $|G_i : G_{i+1}| = p$ for all $i = 0, \ldots, n-1$.

Another important filtration of a p-group G is the derived series $G^{(i)}$ ($i \geqslant 0$), defined by $G^{(0)} = G$ and $G^{(i+1)} = [G^{(i)}, G^{(i)}]$, the commutator subgroup of $G^{(i)}$. The derived length of G, which we denote by $\mathrm{dl}(G)$, is the minimal i such that $G^{(i)} = 1$.

It is easy to see that the maximal subgroups of a p-group G are exactly its subgroups of index p, which are all normal in G. The intersection of all the maximal subgroups of G is denoted by $\Phi(G)$, the Frattini subgroup of G. Since for each maximal subgroup $M \subset G$ the quotient G/M is abelian of exponent p, the same is true for $G/\Phi(G)$. Thus $G/\Phi(G)$ is elementary abelian; it is in fact the largest elementary abelian quotient of G. In other words we have $\Phi(G) = G'G^p$ where G^p is the group generated by all p-th powers in G. In general we use G^n to denote the subgroup generated by all n-th powers in G. The exponent of G, denoted by $\exp(G)$, is defined to be the minimal n such that $G^n = 1$. We let $\Omega_i(G)$ denote the subgroup generated by the elements of G whose order divides p^i.

The Frattini subgroup is quite useful in the study of p-groups because of its relation to generating sets. Let X be a subset of a p-group G. Then X generates G if and only if there is no maximal subgroup $M \subset G$ containing X. This shows that X generates G if and only if the image of X in $G/\Phi(G)$ generates $G/\Phi(G)$. Since $G/\Phi(G)$ is elementary abelian, it can be viewed as a finite-dimensional linear space V over the field $\mathbb{F}_p$ with p elements. A subset $X \subseteq G$ is a minimal generating set for G if and only if its image in V forms a basis for V. Thus all minimal generating sets for G have the same size, which is denoted by $d(G)$ and coincides with the dimension of V over $\mathbb{F}_p$.

In some contexts it is useful to associate additional filtrations with a p-group G. The *lower p-series* $P_i(G)$ of G is defined inductively by

$$P_1(G) = G \quad \text{and} \quad P_{i+1}(G) = P_i(G)^p[P_i(G), G].$$

Thus the sections $P_i(G)/P_{i+1}(G)$ are all elementary abelian and central. Clearly $P_2(G) = \Phi(G)$. It is interesting to note that there is an explicit expression for the subgroups $P_i(G)$:

$$P_i(G) = \prod_{j+k=i} \gamma_j(G)^{p^k}.$$

See [31, p. 242].

Another important filtration is the so-called Zassenhaus-Jennings-Lazard filtration $D_i(G)$ given by $D_1(G) = G$ and $D_{i+1}(G) = D_j(G)^p[D_i(G), G]$, where j is the minimal integer which is greater than or equal to $(i+1)/p$. This filtration plays an important rôle in the study of modular p-group algebras, since it coincides with the dimension subgroup filtration. More specifically, if K is a field of characteristic p and Δ is the augmentation ideal of the group algebra KG, then $G \cap (1+\Delta^i) = D_i(G)$ for all i. See Jennings [34]. The following explicit expression for $D_i(G)$ was derived by Lazard [44].

$$D_i(G) = \prod_{jp^k \geqslant i} \gamma_j(G)^{p^k}.$$

Unlike all previous filtrations, the dimension subgroup filtration need not be strictly descending: we may well have $D_i(G) = D_{i+1}(G)$ while $D_i(G) \neq 1$. Recent studies of dimension subgroups and their applications in p-group theory can be found in [97], [95] and [94].

We say that a filtration $\{G_i\}_{i \geqslant 1}$ of G is an *N-series* if it satisfies $[G_i, G_j] \subseteq G_{i+j}$ for all i, j. In particular an N-series is a central series. An N-series $\{G_i\}$ is called an N_p-*series* if it also satisfies $G_i^p \subseteq G_{pi}$ for all i. The lower central series and the lower p-series are N-series. The dimension subgroup filtration $\{D_i(G)\}$ is an N_p-series.

In what follows the phrase '$a, b, \ldots$-bounded' means bounded above by a certain function of $a, b, \ldots$ alone.

2.2. COLLECTION FORMULAE AND REGULAR p-GROUPS

In his seminal 1933 paper [19], P. Hall developed combinatorial machinery for the study of p-groups, which has a number of important applications. The collection process he developed gave rise to the following formulae which hold in every p-group:

$$(xy)^{p^n} \equiv x^{p^n} y^{p^n} \mod \gamma_2(H)^{p^n} \prod_{r=1}^{n} \gamma_{p^r}(H)^{p^{n-r}}, \text{ where } H = \langle x, y \rangle;$$

$$[x^{p^n}, y] \equiv [x, y]^{p^n} \mod \gamma_2(N)^{p^n} \prod_{r=1}^{n} \gamma_{p^r}(N)^{p^{n-r}}, \text{ where } N = \langle x, [x, y] \rangle.$$

See [31, pp. 240–241].

A p-group is called *regular* if for every $x, y \in G$ we have $(xy)^p \equiv x^p y^p \mod (H')^p$ where $H = \langle x, y \rangle$. Clearly, groups of exponent p are regular. Furthermore, in view of the first collection formula above we see that p-groups of class less than p are regular. Regular p-groups share various important properties with abelian groups. For example they admit a *basis* $a_1, \ldots, a_d$, so that every $g \in G$ is uniquely expressible as $g = a_1^{k_1} \cdots a_d^{k_d}$ where $0 \leqslant k_i < |a_i|$. It is also known that, if G is regular, then a product of p^n-th powers in G is itself a p^n-th power; and the product of elements of order dividing p^n has order dividing p^n. In fact certain non-regular p-groups also have these properties; see Mann [63] for more details.

Regular p-groups served as an important tool in many classical results on p-groups. Nowadays it seems that their rôle has been partially taken up by powerful p-groups (see Chapter 4).

2.3. Associated Lie Rings

Lie rings were associated with p-groups already in the 30s in the context of the Restricted Burnside Problem (see Chapter 5). Let G_i be any N-series of G. Then we can form the direct sum L of the abelian groups $L_i := G_i/G_{i+1}$ and obtain a finite abelian group. We use additive notation in L. Commutation in G induces a well-defined binary operation on L; it is defined on homogeneous elements by

$$[xG_{i+1}, yG_{j+1}] := [x,y]G_{i+j+1} \ (x \in G_i, y \in G_j),$$

and extended to non-homogeneous elements by additivity. It is easy to check that the operation $[\ ,\ ]$ on L is skew-symmetric and additive in each variable; moreover, it satisfies the Jacobi identity

$$[[x,y],z] + [[y,z],x] + [[z,x],y] = 0.$$

Hence $(L, +, [\ ,\ ])$ has the structure of a Lie ring. This Lie ring is $\mathbb{N}$-graded in the sense that

$$L = \bigoplus_{i \in \mathbb{N}} L_i, \text{ and } [L_i, L_j] \subseteq L_{i+j}.$$

(Our natural numbers $\mathbb{N}$ start with 1.) Since $G_i = 1$ for large i, we have $|L| = |G|$ and $L_i = 0$ for large i.

The above procedure can be performed for each N-series G_i, and by choosing G_i appropriately we maintain essential flexibility. There are however some standard choices. If $G_i = \gamma_i(G)$ then the Lie ring L obtained in this way is the classical graded Lie ring associated to G. In this case L is a $\mathbb{Z}$-module, but not necessarily a Lie algebra (over a field). To get a Lie algebra over $\mathbb{F}_p$ we have to require that each section G_i/G_{i+1} is elementary abelian. This is the case if we let $G_i = P_i(G)$, or if we set $G_i = D_i(G)$ (see §2.1). In the latter case the Lie algebra obtained has an additional structure, namely a (formal) p-th power operation (coming from taking p-th powers in G), which makes it a restricted Lie algebra (or a Lie p-algebra); see Jacobson [33], Chapter 5.

For later use we introduce some Lie-theoretic notation. Long Lie products $[x_1, \ldots, x_n]$ are interpreted using the left-normed convention. The i-th term of the lower central series (the derived series) of a Lie ring L is denoted by L^i ($L^{(i)}$). The nilpotency class and the derived length of L (assuming it is nilpotent or soluble) are denoted by $c(L)$ and $\mathrm{dl}(L)$ respectively. The n-th Engel identity is the identity $[x, y, \ldots, y] = 0$ where y occurs n times. An n-Engel Lie ring is a Lie ring satisfying this identity.

Let $L = \bigoplus L_i$ be a graded Lie ring constructed from a filtration G_i of G, and let H be a subgroup of G. Then we can associate to H a subalgebra $K = K(H)$ of L, defined by $K = \bigoplus K_i$, where $K_i = (H \cap G_i)G_{i+1}/G_{i+1} \subseteq L_i$. It is easy to see that $|K| = |H|$ and so $|G : H| = |L : K|$. If H is normal in G, then K is a Lie ideal of L.

The importance of the Lie ring construction is that it enables one to reduce various problems in finite p-groups to questions about Lie algebras, which can then be solved using Lie-theoretic techniques. In Chapters 3, 5, 6 we describe in some detail two main applications of such an approach: the solution of the Restricted Burnside Problem, and the study of automorphisms with few fixed points. Additional recent applications can be found in Chapters 7, 8.

We now briefly mention some other applications. We say that a p-group G has *breadth* b if the maximal size of a conjugacy class in G is p^b. We define the *co-breadth* of a group of order p^n and breadth b to be p^{n-b} (it coincides with the minimal order of a centralizer $C_G(x)$ in G). It has been known for some time that the order of the commutator subgroup of a group G is bounded in terms of the maximal size of a conjugacy class in G. The best such bound in the case of p-groups was obtained by M. R. Vaughan-Lee [111]. Using Lie ring methods he showed that if G is a p-group of breadth b then $|G'| \leqslant p^{b(b+1)/2}$. This inequality is best possible.

The Lie ring method was applied by C. R. Leedham-Green, P. M. Neumann and J. Wiegold in the context of the so-called class-breadth conjecture. They show in [53] that the nilpotency class of a p-group of breadth b is at most $2b$. This is still larger than the conjectured bound $b + 1$, but it is now known that this latter bound does not hold in general.

Finally, let us mention a recent application of the Lie ring method in the study of groups of exponent p. The lower central series of such groups have special properties which do not hold in p-groups of larger exponent. It is shown in [98] that, if G has exponent p, then $[\gamma_i(G), \gamma_j(G)] \supseteq \gamma_k(G)$ where $k = i + (p - 2)(j - 1) + 1$. This extends the classical result of Meier-Wunderli which shows that $G'' \supseteq \gamma_{p+1}(G)$ [76]. The proof uses elementary properties of Engel Lie algebras.

We note that, in addition to this classical construction, there are other ways to associate Lie rings with p-groups. These additional constructions apply only in special situations (e.g. when the underlying group has small nilpotency class, or when it is 'uniform'). See Chapter 6 for more details.

We close this section with a question.

Problem 1 *For which Lie rings L is there a p-group G such that the Lie ring obtained from G using the filtrations γ_i, P_i or D_i is isomorphic to L?*

Of course, we really have here three different questions. In general, the Lie rings L should be finite (of p-power order), $\mathbb{N}$-graded, and generated by their first homogeneous component L_1. If the filtration P_i is used then L should be a Lie algebra over $\mathbb{F}_p$; and, if D_i is used instead, then L should also be a restricted Lie algebra, and its grading should be compatible with its p-th power map (which means $L_i^p \subseteq L_{pi}$). Are these necessary conditions also sufficient? Interesting partial results have recently been obtained by B. Kahn and P. A. Minh.

2.4. RANK AND SUBGROUP STRUCTURE

The study of the subgroup structure of p-groups dates back to Burnside, who classified p-groups which satisfy a strong 'narrowness' condition. To be precise, Burnside showed that if a p-group G has only one subgroup of order p (equivalently, if every abelian subgroup of G is cyclic) then G is cyclic or a generalized quaternion 2-group. This basic result, which plays an important rôle in the study of Frobenius complements, was subsequently extended in many ways.

First, the p-groups in which every abelian *normal* subgroup is cyclic can be classified. For odd p these groups are necessarily cyclic, while for $p = 2$ one obtains three additional infinite families: the generalized quaternion 2-groups, the dihedral 2-groups, and the semi-dihedral 2-groups. Second, one may weaken even further the narrowness condition by assuming that the *characteristic* abelian subgroups of G are cyclic. In this case a result of P. Hall shows that G is a central product of subgroups E and R, where E is extra-special (or trivial) and R is cyclic or $p = 2$ and R is generalized quaternion, dihedral, or semi-dihedral.

Let us say that a p-group G has p-*rank* r if the maximal rank of an elementary abelian subgroup of G is r. Thus, the groups of p-rank 1 were classified by Burnside. The determination of the p-groups of given p-rank is important partially because of Quillen's celebrated theorem [91], showing that the p-rank of G equals the Krull dimension of the cohomology ring $H^*(G, \mathbb{F}_p)$. The groups of p-rank 2 were classified by N. Blackburn for odd p [6].

Problem 2 *Determine the 2-groups of 2-rank two.*

See Rusin [93] for some recent results.

We mention below some additional useful results on the subgroup structure of p-groups. Let G be a p-group and let $A \lhd G$ be an abelian normal subgroup of G which is maximal (with respect to inclusion). Then A is self-centralizing, namely $C_G(A) = A$. This can easily be proved by looking at the intersection of $C_G(A)$ with the inverse image of $Z(G/A)$ in G. Note that this result implies that, under the above assumptions, G/A can be embedded in $\mathrm{Aut}(A)$; this is the form in which this basic fact is often used. As a corollary one concludes that the order of a p-group G is bounded above in terms of the order of any maximal abelian normal subgroup A of G. In fact it follows from this argument that, if $|A| = p^a$, then $|G| \leqslant p^{a(a+1)/2}$. This bound is best possible. For the construction of p-groups in which all abelian subgroups are small, see Ol'shanskiĭ [86].

Define the *rank* of G (denoted by $\mathrm{rk}(G)$) to be the minimal integer r such that $d(H) \leqslant r$ for all subgroups $H \subseteq G$. Using the existence of a self-centralizing abelian normal subgroup it follows at once that p-groups of p-rank s have rank at most $s(s+1)/2$; this bound can be improved by (roughly) a factor of two.

The fact that maximal abelian normal subgroups are self-centralizing was extended by Alperin [2] as follows. Let G be a p-group and let A be an abelian normal subgroup of G, maximal with respect to having exponent p^n. Suppose further that $p^n \neq 2$. Then the elements of G of order at most p^n which centralize A lie in A. The particular case $n = 1$ of this result was obtained by Feit and Thompson and plays a rôle in their Odd Order paper.

Burnside's result with which we started this section can be viewed as a subgroup counting theorem. Many more counting theorems have been proved since then, and various congruences related to the number of subgroups of given type in a p-group

G have been established. We refer the reader to Berkovich's papers [3], [4] and their reference lists for up-to-date results.

2.5. p-GROUPS OF MAXIMAL CLASS

The nilpotency class of a p-group G of order p^n $(n \geqslant 2)$ is at most $n - 1$. This is because $|G/\gamma_2| \geqslant p^2$ for G non-cyclic. If $|G| = p^n$ and $c(G) = n - 1$, G is said to be *of maximal class*. The 2-groups of maximal class form three infinite families: the dihedral groups, the semi-dihedral groups, and the generalized quaternion groups. Note that all these groups have a cyclic subgroup of index 2. In his seminal paper [5], N. Blackburn determines all the 3-groups of maximal class. He also obtains interesting results for arbitrary p. For example, it is shown in [5] that a p-group G is of maximal class if and only if there exists an element $x \in G$ of order p such that $|C_G(x)| = p^2$; it is also shown there that, for a p-group G, if G/γ_{p+1} (or G/γ_2 if $p = 2$) is of maximal class, then so is G.

The investigation of p-groups of maximal class continued with works by Shepherd [109], Leedham-Green and McKay [48, 49, 50], and others. It has resulted in a deep theory with somewhat surprising corollaries. Some of the highlights of this theory are surveyed in Hartley [23], §1.2. For example, by the results of [109] and [48], p-groups of maximal class have class two subgroups of p-bounded index. The coclass theory, described in Chapter 7, establishes results of this type in greater generality.

The following problem seems to be very difficult.

Problem 3 *Classify the 5-groups of maximal class.*

A detailed analysis, using computers, has been carried out by various mathematicians, including C. R. Leedham-Green, S. McKay and M. F. Newman. Consequently, some interesting conjectures can be formulated. See Newman [80]. Finally, let us mention Héthelyi's work [26] in which some extensions of p-groups of maximal class are considered.

2.6. CONJUGACY CLASSES, I

Let $k(G)$ denote the number of conjugacy classes in a group G. Since a group G of order p^n has a normal series of length n, it follows immediately that such a group satisfies $k(G) \geqslant n$, and even $k(G) \geqslant (p - 1)n$ (since $x, x^2, \ldots, x^{p-1}$ lie in distinct conjugacy classes).

A more detailed analysis yields sharper bounds. By a result of P. Hall (see Huppert [30], p. 549), if $|G| = p^n$ and $n = 2m + e$ where $e = 0, 1$, then there exists a non-negative integer $l = l(G)$, which we call the *abundance* of G, such that

$$k(G) = (p^2 - 1)m + (p - 1)(p^2 - 1)l + p^e.$$

In particular $k(G) \geqslant (p^2 - 1)[n/2]$. Very little seems to be known about the relation between $l(G)$ and the structure of G, except when $l(G) = 0$. In this latter case it was shown by Poland [88] that G is a group of maximal class whose order is at most p^{p+2}. More detailed results on groups for which $l(G) = 0$ were subsequently obtained by Vera-López et al. (see [115], [116]).

Computer calculations carried out independently by E. A. O'Brien and by J. Neubüser and his student B. Rothe show that, for small primes p, there is a bound to the order of p-groups G with $l(G) = 1$. No general result of this type seems to be known. In [85] it is shown that there are only finitely many p-groups G of given abundance and given co-breadth. This result also follows from Proposition 6.4 below.

Problem 4 *Show that there are only finitely many p-groups G of any given abundance l.*

The question how small $k(G)$ can be, given $|G|$, has already been posed by R. Brauer. A construction due to Kovács and Leedham-Green [42] establishes the existence of groups of order p^p with less than p^3 conjugacy classes. These groups occur as quotients of the Nottingham group (see §2.8). The search for p-groups with a small number of conjugacy classes is still on. In this context, it is natural to consider the following problem, which is posed by Pyber in [89].

Problem 5 *Is there a constant c such that there are infinitely many p-groups G (with p fixed) for which $k(G) \leqslant c \log |G|$?*

It is a common belief that the finite quotients of the Nottingham group have this property; this has not been verified. On the other hand, the following provocative question may also be considered.

Problem 6 *Given p, is there $\epsilon(p) > 0$ such that $k(G) \geqslant |G|^{\epsilon(p)}$ for all p-groups G?*

If true, this would be rather unexpected; still, it seems that such an implausible conjecture is consistent with our current knowledge of p-groups.

As for upper bounds on $k(G)$, suppose that $|G| = p^n$. Then it is obvious that $k(G) \leqslant p^n$ with equality if and only if G is abelian. In general one expects groups with many conjugacy classes to be close to abelian. An asymptotic result of this type, which holds for arbitrary finite groups, is proved by P. M. Neumann in [79]. In the case of p-groups, it is easy to see that $k(G) \geqslant 2p^{n-2}$ if and only if $|G'| \leqslant p$. A stronger result has recently been proved by Mann and Scoppola [67]. They show that, if $|G| = p^n$ and p is odd, then $k(G) \geqslant p^{n-2}$ if and only if G has breadth at most 2 or G has an abelian maximal subgroup.

Much work has been done on $k(G)$ for particular p-groups G, such as the Sylow p-subgroups of Sym_n and $\mathrm{GL}_n(p)$. These p-groups turn out to have many conjugacy classes. See Pálfy and Szalay [87], Higman [29], as well as Vera-Lopéz and Arregi [114]. For the best known lower bounds on $k(G)$ for arbitrary finite groups, see Pyber [89].

2.7. ENUMERATION PROBLEMS

Let $f(m)$ denote the number of groups of order m up to isomorphism. What can be said about the growth of $f(m)$, particularly for prime powers $m = p^n$? This fundamental question was addressed by many authors. By the results of Higman [29] and Sims [110], there are constants c_1 and c_2 such that

$$p^{\frac{2}{27}n^3 - c_1 n^2} \leqslant f(p^n) \leqslant p^{\frac{2}{27}n^3 + c_2 n^{8/3}}. \tag{1}$$

In the proof of the upper bound one actually counts power-commutator representations with respect to a normal series of length n. The lower bound is proved by counting p-groups of class two. So in some (possibly weak) sense one may say that most p-groups have class two.

For small values of n and p, $f(p^n)$ has been determined explicitly. The most striking result in this direction is O'Brien's enumeration of the groups of order 2^8 [84], according to which

$$f(256) = 56,092.$$

There have been various attempts to improve the error term in the upper bound (1) on $f(p^n)$ given above. A recent (yet unpublished) work by Mike Newman and Craig Seeley [83] shows that

$$f(p^n) \leqslant p^{\frac{2}{27}n^3 + c_3 n^{5/2}}. \tag{2}$$

We note that it follows from a recent result of Pyber [90] that $f(m) \leqslant 2^{(\frac{2}{27}+o(1))(\log m)^3}$ for arbitrary m. Pyber's arguments, which apply the Classification Theorem and the theory of finite permutation groups, actually reduce the enumeration of arbitrary finite groups to p-group enumeration.

The following new result, which relies on [83] and [90], shows that a random finite group has a 2-subgroup of very small index.

Proposition 2.1   *There exists a constant c with the following property: if n is a positive integer and G is a randomly chosen group of order at most n, then the probability that G has a 2-subgroup H of order at least $2^{\log n - c\sqrt{\log n}}$ tends to 1 as $n \to \infty$.*

Proof.   We should show that, for some c, the proportion of groups not having such a large 2-subgroup tends to 0 as $n \to \infty$. We describe the proof, leaving some easy details for the reader. First note that, by (1), if 2^k and p^l ($p > 2$) are of the same order of magnitude, then $f(p^l) << f(2^k)$, where $<<$ stands for 'much smaller' (in this case $f(p^l) < f(2^k)^{c_4}$ for some $c_4 < 1$). Furthermore, it follows from the lower bound (1) and the upper bound (2) above that for large c_5

$$f(2^{k-c_5\sqrt{k}}) << f(2^k). \tag{3}$$

More specifically we can say that, given c_6, we can choose c_5 such that the left-hand side divided by the right-hand side of (3) is at most $2^{-c_6 k^{5/2}}$.

Now, given $m \leqslant n$, let $m = \prod_p p^{k(m,p)}$ be its prime factorization. By [90] we have

$$f(m) \leqslant m^{c_7 k} \cdot \prod_p f(p^{k(m,p)}),$$

where $k = \max\{k(m,p) : p\}$. Putting everything together it is now easy to verify that, for large c,

$$\sum_{\{m \leqslant n : k(m,2) < \log n - c\sqrt{\log n}\}} f(m) << \sum_{m \leqslant n} f(m). \qquad \square$$

Recent results of Simon Blackburn shed more light on the subject. In his D. Phil. thesis [7] Blackburn proves that there are $p^{\frac{2}{27}n^3+O(n^2)}$ groups of order p^n and class 3 (thus extending known results for class two p-groups). He also enumerates groups of order p^n inside a given isoclinism class [8], showing that their number cannot exceed $p^{n^2/3+O(\sqrt{n})}$. It follows that there are many isoclinism classes of groups of order p^n. Roughly speaking, p-groups are called isoclinic if they have the same commutator structure. For more details see P. Hall's 1940 paper [20] (where the term family is used for isoclinism class).

We close this section with enumeration results for particular types of p-groups. It turns out that the number of d-generated p-groups of order p^n is bounded by p^{cn^2} (where c depends on d), so it is significantly smaller than the number of all p-groups of that order. This result is contained in Peter Neumann's 1969 paper [78]. Furthermore, it has recently been observed by Pyber that there are at most p^{cn} isomorphism types of groups of rank r and order p^n (where c depends on r). This result is one of the many consequences of the theory of powerful p-groups.

2.8. PRO-p GROUPS

A pro-p group is an inverse (i.e. projective) limit of finite p-groups. Thus, to define a pro-p group, one starts with an inverse system $\{G_i : i \in I\}$ of finite p-groups with compatible homomorphisms $\alpha_{ij} : G_j \to G_i$ ($i < j$). If the ordered set I is countable (as is usually the case) we can replace it with a chain of type ω where the respective homomorphisms are all surjective, and get the same inverse limit. Equivalently, pro-p groups can be defined as topological groups which are compact, Hausdorff, totally disconnected, and in which every open normal subgroup has a p-power index. In analyzing properties of infinite families of finite p-groups, it is often useful to form inverse limits and to study the resulting pro-p groups. Leedham-Green's paper [47] is a good example.

Note that, formally, finite p-groups are also pro-p groups (though by 'pro-p groups' people usually mean 'infinite pro-p groups'). When dealing with pro-p groups G, all notions are interpreted topologically. For example, $X \subseteq G$ is said to be a generating set if the minimal closed subgroup containing X is G. The cardinality of a minimal generating set is denoted by $d(G)$. Similarly the various characteristic groups G', $\gamma_n(G)$, G^{p^k}, $\Phi(G) = G'G^p$ are interpreted as closed subgroups.

In this respect it is worthwhile mentioning a result of Serre, according to which the *abstract* commutator group G' of a finitely generated pro-p group G is closed. This implies that every abstract finite index subgroup of a finitely generated pro-p group is open. It is still not known whether a similar result holds for arbitrary finitely generated profinite groups.

Pro-p groups are equipped with natural filtrations such as $\gamma_i(G)$, $P_i(G)$, $D_i(G)$, etc. (see §2.1). Of course, these filtrations are usually of infinite length. Thus, in pro-p groups, instead of requiring that filtrations $\{G_i\}$ end up with the trivial group 1, we require that $\bigcap_{i\geqslant 1} G_i = 1$. To an N-series $\{G_i\}_{i\geqslant 1}$ in a pro-p group G we can associate a graded Lie ring L exactly as in §2.3. This Lie ring is infinite (or infinite dimensional over $\mathbb{F}_p$) if G is infinite. In fact we usually take L to be the *Cartesian* product of the quotients G_i/G_{i+1}, instead of the direct product. Thus L can be regarded as a topological Lie ring, and as such it is compact. Given L, we can associate a subalgebra $K = K(H) \subseteq L$ to a (closed) subgroup $H \subseteq G$. This is done exactly as in the p-group case (see §2.3). We then have $|L : K| = |G : H|$ (where both sides may be infinite).

We give below some natural examples of pro-p groups. The most basic example is the additive group of the p-adic integers, denoted by $\mathbb{Z}_p$; it can be thought of as an inverse limit of the cyclic p-groups C_{p^i} (with obvious epimorphisms). By forming direct products of finitely many copies of $\mathbb{Z}_p$ with a suitable abelian p-group P we obtain the finitely generated abelian pro-p groups. The simplest example of a pro-p group which is not finitely generated is the Cartesian product of countably many copies of C_p. It can be identified with the additive group of the power series ring $\mathbb{F}_p[[t]]$.

A somewhat less trivial example of a pro-p group is the dihedral pro-2 group. It can be constructed as a split extension of $\mathbb{Z}_2$ by the involution $\alpha \in \mathrm{Aut}\mathbb{Z}_2$ sending 1 to -1. This can be generalized as follows: let G be a pro-p group containing an open subgroup T which is a finitely generated free p-adic module; suppose G/T acts irreducibly on $T \otimes \mathbb{Q}_p$; then G is called a *p-adic pre space group*, or a *p-adic space group* if the action is faithful. In such a situation we say that T is the *translation group* and G/T is the *point group*. The p-adic space groups play an essential rôle in the coclass theory (see Chapter 7).

Certain pro-p groups occur naturally as matrix groups. For example, let $\mathrm{GL}_n^1(\mathbb{Z}_p)$ ($n \geqslant 2$) denote the first congruence subgroup in $\mathrm{GL}_n(\mathbb{Z}_p)$, namely, the kernel of the natural epimorphism of the latter group onto $\mathrm{GL}_n(\mathbb{F}_p)$. Then $\mathrm{GL}_n^1(\mathbb{Z}_p)$ is a pro-p group; it is powerful if $p > 2$, and in any case it is p-adic analytic and non-soluble (see Chapter 4 for the terminology). In a similar manner one may consider $\mathrm{GL}_n^1(\mathbb{F}_p[[t]])$ which is not p-adic analytic (but may be regarded as $\mathbb{F}_p[[t]]$-analytic; see Chapter 8).

It seems that even in such a quick excursion into the universe of pro-p groups it is necessary to pay some attention to one of the most striking of them all, namely the so-called *Nottingham group*. This group, defined and studied by D. L. Johnson [35] and I. York [121] in Nottingham, can be regarded as a 'congruence subgroup' in $\mathrm{Aut}\mathbb{F}_p[[t]]$, namely the centralizer of $(t)/(t)^2$ in the latter group. Alternatively, it can be described as the group of all power series of the form $t + a_2 t^2 + a_3 t^3 + \cdots$ ($a_i \in \mathbb{F}_p$) under substitution. One of its most striking properties is that it contains an isomorphic copy of every finite p-group. See Chapter 8 as well as [57] for more details.

It is possible to form the *free* pro-p group on a given number of generators, and to consider pro-p presentations of pro-p groups (in terms of generators and relators). Finally, it is worth mentioning that pro-p groups arise naturally in field arithmetic (as Galois groups of certain infinite field extensions).

3. Fixed Point Theorems, I

This chapter, as well as Chapter 6, is devoted to the study of fixed points of auto-morphisms of p-groups. We shall describe some rather deep results, whose proofs form beautiful examples of the so-called *Lie ring method*. In this chapter we study automorphisms of prime order, while Chapter 6 is devoted to automorphisms of composite order. Although the main object of our study is finite p-groups, some of the results apply for a larger class of groups, such as (possibly infinite) nilpotent groups.

3.1. FIXED-POINT-FREE AUTOMORPHISMS: EARLY RESULTS

Let G be a group and let $\alpha \in \text{Aut}\,G$. We say that α is a *fixed-point-free* automorphism (f.p.f. for short) if the identity element of G is the only fixed point of α. What can be said of the structure of a group G which admits a fixed-point-free automorphism α?

An old result of Burnside shows that finite groups which admit a f.p.f. automorphism of order 2 are necessarily abelian. The (easy) proof is purely group-theoretic and makes use of properties of dihedral groups. A similar result also holds for periodic groups (with basically the same proof). Note however that the free group on two generators x, y admits a f.p.f. involutory automorphism interchanging x and y.

Burnside also showed that finite groups which admit a f.p.f. automorphism of order 3 are nilpotent of class at most two. Fixed-point-free automorphisms of order 5 have been studied by B. H. Neumann [77].

3.2. A REDUCTION

While the study of fixed-point-free automorphisms of very small order can be carried out within group theory proper, it seems necessary to use Lie-theoretic methods in order to derive more general results. Let G be a finite group admitting a fixed-point-free automorphism α of prime order p. In view of the results stated above, it makes sense to ask whether G is nilpotent, and whether there is a bound on its nilpotency class in terms of p. Of these two questions, only the second is really about groups of prime power order. An affirmative answer to the first question was given by Thompson in 1959. As for the second question, the following procedure can be carried out. Let $L = L(G)$ be the graded Lie ring which is constructed from G using its lower central series (see §2.3). Note that G and L have the same nilpotency class. Since α acts fixed-point-freely on every α-invariant section of G, it induces on L a fixed-point-free automorphism, which is also of order p. Suppose we can deduce that the nilpotency class of the Lie ring L is at most $h(p)$ (for a suitable function h). Then the same will hold for the nilpotency class of G.

3.3. HIGMAN'S THEOREM

The above reduction was used by G. Higman in [28]. In the same paper Higman solves the resulting Lie-theoretic problem, proving:

Theorem 3.1 *Let L be a Lie ring which admits a fixed-point-free automorphism of prime order p. Then L is nilpotent of class at most $h(p)$ (for a suitable function h).*

The basic idea in Higman's proof of this theorem (which was later adopted in other applications) is the use of eigenvalues and eigenspaces for the automorphism α. Consider L as a $\mathbb{Z}$-module and let $\widetilde{L} = L \otimes_{\mathbb{Z}} \mathbb{Z}[\omega]$, where ω is a primitive p-th root of unity. The automorphism α is uniquely extended to an automorphism $\widetilde{\alpha}$ of

$\tilde{L}$, acting trivially on ω. It is clear that $\tilde{\alpha}$ is also fixed-point-free. For $0 \leqslant i < p$ let $H_i = \{x \in \tilde{L} \mid \tilde{\alpha}(x) = \omega^i x\}$, and $H = H_0 \oplus \cdots \oplus H_{p-1}$. Then $H_0 = 0$ and the ring H is $\mathbb{Z}/p\mathbb{Z}$-graded (i.e. $[H_i, H_j] \subseteq H_{i+j}$ where the sum of the indices is mod p). Moreover, it is easy to verify that $pL \subseteq p\tilde{L} \subseteq H$. At this stage, bounding the class of L becomes essentially a combinatorial problem. We refer the reader to [28] and Chapter 8 of [31] for more details.

Theorem 3.1 gives rise to:

Theorem 3.2   *The nilpotency class of a nilpotent group which admits a fixed-point-free automorphism of prime order p is at most $h(p)$.*

Kostrikin and Kreknin gave explicit bounds on Higman's function $h(p)$, showing that

$$h(p) \leqslant \frac{(p-1)^{2^{p-1}-1} - 1}{p-2}.$$

Problem 7   *Find better estimates for Higman's function h.*

The exact value of $h(p)$ is only known for small primes $p = 2, 3, 5, 7$. A general lower bound of the form $(p^2 - 1)/4$ is obtained in [28]. The gap between this bound and the super-exponential upper bound of Kostrikin and Kreknin is enormous, and it would be nice to reduce it.

3.4. ALPERIN'S THEOREM

Alperin was the first to realize that Higman's method can be used in investigating automorphisms which are only 'almost' fixed-point-free. His paper [1] deals with p-groups G which admit an automorphism of order p with few fixed points. In this case it is impossible to bound the nilpotency class of G. Indeed, p-groups of maximal class admit an automorphism of order p with p^2 fixed points, though their class can be arbitrarily large (see §2.5). However, Alperin showed that it is possible to bound the derived length of G in such circumstances.

Theorem 3.3   *There exists a function f such that the derived length of a p-group admitting an automorphism of order p with m fixed points is at most $f(p, m)$.*

In view of the remark preceding the theorem we obtain the following result from [1]:

Corollary 3.4   *The derived length of a p-group of maximal class is bounded above in terms of p alone.*

Indeed, if G has maximal class and f is as above, then $\mathrm{dl}(G) \leqslant f(p, p^2)$. Conjecture B of Leedham-Green and Newman (which is now a theorem) can be regarded as a generalization of Corollary 3.4. See Chapter 7 for more details. We note that there is no absolute bound on the derived length of p-groups of maximal class. Indeed, it is shown in [42] that a suitable factor group of the Nottingham group has maximal class and derived length $\sim \log p$.

Theorem 3.3 is proved using a tricky application of Higman's method. The proof actually gives a bit more, namely: there exists a subgroup $H \lhd G$ such that the class of G/H is p, m-bounded, and the class of H is at most $h(p) + 1$.

3.5. MORE RECENT DEVELOPMENTS

In the late 80s Khukhro was able to extend Higman's method and to study systematically groups and Lie rings with almost fixed-point-free automorphisms of prime order. His main group-theoretic result is as follows.

Theorem 3.5 *Let G be a nilpotent torsion group which admits an automorphism of prime order p with m fixed points. Then there is a subgroup $H \lhd G$ of p, m-bounded index whose nilpotency class is p-bounded.*

If the group G above happens to be a p-group (as in Alperin's theorem) we can require further that the class of H is at most $h(p) + 1$. However, in this case the theorem is rather easy. The deeper part of Khukhro's result is in the 'coprime' case, where a q-group G (q a prime different from p) admits an automorphism of order p with m fixed points.

Theorem 3.5 has just been generalized by Yu. Medvedev [73] to arbitrary nilpotent groups. See also Khukhro's recent monograph [36].

3.6. PRO-p GROUPS

We note that pro-p groups behave quite differently with respect to almost f.p.f. automorphisms, and results of the type described above are not valid for them. For example, the non-abelian free pro-p group on $x_1, \ldots, x_p$ has a fixed-point-free automorphism of order p inducing a cyclic permutation on the generators x_i. Automorphisms of free pro-p groups have been studied in various papers by Herfort, Ribes and Zalesskiĭ; see, for instance, [25], where the following problem is posed:

Problem 8 *Let G be a finitely generated free pro-p group, and let $\alpha \in \operatorname{Aut} G$ be an automorphism of order p.*

(i) *Show that the centralizer $C_G(\alpha)$ is finitely generated.*

(ii) *Show that $d(C_G(\alpha)) \leqslant d(G)$.*

4. The Importance of Being Powerful

4.1. SOME DEFINITIONS

A p-group G is called *powerful* if $p > 2$ and G/G^p is abelian, or if $p = 2$ and G/G^4 is abelian. Powerful p-groups were studied long ago by C. Hobby, B. King, and others. It was M. Lazard who established the fundamental rôle of these objects in the theory of p-adic Lie groups [45] (note that Lazard used the term 'p-saturated' instead of powerful). Twenty years later Lubotzky and Mann carried out a systematic study of powerful groups, and derived several new applications to p-adic Lie groups.

We now proceed with some more definitions. A normal subgroup $N \lhd G$ is said to be *powerfully embedded* in G if $[N, G] \subseteq N^p$ (N^4 if $p = 2$). Note that a powerfully embedded subgroup is itself powerful. Also, for odd p, every cyclic normal subgroup of a p-group G is powerfully embedded in G. It is known that 'natural' subgroups of a powerful p-group G are powerfully embedded (hence powerful). These include $\gamma_i(G)$, G^{p^i}, $G^{(i)}$, etc. We also say that $N \lhd G$ is *hereditarily powerful* in G if every normal subgroup of G which is contained in N is powerful. Finally, a powerful p-group G is said to be *uniform* if $d(G^{p^i}) = d(G)$ unless $G^{p^i} = 1$.

All these definitions apply for pro-p groups (where we only deal with closed subgroups). Unless otherwise stated, all the results in this chapter are taken from Lubotzky and Mann [58]. See also the recent book [12].

4.2. BASIC PROPERTIES

Obviously, all abelian p-groups are powerful. Moreover, powerful groups can be thought of as groups which are abelian in a graded sense. For example, if $L(G)$ is the restricted Lie algebra constructed from the dimension subgroup filtration of G (see §2.3), then $L(G)$ is abelian if and only if G is powerful (as noted in [104]). This observation can be used to provide a uniform definition of powerful groups, which does not distinguish between odd and even primes.

The similarity between powerful groups and abelian groups is rather striking. For example, we have:

Proposition 4.1   *Let G be a powerful p-group. Then*

 (i) $\Phi(G) = G^p$.

 (ii) *The sections $G^{p^i}/G^{p^{i+1}}$ are elementary abelian and central.*

 (iii) *The series of orders $|G^{p^i}/G^{p^{i+1}}|$ is (weakly) monotonically decreasing.*

 (iv) *Every element of G^{p^i} is of the form x^{p^i} for some x.*

 (v) *Suppose G is uniform of exponent at least p^j, and let $i \leqslant j$. Then $x^{p^i} \in G^{p^j}$ if and only if $x \in G^{p^{j-i}}$.*

 (vi) *If $G = \langle a_1, \ldots, a_d \rangle$, then $G = \langle a_1 \rangle \cdots \langle a_d \rangle$.*

 (vii) *If $G = \langle a_1, \ldots, a_d \rangle$, then $G^{p^i} = \langle a_1^{p^i}, \ldots, a_d^{p^i} \rangle$.*

 (viii) *We have $d(H) \leqslant d(G)$ for all subgroups $H \subseteq G$.*

Note that part (ii) above yields $P_i(G) = G^{p^{i-1}}$ for all $i \geqslant 1$, where P_i denotes the lower p-series. Part (vi) gives rise to the following bound on the order of G in terms of the exponent of G and its number of generators.

Corollary 4.2   *Let G be a d-generated powerful p-group of exponent $n = p^k$. Then $|G| \leqslant n^d$.*

The investigation of such bounds for arbitrary p-groups is the subject of the next chapter.

The following result from [101] demonstrates the 'linearity' of powerful p-groups.

Lemma 4.3   *Let G be a p-group and let $M, N \lhd G$ be powerfully embedded subgroups. Then $[M^{p^i}, N^{p^j}] = [M, N]^{p^{i+j}}$.*

In particular, if G is powerful then $\gamma_k(G^{p^i}) = \gamma_k(G)^{p^{ki}}$, and $(G^{p^i})^{(k)} = G^{(k)^{p^{2^k i}}}$. Results of this type play some rôle in the study of almost f.p.f. automorphisms (see Chapter 6).

4.3. FINDING POWERFUL SUBGROUPS

How can we find powerful subgroups in a given p-group G? The ways to do this are usually based on the following result, which is implicit in [58].

Lemma 4.4 *Let G be a p-group and let $M \lhd G$ be a normal subgroup which centralizes every normal elementary abelian section of G. Then M (or M^2 if $p = 2$) is powerful.*

An explicit proof of the lemma can be found in [99]. Note that any normal subgroup of G which is contained in M satisfies the assumption on M. It follows that, under the conditions of the lemma, the subgroups M, M^2 (respectively) are hereditarily powerful.

According to Lemma 4.4, the intersection $P(G)$ of the centralizers of the elementary abelian normal sections of a p-group G is always powerful (or has a powerful square power). Note that if G has rank r, then $G/P(G)$ can be embedded in a certain power of $\mathrm{GL}_r(p)$. This enables one to bound the exponent and the derived length of the factor group $G/P(G)$, and eventually to bound its order.

We thus obtain the following important result from [58]. (For a real number x we let $\lceil x \rceil$ denote the smallest integer which is greater than or equal to x.)

Theorem 4.5 *Let G be a p-group of rank r. Then G has a characteristic powerful subgroup N of p, r-bounded index. In fact we have $|G : N| \leqslant p^{rk}$, where $k = \lceil \log_2 r \rceil$ if $p > 2$ and $k = \lceil \log_2 r \rceil + 1$ if $p = 2$.*

The same conclusion holds even if we only assume that the *characteristic rank* of G is at most r, namely, that $d(H) \leqslant r$ for all characteristic subgroups H of G.

Problem 9 *Can the bound in Theorem 4.5 be improved to p^{cr} for some absolute constant c?*

This is closely related to finding the maximal order of rank r p-subgroups of $\mathrm{GL}_r(p)$.

4.4. p-ADIC ANALYTIC GROUPS

One of the main motivations for the study of powerful groups is their special rôle in the theory of p-adic analytic groups. A topological group is said to be p-*adic analytic* if it can be given the structure of a p-adic manifold, with the group operations being analytic functions. In his fundamental work of 1965 [45], Lazard develops a beautiful theory of such groups. It turns out that p-adic analytic groups have open finitely generated pro-p subgroups, so it is natural to focus on p-adic analytic pro-p groups. Which pro-p groups are p-adic analytic? This question may be viewed as a p-adic analogue of Hilbert's fifth problem. The following theorem summarizes various solutions which have been given to this question over the years.

Theorem 4.6 *Let G be a finitely generated pro-p group. Then each of the following conditions is equivalent to G being p-adic analytic.*

(i) *G is virtually powerful.*

(ii) *The restricted Lie algebra $L(G)$ corresponding to the dimension subgroup filtration is nilpotent.*

(iii) *G has finite rank.*

(iv) *For some c and for all n, G has at most n^c subgroups of index n.*

(v) *For some $c < 1/8$ and for all large n, G has at most $n^{c \log_p n}$ subgroups of index n.*

(vi) *The graded group ring of G over $\mathbb{F}_p$ satisfies a non-trivial polynomial identity.*

Parts (i) and (ii) are due to Lazard. Parts (iii) and (iv), which are useful in many applications, are due to Lubotzky and Mann [58, 59]. Parts (v) and (vi) are taken from [100] and [104] respectively.

As shown in [45], p-adic analytic pro-p groups G have many nice properties. They are linear (in characteristic zero), and they can be associated with a finite-dimensional p-adic Lie algebra Lie(G) which reflects their structure rather closely. For example, G is soluble if and only if Lie(G) is soluble. As in classical Lie groups, G acts on Lie(G) via the adjoint action. A very recent work by Thomas Weigel [117] on the Exp and Log functors examines the relations between (certain) finite powerful p-groups and their Lie algebras.

Over the past decade p-adic analytic groups have proved instrumental in the study of residually finite groups. In many cases one starts with an abstract group G satisfying a certain condition and concludes that the pro-p completion of G is p-adic analytic. The linearity of p-adic analytic pro-p groups can then be used to deduce the linearity of the original group G. Once we are in the domain of linear groups, special tools and results become available. See §5.8 below for one particular example of such an application, as well as [12] for a much more thorough discussion.

We close this section with some problems which may lead to new characterizations of p-adic analytic groups.

By an old result of Serre, p-adic analytic pro-p groups are Noetherian (namely, they satisfy the ascending chain condition on closed subgroups; equivalently, their closed subgroups are finitely generated). Note that this also follows from part (iii) of Theorem 4.6. It would be interesting to know whether the converse is true.

Problem 10   *Is every Noetherian pro-p group p-adic analytic?*

This problem is posed in [58] and in [64].
The next question is somewhat related.

Problem 11   *Let G be a pro-p group, and suppose every closed subgroup $H \subset G$ of infinite index is p-adic analytic; does it follow that G is p-adic analytic?*

It is easy to see that, if G is as above, then G is Noetherian. Hence an affirmative solution to Problem 10 implies an affirmative solution to Problem 11.

Note that, if G is p-adic analytic, then there is a bound on the length of chains of closed subgroups $1 = G_0 \subset G_1 \subset \ldots \subset G_n = G$ with the property that the indices $|G_{i+1} : G_i|$ are all infinite (as G_i are all p-adic analytic and dim G_i is strictly increasing). A positive answer to Problem 11 would imply the converse: namely, that the existence of such a bound is equivalent to G being p-adic analytic.

Let us now consider the power structure of pro-p groups. It follows from previously mentioned results that p-adic analytic pro-p groups have many p-th powers, in the sense that they have open subgroups consisting entirely of p-th powers. It would be interesting to know whether the converse is true.

Problem 12 *Let G be a pro-p group which has an open subgroup H consisting entirely of p-th powers; does it follow that G is p-adic analytic?*

We conjecture a positive answer. Actually, even the following is not known:

Problem 13 *Let G be a finitely generated pro-p group, and suppose that for each $x, y \in G$ there is $z \in G$ such that $x^p y^p = z^p$; does it follow that G is p-adic analytic?*

4.5. CONJUGACY CLASSES, II

We conclude this chapter with a new result on conjugacy classes in finite p-groups. While the general lower bounds on $k(G)$ are of the form $c \log |G|$, we use powerful p-groups to obtain sharper bounds for groups of bounded rank.

Lemma 4.7 *Let G be a powerful p-group and let $c = 1 - p^{-1}$. Then*

(i) $k(G) \geqslant c \cdot \exp(G)$.

(ii) $k(G) \geqslant c \cdot |G|^{1/d}$ *where* $d = d(G)$.

Proof. Set $\exp(G) = n = p^k$ and let $S \subset G$ be the set of elements of order n in G. We claim that $|S| \geqslant c|G|$. To show this let $T = G \smallsetminus S$, the set of elements of order dividing p^{k-1}. It is clear that T does not generate G, otherwise it would follow from part (vii) of Proposition 4.1 that $G^{p^{k-1}} = 1$. Thus T is contained in a maximal subgroup $M \subset G$, and so

$$|S| \geqslant |G \smallsetminus M| = c|G|.$$

Note that $|C_G(x)| \geqslant |\langle x \rangle| = n$ for all $x \in S$. Hence, if $\mathrm{Cl}(x)$ denotes the conjugacy class of x in G, we have $|\mathrm{Cl}(x)| \leqslant |G|/n$ for all $x \in S$. As a normal subset of G, S is a union of, say, l conjugacy classes. It follows that

$$l \geqslant \frac{|S|}{|G|/n} \geqslant \frac{c|G|}{|G|/n} = cn.$$

Since $k(G) \geqslant l$, part (i) follows.

Part (ii) is a consequence of part (i) and Corollary 4.2. $\qquad\square$

Proposition 4.8 *Let G be a p-group of rank r. Then $k(G) \geqslant |G|^{\epsilon}$ where the number $\epsilon = \epsilon(r) > 0$ depends only on r.*

Proof. Let $N \lhd G$ be the powerful subgroup obtained in Theorem 4.5, and let $f(p, r)$ be the bound obtained on its index. It is easy to see using Lemma 4.7 that the number of G-classes of N is at least $k(N)/|G : N| \geqslant c|N|^{1/r}/f(p, r)$. The result now readily follows. $\square$

In fact, combining the above proposition with the Classification Theorem and some additional results, one can deduce that $k(G) \geqslant |G|^{\epsilon(r)}$ for all finite groups G of rank r.

5. Around Zelmanov

5.1. THE RESTRICTED BURNSIDE PROBLEM

The famous Burnside problems may be stated as follows.

1. *Is every finitely generated torsion group finite?*

2. *Is every finitely generated group of bounded exponent finite?*

3. *Is there a bound on the order of d-generated finite groups of exponent n?*

The first two questions were posed by W. Burnside (who provided positive answers for exponents $\leqslant 3$) in 1902 [10]. The study of the third question dates back to the late 30s, following many years of unsuccessful attempts to solve problems 1 and 2. Later (thanks to Magnus [61]) question 3 became known as the Restricted Burnside Problem (RBP for short).

In spite of positive solutions for exponents 4 and 6, the answers to questions 1 and 2 turned out to be negative, as shown by Golod and by Novikov-Adian respectively. Gradually it became clear how wild infinite p-groups can be: for example, for large enough p there exist infinite 2-generated p-groups G all of whose non-trivial proper subgroups have order p (this construction is due independently to Ol'shanskiĭ and to Rips). Other constructions, by Golod, Gupta and Sidki, and Grigorchuk gave rise to infinite finitely generated p-groups which are residually-p.

The history of question 3 developed rather differently. A reduction theorem proved by Hall and Higman in 1956 shows that, assuming the outer automorphism groups of the finite simple groups are soluble and that there are only finitely many finite simple groups of any given exponent, it suffices to provide a positive answer to the RBP for prime power exponents $n = p^k$ [21]. Of course, these assumptions now follow from the Classification Theorem. Consequently, the RBP is reduced to the case of p-groups.

5.2. PRIME EXPONENT: REDUCTION TO LIE RINGS

Let G be a d-generated finite group of prime exponent p. Let $L = L(G)$ be its associated graded Lie ring, which is formed using the lower central series of G (see §2.3). Then L is a Lie algebra over $\mathbb{F}_p$ and we have $|L| = |G|$. It is also easy to see that the Lie algebra L is d-generated. A crucial observation, due to W. Magnus, shows that L is $(p-1)$-Engel (i.e. it satisfies the identity $\mathrm{ad}(y)^{p-1} = 0$).

To bound the order of G it therefore suffices to show that the dimension of a finite d-generated $(p-1)$-Engel Lie algebra L over $\mathbb{F}_p$ is d, p-bounded. Since the dimension of d-generated nilpotent Lie algebras of class c is easily seen to be d, c-bounded, it suffices to show that the class of L is d, p-bounded. Let $\widetilde{L}$ be the relatively free Lie algebra on d-generators satisfying the $(p-1)$-Engel identity. Then L is an image of $\widetilde{L}$. If we show that $\widetilde{L}$ is nilpotent we may immediately conclude that $c(L) \leqslant c(\widetilde{L})$ is d, p-bounded. This reduces the RBP for exponent p to the following question:

Are $(p-1)$-Engel Lie algebras of characteristic p locally nilpotent?

5.3. Kostrikin's Theorem

By Engel's classical result, m-Engel finite dimensional Lie algebras are nilpotent. This result cannot be applied in our case, as we are dealing with infinite-dimensional objects. However, a deep theorem of Kostrikin [38], [39] provides the desired answer to the Lie-theoretic question posed above.

Theorem 5.1

(i) *$(p-1)$-Engel Lie algebras of characteristic p are locally nilpotent.*

(ii) *Consequently, the order of a d-generated p-group of exponent p is d, p-bounded.*

The particular case $p = 5$ was also treated by Higman [27].

Kostrikin's Theorem has a number of applications in the theory of p-groups. For example, it implies that, for some function f, if all sections of class at most $f(p)$ of a p-group G are regular, then G itself is regular; see Mann [62].

5.4. Prime Power Exponents: Reduction to Lie Rings

One of the difficulties in extending Kostrikin's approach to prime power exponents $n = p^k$ lies in the fact that the following problem is still open.

Problem 14 *Given $n = p^k$, is there an integer m, depending only on n, such that the associated Lie algebra of every p-group of exponent n is m-Engel?*

Consequently, there was no reduction to a natural Lie-theoretic question. The first significant step in Zelmanov's work on the Restricted Burnside Problem was to establish such a reduction.

Let G be a group of exponent p^k. Then the lower central series of G gives rise to a Lie ring L which is not necessarily a Lie algebra over $\mathbb{F}_p$. This is why it is convenient to replace the lower central series by the lower p-series $P_i(G)$ (see §2.1). So let $L = L(G)$ be the graded Lie algebra over $\mathbb{F}_p$ which is constructed from the filtration $P_i(G)$. If $d = d(G)$ then L has d generators, say $a_1, \ldots, a_d$. By results of Higman and Sanov, L satisfies the linearized Engel identity

$$\sum_{\sigma \in \mathrm{Sym}_{p^k - 1}} \mathrm{ad}(x_{\sigma(1)}) \cdots \mathrm{ad}(x_{\sigma(p^k - 1)}) = 0,$$

as well as the identity

$$\mathrm{ad}(y)^{p^k} = 0 \text{ for all } y \in B,$$

where B is the set of all Lie products in the generators a_i.

It is thus clear that to solve the RBP it suffices to show that Lie algebras L satisfying the above conditions are nilpotent. Using combinatorics of words in the spirit of Shirshov, Zelmanov carried out the Lie-theoretic reduction much further. He proved in [122] that, in order to provide an affirmative solution, it suffices to show that a Lie algebra over an infinite field of characteristic p which satisfies an Engel identity is locally nilpotent.

5.5. ZELMANOV'S THEOREM

Having completed the reduction, Zelmanov set out to prove the local nilpotency of Engel Lie algebras of characteristic p. His ingenious proof, published in [123] for $p > 2$ and in [124] for $p = 2$ (see also [125]), uses a deep structure theory for Jordan algebras which he had previously developed [70] as well as divided powers and other tools. The proof also relies on the joint work of Kostrikin and Zelmanov [40], which establishes the local nilpotency of so-called sandwich algebras. A description of the proof is beyond the scope of this survey. A more elementary proof of the local nilpotency of Engel Lie algebras can be found in Vaughan-Lee's recent monograph [112].

Zelmanov's Theorem can be stated as follows.

Theorem 5.2   *The orders of d-generated finite groups of exponent $n = p^k$ are d, n-bounded.*

Using the Hall-Higman reduction and the Classification Theorem, a similar result follows for arbitrary exponents n.

5.6. BOUNDS

The question of explicit bounds in the RBP (even for prime exponents) has been open for a long time. Let $B(d,n)$ denote the largest order of a finite d-generated group of exponent n. Then $B(d,2) = 2^d$, and it is not difficult to check that

$$B(d,3) = 3^{d + \binom{d}{2} + \binom{d}{3}}.$$

The value of $B(d,6)$ is also known (see [112], Chapter 5).

The case of exponent 4 is harder. Still, rather precise bounds were obtained by A. J. Mann in 1982 [68]; he showed that

$$2^{4^d/2} < B(d,4) < 2^{(4+2\sqrt{2})^d/2},$$

for all $d > 1$.

In the recent work [113], M. R. Vaughan-Lee and Zelmanov obtain explicit upper bounds on $B(d,n)$ for all prime power exponents n.

Theorem 5.3

(i) *Let $n = p^k$. Then*

$$B(d,n) \leqslant d^{d^{\cdot^{\cdot^{\cdot^{d}}}}},$$

where d occurs n^{n^n} times.

(ii) *For $n = p$ we have*

$$B(d, n) \leqslant d^{d^{.^{.^{.^{d}}}}},$$

where d occurs 3^n times.

We remark that the known lower bounds on $B(d, n)$ are much smaller.

Problem 15 *What is the real order of magnitude of the Burnside numbers $B(d, n)$?*

Some related conjectures are posed in [113].

5.7. Application to Analytic Groups

Zelmanov's Theorem can be used to obtain the following characterization of p-adic analytic pro-p groups [99].

Theorem 5.4 *Let G be a finitely generated pro-p group. Then G is p-adic analytic if and only if there exists k such that the wreath product $C_p \wr C_{p^k}$ is not isomorphic to an upper section of G.*

Recall that an upper section is a finite quotient of a finite index subgroup.

The 'only if' part follows easily from part (iii) of Theorem 4.6, as the above wreath products have unbounded rank. So the main content of the theorem is the 'if' part. The proof combines Zelmanov's Theorem and the theory of powerful p-groups with the well-known structure of $\mathbb{F}_p C_{p^k}$-modules.

As an application we obtain a characterization of p-adic analytic pro-p groups in terms of p-th powers. It is easy to see (using the results of the preceding chapter) that if G is a p-adic analytic pro-p group then, for some c and for every upper section H of G, a product of p^{i+c}-th powers in H is a p^i-th power in H. Groups with such a property are termed c-*power closed*. Using the theorem above it is easy to deduce the converse, so we have:

Corollary 5.5 *A finitely generated pro-p group is p-adic analytic if and only if it is c-power closed for some c.*

It is unlikely that Theorem 5.4 and Corollary 5.5 could be given elementary proofs (not relying on Zelmanov's Theorem); indeed, each of these result is actually equivalent to Zelmanov's solution to the RBP. For example, suppose Corollary 5.5 holds and let G be a finitely generated pro-p group of finite exponent p^k. Then G is clearly k-power closed. Thus we can conclude that G is p-adic analytic. However, it is easy to see that a p-adic analytic pro-p group G of finite exponent is necessarily finite (apply Corollary 4.2 for an open powerful subgroup of G, or simply use the fact that G is linear). It follows that finitely generated pro-p groups of finite exponent are finite. By considering the relatively free pro-p group on d generators with exponent $n = p^k$, Theorem 5.2 easily follows.

It is tempting to try to obtain stronger versions of Theorem 5.4.

Problem 16 *Let G be a finitely generated pro-p group, and suppose G has no closed sections isomorphic to the infinite wreath product $C_p \wr \mathbb{Z}_p$. Does it follow that G is p-adic analytic?*

A positive answer would imply that Noetherian pro-p groups are p-adic analytic (see Problem 10 above).

5.8. The Engel Identity for p-Groups

The study of the Engel identity $[x, y, \ldots, y] = 1$ in groups is somewhat related to the study of the Burnside identity $x^n = 1$. A classical result of Zorn shows that finite groups satisfying an Engel identity are nilpotent. The analogue of the RBP in this case would be:

Is the nilpotency class of a finite d-generated group satisfying the n-th Engel identity d, n-bounded?

Obviously, this is really a question about finite p-groups. There are several ways to solve this problem: see Zelmanov [122] and Wilson [119]. Here we shall describe Wilson's approach, which is more group-theoretic.

By considering the relatively free n-Engel pro-p group on d generators, we see that it suffices to prove that finitely generated n-Engel pro-p groups are nilpotent. Let G be such a group. Then every upper section of G also satisfies the n-th Engel identity. However, the wreath products $C_p \wr C_{p^k}$ do not satisfy a fixed Engel indentity. It follows that not all of them are obtained as upper sections of G. Applying Theorem 5.4 above, we conclude that G is p-adic analytic. Thus G is linear (see §4.4). Implementing the well-known Tits Alternative, we conclude that either G is virtually soluble or it has a nonabelian free subgroup. The latter is impossible, since nonabelian free groups do not satisfy an Engel identity. It follows that G is soluble (as virtually soluble pro-p groups are soluble). However, by a result of K. W. Gruenberg [18], n-Engel soluble groups are nilpotent. We have established the following:

Theorem 5.6 *The nilpotency class of d-generated n-Engel p-groups is d, n-bounded.*

A similar result follows for arbitrary finite, and residually finite, groups. It is not known whether finitely generated n-Engel groups are always nilpotent.

5.9. Semigroup Identities

We note that some other group identities can be studied by combining Zelmanov's work with the theory of powerful p-groups. A *semigroup identity* is an identity of the form $u = v$, where u, v are words without inverses in a suitable free group. There are two important examples: the Burnside identity $x^n = 1$, and the Morse identity $u_m = v_m$ defined by $u_1 = x, v_1 = y, u_{m+1} = u_m v_m$ and $v_{m+1} = v_m u_m$. The following result is a particular case of [102], Theorem B.

Theorem 5.7 *For every semigroup identity $u = v$ there is a Burnside identity $x^n = 1$ and a Morse identity $u_m = v_m$ such that any finite group G satisfying the identity $u = v$ is an extension of a nilpotent group N satisfying $u_m = v_m$ by a group H satisfying $x^n = 1$.*

It can then be shown that, for some function f (depending on u, v), the order of d-generated subgroups of H, and the nilpotency class of d-generated subgroups of N, are bounded above by $f(d)$. As a consequence we obtain the following result:

Corollary 5.8 *A d-generated finite group satisfying a semigroup identity $u = v$ has a nilpotent normal subgroup whose index and nilpotency class are both d, u, v-bounded.*

Zelmanov's Theorem (which is essential in the proof) is obtained as a special case (take $u = x^n$, $v = 1$).

5.10. EXTENSIONS FOR PRO-p GROUPS

Zelmanov's Theorem is tantamount to saying that pro-p groups of bounded exponent are locally finite. In a subsequent work [126], Zelmanov proves an even stronger result.

Theorem 5.9 *Torsion pro-p groups are locally finite.*

As an application he obtains the following.

Corollary 5.10 *Every infinite pro-p group has an infinite abelian subgroup.*

Indeed, let us show how the corollary follows by combining Theorem 5.9 with Alperin's Theorem (see §3.4). Let G be a counter-example. Then G is a torsion group, otherwise it has an infinite cyclic subgroup. Hence G is locally finite. Clearly, G is non-soluble, since infinite soluble groups have infinite abelian subgroups. It follows from Theorem 3.3 that the centralizer of every element of order p in G is infinite. Moreover, since every infinite subgroup H of G is also a counter-example to Corollary 5.10, we see that $|C_H(x)| = \infty$ for every $x \in H$ of order p. This gives rise to the following process: set $H_1 = G$ and pick $x_1 \in H_1$ of order p; now set $H_2 = C_{H_1}(x_1)$ and let $x_2 \in H_2$ with $x_2 \neq x_1$ be of order p. In general, if H_i, x_i are defined for $i \leqslant n$, we let $H_{n+1} = C_{H_n}(x_n)$ and pick $x_{n+1} \in H_{n+1}$ of order p such that $x_{n+1} \neq x_1, \ldots, x_n$. Then the group generated by $x_1, x_2, \ldots, x_n, \ldots$ is an infinite abelian subgroup of G.

Recall that, by the classical result of Hall and Kulatilaka [22], every infinite locally finite group has an infinite abelian subgroup (in fact Alperin's Theorem is not used in their proof).

While Theorem 5.9 extends Zelmanov's solution to the RBP, the following interesting result of Wilson and Zelmanov extends Theorem 5.6 in a similar direction. We say that G is a *weakly Engel group* if for every $x, y \in G$ there exists $n = n(x, y)$ (depending on x and y) such that $[x, y, \ldots, y] = 1$, where y occurs n times. Of course, the numbers $n(x, y)$ may not be bounded.

Theorem 5.11 *Weakly Engel pro-p groups are locally nilpotent.*

See [120] for more details.

We note that the following problem, posed by Platonov, is still open.

Problem 17 *Is every torsion pro-p group a group of finite exponent?*

A result of Herfort [24] shows that a positive answer to Problem 17 would imply that every profinite torsion group has finite exponent.

5.11. CENTRALIZERS IN PRO-p GROUPS

In this section we discuss the following application of Theorem 5.9 (see [106]).

Theorem 5.12 *Let G be a pro-p group and suppose any centralizer $C_G(x)$ $(x \in G)$ is either finite or of finite index. Then*

(i) *G is finite-by-abelian-by-finite.*

(ii) *If G is just-infinite, then G is a p-adic space group whose point group is cyclic or generalized quaternion.*

The rôle of Theorem 5.9 in the proof is to reduce to the locally finite case. To see how this is done, let G be as in Theorem 5.12 and let Δ be its FC-centre, namely

$$\Delta = \{x \in G \mid |G : C_G(x)| < \infty\}.$$

Note that, in general, Δ may not be closed in G. We claim that $G/Z(\Delta)$ is a torsion group. Indeed, elements outside Δ have finite centralizers (by the assumption on G), so they all have finite order. In particular G/Δ is a torsion group. However, it is well known that $\Delta/Z(\Delta)$ is always a torsion group. Hence $G/Z(\Delta)$ is a torsion group, as claimed.

Suppose now that G is not locally finite. By Theorem 5.9 it follows that G is not a torsion group. In view of the claim just proved we conclude that there is an element $x \in Z(\Delta)$ which has infinite order. We claim that $\Delta = C_G(x)$. The inclusion $\subseteq$ follows from the fact that x is central in Δ. To verify the reverse inclusion, note that every element $y \in C_G(x)$ has infinite centralizer (since $\langle x \rangle$ is infinite), so $C_G(y)$ has finite index and $y \in \Delta$ by the assumption on G.

Having proved that $\Delta = C_G(x)$ it follows that $|G : \Delta| < \infty$. It also follows that Δ is a *closed* subgroup of G (centralizers are always closed). Consequently, Δ is an open subgroup of G. We see that Δ is an FC pro-p group. It is not difficult to verify, using a Baire category argument, that FC pro-p groups have finite commutator subgroups. Thus

$$|G : \Delta| < \infty \text{ and } |\Delta'| < \infty,$$

so part (i) of the theorem follows in this case. Part (ii) can be established with some extra work.

We are left with the case where G is locally finite. This case can be handled using Khukhro's Theorem 3.5 and some other tools. We omit the details.

5.12. MORE ON POWERS

What can be said about finitely generated pro-p groups with 'many' elements of bounded order? The following result of Khukhro gives a partial solution.

Theorem 5.13 *Let G be a finitely generated pro-p group which has an open subset consisting of elements of order p. Then G is virtually nilpotent.*

This result applies Kostrikin's Theorem and is strongly related to the notion of splitting automorphisms (see [36], Chapter 7). No parallel result is known for elements whose order is a proper power of p.

Problem 18   *Let G be a finitely generated pro-p group which has an open subset consisting of elements of order p^k. Does it follow that G is soluble?*

We conjecture that the answer is positive.

Finally, consider the following problem: let G be a finitely generated pro-p group; is the *abstract* subgroup G^p closed in G? If we replace G^p by G' then the answer is positive, as we have already noted (see §2.8). Since $G^p \supseteq G'$ in powerful groups, it easily follows that G^p is closed if G is p-adic analytic. But what is the answer in general?

A standard Baire Category argument shows that G^p is closed if and only if every product of p-th powers in G can be represented by a product of bounded length (where the bound may depend on p and $d(G)$). A recent (yet unpublished) result of C. Martinez, which relies on Zelmanov's work, establishes this property.

Theorem 5.14   *For every d, p^k there exists an integer $f(d, p^k)$ with the following property: if G is a d-generated p-group, then every element of G^{p^k} can be expressed as a product of at most $f(d, p^k)$ p^k-th powers in G.*

Corollary 5.15   *Let G be a finitely generated pro-p group and let $k \geqslant 1$. Then the abstract subgroup G^{p^k} is closed.*

This result, proved in [69], provides a partial solution to the following.

Problem 19   *Let G be any finitely generated profinite group. Does it follow that the abstract subgroup G^n is closed for all n?*

We conjecture a positive answer. This would imply that a finite index subgroup of a finitely generated profinite group is necessarily open, thus solving a longstanding problem, studied by Hartley and many others.

Let us explain the implication, as it makes essential use of Zelmanov's Theorem. Let G be a finitely generated profinite group, and let $H \subseteq G$ be a finite index subgroup. To show that H is open it suffices to show that some subgroup N of H is open (as H is a union of cosets of N). Let H_G be the core of H in G and let $n = |G : H_G|$. Then $H \supseteq H_G \supseteq G^n$, so it suffices to show that the (abstract) subgroup G^n is open. Now, suppose Problem 19 has a positive solution, so that G^n is closed. Then $\overline{G} = G/G^n$ is a finitely generated profinite group of bounded exponent. Hence $\overline{G}$ is finite by the positive solution to the RBP. It follows that G^n is closed of finite index, hence open. Therefore H is open, as required.

6. Fixed Point Theorems, II

In this chapter we describe how the Lie ring method can be adapted to the study of certain almost fixed-point-free automorphisms of composite order. We shall mainly be concerned with automorphisms of p-groups whose order is a (proper) power of p. The proofs of all the results below apply the theory of powerful p-groups in an essential way. Some of the proofs are also based on a new Lie ring construction (see [101], §3) which follows ideas from the theory of p-adic Lie groups. In a way, we may view the results of this chapter as yet another triumph of p-adic ideas and the theory of powerful p-groups in particular.

6.1. The Problem

Let G be a p-group admitting an automorphism α of order p^k with p^m fixed points. If $k = 1$ then Alperin's theorem shows that the derived length of G is bounded in such a situation. What can be said of the structure of G in the general case? In particular, is $\mathrm{dl}(G)$ still bounded?

The situation was unclear for a long time. The main problem was in the translation of the question into Lie-theoretic language. To be specific, let $L(G)$ be the usual graded Lie ring associated with G (see §2.3). The difficulty lies in the fact that there is no general (two-sided) connection between the derived length of G and the derived length of $L(G)$. Thus, even if we prove that $L(G)$ has bounded derived length, it is not clear how to derive a similar conclusion for G. In the next sections we show how the theory of powerful p-groups can be applied to overcome this difficulty. Our arguments follow [101].

6.2. Reduction to Powerful Groups

Let G, α, p, m, k be as in 6.1. Let E be an elementary abelian characteristic section of G. Then α acts on E with at most p^m fixed points. Consider E as an $\mathbb{F}_p\langle\alpha\rangle$-module and write $E = \bigoplus_{i=1}^l V_i$ where the V_i are indecomposable. Since $\langle\alpha\rangle \cong C_{p^k}$, every V_i is uniserial of dimension at most p^k. In particular $|C_E(\alpha)| = p^l$ so $l \leqslant m$. We conclude that $\dim E \leqslant lp^k \leqslant mp^k$. It follows that the characteristic rank of G is at most mp^k. By the remark following Theorem 4.5, G has a characteristic powerful subgroup P such that $|G : P|$ is p, m, k-bounded. Obviously, $d(P) \leqslant mp^k$ (and therefore $\mathrm{rk}(G)$ is p, m, k-bounded). Clearly, α acts on P with at most p^m fixed points. Assuming $\mathrm{dl}(P)$ is p, m, k-bounded, we would obtain a bound on $\mathrm{dl}(G)$. Thus, in order to bound $\mathrm{dl}(G)$ we may assume that G is powerful with at most mp^k generators.

6.3. Reduction to Uniform Groups

Suppose G is powerful with $d \leqslant mp^k$ generators, and let α be as above. It is easy to see that G has a filtration of characteristic subgroups

$$G = G_0 \supset G_1 \supset \ldots \supset G_l = 1,$$

where $l \leqslant d$ and where each section G_i/G_{i+1} is uniform. Indeed, for a suitable choice of indices j_i, $G_i = G^{p^{j_i}}$ will do.

Clearly α acts on each of the uniform sections G_i/G_{i+1} with at most p^m fixed points, and bounding $\mathrm{dl}(G_i/G_{i+1})$ for all i amounts to bounding $\mathrm{dl}(G)$. Therefore we may assume that G is uniform.

6.4. A Lie Ring Construction

A uniform group is sufficiently linear so that special methods and constructions which do not apply in general do apply here. For our purpose, the following construction will be useful. Let G be a uniform p-group of exponent p^e with d generators. Fix $i \leqslant e/4$ and consider the sections $A = G^{p^i}/G^{p^{2i}}$ and $B = G^{p^{2i}}/G^{p^{3i}}$. Then A, B are abelian (by Lemma 4.3); furthermore, A and B are both isomorphic to the direct product of d copies of C^{p^i}. The p-th power map in G gives a well-defined isomorphism $q : A \longrightarrow B$, and commutation in G induces a well defined skew-symmetric bilinear operation $c : A \times A \longrightarrow B$. Using q^{-1} we can pull back and obtain a well-defined operation $[\,,\,] : A \times A \longrightarrow A$, given by $[x,y] = q^{-1}(c(x,y))$. It can be shown that A, equipped with the induced group-multiplication (considered as $+$) and with the operation $[\,,\,]$ (considered as Lie product), has the structure of a Lie ring, which we call L. In fact L is *uniform*, in the sense that its additive group is a homocyclic p-group and L/pL is commutative.

It is clear that α induces on L an automorphism of order dividing p^k with at most p^m fixed points. The crucial point here is that there is a strong connection between the derived length of G and the derived length of L. To be precise, it can be shown (using Lemma 4.3) that, if the derived length of each Lie ring L obtained in this manner (when i varies) is at most l, then $\mathrm{dl}(G) \leqslant 2l + 1$. This establishes a reduction of our question to a Lie-theoretic one:

Is the derived length of a uniform Lie ring L admitting an automorphism α of order p^k with p^m fixed points p^k, p^m-bounded?

6.5. A RESULT OF KREKNIN

The following result of Kreknin can be used to solve the resulting Lie-theoretic question.

Proposition 6.1 *Let L be a Lie ring and let $\alpha \in \mathrm{Aut}\, L$ be an automorphism of finite order n. Let I be an ideal of L which contains all fixed points of α. Then*

$$n^{n-1}(nL)^{(2^{n-1}-1)} \subseteq n^{n-1}I.$$

This result had been used by Kreknin to show that a Lie ring which admits a f.p.f. automorphism of finite order n is soluble of derived length $\leqslant 2^n - 2$ [43]. In our case we can use Proposition 6.1 to provide an affirmative answer to the question concluding section 6.4. The idea is to take for I all elements $x \in L$ satisfying $p^m x = 0$. Then I, α satisfy the conditions of Proposition 6.1 with $n = p^k$. It is now straightforward to deduce that $\mathrm{dl}(L)$ is bounded.

6.6. EXTENSION OF ALPERIN'S THEOREM

Having outlined the reduction and the solution to the Lie-theoretic problem, we can formulate the resulting group-theoretic theorem.

Theorem 6.2 *There exists an integer $f(p^k, p^m)$ such that, if G is a p-group admitting an automorphism α of order p^k with p^m fixed points, then $\mathrm{dl}(G) \leqslant f(p^k, p^m)$.*

Of course, for $k = 1$ this is exactly Alperin's Theorem 3.3. In general the order of magnitude of the function f obtained from the proof is approximately $mp^k 2^{p^k}$ (see [101], p. 281 for the exact expression).

Theorem 6.2 gives rise to the following somewhat counter-intuitive result, showing that p-groups with elements whose centralizer is small are in some sense close to being abelian.

Corollary 6.3 *The derived length of a p-group G is bounded above in terms of the order of the centralizer of any element $x \in G$.*

Indeed, let $x \in G$ and set $|C_G(x)| = p^m$. Then the inner automorphism of G induced by x has order dividing p^m and p^m fixed points. Thus $\mathrm{dl}(G) \leqslant f(p^m, p^m)$ where f is as above.

Equivalently, one may say that if G has a conjugacy class of size at least $|G|/q$ then $\mathrm{dl}(G)$ is q-bounded.

6.7. CONJUGACY CLASSES, III

By combining previous results we can now obtain the following.

Proposition 6.4   *Let G be a p-group and let $x \in G$ with $|C_G(x)| = p^m$. Then $k(G) \geqslant |G|^\epsilon$ where $\epsilon = \epsilon(p^m) > 0$ depends only on p^m.*

Proof.   We have observed (see §6.2) that $\mathrm{rk}(G)$ is p^m-bounded. Hence the result follows from Proposition 4.8.   □

This result shows that p-groups with a very large conjugacy class must have many conjugacy classes. In particular it follows that only finitely many p-groups G can have given abundance and given co-breadth (a result already proved in [85]).

6.8. STRONGER BOUNDS: KHUKHRO'S THEOREM

Some extensions of Theorem 6.2 have recently been obtained. The idea is to bound the derived length in terms of fewer parameters. It turns out that this is possible, if one restricts to a suitable subgroup H of G. The first result of this type has been obtained by Khukhro (see [36], p. 237).

Theorem 6.5   *Let G be a p-group admitting an automorphism α of order p^k with p^m fixed points. Then G has a subgroup H of p^k, p^m-bounded index such that $\mathrm{dl}(H)$ is p^k-bounded.*

We note that the proof of Khukhro's theorem involves reductions to powerful groups and uniform groups, but avoids the use of our p-adic Lie ring L. Another ingredient of Khukhro's proof is the Mal'cev correspondence and the Baker-Campbell-Hausdorff formula.

6.9. STRONGER BOUNDS: MEDVEDEV'S THEOREM

A somewhat different generalization of Theorem 6.2 has just been announced by Medvedev.

Theorem 6.6   *Let G be a p-group admitting an automorphism of order p^k with p^m fixed points. Then G has a subgroup H of p^m, p^k-bounded index whose derived length is m-bounded.*

Medvedev also shows that, if $k = 1$, then G has a subgroup H of p^m-bounded index whose nilpotency class is m-bounded. These results were conjectured by Khukhro. The case $m = 1$ of the above result (where $c(H) \leqslant 2$) is due independently to S. McKay [71] and to I. Kiming [37]. Medvedev's arguments implement all the machinery used in proving Theorem 6.2, including the p-adic Lie ring construction. See [74], [75].

6.10. GROUPS OF BOUNDED RANK

Let G be a p-group admitting an almost f.p.f. automorphism α. Can we obtain bounds on the derived length of G which do not depend on the order of α? As shown in [103], this can be done in some situations where the rank of the group is bounded.

Theorem 6.7 *Let G be a p-group of rank r which admits a p'-automorphism α with m fixed points. Then the derived length of G is r, m-bounded.*

Recall that a p'-automorphism is an automorphism whose order is prime to p. If $m = 1$ (namely, α is f.p.f.), the bound on $\mathrm{dl}(G)$ is approximately 2^{r+1}.

The proof of Theorem 6.7 consists of reductions similar to those described above. The p-adic Lie ring construction is also applied. However, at the Lie-theoretic level Kreknin's theorem is not enough, and so the following result has to be proved.

Proposition 6.8 *Let L be a Lie ring of finite rank r admitting a semisimple fixed-point-free automorphism α. Then $\mathrm{dl}(L)$ is r-bounded. In fact we have $\mathrm{dl}(L) \leqslant 2^{r-1}$.*

Here the rank of L is the rank of its additive group. An automorphism is called *semisimple* if it can be represented by a diagonal matrix (over a suitable integral extension of $\mathbb{Z}$). This result, but with the bound $2^{r-1} - 1$, appears in [103], Proposition. However, the proof given there only establishes the bound $2^r - 1$. I am grateful to A. Turull for pointing out this correction. In fact by slightly modifying the arguments in [103] it is possible to obtain the inequality $\mathrm{dl}(L) \leqslant 2^{r-1}$ stated above.

We note that Theorem 6.7 can be extended to arbitrary finite groups (see the main result of [103]).

Problem 20 *In the statement of Theorem 6.7, can we omit the restriction that α is a p'-automorphism?*

To obtain this, it would suffice to extend Proposition 6.8 to general automorphisms α.

6.11. p-GROUPS ACTING ALMOST FIXED-POINT-FREELY

In addition to studying p-groups admitting almost f.p.f. automorphisms, there is some interest in the structure of p-groups G which can act almost fixed-point-freely on some finite group H, or on a linear space V. Let V be a finite dimensional complex linear space acted on faithfully by a finite group G. Define the *fixity* of this action to be the maximal dimension of a space of fixed points $C_V(g)$ of a nontrivial element $g \in G$. Thus fixed-point-free actions are actions of fixity 0, and the groups G which can act in such a way are exactly the Frobenius complements. By a classical result of Burnside, Frobenius complements of prime power order are cyclic or generalized quaternion. This result is extended in [82], where, in particular, the p-groups which can act with fixity 1 are determined. We omit the details. Let us just point out an elementary—yet useful—fact: p-groups which can act with fixity f have p-rank at most $f + 1$.

Problem 21 *Determine the p-groups which can act with fixity p on a complex linear space.*

This is related to classifying the irreducible p-subgroups of $\mathrm{GL}_{p^2}(\mathbb{C})$. Some progress has recently been made for $p = 2$ by D. L. Flannery [14].

6.12. CONCLUDING REMARKS

In spite of the recent progress in the study of almost f.p.f. automorphisms, the following fundamental question is still open:

Problem 22 *Let G be a p-group admitting a fixed-point-free automorphism of order n. Is* $\mathrm{dl}(G)$ *n-bounded?*

This question, which appears in the Kourovka Notebook, has an affirmative answer if n is prime (by Higman's Theorem 3.2) and if $n = 4$ (see Kovács [41]). It seems to be open in all other cases. Theorem 6.7 shows that it suffices to bound the rank of G in terms of n. However, the rank of G need not be n-bounded, as is easily demonstrated by elementary abelian 3-groups acted on by an involution inverting all elements.

7. The By-Products of Being Thin

7.1. THE COCLASS CONJECTURES

This chapter is devoted to the coclass theory for p-groups.

In 1980 C. R. Leedham-Green and M. F. Newman proposed a program for classifying finite p-groups using coclass as the primary invariant [54]. The *coclass* of a group of order p^n and nilpotency class c is defined to be $n - c$; a pro-p group of coclass r is an inverse limit of finite p-groups of coclass r. Thus, groups of coclass 1 are exactly the groups of maximal class studied by N. Blackburn [5] and many others (see §2.5).

Let us formulate the coclass conjectures made in [54], ordered by decreasing strength (that is, each conjecture implies the next one).

Conjecture A *For some function f, every p-group of coclass r has a normal subgroup N of class at most 2 (or 1, if $p = 2$) and index at most* $f(p, r)$.

Conjecture B *For some function g, every p-group of coclass r has derived length at most* $g(p, r)$.

Conjecture C *Pro-p groups of finite coclass are soluble.*

Conjecture D *Given p and r, there are only finitely many isomorphism types of infinite pro-p groups of coclass r.*

Conjecture E *Given p and r, there are only finitely many isomorphism types of infinite soluble pro-p groups of coclass r.*

There is a certain counter-intuitive aspect in some of the conjectures, in that they assert that groups of small coclass, that is, of very large nilpotency class, are in a sense almost abelian.

At the time the coclass conjectures were posed, there was no real evidence that even the weakest of them holds. However, these conjecture have all become theorems; we describe below some stages in their proof.

7.2. UNISERIALITY

A fundamental notion in the context of coclass is that of uniserial actions. Let G be a p-group acting on a p-group H. We say that this action is *uniserial* if all G-invariant sections of H on which G acts trivially have order p (or 1). This is tantamount to saying that $|H : [H, G]| \leqslant p$, $|[H, G] : [H, G, G]| \leqslant p$, etc.

A similar definition applies for pro-p groups. Let G be a p-adic (pre) space group with translation group T. We say that G is a uniserial p-adic (pre) space group if G (equivalently, G/T) acts uniserially on T. This implies that G is a pro-p group of finite coclass. Conversely, all known examples of infinite pro-p groups of finite coclass consist of uniserial (pre) space groups. As we shall see below, this is not a coincidence.

We shall also deal with uniserial actions in a Lie-theoretic context. As the definitions are completely analogous we omit the details.

7.3. CONJECTURE E

Let G be an infinite pro-p group of finite coclass. Suppose G is soluble. Then it was shown in [54] that G is a p-adic pre space group, so in particular it is abelian-by-finite. Let $T \subseteq G$ be the translation group. Fixing T, it follows that G/T is a finite p-group of bounded order acting uniserially on T. Homological arguments show that, given T and G/T (up to isomorphism), there are only finitely many extensions of T by G/T. Thus to prove Conjecture E it suffices to bound the number of possibilities for T. This reduces Conjecture E to

Conjecture E′ *There exists a function $e = e(p, r)$ such that the dimension of a p-adic (pre) space group of coclass r is at most $e(p, r)$.*

Using results on the Sylow structure of $\mathrm{GL}_n(\mathbb{Z}_p)$, some representation theory and elementary (yet rather long) computations, Leedham-Green, McKay and Plesken proved Conjecture E′ in [51], [52]. Their bounds on the function e were not optimal. Recently, S. McKay obtained the best possible bounds, namely

$$e(p, r) = (p - 1)p^{r-1}.$$

Using the theory of powerful p-groups, much simpler proofs of Conjecture E can be given (see [65], [100]). The idea is to bound the rank of a finite p-group of coclass r; the existence of such a bound readily implies Conjecture E′. The shortest proof of this type (for odd p) is given in Mann's paper [65], where the following result is proved.

Proposition 7.1 *Let G be a p-group of coclass r $(p > 2)$. Then there exists a powerful normal subgroup $N \lhd G$ such that $d(N) \leqslant p^r$ and $|G : N| \leqslant p^{p^r+r}$. In particular, the rank of G is p, r-bounded.*

In view of part (iii) of Theorem 4.6 we obtain the following.

Corollary 7.2   *Pro-p groups of finite coclass are p-adic analytic.*

This corollary was first proved by Leedham-Green [46] in the mid 80s. It paved the way for most of the attacks on the more difficult conjectures.

7.4. CONJECTURE C

The following result of S. Donkin [13], combined with Corollary 7.2, implies Conjecture C for all primes $p \geqslant 5$.

Theorem 7.3   *Let G be a p-adic analytic pro-p group of finite coclass ($p \geqslant 5$). Then G is soluble.*

Donkin's proof, which was given in the mid 80s, is based on the theory of semisimple algebraic groups. Let G be a counter-example to the theorem and let $L = \mathrm{Lie}(G)$ be its associated p-adic Lie algebra. Since G is non-soluble (by assumption) it easily follows that L is non-soluble. The fact that G has finite coclass implies that the adjoint action of G on L is uniserial. We have thus obtained a non-soluble p-adic Lie algebra L acted on uniserially by a pro-p group. The next stage is to reduce to the case where L (tensored with the p-adic field) is simple (nonabelian). This leads to the following question:

Can a pro-p group act uniserially on a simple p-adic Lie algebra?

Using the classification of simple p-adic Lie algebras and Iwahori and Matsumoto's theory of p-adic Chevalley groups, Donkin was able to obtain a negative answer, thus proving the theorem.

Donkin's proof is rather complicated and highly non-elementary. Another non-elementary approach, which applies the theory of buildings and might handle smaller primes, has been suggested by J. Tits (see [13, p. 341]).

In 1991 Zelmanov and myself obtained a short proof of Conjecture C for all primes p [108]. Our approach was based on the theory of almost f.p.f. automorphisms. We first observed that a p-adic analytic pro-p group with an element whose centralizer is finite is soluble. This can be deduced from Kreknin's result [43] on Lie rings with fixed-point-free automorphisms (see §6.5). At this stage one is led to the following question:

Does every pro-p group of finite coclass have an element whose centralizer is finite?

Clearly, a positive answer would imply Conjecture C. By making additional reductions to questions on modular Lie algebras we gave an affirmative answer to this question, and deduced Conjecture C for all primes p (the case $p = 2$ is in fact the simplest). Note that, if $x \in G$ and the centralizer $C_G(x)$ is finite, then x is automatically of finite order. As a by-product it follows that pro-p groups of finite coclass cannot be torsion-free. We are not aware of any direct proof of this somewhat curious fact (which of course follows from Conjecture C).

Note that, combining [54] with Conjecture (i.e. Theorem) C, it follows that every infinite pro-p group G of finite coclass has the structure of a uniserial p-adic pre space group; moreover, if G happens to be just-infinite, then G is a p-adic space group.

We close this section with a problem. For an infinite pro-p group G define a series $c_n = c_n(G) \leqslant \infty$ $(n \geqslant 1)$ by

$$|G : \gamma_n| = p^{c_n}.$$

Note that G has coclass r if and only if $c_n(G) = n + r - 1$ for all sufficiently large n; thus G has finite coclass if and only if the series $\{c_n(G) - n\}$ is bounded. However, one may try to impose weaker conditions on the series $\{c_n\}$. For example, in [100] it is shown that the inequality $c_n < n + [\log_p(n/2)]$ for a single value of n already implies that G has finite coclass. We therefore ask:

Problem 23 *Given n, find the maximal integer $f(n)$ such that $c_n(G) \leqslant f(n)$ implies that G has finite coclass.*

It can be shown, using the example of the Nottingham group, that

$$f(n) < n + [(n-2)/(p-1)].$$

This still leaves open a fairly large gap.

7.5. CONJECTURE A

Let us now discuss the strongest of the coclass conjectures, namely Conjecture A. This conjecture was deduced from Conjecture C by Leedham-Green in [47], shortly after the work of Donkin on Conjecture C. The deduction is non-trivial, and, curiously enough, applies Conjecture C twice (for odd primes). Thus a proof of Conjecture A for $p \geqslant 5$ was available in the second half of the last decade. A proof for all primes is obtained by combining [108] and [47]. Leedham-Green's work makes use of various limit arguments; consequently, though the existence of a function $f(p,r)$ with the required property is established, no explicit bounds are obtained.

Can we find an effective proof of Conjecture A which yields explicit bounds on f? Such a proof has recently been given in [105], where the following results are established.

Theorem 7.4 *Let G be a finite p-group of coclass r.*

(i) *Suppose $p > 2$. Then $\gamma_{2(p^r - p^{r-1} - 1)}$ has class at most 2.*

(ii) *Suppose $p = 2$ and $|G| \geqslant 2^{2^{2r+5}}$. Then $\gamma_{7.2^r - 2}$ is abelian.*

For odd primes this confirms Conjecture A with

$$f(p,r) \leqslant p^{2(p^r - p^{r-1} - 1) + r - 1} < p^{2p^r},$$

a bound which is quite realistic. The bound for $p = 2$ obtained from part (ii) of Theorem 7.4 is probably less accurate.

Let us briefly outline the main stages in the proof of part (i) of Theorem 7.4. We first study uniserial actions on powerful p-groups in some detail. We then show that, if G is as in part (i) of the theorem and $m = (p-1)p^{r-1}$, then (unless G is very small) γ_m is powerful with $d = (p-1)p^s$ generators for some $s < r$, and G acts uniserially on γ_m. We also show that $\gamma_i^p = \gamma_{i+d}$ for $i \geqslant m$, so the p-th power operation in γ_m is well-behaved. This establishes various known properties of groups of maximal class in greater generality.

The next stage consists of Lie-theoretic constructions. We let $L = \bigoplus_{i \geqslant 1} \gamma_i / \gamma_{i+1}$ be the usual graded Lie ring associated with G, and set $M = L^m = \bigoplus_{i \geqslant m} \gamma_i / \gamma_{i+1}$. Then M can be regarded as a Lie algebra over $\mathbb{F}_p$ which is also an $\overline{L}$-module, where $\overline{L} = L/pL$. It is clear that $\overline{L}$ acts uniserially on M. We use the p-th power map in G and its nice properties to turn M into an $\mathbb{F}_p[t]$-Lie algebra. It then turns out that the action of t on M commutes with the action of $\overline{L}$, so M may be regarded as an $\overline{L}$-module over $\mathbb{F}_p[t]$.

Now, assuming the conclusion in part (i) does not hold, we show that M can be lifted to a Lie algebra N which is a *free* $\mathbb{F}_p[t]$-module of rank $d = (p-1)p^s$. We also have $|N : N'| < \infty$. Moreover, the elements of $\overline{L}_1$ can be viewed as derivations of N and the resulting action is uniserial. Tensoring N and $\overline{L}_1$ with the function field $\mathbb{F}_p(t)$ (or with a suitable algebraic extension of it) we obtain a *perfect* Lie algebra K of dimension d and a linear space V of derivations of K, most of which are *non-singular* (as linear transformations). In fact we show that, for some derivation $D \in V$, we have $D^d = 1$.

The question now arises:

> *Does there exist a finite-dimensional perfect Lie algebra K in characteristic p admitting a non-singular derivation D satisfying $D^{(p-1)p^s} = 1$?*

A negative answer is required in order to complete the proof of part (i) of Theorem 7.4. So suppose, in contradiction, that there exists such a Lie algebra K and such a derivation D. Replacing D by D^{p^s} (which is also a derivation) we may assume that $D^{p-1} = 1$. This implies that D is semisimple and that all its eigenvalues are in $\mathbb{F}_p^*$. For $\alpha \in \mathbb{F}_p$ let K_α be its eigenspace. Then $K = \sum_\alpha K_\alpha$ is a grading of K over the additive group of $\mathbb{F}_p$ such that $K_0 = 0$. It is now straightforward to verify that K satisfies the $(p-1)$-Engel identity for homogeneous elements, and is thus nilpotent (by a variation of Jacobson on Engel's theorem [32]). Since K is nilpotent, it cannot be perfect. This contradiction completes the sketch of the proof.

The proof of part (ii) requires some more ideas and will not be sketched here.

It is worth mentioning that, while the proof of Theorem 7.4 does fall under the category of Lie ring methods, we have here an unusual case where the reduction to the Lie-theoretic problem is significantly more difficult than the solution of the resulting problem.

7.6. CONJUGACY CLASSES, IV

In the course of the proof of the effective version of Conjecture A, we obtain the following result, which has some independent interest.

Proposition 7.5   *The co-breadth of p-groups of coclass r is p, r-bounded.*

Thus every p-group G of coclass r possesses an element x whose centralizer is small. The result is effective: for odd p we have the bound $|C_G(x)| \leqslant p^{p^r + r - 1}$, while for $p = 2$ the bound is slightly larger.

Note that the above proposition, combined with Corollary 6.3, immediately implies Conjecture B with explicit bounds. However, much better bounds on $\mathrm{dl}(G)$ already follow from Theorem 7.4.

The above proposition shows that p-groups of small coclass have a large conjugacy class. We now make some more remarks on conjugacy classes in p-groups of given coclass.

Proposition 7.6 *Let G be a p-group of coclass r. Then $k(G) > |G|^\epsilon$ for some $\epsilon = \epsilon(p, r) > 0$ depending only on p and r.*

Proof. By Proposition 7.1, $\mathrm{rk}(G)$ is p, r-bounded. Therefore the result follows from Proposition 4.8. □

Our next result deals only with 2-groups. It shows that if the coclass of G is bounded then conjugacy classes in G are either very large or very small.

Proposition 7.7 *There exist functions $\epsilon(r) > 0$ and $c(r)$ with the following property: if G is a 2-group of coclass r and $C \subset G$ is a conjugacy class in G, then either $|C| > |G|^{\epsilon(r)}$ or $|C| < c(r)$.*

Proof. We rely on part (ii) of Theorem 7.4 (and on some other results from [105]). We may assume that $|G| \geqslant 2^{2^{2r+5}}$, and let $A = \gamma_{7.2^r - 2}$. Then A is abelian of r-bounded index, and $d(A)$ is also r-bounded. Applying [105], Theorem 1.5(ii), it is easy to see that A is almost homocyclic, namely, it becomes homocyclic after factoring out an elementary abelian subgroup $B \subseteq A$ which is normal in G. Note that $|B|$ is r-bounded. Replacing G by G/B (as we may) we can assume that A is homocyclic. Denote its exponent by p^e. Let $H = C_G(A/A^4)$ and let P be the centralizer of A/A^4 in $\mathrm{Aut} A$. It is well-known that P is a uniform 2-group of exponent 2^{e-2}. Moreover, setting $P_i = C_P(A/A^{2^{i+2}})$ ($0 \leqslant i \leqslant e - 2$), we obtain $P_i = P^{2^i}$. By uniformity we have $\Omega_j(P) = P_{e-2-j}$. Let $\alpha : H \to P$ be the obvious homomorphism (attaching to $x \in H$ its conjugation action on A). Fix $x \in G$.

Case 1. $x \in H$. Since $|G/A|$ is r-bounded, an r-bounded power of x lies in A. This shows that $\alpha(x)$ has r-bounded order, say $\alpha(x)^{2^j} = 1$ for some r-bounded j. Then $\alpha(x) \in \Omega_j(P) = P_{e-2-j}$, so x acts trivially on $A/A^{2^{e-j}}$. This implies that $C_A(x)$ has bounded index in A. Since

$$|G : C_G(x)| \leqslant |G : A||A : C_A(x)|,$$

we conclude that the conjugacy class of x in G has r-bounded size.

Case 2. $x \notin H$. Then x acts non-trivially on A/A^4. Choose $a \in A$ satisfying $a^x \not\equiv a \bmod A^4$. Then $[a, x] \notin A^4$ and so $[a, x]$ has order at least 2^{e-1}. Note that $[a^k, x] = [a, x]^k$. It therefore follows that the set $[G, x]$ of all commutators $[g, x]$ ($g \in G$) has size at least 2^{e-1}. Thus x has at least 2^{e-1} conjugates. Finally, note that $|G| \leqslant |G/A|(2^e)^{d(A)}$, where $|G/A|$ and $d(A)$ are r-bounded. This yields $2^{e-1} \geqslant |G|^{\epsilon(r)}$ for a suitable $\epsilon(r) > 0$. The result follows. □

It is easy to obtain explicit (and realistic) estimates for the functions $c(r), \epsilon(r)$ above.

Problem 24 *Is Proposition 7.7 valid for odd primes?*

A result of Vera-López provides a positive answer in the case $r = 1$.
Recall that p-groups of abundance 0 are of maximal class (see §2.6).

Problem 25 *Is the coclass of a p-group G bounded above in terms of its abundance*
l?

This is open even for $l = 1$. However, in this case computer calculations carried
out by E. A. O'Brien and by J. Neubüser and B. Rothe provide positive results for
small primes p.

7.7. LEEDHAM-GREEN'S THEOREM A*

A result of Leedham-Green reveals rather surprising relations between finite p-groups
and p-adic space groups. It is shown in [47] that there exists a function h such that
every finite p-group G of coclass r has a normal subgroup N whose order is at most
$h(p, r)$ such that G/N is obtained from a p-adic space group of coclass at most r
via a series of 'elementary operations' (such finite p-groups are called *constructible*).
While the precise result for odd p is not easily formulated, for $p = 2$ this simply
means that G/N is a *quotient* of some 2-adic space group of coclass at most r.

Problem 26 *Find explicit bounds on the function $h(p, r)$.*

No such bounds are known up to now. For $p = 2$ this problem has a geometric
interpretation concerning the twig lengths in the graph $T_{2,r}$ (see below).

7.8. GRAPHS; EXPERIMENTAL EVIDENCE

Some of the coclass conjectures may be reformulated as combinatorial statements on
the structure of certain graphs. Thus, let $T_{p,r}$ be a (directed) graph whose vertices
are all p-groups of coclass r, where (H, G) is an edge if there is an epimorphism from
H to G whose kernel has order p. Then Conjecture D is tantamount to saying that
there are only finitely many infinite chains in $T_{p,r}$.

Recently M. F. Newman and E. A. O'Brien [81] studied the graph $T_{2,3}$ (representing the 2-groups of coclass 3) in great detail, using the p-group generation
algorithm. Apart from determining the infinite chains, they obtained information
on the length of twigs splitting from an infinite chain, and about various periodic
patterns. This has led them to formulate several delicate conjectures, such as:

Problem 27 *Show that the number of 2-groups of given coclass r and order 2^n is*
virtually periodic in n, with period 2^{r-1}.

It is also conjectured in [81] that the 2-groups of given coclass can be divided
into finitely many 'one-parameter families' and finitely many 'sporadics'.

7.9. COCLASS THEORY FOR LIE ALGEBRAS

The remarkable accuracy of the coclass conjectures for p-groups suggests the possibility that a coclass theory could be developed for other objects, such as Lie algebras. The coclass of a nilpotent or a residually nilpotent Lie algebra L is defined by $\sum_{i \geqslant 1}(\dim(L^i/L^{i+1}) - 1)$ (possibly infinity). A coclass theory for Lie algebras may be relevant for the structure of certain infinite groups, such as torsion-free nilpotent groups (see Corollary 7.11 below).

In this section we record the known results on this subject (only a small part of which has already been published). All Lie algebras below are assumed to be residually nilpotent. We start with the case of restricted Lie algebras. The following result is due to D. M. Riley and J. F. Semple [92].

Theorem 7.8 *Let L be a Lie algebra of coclass r over a field of characteristic p. Then L is finite-dimensional and its dimension is p, r-bounded. In fact we have $\dim(L) \leqslant 2p^r + r + 1$ if $p > 2$, and $\dim(L) \leqslant 6 \cdot 2^r + r + 1$ if $p = 2$.*

Thus we may say that the coclass conjectures hold in a stronger form for restricted Lie algebras. As shown below, non-restricted Lie algebras in characteristic p behave much more wildly. In fact even Conjecture C is not valid for them.

Theorem 7.9 *For every prime p there are non-soluble $\mathbb{N}$-graded Lie algebras of coclass 1 over $\mathbb{F}_p$.*

This result is proved in [107]. In fact, for each p, infinitely many (pairwise non-isomorphic) Lie algebras of this type are constructed. These Lie algebras have the form $L = \bigoplus_{i \geqslant 1} L_i$ where $\dim L_1 = 2$, $\dim L_i = 1$ for all $i > 1$, and L is generated by L_1. It is interesting to note that our construction (as the proof of Conjecture A for p-groups) is strongly related to the notion of nonsingular derivations. In fact we first show that there are finite-dimensional simple cyclically graded Lie algebras $S = \bigoplus_{i \in \mathbb{Z}/n\mathbb{Z}} S_i$ over $\mathbb{F}_p$ with 1-dimensional homogeneous components S_i and with a nonsingular derivation D acting cyclically by $D(S_i) = S_{i+1}$. We then obtain the required Lie algebras L by 'untying' S. To be precise, L is generated by $L_1 \otimes t$ and $D \otimes t$ inside $S.\langle D \rangle \otimes \mathbb{F}_p[t]$.

The next result, which is contained in a joint (yet unpublished) work of Zelmanov and myself, deals with Lie algebras of characteristic zero.

Theorem 7.10 *Let L be an $\mathbb{N}$-graded Lie algebra of finite coclass r over a field of characteristic zero, and suppose L is generated by L_1. Then*

(i) *L has an abelian ideal of finite codimension; in particular L is soluble.*

(ii) *The derived length of L is r-bounded.*

While Theorem 7.8 above is proved by adapting the methods of [105], the proof of Theorem 7.10 uses completely different methods. As a consequence we obtain the following new result in the theory of infinite nilpotent groups.

Corollary 7.11 *There exists a function f such that if G is a torsion-free nilpotent group of Hirsch rank h and class c, then $\mathrm{dl}(G) \leqslant f(h - c)$.*

This answers a question of P. H. Kropholler. It is an easy exercise to deduce the corollary from Theorem 7.10, so we leave it to the reader. Let us just point out that the function $f(r)$ does not grow fast; in fact it is bounded by $3r$. However, its exact behaviour remains to be determined.

We conclude with the following plausible conjecture on the structure of modular Lie algebras of finite coclass.

Problem 28 *Let L be an $\mathbb{N}$-graded Lie algebra of finite coclass over $\mathbb{F}_p$ (generated by L_1). Show that L has a subalgebra of finite co-dimension which is finite-dimensional over $\mathbb{F}_p[t]$.*

7.10. OTHER NARROWNESS CONDITIONS

Restricting the coclass of a group may be regarded as a narrowness condition. Some authors have considered other narrowness conditions. For example, R. Brandl, A. Caranti and C. M. Scoppola call a p-group G *thin* if every anti-chain of normal subgroups of G has size at most $p + 1$; note that all p-groups of maximal class have this property; thus, to avoid trivialities, one requires further that G not be of maximal class.

Rather wild thin p-groups can be obtained by forming quotients of the Nottingham group. However, metabelian thin groups are quite well understood (see [9]); in particular, for fixed p, there are only finitely many metabelian thin p-groups. Various results on the lattice of normal subgroups of a thin group can be found in the recent preprint [11] by Caranti, Mattarei, Newman and Scoppola.

The following question is posed in [9].

Problem 29 *Given p and k, are there only finitely many thin p-groups of derived length k?*

The notion of a thin group applies equally well to pro-p groups. There is also some interest in pro-p groups G which are 'virtually thin' in the sense that anti-chains of normal subgroups of large index have size $\leqslant p + 1$; to avoid trivialities it is natural to require further that G not be of finite coclass. It turns out that, while infinite metabelian thin pro-p groups do not exist, certain (non-uniserial) p-adic space groups provide examples of infinite metabelian virtually thin pro-p groups.

The lower central factors of thin p-groups are elementary abelian of order p or p^2. Leedham-Green proposed the study of a larger class of groups, those of *bounded width*. Here the width of a pro-p group G can be defined as the supremal value of $|\gamma_i(G)/\gamma_{i+1}(G)|$. It is known that the Nottingham group has width p^2, so groups of bounded width may be fairly complicated.

The following challenging problem was suggested by Leedham-Green.

Problem 30 *Let G be a just-infinite pro-p group of finite width. Show that one of the following holds:*

(i) G *is linear over $\mathbb{Z}_p$.*

(ii) G *is linear over $\mathbb{F}_p[[t]]$.*

(iii) G *is a closed subgroup of the Nottingham group.*

Methodologically it makes sense to start the examination of pro-p groups of bounded width by considering the analogous notion for Lie algebras. Various problems and conjectures can be posed. However, the structure of pro-p groups and Lie algebras of given width is still mysterious, in spite of several recent attempts to clarify the situation.

8. Just-Infinite Pro-p Groups: an Inventory

8.1. ROUGH PLAN

Recall that an infinite pro-p group G is termed *just-infinite* if all non-trivial closed normal subgroups of G have finite index. Clearly, just-infinite pro-p groups are finitely generated (otherwise $G/\Phi(G)$ is elementary abelian of infinite rank, so it has non-trivial closed normal subgroups of infinite index). It is also easy to see (using Zorn's Lemma) that every infinite finitely generated pro-p group has a just-infinite quotient.

Classifying the just-infinite pro-p groups (at least in the weak sense of dividing into well-understood classes) is one of the greatest challenges in the theory of pro-p groups. While in classifying finite simple groups one could use induction (i.e. consider a minimal counter-example), it is not clear that even such a basic idea could apply here. In any case, it seems that we are still at the stage of discovering new objects, so classification attempts may be somewhat premature. Indeed, some new examples of just-infinite pro-p groups will be presented in Theorem 8.5 below.

Roughly speaking, we shall try to draw a 'map' of the universe of the *known* just-infinite pro-p groups, starting with relatively tame groups, and passing gradually to wilder ones. It turns out that the known just-infinite pro-p groups can be divided into 4 natural families: the soluble ones, the non-soluble p-adic analytic ones, the groups which are analytic over $\mathbb{F}_p[[t]]$, and the pro-p groups of *Cartan type*. The last family, introduced by Leedham-Green and myself in [56], includes the Nottingham group (among other examples).

We point out that some of the results in section 8.5 should be treated with some caution, as they are not fully written up yet.

8.2. SOLUBLE GROUPS

Determining the just-infinite soluble pro-p groups is rather easy.

Lemma 8.1 *Let G be a soluble pro-p group. Then G is just-infinite if and only if G is a p-adic space group.*

Proof. Let G be a just-infinite soluble pro-p group. Then G has a non-trivial (closed) abelian normal subgroup A. It follows that A has finite index in G. Replacing A by a larger subgroup if possible, we can assume that A is a maximal abelian normal subgroup in G. Standard arguments (borrowed from the finite p-group case) now show that A is self-centralizing in G, so G/A acts faithfully on A. It is clear that A is finitely generated (as G is). Note that A is torsion-free, since its torsion subgroup is normal in G and of infinite index. Thus A has the structure of a finitely generated free p-adic module. Finally, all G-invariant subgroups of A have finite index in A, and this implies that $A \otimes \mathbb{Q}_p$ is an irreducible (faithful) $\mathbb{Q}_p P$-module, where $P = G/A$. We see that G is a p-adic space group with translation group A and point group P.

The other implication is easier and is left to the reader.   □

Note that just-infinite soluble pro-p groups are always p-adic analytic.

8.3. p-ADIC ANALYTIC GROUPS

Non-soluble examples of just-infinite pro-p groups can easily be found among the p-adic analytic groups. A typical example is given by the first congruence subgroup $G = \mathrm{SL}_2^1(\mathbb{Z}_p)$ (see §2.8). There are several ways to show that G is just-infinite; for example, we can rely on the following useful observation.

Lemma 8.2 *Let G be a pro-p group with an N-series $\{G_i\}_{i \geqslant 1}$, and let L be the graded Lie ring associated with that filtration. Suppose L is just-infinite (namely, every non-zero graded Lie ideal of L has finite index in L). Then the group G is just-infinite.*

Proof. Arguing by contradiction, let $1 \neq N \lhd G$ be a closed normal subgroup of infinite index in G. Then the Lie subalgebra $K(N) \subseteq L$ corresponding to N is a non-zero graded Lie ideal of infinite index in L (see §2.3). This contradicts the assumption that L is just-infinite.   □

Corollary 8.3 *Under the above conditions, suppose $L = S \otimes t\mathbb{F}_p[[t]]$ for some finite-dimensional simple Lie algebra S over $\mathbb{F}_p$. Then G is just-infinite.*

Proof. It is straightforward to verify, using the simplicity of S, that the $\mathbb{N}$-graded Lie algebra $S \otimes t\mathbb{F}_p[[t]]$ is just-infinite. Now apply Lemma 8.2.   □

To complete the proof that $\mathrm{SL}_2^1(\mathbb{Z}_p)$ is just-infinite, one checks that, if $G_i = \mathrm{SL}_2^i(\mathbb{Z}_p)$ ($i \geqslant 1$) is the congruence filtration, then $[G_i, G_j] \subseteq G_{i+j}$, and the resulting Lie algebra is isomorphic to $\mathfrak{sl}_2(\mathbb{F}_p) \otimes t\mathbb{F}_p[[t]]$. For odd p, the Lie algebra $\mathfrak{sl}_2(\mathbb{F}_p)$ is simple, so the claim follows. A slight variation is needed to take care of the case $p = 2$. Furthermore, it can be shown that all open subgroups of $\mathrm{SL}_2^1(\mathbb{Z}_p)$ are also just-infinite.

In general, congruence subgroups in p-adic Chevalley groups form a rich source of examples of non-soluble p-adic analytic just-infinite pro-p groups.

8.4. $\mathbb{F}_p[[t]]$-ANALYTIC GROUPS

Similar constructions can be carried out when the p-adic integers are replaced with the power series ring $\mathbb{F}_p[[t]]$. Thus we can consider, for example, $G = \mathrm{SL}_2^1(\mathbb{F}_p[[t]])$ and show that it is just-infinite in exactly the same way (i.e. using Corollary 8.3). This group is non-soluble and not even p-adic analytic (since it has infinite rank). This basic example can be generalized. In fact, in [60] Lubotzky and I consider pro-p groups which are analytic over $\mathbb{F}_p[[t]]$ (or over more general pro-p rings Λ). To some extent, these objects were already studied by Serre in [96]. We show in [60] that there is a continuous transition between p-adic analytic pro-p groups and (certain) $\mathbb{F}_p[[t]]$-analytic groups. For example, while p-adic analytic pro-p groups have polynomial subgroup growth, the subgroup growth of $\mathbb{F}_p[[t]]$-analytic pro-p groups is usually of type $n^{\log n}$ (which is the minimal growth type of non p-adic analytic pro-p groups [100]); and, while p-adic analytic groups have finite rank, $\mathbb{F}_p[[t]]$-analytic groups (usually) have finite lower rank, i.e. they have a series H_i of open subgroups converging to 1 such that $d(H_i)$ is uniformly bounded. Analytic groups over $\mathbb{F}_p[[t]]$ are useful in the study of arithmetic groups in characteristic p and in some number-theoretic contexts. We shall not elaborate on these points here, but refer the interested reader to [60] for more details.

We close this section with the following intriguing problem.

Problem 31 *Does* $\mathrm{SL}_n(\mathbb{F}_p[[t]])$ *have a closed subgroup which is isomorphic to the free pro-p group on two generators?*

For $n = 2$ and p odd the answer is negative, as proved by Zubkov [128]. This question is closely related to the notion of topological identities for profinite groups and pro-p identities in particular.

8.5. Pro-p Groups of Cartan Type

The pro-p groups of Cartan type have recently been defined in [56]. The motivation is three-fold: first, to show that the Nottingham group which was considered 'sporadic' actually lies in a natural infinite family of pro-p groups; second, to find pro-p analogues of the simple modular Lie algebras of Cartan type; and third, to obtain new examples of just-infinite pro-p groups. As a by-product we shall construct what seems to be the first example of a just-infinite pro-p group of infinite width.

We start with a brief discussion of the Nottingham group. Recall that this group consists of normalized automorphisms of $\mathbb{F}_p[[t]]$, namely those acting trivially on $(t)/(t)^2$. This group has a natural congruence filtration $\{G_i\}_{i \geqslant 1}$, G_i being the centralizer of $(t)/(t)^{i+1}$ in G. It is easy to see that $|G_i : G_{i+1}| = p$ for all i, and that $\{G_i\}$ is an N_p-series. It is natural to choose representatives $e_i \in G_i \smallsetminus G_{i+1}$, where e_i is represented by the power series $t + t^{i+1}$. We then have

$$[e_i, e_j] \equiv e_{i+j}^{i-j} \mod G_{i+j+1}.$$

This shows that the graded Lie algebra L corresponding to the filtration $\{G_i\}$ can be thought of as spanned by $e_1, e_2, e_3, \ldots$ subject to $[e_i, e_j] = (i-j)e_{i+j}$. This Lie algebra coincides with the annihilator of $(t)/(t^2)$ in the derivation algebra $\mathrm{Der}(\mathbb{F}_p[[t]])$. It is easy to see that L is a just-infinite Lie algebra, and so it follows (using Lemma 8.2) that the Nottingham group is just-infinite. For more information see [121] and [57].

The following problems concerning the Nottingham group are still unsettled.

Problem 32 *Does the Nottingham group have a closed subgroup (or section) isomorphic to $C_p \wr \mathbb{Z}_p$?*

A negative answer would imply a negative answer to Problem 16. It has been shown by A. Weiss, using Galois Theory, that the Nottingham group has an elementary abelian subgroup of infinite rank.

Problem 33   *Does the Nottingham group have (non-abelian) free (discrete) subgroups?*

The answer ought to be 'yes'. There have been various attempts to solve this problem in the past few years, but as far as I know it is still open.

It turns out that the construction of the Nottingham group can be generalized as follows. Consider the power series ring $R = \mathbb{F}_p[[x_1, \ldots, x_r]]$ in r variables, and let $M = (x_1, \ldots, x_r)$ be its unique maximal ideal. For $i \geqslant 1$ let

$$\mathrm{Aut}_i(R) = C_{\mathrm{Aut}R}(M/M^{i+1}).$$

We are particularly interested in the group $\mathrm{Aut}_1(R)$ (which, for $r = 1$ coincides with the Nottingham group). It can be shown that $\{\mathrm{Aut}_i(R)\}$ is an N_p-series in $\mathrm{Aut}_1(R)$. Let $L(\mathrm{Aut}_1(R))$ be the associated Lie algebra, and set $\mathrm{Der}_1(R) = \mathrm{Ann}_{\mathrm{Der}R}(M/M^2)$.

Proposition 8.4   *With the above notation we have*

$$L(\mathrm{Aut}_1(R)) \cong \mathrm{Der}_1(R).$$

This enables us to reduce some questions about the group $\mathrm{Aut}_1(R)$ to Lie-theoretic questions regarding the Lie algebra $\mathrm{Der}_1(R)$. In fact it is easy to see that

$$\mathrm{Der}_1(R) \cong M^2 \partial/\partial x_1 \oplus \cdots \oplus M^2 \partial/\partial x_r,$$

the free M^2-module on the set of partial derivatives $\partial/\partial x_1, \ldots, \partial/\partial x_r$. This isomorphism facilitates explicit calculations in the Lie algebra $\mathrm{Der}_1(R)$.

Theorem 8.5   *With the above notation, let $G = \mathrm{Aut}_1(R)$, $G_i = \mathrm{Aut}_i(R)$, and suppose $r > 1$. Then*

(i) $\{G_i\}$ *coincides with the lower central series of G.*

(ii) G *is finitely generated; in fact $d(G) = r\binom{r+1}{2}$.*

(iii) G_i/G_{i+1} *is elementary abelian of rank $r\binom{i+r}{r-1}$. In particular G has infinite width.*

(iv) G *is just-infinite.*

For example, for $r = 2$ we have $d(G) = 6$ and $|\gamma_i/\gamma_{i+1}| = p^{2i+4}$.

While the proof of parts (i)–(iii) is based on reductions to Lie-theoretic questions, this is not the case with respect to part (iv). The reason for this is that for $r \geqslant 2$ the Lie algebra $\mathrm{Der}_1(R)$ is *not* just-infinite; thus Lemma 8.2 does not apply, and more delicate arguments should be used in order to show that the group is just-infinite.

The groups $\mathrm{Aut}_1(R)$ may be regarded as 'generalized Nottingham groups'; they bear a close relation to the Witt Lie algebras W_r. We shall not elaborate on this point here. However, it turns out that one can associate pro-p groups not only with the Witt algebras but also with the other simple restricted Lie algebras of Cartan type (of type S, H and K). Roughly speaking, while Lie algebras of Cartan type arise as collections of derivations annihilating a given differential form, their corresponding groups consist of automorphisms preserving that form. For example, to the so-called special algebra

$$S_r = \text{Ann}_{W_r}(dx_1 \wedge \cdots \wedge dx_r),$$

we associate the pro-p group

$$C_{\text{Aut}_1(R)}(dx_1 \wedge \cdots \wedge dx_r),$$

which consists of those elements of $\text{Aut}_1(R)$ whose Jacobian matrix has determinant 1. We call the groups obtained in this way pro-p groups of *Cartan type*. Whether these groups are always just-infinite is still unclear.

8.6. CONCLUDING REMARKS

Let G be a finitely generated pro-p group, and let $r(G)$ denote the number of relators in a *minimal* pro-p presentation of G (namely, a presentation which has the form $G = F/R$, where F is a free pro-p group on $d(G)$ generators). It is known that $r(G) = \dim H_2(G, \mathbb{F}_p)$ so it does not depend on the particular presentation (as long as it is minimal). Let us say that a pro-p group G satisfies the *Golod-Šafarevič Inequality* (GSI for short) if it satisfies

$$r(G) \geqslant d(G)^2/4.$$

In this case we shall say that G is a GS *group*. Here the case $r(G) = \infty$ is allowed (so finitely generated pro-p groups which are not finitely presented are included among the GS groups). That finite pro-p groups are GS groups was proved by Golod and Šafarevič in 1964 [16]. Gradually it became clear that some infinite pro-p groups also satisfy the GSI. These include p-adic analytic groups and certain other analytic groups (discussed in §8.4). Important results on the Golod-Šafarevič Inequality for other types of pro-p groups have recently been established by Wilson and by Zelmanov [118], [120], [127]. In [127] it is shown that a non GS pro-p group has a closed subgroup which is isomorphic to the free pro-p group on two generators.

Here we make the following observation.

Lemma 8.6   *Let G be a just-infinite pro-p group. Then G satisfies the Golod-Šafarevič Inequality.*

Though it seems that this result has not been formulated before as such, it is in essence known. The idea is that, if G is not a GS group, then, by adding a relator lying sufficiently deep in the dimension subgroup filtration of G, we still obtain an infinite group. Hence such a pro-p group G must have proper infinite quotients. In fact it was also observed by J. S. Wilson and by myself that a non GS pro-p group has continuously many closed normal subgroups, so it is very far from being just-infinite.

We end this section with a recent, yet unpublished, result of Leedham-Green, Plesken and Ziegler [55].

Theorem 8.7   *There are $2^{\aleph_0}$ many isomorphism types of just-infinite pro-p groups of finite width.*

At a first glance, this result can be regarded as rather disappointing, hampering the attempt to classify just-infinite groups. However, there are continuously many isomorphism types of p-adic analytic pro-p groups, and still our understanding of these groups is rather satisfactory. More importantly, the groups constructed in [55] are all finite extensions of one particular pro-p group G (which happens to be $\mathbb{F}_p[[t]]$-analytic). Thus, up to commensurability, they all coincide.

It would be interesting to know how wide just-infinite pro-p groups can be. This may establish an interesting connection between the 'bottom' and the 'top' of pro-p groups.

Problem 34 *Let G be a just-infinite pro-p group and let*

$$p^{d_i} = |D_i(G)/D_{i+1}(G)|.$$

How fast can the sequence $\{d_i\}$ grow? Can it grow exponentially with i?

We note that in free (nonabelian) pro-p groups the series $\{d_i\}$ grows exponentially. In all known just-infinite pro-p groups $\{d_i\}$ grows at most polynomially.

Problem 35 *How many just-infinite pro-p groups are there up to commensurability?*

Problem 36 *Does every just-infinite pro-p group belong (up to commensurability) to one of the families described above?*

It would be over-optimistic at this stage to conjecture a positive answer. We therefore conclude these notes with the following problem.

Problem 37 *Find new types of just-infinite pro-p groups.*

References

1. J. L. Alperin, Automorphisms of solvable groups, *Proc. Amer. Math. Soc.* 13 (1962), 175–180.
2. J. L. Alperin, Centralizers of abelian normal subgroups of p-groups, *J. Algebra* 1 (1964), 110–113.
3. Y. Berkovich, Counting theorems for p-groups, *Arch. Math.* 59 (1992), 215–222.
4. Y. Berkovich, On the number of subgroups of given order and exponent p in a finite irregular p-group, *Bull. London Math. Soc.* 24 (1992), 259–266.
5. N. Blackburn, On a special class of p-groups, *Acta Math.* 100 (1958), 49–92.
6. N. Blackburn, Generalizations of certain elementary theorems on p-groups, *Proc. London Math. Soc.* (3) 11 (1961), 1–22.
7. S. R. Blackburn, D. Phil. thesis, Oxford, 1992.
8. S. R. Blackburn, Enumeration within isoclinism classes of groups of prime power order, *J. London Math. Soc.* 50 (1994), 293–304.
9. R. Brandl, A. Caranti and C. M. Scoppola, Metabelian thin p-groups, *Quart. J. Math. Oxford* (2) 43 (1992), 157–173.
10. W. Burnside, On an unsettled question in the theory of discontinuous groups, *Quart. J. Pure Appl. Math.* 33 (1902), 230–238.
11. A. Caranti, S. Mattarei, M. F. Newman and C. M. Scoppola, *Thin groups of prime power order and thin Lie algebras*, Preprint, 1994.
12. J. Dixon, M. P. F. Du Sautoy, A. Mann and D. Segal, *Analytic Pro-p Groups*, London Math. Soc. Lecture Note Series 157, Cambridge University Press, Cambridge, 1991.

13. S. Donkin, Space groups and groups of prime power order. VIII. Pro-p groups of finite coclass and p-adic Lie algebras, *J. Algebra* 111 (1987), 316–342.

14. D. L. Flannery, Finite irreducible linear 2-groups of degree 4, Ph. D. thesis, Australian National University, 1992.

15. E. S. Golod, On nil algebras and finitely approximable groups, *Izv. AN SSSR* 28 (1964), 273–276 (in Russian).

16. E. S. Golod and I. Šafarevič, On the class field tower, *Izv. AN SSSR* 28 (1964), 261–272 (in Russian).

17. D. Gorenstein, *Finite Groups*, 2nd edition, Chelsea, New York, 1980.

18. K. W. Gruenberg, Two theorems on Engel groups, *Proc. Cambr. Phil. Soc.* 49 (1953), 377–380.

19. P. Hall, A contribution to the theory of groups of prime power orders, *Proc. London Math. Soc.* 36 (1933), 29–95.

20. P. Hall, The classification of prime-power groups, *J. Reine Angew. Math.* 182 (1940), 130–141.

21. P. Hall and G. Higman, On the p-length of p-solvable groups and reduction theorems for Burnside's problem, *Proc. London Math. Soc.* 6 (1956), 1–42.

22. P. Hall and C. R. Kulatilaka, A property of locally finite groups, *J. London Math. Soc.* 39 (1964), 235–239.

23. B. Hartley, Topics in the theory of nilpotent groups, in *Group Theory—Essays for Philip Hall*, eds: K. W. Gruenberg and J. E. Roseblade, Academic Press, London, 1984.

24. W. N. Herfort, Compact torsion groups and finite exponent, *Arch. Math.* 33 (1979), 404–410.

25. W. N. Herfort, L. Ribes and P. A. Zalesskiĭ, Fixed points of automorphisms of free pro-p groups of rank two, *Canad. J. Math.*, to appear.

26. L. Héthelyi, On subgroups of p-groups having soft subgroups, *J. London Math. Soc.* (2) 41 (1990), 425–437.

27. G. Higman, On finite groups of exponent 5, *Proc. Cambr. Phil. Soc.* 52 (1956), 381–390.

28. G. Higman, Groups and Lie rings having automorphisms without non-trivial fixed points, *J. London Math. Soc.* 32 (1957), 321–334.

29. G. Higman, Enumerating p-groups, I: inequalities, *Proc. London Math. Soc.* 10 (1960), 24–30.

30. B. Huppert, *Endliche Gruppen I*, Springer, Berlin-Heidelberg-New York, 1967.

31. B. Huppert and N. Blackburn, *Finite Groups II*, Springer, Berlin-Heidelberg-New York, 1982.

32. N. Jacobson, A note on automorphisms and derivations of Lie Algebras, *Proc. Amer. Math. Soc.* 6 (1955), 281–283.

33. N. Jacobson, *Lie Algebras*, Wiley-Interscience, New York, 1962.

34. S. A. Jennings, The structure of the group ring of a p-group over a modular field, *Trans. Amer. Math. Soc.* 50 (1941), 175–185.

35. D. L. Johnson, The group of formal power series under substitution, *J. Austral. Math. Soc.* (Ser. A) 45 (1988), 296–302.

36. E. I. Khukhro, *Nilpotent Groups and their Automorphisms*, de Gruyter, Berlin, 1993.

37. I. Kiming, Structure and derived length of finite p-groups possessing an automorphism of p-power order having exactly p fixed points, *Math. Scand.* 62 (1988), 153–172.

38. A. I. Kostrikin, On the Burnside problem, *Izv. AN SSSR* 23 (1959), 3–34 (in Russian).

39. A. I. Kostrikin, Sandwiches in Lie algebras, *Mat. Sb.* 110 (1979), 3–12 (in Russian).

40. A. I. Kostrikin and E. I. Zelmanov, A theorem on sandwich algebras, *Proc. Steklov Inst. of Math.* no. 4 (1991), 121–126.

41. L. G. Kovács, Groups with regular automorphisms of order four, *Math. Z.* 75 (1961), 277–294.

42. L. G. Kovács and C. R. Leedham-Green, Some normally monomial p-groups of maximal class and large derived length, *Quart. J. Math. Oxford* (2) 37 (1986), 49–54.

43. V. A. Kreknin, Solvability of Lie algebras with a regular automorphism of finite period, *Soviet Math. Dokl.* 4 (1963), 683–685.

44. M. Lazard, Sur les groupes nilpotents et les anneaux de Lie, *Ann. Sci. École Norm. Sup.* 71 (1954), 101–190.

45. M. Lazard, Groupes analytiques p-adiques, *Publ. Math. I.H.E.S.* 26 (1965), 389–603.

46. C. R. Leedham-Green, Pro-p groups of finite coclass, *J. London Math. Soc.* 50 (1994), 43–48.

47. C. R. Leedham-Green, The structure of finite p-groups, *J. London Math. Soc.* 50 (1994), 49–67.

48. C. R. Leedham-Green and S. McKay, On p-groups of maximal class I, *Quart. J. Math. Oxford* (2) 27 (1976), 297–311.

49.   C. R. Leedham-Green and S. McKay, On p-groups of maximal class II, *Quart. J. Math. Oxford* (2) 29 (1978), 175–186.
50.   C. R. Leedham-Green and S. McKay, On p-groups of maximal class III, *Quart. J. Math. Oxford* (2) 29 (1978), 281–299.
51.   C. R. Leedham-Green, S. McKay and W. Plesken, Space groups and groups of prime power order. V. A bound to the dimension of space groups with fixed coclass, *Proc. London Math. Soc.* 52 (1986), 73–94.
52.   C. R. Leedham-Green, S. McKay and W. Plesken, Space groups and groups of prime power order. VI. A bound to the dimension of a 2-adic group with fixed coclass, *J. London Math. Soc.* 34 (1986), 417–425.
53.   C. R. Leedham-Green, P. M. Neumann and J. Wiegold, The breadth and the class of a finite p-group, *J. London Math. Soc.* (2) 1 (1969), 409–420.
54.   C. R. Leedham-Green and M. F. Newman, Space groups and groups of prime power order I, *Arch. Math.* 35 (1980), 193–202.
55.   C. R. Leedham-Green, W. Plesken and G. Ziegler, Pro-p groups of finite width, in preparation.
56.   C. R. Leedham-Green and A. Shalev, Pro-p groups of Cartan type I, in preparation.
57.   C. R. Leedham-Green, A. Shalev and A. Weiss, Reflections on the Nottingham group, in preparation.
58.   A. Lubotzky and A. Mann, Powerful p-groups. I, II, *J. Algebra* 105 (1987), 484–515.
59.   A. Lubotzky and A. Mann, On groups of polynomial subgroup growth, *Invent. Math.* 104 (1991), 521–533.
60.   A. Lubotzky and A. Shalev, On some Λ-analytic pro-p groups, *Israel J. Math.* 85 (1994), 307–337.
61.   W. Magnus, Über Gruppen und zugeordnete Liesche Ringe, *J. Reine Angew. Math.* 182 (1940), 142–159.
62.   A. Mann, Regular p-groups, *Israel J. Math.* 10 (1971), 471–477.
63.   A. Mann, The power structure of p-groups, I, *J. Algebra* 42 (1976), 121–135.
64.   A. Mann, Some applications of powerful p-groups, in *Groups—St Andrews 1989*, vol. 2, London Math. Soc. Lecture Note Series 160, Cambridge University Press, Cambridge, 1990.
65.   A. Mann, Space groups and groups of prime power order. VII. Powerful p-groups and uncovered p-groups, *Bull. London Math. Soc.* 24 (1992), 271–276.
66.   A. Mann, *Finite p-Groups*, to be published by Oxford University Press.
67.   A. Mann and C. M. Scoppola, Groups of prime power order with many conjugacy classes, Preprint, 1994.
68.   A. J. Mann, On the order of groups of exponent 4, *J. London Math. Soc.* 26 (1982), 64–76.
69.   C. Martinez, On power subgroups of profinite groups, *Trans. Amer. Math. Soc.*, to appear.
70.   K. McCrimmon and E. I. Zelmanov, The structure of strongly prime quadratic Jordan algebras, *Adv. Math.* 69 (1988), 133–222.
71.   S. McKay, On a special class of p-groups, *Quart. J. Math. Oxford* (2) 38 (1987), 489–502.
72.   S. McKay, On a special class of p-groups II, *Quart. J. Math. Oxford* (2) 41 (1990), 431–448.
73.   Yu. Medvedev, Groups and Lie rings with almost regular automorphisms, *J. Algebra* 164 (1994), 877–885.
74.   Yu. Medvedev, p-Groups, Lie p-rings and p-automorphisms, Preprint, 1994.
75.   Yu. Medvedev, p-Divided Lie rings and p-groups, Preprint, 1994.
76.   H. Meier-Wunderli, Über endliche p-Gruppen deren Elemente der Gleichung $x^p = 1$ genügen, *Comment. Math. Helv.* 24 (1950), 18–45.
77.   B. H. Neumann, Groups with automorphisms that leave only the identity element fixed, *Arch. Math.* 7 (1956), 1–5.
78.   P. M. Neumann, An enumeration theorem for finite groups, *Quart. J. Math.* 20 (1969), 395–401.
79.   P. M. Neumann, On two combinatorial problems in group theory, *Bull. London Math. Soc.* 21 (1989), 456–458.
80.   M. F. Newman, Groups of prime-power order, in *Groups—Canberra 1989*, Lecture Notes in Math. 1456, Springer, Berlin, 1990, pp. 49–62.
81.   M. F. Newman and E. A. O'Brien, Classifying 2-groups by coclass, in preparation.
82.   M. F. Newman, E. O'Brien and A. Shalev, The fixity of groups of prime power order, *Bull. London Math. Soc.*, to appear.
83.   M. F. Newman and C. Seeley, in preparation.

84. E. A. O'Brien, The groups of order 256, *J. Algebra* 143 (1991), 219–235.

85. E. A. O'Brien and A. Shalev, Conjugacy classes in p-groups, in preparation.

86. A. Yu. Ol'shanskiĭ, The number of generators and orders of abelian subgroups of finite p-groups, *Math. Notes* 23 (1978), 183–185.

87. P. P. Pálfy and M. Szalay, The distribution of the character degrees of the symmetric p-groups, *Acta Math. Hung.* 41 (1983), 137–150.

88. J. Poland, Two problems on finite groups with k conjugate classes, *J. Austral. Math. Soc.* (Ser. A) 8 (1968), 49–55.

89. L. Pyber, Finite groups have many conjugacy classes, *J. London Math. Soc.* (2) 46 (1992), 239–249.

90. L. Pyber, Enumerating finite groups of given order, *Ann. Math.* 137 (1993), 203–220.

91. D. Quillen, The spectrum of an equivariant cohomology ring, I-II, *Ann. Math.* 94 (1971), 549–602.

92. D. M. Riley and J. F. Semple, The coclass conjecture for restricted Lie algebras, *Bull. London Math. Soc.*, to appear.

93. D. J. Rusin, The 2-groups of rank 2, *J. Algebra* 149 (1992), 1–31.

94. C. M. Scoppola, Groups of prime power order as Frobenius-Wielandt complements, *Trans. Amer. Math. Soc.* 325 (1991), 855–874.

95. C. M. Scoppola and A. Shalev, Applications of dimension subgroups to the power structure of p-groups, *Israel J. Math.* 73 (1991), 45–56.

96. J.-P. Serre, *Lie Groups and Lie Algebras* (new edition), Lecture Notes in Math. 1500, Springer, Berlin, 1991.

97. A. Shalev, Dimension subgroups, nilpotency indices, and the number of generators of ideals in p-group algebras, *J. Algebra* 129 (1990), 412–438.

98. A. Shalev, A note on groups of prime exponent, *Quart. J. Math. Oxford* (2) 42 (1991), 213–217.

99. A. Shalev, Characterization of p-adic analytic groups in terms of wreath products, *J. Algebra* 145 (1992), 204–208.

100. A. Shalev, Growth functions, p-adic analytic groups, and groups of finite coclass, *J. London Math. Soc.* 46 (1992), 111–122.

101. A. Shalev, On almost fixed point free automorphisms, *J. Algebra* 157 (1993), 271–282.

102. A. Shalev, Combinatorial conditions in residually finite groups, II, *J. Algebra* 157 (1993), 51–62.

103. A. Shalev, Automorphisms of finite groups of bounded rank, *Israel J. Math.* 82 (1993), 395–404.

104. A. Shalev, Polynomial identities in graded group rings, restricted Lie algebras and p-adic analytic groups, *Trans. Amer. Math. Soc.* 337 (1993), 451–462.

105. A. Shalev, The structure of finite p-groups: effective proof of the coclass conjectures, *Invent. Math.* 115 (1994), 315–345.

106. A. Shalev, Profinite groups with restricted centralizers, *Proc. Amer. Math. Soc.* 122 (1994), 1279–1284.

107. A. Shalev, Simple Lie algebras and Lie algebras of maximal class, *Arch. Math.* 63 (1994), 297–301.

108. A. Shalev and E. I. Zelmanov, Pro-p groups of finite coclass, *Math. Proc. Cambr. Phil. Soc.* 111 (1992), 417–421.

109. R. T. Shepherd, Ph. D. thesis, University of Chicago, 1971.

110. C. C. Sims, Enumerating p-groups, *Proc. London Math. Soc.* 15 (1965), 151–166.

111. M. R. Vaughan-Lee, Breadth and commutator subgroups of p-groups, *J. Algebra* 32 (1976), 278–285.

112. M. R. Vaughan-Lee, *The Restricted Burnside Problem*, 2nd edition, Oxford University Press, Oxford, 1993.

113. M. R. Vaughan-Lee and E. I. Zelmanov, Upper bounds in the Restricted Burnside Problem, *J. Algebra* 162 (1993), 107–145.

114. A. Vera-López and J. M. Arregi, Conjugacy classes in Sylow p-subgroups of $GL(n, q)$, *J. Algebra* 152 (1992), 1–19.

115. A. Vera-López and G. A. Fernández-Alcober, On p-groups of maximal class, I, *J. Algebra* 143 (1991), 179–207.

116. A. Vera-Lopéz and B. Larrea, On p-groups of maximal class, II, *J. Algebra* 137 (1991), 77–116.

117. T. S. Weigel, Exp *and* Log *Functors for the Categories of Powerful p-central Groups and Lie Algebras*, Habilitationsschrift, Freiburg, 1994.

118. J. S. Wilson, Finite presentations of pro-p groups and discrete groups, *Invent. Math.* 105 (1991), 177–183.

119. J. S. Wilson, Two-generator conditions for residually finite groups, *Bull. London Math. Soc.* 23 (1991), 239–248.

120. J. S. Wilson and E. I. Zelmanov, Identities for Lie algebras of pro-p groups, *J. Pure and Appl. Algebra* 81 (1992), 103–109.

121. I. O. York, *The Group of Formal Power Series under Substitution*, Ph. D. thesis, Nottingham, 1990.

122. E. I. Zelmanov, On some problems of group theory and Lie algebras, *Math. USSR-Sb.* 66 (1990), 159–168.

123. E. I. Zelmanov, The solution of the restricted Burnside problem for groups of odd exponent, *Math. USSR-Izv.* 36 (1991), 41–60.

124. E. I. Zelmanov, The solution of the restricted Burnside problem for 2-groups, *Math. USSR-Sb.* 72 (1992), 543–565.

125. E. I. Zelmanov, *The solution of the Restricted Burnside Problem for Groups of Prime Power Exponent*, Yale University Notes, 1990.

126. E. I. Zelmanov, On periodic compact groups, *Israel J. Math.* 77 (1992), 83–95.

127. E. I. Zelmanov, in preparation.

128. A. N. Zubkov, Non-abelian free pro-p groups cannot be represented by 2-by-2 matrices, *Siberian Math. J.* 28 (1987), 742–747.

INDEX

451